Paleozoic Plays of NW Europe

Geological Society books refereeing procedures

The Society makes every effort to ensure that the scientific and production quality of its books matches that of its journals. Since 1997, all book proposals have been refereed by specialist reviewers as well as by the Society's Books Editorial Committee. If the referees identify weaknesses in the proposal, these must be addressed before the proposal is accepted.

Once the book is accepted, the Society Book Editors ensure that the volume editors follow strict guidelines on refereeing and quality control. We insist that individual papers can only be accepted after satisfactory review by two independent referees. The questions on the review forms are similar to those for *Journal of the Geological Society*. The referees' forms and comments must be available to the Society's Book Editors on request.

Although many of the books result from meetings, the editors are expected to commission papers that were not presented at the meeting to ensure that the book provides a balanced coverage of the subject. Being accepted for presentation at the meeting does not guarantee inclusion in the book.

More information about submitting a proposal and producing a book for the Society can be found on its website: www.geolsoc.org.uk.

It is recommended that reference to all or part of this book should be made in one of the following ways:

Monaghan, A. A., Underhill, J. R., Hewett, A. J. & Marshall, J. E. A. (eds) 2019. *Paleozoic Plays of NW Europe*. Geological Society, London, Special Publications, **471**.

Arsenikos, S., Quinn, M., Kimbell, G., Williamson, P., Pharaoh, T., Leslie, G. & Monaghan, A. 2018. Structural development of the Devono-Carboniferous plays of the UK North Sea. *In*: Monaghan, A. A., Underhill, J. R., Hewett, A. J. & Marshall, J. E. A. (eds) 2019 *Paleozoic Plays of NW Europe*. Geological Society, London, Special Publications, **471**, 65–90, https://doi.org/10.1144/SP471.3

GEOLOGICAL SOCIETY SPECIAL PUBLICATION NO. 471

Paleozoic Plays of NW Europe

EDITED BY

A. A. MONAGHAN
British Geological Survey, The Lyell Centre,
Research Avenue South, Edinburgh, EH14 4AP, UK

J. R. UNDERHILL
Shell Centre for Exploration Geoscience,
Heriot-Watt University, Edinburgh, EH14 4AS, UK

A. J. HEWETT
10 Murrells Walk, Great Bookham, Surrey, KT23 3LP, UK

and

J. E. A. MARSHALL
Ocean & Earth Science, University of Southampton,
National Oceanography Centre, European Way,
Southampton, SO14 3ZH, UK

2019
Published by
The Geological Society
London

THE GEOLOGICAL SOCIETY

The Geological Society of London (GSL) was founded in 1807. It is the oldest national geological society in the world and the largest in Europe. It was incorporated under Royal Charter in 1825 and is Registered Charity 210161.

The Society is the UK national learned and professional society for geology with a worldwide Fellowship (FGS) of over 10 000. The Society has the power to confer Chartered status on suitably qualified Fellows, and about 2000 of the Fellowship carry the title (CGeol). Chartered Geologists may also obtain the equivalent European title, European Geologist (EurGeol). One fifth of the Society's fellowship resides outside the UK. To find out more about the Society, log on to www.geolsoc.org.uk.

The Geological Society Publishing House (Bath, UK) produces the Society's international journals and books, and acts as European distributor for selected publications of the American Association of Petroleum Geologists (AAPG), the Indonesian Petroleum Association (IPA), the Geological Society of America (GSA), the Society for Sedimentary Geology (SEPM) and the Geologists' Association (GA). Joint marketing agreements ensure that GSL Fellows may purchase these societies' publications at a discount. The Society's online bookshop (accessible from www.geolsoc. org.uk) offers secure book purchasing with your credit or debit card.

To find out about joining the Society and benefiting from substantial discounts on publications of GSL and other societies worldwide, consult www.geolsoc.org.uk, or contact the Fellowship Department at: The Geological Society, Burlington House, Piccadilly, London W1J 0BG: Tel. +44 (0)20 7434 9944; Fax +44 (0)20 7439 8975; E-mail: enquiries@geolsoc.org.uk.

For information about the Society's meetings, consult *Events* on www.geolsoc.org.uk. To find out more about the Society's Corporate Affiliates Scheme, write to enquiries@geolsoc.org.uk.

Published by The Geological Society from:
The Geological Society Publishing House, Unit 7, Brassmill Enterprise Centre, Brassmill Lane, Bath BA1 3JN, UK

The Lyell Collection: www.lyellcollection.org
Online bookshop: www.geolsoc.org.uk/bookshop
Orders: Tel. +44 (0)1225 445046, Fax +44 (0)1225 442836

British Library Cataloguing in Publication Data

A catalogue record for this book is available from the British Library.
ISBN 978-1-78620-395-3
ISSN 0305-8719

Distributors
For details of international agents and distributors see:
www.geolsoc.org.uk/agentsdistributors

Typeset by Nova Techset Private Limited, Bengaluru & Chennai, India
Printed and bound by CPI Group (UK) Ltd, Croydon CR0 4YY

Contents

Acknowledgements

The Geological Society thanks the following companies for their generous sponsorship of the Petroleum Group and the conference that led to this volume.

Petroleum Group corporate sponsors at the time of the conference:

Sponsors of Paleozoic Plays of Northwest Europe conference:

Paleozoic plays of NW Europe: an introduction

A. A. MONAGHAN[1]*, J. R. UNDERHILL[2], J. E. A. MARSHALL[3] & A. J. HEWETT[4]

[1]*British Geological Survey, Research Avenue South, Edinburgh, EH14 4AP, UK*

[2]*Centre for Exploration Geoscience, Applied Geoscience Unit, Institute of Petroleum Engineering, Heriot–Watt University, Edinburgh, EH14 4AS, UK*

[3]*Ocean & Earth Science, University of Southampton, National Oceanography Centre, European Way, Southampton, SO14 3ZH, UK*

[4]*10 Murrells Walk, Great Bookham, Surrey, KT23 3LP, UK*

A.A.M., 0000-0003-2147-9607

**Correspondence: als@bgs.ac.uk*

Abstract: Despite successful production from Carboniferous and Permian reservoirs in the Southern North Sea and onshore Netherlands and Germany, Paleozoic hydrocarbon plays across parts of NW Europe remain relatively under-explored onshore and offshore. This volume brings together new and previously unpublished knowledge about the Paleozoic plays of NW Europe. Improvements in seismic data quality and availability tied to previously unpublished well datasets form the basis for improved understanding of local to regional structural interpretations, depositional environments and basin history. New interpretations move significantly away from generalized basin development models, with improved definition of structural traps and source rock basins feeding to better constrained, locally variable burial, uplift, maturation and migration models. Particularly notable are the significant mapped extents and thickness of Paleozoic source, reservoir and seal rocks. Areas previously dismissed as regional highs and platforms are dissected by Paleozoic basins with evidence for mature source rocks into basin centres. Numerous potential Paleozoic plays or play elements result within thick organic-rich and variably mature successions. Outside or below existing Jurassic and Southern North Sea to onshore Netherlands and German Permian-Carboniferous plays, Paleozoic plays in frontier areas offer significant additional exploration opportunities.

In the 2010s, Government-level policy in countries with North Sea oil and gas fields increased the focus on extending the life of mature basins and opening up new plays, before ageing infrastructure is decommissioned. The aim has been to maximize economic recovery and maintain the energy security of individual countries. In the mature North Sea petroleum province and surrounding onshore plays of NW Europe, Paleozoic petroleum plays offer new exploration targets in frontier areas in shallow waters, either below or close to producing fields.

Carboniferous source rocks and Carboniferous, Permian and Triassic reservoirs and seals have formed a highly productive gas province in the East Irish Sea, the Southern North Sea (SNS) and onshore into the Netherlands and Germany since the 1960s (e.g. Meadows *et al.* 1997; Glennie & Underhill 1998; Fraser & Gawthorpe 2003; Underhill 2003; Cameron *et al.* 2005; Breunese *et al.* 2010; Gast *et al.* 2010; Kombrink *et al.* 2010; Pletsch *et al.* 2010; **Pharaoh et al. 2018**; Figs 1 & 2). Substantial quantities of gas have been produced, e.g. *c.* 3400 bcm at 2005 from fields within the Anglo-Dutch and North German basins (Breunese *et al.* 2010), or are estimated, e.g. estimated recoverable volume of 102 bcm (3.6 tcf[1]) of gas in Carboniferous SNS gas fields in the UK sector at 2015 (**Besly 2018**). The discovery and development of fields such as Breagh with its lower Carboniferous reservoir and the Cygnus (Catto *et al.* 2017) Rotliegend Group, Leman Sandstone reservoir have extended beyond the margins of the established plays (Fig. 1) and within a wider stratigraphical range (Fig. 2). The exploration and development success at these fields exemplifies significant remaining Paleozoic play opportunities in adjacent 'frontier' areas. In addition, unconventional and tight gas resources are known and/or under active exploration within Carboniferous rocks across

[1]bcm = billion cubic metres, tcf = trillion cubic feet.

From: MONAGHAN, A. A., UNDERHILL, J. R., HEWETT, A. J. & MARSHALL, J. E. A. (eds) *Paleozoic Plays of NW Europe*. Geological Society, London, Special Publications, **471**, 1–15.
First published online December 19, 2018, https://doi.org/10.1144/SP471.13

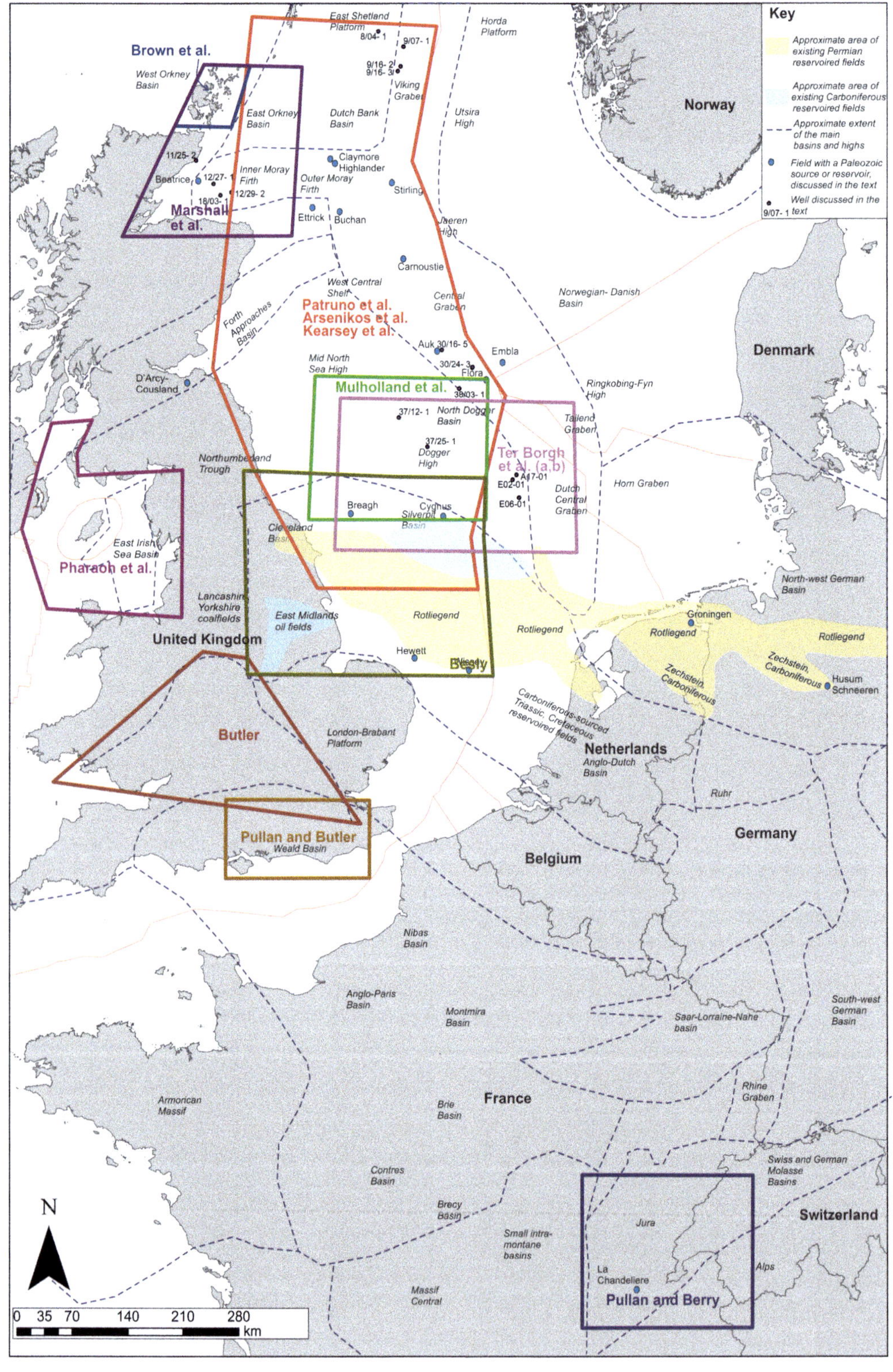
Key
Approximate area of existing Permian reservoired fields
Approximate area of existing Carboniferous reservoired fields
Approximate extent of the main basins and highs
Field with a Paleozoic source or reservoir, discussed in the text
Well discussed in the text
Brown et al.
Marshall et al.
Patruno et al.
Arsenikos et al.
Kearsey et al.
Mulholland et al.
Ter Borgh et al. (a,b)
Pharaoh et al.
Besly
Butler
Pullan and Butler
Pullan and Berry
Norway
Denmark
United Kingdom
Netherlands
Belgium
Germany
France
Switzerland
East Shetland Platform
Horda Platform
West Orkney Basin
East Orkney Basin
Dutch Bank Basin
Viking Graben
Utsira High
Inner Moray Firth
Outer Moray Firth
Claymore
Highlander
Stirling
Beatrice
Ettrick
Buchan
Jaeren High
Carnoustie
West Central Shelf
Central Graben
Norwegian- Danish Basin
Forth Approaches Basin
Mid North Sea High
Auk
Embla
Flora
Ringkobing-Fyn High
D'Arcy-Cousland
North Dogger Basin
Dogger High
Tail End Graben
Northumberland Trough
Horn Graben
Dutch Central Graben
Breagh
Cygnus
Silverpit Basin
Cleveland Basin
East Irish Sea Basin
Lancashire Yorkshire coalfields
East Midlands oil fields
Rotliegend
Groningen
North-west German Basin
Hewett
Wissey
Zechstein, Carboniferous
Husum Schneeren
Carboniferous-sourced Triassic, Cretaceous reservoired fields
Anglo-Dutch Basin
London-Brabant Platform
Ruhr
Weald Basin
Nibas Basin
Anglo-Paris Basin
Montmira Basin
Saar-Lorraine-Nahe basin
South-west German Basin
Rhine Graben
Armorican Massif
Brie Basin
Swiss and German Molasse Basins
Contres Basin
Brecy Basin
Small intra-montane basins
Jura
Alps
La Chandeliere
Massif Central
N
0 35 70 140 210 280 km

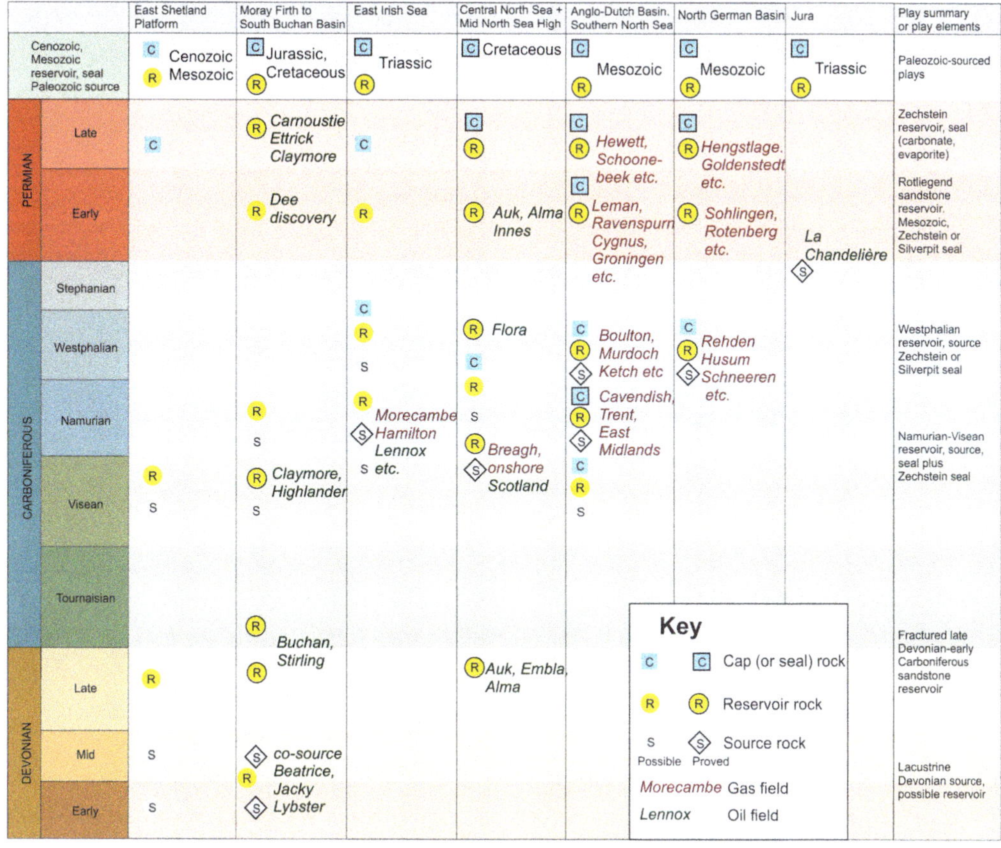

Fig. 2. Summary of proved and possible upper Paleozoic petroleum play elements across NW Europe. Information synthesized and modified from wide range of sources including Bruce & Stemmerik (2003), Marshall & Hewett (2003), Breunese *et al.* (2010), Gast *et al.* (2010), Kombrink *et al.* (2010), Pletsch *et al.* (2010), BGS (2016) and reports therein; Monaghan *et al.* (2016), Patruno & Reid (2016, 2017), BGS (2017), IGI Ltd (2017) and Pullan & Berry (2018).

NW Europe (e.g. USEIA 2013; Clarke *et al.* 2014; Hennissen *et al.* 2017; OGA 2017*a*, *b*).

In the Central–Northern North Sea, lacustrine Devonian source rocks have long been recognized (Andrews *et al.* 1990; Hillier & Marshall 1992; Duncan & Buxton 1995; Marshall 1998; Marshall & Hewett 2003) with renewed exploration interest in their charge to Mesozoic- and Cenozoic-hosted fields and prospects (e.g. Richardson *et al.* 2005; Patruno & Reid 2016, 2017). More widely across NW Europe, Paleozoic source rocks have been documented in the Northern Permian Basin (Pedersen *et al.* 2006; Ohm *et al.* 2012); Permo-Carboniferous source and reservoir intervals are proved in the developing Barents Sea plays (e.g. Van Koeverden *et al.* 2010), as well as forming reservoirs and providing source rock contributions to fields west of Shetland (e.g. Coney *et al.* 1993; Mark *et al.* 2008).

However, in comparison with the well-characterized Jurassic-sourced petroleum systems of the North Sea a number of challenges exist in Paleozoic plays:

- data quality – e.g. seismic below Zechstein evaporites, quality of old well logs; biostratigraphic control for accurate subsurface calibration;

Fig. 1. Approximate areas of interest of the papers in this volume with extent of main basins and highs, selected fields with a Paleozoic involvement and extent of established, producing Paleozoic play systems. Information synthesized and modified from wide range of sources including USGS (1997), Bruce & Stemmerik (2003), Marshall & Hewett (2003), Breunese *et al.* (2010), Gast *et al.* (2010), Kombrink *et al.* (2010), Pletsch *et al.* (2010), BGS (2016) and BGS (2017).

- data availability – e.g. limited well penetrations owing to burial depth, limited sample analyses;
- variable source, reservoir and seal quality over large geological time spans and widespread geographical areas;
- complex structural and maturation/migration history.

As a result, Paleozoic plays can be perceived as complex and risky, with expert knowledge built up through years of experience, and collaboration being particularly important. Historically, the loss of knowledge from company reorganizations related to fluctuations in the oil price has significantly hampered wider understanding of the plays (**Besly 2018**).

Various leading explorationists from industry and academia have suggested that a paradigm shift is required in knowledge and play concepts to open up major new Paleozoic plays and reduce perceived risk. In this volume, **Besly (2018)** discusses the history of Carboniferous gas exploration and production in the UK sector of the SNS and the challenges faced. Five areas of Carboniferous petroleum geology in which the currently accepted status quo is open to question, named the 'founding myths', are described; (1) Westphalian coals as the dominant source rock; (2) lack of reservoir intervals; (3) pessimistic view of intra-formational seals; (4) sub-basin depocentre distribution and migration pathways; and (5) oversimplified tectonic models and burial histories. **Besly (2018)** provides a summary of evidence to question these 'founding myths', which provide new positive insight. Together with papers in this volume and other recently published papers, reports and datasets, evidence is growing of potentially prospective Paleozoic plays.

In the Netherlands, EBN and TNO have been active in understanding Paleozoic plays in the northern offshore Dutch sector (Schroot *et al.* 2006; de Bruin *et al.* 2015; EBN 2015*a*, *b*; **Ter Borgh *et al.* 2018*a*, *b***). Despite the challenge of limited well datasets and a complex structural and burial history, evidence is growing for Rotliegend and Carboniferous (Visean–Namurian) plays (de Bruin *et al.* 2015; **Ter Borgh *et al.* 2018*a*, *b***).

In the UK, following a major review of the oil and gas industry (Wood 2014), the UK Government (the then Department for Energy and Climate Change), Oil and Gas UK and the British Geological Survey (BGS) consulted with industry as to their priorities for future maximizing economic recovery on the UK continental shelf. The 'frontier' plays of the Paleozoic were ranked as top priority for new datasets, regional interpretations and collaborative working. A joint industry project, the 21st Century Exploration Roadmap (21CXRM) Paleozoic Project, was initiated, undertaken by BGS in 2014–16, which included a focus on collaboration, data sharing and release (products at BGS 2016). In tandem, the newly formed Oil and Gas Authority (OGA) acquired and released the results of a £20 million Government-funded seismic survey across the Paleozoic frontier area of the Mid North Sea High, and initiated a number of innovation projects and university research studies to further stimulate exploration, culminating in the 29th Offshore Licensing Round in 2016/17. The 2018 31st Offshore Licensing Round also includes frontier Paleozoic plays of the East Shetland Platform and to the west of Britain, with an additional set of seismic data release, background datasets and interpretative reports released in 2017 (BGS 2017; Frogtech Geoscience 2017; Getech 2017; IGI Ltd 2017; OGA 2017*c*).

Paleozoic Plays of NW Europe conference and aim of this volume

The renewed interest in Paleozoic prospectivity led to a Geological Society of London 'Paleozoic Plays of NW Europe' conference convened by industry, academic, BGS and OGA representatives in May 2016.

This volume contains some of the key papers presented at that conference and aims to bring new insights to Paleozoic petroleum play systems and fill a gap between academic and industry studies. It aims to bring previously unpublished literature into the public domain. Most of the papers presented fall within a scale intermediate between that of Millennium Atlas (Bruce & Stemmerik 2003; Glennie *et al.* 2003; Marshall & Hewett 2003)/Southern Permian Basin Atlas (Breunese *et al.* 2010; Gast *et al.* 2010; Kombrink *et al.* 2010; Peryt *et al.* 2010; Pletsch *et al.* 2010) and field-scale studies and provide a regional overview, forming a data-evidenced framework for future exploration. Several papers are focused on Paleozoic plays outside the extent of mature Jurassic (Kimmeridge) North Sea source rocks (e.g. **Besly 2018**; **Mulholland *et al.* 2018**; **Pullan & Butler 2018**; **Ter Borgh *et al.* 2018*a*, *b***). In some areas, Paleozoic play elements such as those described in papers by **Marshall *et al.* (2018)**, **Patruno *et al.* (2018)** and others provide an additional source beneath existing Jurassic and younger plays (e.g. Moray Firth) or act as a reservoir for migrated Jurassic oil (e.g. Auk-Flora Ridge, Buchan Field). The focus of prospective play systems described in this volume is dominantly upper Paleozoic (Devonian, Carboniferous, Permian). Western European substage timescale names are commonly used (e.g. Heckel & Clayton 2006; Holliday & Molyneux 2006) throughout this volume owing to their basis in biostratigraphical subdivision.

This introductory paper begins with a summary of exploration history of Paleozoic plays. Particular detail has been included on Devonian exploration as this is not documented elsewhere. The increasing knowledge of the structure, stratigraphy, petroleum systems and basin and migration histories of Paleozoic plays across NW Europe is then discussed, placing the new papers in this volume into context. The paper concludes with a summary of current and future exploration opportunities.

Exploration history: Devonian

It is worth taking the opportunity at this time to review the thinking that lay behind the exploration of the Devonian, as both a source and a reservoir. The generation of geologists who led much of this exploration is now passing into retirement and much of the information is anecdotal oral information with the relevant company files long shredded.

As regards Devonian source rocks, it was long known to the onshore geological community that the lacustrine sediments of the Orcadian Basin were rich in organic matter. There had been an abandoned attempt in the 17th or 18th centuries to mine for 'coal' (Miller 1841) at a Cromarty Fish Bed locality (Coal Heugh) based on the assumption that these black shales were related to coals. This was a prediction that was not entirely unwarranted given the existence of the Jurassic Brora Coal further north along the coast and other isolated 'Highland' occurrences of Carboniferous coals (e.g. Inninmore, Morvern). Following the primary survey of Caithness (Crampton & Carruthers 1914), a bituminous shale known locally as the 'black man' was reported. In addition, there have been many records of solid bitumens from onshore Orkney (**Brown *et al.* 2018**) that were numerous enough to be named the mineral Cloustonite. Beds that 'resemble impure oil shales' were also described in Orkney (Wilson *et al.* 1935). Many of these records have been reviewed by Parnell (1983), including a recent account (Parnell *et al.* 2017) of the extensive networks of basement-hosted bitumen veins from the Inverness area that were economically mined from at least the 18th century. These fractures represent a reservoir analogue for the Jurassic sourced hydrocarbons found within 'basement' reservoirs of both the Clair and Lancaster fields and related discoveries/prospects under development by Hurricane Energy. Despite this long record of occurrences of Orcadian bitumens, it proved very difficult for some groups to accept that there could be any source rock other than the Jurassic Kimmeridge Clay. For example, J.G.C.M. Fuller – a senior industry figure in the 1970s development of the North Sea – was an arch proponent of this single Kimmeridge source rock model (see review in Jones & Scott 1980) and regarded it as the only source rock capable of generating significant volumes of North Sea oil. He had a lecture party piece of setting fire to a piece of Kimmeridge Clay to show its richness as a source rock. Whilst very entertaining, this did nothing to disprove the existence of other source rock systems. This was at a time when the Devonian was often regarded as economic basement within the North Sea area. However, there was increasing knowledge from the industry of the source potential of the onshore Devonian which was incorporated in the widely attended fieldtrips to the Moray Firth run by Robertson Research. In addition, there was continuing research from UK universities that was showcased at the Orcadian Basin meeting in Cambridge in September 1984 that was attended by many industrial participants and published as an Orcadian Basin thematic set in the *Scottish Journal of Geology* (Trewin 1985), including three source rock papers.

As documented by **Marshall *et al.* (2018)**, there was a significant inability to recognize top Devonian beneath the Rotliegend Group and this resulted in many long Devonian sections being drilled. Top Devonian was often only recognized when obvious dark-coloured lacustrine mudstones were finally penetrated. However, the offshore section that drew the most attention from exploration geologists was in well 12/27-1. Here in 1982 Burmah Oil were drilling a Jurassic play on a fault block in the Inner Moray Firth. The well was drilled to basement as the agreed target and in doing so revealed (Richards 1985) a kilometre of Early Devonian lacustrine mudstones rich in organic matter. These were the equivalents of the rather poorly exposed and presumed much thinner onshore equivalents around the Strathpeffer area in Easter Ross. This discovery subsequently made the Devonian both a primary and a secondary target. For example, in 1985 Kerr-McGee drilled 12/29-2, which penetrated both a Lower Devonian source rock section and overlying alluvial fan deposits that included a thick interval (*c.* 200 m) of high-porosity (15–21%) sandstone. This was followed by similar wells such as 18/3-1 (1992) and 11/25-2ST1 (1986), the latter being drilled into the Great Glen Fault Zone. The Eday Group equivalent sandstones were tested in Quadrant 13, largely in the 1980s and generally identified at that time as Rotliegend sandstone.

A key Devonian well in the southern part of the North Sea was 30/24-3, drilled in 1972 on what was to become the Argyll Field with its combined Jurassic, Zechstein, Rotliegend and Devonian reservoir section. Well 30/24-3 penetrated an interval of what proved to be a Mid Devonian marine limestone (Pennington 1975) that was considerably distant from any known marine margin and in what was generally believed to be well within the Old Red Sandstone continent. This discovery caused considerable discussion with various counter claims, including that the Devonian stromatoporoid was not truly restricted to marine environments or that it was merely a thick calcrete. Subsequent wells with much better cuttings recovery (Marshall *et al.* 1996), together with core, proved the existence of what was to become known as the Kyle Limestone. This Kyle Limestone has been the target for subsequent exploration, including well 37/25-1 drilled on the Corbenic Prospect in 2009.

The thickest Central North Sea Devonian section was drilled by Mobil in 1975 in well 38/03-1, which penetrated over 1500 m of sandstone before terminating in the Kyle Limestone. In part, this could be palynologically dated as Frasnian to Givetian in age. It was equally unusual in that Devonian 'coal seams' were apparently identified. However, investigation of wireline and mud logs together with analyses (atomic H/C ratio, vitrinite reflectivity, palynology) of hand-picked coal fragments from cuttings shows them to be a drilling mud additive. This additive was found to be coincident with hole rugosities that were of sufficient size to cause cycle skipping on the sonic log and give the appearance of interbedded coals. We can only speculate that the operator continued drilling such a long section based on the presumption that it was drilling through Carboniferous Coal Measures.

There have been rare longer Devonian penetrations on the East Shetland Platform and particularly towards the graben margins. These include wells 9/7-1 (1974), 9/16-2 (1987), 9/16-3 (1991) and 8/4-1 (1994). These are particularly important in that they reveal a stratigraphy and sediment infill that is common to other parts on the Orcadian Basin, including that of SE Shetland. The key stratigraphic markers can be recognized and the infill includes organic matter rich lacustrine sediment. They provide the essential information for focused sub-Permian exploration (**Patruno *et al.* 2018**) on the East Shetland Platform.

Throughout this exploration of the Devonian, there was a continuing debate about the origin of oil from the Beatrice Field (1976; Stevens 1991). This was the considerable oilfield that lies close inshore in the Inner Moray Firth and contained nearly 500 mmboe (millions of barrels of oil equivalent) of heavy degraded oil within a Jurassic reservoir. There has been a continuing history of biomarker analysis of this oil that has paralleled technological advances in organic geochemistry. This has seen the Beatrice oil go from a Jurassic source to a mixed Devonian–Jurassic source (e.g. Stevens 1991) to a wholly Devonian source (Bailey *et al.* 1990), with the final argument based on $\delta^{13}C$ values (Marshall & Hewett 2003). This has been paralleled by changes in our understanding of its geological context with claims for long-distance migration (from what proved to be a thermally immature Jurassic section) to more direct evidence of bitumen being sourced from Devonian intervals as in 12/27-1 (Marshall 1998).

As regards the Devonian as a reservoir interval, there were initial early discoveries of Devonian sections within tilted fault blocks and charged from the Mesozoic. For example, the Buchan Field (Quadrants 20/21, fractured Upper Devonian–lower Carboniferous reservoir) was discovered in 1974 (Edwards 1991) and the Stirling Field (Block 16/21) produces from an Upper Devonian reservoir (Gambaro & Currie 2003).

Devonian play elements have been tested by a handful of wells further south, for example penetrations of the Kyle Limestone Group and Buchan Formation (Mid-Upper Devonian) in the Auk–Flora Ridge area (UK Quadrant 30, e.g. wells 30/16-5, 30/24-3, 30/25a-2). The Auk and Argyll (renamed Alma) fields (UK sector blocks 30/16, 30/24 and 30/25) and Embla Field (Norwegian sector, block 2/7) contain proved Devonian fluvial sandstone reservoirs, in fault blocks updip of a Jurassic source (Marshall & Hewett 2003; Ohm *et al.* 2012).

In central-southern parts of the North Sea, a Mid Devonian play system comprising karstified carbonate reef and platform reservoirs fringed by mudstone source rocks and a mudstone seal has been advocated by analogy to Mid Devonian strata in Belgium and the West Canadian Basin (Bełka *et al.* 2010; De Jong *et al.* 2016). However, limited numbers of tests of the Kyle Limestone Group on structural highs either were dry (37/12-1, in 1985) or the target strata were absent (37/25-1, in 2009; ExxonMobil 2010).

In the northern Dutch offshore sector, three wells (A17-1, E06-1 and E02-1) penetrate the Middle–Upper Devonian (Bełka *et al.* 2010). The hydrocarbon potential of the Devonian onshore NW Europe is limited by sparse data and high thermal maturation of source rock intervals, with Bełka *et al.* (2010) stating that there was no hydrocarbon production in onshore areas.

Exploration history: Carboniferous

Onshore UK oil and gas exploration of the Carboniferous sequence in the UK began as early as 1919, with small fields producing in the East Midlands and Yorkshire (1940s to present day; Pharaoh *et al.* 2011) and Midland Valley of Scotland (1960s, Hallett *et al.* 1985). In the early days of North Sea exploration, 1965–70, wells were drilled across the Central North Sea/Mid North Sea High, with many of these wells remaining the only penetrations of the Carboniferous. Bruce & Stemmerik (2003) summarize fields with a Carboniferous component in the Central-Northern North Sea, with details given by Edwards (1991; Buchan) and Harker *et al.* (1991; Claymore).

In the East Irish Sea, the Carboniferous-sourced, Triassic-reservoir Morecambe fields were first drilled in 1969 and discovered in the 1970s (Cowan 1996). Renewed drilling activity in the East Irish Sea from 1989 resulted in the discovery of oil and gas fields such as Hamilton, Douglas and Lennox (e.g. Haig *et al.* 1997; Yaliz 1997).

Many significant gas discoveries in the Carboniferous Southern North Sea play were made between 1984 and 2000 (Cameron *et al.* 2005). The exploration history within the UK Southern North Sea gas basin is further summarized by **Besly (2018)** and references therein. Outside of the Westphalian and Namurian play, the Visean–Namurian Breagh gas field came onstream in 2013, following re-appraisal of gas shows in well 42/13- 2 first identified in 1997.

Kombrink *et al.* (2010), Breunese *et al.* (2010) and Pletsch *et al.* (2010) give details of Carboniferous exploration from onshore UK to the Netherlands–North German Basin, with Schroot *et al.* (2006) and De Bruin *et al.* (2015) giving additional detail on the northern Dutch offshore. Exploration of the first onshore Carboniferous fields (e.g. Coevorden gas field 1951) was followed by discoveries in the gas province of the Cleaver Bank High area, e.g. the 1974 K4-FA field of the Netherlands sector, with discoveries such as Husum Schneeren gas field in Germany from the 1980s (Van Buggenum & Den Hartog Jager 2007; Kombrink *et al.* 2010).

Exploration success based on Westphalian–Stephanian Coal Measures to Autunian (Permian) bituminous shales source rocks within the Variscan basins of France, Switzerland and Germany has been limited (literature summary in BGS 2017) with a case history given by **Pullan & Berry (2018)**.

Exploration history: Permian

Exploration of Permian reservoirs was ignited by the discovery of the major Groningen Field in the northern Netherlands in the late 1950s. Recognition that similar aeolian sandstones were exposed in NE England led to speculation that the Rotliegend Group reservoir may extend beneath the North Sea (Glennie 1998). Exploration drilling during the late 1960s and early 1970s not only confirmed the continuous nature of the sandstone play fairway, but also opened up a major gas province throughout the Southern Permian Basin in which the aeolian and fluvial sandstones ascribed to the Schlochteren or Leman Sandstone Formation were charged by Carboniferous source rocks beneath and sealed by evaporites belonging to the Zechstein Supergroup (Glennie 1998; Glennie & Underhill 1998; Underhill 2003; Gast *et al.* 2010).

By the 1980s, it was clear that the northern limit of the Rotliegend play fairway was defined by the deposits of a large, intracontinental saline Silverpit Lake, the southern margin of which comprised a continental sandy sabkha forming a 'waste zone' (i.e. an area that was neither the high-quality reservoir of the Rotliegend or a complete Silverpit seal; Alberts & Underhill 1991; Glennie 1998; Glennie & Underhill 1998; Underhill 2003). Recognition that the Silverpit Claystone Formation could act as a seal for erosionally truncated Carboniferous sequences lying beneath the Base Permian (Variscan) Unconformity led to an exploration campaign for Westphalian and Namurian reservoirs described in the previous section. The recent development of the Cygnus field (Catto *et al.* 2017) has extended the Rotliegend play to the northern edge of the Silverpit Lake.

Some notable, but more limited, exploration success in the Permian Rotliegend Group also occurred in the Northern Permian Basin, most notably at Auk, where oil derived from the Kimmeridge Clay Formation migrated from its kitchen in the Central Graben to fill traps on the rift flanks (e.g. Glennie *et al.* 2003).

The depositional extent of the Zechstein evaporites forms an effective limit to the regional seal. However, whilst that is detrimental for the underlying Rotliegend Group reservoir plays, it has led to gas migrating into Zechstein carbonates (e.g. Hewett field, Cooke-Yarborough & Smith 2003; Underhill & Hunter 2008; Wissey field, Duguid & Underhill 2010; Peryt *et al.* 2010) and Triassic reservoirs belonging to the Hewett and Bunter Sandstone formations (Underhill 2003). The Zechstein and Triassic play continues into the Netherlands sector (e.g. P6 gas field) and onshore into the Netherlands and northern Germany (e.g. Schoonebeek gas field; see Peryt *et al.* 2010). The Zechstein carbonate play has proved to be successful in onshore areas of the Cleveland Basin, UK and the acquisition of new seismic data over the Mid North Sea High has suggested that some similar opportunities may exist there too (Patruno *et al.* 2017; **Mulholland *et al.* 2018**). Farther north, the Zechstein is a reservoir in the Carnoustie, Ettrick and Claymore fields (Glennie *et al.* 2003).

Structural complexity of the Paleozoic

As a result of syn- and post-depositional tectonism, Paleozoic plays are generally faulted, folded and bound by unconformities. Structural traps are common. For example:

- Late Carboniferous–early Permian Variscan inversion, uplift and folding along with NW–SE/WNW–ESE and NE–SW faults formed the many structural traps of the Westphalian UK SNS play (e.g. Cameron *et al.* 2005). Dip and fault closures adjacent to the base Permian unconformity (Variscan) and Permian seal rocks form a major trap type (Pletsch *et al.* 2010).

- The Groningen field onshore the Netherlands, a large structural high, is dissected by east–west- and NNW–SSE-trending faults (Gast *et al.* 2010).
- In the Moray Firth, Mesozoic rifting and Cenozoic uplift control trapping geometries, and in areas lacking Zechstein evaporite seal, faults and tilted saline aquifers that rise to subcrop the seabed may act as key migration pathways (Marshall 1998; Underhill 1991; Hillis *et al.* 1994; Richardson *et al.* 2005; Guariguata-Rojas & Underhill 2017).

High-quality seismic data of kilometres-deep, faulted and folded Paleozoic strata are essential to unravel the structural evolution and its impact on hydrocarbon prospectivity. Seismic data quality can be challenging, especially beneath Zechstein evaporites. Taken together with the cost and availability of seismic data, many studies had, until recently, tended to be localized to block- or field-scale interpretations, lacking semi-regional context. Prior exploration had focussed on significant regional highs observed in two-way travel time, with little widespread mapping of Paleozoic structures or detailed depth conversion.

Significant improvements in seismic data quality and image fidelity of Paleozoic successions have been made in recent years along with a more widespread acquisition of 3D surveys (e.g. Rodriguez *et al.* 2014; Patruno & Reid 2016, 2017; **Patruno *et al.* 2018**; **Ter Borgh *et al.* 2018*b***). Together with the acquisition and release of Government-funded seismic data in the UK sector and regional studies (e.g. De Bruin *et al.* 2015; BGS 2016; **Arsenikos *et al.* 2018**), recent studies greatly improve our understanding of the extent of play fairways along the northern margin of the Southern Permian Basin, the structural complexity of the basin's Paleozoic succession and their combined influence on hydrocarbon prospectivity.

In this volume, **Arsenikos *et al.* (2018)** present regional structure maps of Devonian and lower Carboniferous seismic reflectors based on released and unreleased seismic datasets, re-interpreted well ties and gravity and magnetic studies from the East Orkney Basin and Central North Sea (Fig. 1). Basement inheritance, stress partitioning and the presence of granite-cored blocks are believed to influence the variety of observed fault trends, with sub-basin depocentres cutting across 'platform' areas, and relative highs. Extensive lower Carboniferous and Lower-Middle Devonian source rock intervals are interpreted at depths of around 4–5 km.

Patruno *et al.* (2018) present seismic interpretations from the East Shetland Platform to the Greater Mid North Sea High, highlighting Paleozoic to recent polyphase subsidence and inversion on a variety of trends and linked to variable prospectivity. The lack of a Permo-Triassic rifting phase in the Mid North Sea High platform area and its influence on maturation and migration is contrasted with interpretations of a deep, mature middle Devonian source rock across the East Shetland Platform.

The continuation of Devono-Carboniferous basins and highs (e.g. Elbow Spit High, North Elbow Basin/Outer Rough Basin) and of WNW–ESE and Carboniferous-Permian NE–SW faults from the UK sector of the Southern-Central North Sea (e.g. Cameron *et al.* 2005; Milton-Worssell *et al.* 2010; **Arsenikos *et al.* 2018**) into the Dutch sector is mapped by **Ter Borgh *et al.* (2018*b*)**. Faults with a NE–SW trend are known to cause reservoir compartmentalization (Oudmayer & De Jager 1993; Van Hulten 2010) and both fault sets have potential to generate intra-Carboniferous closures (**Ter Borgh *et al.* 2018*b***).

Paleozoic-sourced potential trapped within inversion structures is highlighted by a number of papers. In the Irish Sea, **Pharaoh *et al.* (2018)** document Variscan inversion structures superimposed on Carboniferous fault blocks and highlight potential prospectivity within the deeply buried Carboniferous–Permian succession and around the margins of the oil- and gas-producing East Irish Sea Basin. **Butler (2018)** documents a regional seismostratigraphic framework of central-southern England that also includes a significant history of inversion and unconformity development within deeply buried sequences, including Variscan inversion over the Midlands Microcraton. Paleozoic to Cenozoic inversion structures also play a role in likely Paleozoic source rock distribution, maturity and migration for a proposed gas source in the Weald Basin of southern England (**Pullan & Butler 2018**). Finally, **Pullan & Berry (2018)** describe a Permo-Carboniferous source and Triassic reservoir proven in a number of small discoveries in thrust-bound structural highs in the Jura fold belt of France and Switzerland.

In summary, high-quality 2D and 3D seismic datasets have allowed detailed mapping of structurally complex Paleozoic successions. Prospective faulted and folded Paleozoic basins and highs are evident across wide areas of the North Sea and onshore. A range of structural orientations, styles and timings evidence polyphase and partitioned extension, strike-slip and inversion applied to an inherited structural fabric and dissected by younger events. Areas previously considered as regional highs (Mid North Sea High, East Shetland Platform) are dissected by Paleozoic basins. New interpretations move significantly away from generalized and simplified basin development models, to better pinpoint structural traps, source rock kitchens and their implications for maturation and migration.

Stratigraphical and facies complexity

High-level summaries of basin and national stratigraphical nomenclature are given in the Millennium Atlas and Southern Permian Basin Atlas (e.g. Bruce & Stemmerik 2003; Marshall & Hewett 2003; Gast *et al.* 2010; Kombrink *et al.* 2010; Pletsch *et al.* 2010) drawing together formal stratigraphic treatise such as the UKOOA volumes (e.g. Cameron *et al.* 1992*a*, *b*; Cameron 1993*a*, *b*), Waters *et al.* (2011), Van Adrichem Boogaert & Kouwe (1993–97) and updated interpretations. However, challenges in Paleozoic stratigraphic correlation and interpretation exist where large, laterally variable volumes of rock are poorly constrained by well and biostratigraphic datasets, and where stratigraphical nomenclature changes onshore to offshore and across basin or country borders. This can hinder facies interpretation and create confusion for prospectivity assessments.

Kearsey *et al.* (2018) present a simplified stratigraphy for the pre-Westphalian Carboniferous rocks of the UK sector from the Outer Moray Firth to Southern North Sea based on revised well and seismic interpretations. Biostratigraphically and lithostratigraphically constrained time slices highlight lateral facies variability within a regionally extensive delta system, mapping more specifically a regional understanding on the distribution of stacked source and reservoir intervals (e.g. Collinson 2005; Kombrink *et al.* 2010) to that farther north (Leeder & Boldy 1990).

Many challenges remain in the detailed understanding of correlation and distribution of Paleozoic strata, based on sparse data and in sequences where facies distribution, extent and thickness exert a strong control on the petroleum system. The impact of facies variations in Zechstein carbonates and anhydrites over the Mid North Sea High to reservoir and seal intervals is described by **Mulholland *et al.* (2018)**. The importance of Rotliegend-age sandstones preserved in areas of low palaeorelief is also highlighted and the predicted extent of each play type is discussed. This adds to a growing interest in the prospectivity of Rotliegend and Zechstein reservoirs to the north of the Silverpit and Southern Permian Basin and above or updip of Visean–Namurian source rocks (see also De Bruin *et al.* 2015; Patruno *et al.* 2017).

The challenges of stratigraphical correlation with largely barren sandstone reservoir intervals have been studied previously in the Westphalian D and Rotliegend of the Southern Permian Basin (e.g. Besly 2005; Rotliegend summary in Gast *et al.* 2010). **Marshall *et al.* (2018)** present revised stratigraphical interpretations of Devonian and Permian Rotliegend sandstones in the Moray Firth that significantly change the understanding of their thickness and distribution. A field analogue in Greenland provides further insight into the nature of the unconformable contact and interpretation of the sandstone units.

New insights into source (distribution, quality, maturity), reservoir and seal

A number of entirely Paleozoic plays are documented across NW Europe, such as the Westphalian-sourced, Rotliegend-reservoired gas play (e.g. Leman, Groningen fields), the Visean–Namurian play (e.g. Breagh) and the Carboniferous-sourced, Zechstein-reservoired play (e.g. Hewett, Schoonebeek fields; Fig. 2). Many more fields and plays have a Paleozoic source rock (e.g. Triassic-reservoired fields of the East Irish Sea and Anglo-Dutch Basin, Devonian co-sourced, Jurassic-reservoired fields in the Moray Firth) with migration into Mesozoic and Cenozoic reservoirs. Given the thickness and diversity of organic-rich successions, it is unsurprising that Paleozoic plays of NW Europe comprise a diverse range of petroleum system elements in a variety of play fairways. New work presented in this volume increases understanding of stratigraphical, facies, structural, thermal and burial history variability within the Paleozoic successions, which highlights some of the future opportunities and plays.

In a detailed examination of the Visean–Namurian play proved by the Breagh field, changes in fluvio-deltaic to basinal shale depositional facies, NE to SW across the southern part of the Central North Sea, are shown by **Ter Borgh *et al.* (2018*a*)** and **Kearsey *et al.* (2018)** to influence reservoir and source rock prospectivity. Post-well analysis in the Dutch sector by **Ter Borgh *et al.* (2018*a*)** demonstrates invalid tests, drilled off-structure by the majority of existing wells, leading to the suggestion of significant hydrocarbon potential remaining untested within this play.

Brown *et al.* (2018) summarize the evidence for an exhumed Devonian petroleum system on Orkney, sourced by lacustrine laminites. Oil seeps are associated with faults and bitumen-bearing sandstones crop out within a broad anticlinal structure. Onshore observations provide insight to Devonian-sourced hydrocarbons, potential Devonian reservoir intervals and migration histories in the Moray Firth to East Shetland Platform offshore to the east.

Possibilities for further Paleozoic prospectivity onshore related to relatively structurally complex settings include a marine Devonian shale source for dry gas in Mesozoic reservoirs of the Weald Basin of southern UK advocated by **Pullan & Butler (2018)** based on isotopic, gas composition and maturity data. The distribution of Paleozoic rocks beneath the Variscan and Acadian unconformities is poorly defined by data and affected by compressional deformation, but this deeper gas source could open up Paleozoic and Triassic plays beneath the traditional Jurassic play of the Weald Basin.

The small gas fields and oil and gas shows in the Jura documented by **Pullan & Berry (2018)** are typical of late and post-orogenic piedmont and intramontane basins of the internal Variscides in containing coals, high-quality algal-rich lacustrine Autunian and bituminous Stephanian shales, in this case sourcing a Triassic reservoir and seal system. Many other similar, small basins exist across France and Germany, with thick coal seams common in Westphalian and Stephanian sequences (e.g. Saar-Lorraine-Nahe basin and coalfield; references summarized in BGS 2017).

Burial and uplift history linked to prospectivity

Increased understanding of structural evolution, source rock distribution and maturity highlighted by regional studies has led to an appreciation of the variability in timing and amount of thermal maturation, source rock productivity and resultant oil and gas migration and accumulation. **Besly (2018)** documents variability in basin modelling studies and that simple rift and sag models of basin evolution are oversimplified for the Carboniferous of the Southern North Sea and onshore UK. Variability in basin evolution, maturation and migration from the Outer Moray Firth, Forth Approaches and to south of the Mid North Sea High has been documented by Monaghan *et al.* (2015, 2016, 2017), Vincent (2015, 2016), IGI Ltd (2017) and **Patruno *et al.* (2018)**. Ter Borgh *et al.* (2018*a*, *b*), De Bruin *et al.* (2015) and Schroot *et al.* (2006) give basin modelling examples from the Dutch sector.

Onshore, **Pullan & Butler (2018)** report on basin modelling that indicates gas-mature Paleozoic source rocks in the centre of the Weald Basin from Jurassic times. **Pullan & Berry (2018)** incorporated fold and thrust models in their basin modelling of a Jura petroleum system that is mature for gas from the Jurassic–Cretaceous with maturity progressively decreasing to the west and SW.

The papers and reports summarized in this section above illustrate how detailed local studies constrained by observations of semi-regional basin history allow improved understanding of Paleozoic or Paleozoic-sourced oil and gas accumulations.

Current exploration of Paleozoic plays

In the 18 months of 2016–17 since this volume was initiated, and despite company reorganizations owing to a sustained low oil price, research, exploration and development of Paleozoic plays of NW Europe have returned some successes. For example, in the UK sector the Cygnus gas field came onstream and the Ruby discovery (Rotliegend sandstone) was announced straddling the Dutch/German sectors. Exploration drilling was active in the Barents Sea, including successful appraisal of the Permian–Triassic reservoir in the Alta discovery.

In the UK sector, the OGA has actively promoted a Southern North Sea tight gas strategy that applies to Carboniferous and Permian sandstones (OGA 2017*a*, *b*). An exploration/appraisal well to test the deep Namurian tight gas play has been drilled by BP at Ravenspurn, Block 43/26, and the Grove Deep prospect by Spirit Energy. The UK 29th offshore licensing round in 2016/17 focussed on frontier areas and, of the licences awarded, the focus for Paleozoic plays was on the southern side of the Mid North Sea High (the north of Quadrants 41–44, the south of Quadrants 35–38). Onshore UK, permissions have been granted for testing of unconventional Namurian shale gas prospects in central England (e.g. Yorkshire, Kirby Misperton; Lancashire, Preston New Road).

Future opportunities

Recently published papers, reports and datasets show progress towards a greater understanding and de-risking of Paleozoic plays in NW Europe. The variety of play fairways through thick successions offering numerous source, reservoir and seal intervals and with variable structural and thermal histories means that a single paradigm shift for Paleozoic exploration is not appropriate. There are, however, a number of emerging, promising plays for which there is now greater data availability and semi-regional knowledge to stimulate and de-risk exploration. Promising plays include:

- The Visean–Namurian play on the southern side of the Mid North Sea High and northern side of the Anglo-Dutch basin in the UK and Netherlands sectors, extending southwards to tight gas plays beneath existing gas fields.
- The Rotliegend play on the northern side of the Silverpit Basin, i.e. extending the Cygnus reservoir interval.
- The Zechstein plays both north and south of the Southern North Sea to onshore gas fields.
- Devonian and Carboniferous sourced plays around the Orcadian Basin, from the south Buchan Basin to the East Shetland Platform, in Paleozoic and younger reservoirs.
- Smaller onshore and structurally complex plays, for example the Westphalian–Autunian late- to post-orogenic basins of the internal and northern

external Variscides (France, Germany, Switzerland, southern Britain).
- Basement plays that involve Paleozoic elements or 'basement' reservoirs, e.g. Clair, Johan Sverdrup.

In addition to conventional Paleozoic plays, exploration of tight oil and gas and shale gas plays will probably form a future opportunity. The long-term use of oil and gas fields and possibly co-development of wells and fields for more widespread energy storage, as well as the use of depleted fields for carbon storage (e.g. IEAGHG 2009; Underhill *et al.* 2009; SCCS 2015) and geothermal energy (e.g. Gluyas *et al.* 2016) may also become economically viable and socially desirable.

Acknowledgements The conference from which this volume originated was run by the Geological Society of London and the Petroleum Group and sponsored by Origo Exploration, EBN, Cairn Energy, Draupner Energy and Nexen. The original additional conference convenors were Hugh Dennis, Henry Allen, Andrea James, Ian Roche, Nick Richardson and Paul Herrington. The Geological Society Publishing House staff are thanked for dealing with numerous queries, particularly Angharad Hills, Tamzin Anderson and Samuel Lickiss. Nick Richardson and Teresa Sabato Ceraldi are thanked as Society Book Editors for this volume.

Numerous industry, academic and retired paper reviewers are thanked for their time and constructive input to the papers within this volume. A. Monaghan publishes this introductory paper with the permission of the Executive Director, British Geological Survey (NERC).

Funding The authors acknowledge financial support from the Natural Environment Research Council (NERC).

References

Alberts, M.A. & Underhill, J.R. 1991. The effect of Tertiary structuration on Permian gas prospectivity, Cleaver bank area, southern North Sea, UK. *In*: Spencer, A.M. (ed.) *Generation, Accumulation and Production of Europe's Hydrocarbons*. Special Publications of the European Association of Petroleum Geoscientists, Memoirs, Oxford University Press, Oxford, **1**, 161–173.

Andrews, I.J., Long, D., Richards, P.C., Thomson, A.R., Brown, S., Chesher, J.A. & Mccormac, M. 1990. *United Kingdom Offshore Regional Report*. The Geology of the Moray Firth, HMSO for the British Geological Survey, London.

Arsenikos, S., Quinn, M., Kimbell, G., Williamson, P., Pharaoh, T. & Monaghan, A. 2018. Structural development of the Devono-Carboniferous plays of the UK North Sea. *In*: Monaghan, A.A., Underhill, J.R., Marshall, J.E.A. & Hewett, A.J. (eds) *Paleozoic Plays of NW Europe*. Geological Society, London, Special Publications, **471**. First published online March 20, 2018, https://doi.org/10.1144/SP471.3.

Bailey, N.J., Burwood, R. & Harriman, G.E. 1990. Application of pyrolysate carbon isotope and biomarker technology to organofacies definition and oil correlation problems in North Sea basins. *Organic Geochemistry*, **16**, 1157–1172.

Bełka, Z., Devleeschouwer, X., Narkiewicz, M., Piecha, M., Reijers, T.J.A., Ribbert, K.-H. & Smith, N.J.P. 2010. Devonian. *In*: Doornenbal, J.C. & Stevenson, A.G. (eds) *Petroleum Geological Atlas of the Southern Permian Basin Area*. EAGE, Houten, 71–79.

Besly, B. 2018. Exploration and development in the Carboniferous of the southern North Sea: a 30year retrospective. *In*: Monaghan, A.A., Underhill, J.R., Marshall, J.E.A. & Hewett, A.J. (eds) *Paleozoic Plays of NW Europe*. Geological Society, London, Special Publications, **471**. First published online May 03, 2018, https://doi.org/10.1144/SP471.10

Besly, B.M. 2005. Late Carboniferous redbeds of the UK southern North Sea, viewed in a regional context. *In*: Collinson, J., Evans, D., Holliday, D. & Jones, N. (eds) *Carboniferous Hydrocarbon Geology: The Southern North Sea and Surrounding Onshore Areas*. Yorkshire Geological Society, Occasional Publications, **7**, 225–226.

BGS 2016. 21CXRM Palaeozoic project reports (9 Central North Sea/Mid North Sea High, 9 Orcadian Basin to Forth Approaches, 6 Greater Irish Sea, 1 overview) and datasets can be downloaded from http://www.bgs.ac.uk/research/energy/petroleumGeoscience/explorationRoadmap.html Also in the open access repository http://nora.nerc.ac.uk/

BGS 2017. *South-West Approaches Study: A Review of Late- and Post- Variscan Basins and Source Potential in Western Europe*. British Geological Survey Commissioned Report, **CR/17/031**. Report produced for the OGA as part of the 21st Century Exploration Road Map (21CXRM), https://www.ogauthority.co.uk/data-centre/interactive-maps-and-tools/

Breunese, J., Anderson, J. *et al.* 2010. Appendix 3. *In*: *Petroleum Geological Atlas of the Southern Permian Basin Area*. EAGE, Houten, 305–314.

Brown, F.J., Astin, T.R. & Marshall, J.E.A. 2018. The Paleozoic petroleum system in the north of Scotland-outcrop analogues. *In*: Monaghan, A.A., Underhill, J.R., Marshall, J.E.A. & Hewett, A.J. (eds) *Paleozoic Plays of NW Europe*. Geological Society, London, Special Publications, **471**. First published online December 19, 2018, https://doi.org/10.1144/SP471.14.

Bruce, D.R.S. & Stemmerik, L. 2003. Carboniferous. *In*: Evans, D., Graham, C., Armour, A. & Bathurst, P. (eds and co-ordinators) *The Millennium Atlas: Petroleum Geology of the Central and Northern North Sea*. Geological Society, London, 83 89.

Butler, M. 2018. Seismostratigraphic analysis of Paleozoic sequences of the Midlands Microcraton. *In*: Monaghan, A.A., Underhill, J.R., Marshall, J.E.A. & Hewett, A.J. (eds) *Paleozoic Plays of NW Europe*. Geological Society, London, Special Publications, **471**. First published online April 30, 2018, https://doi.org/10.1144/SP471.6

Cameron, T.D.J. 1993*a*. 5. Carboniferous and Devonian of the southern North Sea. *In*: Knox, R. & Cordey, W. (eds) *Lithostratigraphical Nomenclature of the UK North Sea*. HMSO for the British Geological Survey, London, 1–118.

CAMERON, T.D.J. 1993*b*. 4. Triassic, Permian and Pre-Permian (Central and Northern North Sea). *In*: KNOX, R. & CORDEY, W. (eds) *Lithostratigraphical Nomenclature of the UK North Sea*. HMSO for the British Geological Survey, London, 1–196.

CAMERON, T.D.J., CROSBY, A., BALSON, P.S., JEFFERY, D.H., LOTT, G.K., BULAT, J. & HARRISON, D.J. 1992*a*. Upper Permian. *In*: *United Kingdom Offshore Regional Report: The Geology of the Southern North Sea*. HMSO for the British Geological Survey, London, 43–54.

CAMERON, T.D.J., CROSBY, A., BALSON, P.S., JEFFERY, D.H., LOTT, G.K., BULAT, J. & HARRISON, D.J. 1992*b*. Lower Permian. *In*: *United Kingdom Offshore Regional Report: The Geology of the Southern North Sea*. HMSO for the British Geological Survey, London, 38–42.

CAMERON, T.D.J., MUNNS, J. & STOKER, S. 2005. Remaining hydrocarbon exploration potential of the Carboniferous fairway. *In*: COLLINSON, J., EVANS, D., HOLLIDAY, D. & JONES, N. (eds) *Carboniferous Hydrocarbon Geology: The Southern North Sea and Surrounding Onshore Areas*. Yorkshire Geological Society, Occasional Publications, **7**, 209–224.

CATTO, R., TAGGART, S. & POOLE, G. 2017. Petroleum geology of the Cygnus gas field, UK North Sea: from discovery to development. *In*: BOWMAN, M. & LEVELL, B. (eds) *Petroleum Geology of NW Europe: 50 Years of Learning – Proceedings of the 8th Petroleum Geology Conference*. Geological Society, London, Petroleum Geology Conference Series, **8**, 307–318, 23 March 2017, https://doi.org/10.1144/PGC8.39

CLARKE, H., EISNER, L., STYLES, P. & TURNER, P. 2014. Felt seismicity associated with shale gas hydraulic fracturing: the first documented example in Europe. *Geophysical Research Letters*, **41**, 8308–8314.

COLLINSON, J. 2005. Dinantian and Namurian depositional systems in the southern North Sea. *In*: COLLINSON, J., EVANS, D., HOLLIDAY, D. & JONES, N. (eds) *Carboniferous Hydrocarbon Geology: The Southern North Sea and Surrounding Onshore Areas*. Yorkshire Geological Society, Occasional Publications, **7**, 35–56.

CONEY, D., FYFE, T.B., RETAIL, P. & SMITH, P.J. 1993. Clair appraisal: the benefits of a co-operative approach. *In*: PARKER, J.R. (ed.) *Petroleum Geology of Northwest Europe: Proceedings of the 4th Conference*. Geological Society, London, Petroleum Geology Conference Series, **4**, 1409–1420, https://doi.org/10.1144/0041409

COOKE-YARBOROUGH, P. & SMITH, E. 2003. The Hewett Fields: Blocks 48/28a, 48/29, 48/30, 52/4a, 52/5a, UK North Sea: Hewett, Deborah, Big Dotty, Little Dotty, Della, Dawn and Delilah Fields. *In*: GLUYAS, J.G. & HICHENS, H.M. (eds) *United Kingdom Oil and Gas Fields, Commemorative Millennium Volume*. Geological Society, London, Memoirs, **20**, 731–739, https://doi.org/10.1144/GSL.MEM.2003.020.01.60

COWAN, G. 1996. The development of the North Morecambe gas field, East Irish Sea Basin, UK. *Petroleum Geoscience*, **2**, 43–52, https://doi.org/10.1144/petgeo.2.1.43

CRAMPTON, C.B. & CARRUTHERS, R.G. 1914. *The Geology of Caithness (sheets 110 and 116 with parts of 109, 115 and 117)*. Memoirs of the Geological Survey, Scotland, HMSO, Edinburgh.

DE BRUIN, G., BOUROULLEC, R. *ET AL*. 2015. *New Petroleum Plays in the Dutch Northern Offshore*. TNO Report, **R10920**.

DE JONG, K., JAARSMA, B., REIJMER, J. & TER BORGH, M. 2016. *Assessing the petroleum potential of Devonian Carbonates in the Mid North Sea area*. Presentation from Msc thesis, https://www.ebn.nl/wp-content/uploads/2016/12/Conference-K.-de-Jong-et-al.-2016.pdf

DUGUID, C. & UNDERHILL, J.R. 2010. Geological Controls on Upper Permian Plattendolomite Formation reservoir propsectivity, Wissey Field, UK Southern North Sea. *Petroleum Geoscience*, **16**, 331–348, https://doi.org/10.1144/1354-0793/10-021

DUNCAN, A.D. & BUXTON, N.W.K. 1995. New evidence for evaporitic Middle Devonian lacustrine sediments with hydrocarbon source potential on the East Shetland Platform, North Sea. *Journal of the Geological Society, London*, **152**, 251–258, https://doi.org/10.1144/gsjgs.152.2.0251

EBN 2015*a*. Poster from EAGE downloaded from the website, https://www.ebn.nl/wp-content/uploads/2014/11/Poster4_EBN_EAGE2015_Dinant.pdf [last accessed 14 December 2015].

EBN 2015*b*. Poster from EAGE downloaded from the website, https://www.ebn.nl/wp-content/uploads/2014/11/Poster6_EBN_EAGE2015_TectD.pdf [last accessed 14 December 2015].

EDWARDS, C.W. 1991. The Buchan Field, Blocks 20/5a and 21/1a, UK North Sea. *In*: ABBOTTS, I.L. (ed.) *United Kingdom Oil and Gas Fields, 25 Years Commemorative Volume*. Geological Society, London, Memoirs, **14**, 253–259.

EXXONMOBIL 2010. Relinquishment Report for Licence P1259 Blocks: 36/15, 36/20, 36/25, 37/11, 37/16, 37/17, 37/18, 37/19, 37/20, 37/21, 37/22, 37/23, 37/24, 37/25, 38/16, 38/21, 38/22, 38/26, 38/27 and 38/28. https://itportal.decc.gov.uk/web_files/relinqs/p1259.pdf

FRASER, A.J. & GAWTHORPE, R.L. 2003. *An Atlas of Carboniferous Basin Evolution in Northern England*. Geological Society, London, Memoirs, **28**, https://doi.org/10.1144/GSL.MEM.2003.028.01.01

FROGTECH GEOSCIENCE 2017. 21XCRM East Shetland Platform Project Phase 1 – SEEBASE Study. Report produced for the OGA as part of the 21st Century Exploration Road Map (21CXRM), https://www.ogauthority.co.uk/data-centre/interactive-maps-and-tools/

GAMBARO, M. & CURRIE, M. 2003. The Balmoral, Glamis and Stirling Fields, Block 16/21, UK Central North Sea. *In*: GLUYAS, J.G. & HICHENS, H.M. (eds). *United Kingdom Oil and Gas Fields Commemorative Millennium Volume*. Geological Society, London, Memoirs, **20**, 395–413.

GAST, R.E., DUSAR, M. *ET AL*. 2010. Rotliegend. *In*: DOORNENBAL, J.C. & STEVENSON, A.G. (eds) *Petroleum Geological Atlas of the Southern Permian Basin Area*. EAGE, Houten, 101–121.

GETECH 2017. *Gravity and Magnetic Data: SW Approaches and Western Europe*. Report produced for the OGA as part of the 21st Century Exploration Road Map (21CXRM), https://www.ogauthority.co.uk/data-centre/interactive-maps-and-tools/

GLENNIE, K., HIGHMAN, J. & STEMMERIK, L. 2003. Permian. *In*: EVANS, D., GRAHAM, C., ARMOUR, A. & BATHURST,

P. (eds and co-ordinators) *The Millennium Atlas: Petroleum Geology of the Central and Northern North Sea*. Geological Society, London, 8–1, 8–44.

Glennie, K.W. 1998. Permian. *In*: Glennie, K. (ed.) *Introduction to the Petroleum Geology of the North Sea*. Blackwell Science, Oxford, 42–84.

Glennie, K.W. & Underhill, J.R. 1998. The development and evolution of structural styles in the North Sea. *In*: Glennie, K. (ed.) *Introduction to the Petroleum Geology of the North Sea*. Blackwell Science, Oxford, 42–84.

Gluyas, J., Mathias, S. & Goudarzi, S. 2016. North Sea – next life: extending the commercial life of producing North Sea fields. *In*: Bowman, M. & Levell, B. (eds) *Petroleum Geology of NW Europe: 50 Years of Learning*. Geological Society, London, Petroleum Geology Conference Series, **8**, 27 October 2016, https://doi.org/10.1144/PGC8.30

Guariguata-Rojas, G.J. & Underhill, J.R. 2017. Implications of Early Cenozoic uplift and fault reactivation for carbon storage in the Moray Firth Basin. *Interpretation*, **5**, 1–21.

Haig, D.B., Pickering, S.C. & Probert, R. 1997. The Lennox oil and gas Field. *In*: Meadows, N.S., Trueblood, S.P., Hardman, M. & Cowan, G. (eds) *Petroleum Geology of the Irish Sea and Adjacent Areas*. Geological Society, London, Special Publications, **124**, 417–436, https://doi.org/10.1144/GSL.SP.1997.124.01.25

Hallett, D., Durant, G.P. & Farrow, G.E. 1985. Oil exploration and production in Scotland. *Scottish Journal of Geology*, **21**, 547–570.

Harker, S.D., Green, S.C.H. & Romani, R.S. 1991. The Claymore Field, Block 14/19, UK North Sea. *In*: Abbotts, I.L. (ed.). *United Kingdom Oil and Gas Fields, 25 Years Commemorative Volume*. Geological Society, London, Memoirs, **14**, 269–278, https://doi.org/10.1144/GSL.MEM.1991.014.01.33

Heckel, P.H. & Clayton, G. 2006. The Carboniferous System. Use of the new official names for the subsystems, series and stages. *Geologica Acta*, **4**, 403–407.

Hennissen, J.A.I., Hough, E., Vane, C.H., Leng, M.J., Kemp, S.J., & Stephenson, M.H. 2017. The prospectivity of a potential shale gas play: an example from the southern Pennine Basin (central England, UK). *Marine and Petroleum Geology*, **86**, 1047–1066, https://doi.org/10.1016/j.marpetgeo.2017.06.033

Hillier, S. & Marshall, J.E.A. 1992. Organic maturation, thermal history and hydrocarbon generation in the Orcadian Basin, Scotland. *Journal of the Geological Society, London*, **149**, 491–502, https://doi.org/10.1144/gsjgs.149.4.0491

Hillis, R.R., Thomson, K. & Underhill, J.R. 1994. Quantification of Tertiary Erosion in the Inner Moray Firth using sonic velocity data from the Chalk and the Kimmeridge Clay. *Marine and Petroleum Geology*, **11**, 283–293.

Holliday, D.W. & Molyneux, S.G. 2006. Editorial statement: new official names for the subsystems, series and stages of the Carboniferous System – some guidance for contributors to the Proceedings. *Proceedings of the Yorkshire Geological Society*, **56**, 57–58, https://doi.org/10.1144/pygs.56.1.57

IEAGHG 2009. *CO_2 Storage in Depleted Gas Fields*. Report number **2009/01**, http://hub.globalccsinstitute.com/sites/default/files/publications/95786/co2-storage-depleted-gas-fields.pdf

IGI Ltd 2017. Source rock potential of the East Shetland Platform Area. Report produced for the OGA as part of the 21st Century Exploration Road Map (21CXRM), https://www.ogauthority.co.uk/data-centre/interactive-maps-and-tools/

Jones, J.M. & Scott, P.W. 1980. Progress report on fossil fuels – exploration and exploitation. *Proceedings of the Yorkshire Geological and Polytechnic Society*, **42**, 581–593.

Kearsey, T.I., Millward, D., Ellen, R., Whitbread, K. & Monaghan, A.A. 2018. Revised stratigraphic framework of pre-Westphalian Carboniferous petroleum system elements from the Outer Moray Firth to the Silverpit Basin, North Sea, UK. *In*: Monaghan, A.A., Underhill, J.R., Marshall, J.E.A. & Hewett, A.J. (eds) *Paleozoic Plays of NW Europe*. Geological Society, London, Special Publications, **471**. First published online May 03, 2018, https://doi.org/10.1144/SP471.11.

Kombrink, H., Besly, B.M. et al. 2010. Carboniferous. *In*: Doornenbal, J.C. & Stevenson, A.G. (eds) *Petroleum Geological Atlas of the Southern Permian Basin Area*. EAGE, Houten, 81–99.

Leeder, M.R. & Boldy, S.A.R. 1990. The Carboniferous of the Outer Moray Firth Basin, quadrants 14 and 15, Central North Sea. *Marine and Petroleum Geology*, **7**, 29–37.

Mark, D.F., Green, P.F., Parnell, J., Kelley, S.P., Lee, M.R. & Sherlock, S.C. 2008. Late Palaeozoic hydrocarbon migration through the Clair field, West of Shetland, UK Atlantic margin. *Geochimica et Cosmochimica Acta*, **72**, 2510–2533, https://doi.org/10.1016/j.gca.2007.11.037

Marshall, J.E.A. 1998. The recognition of multiple hydrocarbon generation episodes: an example from Devonian lacustrine sedimentary rocks in the Inner Moray Firth, Scotland. *Journal of the Geological Society, London*, **155**, 335–352, https://doi.org/10.1144/gsjgs.155.2.0335

Marshall, J.E.A. & Hewett, A.J. 2003. Devonian. *In*: Evans, D., Graham, C., Armour, A. & Bathurst, P. (eds and co-ordinators) *The Millennium Atlas: Petroleum Geology of the Central and Northern North Sea*. Geological Society, London, 65–81.

Marshall, J.E.A., Rogers, D.A. & Whiteley, M.J. 1996. Devonian marine incursions into the Orcadian Basin, Scotland. *Journal of the Geological Society, London*, **153**, 451–466.

Marshall, J.E.A., Glennie, K.W., Astin, T.R. & Hewett, A.J. 2018. The Old Red Group (Devonian) – Rotliegend (Permian) unconformity in the Inner Moray Firth. *In*: Monaghan, A.A., Underhill, J.R., Marshall, J.E.A. & Hewett, A.J. (eds) *Paleozoic Plays of NW Europe*. Geological Society, London, Special Publications, **471**. First published online June 22, 2018, https://doi.org/10.1144/SP471.12.

Meadows, N.S., Trueblood, S., Hardman, M. & Cowan, G. 1997. *Petroleum Geology of the Irish Sea and Adjacent Areas*. Geological Society, London, Special Publications, **124**, https://doi.org/10.1144/GSL.SP.1997.124.01.26

Miller, H. 1841. *The Old Red Sandstone, or New Walks in an Old Field*. Johnstone, Edinburgh.

Milton-Worssell, R., Smith, K., Mcgrandle, A., Watson, J. & Cameron, D. 2010. The search for a Carboniferous petroleum system beneath the Central North Sea. *In*: Vining, B.A. & Pickering, S.C. (eds) *Petroleum Geology: From Mature Basins to New Frontiers – Proceedings of the 7th Petroleum Geology Conference*, The Geological Society, London, 57–75, https://doi.org/10.1144/0070057

Monaghan, A.A., Arsenikos, S. *et al.* 2015. *Palaeozoic Petroleum Systems of the Central North Sea/Mid North Sea High.* British Geological Survey Commissioned Report, **CR/15/124**, http://nora.nerc.ac.uk/516766/

Monaghan, A.A., Johnson, K. *et al.* 2016. *Palaeozoic Petroleum Systems of the Orcadian Basin to Forth Approaches, Quadrants 6–21, UK.* British Geological Survey Commissioned Report, **CR/16/038**, http://nora.nerc.ac.uk/id/eprint/516781/

Monaghan, A.A., Arsenikos, S. *et al.* 2017. Carboniferous petroleum systems around the Mid North Sea High. *Journal of Marine and Petroleum Geology*, **88C**, 282–302, https://doi.org/10.1016/j.marpetgeo.2017.08.019

Mulholland, P., Esestime, P., Rodriguez, K. & Hargreaves, P.J. 2018. The role of paleo-relief in the control of Permian facies distribution over the Mid North Sea High, UKCS. *In*: Monaghan, A.A., Underhill, J.R., Marshall, J.E.A. & Hewett, A.J. (eds) *Paleozoic Plays of NW Europe*. Geological Society, London, Special Publications, **471**. First published online May 03, 2018, https://doi.org/10.1144/SP471.8

OGA 2017*a*. https://www.ogauthority.co.uk/media/3770/oil-gas-southern-north-sea-tight-gas-strategy.pdf

OGA 2017*b*. https://www.ogauthority.co.uk/media/4521/oga-sns-tight-gas-stimulation-december-2017.pdf

OGA 2017*c*. OGA Regional Geological Mapping Project - Southern North Sea, Central North Sea and Moray Firth, 2017. Maps produced by Lloyds Register under contract to the OGA, https://www.ogauthority.co.uk/data-centre/interactive-maps-and-tools/

Ohm, S.E., Karlsen, D.A., Phan, N.T., Strand, T. & Iversen, G. 2012. Present Jurassic petroleum charge facing Paleozoic biodegraded oil: geochemical challenges and potential upsides, Embla field, North Sea. *AAPG Bulletin*, **96**, 1523–1552.

Oudmayer, B.C. & De Jager, J. 1993. Fault reactivation and oblique-slip in the Southern North Sea. Geological Society, London, Petroleum Geology Conference Series, **4**, 281–1290, 1 January 1993, https://doi.org/10.1144/0041281

Parnell, J. 1983. The distribution of hydrocarbon minerals in the Orcadian Basin. *Scottish Journal of Geology*, **19**, 205–213, https://doi.org/10.1144/sjg19020205

Parnell, J., Baba, M. & Bowden, S. 2017. Emplacement and biodegradation of oil in fractured basement: the 'coal' deposit in Moinian gneiss at Castle Leod, Ross-shire. *Transactions of the Royal Society of Edinburgh, Earth and Environmental Science*, **107**, 23–32.

Patruno, S. & Reid, W. 2016. New plays on the Greater East Shetland Platform (UKCS Quadrants 3, 8–9, 14–16) – part 1: regional setting and a working petroleum system. *First Break*, **34**, 33–45.

Patruno, S. & Reid, W. 2017. New plays on the Greater East Shetland Platform (UKCS Quadrants 3, 8–9, 14–16) – part 2: newly reported Permo-Triassic intra-platform basins and their influence on the Devonian-Paleogene prospectivity of the area. *First Break*, **35**, 59–69.

Patruno, S., Reid, W., Jackson, C.A.-L. & Davies, C. 2017. New insights into the unexploited reservoir potential of the Mid North Sea High (UKCS quadrants 35–38 and 41–43): a newly described intra-Zechstein sulphate–carbonate platform complex. *In*: Bowman, M. & Levell, B. (eds) *Petroleum Geology of NW Europe: 50 Years of Learning*. Geological Society, London, Petroleum Geology Conference Series, **8**, 87–124, 13 June 2017, https://doi.org/10.1144/PGC8.9

Patruno, S., Reid, W., Berndt, C., Feuilleaubois, L. 2018. Polyphase tectonic inversions and hydrocarbon prospectivity: a comparative synopsis for the under-explored platforms of the eastern UK Continental Shelf (Greater East Shetland Platform and Mid North Sea High). *In*: Monaghan, A.A., Underhill, J.R., Marshall, J.E.A. & Hewett, A.J. (eds) *Paleozoic Plays of NW Europe*. Geological Society, London, Special Publications, **471**. First published online May 04, 2018, https://doi.org/10.1144/SP471.9.

Pedersen, J.H., Karlsen, D.A., Lie, J.E., Brunstad, H. & Di Primio, R. 2006. Maturity and source-rock potential of Palaeozoic sediments in the NW European Northern Permian Basin. *Petroleum Geoscience*, **12**, 13–28, https://doi.org/10.1144/1354-079305-666

Pennington, J.J. 1975. The geology of the Argyll Field. *In*: Woodland, A.W. (ed.) *Petroleum Geology of the Continental shelf of NW Europe*. Applied Science, Barking, 285–291.

Peryt, T.M., Geluk, M., Mathiesen, A., Paul, J. & Smith, K. 2010. Zechstein. *In*: Doornenbal, J.C. & Stevenson, A.G. (eds) *Petroleum Geological Atlas of the Southern Permian Basin Area*. EAGE, Houten, 123–148.

Pharaoh, T.C., Vincent, C.J., Bentham, M.S., Hulbert, A.G., Waters, C.N. & Smith, N.J. 2011. *Structure and Evolution of the East Midlands Region of the Pennine Basin*. The British Geological Survey, Keyworth, Nottingham, Subsurface Memoir.

Pharaoh, T.C., Gent, C.M.A. *et al.* 2018. An overlooked play? Structure, stratigraphy and hydrocarbon prospectivity of the Carboniferous in the East Irish Sea-North Channel basin complex. *In*: Monaghan, A.A., Underhill, J.R., Marshall, J.E.A. & Hewett, A.J. (eds) *Paleozoic Plays of NW Europe*. Geological Society, London, Special Publications, **471**. First published online March 22, 2018, https://doi.org/10.1144/SP471.7.

Pletsch, T., Appel, J. *et al.* 2010. Petroleum generation and migration. *In*: Doornenbal, J.C. & Stevenson, A.G. (eds) *Petroleum Geological Atlas of the Southern Permian Basin Area*. EAGE, Houten, 225–253.

Pullan, C. & Berry, M. 2018. Paleozoic sourced oil play in the Jura mountains of France and Switzerland. *In*: Monaghan, A.A., Underhill, J.R., Marshall, J.E.A. & Hewett, A.J. (eds) *Paleozoic Plays of NW Europe*. Geological Society, London, Special Publications, **471**. First published online March 20, 2018, https://doi.org/10.1144/SP471.2.

Pullan, C. & Butler, M. 2018. Paleozoic Gas Potential in the Weald Basin of Southern England. *In*: Monaghan, A.A., Underhill, J.R., Marshall, J.E.A. & Hewett,

A.J. (eds) *Paleozoic Plays of NW Europe*. Geological Society, London, Special Publications, **471**. First published online March 22, 2018, https://doi.org/10.1144/SP471.1.

Richards, P.C. 1985. A Lower Old Red Sandstone lake in the offshore Orcadian Basin. *Scottish Journal of Geology*, **21**, 381–383, https://doi.org/10.1144/sjg21030381.

Richardson, N.J., Allen, M.R. & Underhill, J.R. 2005. Role of Cenozoic fault reactivation in controlling pre-rift plays, and the recognition of Zechstein Group evaporite-carbonate lateral facies transitions in the East Orkney and Dutch Bank basins, East Shetland Platform, UK North Sea. *In*: Doré, A.G. & Vining, B.A. (eds) *Petroleum Geology: North-West Europe and Global Perspectives – Proceedings of the 6th Petroleum Geology Conference*. Geological Society, London, 337–348.

Rodriguez, K., Wrigley, R., Hodgson, N. & Nicolls, H. 2014. Southern North Sea: unexplored multi-level exploration potential revealed. *First Break*, **32**, 107–113.

SCCS 2015. CO2-EOR Joint Industry Project Reports. University of Edinburgh, http://www.sccs.org.uk/expertise/reports/co2eor-joint-industry-project

Schroot, B.M., v. Bergen, F., Abbink, O.A., David, P., v. Eijs, R. & Veld, H. 2006. *Hydrocarbon Potential of the Pre-Westphalian in the Netherlands on- and Offshore – Report of the Petroplay Project*. TNO Report, **NITG 05-155-C**.

Stevens, V. 1991. The Beatrice Field, Block 11/30a, UK North Sea. *In*: Abbotts, I.L. (ed.) *United Kingdom Oil and Gas Fields, 25 Years Commemorative Volume*. Geological Society, London, Memoirs, **14**, 245–252.

Ter Borgh, M.M., Eikelenboom, W. & Jaarsma, B. 2018*a*. Hydrocarbon potential of the Visean and Namurian in the northern Dutch offshore. *In*: Monaghan, A.A., Underhill, J.R., Marshall, J.E.A. & Hewett, A.J. (eds) *Paleozoic Plays of NW Europe*. Geological Society, London, Special Publications, **471**. First published online March 20, 2018, https://doi.org/10.1144/SP471.5.

Ter Borgh, M.M., Jaarsma, B. & Rosendaal, E.A. 2018*b*. Structural development of the Mid North Sea Area, NL North Sea: paleozoic to present. *In*: Monaghan, A.A., Underhill, J.R., Marshall, J.E.A. & Hewett, A.J. (eds) *Paleozoic Plays of NW Europe*. Geological Society, London, Special Publications, **471**. First published online March 20, 2018, https://doi.org/10.1144/SP471.4.

Trewin, N.H. 1985. Orcadian Basin issue. Editorial introduction. *Scottish Journal of Geology*, **21**, 225–226, https://doi.org/10.1144/sjg21030225

Underhill, J.R. 1991. Implications of Mesozoic–Recent basin development in the western Inner Moray Firth, UK. *Marine and Petroleum Geology*, **8**, 359–369.

Underhill, J.R. 2003. The tectonic and stratigraphic framework of the United Kingdom's oil and gas fields. *In*: Gluyas, J.G. & Hichens, H.M. (eds) *United Kingdom Oil and Gas Fields, Commemorative Millennium Volume*. Geological Society, London, Memoirs, **20**, 17–59, https://doi.org/10.1144/GSL.MEM.2003.020.01.04

Underhill, J.R. & Hunter, K.L. 2008. Effect of Zechstein Supergroup (Z1 Cycle) Werrahalit pods on Prospectivity in the Southern North Sea. *AAPG Bulletin*, **92**, 827–851.

Underhill, J.R., Lykakis, N. & Shafique, S. 2009. Turning exploration risk into a carbon storage opportunity in the UK Southern North Sea. *Petroleum Geoscience*, **15**, 291–304, https://doi.org/10.1144/1354-079309-839

US Energy Information Administration (USEIA) 2013. Technically recoverable shale oil and shale gas resources: an assessment of 137 shale formations in 41 countries outside the United States. Report prepared by Advanced Resources International Inc., https://www.eia.gov/analysis/studies/worldshalegas/ [last accessed October 2016]

USGS 1997. *Map Showing Geology, Oil and Gas Fields, and Geologic Provinces of Europe including Turkey*. Compiled by Mark J. Pawlewicz, Douglas W. Steinshouer and Donald L. Gautier. Open file report **97-470I**, https://pubs.usgs.gov/of/1997/ofr-97-470/OF97-470I/

Van Adrichem Boogaert, H.A. & Kouwe, W.F.P. 1993–97. *In*: *Stratigraphic Nomenclature of the Netherlands*, https://www.dinoloket.nl/nomenclature-deep

Van Buggenum, J.M. & Den Hartog Jager, D.G. 2007. Silesian. *In*: Wong, T.E., Batjes, D.A.J. & de Jager, J. (eds). *Geology of the Netherlands*. Royal Netherlands Academy of Arts and Sciences, Amsterdam, 43–62.

Van Hulten, F.F.N. 2010. Geological factors effecting compartmentalization of Rotliegend gas fields in the Netherlands. *In*: Jolley, S.J., Fisher, Q.J., Ainsworth, R.B., Vrolijk, P.J. & Delisle, S. (eds). *Reservoir Compartmentalization*. Geological Society, London, Special Publications, **347**, 301–315.

Van Koeverden, J.H., Karlsen, D.A., Schwark, L., Chpitsglouz, A. & Backer-Owe, K. 2010. Oil-prone Lower Carboniferous coals in the Norwegian Barents Sea: implications for a Palaeozoic petroleum system. *Journal of Petroleum Geology*, **33**, 155–182, https://doi.org/10.1111/j.1747-5457.2010.00471.x

Vincent, C.J. 2015. *Maturity Modelling of Selected Wells in the Central North Sea*. British Geological Survey Commissioned Report, **CR/15/122**, http://nora.nerc.ac.uk/516764/

Vincent, C.J. 2016. *Maturity Modelling of Selected Wells in the Orcadian Basin*. British Geological Survey Commissioned Report, **CR/16/036**, http://nora.nerc.ac.uk/id/eprint/516743/

Waters, C.N., Somerville, I.D. *et al.* (eds) 2011. *A Revised Correlation of Carboniferous Rocks in the British Isles*. Geological Society, London, Special Report, **26**.

Wilson, G.V., Edwards, W., Knox, J., Jones, R.C.B. & Stephens, J.V. 1935. *The Geology of the Orkneys*. Memoirs of the Geological Survey Scotland. HMSO, Edinburgh.

Wood, I. 2014. UKCS Maximising Economic Recovery Review: Final Report 'The Wood Review', https://www.gov.uk/government/uploads/system/uploads/attachment_data/file/471452/UKCS_Maximising_Recovery_Review_FINAL_72pp_locked.pdf [last accessed December 2016]

Yaliz, A.M. 1997. The Douglas Oil Field. *In*: Meadows, N.S., Trueblood, S.P., Hardman, M. & Cowan, G. (eds) *Petroleum Geology of the East Irish Sea and Adjacent Areas*. Geological Society, London, Special Publications, **124**, 399–416, https://doi.org/10.1144/GSL.SP.1997.124.01.24

Exploration and development in the Carboniferous of the Southern North Sea: a 30-year retrospective

BERNARD BESLY

School of Geography, Geology and the Environment, University of Keele, Keele, Staffordshire ST5 5BG, UK

b.besly@keele.ac.uk

Abstract: A review is presented of the progress of exploration for, and development of, gas fields in the Carboniferous of the UK Southern North Sea in the period since the first significant discoveries were made in 1984. The outcomes of such exploration have generally failed to live up to high initial expectations and exploration targeting of the Carboniferous has declined, the objective having come to be seen by many as difficult and risky. This review includes a summary of the published consensus regarding elements of the Carboniferous petroleum system and discusses the reasons for the decline in interest, which encompass geological complexity, interpretational and operational problems and other non-technical factors. Five areas of Carboniferous petroleum geology are identified in which the currently accepted status quo is open to challenge. More detailed discussion of these leads to the following general conclusions: (1) the distribution of source rocks and their maturation history remains poorly understood, largely as a result of the hitherto unquestioned acceptance that Westphalian coals have acted as the dominant gas source; (2) in many early wells the combination of formation damage and shortcomings in petrophysical data acquisition and evaluation has resulted in a failure to identify potential pay in low permeability formations and an overemphasis on the importance of channel sand bodies as reservoir objectives; (3) the controls on seal capacity and integrity within the Carboniferous succession have been little studied and, as a result, an unduly pessimistic view of intra-Carboniferous sealing potential has prevailed; (4) the distribution of sub-basin depocentres, and thus of basinal shale source rocks and potential hydrocarbon migration paths, remains poorly understood; and (5) conceptual models of the large-scale tectonic history of the Carboniferous basin complex have failed to evolve from early and simplistic rift and sag models, which do not adequately explain the observed distribution of stratigraphic thicknesses and are inconsistent with some published burial histories.

When exploration moved offshore into the Southern North Sea in the early 1960s, the impetus was provided by the discovery of Carboniferous-sourced gas in the lower Permian aeolian sandstone reservoirs of the Rotliegend. At this time there had been a 20-year history of exploration success targeting Carboniferous sandstone reservoirs in the English East Midlands; there were also well-known occurrences of potentially rich oil source rocks in the lower Carboniferous of the Midland Valley of Scotland and in the basal Namurian of northern England (Falcon & Kent 1960; Corfield in press). It is therefore not surprising that a number of the earliest wells drilled in the UK offshore sector (38/29-1, 44/2-1 and 53/10-1, all in 1965; and 41/18-2 in 1966) targeted Carboniferous objectives where these were accessible at the margins of the Mesozoic and Cenozoic basin. These proved disappointing: the sandstone reservoirs encountered were tight and early indications suggested the presence of only immature gas-prone source rock. Only in well 41/18-2 was any flow obtained from a Carboniferous reservoir, and that was at trivial rates. With the discovery of the major Rotliegend gas accumulations in the West Sole, Viking and Leman fields, the Carboniferous ceased to be an exploration target, although significant penetrations continued to be made in bottom-hole sections, which generally showed a lack of reservoir development. Between the end of the 1960s and the early 1980s, the consensus in the exploration and production community was that, although the Carboniferous provided a world-class, regionally developed source rock, it formed economic basement as far as exploration objectives were concerned.

Renewed interest in the prospectivity of the Carboniferous in the early 1980s was driven by a number of factors. First, it was recognized that a number of gas-bearing sandstones in supposed Rotliegend penetrations in Quadrants D and F of the Dutch offshore area were in fact of late Carboniferous age. Second, few obvious dip closures remained to be drilled in the area of Rotliegend reservoir development. Third, academic research in the UK highlighted the presence of sand-rich potential reservoir objectives in the red bed successions in the youngest Carboniferous. Fourth, and most importantly, the economic framework for gas exploration in the

From: Monaghan, A. A., Underhill, J. R., Hewett, A. J. & Marshall, J. E. A. (eds) 2018. *Paleozoic Plays of NW Europe*. Geological Society, London, Special Publications, **471**, 17–64.
First published online May 3, 2018, https://doi.org/10.1144/SP471.10

Southern North Sea changed: it had become evident that a UK gas supply shortfall might occur from the year 2000 and the newly elected Conservative government decided to pursue a policy of deregulation and, ultimately, denationalization. These actions led to potentially more competitive gas prices and to the opening up of new exploration acreage, acting as a spur for a renewed drilling campaign targeting the Carboniferous. After the award of licences in the 8th Round, a flurry of drilling led to the announcement of the discovery of the first four significant Carboniferous fields (Chiswick, Murdoch, Boulton and Ketch) in the space of only 19 days in November 1984, initiating a scramble for acreage in the ensuing 9th and 10th Rounds.

The objective of this paper is to review what has happened since then, both in terms of the operational experience gained and of the understanding of plays within the Carboniferous. Although mention will be made of results in the Netherlands and Germany, the paper will concentrate on UK acreage. A brief review of the play elements and exploration results is followed by some remarks on the production performance of the early fields and on the combination of unfavourable circumstances that led to delays in the commercialization of Carboniferous discoveries and the growth of a general perception that the Carboniferous represents a risky and 'difficult' play. In the following part of the paper a number of general beliefs that have grown up about the Carboniferous geology of the Southern North Sea are identified and discussed. Because of the wide informal currency these beliefs have gained, they are referred to here as 'Founding Myths'. It is the author's belief that the correct analysis of the subjects of these myths will lead to the identification of renewed prospectivity and exploration success in what are currently regarded as fairly mature to mature areas of the UK Southern North Sea.

Since the initial version of this paper was written, a large number of studies carried out by the British Geological Survey been published and significant compilations of data have been released into the public domain by the UK Oil and Gas Authority (OGA). This work, carried out under the OGA's 21st Century Exploration Road Map initiative, provides an extensive dataset and associated interpretations relating principally to the Mid North Sea High area, but extending to cover the northern edge of the Southern North Sea basin. A summary with extensive bibliography is provided by Monaghan *et al.* (2017). Some elements of this work and the newly released data have been incorporated into this paper during post-review revision. To fully incorporate this work would have been neither feasible nor desirable, providing as it does a wide-ranging overview that starts to address some of the issues raised here.

Exploration history from the mid-1980s

Large areas of acreage in which the Carboniferous was seen as the main objective were made available in the 8th to 10th Rounds of UK offshore licencing (1982–1987). This acreage lay principally to the north of the established Rotliegend fairway in UK Quadrants 42–44 and the northern part of Quadrant 49. In this area it was already known that the Leman Sandstone reservoir passes laterally into the mudstone and halite succession of the Silverpit Formation, providing a potential topseal to reservoirs in the Carboniferous. Carboniferous exploration was not limited to these areas. A combination of government initiatives and the desire to identify additional reserves close to the existing infrastructure also led to the drilling of a number of deep Carboniferous penetrations below the existing Rotliegend fields. In the period between 1984 and 2014 *c.* 143 exploration wells were drilled to test Carboniferous objectives in the UK sector (Fig. 1). These resulted in 37 discoveries, of which 27 named fields have been placed on production with an estimated recoverable volume of 3.6 TCF of gas (Table 1). The results of drilling, combined with those of bottom-hole penetrations of earlier wells and regional seismic interpretation, showed that the Southern North Sea area was underlain by a thick Carboniferous succession, folded into a series of NW- to SE-trending major anticlines and synclines (Fig. 1). Valid traps with Carboniferous reservoirs occur in faulted blocks and small-scale folds superimposed on these major structures, particularly in areas where these have been modified by Mesozoic or Cenozoic transpression (Fig. 2). In the majority of cases the ultimate topseal is indeed provided by the Silverpit Formation. However, in a number of areas pay has been encountered in Carboniferous reservoirs even where Rotliegend sandstones are present. In some of these (Wollaston, Ravenspurn, Johnston), this is the result of a gas column continuing downwards where the Rotliegend reservoir is very thin. However, in the onshore Saltfleetby Field (Hodge 2003) and the offshore discoveries in wells 48/23-3 (Blythe) and 48/24a-3 (Wherry), it is clear that the Carboniferous contains seals that are able to isolate a deep reservoir layer from the regional Rotliegend sand blanket.

Stratigraphy

The Carboniferous succession encountered in the Southern North Sea (Fig. 3) is similar to that present in adjacent onshore areas (Besly 1998; Kombrink *et al.* 2010; northern England, Fraser & Gawthorpe 2003; Netherlands, Geluk *et al.* 2007; van Buggenum & den Hartog Jager 2007; Kombrink 2008).

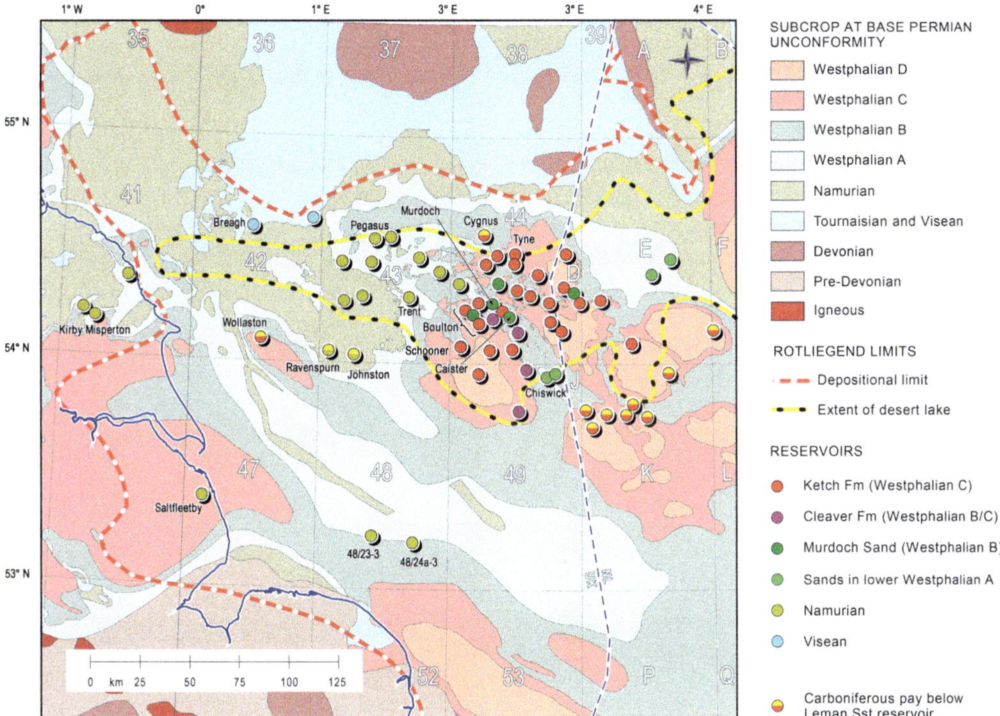

Fig. 1. Carboniferous subcrop below the base Permian unconformity in the UK Southern North Sea and adjacent areas showing established gas fields, some key discoveries in Carboniferous reservoirs and the limits of Rotliegend facies belts that control the presence of a topseal at the unconformity. Named wells and fields are those mentioned in the text. Subcrop map reproduced with permission from Kombrink (2008).

This major basin fill, deposited over a period of *c.* 60 Ma between *c.* 360 and 300 Ma, consists of three stratigraphic elements: (1) the infill in early Carboniferous times of a series of major rift basins initiated during the late Devonian; (2) a thick and broad regional drape, generally interpreted as a post-rift sag succession, deposited in mid-Carboniferous times; and (3) a dominantly red bed succession associated with the early stages of Variscan basin inversion.

In this paper the footwall highs of the rift basins are generally referred to as ‘blocks’ and the rift depocentres as ‘basins’. This is in keeping with a widely used informal classification introduced by Miller & Grayson (1982), which usefully skirts around the problem of the exact structural description of complex basin geometries that are not known in detail at depth, and in which the shapes of shallow water areas associated with the footwall highs were subject to change as carbonate ramps evolved into flat-topped platforms in the later Visean. The aggregate thickness of the Carboniferous succession originally deposited in some parts of the basins may have been >6 km. Although large thicknesses have been removed during end-Carboniferous inversion, very thick successions remain. As a result, individual well penetrations seldom prove a significant proportion of the entire basin fill and the base of the Carboniferous has only been penetrated in a few places offshore, all on the margins of the Mid North Sea High (e.g. wells 42/10b-2, 44/2-1, E2-1 and E6-1). The nature of the deeper parts of the basin fills remains virtually unknown, even in the onshore areas (cf. Hughes *et al.* 2018; Clarke *et al.* in press).

This fairly simple model of basin evolution, involving rift, post-rift and deformation phases, is widely accepted and forms the basis for the following generalized description of the stratigraphic evolution. It should be noted, however, that basin formation in the southern parts of the UK and the Netherlands has unambiguously been shown to involve flexural subsidence and the formation of a foreland basin (Burgess & Gayer 2000; Kombrink *et al.* 2008). It proves difficult to reconcile these subsidence histories with those observed in northern England – the classic rift and sag area – where the unconformities associated with foreland bulge migration predicted by Burgess & Gayer (2000)

Table 1. *Carboniferous gas fields in the UK Southern North Sea*

Field	Discovery year	Discovery well	Start of production	Published account(s)
Boulton B	1984	44/21-2	1997	Conway & Valvatne (2003*a*)
Boulton H	1989	44/21a-4	2004	Cooper *et al.* (2005)
Breagh	1997	42/13-2	2013	Documents at www.sterling-resources.com; [McPhee *et al.* (2008)]
Caister C	1985	44/23-4	1993	Ritchie & Pratsides (1993)
Cavendish	1989	43/19-1	2007	[Kersten *et al.* (2013)]
Chiswick	1984	49/04-1	2007	Nesbit & Overshott (2010)
Hawksley	1998	44/17a-4	2002	Cooper *et al.* (2005)
Katy	2007	44/19b-6	2013	
Kelvin	2005	44/23b-11	2007	
Ketch	1984	44/28-1	1999	
Kilmar	1992	43/22-1	2006	
McAdam	1990	44/17-1	2003	Cooper *et al.* (2005)
Minke	1993	44/24-4	2007	
Munro	2004	44/17b-7	2005	
Murdoch	1984	44/22-1	1993	Conway & Valvatne (2003*b*)
Murdoch K	2002	44/22a-10Z	2003	Cooper *et al.* (2005)
Orca	1993	44/29b-4	2013	
Rita	2008	44/22c-9	2009	
Saltfleetby	1996	Saltfleetby-1Z	1999	Hodge (2003)
Schooner	1986	44/26-2	1996	Moscariello (2003), Mijnssen (1997)
Topaz	1987	49/01-3	2009	
Trent	1991	43/24-1	1996	O'Mara *et al.* (1999, 2003*a*)
Tyne North	1993	44/18-2	1996	O'Mara *et al.* (2003*b*)
Tyne South	1992	44/18-1	1996	O'Mara *et al.* (2003*b*)
Tyne West	1994	44/18-4A	1996	O'Mara *et al.* (2003*b*)
Watt	1990	44/22b-8	2003	Cooper *et al.* (2005)
Wingate	2008	44/24b-7	2012	

Publications in square brackets describe specific technical issues without providing a full description of the field.

are clearly not developed. Thus the true history of basin development remains unclear, a question that is returned to at the end of the paper.

Generalized stratigraphic evolution

These three main stratigraphic elements were formalized and named as five stratigraphic megasequences in northern England by Fraser & Gawthorpe (2003). Although the applicability of parts of this scheme in the North Sea can be questioned, it forms a good first approximation for a brief summary of the stratigraphic history. Of these five megasequences, the first two correspond to the formation and infill of the early Carboniferous rift systems, and the third and fourth to the mid-Carboniferous post-rift sag phases.

(1) EC1–EC2 (Fammenian–Chadian). In northern England, the earliest parts of the rift basin fills are generally inferred to consist of red beds of Old Red Sandstone facies. Marine transgression during the Tournaisian led to the widespread establishment of carbonate facies as far north as present day latitude 55° N. Shallow water carbonate platforms became established on topographic highs formed at the crests of extensional fault blocks, while the deeper water areas of the rifts developed deep ramp and basinal successions of shale, carbonate mud, deep water mud mounds and carbonate turbidites. To the north of present day latitude 55° N, fluvial and lacustrine conditions prevailed during the lower part of this succession in both block and basinal areas. Little is known of the equivalent succession in the offshore area, except that extremely thick (2–4 km) Upper Devonian Old Red Sandstone successions are seismically imaged in the Mid North Sea High (S. Corfield, 2016, pers. comm.). It is not known whether these late Devonian thicknesses are also present where the very thick Carboniferous successions occur under the main part of the Southern North Sea complex of basins.

(2) EC4–EC6 (Arundian–Brigantian). A major northerly sourced fluvial system became established at some time during the middle part of the Visean, following which deep water sub-basin areas were progressively infilled from north to south.

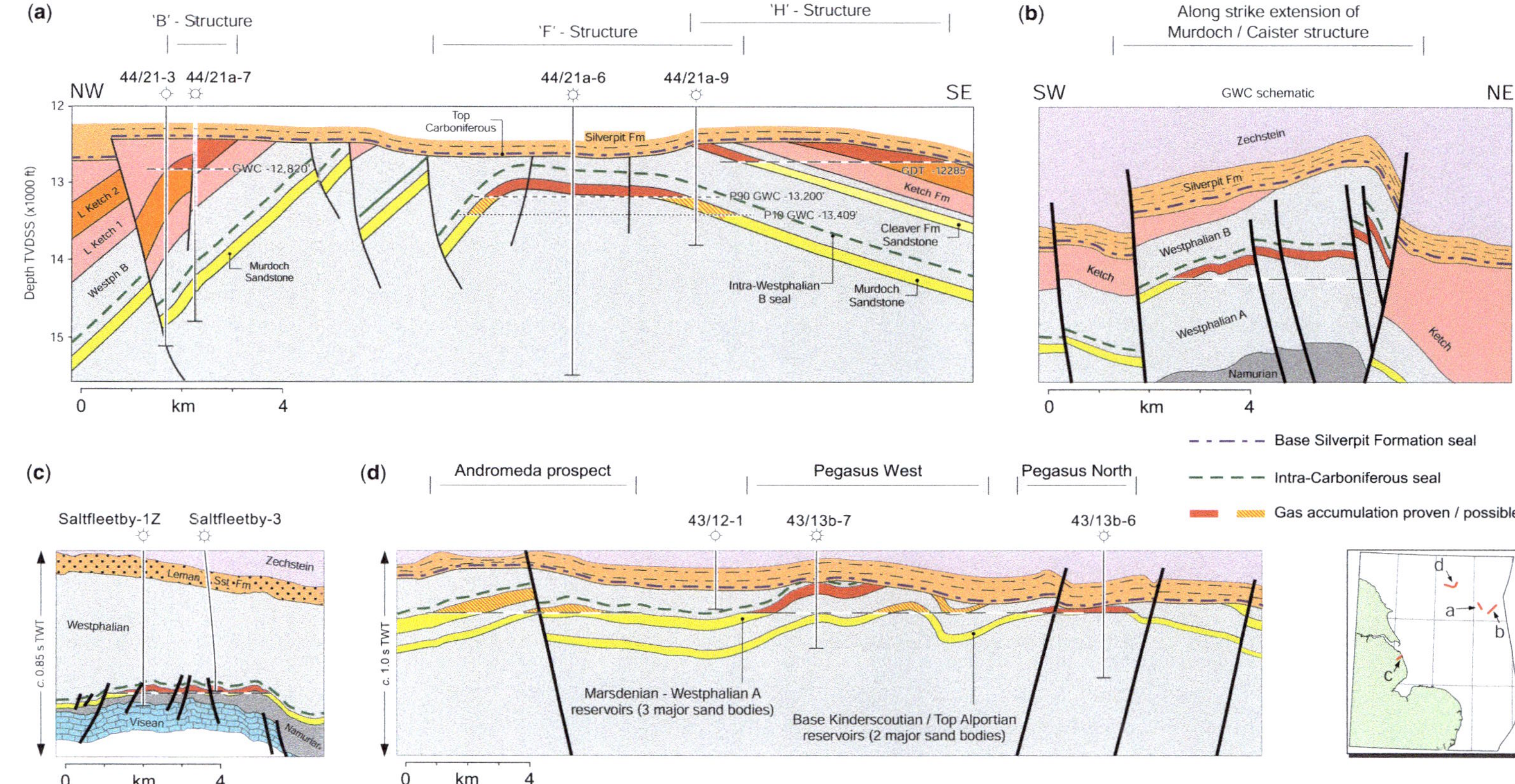

Fig. 2. Examples of Carboniferous trap configurations. (**a**) Boulton Field: Variscan anticlinal trap, structurally modified during Mesozoic–Cenozoic transpression, containing both intra-Carboniferous dip closure (Boulton F) and truncation traps sealed by the Silverpit Formation (Boulton B and H). (**b**) Murdoch and Caister Fields: Variscan anticline modified by the development of a positive flower structure during Mesozoic–Cenozoic transpression; gas in Murdoch Sandstone trapped by a combination of intra-Carboniferous and base Permian seals and lateral fault seal. (**c**) Saltfleetby Field: Variscan drape fold over Lower Carboniferous footwall high, relying entirely on intra-Carboniferous (base Westphalian A) topseal. (**d**) Pegasus Fields: set of Variscan anticlines truncated by, but lacking structural closure at, the base Permian unconformity and reliant on a combination of base Permian and intra-Carboniferous seals. *Sources:* (a) partly after Conway & Valvatne (2003*a*), reproduced with permission of the Petroleum Exploration Society of Great Britain; (b) after Corfield *et al.* (1996); (c) after Hodge (2003); (d) adapted by permission of Centrica from presentation at DEVEX Aberdeen, 20–21 May, 2015.

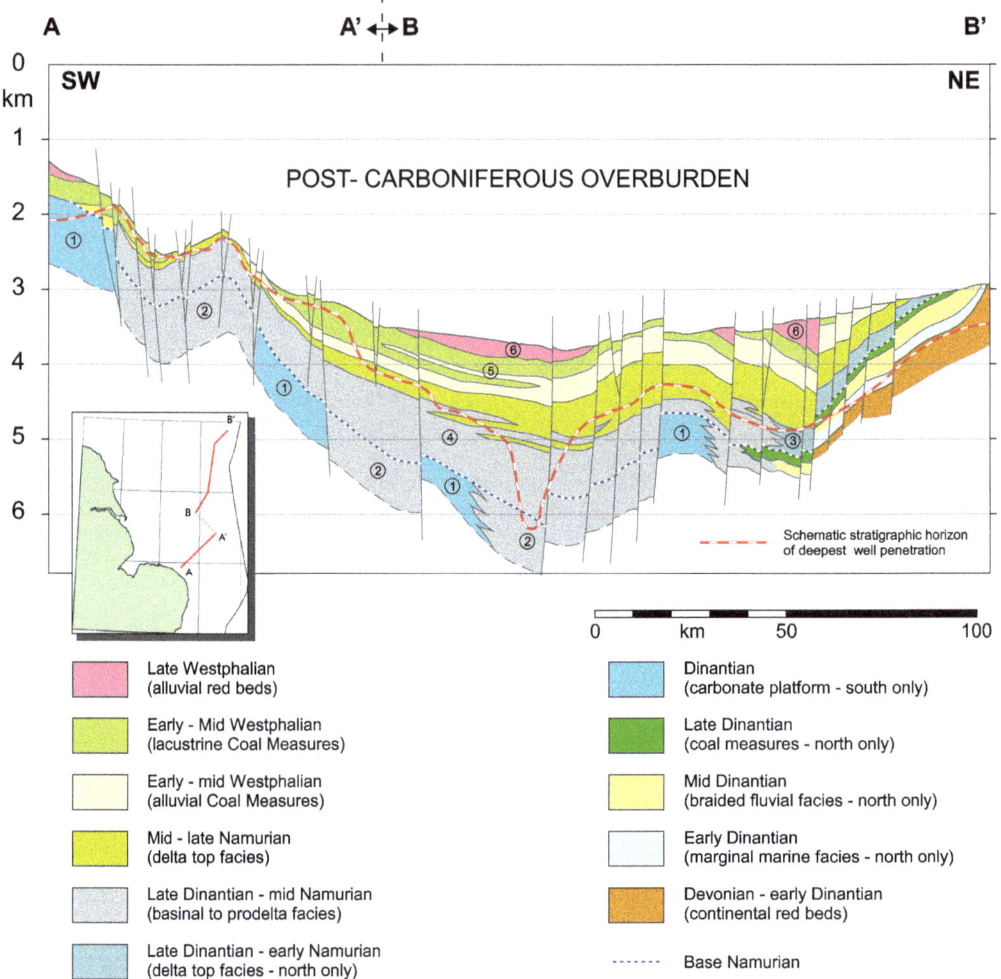

Fig. 3. Generalized stratigraphic cross-section through the Carboniferous succession underlying the Permian and later fill of the Southern North Sea basin. 1, Lower Carboniferous platform carbonate facies developed at basin margins and on intra-basinal footwall highs; except in the extreme south, its distribution is speculative. 2, Lower Carboniferous basinal facies developed in rifted depocentres; distribution speculative. 3, Exact geographical location of deep to shallow water transition in early Namurian not accurately known (see also Fig. 20). 4, Stratigraphic position of deep to shallow water transition in mid- and late Namurian only locally known in a few well penetrations in Quadrant 43. 5, Pronounced proximal to distal decrease in sand content in both Namurian and early to mid-Westphalian from NE to SW. 6, Late Westphalian alluvial red beds restricted to inliers in cores of eroded Variscan synclines.

(3) LC1 (Pendleian–early Westphalian A). In the UK onshore area, active rifting ceased at the end of the Visean: deep water successions deposited after this time represent the infilling of residual bathymetry. The end of the Visean was also marked by a pronounced reduction in carbonate production; carbonate platform sedimentation ceased and shallow marine limestones were thereafter only deposited as a volumetrically insignificant facies associated with sea-level highstands in shoal water delta successions to the north of present day latitude 54° N (the so-called Yoredale facies). Between the mid-Namurian and the early Westphalian, repeated phases of major delta progradation resulted in the elimination of inherited basin bathymetry. These deltas deposited the classic Millstone Grit succession of northern England. During this period, pronounced glacio-eustatic sea-level fluctuations resulted in frequent

short-lived marine flooding events (the so-called 'Marine Bands'). Rapid evolutionary change in goniatite faunas allows exceptionally precise stratigraphic correlation in this succession if suitable macrofossil material can be collected. Infill of deep water areas continued to take place from north to south (Collinson 2005) and the final deep water area was infilled in the early Westphalian A in UK Quadrants 48 and 49 and the southern part of the Netherlands offshore and onshore areas (see well sections in van Adrichem Boogaert & Kouwe 1995).

(4) LC2 (Early Westphalian A–late Westphalian C). Following this shallowing episode a widespread shallow water to subaerial depositional surface became established, consisting of shallow lakes, lacustrine deltas, fluvial channels and mires in which the thick and extensive coal seams of the productive coal measures were deposited. Glacio-eustatic Marine Bands continue to allow precise correlation until the Westphalian B/C boundary. A dominance of northern sediment input led to a marked decrease in sand content from north to south (Quirk 1997; Cole *et al.* 2005), reversed only in the late Westphalian C by the appearance of southerly derived sandstones sourced from uplifts in the approaching Variscan Front (Morton *et al.* 2005).

(5) Inversion megasequence (Late Westphalian C–Stephanian). From earliest Westphalian times fluvial red beds accumulated along some margins of the basin complex (Besly 1988; Mitchell & Owens 1990). These became more extensive from mid-Westphalian B time in response to basin margin uplift associated with the northwards progression of Variscan deformation (Besly 1988) and by late Westphalian C time were the dominant facies in the basin. Red bed deposition, with the development of minor local coal-bearing facies, continued in the UK onshore area until at least the earliest Permian and in north Germany until at least the Stephanian. The stratigraphic relationships in the youngest part of the succession are complicated by poor biostratigraphic resolution and the presence of complex and poorly documented unconformities associated with the early development of Variscan folds.

Stratigraphic problems

As has already been mentioned, even the longest well penetrations of Carboniferous strata only intersect a small proportion of the entire basin fill. With the exception of a few key wells, most penetrations are <300 m. The correct identification of absolute horizon and marker beds in such short sections of thick and usually monotonous clastic successions is fraught with difficulty and ambiguity. In onshore areas of the UK, subdivision and correlation of the deltaic and alluvial successions of pre-mid-Westphalian C age has for many years been based on thin marine flooding surfaces (Marine Bands in the Namurian and Westphalian; widespread thin limestones in the Visean and early Namurian). As these can only be unambiguously identified on the basis of macrofauna, it has proved difficult to identify these in the offshore wells, even where core is available. Any attempt to correlate or identify absolute stratigraphic horizons on the basis of wireline response alone is subject to almost insuperable problems of aliasing in generally repetitive cyclic sequences.

The resolution of the miospore biostratigraphy – the principal useful fossil group – has improved substantially (McLean *et al.* 2005) and this has led to better relative age determinations, but most of the older well penetrations lack palynological analyses of sufficient number, quality and consistency for reliable biostratigraphically-based correlations to be made, and palynomorphs become degraded and difficult to identify once the rocks reach gas-generating maturities. Subdivision and classification in the late Westphalian to Stephanian red bed sequences still relies exclusively on lithostratigraphy. Unless based on extensive mineralogical, petrographic and chemostratigraphic data and clearly linked to regional provenance patterns (Besly 2005), lithostratigraphic units named in the red bed successions may be meaningless.

Early hopes that a combination of fairly simple geochemical indicators used in combination with spectral gamma ray logs (Archard & Trice 1990; Leeder *et al.* 1990) have largely been disappointed (e.g. Bristow & Williamson 1998). However, statistical analysis of large and closely sampled multi-element geochemical datasets has proved to be a powerful tool for local correlation, particularly in the generally unfossiliferous red beds of the youngest Westphalian (Pearce *et al.* 2005*a*). The effectiveness of this technique is enhanced by careful integration with palynological, trace element and isotope geochemical studies, leading to possibilities of regional genetic stratigraphic subdivision (Pearce *et al.* 2005*b*). As yet, there is insufficient chemostratigraphic data in the public domain and this, when coupled with the inadequate availability of reliable palynological data, means that the technique does not yet provide a basis for a regional genetic stratigraphic framework.

In the absence of regional chronostratigraphic markers, attempts to set up a widely applicable genetic stratigraphy that might act as a basis for robust play fairway analysis have been patchy and inconsistent. Preliminary sequence stratigraphic schemes (Quirk 1997; Cole *et al.* 2005) have been restricted to the Westphalian, producing inconsistent results from non-overlapping or incompletely documented well datasets. Other stratigraphic nomenclature schemes have largely – it would seem – been designed to bring some consistency into the nomenclature used in well completion logs (Cameron 1993 for the UK sector; van Adrichem Boogaert & Kouwe 1995 for the Netherlands sector). These schemes have resulted in a hotch-potch of stratigraphic nomenclature, some of it based on age considerations, some on the extrapolation of onshore lithostratigraphic names that are, in practice, difficult to apply away from their well-defined onshore type areas. The result is a stratigraphic framework that lacks precision, fails to account meaningfully for some major observed lateral variations (e.g. the stratigraphic identity in UK well 41/10-1 in fig. 3 of Collinson 2005) and becomes inoperable across the national median lines (Fig. 4).

Major play elements in the Carboniferous

The main elements of the Carboniferous play are summarized in a series of annotated cross-sections (Fig. 5). Apart from sealing and the nature and timing of trap formation, most of the play elements in the offshore area are the same as those in the adjacent UK onshore areas (Fraser & Gawthorpe 2003). The difference in sealing and trapping is the result of the presence of a topseal at the base Permian unconformity, which is absent in large areas of the UK onshore, but widely present in the Dutch and German onshore areas (Kombrink *et al.* 2010). This section of this review summarizes the consensus view provided by previous publications; some of these views are open to challenge and further discussion is provided later in the paper.

Source rocks

Source rocks are documented at all horizons in the Carboniferous succession from the middle part of the Visean up to the base of the red bed succession in the middle Westphalian (Fig. 5a). The thick coal seams in the Westphalian are the best known of these and have widely been assumed to be the main gas source rock in the Southern North Sea (e.g. Cornford 1998; Pletsch *et al.* 2010). Thin coals are, however, present throughout the delta and alluvial plain facies of the upper Visean and Namurian, occurring in progressively older strata to the north, reflecting the earlier onset of shallow water deposition in this area. Oil shales of basinal facies are well known in the English East Midlands, occurring principally in the basal Namurian, but also present at shallower horizons up to the Marsdenian (Fraser *et al.* 1990). Minor oil source potential is also present in the Marine Bands in the Westphalian. Further oil shales of shallow lacustrine facies are present in the upper Visean and lower Namurian, where these are developed in the shoal water Yoredale facies. In general, these are thin and probably insignificant, but, where exceptional local palaeogeographical conditions obtained, thick oil-prone lacustrine shales accumulated, which, in the past, formed the basis for the oil retorting industry in the eastern Midland Valley of Scotland (Loftus & Greensmith 1988; Parnell 1988). It is not known whether any similar exceptional development of lacustrine oil shales is present in the offshore, although such developments have been suggested in UK Quadrant 20 (Bruce & Stemmerik 2003).

Although there is an extensive literature indicating the presence of source rocks in the Carboniferous, there is, with one exception, little published on the quantitative aspects of source rock development. Some indication of organic richness and yield in the basinal shales is given by Fraser *et al.* (1990) and by Pletsch *et al.* (2010), and a compilation of measured and log-derived total organic carbon (TOC) from 31 wells is given by Gent (2015). In general, it seems to have been assumed that the gas-generative potential of coal seams has always been more than sufficient to charge conventional traps and little attention has been paid to the origin of hydrocarbons in areas where the Westphalian coals are absent.

The exception to the lack of detailed study of quantitative aspects of source rock development is provided by a study of the Pendleian to Arnsbergian Bowland Shale Formation in the Widmerpool Basin in the English Midlands (Gross *et al.* 2015). Here, in rocks that have not reached the oil window, the TOC content of basin mudstones and of fine-grained materials intercalated with turbidites varies between 1.3 and 9.1% (average 4.2%), with the hydrogen index ranging from 50 to 425 mg hydrocarbons/g TOC, attesting to the presence of kerogens of both types II and III. The latter type is formed by detrital plant material associated with turbidite inputs inferred to be related to the development of lowstand systems tracts. The comparative lack of oil shows in wells drilled in the offshore successions in the North Sea suggests that such oil-prone source rock developments may be the exception rather the norm, reflecting an exceptional combination of palaeogeographical circumstances (see also Fig. 20 in Fraser *et al.* 1990 and associated discussion therein).

It should be noted that disseminated plant material is widely present throughout the Carboniferous succession. Barnard & Cooper (1983) have suggested that this may be as important as the organic material contained within coal seams as a potential source material. In-seam coal is roughly twice as

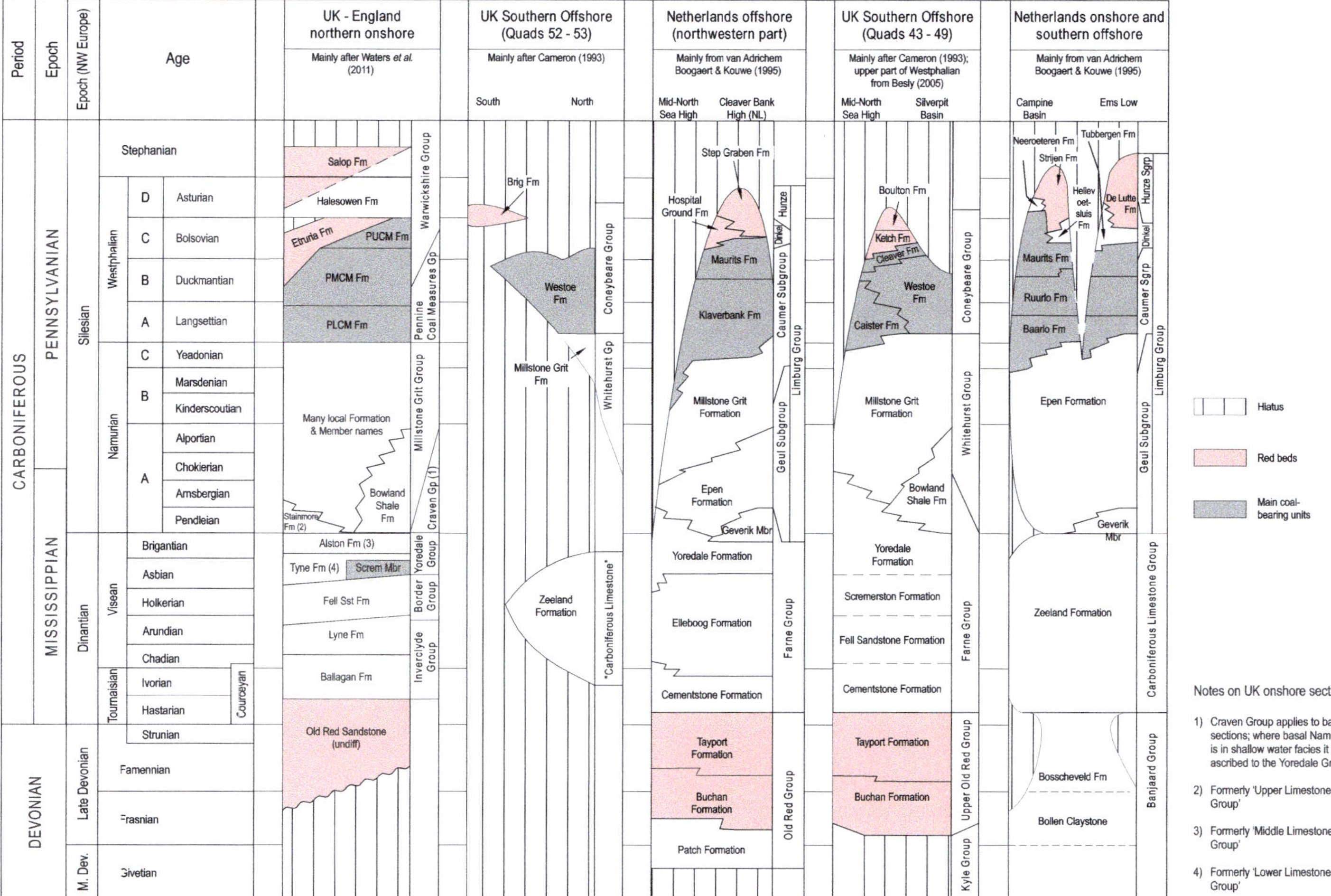

Fig. 4. Summary of Group and Formation names currently used in onshore and offshore areas of the UK and the Netherlands, illustrating the complexity and ambiguity of lithostratigraphic usage. Modified from regional correlation panel in www.dinoloket.nl/carboniferous, with additions and modifications as detailed in the individual columns.

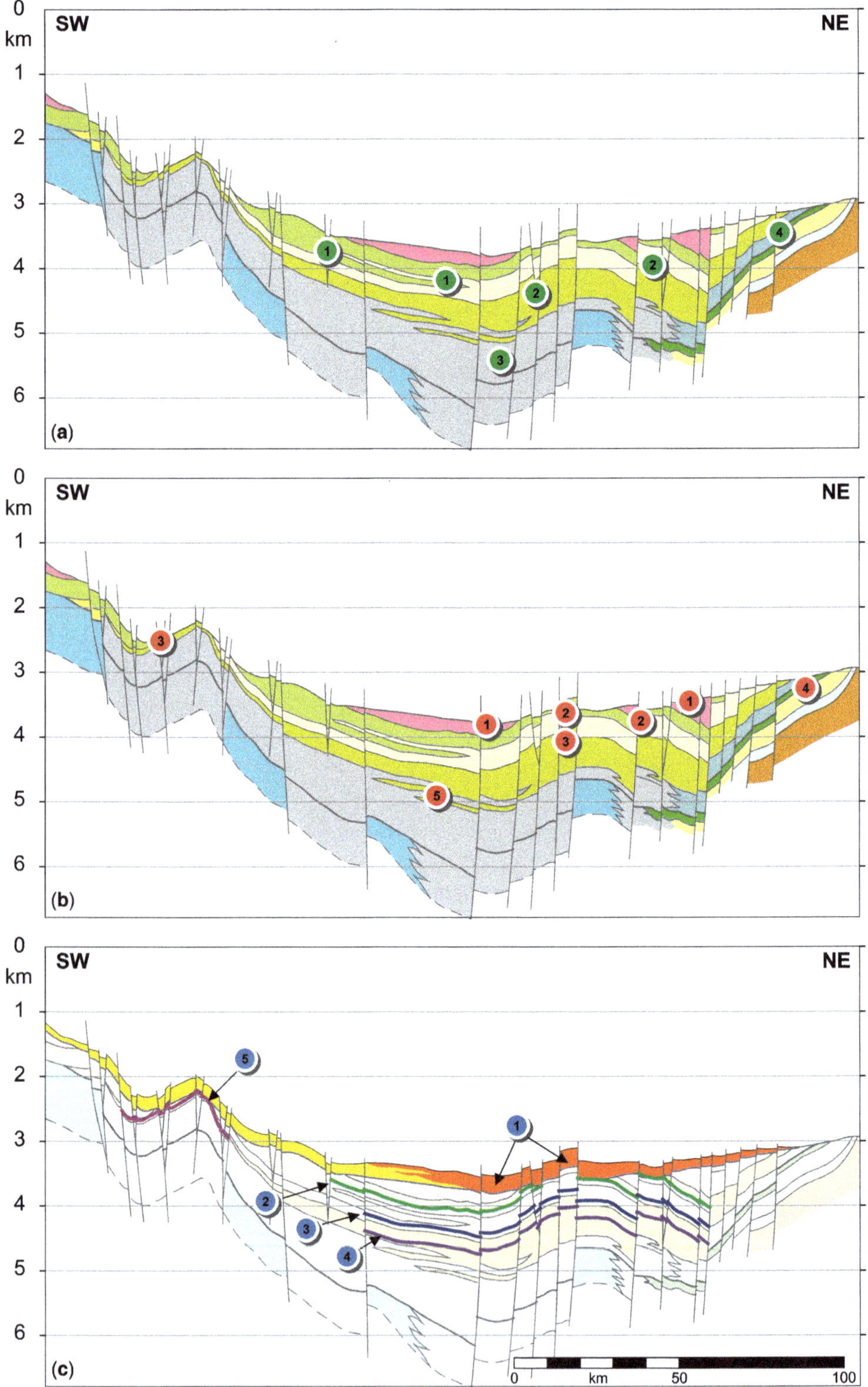

0
km
1
2
3
4
5
6
SW
NE
(a)
(b)
(c)
0
km
50
100

abundant as disseminated plant material in the Namurian C and Westphalian in the Ruhr coalfield (Scheidt & Littke 1994), suggesting that coal seams should dominate gas production in the Westphalian. However, the occurrence of dry gases and the almost total absence of oil in parts of the stratigraphy that lack bedded coal require disseminated plant material to have acted as the dominant source rock in some areas (e.g. North Yorkshire, UK onshore; Hughes *et al.* 2018). Both TOC measurements and log-derived profiles in Gent (2015) suggest the presence of large thicknesses of shales with significant TOC contents in the Visean to early Namurian Scremerston and Yoredale formations and some horizons in the late Namurian to early Westphalian Millstone Grit and Caister formations. Comparison with studied onshore sections suggests that much of this organic content is likely to be disseminated plant material. The question of the relative contribution of Westphalian coals to the generation of currently reservoired gas is returned to in a later section of the paper.

Reservoirs

The most important reservoirs encountered in the Carboniferous of the Southern North Sea occur in fluvial channel sand bodies in the red bed succession in the upper part of the Westphalian (the Ketch Formation in the UK and the Hospital Ground Formation in the Netherlands) and in the basal Westphalian B and Westphalian A (Murdoch Sandstone Member and other unnamed sandstone units of the Caister Formation in UK and the Klaverbank Formation in the Netherlands) (Fig. 5b; Table 1). Further important reservoirs are present in stacked fluvial channel sand bodies and in incised valley fills and associated facies in the Namurian and Visean.

In the early stages of exploration, the prognosis of reservoir development was based on the exposed succession in central and northern England. In retrospect, this analogy turns out to have been misleading. The occurrence of a high net : gross sandstone reservoir in the late Carboniferous red bed succession was predicated on the occurrence of such facies in the Westphalian D to early Permian Enville Formation and related units in the English Midlands. The general log shape of the red bed reservoirs encountered in early wells suggested that this succession had indeed been found offshore (e.g. compare the log shapes of well 44/28-1 in Besly *et al.* 1993 with the Enville Formation logs illustrated in Besly & Cleal 1997), but the drilling of well 44/21-3 demonstrated that the reservoir succession that is now called the Ketch Formation was a sand-rich unit of Westphalian C age, derived from a previously unsuspected northeastern source area and not present in the onshore area. Similarly, the most significant reservoir in the Westphalian – the early Westphalian B Murdoch Sandstone – is not developed onshore. The targeting of sand-rich potential reservoir successions in the upper part of the Namurian was based on the occurrence of thick and extensive sandstones (the Millstone Grit) in the exposed succession of this age onshore. In practice, Millstone Grit sandstones have proved to have relatively poor reservoir properties offshore, but commercially producible reservoirs have been encountered at this stratigraphic horizon, in the Trent and Pegasus fields, in a quartzite facies for which no clear onshore analogue is known.

Seals and traps

Exploration for Carboniferous objectives in the Southern North Sea has hitherto almost exclusively relied on the identification of traps that ultimately have the halite and shale facies of the lower Permian as their topseal (Silverpit Formation in the UK; the Ten Boer Member in the Netherlands) (Figs 2, 5c & 6). There is no UK onshore analogue for this play. The concentration on exploration for traps of this type has largely been driven by the comparative ease with which structures at the base Permian can be mapped and the comparatively low risk attached to this topseal. Although traps of this type may contain elements of Variscan structure, they are essentially formed by folding and the development of positive flower structures associated with Mesozoic or Cenozoic tectonic events. Even in these traps, careful examination of the structure shows the possibility of elements of topseal provided by intra-Carboniferous shales and/or lateral seal provided by fault juxtaposition against the Silverpit Formation or against Zechstein salts (e.g. the Murdoch Field, Conway & Valvatne 2003*b*; Fig. 2b).

Fig. 5. Generalized stratigraphic cross-section annotated with elements of the petroleum system. (**a**) Major source intervals: 1, coals and carbonaceous mudrocks in Westphalian; 2, minor coals and disseminated humic material in middle–upper Namurian; 3, basinal shales in lower Namurian and Visean; 4, coals and oil shales in Visean. (**b**) Main reservoirs: 1, Ketch Formation; 2, Murdoch Sandstone; 3, sands in Caister Formation and upper Namurian; 4, fluvial channel sands in upper Visean; 5, basinal facies in lower Namurian. (**c**) Seals: 1, sub-regional topseal formed by Rotliegend where developed in lacustrine facies of Silverpit Formation; 2–4, mudrocks in Westphalian B (2), basal Westphalian A (3) and Alportian (4), all known to form effective seals in UK Quadrants 43, 44 and 49; and 5, basal Westphalian or uppermost Namurian forming effective seal in UK Quadrant 48 and onshore. Stratigraphic legend as in Figure 3. See text for further discussion.

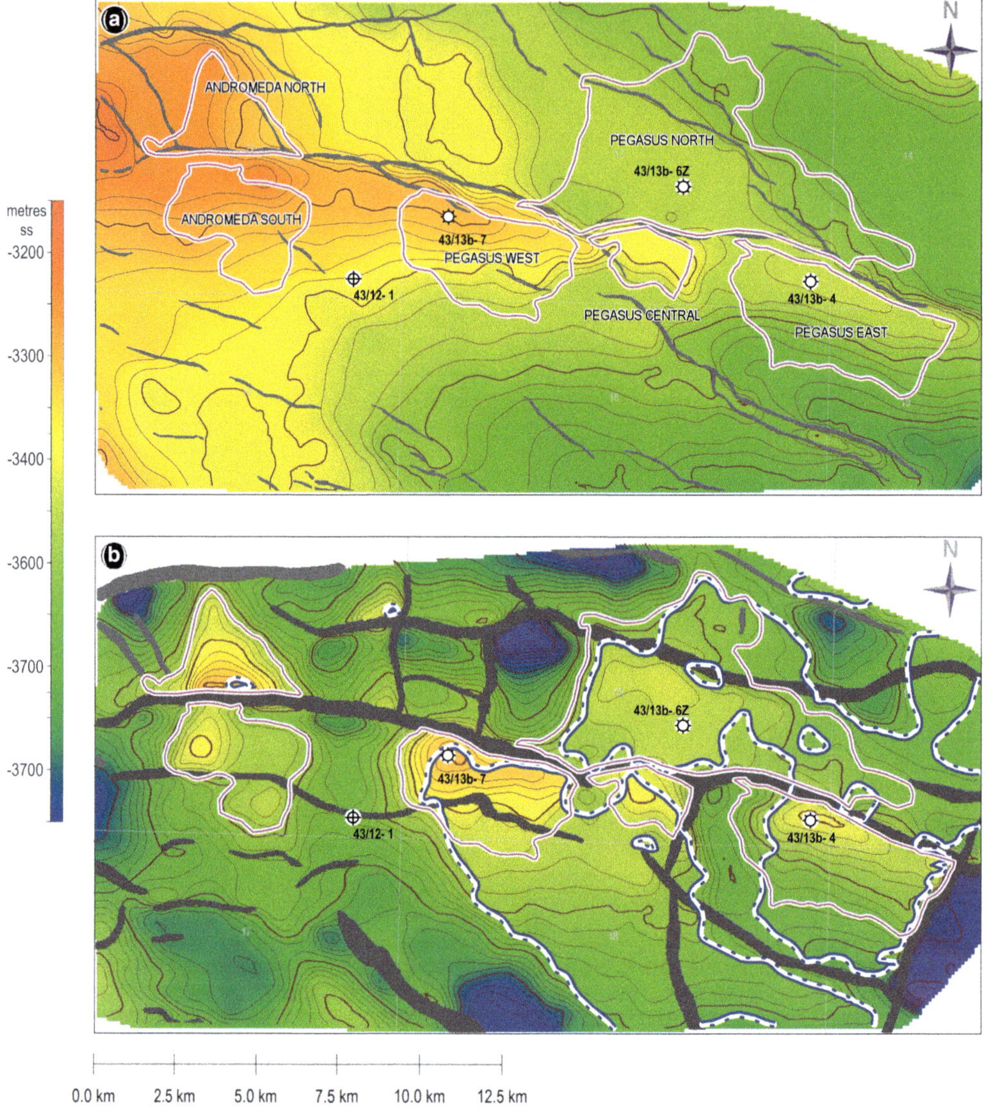

Fig. 6. Example of a gas accumulation requiring a combination of base Permian and intra-Carboniferous seals, Pegasus Field, UK block 43/13. (**a**) Base Permian depth map showing a lack of closure of field and prospect outline polygons. (**b**) Top reservoir map: blue line shows subcrop of Westphalian A seal at base Permian unconformity, with ticks indicating the parts of the accumulations and prospects for which dip closure is provided by the intra-Carboniferous seal. Adapted with permission from 2016 PESGB presentation by Centrica.

Intra-Carboniferous seals are also demonstrable in fields containing multiple pay horizons with different gas–water contacts (e.g. the Boulton Field, Conway & Valvatne 2003*a*: Fig. 2a).

In a number of accumulations, however, it is clear that the Variscan structure provides the dominant or only control on trap geometry and that effective seals must be present within the Carboniferous. In neither the Cavendish Field (Block 43/19) nor the recent Pegasus discovery (Block 43/13) is there a mappable closure at base Permian level and the trapping mechanism must involve a combination of base Permian and intra-Carboniferous seals (Figs 2d & 6), in which a significant element of dip closure is provided by Variscan folding. Variscan dip closures may involve folding related to Variscan shortening (e.g. the Boulton F and Pegasus structures, Figs 2a, d), or may be related to compactional drape over basement highs (e.g. Saltfleetby Field, Fig. 2d). The Kepler accumulation (Block 43/20b) is a Variscan structure entirely sealed by a Carboniferous shale seal (Cameron *et al.* 2005). Intra-Carboniferous seals must also be present in some stratigraphic traps that rely on seat seal (e.g. the Ketex discovery in well 49/3-3).

Given the plentiful occurrence of fine-grained clastics in the Carboniferous and the ample burial they have undergone, there is absolutely no reason why viable seals should not exist throughout the Carboniferous section. This is supported by the obvious occurrence of some very long gas columns in wells that have not been regarded as discoveries. The question of viability and risking of intra-Carboniferous seals is further discussed later in this paper.

Maturation and timing

The consensus view in the literature appears to suggest that the maturation history of Carboniferous source rocks in the Southern North Sea is well known (e.g. Cornford 1998). The deeper parts of the Carboniferous basin fills first entered the hydrocarbon generation window during Carboniferous burial. Any oil and gas that had already migrated during this early burial phase is likely to have escaped during a major phase of fluid expulsion that accompanied the end-Carboniferous Variscan deformation episode (e.g. Hollis 1998). In most of the Southern North Sea area the Carboniferous succession reached its maximum burial during the Mesozoic and Cenozoic (Fig. 7a–c; Cornford 1998,

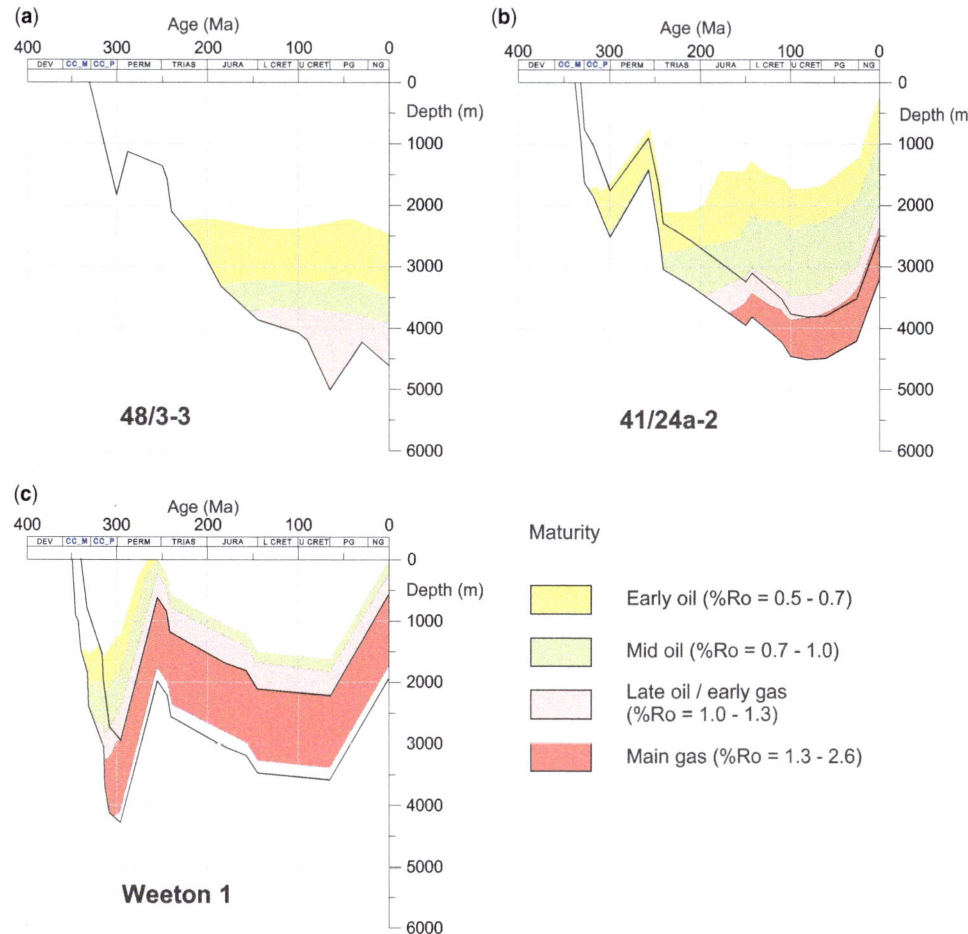

Fig. 7. Burial history plots illustrating varied maturation histories in different parts of the UK Southern North Sea and adjoining onshore areas. (**a**) Sole Pit Basin, after Leeder & Hardman (1990). (**b**) Offshore Cleveland Basin, redrawn with permission from non-proprietary report by Collinson Jones Consultants. (**c**) Onshore Leeds Basin, after Kerr McGee (2002). CC_M, Mississippian; CC_P, Pennsylvanian. See text for discussion.

fig. 11.41; Leeder & Hardman 1990; Pearson & Russell 2000). Source intervals entered the gas window during the Jurassic or Cretaceous and are presently at or near maximum burial over wide areas. Over most of the area neither the volume of generated gas nor the relationship of generation and migration to trap timing have been perceived as risks.

This simple picture is complicated by three factors: (1) the inversion of source kitchens and deeply buried reservoirs in a number of tectonic episodes during the late Jurassic to late Cretaceous; (2) widespread regional tilting in the Neogene, mainly affecting the west of the UK area; and (3) the progressive thinning of the post-Variscan overburden to the north and west.

Inversion episodes of Mesozoic depocentres in the late Jurassic and late Cretaceous (the 'late Cimmerian' and 'sub-Hercynian/Laramide' phases of Glennie & Boegner 1981) are interpreted to have led to multiple episodes of migration and remigration in the Rotliegend reservoirs. During initial deep burial, primary migration of gas occurred from deeply buried source kitchens into marginal areas. Subsequent inversion phases led to remigration into newly formed structures in the cores of the inversion axes. This process appears not to have been documented in detail, but is widely accepted (see summaries in Glennie 1998, pp.164–167; Johnson & Fisher 1998, pp. 516–519).

Neogene regional tilting is well documented, but its causes remain poorly understood (Japsen 1997; Japsen & Chalmers 2000). It is clearly differentiated from the Paleogene regional uplift related to the Iceland thermal plume (White & Lovell 1997). Although the earlier phases of tilting may be related to widespread regional late Miocene ('Alpine') inversion, the later phases cannot be so explained and remain incompletely understood (Blundell 2002). Plio-Pleistocene isostatic uplift may be involved, as may be flexural responses to more localized patterns of denudation (Watts *et al.* 2000). The tilting has the effect of bringing highly mature source material to shallow depths in the western part of the UK sector and the UK onshore area. Various approaches to the quantification of this uplift have been attempted (analysis of sonic velocities, Whittaker *et al.* 1985; Hillis 1995*a*, *b*; vitrinite reflectance, Pearson & Russell 2000; apatite fission track analysis (AFTA), Bray *et al.* 1992; Green 2005), but it proves difficult to reconcile them, partly because of the non-linear nature of some of the parameters involved (Japsen 2000) and partly because the uplift and burial histories are probably more complex than was originally suspected (e.g. Japsen 1997).

The regional tilting has undoubtedly had major consequences for the integrity and geometry of gas accumulations. It may be speculated that uplift may have led to seal failure and further large-scale remigration may be inferred. In this case, as in the case of remigration associated with late Cretaceous or earlier Cenozoic inversion, a laterally extensive blanket of permeable Rotliegend reservoir/aquifer is required. The likely impacts of the switching off of source kitchens and remigration in the less well-connected and lower net : gross Carboniferous successions have not as yet been studied.

The effects of progressive thinning of the post-Variscan overburden can be inferred from the burial history curve for the UK onshore well Weeton-1 (Fig. 7c). Here, maximum burial of the base Namurian source interval (and by implication all source intervals in at least the Tournaisian, Visean and Namurian successions) occurred during Carboniferous burial. It may be inferred that some of the hydrocarbons generated were lost during Variscan deformation and there has been little or no subsequent hydrocarbon generation. This combination of deep pre-Variscan and modest post-Variscan burial must occur over a wide zone in the Mid North Sea High and the UK onshore area and forms a northern and western limit beyond which prospectively is much reduced. This area has yet to be delineated, but a number of wells in which this effect may have been produced are documented by Vincent (2015). The same pattern is general in the more southerly parts of the basin that form an obvious Variscan foreland downwarp, where peak maturity is documented to have been reached prior to Variscan inversion throughout the Ruhr coalfield and adjacent areas (Littke *et al.* 2000), in the UK Oxfordshire coalfield (Green *et al.* 2001) and in the South Wales coalfield (review in Hower & Gayer 2002).

Gas composition

Gas derived from the Carboniferous – in both Carboniferous and younger reservoirs – is generally regarded as dry. This is largely true, with condensate to gas ratios of <5 bbls/MMscf being widely recorded in the central part of the Rotliegend reservoir fairway (figures compiled from papers in Abbotts 1991 and Gluyas & Hichens 2003). However, there are significant variations, with condensate contents of >10 bbls/MMscf present in the Cavendish, Boulton, Schooner and Tyne fields (Conway & Valvatne 2003*a*, *b*; Moscariello 2003; Kersten *et al.* 2013) and >20 bbls/MMscf in the Amethyst West and Saltfleetby fields (Garland 1991; Hodge 2003). In the Cavendish Field, fluid banking of retrograde condensate has proved a productivity impairment mechanism (Kersten *et al.* 2013).

Other compositional factors that have impeded the development of Carboniferous gas accumulations have been the local occurrence of high contents of nitrogen and CO_2 (Corbin *et al.* 2005). The high nitrogen contents conform to a gross regional pattern of increasing nitrogen content in both Carboniferous and Rotliegend reservoirs towards the east (Lokhorst 1998; Gerling *et al.* 1999). There is a consensus that the nitrogen derives from two main sources (Gerling *et al.* 1998), one related to the release of nitrogen from coals at very high maturities (Krooss *et al.* 1995; Gerling *et al.* 1997) and one resulting from the thermal degradation of both organic and inorganic nitrogen-bearing components in the mainly pre-Westphalian marine source rocks (Mingram *et al.* 2005). Both of these processes are encouraged by the generally increased burial of all Carboniferous source intervals moving eastwards from the UK towards the NW German plain. For high nitrogen fields to be commercialized, it has been necessary either to export gas to the European network via the Netherlands (to exploit the higher accepted nitrogen content in the European transmission systems) or to blend the gas with low nitrogen gas to meet UK specifications (Corbin *et al.* 2005).

The occurrence of gases with high CO_2 contents is much more patchy, with isolated accumulations having anomalously high CO_2 (e.g. the undeveloped accumulations in UK well 43/20b-2 (Kepler) and 49/5a-5 ('Coca Cola'), see fig. 9 in Corbin *et al.* 2005). Studies on the controls on CO_2 occurrence have yet to be published, although it may be speculated that these anomalously high CO_2 bullseyes are related to Cenozoic magmatic activity, the two occurrences cited here both being located near areas known to be affected by Paleogene dyke intrusion. Isotope data that might allow a more precise determination of the origin or origins of the CO_2 isotope have yet to be published. As with nitrogen-rich gases, development has been predicated by the ability to blend the gas to a marketable specification. This has enabled the development of the Breagh Field (initial CO_2 content *c.* 7%; DECC 2013), which is blended with gas from the Central North Sea at its landing point in Teeside to reduce the CO_2 content to transmission grid specification.

Exploration challenges presented by the Southern North Sea Carboniferous

Difficulty of prospect definition

Exploration for Carboniferous objectives in the Southern North Sea is bedevilled by a complex set of interpretation problems that make it difficult to define prospects and calculate unambiguous volumetric estimates. Most of these are summarized by Corbin *et al.* (2005).

The principal problems are related to the exceptionally complex overburden, which leads to severe imaging problems (Fig. 8). These result from extensive mobilization of the Zechstein salt (Stewart & Coward 1995), which has led to salt pillows and diapirs – the latter in places having overhang geometries – and to areas of more or less complete salt withdrawal. The presence of the Zechstein salts results in partitioning of the effects of Mesozoic and Cenozoic extension and inversion. Below the salt, post-Mesozoic structuration involves the movement of rigid blocks, while detachment in the salt leads to thin-skinned extensional and wrench-generated structures in the younger Mesozoic and Cenozoic. The resulting ray path complexity causes severe imaging difficulties. Interpretation is compounded by the problems of depth conversion in such a structural setting. Apart from the complex ray paths, which can only be satisfactorily managed by several phases of pre-stack depth migration (pre-SDM), differential burial in the Mesozoic succession leads to a complex velocity structure, particularly in the Upper Cretaceous Chalk (Davis 1987). A graphic example of the range of uncertainty of prospect definition that can result from salt in the overburden is provided by Corbin *et al.* (2005; Fig. 9) and an example of the application of pre-SDM is given by Jones *et al.* 2005.

The imaging problems caused by overburden heterogeneity are compounded by two major problems inherent to the stratigraphy of the base Permian unconformity and subjacent Carboniferous. Local facies variation at the base of the Permian – the presence or absence of a basal sand facies and/or salt layer – leads to marked changes in acoustic impedance at the base Permian unconformity, which may, as a result, be seismically invisible (Fig. 10a; see also discussion in Cooper *et al.* 2005, pp. 323–325). Mapping of structure at the base Permian unconformity level may require an indirect approach involving the picking of a horizon some way above the base Permian unconformity and the construction of isopachs to map the unconformity itself.

Similar difficulties arise in identifying the stratigraphic horizon in the Carboniferous that subcrops beneath the unconformity, a key uncertainty in prospect evaluation because it controls the presence or absence of reservoir within closure. Seismic character within the Carboniferous succession varies markedly over short distances and some key objectives – particularly in the upper part of the Namurian and in the late Westphalian Ketch Formation – are seismically featureless (Fig. 10b). When combined with the stratigraphic difficulties already mentioned and the presence of significant faulting in the succession, these features lead to dangers of aliasing the interpreted stratigraphic horizon. The discovery of numerous small fields in the Ketch reservoir in the CMSIII development (Cooper *et al.* 2005) and the delineation of complex stratigraphic trapping geometries in the subordinate Ketch reservoir in the Cygnus Field (Catto *et al.* 2018) has required detailed

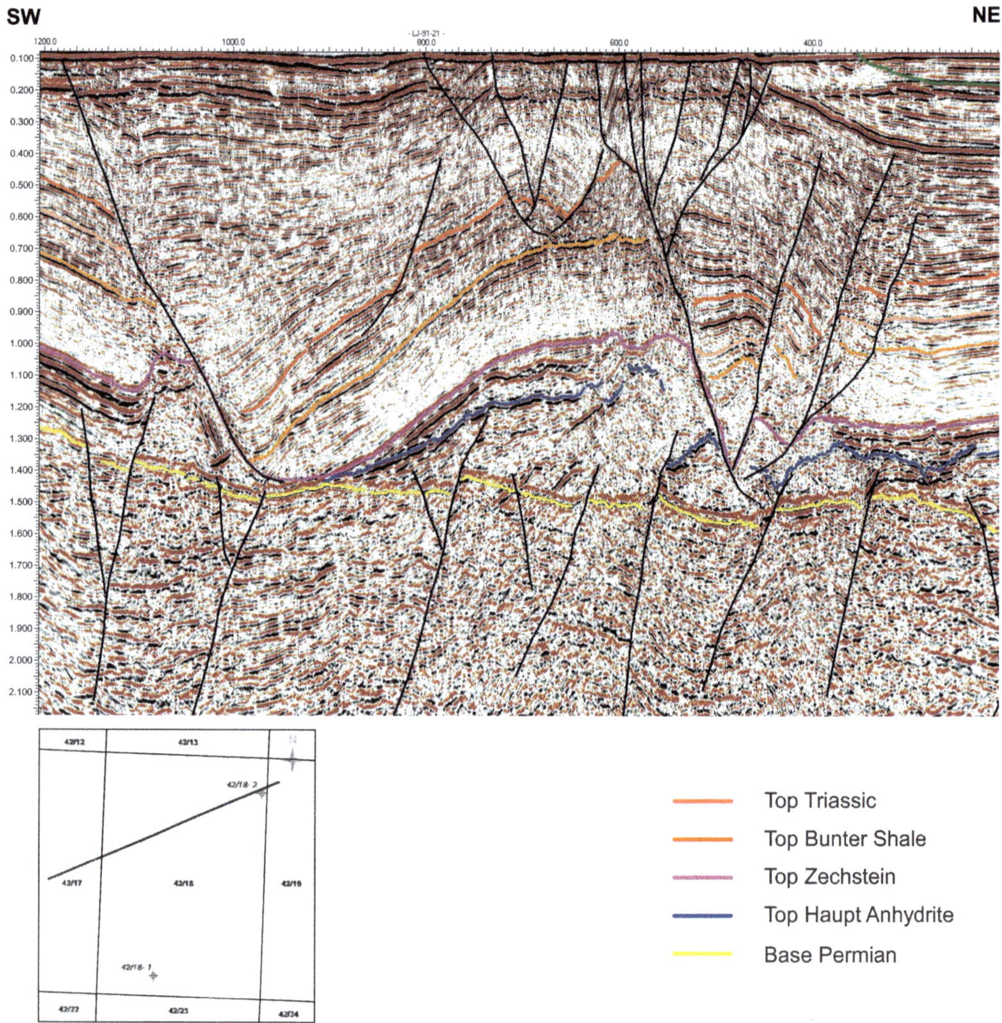

Fig. 8. Seismic line illustrating some of the problems encountered when imaging the Carboniferous beneath the complex overburden in the UK Southern North Sea. *Note:* faults cutting the post-Permian succession sole out in the Zechstein salt; thick localized Mesozoic sediments occupying a collapse graben immediately adjacent to a residual salt swell, producing complex lateral velocity distributions; lateral impersistence and changes in seismic character in Carboniferous associated with heterogeneity in overburden.

and accurate mapping of the subcrop to the base Permian unconformity. This has involved careful multidisciplinary stratigraphic analysis of offset wells to determine reliable seismic markers in the Carboniferous and the construction of local isopach maps to project the seismic interpretation into stratigraphic sections that lack seismic resolution. Even with such studies, wells can miss their reservoir targets, either because of seismic mis-picks or lateral facies variation (O'Mara *et al.* 1999: p. 814). Where seismic quality is poor and well control is distant, well results have differed markedly from prognosis – for instance, in well 41/10-1 the top Visean came in 3000 ft shallower than predicted pre-drill (released operator's completion report).

Drilling problems and difficulties in formation evaluation

A key characteristic of the Carboniferous objective is that the wells needed to target them are generally deep, complex and expensive (e.g. O'Mara *et al.* 1999, 2003*b*). Except around the margins of the basin (Mid North Sea High, Quadrants 41, 42

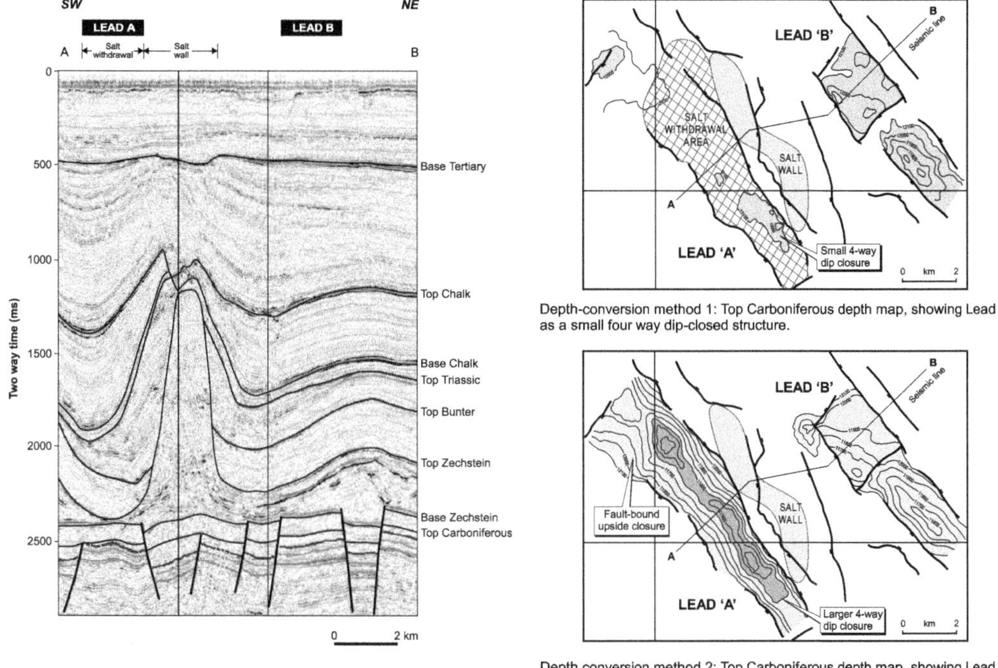

Seismic line through salt wall and salt withdrawal areas.

Depth-conversion method 1: Top Carboniferous depth map, showing Lead A as a small four way dip-closed structure.

Depth conversion method 2: Top Carboniferous depth map, showing Lead A as a large four way dip-closed structure with upside fault-closed potential.

Fig. 9. Examples of ambiguity in structural mapping produced by alternative depth conversion methods (from Corbin *et al.* 2005). The location of the structures is not specified in the original publication; the Zechstein salt diapir map forming fig. 3.30 of Pharaoh *et al.* (2010) suggests a location in the eastern part of UK Quadrant 44.

and 53), the top Carboniferous is at >10 000 ft TVDSS and the overburden consists of a complex succession, including salts and highly overpressured dolomite stringers in the Zechstein and bedded evaporites in the Silverpit Formation facies of the Rotliegend. The latter are prone to flow, causing casing deformation or collapse.

If the complications and cost of reaching the Carboniferous were not enough, drilling good quality well sections within the Carboniferous presents a set of challenges that differ entirely from those to which the industry is accustomed in the Southern North Sea. The highly heterolithic nature of the Carboniferous clastic successions, in which fairly thin-bedded units of mechanically strong sandstone alternate with mechanically weak and/or brittle shale and coal, leads to extensive caving and borehole collapse unless drilling is carefully managed. Many of the early wells were drilled with heavily weighted, salt-saturated, water-based mud, which resulted in severe caving and the development of ledges. The resulting wireline logs were in many cases of poor quality (Fig. 11a). Salt-saturated muds were the inevitable result of leaving openhole sections that had exposed Rotliegend and/or Zechstein salts between the deepest casing point and the drilled Carboniferous section. In at least some cases this resulted from the enforced choice of casing points at depths shallower than had been planned following well control incidents in the Zechstein (McPhee & Byrne 2009). More careful mud formulation and/or the use of oil-based fluids can largely eliminate this problem (Fig. 11b).

Apart from problems caused by poor log quality, the use of significantly overweight mud has led to ambiguities in both petrophysical evaluation and the interpretation of test results. Where the petrophysical method is detailed in well completion reports, it appears that, in some cases, evaluation has relied on standard methods with no consideration given to the likelihood of deep invasion resulting from the combination of significant fluid pressure overbalance and low permeability sandstone reservoirs. A particularly striking illustration of the detrimental effects that these have is provided by McPhee *et al.* (2008). In their careful study of the discovery well of the then-undeveloped Breagh Field, they used SCAL to demonstrate that the use of excess overbalance (400 psi) while drilling had led to up to 60 inches of invasion of filtrate from the brine water-based mud. Well test analysis showed a skin ranging from +24 to +175. Redrilling of the well

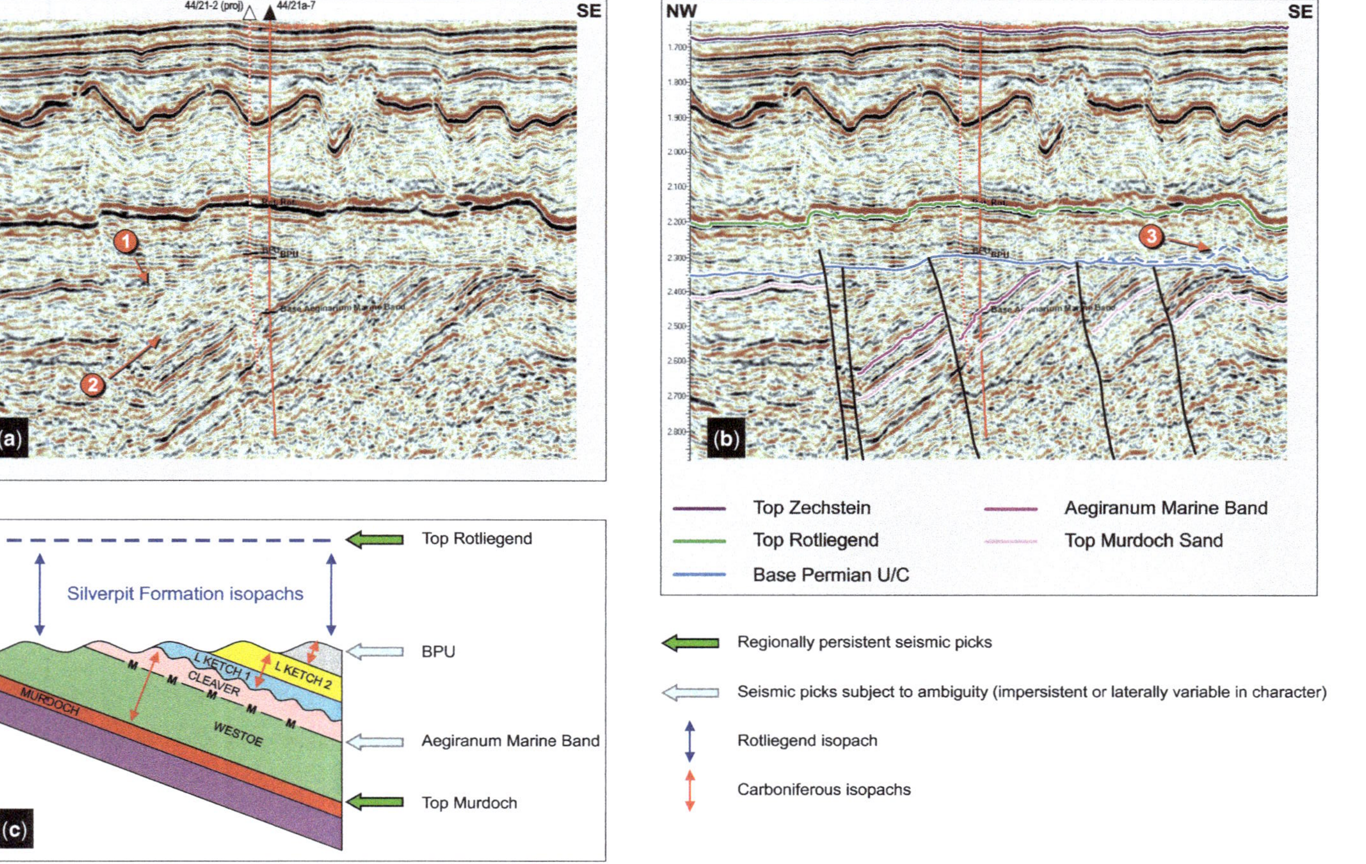

Fig. 10. Base Permian unconformity and intra-Carboniferous imaging problems: example from the Boulton B Field. (**a**) Uninterpreted seismic line; (**b**) interpreted seismic line; and (**c**) summary of isopach methods for the reconstruction of base Permian geometry and mapping of subcrop. Section line corresponds to left-hand half of Figure 2a. 1, Lateral variability and local discontinuity of base Permian reflector; 2, acoustically transparent zone incorporating the Ketch and Bouton formations; and 3, possible ambiguities introduced in base Permian mapping by combination of seismic and isopach derived picks. Images supplied by Engie E&P UK Ltd (*Source:* PESGB Course Manual – Petroleum Geology of the North Sea 2013). Published with permission of the Petroleum Exploration Society of Great Britain.

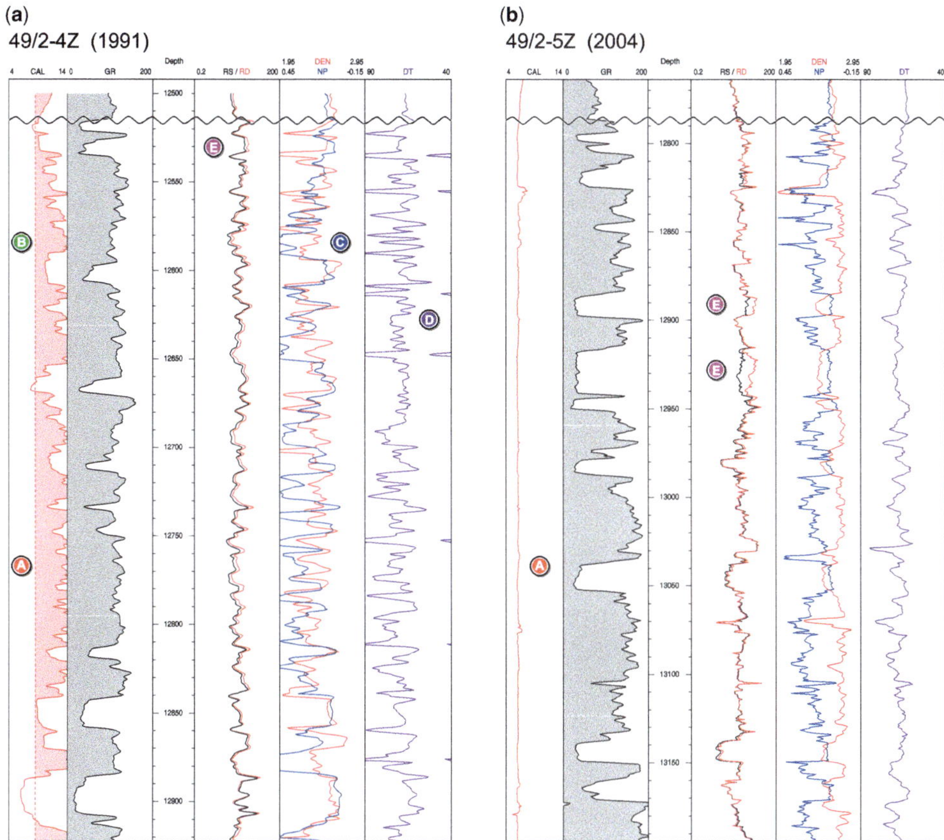

Fig. 11. Contrast in quality of wireline logs in similar stratigraphic penetrations in adjacent boreholes drilled (**a**) without and (**b**) with careful attention to mud formulation. Sections are in the Ketch Formation and are hung on a datum at the base Permian unconformity. *Note:* A, severe caving and B, development of ledges in earlier well; C, degradation of pad-mounted density/neutron logs and D, noisy sonic with frequent cycle-skipping in oversize rugose hole sections; and E, suppression of true resistivity in gas-bearing sandstone owing to deep invasion.

using oil-based mud with minimal overbalance led to negligible invasion and eliminated the skin. The resulting nearly six-fold increase in flow rate demonstrated the commercial viability of the field (Fig. 12).

Development challenges and unpredictable field performance

Development of Carboniferous accumulations in the UK sector of the Southern North Sea has been characterized by generally long lead times. Of the first ten fields discovered, the shortest interval between discovery and production start-up was eight years (Caister Field, discovered 1985, start-up 1993), while the longest so far has been 23 years (Chiswick Field, discovered 1984, start-up 2007).

This delay was partly due to the lack of infrastructure to the north of the Rotliegend pinchout line, but was also a consequence of the dramatic collapse in oil and gas prices in 1985, immediately after the discovery of the first Carboniferous fields (Fig. 13). Apart from the restriction in budgets that this caused, widespread redundancies and the wholesale relocation of exploration departments from London to Aberdeen led to the break-up of teams that had built competence in the Carboniferous and to the loss of experienced individuals. The modest revival of the mid-1990s was followed by a second price slump that led to another large-scale restructuring in the industry at the end of 1998, which was accompanied by a similar fragmentation and loss of expertise. The same period saw major changes in the organization and priorities of both academic and statutory organizations, such as the British Geological Survey, which might otherwise have provided research support to an emerging Carboniferous play. Research groups that had long histories of

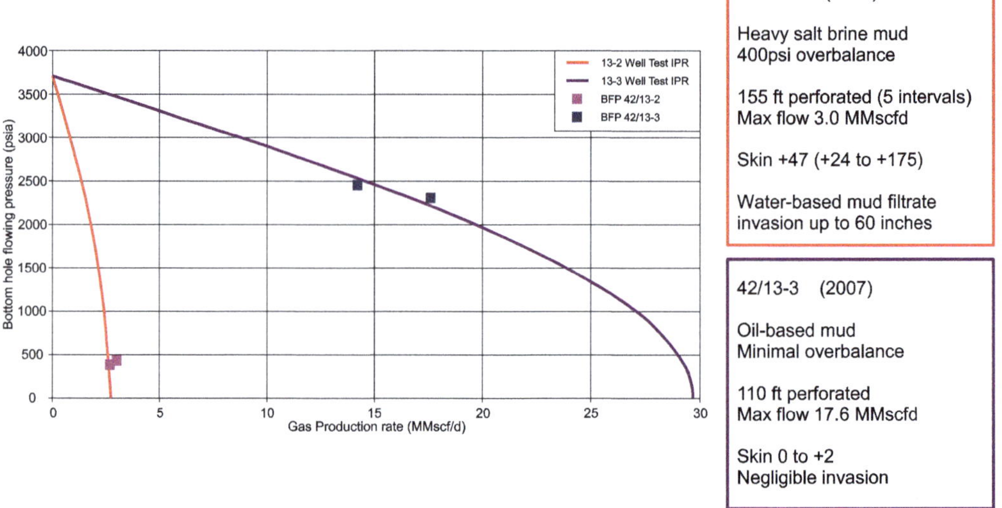

Fig. 12. Comparison of well test results and interpretation in Breagh Field discovery well 42/12-2 and appraisal well 42/12-3. Contrasting bottom-hole flowing pressures and gas production rates between the two wells drilled with different mud compositions and weights define sharply differing inflow performance relationships. Use of correctly weighted oil-based mud leads to an increase in productivity of a factor of ten. See text for discussion. Inflow performance relationships plot from McPhee & Byrne 2009. Copyright 2009, Society of Petroleum Engineers. Reproduced with permission of SPE. Further reproduction prohibited without permission.

Carboniferous studies were downgraded and expertise was lost through retirement. The intense research effort that had accompanied the opening up of the Rotliegend and Jurassic plays in the North Sea was not replicated.

A telling example is provided by the development of stratigraphic concepts. From a start of virtually no knowledge in the early 1970s, understanding of Jurassic stratigraphy in the Central and Northern North Sea had advanced to a fully documented genetic stratigraphy including the latest sequence stratigraphic concepts by the early 1990s (Partington *et al.* 1993). By contrast, there was still, in 2017, no agreed genetic stratigraphy for the

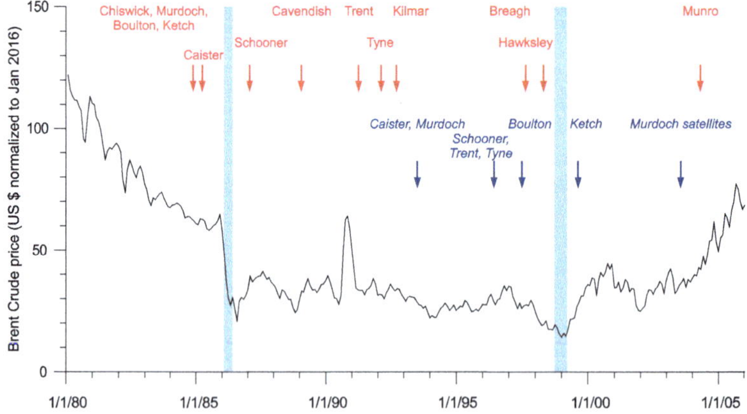

Fig. 13. Plot of oil price (normalized to January 2016) v. timing of discovery (red upright text) and production start (blue italic text) of principal Carboniferous gas fields in the UK Southern North Sea. Vertical blue bars mark the timing of significant industry retrenchment and re-organization. Major retrenchments followed initial discoveries and the start-up of most production, leading to fragmentation of the knowledge base and loss of expertise.

Carboniferous succession in the UK offshore – let alone the wider North Sea area – despite more than 100 years of intense litho- and biostratigraphic research in the adjoining UK and Netherlands onshore areas.

The slow pace of development was not only driven by commodity price and infrastructure issues. When targeted exploration for Carboniferous objectives started in 1984, there was already a large volume of undeveloped reserves in the Rotliegend (Fig. 14). These had remained undeveloped for a combination of reasons that rendered them uncommercial at the low prevailing gas price. Many accumulations were in reservoirs that were tight, either as a result of diagenetic modification or unfavourable facies development. These would not produce at viable rates using the drilling and completion methods that were then standard. Projects were also rendered unviable by lack of ullage in the evacuation systems and high tariffs demanded for pipeline access. With more favourable gas prices, decline in the first generation fields and improved subsurface technology (identification of fracture systems, underbalanced drilling) these accumulations, located in shallow water near existing infrastructure, became the main focus for development in a low-price environment. At the same time, improvements in imaging and depth conversion brought about by the universal adoption of three-dimensional seismic, and the improvement in pre-stack seismic processing and interpretation, led to the delineation and discovery of additional reserves in the main Rotliegend fairway (Fig. 14). Viewed in this context, the volumes discovered in the Carboniferous are relatively small and, given their remote location and the other technical difficulties in developing them, it is not surprising that development was so slow.

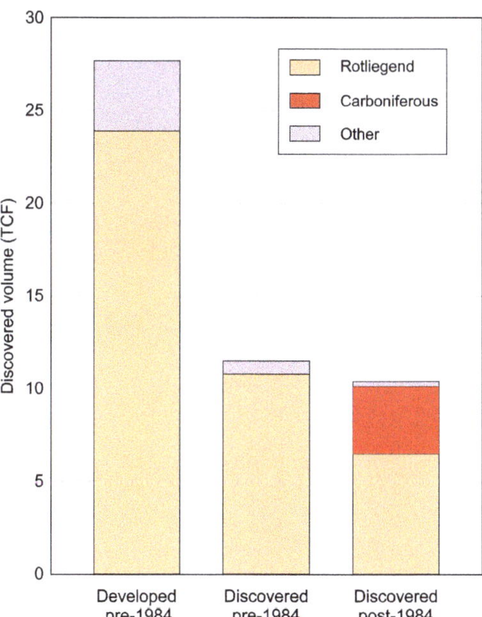

Fig. 14. Context of discovered volumes of gas in Carboniferous reservoirs before and after beginning of targeted exploration in 1984. Note that Carboniferous volumes are dwarfed by volumes in the Rotliegend that were undeveloped in 1984 and are significantly smaller than Rotliegend volumes discovered since that date.

Even when development has proceeded, it has been accompanied by much greater uncertainties than would have been the case in most Rotliegend fields. The primary uncertainty relates to the distribution of the reservoir, particularly in undrilled fault blocks. This arises from a combination of two factors. It is often difficult to predict the occurrence of a particular reservoir horizon at the erosionally truncated subcrop to the base Permian unconformity surface and the channelized nature of the reservoir sands means that they are not developed in a predictable manner over the area of a field. The overall low net : gross ratio in most of the succession introduces uncertainty regarding connectivity, which may be either enhanced or further degraded by faulting, much of which is below seismic resolution (Bailey *et al.* 2002). In general, low porosity and permeability, together with uncertain connectivity, give rise to uncertainties as to whether sufficient volumes of gas are connected to the wellbore to achieve economic flow rates and volumes. All of these difficulties are embedded in a situation where uncertainty is introduced into the overall structure by the complex overburden and difficulties in depth conversion.

The question of well cost has already been mentioned as an exploration challenge, but this challenge becomes even more acute when a development plan has to access all parts of an accumulation beneath the complex Zechstein overburden. Finding windows through the potentially overpressured Plattendolomite leads to long and complex well paths and the alternative requires drilling and casing contingencies to manage the possible overpressure zones. Well cost has been further increased by the need to employ heavy casing in the Silverpit Formation, where, in many of the early development wells, there has been widespread occurrence of casing crimps or collapse as a result of movement in the Rotliegend salts.

The production histories of some of the early Carboniferous fields reflect all these uncertainties and technical issues. Field performance has varied widely, with large divergences between predicted recoverable volumes and actual performance (Table 2). Without access to detailed subsurface and production data, any analysis of this variation is speculative, but some general trends emerge, by inference mainly related to issues of reservoir connectivity. Fields relying on the Murdoch reservoir

Table 2. *Comparison between volumes produced up to September 2015 and previously published estimates of recoverable reserves for selected Carboniferous gas fields*

Field	Start of production	Reservoir	Status (April 2017)	Published volumes (BCF)			Source	Production volume to April 2017	Fraction of published recoverable volume (%)
				Recoverable	GIIP	RF (%)			
Boulton B	1997	Ketch Formation	Producing	142	206	69	Conway & Valvatne (2003*a*)	279	196
Hawksley	2002	Ketch Formation	Ceased	*42*	60	*71*	Cooper *et al.* (2005)	52	124
Murdoch K	2003	Ketch Formation	Producing	*94*	134	*71*	Cooper *et al.* (2005)	223	237
Schooner	1996	Ketch Formation	Producing	612	1059	58	Moscariello (2003)*	308	50
Tyne North	1996	Ketch Formation	Ceased	82	163	50	O'Mara *et al.* (2003*b*)	39	48
Tyne South	1996	Ketch Formation	Ceased	54	61	89	O'Mara *et al.* (2003*b*)	91	169
Watt	2003	Cleaver Formation	Ceased	*40*	57	*71*	Cooper *et al.* (2005)	2	5
Murdoch	1993	Murdoch Sandstone	Producing	348	478	73	Conway & Valvatne (2003*b*)	378	109
McAdam	2003	Caister Formation/ Murdoch Sandstone	Producing	*137*	196	*71*	Cooper *et al.* (2005)	140	102
Trent	1996	Millstone Grit Formation/Caister Formation	Ceased	92	111	83	O'Mara *et al.* (2003*a*)	113	123
Saltfleetby	1999	Millstone Grit Formation	Producing	72.5	114	63.25	Hodge (2003)	67	92

*Operator has subsequently redetermined GIIP as 654 BCF (see text).
Recoverable volume figures printed in italics are calculated from published GIIP figures where no recoverable volume was published, calculated using an average of all published recovery factors. Cumulative production figures from UK Oil and Gas Authority website.
GIIP, gas initially in place; RF, recovery factor.

have performed more or less as predicted, reflecting the laterally extensive sheet-like nature of this unit where developed (Conway & Valvatne 2003*b*; Cameron *et al.* 2005). Fields with reservoirs in the Ketch Formation have, by contrast, shown highly variable performances, the Boulton B and Murdoch K fields having produced about twice their published estimated recoverable volumes with two and one development wells, respectively, while the Schooner Field has managed less than half with 11 development wells (Fig. 15a, b). Some fields producing from the Namurian (Trent, Cavendish) appear to have produced more than their gas initially in place (GIIP), suggesting drainage from surrounding reservoir sections that were not initially considered as part of the net reservoir. Another example – the Watt Field – marks the only attempt to date to develop a reservoir in the Cleaver Formation. Although the sand bodies in this unit look sedimentologically similar to those in the Ketch Formation, the Cleaver clearly has a much lower net : gross ratio and lacks large-scale connectivity. The Watt development accordingly failed after draining a very restricted volume of gas.

The widely varying performances of the Ketch Formation reservoirs deserve further comment. In Schooner, increased understanding of the reservoir geology during the long development history has led to a downgrading by the present operator of the GIIP from the published 1059 BCF (Moscariello 2003) to 654 BCF (Faroe Petroleum, unpublished data). The disparity between estimated and actual recovery in the Ketch reservoir reflects the difficulty of extrapolating and averaging net : gross ratios in fluvial deposits and the much better than assumed connectivity between fluvial channel sand bodies. In the Boulton B and Murdoch K cases, better than assumed connectivity in a fairly high net : gross, more proximal position in the Ketch depositional system has been a positive feature, allowing ready access to all of the gas in the accumulation from a small number of wells. Schooner, by contrast, is located in a more distal position in the depositional system and has lower net : gross ratio (Stone & Moscariello 1999). Here, the large well count results in part from major problems of well stability and scale development, but reservoir connectivity has also been an important factor. The assumptions made regarding sedimentological compartmentalization (cf. Mijnssen 1997) proved to be ill-founded, successive development wells encountering progressively more depleted reservoir, even when located far from the initial core production area. It is not

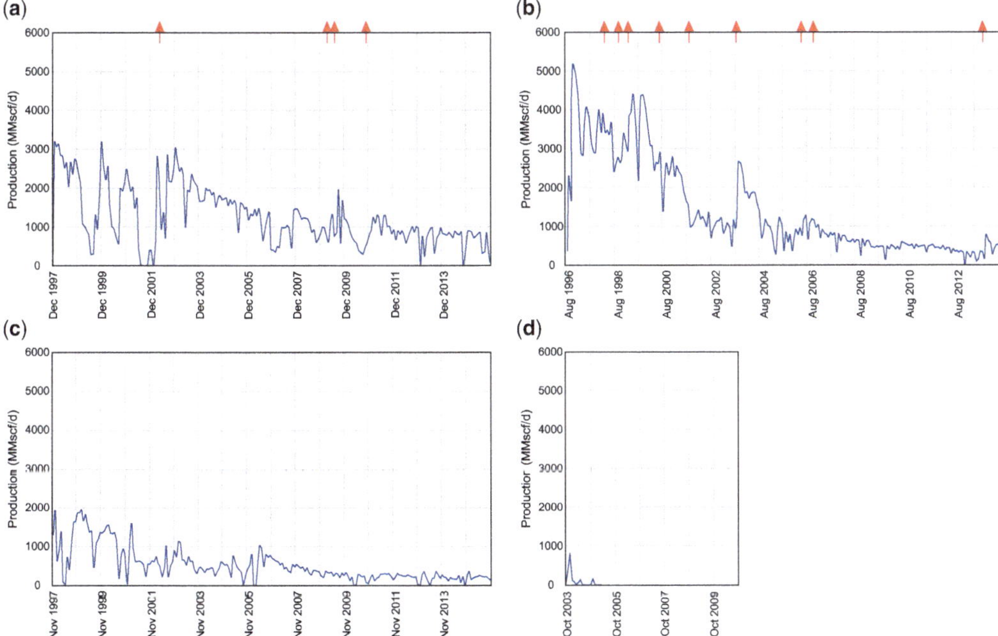

Fig. 15. Field decline curves showing first 6500 days of production for four selected UK Carboniferous fields: (**a**) Boulton B; (**b**) Schooner; (**c**) Trent; and (**d**) Watt. Production commenced through 1, 3, 2 and 1 wells, respectively; triangular symbols mark start-up of additional wells. Data from UK Oil and Gas Authority website (www.ogauthority.co.uk/data-centre/data-downloads-and-publications/production-data/ [last accessed 5 December 2017]). See text for discussion.

clear whether the greater than anticipated connectivity in this case is the result of the sedimentological/stratigraphic juxtapositions of sand bodies, or of greater than anticipated fault juxtaposition. It was known at the time of the initial development that three-dimensional seismic revealed a much greater degree of minor faulting than was initially apparent from the two-dimensional surveys on which the first generation of Carboniferous discoveries had been initially mapped (Oudmayer & de Jager 1993). More recent studies in the Yorkshire mining area in the UK onshore (Bailey *et al.* 2002) have shown that sub-seismic faulting is likely to bring about complete connectivity in channelized sandstone bodies in the Coal Measure succession in that area, so, in retrospect, it is likely that early assumptions regarding sedimentological heterogeneity were overplayed.

Taken overall, all of the Carboniferous fields show initially rapid rates of decline, followed by long tails in which production remains more or less steady over a period of years (Fig. 15). This pattern replicates that seen in Paleozoic-reservoired oil fields in the Central North Sea (Argyll, Buchan), where rapid early decline has been followed by much longer than anticipated periods of low, but consistent, flow rates, reflecting the slow depletion of a labyrinthine reservoir of generally low permeability.

Understanding the Carboniferous petroleum system: recognizing and challenging the founding myths

An understanding of the petroleum geology of a basin or sub-basin involves the collation of lines of evidence to formulate a play concept that acts as the basis for continuing exploration. Plays, thus defined, exhibit diminishing returns, usually expressed through the creaming curve, unless new ideas come forward (the 'paradigm shifts' of Kuhn 1962) to allow new insights and rejuvenate exploration through the identification of new play fairways.

The Carboniferous play in the Southern North Sea is no exception to this generalization, but, through a combination of circumstances, the paradigm shifts that might allow innovative thinking have not occurred. As a result, attitudes to the petroleum geology have stagnated. In this section, I examine five fundamental areas of Carboniferous petroleum geology that merit re-examination. Attitudes in these areas have largely remained frozen in time since the early days of exploration of the Carboniferous fairway. I here refer to them as the 'Founding Myths' of North Sea Carboniferous geology, in reference to their resemblance to the immutable accounts found in traditional societies that describe the absolute truth about events that made the natural world the way it is (cf. Eliade 1968, p. 23).

Founding Myth 1: 'The gas comes from Westphalian coals'

From very early in the exploration and development of the Southern North Sea basin it has been generally accepted that the gas in Rotliegend and younger reservoirs is derived from the coals in the Westphalian. This interpretation was first articulated by Patijn (1964) and has since been repeated without question in all general accounts of the petroleum system in the basin (e.g. Glennie 1998, p. 140). It is not difficult to see why this interpretation has been so widely accepted. Cornford (1998, pp. 428–430) summarizes the coal thicknesses and published maturities in the Westphalian and it is obvious from source rock/reservoir mass balance calculations that the known volumes of reservoired gas represent only a fraction of the total gas volume that has been potentially generated. Thus, although other sourcing possibilities have been identified (e.g. Cornford 1998), the Westphalian is still cited as the dominant source (e.g. de Jager & Geluk 2007), even in areas requiring long and improbable migration paths (e.g. Rodriguez *et al.* 2014).

Two features of the gas occurrences do, however, suggest that the pattern of gas sourcing is more complex, or even that in many areas the assumption of sourcing from Westphalian coals is incorrect.

Maturity patterns. Published maturity data for Carboniferous source rocks in the UK are sparse, with large areas – notably the Carboniferous heartland in Quadrants 44 and 49 – having no published data. Data that have been published are scattered through a wide variety of literature sources, with much data, although nominally in the public domain, having hitherto been difficult to access. Where data have been published, for instance in the *Petroleum Geological Atlas of the Southern Permian Basin Area* (Kombrink *et al.* 2010) or for the reference well 48/3-3 (Leeder *et al.* 1990; see Fig. 7a), they may be difficult to interpret, either consisting of only a small number of data points, covering a limited depth range, or lacking accompanying information on data quality. In the Netherlands the situation is very different, the data release policy being such that vitrinite reflectance and pyrolysis data for a large number of released wells are in the public domain and can be freely downloaded from the Netherland Geological Survey portal (www.nlog.nl; e.g. Fermont & Jegers 1991). This source does not always contain information on the quality of the data, some of which is clearly poor.

Pending the availability of more complete data of better quality, a degree of caution is required in

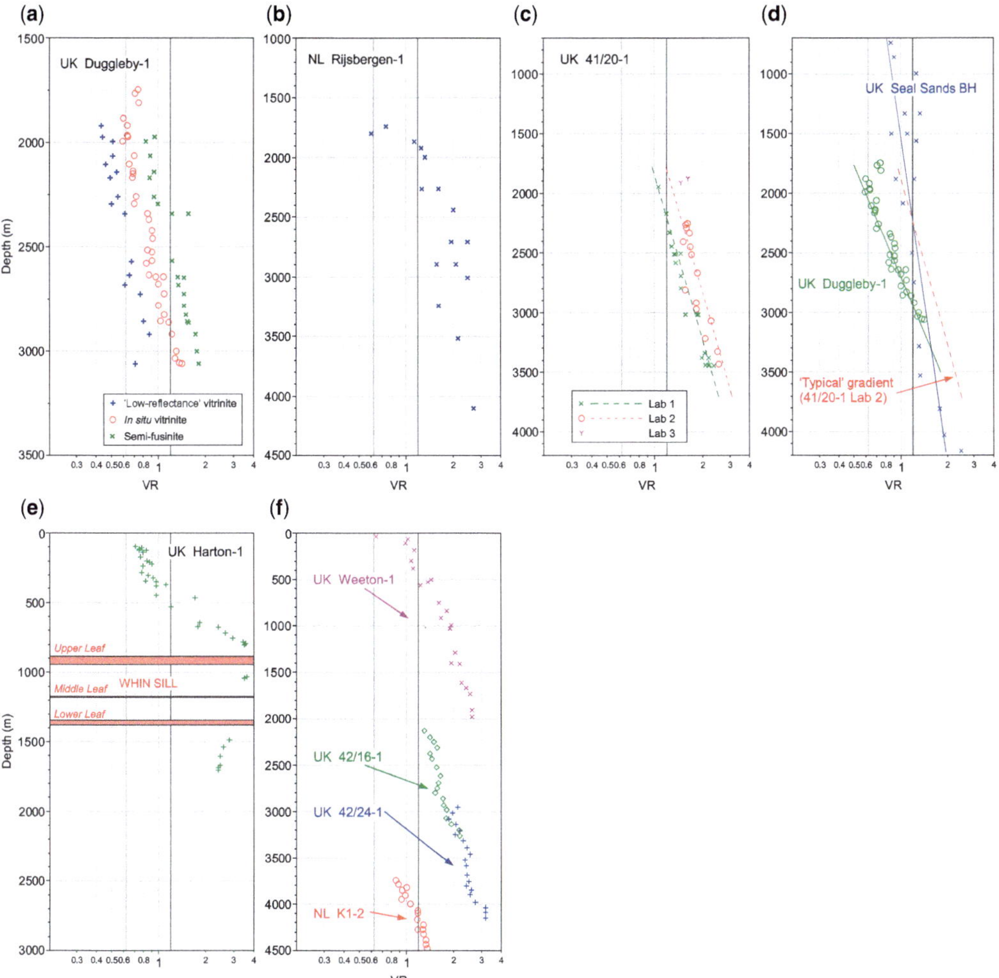

Fig. 16. Examples of vitrinite reflectance data from the Southern North Sea and adjoining areas. (**a**) Wide dispersal of data resulting from presence of multiple vitrinite populations. (**b**) Ambiguous apparent maturation gradients in a dataset for which no quality control information is available. (**c**) Inconsistent maturation profiles in analyses of samples from the same well by multiple laboratories. (**d**) Different maturation gradients in wells located on a granite-cored high (Duggleby-1) and in a deep Lower Carboniferous graben (Seal Sands borehole). (**e**) Thermal anomalies associated with late Carboniferous tholeiitic intrusion. (**f**) Effect of Neogene uplift. Note that the scaling of vertical axis varies. The Carboniferous section in well K1-2 (plot (f)) is currently at maximum burial; all other wells in these plots have been subject to some degree of Neogene uplift. *Sources:* UK 41/20-1 (Lab 1) and Seal Sands borehole, courtesy of CGG Robertson; Duggelby-1, 41/20-1 (Lab 2), 42/16-1 and 42/24-1 from released operators well completion reports; 41/20-1 (Lab 3) from Kombrink *et al.* (2010); Rijsbergen-1 and K1-2 from www.nlog.nl; Harton-1 from Ridd *et al.* (1970); Weeton-1 from Kerr McGee (2002).

interpreting patterns of Carboniferous maturity from vitrinite reflectance data.

(1) Vitrinites in the Carboniferous can be divided into at least three populations (Fig. 16a): the so-called low-reflectance vitrinite, *in situ* vitrinite and semi-fusinite or partially oxidized vitrinite. The first of these may be related to the presence of liptinitic maceral components in coal source material (Murchison & Pearson 2000; Murchison 2004), but may also simply result from caving. The third is the result of penecontemporaneous oxidation, possibly associated with reworking. Clearly only data from the second population can contribute to correct modelling of maturity. Unless vitrinite

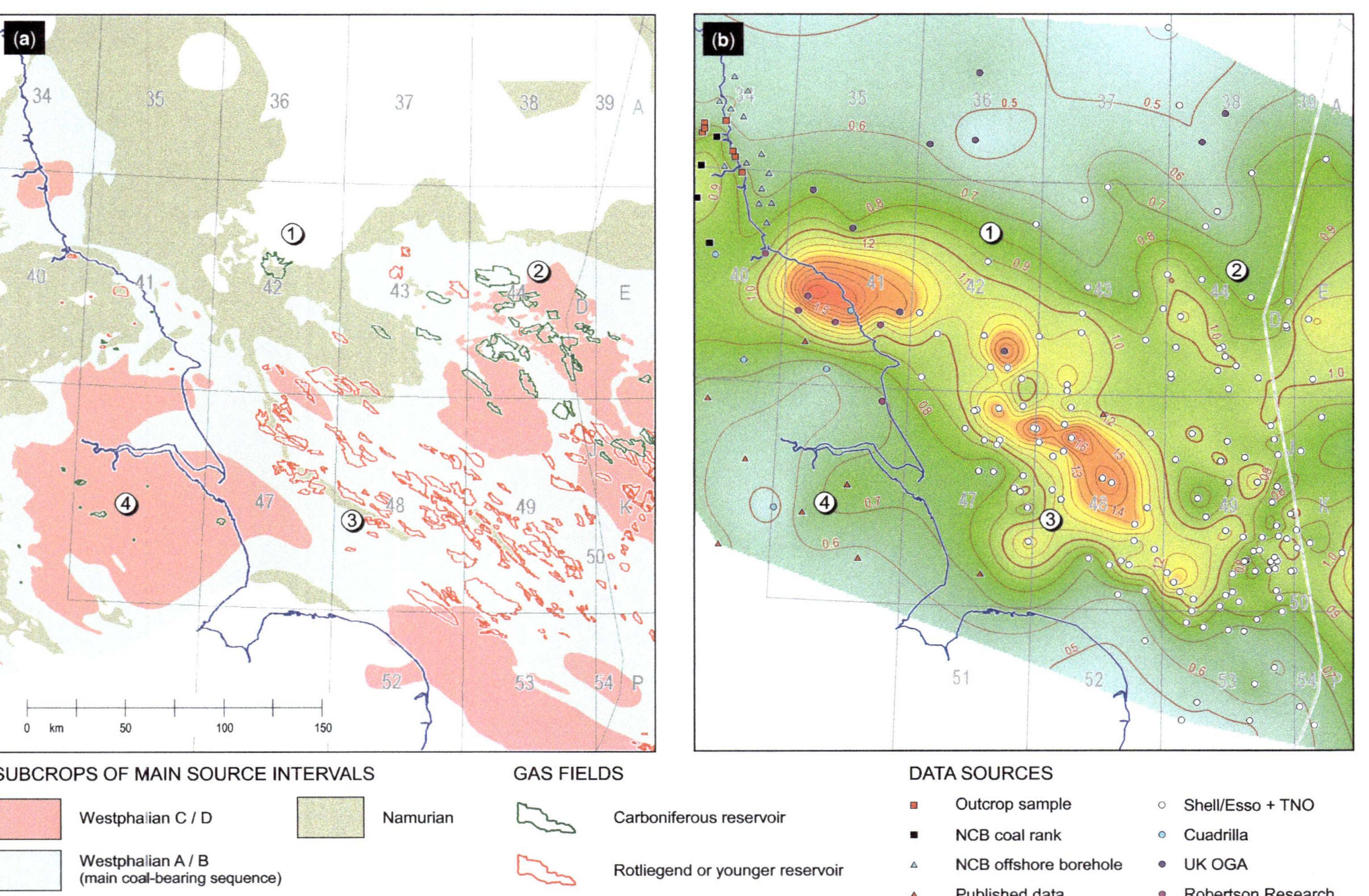
(a)
(b)
0 km 50 100 150
SUBCROPS OF MAIN SOURCE INTERVALS
Westphalian C / D
Westphalian A / B
(main coal-bearing sequence)
Namurian
GAS FIELDS
Carboniferous reservoir
Rotliegend or younger reservoir
DATA SOURCES
Outcrop sample
NCB coal rank
NCB offshore borehole
Published data
Shell/Esso + TNO
Cuadrilla
UK OGA
Robertson Research

types are explicitly identified, the apparent maturation profiles obtained may be confusing or misleading (e.g. Fig. 16b).

(2) Vitrinite reflectance is a subjective measurement and there is variability between laboratories and interpreters (Fig. 16c).

(3) Some workers claim that there are marked differences in maturation gradient between different parts of the depositional area, which are interpreted to result from prolonged differences in the thermal gradient (Creaney *et al.* 1985; Murchison 2004). These are generally believed to relate to residual heat flow derived from Devonian granite masses underlying the basement highs that flank the early Carboniferous rift depocentres. The presence of clearly defined differences in maturation gradient in wells that are not obviously near any of the mapped granite bodies (Fig. 16d) suggests that this explanation may be simplistic, although there might be a link to the position relative to the rift depocentres.

(4) The maturities developed in vitrinites in coal seams may be significantly affected by the overlying lithology: sandstone seam roofs, having a higher thermal conductivity, give rise to a lowering of maturity relative to that developed in seams with shale roofs (Murchison 2004). While this should average out over long stratigraphic sections, it is another potential cause of inaccuracy in limited datasets, although it is possibly within the noise of the calibration measurements employed. Assumptions regarding lithology of overburden sections removed by erosion can have major effects on the outcomes of maturity modelling (Pearson & Russell 2000).

(5) In an extensive study in the Ruhr coalfield, there is a consistent mismatch between reflectance results obtained from coal and those obtained from dispersed vitrinite in clastic sediments, the latter consistently showing lower reflectances, particularly in the oil window (Scheidt & Littke 1994). These researchers ascribe this to a combination of differences in diagenetic reactions affecting the vitrinite within or outside the coal seam and to differences in the quality of the reflectance measurement controlled by the lower availability of ideal material for measurement in dispersed vitrinite samples.

(6) There are widespread occurrences of a suite of tholeiitic basalt intrusions of latest Carboniferous age, known at outcrop and in wells in northern England, where they form the Whin Sill and Causey Dyke complexes, and in wells in Quadrant E in the Netherlands. These have major thermal anomalies associated with them (Ridd *et al.* 1970; Pearson 1988; Fig. 16e).

(7) Where good quality data are available, the vitrinite reflectance profiles clearly demonstrate the effects of the Cenozoic uplift alluded to previously (Fig. 16f).

Any or all of these factors need to be considered in interpreting reflectance profiles. In particular, it may be difficult to identify which, if any, points in a small dataset are anomalous, particularly if: (1) there is no quality control report; (2) the data are limited to a small vertical interval; (3) the data are sparse and/or widely separated; or (4) the data are derived from immediately below the base Permian unconformity.

Several sub-regional and regional maps have been published showing maturity at top Carboniferous (Cope 1986; Leeder & Hardman 1990; Bailey *et al.* 1993; Kombrink *et al.* 2010). All concur in showing that, at subcrop, the Carboniferous is at best marginally mature for gas generation in most areas and that many gas fields occur in areas where the Westphalian is either immature or absent (Fig. 17). Clearly charge in these cases is derived from deeper horizons within the Carboniferous, which have always been assumed to be coals in the lower parts of the Westphalian. Where reasonable maturity data are in the public domain, it is by no means certain that this is the case. Thus the map of maturity at the Westphalian A/B boundary in the

Fig. 17. Regional maps showing (**a**) distribution of gas fields in relation to Carboniferous subcrop map of Kombrink (2008) and (**b**) maturity at the top Carboniferous. 1, Breagh Field; 2, Cygnus Field and adjoining fields in northern Silverpit area; 3, southern Sole Pit; and 4, Gainsborough Trough, UK onshore. See text for discussion. The majority of the database for this compilation is derived from released data from the Netherlands oil and gas portal (http://www.nlog.nl) and from the Shell/ExxonMobil Geochemistry Database for the Southern North Sea released by UK Oil and Gas Authority (http://data-ogauthority.opendata.arcgis.com/); onshore outcrop data from Burnett (1987); approximate vitrinite reflectance equivalences from coal rank data in onshore Durham coalfield from National Coal Board (1960); vitrinite reflectance data from National Coal Board offshore boreholes from Pearson (1988); other published data points from Andrews (2013), Bray *et al.* (1992), Fraser *et al.* (1990) and Pearson (1988); other published data from Hughes *et al.* (2018), Kerr McGee (2002), Leeder & Hardman (1990) and Pearson & Russell (2000); data for other offshore wells from UK Oil and Gas Authority well release packages for UK 30th Licensing Round (http://data-ogauthority.opendata.arcgis.com/); remaining onshore data published by kind permission of Cuadrilla Resources and CGG Robertson.

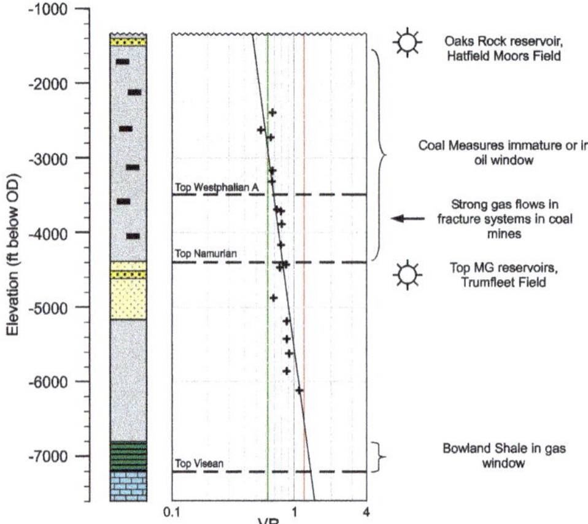

Fig. 18. Source rock maturity profile and reservoired gas horizons, UK Gainsborough Trough. Vitrinite reflectance (VR) data from Pearson & Russell (2000); stratigraphic profile from same source (well Gainsborough-2), supplemented with data from Ward *et al.* (2003).

Netherlands published by TNO (available from www.nlog.nl) shows marginal maturity at depth within the main part of the coal-bearing succession; in addition, the distribution of migrated gas in the Gainsborough Trough area (UK onshore) is clearly incompatible with a Westphalian source (Fig. 18). Where Rotliegend sandstones containing gas fields occur above immature Carboniferous source rocks (for instance, in the southern part of the Sole Pit area, UK Quadrant 48; Fig. 17), this can satisfactorily be explained by remigration within the Rotliegend. However, to charge a field such as Breagh from a Westphalian source in the absence of Rotliegend sandstones involves an improbably complex migration path.

Isotope data. To obtain a better understanding of the origins of gases throughout the NW European Basin, a collaborative research project was undertaken in the mid-1990s by the Geological Surveys of the UK, the Netherlands, Germany, Denmark and Poland, which included the compilation of a large set of gas isotope data (the *Northwest European Gas Atlas*, Lokhorst 1998; Gerling *et al.* 1998). Among the data collected, measurements of carbon isotope ratios in co-occurring methane and ethane allow the identification of both source rock type and source rock maturity using the method of Berner & Faber (1996). As the investigators were expecting to confirm the Westphalian coal origin of the gases, the diversity of the results came as a considerable surprise (Gerling *et al.* 1998, p. 223). These unexpected results do not appear to be of unreliable quality or anomalous. Although the exact provenance of the individual gas samples remains confidential, the number of samples analysed is such as to preclude experimental error (P. Gerling pers. comm. 2016). The reliability of the data is confirmed by an independently generated duplicate dataset from the UK sector only (Fig. 19d).

In the following discussion of the isotope results, the conclusions of the cited authors are taken at face value. It should, however, be stated in advance that the interpretation of such isotopic results should be attempted with caution. Berner & Faber (1996) recognize that the quality of any relationship between hydrocarbon and source depends on the correct specification of the bulk $\delta^{13}C$ value in the source organic material, which is unknown in this case. For humic source rocks, their calibration below maturities of 1.5% R_o is based, in the absence of any experimental data, on an extrapolation of experimental results from source material of higher maturity, backed up by empirical observations of gas isotope compositions. They are at pains to point out that generation of methane at maturities <1.5% R_o involves several reactions that may give rise to inhomogeneities in the isotope content. This may, in part, reflect the different mechanisms involved in methane formation in sapropelic and humic source material (breakage of C–C bonds in the former, condensation of aromatic rings in the latter; Galimov 2006). The approach adopted by Berner & Faber (1996) is strongly criticized by Galimov (2006, p. 1238) on the grounds that it relies on the incorrect assumption that laboratory pyrolysis results can be used as a proxy for maturation processes occurring in natural systems.

In the area in and around the Southern North Sea, three populations of gas can be identified in the *Northwest European Gas Atlas* data. Population 1 (Fig. 19a) consists of gases from Rotliegend and Carboniferous reservoirs in the UK and Netherlands offshore areas. Here, gases uniformly indicate predominant derivation from marine sapropelic source rocks at maturities of R_o between 1.6 and 2.5%. Population 2 (Fig. 19b) consists of gases from Rotliegend and Carboniferous reservoirs in the Netherlands onshore with a gas composition indicative of a mixture of humic and marine sapropelic sources. Here, maturities in the sapropelic source appear higher (R_o = 2.0–2.5%) with admixed gas from the coal source suggesting generally lower maturity (R_o = 0.8–1.6%). The gases in population 3 (Fig. 19c) are from Rotliegend reservoirs in the North German Plain, showing derivation from humic source rocks at generally higher maturities (R_o consistently >1.6%).

Although presented at a conference in London in 1998 and subsequently published in a widely circulated book (Gerling *et al.* 1999), the two radical implications of these results have yet to be fully appreciated.

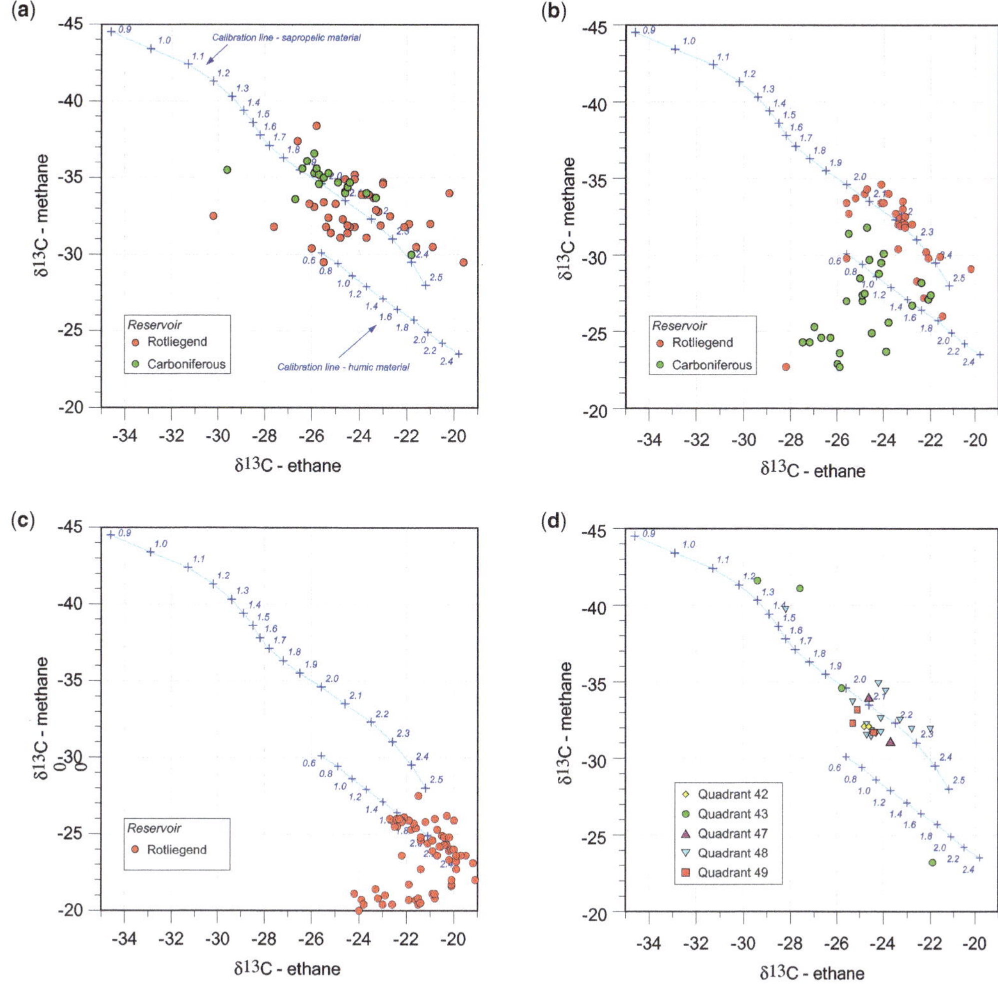

Fig. 19. δ^{13}C isotope data plots for gases in Rotliegend and Carboniferous reservoirs in different areas of the Southern Permian Basin. (**a**) UK and Netherlands offshore; (**b**) Netherlands onshore; (**c**) NW Germany onshore; and (**d**) UK sector only. Data in parts (a–c) from Lokhorst (1998); (d) proprietary data provided by Applied Petroleum Technology (UK) Ltd. Equivalent vitrinite reflectance value calibration lines use precursor δ^{13}C values of −29‰ for sapropelic and −23‰ for humic material (after Gerling *et al.* 1999).

(1) Most, if not all, of the gas discovered in the offshore area of the Southern North Sea (Population 1) is not derived from the Westphalian Coal Measures, but must instead be sourced from the marine shales in the Namurian and possibly Lower Carboniferous (equivalents of the Upper and Lower Bowland Shale of the UK onshore area and Geverik Member in the Netherlands). This is exactly consistent with the inferences that can be made in the UK Gainsborough Trough (Fig. 18) and with the situation in the Breagh Field, where basal Namurian shales accompanied by strong gas shows are encountered in the maturity range $R_o = 1.3$–1.42% in well 42/16-1 located 35 km downflank of the structure to the SW (Geolab UK 1994). Moving eastwards, the gases in the Netherlands onshore (Population 2) appear to consist of a mixture of gas derived from a high maturity (more deeply buried) Namurian marine source rock and less deeply buried, presumably Westphalian, coals. Only in areas of much deeper Mesozoic burial in the north German onshore do the Westphalian coals appear be the dominant source of migrated gas. This corresponds to areas of deeper burial in which the earlier Carboniferous marine shales are post-mature

(Teichmüller *et al.* 1979) and is consistent with the higher nitrogen contents of gases in these areas.

(2) In almost no cases in the offshore area (Fig. 19a) have any of the reservoired gases apparently been derived from source rocks in the early to middle parts of the conventionally accepted gas window. The generation window for gas in the Southern North Sea has generally been taken as between $R_o = \pm 1.1\%$ and $R_o = \pm 2.0\%$. The isotope results suggest that these thresholds are misleading for Carboniferous coals in the Southern North Sea; this question is further discussed in the following.

Source rock distribution and play fairways. If it is the case that most Southern North Sea gas is derived from Lower Carboniferous and Namurian basinal shales, rather than from Westphalian coals, then the entire understanding of the Southern North Sea petroleum systems needs revision. The area of potential source rock development is larger (see subcrop map in Fig. 17). The distribution of effective source rocks is no longer a question of preserved subcrop, but instead will rely on a much better understanding of stratigraphic relationships and basin architecture within the Carboniferous succession. Whereas previously source presence and effectiveness has not been regarded as a high risk, it becomes necessary to map intra-Carboniferous horizons at a regional scale and to define exploration play fairways that relate to both source rock presence (from mapping facies relationships) and source rock quality and maturity within individual source-prone stratigraphic units.

The question of the stratigraphic distribution of potential source rocks other than the Westphalian coals remains insufficiently studied. Lateral facies relationships in the onshore area demonstrate a northern limit to basinal shales in the Namurian (e.g. Fraser & Gawthorpe 2003), which can be replicated in parts of the offshore area where there are well penetrations (Fig. 20). However, in the majority of the main gas-bearing parts of the offshore area, there is no well that penetrates enough of the Namurian to allow such mapping. Coal seams have a wide distribution in the coeval shallow water facies in the basal Namurian and underlying Asbian to Brigantian in the Mid North Sea High and the northern parts of Quadrants 41–44 (the Yoredale and Scremerston formations of Cameron 1993), but the detail of the succession onshore shows significant changes in facies and thickness (e.g. Trueman 1954, pp. 293–297) that are glossed over in Cameron's unilateral extension of the onshore lithostratigraphy into the offshore area. The extent of these coals, particularly in Quadrants 41–44, cannot be taken for granted.

The shallow water facies of the Asbian to Alportian is also known to contain oil shales, some of which are geographically widespread (Archer 1926; Carruthers *et al.* 1927, 1932; Trueman 1954, p. 295). These have not been systematically documented or studied, but have the potential to act as sources of both oil and gas if locally developed in sufficient volume (Powell *et al.* 1976). Although the Asbian to Pendleian succession in northern Northumberland appears to contain few and thin oil shales, it should be borne in mind that the rich oil shales in the succession of the same age in Scotland were developed in an aerially restricted sub-basin of only *c.* 40 × 20 km extent, bounded by a combination of volcanic topography and syndepositional extensional faulting (Loftus & Greensmith 1988). The perceived scarcity of oil shale facies in Northumberland cannot be taken as evidence for their absence in the sparsely drilled areas of the Mid North Sea High or deeper sections that have not yet been penetrated by the drill bit in Quadrants 41–44.

The final and unanswered question concerns the possibility of an effective source in the deeper part of the Lower Carboniferous. Such a source, if present, potentially has the widest geographical distribution of any source rock and the best potential for maturation. It would also underlie, and potentially charge, the widespread sandstone reservoirs in the Chadian to Holkerian Fell Sandstone Formation. The possibility of such a source is hinted at by the occurrence of reservoired gas in the Fell Sandstone in the Northumberland Trough (UK onshore well Errington-1, communicated in the presentation given by Parsons & Trythall 2015). These researchers suggested sourcing of this gas from coals in the stratigraphically overlying Scremerston Formation, but this seems unlikely in view of the low maturities reported at outcrop ($R_o = 0.7\%$ in Namurian coal at outcrop 6 km to NW, Burnett 1987; $R_o = 1.0\%$ reported at 610 m depth in the Stonehaugh Borehole 19 km to the NW, Scott & Colter 1987). These shows must be sourced from deeper horizons in the Visean or Tournaisian sections, which have largely been regarded as devoid of source material on the basis of a few well penetrations in UK Quadrants 41 and 44.

Some encouragement may be offered by the presence of coaly material accompanied by gas shows in the deepest section, of Arundian to Holkerian age, penetrated at the western, most marine-influenced end of the Northumberland Trough in well West Newton-1 (OGA released well data). The dominance of more terrestrial environments in the pre-Asbian succession on the eastern end of the Northumberland Trough might be taken to imply a lower probability of source rock development in this interval; however, coal and mudrocks with moderate TOC contents are recorded in the Chadian to lower Holkerian Cementstone Formation just above total depth in well 41/10-1 (operator's well completion report released by UK OGA; Gent 2015). The

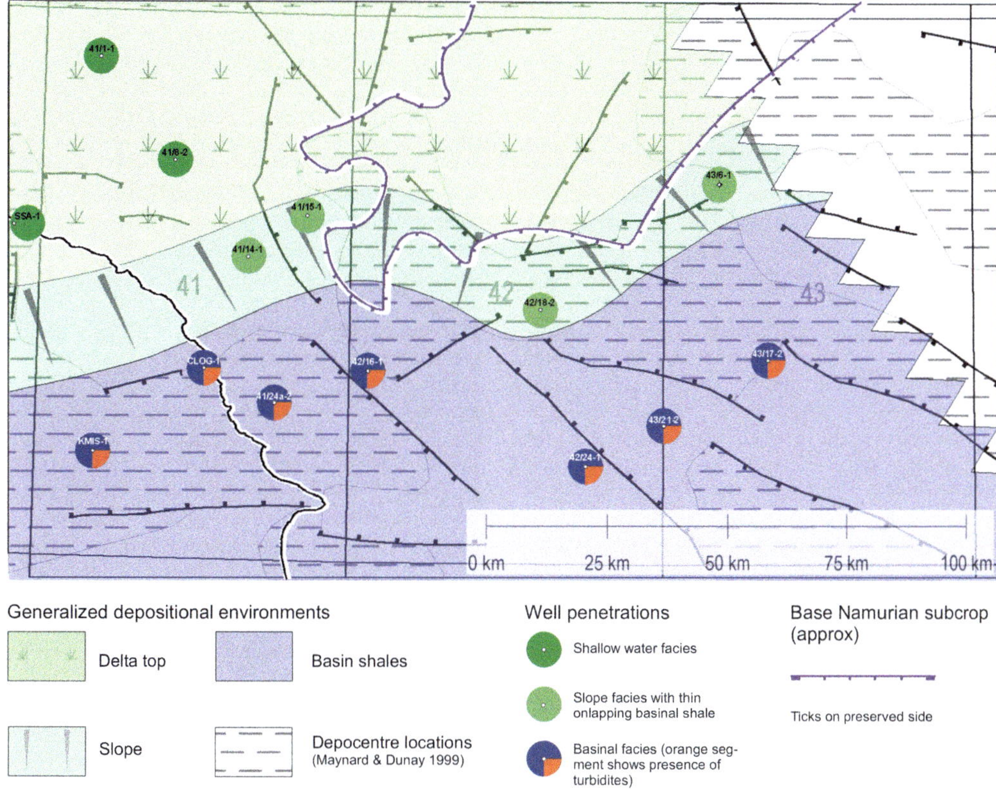

Fig. 20. Palaeogeographical map of the northwestern area of the main gas fairway in the UK Southern North Sea showing the position of the shallow to deep water facies transition at the time of deposition of the main Namurian basinal shale source rock of Pendleian age. Very sparse well control means that boundaries are approximate, particularly given the complex controls on bathymetry arising from the poorly understood framework of late Visean rift depocentres and footwall uplifts. Wells marked as containing slope deposits are those in which the Bowland shale is recognizable by a thin, possibly onlapping, development of high gamma ray shale intercalated within non-basinal deposits. A lack of well penetrations prevents extension of this map to the NE.

occurrence of bitumen throughout the pre-Asbian section penetrated in the Seal Sands borehole (Johnson *et al.* 2011) also requires a deeper, pre-Scremerston source rock.

The 'missing gas' problem. As well as revealing the unexpected patterns of gas sourcing, the isotope data discussed in this paper raise a significant problem. There appears to be little or no reservoired gas in the Southern North Sea area that has been sourced from coals or disseminated terrestrial material in the lower part of the conventionally accepted peak gas generation window (R_o = 1.2–1.6%). This is despite the presence of large volumes of coal in the Westphalian that must be in this maturity window, which conventional wisdom would dictate to be a volumetrically overwhelming source of gas.

Unlike the oil source rocks of the Central and Northern North Sea, in which every aspect of the primary hydrocarbon expulsion mechanism has been intensely studied (see summary in Cornford 1998, pp. 410–414), the conventional oil industry has paid little attention to the processes by which gas is generated in, and expelled from, the coal seams and associated organic matter in the Carboniferous. This makes it difficult to fully analyse the missing gas problem. However, a number of preliminary suggestions can be made.

(1) As has already been mentioned, the basic assumptions underlying the interpretation of the dominance of sapropelic source material may be incorrect. The maturation trend of Berner & Faber (1996) for humic source materials below maturities of R_o = 1.5% is extrapolated, albeit with empirical supporting evidence. The true maturation trend, in combination with different assumptions regarding the initial $\delta^{13}C$ composition of the organic matter, might bring the isotopic ratios of gases derived from

coals at low maturities closer to those of sapropel-derived gas.

(2) The underlying assumption still remains that the gas is derived from the bedded coals in the succession and that those coals are dominantly humic in composition. The reality is that both the make-up and distribution of source material in the Westphalian is much more diverse. Coals of this age in the UK onshore have a wide range of compositions, in many cases containing significant volumes of oil-prone algal and cuticle material (e.g. BCURA 2002). These occur both within the main body of a coal seam dominated by humic material (e.g. Smith 1968) and in intimately associated units of cannel coal and boghead coal formed by the accumulation of algal material in lakes and pools associated with coal swamps (Sullivan 1959). Leaving the bedded coals aside, Gent (2015) identify thick units (up to 15 m) of mudrocks with a TOC of 2–4% in the Westphalian A and upper parts of the Namurian. Evidence from the UK onshore suggests these may correspond either to the fairly abundant Marine Bands known at these levels, or to lacustrine oil shales, which have localized occurrences throughout the English coalfields (e.g. Flintshire, North Staffordshire; Giffard 1938). It is thus likely that maturation of a typical Westphalian succession in the conventional early to peak gas window (R_o = 1.1–1.6) will generate gas from both humic and sapropelic sources. Because of the differences in expulsion efficiency between bedded coals and sapropelic material disseminated in shale (Pepper & Corvi 1995), the sapropel-sourced gas may dominate in the expelled and migrated product.

(3) Most burial history models for the Carboniferous succession in the Southern North Sea assume limited overburden removal during Variscan inversion (Fig. 7a, b) and the onset of significant gas expulsion at the lower threshold of the generally accepted 'gas window' of $R_o > 1.1$ or 1.2%. As a result, it is generally believed that the onset of gas generation postdates Variscan uplift. However, both of these assumptions are open to question and it may be postulated that the generative capacity of the coal seams has been impaired by an early phase of gas loss associated with pre-Variscan burial and Variscan inversion.

The generation and expulsion of gas from coal seams commences at maturity levels significantly below the lower threshold of the generally accepted 'gas window' of $R_o > 1.1$ or 1.2%. Cornford (1998) places the onset of wet gas generation in humic source rocks at *c.* $R_o = 0.8\%$ and the gas yield curves reported by Gaschnitz (2001) for Westphalian coals from the Ruhr coalfield show methane generation starting at levels of *c.* $R_o = 0.65\%$, with the yield exceeding the sorption capacity of the coal (as defined by the curves created by Hildenbrand *et al.* 2006) at a maturity level of $R_o = \pm 0.8\%$. This implies that gas expulsion from coals starts much earlier than is generally envisaged in conventional assessments of bedded coal source rocks, as is suggested by Galimov (2006). The data of Galimov (2006) suggest that by the time more 'typical' gas window maturities are reached, *c.* 40% of the ultimate methane generation has already occurred.

It may therefore be suggested that, even in successions where pre-Variscan maturation failed to reach, or only just reached, the 'conventional' gas window, significant amounts of gas were generated during late Carboniferous burial and expelled during Variscan uplift before the onset of the Mesozoic and Cenozoic burial that is supposed to control the present disposition of gas kitchens. This degassing may have impaired the generative capacity of the coals during subsequent deeper burial. The extent of Variscan degassing can be assessed in the Ruhr area in Germany and the Campine Basin in Belgium, where maximum burial and maturation occurred before the Variscan uplift (Littke *et al.* 2000). Studies undertaken in these areas in the course of coal bed methane evaluation have shown that the changes in gas sorption capacity that accompany post-Variscan uplift and subsequent reburial do indeed result in significant losses of any gas generated during pre-Variscan burial (Juch *et al.* 2004; Hildenbrand *et al.* 2006). Except where in-seam migration has occurred, the coals are significantly undersaturated (Hildenbrand *et al.* 2006; Freudenberg *et al.* 1996), showing that no further gas generation has occurred during post-Variscan reburial. However, it is clear from the data of Gaschnitz (2001) that, even if 40% of the potential gas has been lost from the coals that reached maturities of *c.* $R_o = 1.1\%$ prior to Variscan uplift, more than enough replacement gas can be generated if post-Variscan reburial brings the coals into the higher maturity ranges (R_o = 1.0–1.6%; see Fig. 7a) currently found in the Southern North Sea area.

The loss of early generated gas therefore only appears to form an acceptable explanation for the missing gas problem if burial and maturation during the late Carboniferous were substantially greater than is currently accepted. The general acceptance that maturation and gas generation in the Westphalian coals is the result of post-Variscan burial relies on burial history analysis from various parts of the basin (e.g. Leeder & Hardman 1990; Pearson & Russell 2000), supported by a variety of maturity indicators (e.g. AFTA, sonic velocities; see review by

Green 2005). The main emphasis in all of this work has been to establish the post-Variscan burial history and little attention has been paid to the extent of late Carboniferous burial and maturation.

Pearson & Russell (2000) evaluated several different burial models for the key UK onshore calibration well Up Holland-1 (Fig. 21). One of their models (Model A) envisages maximum burial of Carboniferous source rocks at the end of the Carboniferous or in the early Permian. In this model, the maturity profile in the Carboniferous penetration is identical to that produced in their preferred model (Model D), which involves maximum burial during the late Cretaceous. Despite this inherent ambiguity, their preference for the model involving maximum burial in the Mesozoic was based on the occurrence of one published well section (Bardney-1; Fraser *et al.* 1990), in which the relationship of maturation gradient in both Carboniferous and post-Carboniferous sediments could be observed, and the consensus established from the various AFTA studies (Bray *et al.* 1992; see also Green 2005).

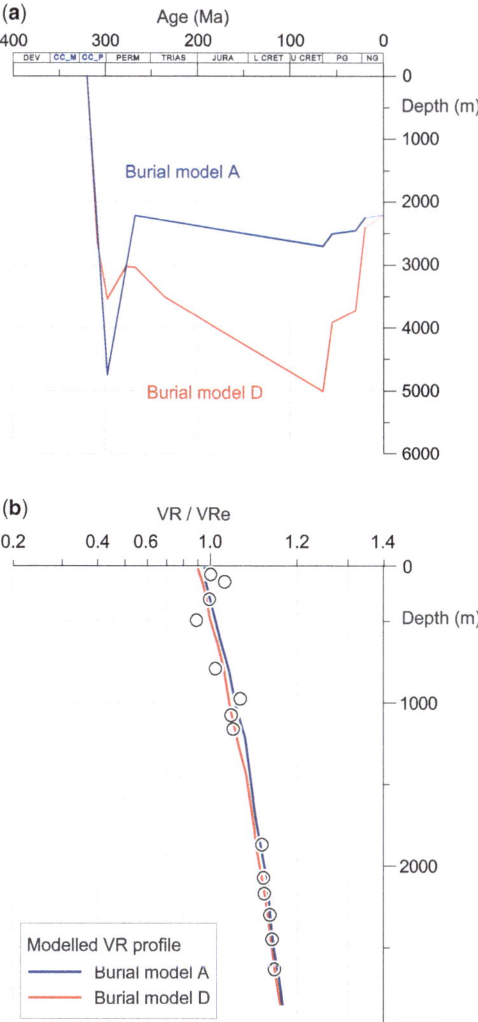

Fig. 21. (**a**) Alternative burial histories and (**b**) resulting modelled vitrinite reflectance (VR) profiles for Westphalian and Namurian section in UK onshore well Up Holland-1. Depth profiles in part (b) have been corrected for offsets of Cenozoic age faulting. Note the similarity of the maturity profile produced by contrasting burial histories. Redrawn from Pearson & Russell (2000). CC_M – Mississippian; CC_P – Pennsylvanian. See text for discussion.

The Up Holland-1 study demonstrates the key failing of this assumption. Multiple burial histories can generate the same observed outcome and no study to date has attempted to differentiate between coals that reached a given level of maturity during Carboniferous burial that has only been equalled or very slightly exceeded following post-Variscan burial from those that have undergone the most significant portion of their maturation during a more recent phase of burial. Although the second group may have reached their present maturity levels with the capacity to generate and expel significant volumes of gas, the generation potential of the first group may have been significantly impaired.

A possible calibration of the generally accepted burial models is provided by Creedy (1991), who summarizes the many years of work done by the British Coal Corporation in their attempts to model gas occurrences as a mining hazard. His analysis of an extensive dataset concludes that the current distribution of gas within the coal seams results from peak maturation having been reached before Variscan uplift, which was accompanied by a significant loss of gas. Fortuitously, the geological cross-section in Creedy (1991) passes close to the well Gainsborough-2, for which maturity indicators and modelled burial history are given by Pearson & Russell (2000). The interpretation of Creedy (1991) directly contradicts the model of Pearson & Russell (2000) and suggests that, in this area at least, a burial history similar to their Model A (Fig. 21) would be more appropriate. Such a burial history would be compatible with the published maturity profile from the Kirby Misperton Field in North Yorkshire, which shows a divergence in maturation gradient between the sections over- and underlying the Variscan unconformity, with a maturation leap at this interface implying maximum burial of the Carboniferous prior to Variscan uplift (Hughes *et al.* 2018).

The extent of end-Carboniferous and early Permian burial is further considered in a later section of this paper. To conclude the present section, it suffices to point out that, in the adjoining East Irish Sea Basin, more extensive burial history studies show that the patterns of burial history are much more diverse than is currently envisaged in the Southern North Sea area (Green *et al.* 1997). Studies of shale gas prospectivity in this area have shown that, at least in some areas, a burial history involving a high degree of burial and uplift in the late Carboniferous is appropriate (Andrews 2013; burial history analysis of wells Thistleton-1 and Hesketh-1). It is thus likely that superficially similar maturity profiles in the Carboniferous succession may have been produced by varied burial and exhumation histories, some entailing peak maturity being reached prior to Variscan inversion and others entailing a history of gas generation and migration related to post-Variscan burial. In the absence of much more extensive AFTA calibration than is presently available, it is difficult to distinguish between these possibilities. It should be remembered that AFTA and other related data can only constrain the extent to which the burial of Carboniferous source rocks is a function of post-Variscan burial.

Why is the missing gas problem important? In areas where there is a demonstrably functioning source rock and migration system – that is, in all of the areas of established discoveries – it is really only of academic interest. However, future success in the Southern North Sea will involve pushing exploration beyond the existing fairways and, in particular, will rely on the successful identification of a play fairway sourced by the abundant coals in the upper part of the Visean and lower Namurian in the shallow water facies flanking the Mid North Sea High. If, for whatever reason, the generative capacity of these coals has been impaired by an early phase of maturation and degassing related to Variscan burial and inversion, then the prospectivity of these areas is greatly downgraded.

Founding Myth 2: 'Carboniferous reservoirs are confined to channel sands'

Almost all production from Carboniferous fields in the Southern North Sea has hitherto come from fluvial channel sandstones. The reservoir quality of these sands appears to be highly variable, with significant degradation due to diagenetic clay (O'Mara *et al.* 2003*a*; Collinson 2005) and pre-Variscan compaction (Bailey *et al.* 1993). They also show variable degrees of post-depositional enhancement, possibly related to weathering beneath the base Permian unconformity (Bailey *et al.* 1993) and/or grain dissolution related to fluid flow associated with fluid migration events during the Mesozoic and unrelated to the base Permian unconformity (Johnson *et al.* 1995). Bailey *et al.* (1993) suggest that reservoir quality decreases with increasing depth in the original stratigraphic thickness as a result of compaction, but that secondary enhancement is present in the topmost 200 m beneath the base Permian unconformity. It is difficult to substantiate this without their original data, but it is evident that the best quality in such supposedly enhanced reservoirs occurs in the coarsest grained deposits and this reinforces the consensus view that effective reservoirs occur in channel sands.

The only productive reservoirs that are not interpreted as fluvial sand bodies occur in the Trent, Cavendish and Pegasus fields, where production comes from clean, quartzitic sandstones in the upper Namurian and basal Westphalian. These were originally interpreted to be the products of reworking by wave or tidal processes (O'Mara *et al.* 2003*a*), but recent work suggests that they may be related to diagenetic enhancement due to the leaching of feldspar and clays in flow paths related to regional seals (J. Collinson, pers. comm. 2017).

The dominance of channel sands can be called into question. Two aspects of petrophysical understanding and exploration concepts combine to demonstrate that a much wider range of reservoirs can be successfully developed, especially when hydraulic fracture stimulation is applied (Chiswick Field, Nesbit & Overshott 2010; Breagh Field, Sterling Resources 2016): (1) the recognition of the extent of formation damage; and (2) the recognition of unconventional plays.

As has already been mentioned, the petrophysical re-evaluation that preceded the successful appraisal of the Breagh Field demonstrated that drilling Carboniferous sections with significantly overbalanced water-based mud leads to extreme formation damage. The damage mechanism has been described by McPhee *et al.* (2008) as involving filtrate retention, fines migration and mud solids invasion in a pore system dominated by micropores as a result of clay diagenesis.

A quick review of a number of the wells included in the OGA public data release in early 2016 shows that drilling with a significant overbalance has been common practice in the UK sector of the Southern North Sea. The difficulty of petrophysical interpretation in such cases and the likelihood of extreme formation damage combine to suggest that producible reservoir sections may have been overlooked in many wells in any sandstone facies. Although the majority of sandstone reservoirs above mid-Namurian levels will of necessity be found in channel sandstone deposits, significant sand bodies of probable turbidite origin have been found in the lower part of the Namurian (Collinson 2005), which may form a viable exploration objective. This play was identified by Cameron & Ziegler (1997), but has not, apparently, been specifically targeted in any exploration well to date, the handful of penetrations being found in deep stratigraphic tests (e.g. 43/17-2, 43/21-2).

The past decade has seen the beginning of exploration for unconventionally reservoired gas in the Carboniferous in the UK onshore area, driven by perceived similarities between British Carboniferous sections and the successful Carboniferous Barnett Shale plays in the USA. Potentially commercial discoveries have been made in the Bowland Basin, Lancashire (Clarke *et al.* in press) and in North Yorkshire, where deeper stratigraphic horizons beneath the small Namurian-reservoired Kirby Misperton Field may, if successfully stimulated and tested, contain a large recoverable resource (Hughes *et al.* 2018). The Bowland Basin discovery is in a shale gas reservoir, which, although providing invaluable additional insights into Carboniferous petroleum systems, is not a realistic exploration target in the offshore area under current commercial conditions. The Greater Kirby Misperton discovery is, however, not a shale gas discovery. The gas is contained in a hybrid play consisting of finely interbedded mudstones and naturally fractured tight siltstones and fine sandstones. The reservoir potential of these sandstones was missed in early log interpretations: their thin-bedded nature makes them invisible on conventional logs and the failure to apply appropriate shaly sandstone methods in petrophysical evaluation led to an underestimation of their porosity and an overestimation of their water saturation. The sandstones have porosities of up to 10% and permeabilities in the microdarcy range. They are probably of pro-delta turbidite origin.

Although no such accumulation has previously been developed offshore, the presence of a natural fracture system suggests an inherently greater permeability than in a pure shale gas play, whereas the dominance of sandstone suggests that more stable completions and longer well lives might be expected. An offshore development in such a reservoir might be feasible. A number of wells in UK Quadrants 41 and 42 have penetrated thick successions that bear a strong resemblance to, and are of roughly the same age as, the newly recognized unconventional reservoirs at Kirby Misperton. At least two of these wells contain significant (±1000 ft) gas columns. Press releases issued while this paper was in press confirm that the first well to test this play (43/26a-E12Z) was abandoned in October 2017 as a technical but not commercial success.

Although channel sandstones will still form the default reservoir when exploring for Carboniferous objectives, a new approach to the choice of drilling fluids and practices and to petrophysical evaluation means that all Carboniferous sand bodies should, in future, be regarded as potential objectives. Diagenetic reservoir degradation need not preclude effective reservoir development and hitherto unrecognized pay in thin-bedded fine-grained turbidites offers a new exploration target.

Founding Myth 3: 'Intra-Carboniferous seals are risky'

Although shales comprise a significant proportion of all Carboniferous clastic successions, their sealing potential is compromised by the silt-rich nature of most of the mudrocks deposited in proximal deltaic and alluvial plain settings (Fraser *et al.* 1990). It is assumed that effective seals are only provided by the shales deposited in marine flooding events, although these are generally only 1–2 m thick and are thus vulnerable to breach by even minor faulting. In the Namurian and basal Westphalian, some specific Marine Bands have been identified as effective seals (Cameron *et al.* 2005; Hodge 2003). In other cases (Cavendish Field, Boulton 'F' Field, Ketex discovery) the exact stratigraphic horizon of the seal is not known, although it is clear in the latter two cases that the seal must be formed by a lacustrine shale because no Marine Band is present at the requisite stratigraphic horizon.

In only one instance has a confident evaluation of the sealing capacity of a Carboniferous shale been published. The Calow Field in the UK onshore is a clearly underfilled dip-closed Variscan inversion anticline with a structural closure in excess of 100 m and containing a gas column of 30–40 m (Fraser *et al.* 1990). This structure is at a very shallow depth in an area of considerable Cenozoic to Recent uplift and stress release. It may be argued that the restricted columns in this example are unlikely to be representative of the situation at depths typically encountered in the offshore area. However, the combination of this published example and the generally thin nature of intra-Carboniferous seals has led to a generally pessimistic view of the possibility of sealing structures within the Carboniferous succession. This may be one of the reasons why comparatively few structures that lack a base Permian closure have been drilled.

A further paradox is introduced by the fact that although shales at base Westphalian and intra-Namurian level have been demonstrated to be effective field-scale seals, there is widespread evidence for the vertical migration of gas through the same stratigraphic horizons (e.g. the isotope data cited earlier in this paper; empirical source to reservoir relationships in the East Midlands, Fraser & Gawthorpe 2003; Ward *et al.* 2003; Fig. 18). There is clearly more going on than simple constraints on seal capacity imposed by the effective seal thickness coupled with typical fault offsets. This remains a subject area that is virtually unstudied. It is evident from coal mine gas studies in the East Midlands that the migration of hydrocarbons derived from the Namurian shale source kitchens is occurring at the present day along faults (Creedy 1985, pp. 238–242; Stevenson 1999). Anecdotal evidence reported by Creedy (1985) suggests that current active migration is associated with small faults rather than with larger structures, a suggestion supported both by Stevenson (1999) and by recent observations made by the author of a major gas and oil influx at Maltby Colliery, South Yorkshire. Here, inflow was associated with subsidiary Riedel shear fractures rotated at *c.* 35° to the major bounding faults of the Gainsborough Trough. These fractures are currently oriented normal to the current minimum horizontal *in situ* stress (i.e. having orientations most likely to create open flow paths).

A model of sealing capacity for intra-Carboniferous shales must therefore take three factors into account: (1) the thickness and capillary properties of an individual candidate shale seal; (2) the extent and magnitude of any faulting affecting the structure; and (3) the *in situ* stress conditions, which may dictate whether or not the faults are likely to form fluid migration conduits. The presence of the effective intra-Carboniferous topseal in the Saltfleetby Field demonstrates that all three criteria can be satisfactorily achieved. It should therefore be possible to identify and map candidate seals that, in suitable structural situations, can form the sealing element of separate play fairways within the Carboniferous succession. The discovery of a 'giant' gas column at Kirby Misperton, greatly exceeding the 1000 ft of mapped structural closure (Hughes *et al.* 2018), requires the presence of multiple seals and may suggest some form of stratigraphic or continuous trapping that expands the spectrum of Carboniferous exploration targets.

Founding Myth 4: 'Basin geometries are fully characterized'

In the past 20 years, studies of the basin history of the Carboniferous in northern England have coalesced around a fairly generally accepted megatectonic framework of fault-bounded Lower Carboniferous highs separated by contemporaneous rift basins. These affect facies development far beyond the active rifting phases, continuing to control depocentre positions and sediment transport pathways well into the Westphalian as a result of minor tectonic accommodation, the infill of relict topography and differential compaction. The megatectonic framework maps (Corfield *et al.* 1996; Fraser & Gawthorpe 2003) have become so familiar and so often republished that it is easy to forget that they are, to a degree, interpretative. Thus, for instance, the pattern of thinning and onlap in the Brigantian to Pendleian within the Alston Block is not at all as would be implied from the generally accepted map, which implies rapid northwards thickening over the Ninety Fathom Fault (the northern bounding fault of the 'Block') rather than the gentle northwards thickening and southwards onlap actually observed (Ridd *et al.* 1970; Chadwick *et al.* 1995). In detail, it is clear that the bounding structures of the highs and lows are highly variable in style, locally consisting of a single large fault, but elsewhere consisting of a complex set of flexural structures and ramps (Fig. 22). The present day expression of the structures at the land surface, or at the sub-Permian subcrop surface, is the result of their response to Variscan inversion and does not always correspond to the position of the major structure at depth (Kimbell *et al.* 1989). Detailed understanding of these structural relationships is key to understanding sediment routing and fluid migration paths.

If the basinal geometries in the onshore area are more complex than generally envisaged, the situation in the offshore is even less clear. Here the deep geometries of the sub-basin areas are not known from drilling; mapping of depocentres has relied entirely on seismic interpretation and gravity modelling, the latter both to locate the positions of granite bodies embedded in basement highs and to delineate deep basin areas. In the UK sector, many publications have recycled the offshore block and basin map of Corfield *et al.* (1996), to the extent that it has effectively become the standard. There are, in fact, four published interpretations of the distribution of highs and basinal areas (Fig. 23). The striking feature of these is the extent of disagreement between them. Even in the comparatively well-known onshore area, major differences of interpretation are apparent, whereas in the offshore there is a lack of agreement as to the location of major basin areas, which fault systems mark major elements in the Carboniferous structural framework and whether bodies interpreted as granite-cored highs on the basis of gravity anomalies might, in fact, be basin depocentres.

Pointing out the differences between these interpretations does not imply any criticism. It is extremely difficult to unravel the deep structure given: (1) the distorting effects of the post-Variscan succession – especially of the mobile Zechstein salt – on the pattern of gravity anomalies; and (2) the difficulty in differentiating between primary data and multiples in the seismic from the deep Carboniferous basin areas (e.g. Cameron & Ziegler 1997, fig. 13). However, the detail of the disposition of the basin segments must play a key part in controlling the source and reservoir facies, source rock maturity gradients and migration paths, especially in the less explored targets in the lower Namurian and Lower Carboniferous. An understanding of these play elements needs an improved definition of the basin geometries.

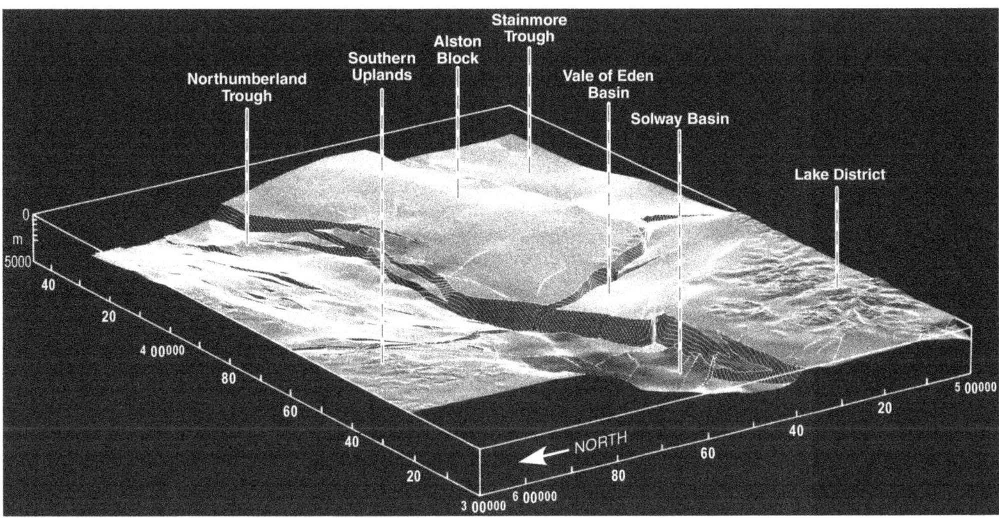

Fig. 22. Variability in structural style of Carboniferous basin depocentres. Northumberland Trough example (Chadwick *et al.* 1995, fig. 7), showing variation in style of extensional fault systems bounding Lower Carboniferous rift basins with the development of complex ramp geometries. Reproduced with permission of the British Geological Survey ©NERC. All rights reserved.

Founding Myth 5: 'Basins formed in rift and sag episode(s)'

The final myth concerns the very fundamentals of understanding the structural and stratigraphic evolution of a sedimentary basin: the gross tectonic environment in which it formed and the stress regimes and resulting subsidence patterns under which sedimentary accommodation space was created.

Because of the extensive stratigraphic and sedimentological work that had taken place on the Carboniferous of northern England in the 1960s and 1970s, it is not surprising that these basins should have been among the first for which subsidence histories were analysed and suggestions made as to their overall plate tectonic setting following the radical changes in basin formation concepts of McKenzie (1978), among others. Leeder (1982) postulated that the block and basin geometry in northern England resulted from a phase of late Devonian and early Carboniferous rifting, followed by a phase of thermal relaxation, possibly accompanied by some wrench faulting, in the Namurian and Westphalian. In a subsequent paper (Leeder and McMahon 1988) undertook a decompaction exercise on four areas, concluding that the Northumberland, Stainmore and Bowland basins had formed in a phase of Visean to early Namurian extension, followed by widespread thermal subsidence in the later Namurian and Westphalian, in which former highs such as the Alston 'Block' were onlapped and incorporated into the depositional area. Subsequent authors have refined this interpretation, modifying the extension factors of Leeder and McMahon (1988), which were generally perceived to be too large (Kimbell *et al.* 1989), identifying timing differences between the major extensional phases in different sub-basins (Fraser & Gawthorpe 2003) and suggesting at least local inversion related to a wrench deformation episode in the Namurian (Chadwick *et al.* 1995). These subtle modifications have not had any impact on the wholesale adoption of a model of Tournaisian–Visean rifting followed by Namurian–Westphalian thermal subsidence, which is now deeply embedded in the British literature.

It was not Leeder's intention to write the tectonic history of the British Carboniferous on tablets of stone. At the time the papers were written he was proposing hypotheses and, in the 1988 paper, it is specifically stated that 'the present knowledge of basin evolution for northern Britain is at a primitive stage of development' (Leeder and McMahon 1988). The latter paper also identifies departures from the ideal model subsidence patterns, with accelerations of subsidence in the Westphalian in Lancashire and Durham that do not fit the simple model. At the time of these studies it was difficult to compile enough data on stratal thicknesses to analyse subsidence patterns throughout the region. It was not always clear whether the compiled data was representative of a basin as a whole, assembled as it was largely from outcrops of the inverted margins of the depocentres.

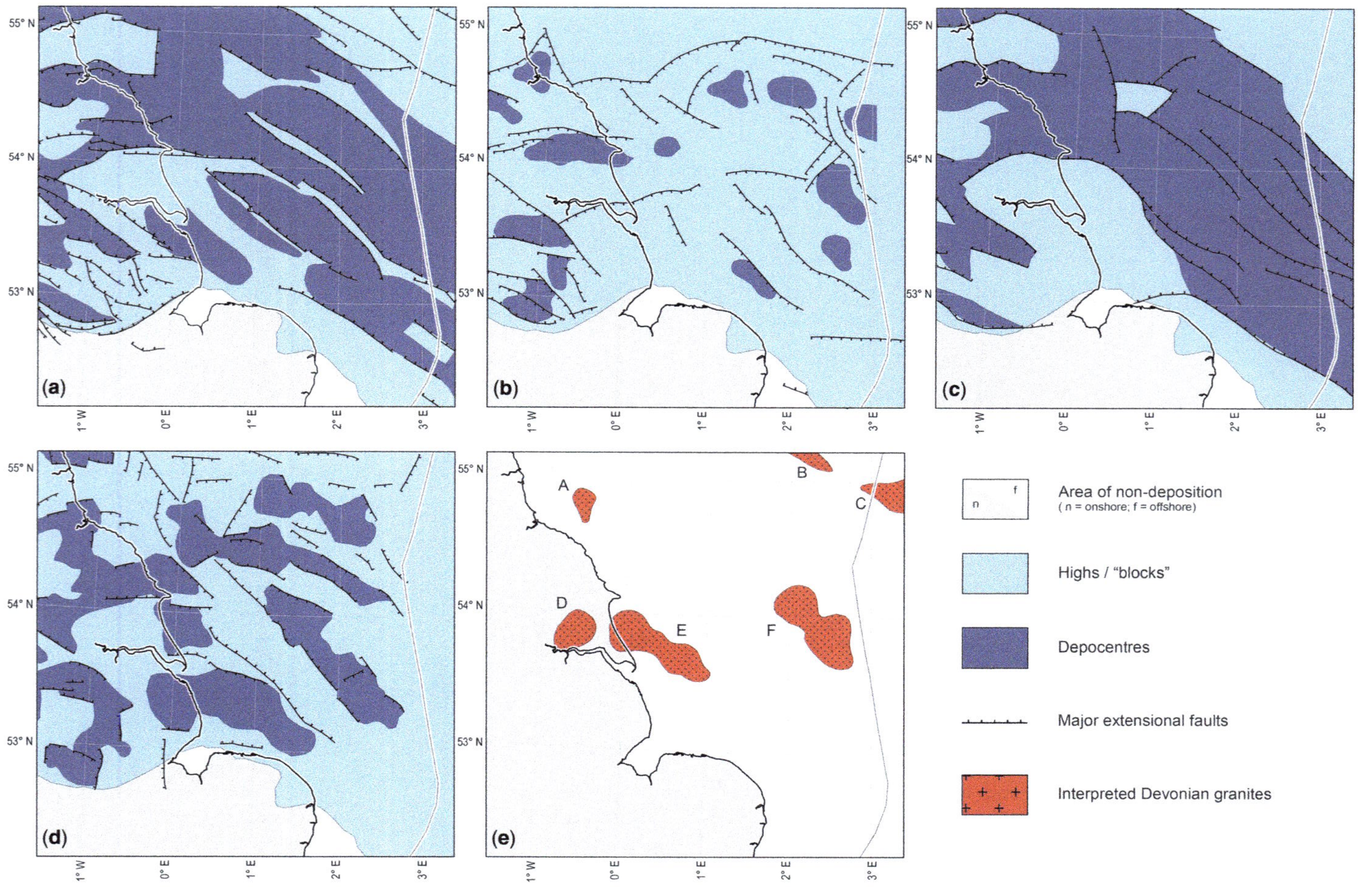

Fig. 23. Comparison of published interpretations of locations of Lower Carboniferous highs and basinal depocentres in the UK offshore area: (**a**) Fraser & Gawthorpe (2003) (UK onshore), Corfield *et al.* (1996) (UK offshore), Kombrink (2008) (the Netherlands); (**b**) Collinson *et al.* (1993); (**c**) Cameron & Ziegler (1997); (**d**) Maynard & Dunay (1999); and (**e**) positions of inferred granites in underlying basement highs (Donato *et al.* 1983; Donato & Megson 1990; Donato 1993; EBN 2015).

The publication of more comprehensive stratigraphic sections by the British Geological Survey (summarized in Waters *et al.* 2011), together with released results of some significant deep onshore wells, allow the compilation of a more extensive set of subsidence curves (Fig. 24). Although no decompaction has been carried out, the patterns are striking. The depocentres studied in Leeder and McMahon (1988) (Fig. 24a) generally show the expected rapid phase of Visean rift subsidence, followed by a slowing that can be equated to a thermal relaxation phase in the Namurian: the anomalous acceleration of subsidence in the Westphalian is confirmed. However, other depocentres within the overall 'Pennine Basin' rift and sag area (Fig. 24b) are dominated by accelerated subsidence in the Westphalian, with little evidence for early Carboniferous rifting. In most cases the overall pattern of accommodation is comparable with that seen in generally accepted foreland flexure areas in South Wales (Kelling 1988) and the Ruhr Basin (Littke *et al.* 2000) (Fig. 24c).

It is not clear what is causing these varied subsidence histories in the basins to the north of the Brabant Massif. Modelling by Kombrink *et al.* (2008) eliminates the possibility that it is caused by foreland flexure, the effects of which do not extend beyond a few hundred kilometres north of the contemporaneous thrust front. The wrench-related deformation and subsidence effects identified by Chadwick *et al.* (1995) may be of much greater importance than hitherto recognized; indeed, it has been suggested that all Carboniferous basin development in northern England and, by implication, the Southern North Sea, should be interpreted in terms of a long-lived regional transtensional regime (de Paola *et al.* 2005). Such a model makes it easier to integrate into a general model the highly diachronous nature of the extension events when viewed over a wider area than just the north of England. A major Namurian rifting event is inferred in the Netherlands by Kombrink *et al.* (2008) and the continuation of extension into the Westphalian is described in the Central Irish Sea Basin by Maddox *et al.* (1995).

The main effect of the adherence to a rift and sag model for the evolution of basins in the UK area is that it has influenced the generally rather small thicknesses of late Carboniferous overburden that are assumed to have been removed by Variscan uplift. Most published burial history models for the UK area involve the removal of <1000 m of upper Carboniferous sediment in the Variscan deformation event (e.g. Fig. 7a, b; Pearson & Russell 2000), reflecting an assumption of gradual decline in thermal subsidence with time. These amounts are clearly inconsistent with the majority of the Carboniferous basin subsidence curves illustrated in Figure 24. The burial history plot for well 48/3-3, compiled before stratigraphic relationships in the Southern North Sea were fully understood, is not even consistent with what is now known about regional stratigraphic thicknesses. The 700 m of removed section

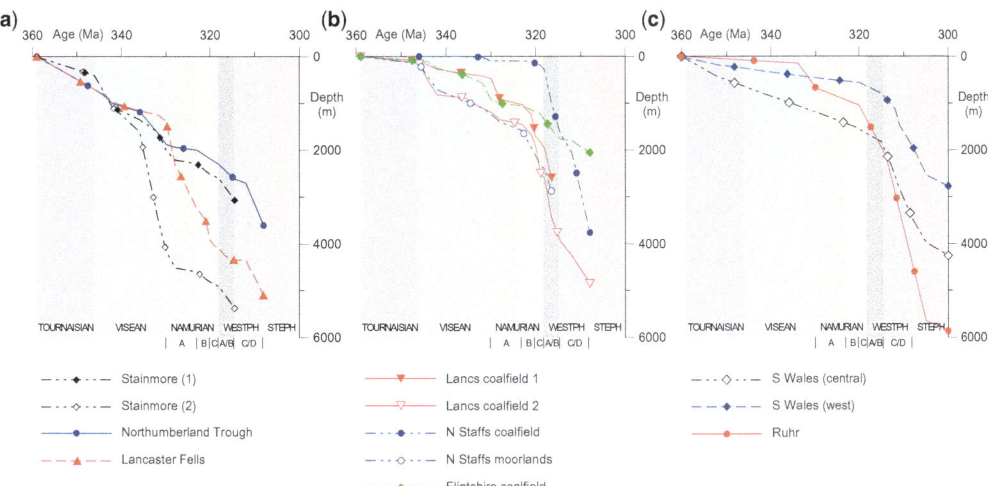

Fig. 24. Compilation of undecompacted Carboniferous subsidence curves for various sub-basin areas in onshore areas surrounding the Southern North Sea: (**a**) basin segments previously interpreted to show the classic rift and sag subsidence history; (**b**) other basin segments in or near the centre of the supposed post-rift thermal sag bulls-eye (cf. fig. 12 in Fraser *et al.* 1990); and (**c**) basin segments in Variscan foredeep area. Data generally from Waters *et al.* (2011) except for Seal Sands borehole (Stainmore 2), South Wales data from Kelling (1988); Ruhr data, Isselburg-3 well from Littke *et al.* (2000).

in Figure 7a compares with a demonstrable removed section of 1800 m (compilation forming fig. 6.12 in Kombrink *et al.* 2010). This should be regarded as a minimum estimate for this area. If the thicknesses of latest Westphalian D and Stephanian observed or inferred in the English Midlands are added, then the removed section might reach thicknesses of between 2600 and 3300 m. End-Carboniferous burial of this extent (Fig. 25), similar to that proposed in Model A by Pearson & Russell (2000), and by Andrews (2013) for wells in Lancashire, has the potential to cause significant pre-erosional maturation and provides a mechanism to explain the lack of coal-derived gas of early mature origin, identified here as the 'missing gas' problem.

In Germany, where thinking has not been dominated by the rift and sag model, extensive thermal, rheological and numerical modelling has led Littke *et al.* (1994, 2000) to conclude that, contrary to previous interpretations (e.g. Ziegler 1990), there was a widespread and thick late Westphalian and Stephanian cover in the Ruhr Basin and NW Germany, with up to 3000 m of erosion associated with the Variscan deformation. In areas proximal to the present Variscan Front, the large thickness results from flexural foreland subsidence. In more distal areas the accommodation may be interpreted to result from a complex subsidence history that includes one or two phases of latest Carboniferous and early Permian extension, with the localized formation of deep pull-apart graben as a result of wrench faulting resulting from regional sinistral transpression (Coward 1993; Schroot & De Haan 2003). It is, however, difficult to create a convincing global interpretation because the time gap between the youngest preserved Carboniferous succession and the base of the overlying Permian is so large (at least 30 Ma) and it is known that this period encompasses a number of phases of basin formation and uplift elsewhere in the basin complex and adjacent intramontane basins (e.g. Roscher & Schneider 2006). A clearer understanding of the latest Carboniferous and early Permian subsidence and burial history patterns will not be obtained until detailed modelling of the type carried out by Littke *et al.* (1994, 2000) for the Ruhr Basin is undertaken throughout the UK onshore and Southern North Sea areas.

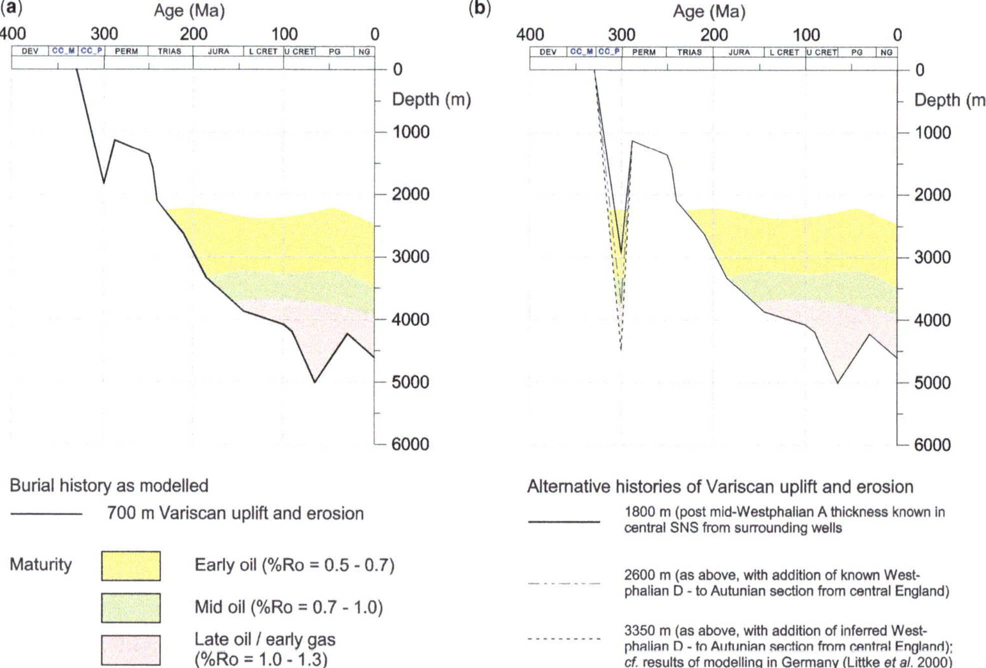

Fig. 25. Alternative possibilities for burial history, UK well 48/3-3 (redrawn after Leeder & Hardman 1990) showing possibilities for enhanced early maturation of the Westphalian succession: (**a**) as published; and (**b**) with realistic alternative values of Variscan denudation. Note that the post-Variscan maturation history has not been remodelled. The combination of greater pre-Variscan burial and loss of early generated hydrocarbon during Variscan uplift result in the present day maturity being similar to, or only slightly greater than, that reached prior to Variscan uplift. CC_M, Mississippian; CC_P, Pennsylvanian.

Discussion and conclusions

Fraser & Gawthorpe (2003) started their monumental *Atlas of Carboniferous Basin Evolution* with the challenge, 'Perhaps all we really know about the Carboniferous is no more than skimming the surface'. Despite decades of research, they concluded – correctly – that a thorough integration of local and regional, surface and subsurface data would yield insights into the structure and stratigraphy of the Carboniferous that would far exceed the sum of the component parts. The same is true for the broader study of the Carboniferous as a petroleum system, or set of petroleum systems, spread across the larger area of the UK and Dutch Southern North Sea. Here, in addition to the challenges of integrating surface and subsurface data, there are the complexities introduced by complex overburden, poor imaging, significant parts of the stratigraphy having little or no well penetration, a range of data-gathering problems induced by technical difficulties in drilling very deep wells, strongly differing local traditions and attitudes to geological interpretation, and language barriers. Add to these the huge stratigraphic thickness of the Carboniferous succession and it becomes less surprising that an integrated basin-wide understanding of the petroleum systems is still in its infancy.

What is evident from this review of the founding myths of Carboniferous geology is that the Carboniferous, far from being a single play, is made up of a number of play fairways consisting of different sources, migration paths, reservoirs, seals and trap configurations. A large number of conventional prospects have been known for some time (Cameron *et al.* 2005), but not drilled. Potential gas discoveries have probably been overlooked in the way that the Breagh Field was, owing to a combination of sub-optimal drilling practice and inadequate petrophysical evaluation. It seems almost certain that hybrid conventional/unconventional accumulations of the type found in the deep extension of the Kirby Misperton Field are present in the offshore area and that, if the technological and cost challenges can be overcome, these will dwarf previous Carboniferous discoveries. The challenge of effectively exploring the immense body of hydrocarbon-bearing rock is daunting, but there is no reason why, with thorough application of the basic principles of play fairway analysis, it should not be successful.

Acknowledgements This paper represents a set of views distilled from many years of working on the Carboniferous in NW Europe as a student, academic, company staff geologist and consultant. The following companies and individuals are thanked for discussion and material assistance at various times: British Geological Survey, Centrica, Conoco Phillips, Cuadrilla Resources, EBN, Engie (Gaz de France), Faroe Petroleum, Esso UK, Shell UK, Third Energy, Tullow Oil; Pat Barnard, Marc Bustin, John Collinson, Steve Corfield, Chris Cornford, Peter Gerling, Colin Jones, Bruce Levell, Duncan McLean, Tim Pearce, David Creedy, Peter Turner. Particular thanks are due to Andy Mortimer and Julian Moore for stimulating discussion that has materially improved the paper, and to the meticulous editing and helpful suggestions of an anonymous referee. Centrica plc and Applied Petroleum Technology (UK) Ltd are thanked for providing images for Figures 2, 6 and 19 specifically for this compilation. Engie E&P UL Ltd are thanked for providing Figure 10. Opinions and interpretations contained herein are solely the responsibility of the author.

References

Abbotts, I. (ed.) 1991. *United Kingdom Oil and Gas Fields, 25 Years Commemorative Volume*. Geological Society, London, Memoirs, **14**, https://doi.org/10.1144/GSL.MEM.1991.014.01.69

Andrews, I.J. 2013. *The Carboniferous Bowland Shale Gas Study: Geology and Resource Estimation*. British Geological Survey for Department of Energy and Climate Change, London.

Archard, G. & Trice, R. 1990. A preliminary investigation into the spectral radiation of the Upper Carboniferous marine bands and its stratigraphic application. *Newsletters on Stratigraphy*, **21**, 167–173.

Archer, A. 1926. *Geology of Berwick-on-Tweed, Norham and Scremerston*. HMSO, London.

Bailey, J.B., Arbin, P., Daffinoti, O., Gibson, P. & Ritchie, J.S. 1993. Permo-Carboniferous plays of the Silver Pit Basin. *In*: Parker, J.R. (ed.) *Petroleum Geology of Northwest Europe: Proceedings of the 4th Conference*. Geological Society, London, 707–715, https://doi.org/10.1144/0040707

Bailey, W.R., Manzocchi, T. *et al.* 2002. The effect of faults on the 3D connectivity of reservoir bodies: a case study from the East Pennine Coalfield, UK. *Petroleum Geoscience*, **8**, 263–277, https://doi.org/10.1144/petgeo.8.3.263

Barnard, P.C. & Cooper, B.S. 1983. A review of geochemical data related to the Northwest European gas province. *In*: Brooks, J. (ed.) *Petroleum Geochemistry and Exploration of Europe*. Geological Society, London, Special Publications, **12**, 19–33, https://doi.org/10.1144/GSL.SP.1983.012.01.04

BCURA 2002. *The BCURA Coal Sample Bank: A Users Handbook*. British Coal Utilisation Research Association, Cheltenham, www.bcura.org/coalbank.html

Berner, U. & Faber, E. 1996. Empirical carbon isotope/maturity relationships for gases from algal kerogens and terrigenous organic matter, based on dry, open-system pyrolysis. *Organic Geochemistry*, **24**, 947–955.

Besly, B.M. 1988. Palaeogeographic implications of late Westphalian to early Permian red beds, Central England. *In*: Besly, B.M. & Kelling, G. (eds) *Sedimentation in a Synorogenic Basin Complex: The Upper Carboniferous of North West Europe*. Blackie, Glasgow, 200–221.

Besly, B.M. 1998. Carboniferous. *In*: Glennie, K.W. (ed.) *Petroleum Geology of the North Sea: Basic Concepts and Recent Advances*. Blackwell, Oxford, 104–136.

BESLY, B.M. 2005. Late Carboniferous redbeds of the UK southern North Sea, viewed in a regional context [extended abstract]. *In*: COLLINSON, J.D., EVANS, D.J., HOLLIDAY, D.W. & JONES, N.S. (eds) *Carboniferous Hydrocarbon Geology of the Southern North Sea and Surrounding Onshore Areas*. Occasional Publications, **7**. Yorkshire Geological Society, 225–226.

BESLY, B.M. & CLEAL, C.J. 1997. Upper Carboniferous stratigraphy of the West Midlands (UK) revised in the light of borehole geophysical logs and detrital compositional suites. *Geological Journal*, **32**, 85–118.

BESLY, B.M., BURLEY, S.D. & TURNER, P. 1993. The late Carboniferous 'Barren Red Bed' play of the Silver Pit area, Southern North Sea. *In*: PARKER, J.R. (ed.) *Petroleum Geology of Northwest Europe: Proceedings of the 4th Conference*. Geological Society, London, 727–740, https://doi.org/10.1144/0040727

BLUNDELL, D.J. 2002. Cenozoic inversion and uplift of southern Britain. *In*: DORÉ, A.G., CARTWRIGHT, J.A., STOKER, M.S., TURNER, J.R. & WHITE, N. (eds) *Exhumation of the North Atlantic Margin: Timing, Mechanisms and Implications for Petroleum Exploration*. Geological Society, London, Special Publications, **196**, 85–101, https://doi.org/10.1144/GSL.SP.2002.196.01.06

BRAY, R.J., GREEN, P.F. & DUDDY, I.R. 1992. Thermal history reconstruction using apatite fission track analysis and vitrinite reflectance: a case study from the UK East Midlands and Southern North Sea. *In*: HARDMAN, R.F.P. (ed.) *Exploration Britain: Geological Insights for the Next Decade*. Geological Society, London, Special Publications, **67**, 3–25, https://doi.org/10.1144/GSL.SP.1992.067.01.01

BRISTOW, C.S. & WILLIAMSON, B.J. 1998. Spectral gamma ray logs: core to log calibration, facies analysis and correlation problems in the Southern North Sea. *In*: HARVEY, P.K. & LOVELL, M.A. (eds) *Core-Log Integration*. Geological Society, London, Special Publications, **136**, 1–7, https://doi.org/10.1144/GSL.SP.1998.136.01.01

BRUCE, D. & STEMMERIK, L. 2003. Carboniferous. *In*: EVANS, D., GRAHAM, C. & BATHURST, P. (eds) *The Millennium Atlas: Petroleum Geology of the Central and Northern North Sea*. Geological Society, London, 83–89.

BURGESS, P.M. & GAYER, R.A. 2000. Late Carboniferous tectonic subsidence in South Wales: implications for Variscan basin evolution and tectonic history in SW Britain. *Journal of the Geological Society, London*, **157**, 93–104, https://doi.org/10.1144/jgs.157.1.93

BURNETT, R.D. 1987. Regional maturation patterns for late Visean (Carboniferous, Dinantian) rocks of northern England based on mapping of conodont colour. *Irish Journal of Earth Sciences*, **8**, 165–185.

CAMERON, D., MUNNS, J. & STOKER, S. 2005. Remaining hydrocarbon exploration potential of the Carboniferous fairway, UK southern North Sea. *In*: COLLINSON, J.D., EVANS, D.J., HOLLIDAY, D.W. & JONES, N.S. (eds) *Carboniferous Hydrocarbon Geology of the Southern North Sea and Surrounding Onshore Areas*. Occasional Publications, **7**. Yorkshire Geological Society, 209–224.

CAMERON, N. & ZIEGLER, T. 1997. Probing the lower limits of a fairway: further pre-Permian potential in the southern North Sea. *In*: ZIEGLER, K., TURNER, P. & DAINES, S.R. (eds) *Petroleum Geology of the Southern North Sea: Future Potential*. Geological Society, London, Special Publications, **123**, 123–141, https://doi.org/10.1144/GSL.SP.1997.123.01.08

CAMERON, T.D.J. 1993. Carboniferous and Devonian of the Southern North Sea. *In*: KNOX, R.W.O'B. & CORDEY, W.G. (eds) *Lithostratigraphic Nomenclature of the UK North Sea*. British Geological Survey, Keyworth.

CARRUTHERS, R.G., DINHAM, C.H., BURNETT, G.A. & MADEN, J. 1927. *Geology of Belford, Holy Island and the Farne Islands*. HMSO, London.

CARRUTHERS, R.G., BURNETT, G.A. & ANDERSON, W. 1932. *Geology of the Cheviot Hills*. HMSO, London.

CATTO, R., TAGGART, S. & POOLE, G. 2018. Petroleum geology of the Cygnus gas field, blocks 44/11 and 44/12, UK North Sea. *In*: BOWMAN, M. & LEVELL, B. (eds) *Petroleum Geology of NW Europe: 50 Years of Learning – Proceedings of the 8th Petroleum Geology Conference* Geological Society, London, 307–318, https://doi.org/10.1144/PGC8.39

CHADWICK, R.A., HOLLIDAY, D.W., HOLLOWAY, S. & HULBERT, A.G. 1995. *The Structure and Evolution of the Northumberland–Solway Basin and Adjacent Areas*. Subsurface Memoirs. British Geological Survey, Keyworth.

CLARKE, H., TURNER, P., BUSTIN, M., RILEY, N.J. & BESLY, B. In press. Shale gas resources of the Bowland Basin, NW England: a holistic study. *Petroleum Geoscience*, https://doi.org/10.1144/petgeo2017-066

COLE, J.M., WHITAKER, M., KIRK, M. & CRITTENDEN, S. 2005. A sequence stratigraphic scheme for the Late Carboniferous, Southern North Sea, Anglo-Dutch sector. *In*: COLLINSON, J.D., EVANS, D.J., HOLLIDAY, D.W. & JONES, N.S. (eds) *Carboniferous Hydrocarbon Geology of the Southern North Sea and Surrounding Onshore Areas*. Occasional Publications, **7**. Yorkshire Geological Society, 75–104.

COLLINSON, J.D. 2005. Dinantian and Namurian depositional systems in the Southern North Sea. *In*: COLLINSON, J.D., EVANS, D.J., HOLLIDAY, D.W. & JONES, N.S. (eds) *Carboniferous Hydrocarbon Geology of the Southern North Sea and Surrounding Onshore Areas*. Occasional Publications, **7**. Yorkshire Geological Society, 35–56.

COLLINSON, J.D., JONES, C.M., BLACKBOURN, G.A., BESLY, B.M., ARCHARD, G.M. & MCMAHON, A.H. 1993. Carboniferous depositional systems of the Southern North Sea. *In*: PARKER, J.R. (ed.) *Petroleum Geology of Northwest Europe: Proceedings of the 4th Conference*. Geological Society, London, 677–687, https://doi.org/10.1144/0040677

CONWAY, A.M. & VALVATNE, C. 2003*a*. The Boulton Field, Block 44/21a, UK North Sea. *In*: GLUYAS, J.G. & HICHENS, H.M. (eds) *United Kingdom Oil and Gas Fields, Commemorative Millennium Volume*. Geological Society, London, Memoirs, **20**, 671–680, https://doi.org/10.1144/GSL.MEM.2003.020.01.53

CONWAY, A.M. & VALVATNE, C. 2003*b*. The Murdoch Gas Field, Block 44/22a, UK Southern North Sea. *In*: GLUYAS, J.G. & HICHENS, H.M. (eds) *United Kingdom Oil and Gas Fields, Commemorative Millennium Volume*. Geological Society, London, Memoirs, **20**, 789–798, https://doi.org/10.1144/GSL.MEM.2003.020.01.66

Cooper, M.M., Easton, S.D.W., Lynch, J.J. & Fozdar, I.M. 2005. The Caister–Murdoch System (CMS) III. Carboniferous cluster development, UK Southern North Sea. *In*: Doré, A.G. & Vining, B.A. (eds) *Petroleum Geology: North-West Europe and Global Perspectives – Proceedings of the 6th Petroleum Geology Conference*. Geological Society, London, 317–326, https://doi.org/10.1144/0060317

Cope, M.J. 1986. An interpretation of vitrinite reflectance data from the Southern North Sea Basin. *In*: Brooks, J., Goff, J.C. & Van Hoorn, B. (eds) *Habitat of Palaeozoic Gas in NW Europe*. Geological Society, London, Special Publications, **23**, 85–98, https://doi.org/10.1144/GSL.SP.1986.023.01.06

Corfield, S.M. In press. The first oil exploration campaign in the UK, 1918-1922. *In*: Craig, J. (ed.) *History of the European Oil and Gas Industry*. Geological Society, London, Special Publications, **465**

Corfield, S.M., Gawthorpe, R.L., Gage, M., Fraser, A.J. & Besly, B.M. 1996. Inversion tectonics of the Variscan foreland of the British Isles. *Journal of the Geological Society, London*, **153**, 17–32, https://doi.org/10.1144/gsjgs.153.1.0017

Corbin, S., Gorringe, S. & Torr, D. 2005. Challenges of developing Carboniferous gas fields in the UK Southern North Sea. *In*: Doré, A.G. & Vining, B.A. (eds) *Petroleum Geology: North-West Europe and Global Perspectives – Proceedings of the 6th Petroleum Geology Conference*. Geological Society, London, 587–594, https://doi.org/10.1144/0060587

Cornford, C. 1998. Source rocks and hydrocarbons of the North Sea. *In*: Glennie, K.W. (ed.) *Petroleum Geology of the North Sea: Basic Concepts and Recent Advances*. Blackwell, Oxford, 376–462.

Coward, M.P. 1993. The effect of Late Caledonian and Variscan continental escape tectonics on basement structure, Paleozoic basin kinematics and subsequent Mesozoic basin development in NW Europe. *In*: Parker, J.R. (ed.) *Petroleum Geology of Northwest Europe: Proceedings of the 4th Conference*. Geological Society, London, 1095–1108, https://doi.org/10.1144/0041095

Creaney, S., Allchurch, D.M. & Jones, J.M. 1985. Vitrinite reflectance variation in northern England. *Comptes rendues 9ème Congres Internationale de Stratigraphie et Géologie du Carbonifère, Urbana, 1979.*, Vol. **4**. Southern Illinois University Press, Carbondale, IL, 583–589.

Creedy, D.P. 1985. *The origin and distribution of firedamp in some British coalfields*. PhD thesis, University of Wales.

Creedy, D.P. 1991. An introduction to geological aspects of methane occurrence and control in British deep coal mines. *Quarterly Journal of Engineering Geology*, **24**, 209–220.

Davis, B.K. 1987. Velocity changes and burial diagenesis in the Chalk of the southern North Sea basin. *In*: Brooks, J. & Glennie, K. (eds) *Petroleum Geology of North West Europe: Proceedings of the 3rd Conference*. Graham & Trotman, London, 307–313.

DECC 2013. Project Pathfinder. Current and Future UKCS Oil & Gas Projects, September 2013, www.decc.gov.uk [last accessed January 2017].

de Jager, J. & Geluk, M.C. 2007. Petroleum geology. *In*: Wong, Th.E., Batjes, D.A.J. & de Jager, J. (eds) *Geology of the Netherlands*. Royal Netherlands Academy of Arts and Sciences, Amsterdam, 241–264.

de Paola, N., Holdsworth, R.E., McCaffrey, K.J.W. & Barchi, M.R. 2005. Partitioned transtension: an alternative to basin inversion models. *Journal of Structural Geology*, **27**, 607–625.

Donato, J.A. 1993. A buried granite batholith and the origin of the Sole Pit Basin, UK Southern North Sea. *Journal of the Geological Society, London*, **150**, 255–258, https://doi.org/10.1144/gsjgs.150.2.0255

Donato, J.A. & Megson, J.B. 1990. A buried granite batholith beneath the East Midland Shelf of the Southern North Sea Basin. *Journal of the Geological Society, London*, **147**, 133–140, https://doi.org/10.1144/gsjgs.147.1.0133

Donato, J.A., Martindale, W. & Tully, M.C. 1983. Buried granites within the Mid North Sea High. *Journal of the Geological Society, London*, **140**, 825–837, https://doi.org/10.1144/gsjgs.140.5.0825

EBN 2015. Paleozoic structural framework in the northern Dutch offshore. Poster from EAGE Conference, Madrid, Spain, 2015, www.ebn.nl

Eliade, M. 1968. *Myths, Dreams and Mysteries: The Encounter Between Contemporary Faiths and Archaic Realities*. Collins, London.

Falcon, N.L. & Kent, P.E. 1960. *Geological Results of Petroleum Exploration in Britain 1945–1957*. Geological Society, London, Memoirs, **2**, https://doi.org/10.1144/GSL.MEM.1960.002.01.01

Fermont, W.J.J. & Jegers, L.F. 1991. *Resultaten van de koolpetrografisch/geochemische prospectiviteitsanalyse aan de offshore DE-blokken* [in Dutch]. Fase 1: Data Summary. Rijks Geologische Dienst, Wetenschappelijk Laboratorium, Vaste en Organische Gesteenten, Rapport **GB2340**, www.nlog.nl

Fraser, A.J. & Gawthorpe, R.L. 2003. *An Atlas of Carboniferous Basin Evolution in Northern England*. Geological Society, London, Memoirs, **26**, https://doi.org/10.1017/S0016756803228788

Fraser, A.J., Nash, D.F., Steele, R.P. & Ebdon, C.C. 1990. A regional assessment of the intra-Carboniferous play of northern England. *In*: Brooks, J. (ed.) *Classic Petroleum Provinces*. Geological Society, London, Special Publications, **50**, 417–440, https://doi.org/10.1144/GSL.SP.1990.050.01.26

Freudenberg, U., Lou, S., Schlüter, R., Schütz, K. & Thomas, K. 1996. Main factors controlling coalbed methane distribution in the Ruhr District, Germany. *In*: Gayer, R. & Harris, I. (eds) *Coalbed Methane and Coal Geology*. Geological Society, London, Special Publications. **109**, 67–88, https://doi.org/10.1144/GSL.SP.1996.109.01.06

Galimov, E.M. 2006. Isotope organic geochemistry. *Organic Geochemistry*, **37**, 1200–1262.

Garland, C.R. 1991. The Amthyst Field, Blocks 47/8a, 47/9a, 47/13a, 47/14a, 47/15a, UK North Sea. *In*: Abbotts, I. (ed.) *United Kingdom Oil and Gas Fields, 25 years Commemorative Volume* Geological Society, London, Memoirs, **14**, 387–393, https://doi.org/10.1144/GSL.MEM.1991.014.01.48

Gaschnitz, R. 2001. Gasgenese und Gasspeicherung im flözführenden Oberkarbon des Ruhr-Beckens. *Berichte des FZ Jülich*, **3859**.

GELUK, M.C., DUSAR, M. & DE VOS, W. 2007. Pre-Silesian. *In*: WONG, TH.E., BATJES, D.A.J. & DE JAGER, J. (eds) *Geology of the Netherlands*. Royal Netherlands Academy of Arts and Sciences, Amsterdam, 27–42.

GENT, C.M.A. 2015. *Total organic carbon calculation using geophysical logs for 31 wells across the Palaeozoic of the Central North Sea*. British Geological Survey Commissioned Report **CR/15/121**, http://nora.nerc.ac.uk/516744/

GERLING, P., IDIZ, E., EVERLIEN, G. & SOHNS, E. 1997. New aspects on the origin of nitrogen in natural gas in northern Germany. *Geologisches Jahrbuch*, **D103**, 65–84.

GERLING, P., LOKHORST, A., NICHOLSON, R.A. & KOTARBA, M. 1998. Natural gas from Pre-Westphalian sources in northwest Europe – a new exploration target. *In*: *Proceedings of the 1998 International Gas Research Conference, San Diego, California, USA. Vol. 1: Exploration and Production*. Gas Research Institute, Chicago, IL, 219–229.

GERLING, P., GELUK, M.C., KOCKEL, F., LOKHORST, A., LOTT, G.K. & NICHOLSON, R.A. 1999. 'NW European Gas Atlas' – new implications for the Carboniferous gas plays in the western part of the Southern Permian Basin. *In*: FLEET, A.J. & BOLDY, S.A.R. (eds) *Petroleum Geology of Northwest Europe: Proceedings of the 5th Conference*. Geological Society, London, 799–808, https://doi.org/10.1144/0050799

GEOLAB UK 1994. A geochemical evaluation of the UKCS well 42/16-1. *Report prepared for Premier Consolidated Oilfields plc*. Released by UK Oil and Gas Authority, April 2016.

GIFFARD, H.P.W. 1938. The former cannel oil industry in North Wales and Staffordshire. *In*: *Oil Shale & Cannel Coal, Proceedings of a Conference held in Scotland*, June 1938. Institute of Petroleum, London, 78–95.

GLENNIE, K.W. 1998. Lower Permian – Rotliegend. *In*: GLENNIE, K.W. (ed.) *Petroleum Geology of the North Sea: Basic Concepts and Recent Advances*. Blackwell, Oxford, 137–173.

GLENNIE, K.W. & BOEGNER, P.L.E. 1981. Sole Pit inversion tectonics. *In*: ILLING, L.V. & HOBSON, G.D. (eds) *Petroleum Geology of the Continental Shelf of North West Europe*. Institute of Petroleum, London, 110–120.

GLUYAS, J.G. & HICHENS, H.M. (eds) 2003. *United Kingdom Oil and Gas Fields, Commemorative Millennium Volume*. Geological Society, London, Memoirs, **20**, https://doi.org/10.1144/GSL.MEM.2003.020.01.01

GREEN, P.F. 2005. Post-Carboniferous burial and exhumation histories of Carboniferous rocks of the southern North Sea and adjacent onshore UK. *In*: COLLINSON, J.D., EVANS, D.J., HOLLIDAY, D.W. & JONES, N.S. (eds) *Carboniferous Hydrocarbon Geology of the Southern North Sea and Surrounding Onshore Areas*. Occasional Publications, **7**. Yorkshire Geological Society, 25–34.

GREEN, P.F., DUDDY, I.R. & BRAY, R.J. 1997. Variation in thermal history styles around the Irish Sea and adjacent areas: implications for hydrocarbon occurrence and tectonic evolution. *In*: MEADOWS, N.S., TRUEBLOOD, S.P., HARDMAN, M. & COWAN, G. (eds) *Petroleum Geology of the Irish Sea and Adjacent Areas*. Geological Society, London, Special Publications, **124**, 73–93, https://doi.org/10.1144/GSL.SP.1997.124.01.06

GREEN, P.F., THOMSON, K. & HUDSON, J.D. 2001. Recognition of tectonic events in undeformed regions: contrasting results from the Midland Platform and East Midlands Shelf, Central England. *Journal of the Geological Society, London*, **158**, 59–73, https://doi.org/10.1144/jgs.158.1.59

GROSS, D., SACHSENHOFER, R.F., BECHTEL, A., PYTLAK, L., RUPPRECHT, B. & WEGERER, E. 2015. Organic geochemistry of Mississippian shales (Bowland Shale Formation) in central Britain: implications for depositional environment, source rock and gas shale potential. *Marine and Petroleum Geology*, **59**, 1–21.

HILDENBRAND, A., KROOSS, B.M., BUSCH, A. & GASCHNITZ, R. 2006. Evolution of methane sorption capacity of coal seams as a function of burial history – a case study from the Campine Basin, NE Belgium. *International Journal of Coal Geology*, **66**, 179–203.

HILLIS, R.R. 1995*a*. Quantification of Tertiary exhumation in the United Kingdom Southern North Sea using sonic velocity data. *American Association of Petroleum Geologists Bulletin*, **79**, 130–152.

HILLIS, R.R. 1995*b*. Regional Tertiary exhumation in and around the United Kingdom. *In*: BUCHANAN, J.G. & BUCHANAN, P.G. (eds) *Basin Inversion*. Geological Society, London, Special Publications, **88**, 167–190, https://doi.org/10.1144/GSL.SP.1995.088.01.11

HODGE, T. 2003. The Saltfleetby Field, Block L 47/16, Licence PEDL 005, onshore UK. *In*: GLUYAS, J.G. & HICHENS, H.M. (eds) *United Kingdom Oil and Gas Fields, Commemorative Millennium Volume*. Geological Society, London, Memoirs, **20**, 911–919, https://doi.org/10.1144/GSL.MEM.2003.020.01.77

HOLLIS, C 1998. Reconstructing fluid history: an integrated approach to timing fluid expulsion and migration in the Carboniferous Derbyshire Platform, England. *In*: PARNELL, J. (ed.) *Dating and Duration of Fluid Flow and Fluid–Rock Interaction*. Geological Society, London, Special Publications, **144**, 153–159, https://doi.org/10.1144/GSL.SP.1998.144.01.12

HOWER, J.C. & GAYER, R.A. 2002. Mechanisms of coal metamorphism: case studies from Paleozoic coalfields. *International Journal of Coal Geology*, **50**, 215–245.

HUGHES, F., HARRISON, D. ET AL. 2018. The unconventional Carboniferous reservoirs of the Greater Kirby Misperton gas field and their potential: North Yorkshire's sleeping giant. In: BOWMAN, M. & LEVELL, B. (eds) *Petroleum Geology of NW Europe: 50 Years of Learning – Proceedings of the 8th Petroleum Geology Conference*. Geological Society, London, 611–625, https://doi.org/10.1144/PGC8.5

JAPSEN, P. 1997. Regional Neogene exhumation of Britain and the western North Sea. *Journal of the Geological Society, London*, **154**, 239–247, https://doi.org/10.1144/gsjgs.154.2.0239

JAPSEN, P. 2000. Investigation of multi-phase erosion using reconstructed shale trends based on sonic data. Sole Pit axis, North Sea. *Global and Planetary Change*, **24**, 189–210.

JAPSEN, P. & CHALMERS, J.A. 2000. Neogene uplift and tectonics around the North Atlantic: overview. *Global and Planetary Change*, **24**, 165–173.

JOHNSON, G.A.L., SOMERVILLE, I.D., TUCKER, M.E. & CÓZAR, P. 2011. Carboniferous stratigraphy and context of the Seal Sands No. 1 Borehole, Teesmouth, NE England: the deepest onshore borehole in Great Britain. *Proceedings of the Yorkshire Geological Society*, **58**, 173–196, https://doi.org/10.1144/pygs.58.3.231

JOHNSON, H.D. & FISHER, M.J. 1998. North Sea plays: geological controls on hydrocarbon distribution. *In*: GLENNIE, K.W. (ed.) *Petroleum Geology of the North Sea: Basic Concepts and Recent Advances*. Blackwell, Oxford, 463–547.

JOHNSON, S.A., TURNER, P., HARTLEY, A. & REY, D. 1995. Palaeomagnetic implications for the timing of hematite precipitation and remagnetization in the Carboniferous Barren Red Measures, UK Southern North Sea. *In*: TURNER, P. & TURNER, A. (eds) *Palaeomagnetic Applications in Hydrocarbon Exploration and Production*, Geological Society, London, Special Publications, **98**, 97–117, https://doi.org/10.1144/GSL.SP.1995.098.01.06

JONES, C.M., ALLEN, P.J. & MORRISON, N.H. 2005. Geological factors influencing gas production in the Tyne Field (Block 44/18a), southern North Sea, and their impact on future infill well planning. *In*: COLLINSON, J.D., EVANS, D.J., HOLLIDAY, D.W. & JONES, N.S. (eds) *Carboniferous Hydrocarbon Geology of the Southern North Sea and Surrounding Onshore Areas*. Occasional Publications, **7**. Yorkshire Geological Society, 183–194.

JUCH, D., GASCHNITZ, R. & THIELEMANN, T. 2004. The influence of geological history on coal mine gas distribution in the Ruhr district – a challenge for the future research and recovery. *Geologica Belgica*, **7**, 191–199.

KELLING, G. 1988. Silesian sedimentation and tectonics in the South Wales Basin: a brief review. *In*: BESLY, B.M. & KELLING, G. (eds) *Sedimentation in a Synorogenic Basin Complex: The Upper Carboniferous of North West Europe*. Blackie, Glasgow, 38–42.

KERR MCGEE 2002. *Carboniferous opportunities in the Bowland–Craven Basin North & West Yorkshire, UK*. PEDL009 Relinquishment Report, www.ukogl.org.uk/industry-reports/

KERSTEN, C., SCHULZE, K., SCHROERS, F., MANDIWALL, D. & JEFFS, P. 2013. Formation damage in the Cavendish gas field – causes, treatment and future measures. Paper presented at the 2013 SPE European Formation Damage Conference, Noordwijk, Netherlands, 5–7 June 2013, **SPE 164185-MS**.

KIMBELL, G.S., CHADWICK, R.A., HOLLIDAY, D.W. & WERNGREN, O.C. 1989. The structure and evolution of the Northumberland Trough from new seismic reflection data and its bearing on modes of continental extension. *Journal of the Geological Society, London*, **146**, 775–787, https://doi.org/10.1144/gsjgs.146.5.0775

KOMBRINK, H. 2008. The Carboniferous of the Netherlands and surrounding areas: a basin analysis. *Geologica Ultraiectina*, **294**.

KOMBRINK, H., LEEVER, K.A., VAN WEES, J.-D., VAN BERGEN, F., DAVID, P. & WONG, T.E. 2008. Late Carboniferous foreland basin formation and Early Carboniferous stretching in north western Europe: inferences from quantitative subsidence analyses in the Netherlands. *Basin Research*, **20**, 377–395.

KOMBRINK, H., BESLY, B.M. *ET AL.* 2010. Carboniferous. *In*: DOORNENBAL, J.C. & STEVENSON, A.G. (eds) *Petroleum Geological Atlas of the Southern Permian Basin Area*. EAGE Publications, Houten, 81–99.

KROOSS, B.M., LITTKE, R., MÜLLER, B., FRIELINGSDORF, J., SCHWOCHAU, K. & IDIZ, E.F. 1995. Generation of nitrogen and methane from sedimentary organic matter: implications on the dynamics of natural gas accumulations. *Chemical Geology*, **126**, 291–318.

KUHN, T.S. 1962. *The Structure of Scientific Revolutions*. University of Chicago Press, Chicago, IL.

LEEDER, M.R. 1982. Upper Palaeozoic basins of the British Isles – Caledonide inheritance v. Hercynian plate margin processes. *Journal of the Geological Society, London*, **139**, 479–491, https://doi.org/10.1144/gsjgs.139.4.0479

LEEDER, M.R. & HARDMAN, M. 1990. Carboniferous geology of the Southern North Sea Basin and controls on hydrocarbon prospectivity. *In*: HARDMAN, R.F.P. & BROOKS, J. (eds) *Tectonic Events Responsible for Britain's Oil and Gas Reserves*. Geological Society, London, Special Publications, **55**, 87–105, https://doi.org/10.1144/GSL.SP.1990.055.01.04

LEEDER, M.R. & MCMAHON, A.H. 1988. Upper Carboniferous (Silesian) basin subsidence in northern Britain. *In*: BESLY, B.M. & KELLING, G. (eds) *Sedimentation in a Synorogenic Basin Complex: The Upper Carboniferous of North West Europe*. Blackie, Glasgow, 43–52.

LEEDER, M.R., RAISWELL, R., AL-BIATTY, H., MCMAHON, A. & HARDMAN, M. 1990. Carboniferous stratigraphy, sedimentation and correlation of well 48/3-3 in the southern North Sea Basin: integrated use of palynology, natural gamma/sonic logs and carbon/sulphur geochemistry. *Journal of the Geological Society, London*, **147**, 287–300, https://doi.org/10.1144/gsjgs.147.2.0287

LITTKE, R., BÜKER, C., LÜCKGE, A., SACHSENHOFER, R.F. & WELTE, D.H. 1994. A new evaluation of palaeoheatflows and eroded thicknesses for the Carboniferous Ruhr basin, western Germany. *International Journal of Coal Geology*, **26**, 155–183.

LITTKE, R., BÜKER, C., HERTLE, M., KARG, H., STROETMAN-HEINEN, V. & ONCKEN, O. 2000. Heat flow evolution, subsidence and erosion in the Rheno-Hercynian orogenic wedge of central Europe. *In*: FRANKE, W., HAAK, V., ONCKEN, O. & TANNER, D. (eds) *Orogenic Processes: Quantification and Modelling in the Variscan Belt*. Geological Society, London, Special Publications, **179**, 231–255, https://doi.org/10.1144/GSL.SP.2000.179.01.15

LOFTUS, G.W.F. & GREENSMITH, J.T. 1988. The lacustrine Burdiehouse Limestone Formation a key to the deposition of the Dinantian oil shales of Scotland. *In*: FLEET, A.J., KELTS, K. & TALBOT, M.R. (eds) *Lacustrine Petroleum Source Rocks*. Geological Society, London, Special Publications, **40**, 219–234, https://doi.org/10.1144/GSL.SP.1988.040.01.19

LOKHORST, A. (ed.) 1998. *Northwest European Gas Atlas Composition and Isotope Ratios of Natural Gases*. NITG-TNO, Haarlem.

MADDOX, S.J., BLOW, R. & HARDMAN, M. 1995. Hydrocarbon prospectivity of the Central Irish Sea Basin with reference to Block 42/12, offshore Ireland. *In*: CROKER, P.F. & SHANNON, P.M. (eds) *The Petroleum Geology of Ireland's Offshore Basins*. Geological Society, London,

Special Publications, **93**, 59–77, https://doi.org/10.1144/GSL.SP.1995.093.01.08

Maynard, J.R. & Dunay, R.E. 1999. Reservoirs of the Dinantian (Lower Carboniferous) play of the Southern North Sea. *In*: Fleet, A.J. & Boldy, S.A.R. (eds) *Petroleum Geology of Northwest Europe: Proceedings of the 5th Conference*. Geological Society, London, 729–745, https://doi.org/10.1144/0050729

McKenzie, D. 1978. Some remarks on the development of sedimentary basins. *Earth and Planetary Science Letters*, **40**, 25–32.

McLean, D., Owens, B. & Neves, R. 2005. Carboniferous miospore biostratigraphy of the North Sea. *In*: Collinson, J.D., Evans, D.J., Holliday, D.W. & Jones, N.S. (eds) *Carboniferous Hydrocarbon Geology of the Southern North Sea and Surrounding Onshore Areas*. Occasional Publications, **7**. Yorkshire Geological Society, 13–24.

McPhee, C. & Byrne, M. 2009. Unlocking hidden reservoir potential through integrated formation damage evaluation. Paper presented at the 2009 SPE European Formation Damage Conference, Scheveningen, Netherlands, 27–29 May 2009, SPE **120964**.

McPhee, C., Judt, M., McRae, D. & Rapach, J. 2008. Maximising gas well potential in the Breagh Field by mitigating formation damage. Paper presented at the 2008 SPE Asia Pacific Oil & Gas Conference, Perth, Australia, 20–22 October 2008, SPE **115690**.

Mijnssen, F.C.J. 1997. Modelling of sandbody connectivity in the Schooner Field. *In*: Ziegler, K., Turner, P. & Daines, S.R. (eds) *Petroleum Geology of the Southern North Sea: Future Potential*. Geological Society, London, Special Publications, **123**, 169–180, https://doi.org/10.1144/GSL.SP.1997.123.01.11

Miller, J. & Grayson, R.F. 1982. The regional context of Waulsortian facies in northern England. *In*: Bolton, K., Lane, H.R. & Lemone, D.U. (eds) *Symposium on the Palaeoenvironmental Setting and Distribution of the Waulsortian Facies*. The El Paso Geological Society and University of Texas at El Paso, 17–30.

Mingram, B., Hoth, P., Lüders, V. & Harlov, D. 2005. The significance of fixed ammonium in Paleozoic sediments for the generation of nitrogen rich natural gases in the North German Basin (NGB). *International Journal of Earth Sciences (Geologische Rundschau)*, **94**, 1010–1022.

Mitchell, W.W. & Owens, B. 1990. The geology of the western part of the Fintona Block, Northern Ireland: evolution of Carboniferous basins. *Geological Magazine*, **127**, 407–426.

Monaghan, A.A., Arsenikos, S. *et al.* 2017. Carboniferous petroleum systems around the Mid North Sea High, UK. *Marine and Petroleum Geology*, **88**, 282–302.

Morton, A., Hallsworth, C. & Moscariello, A. 2005. Interplay between northern and southern sediment sources during Westphalian deposition in the Silverpit Basin, southern North Sea. *In*: Collinson, J.D., Evans, D.J., Holliday, D.W. & Jones, N.S. (eds) *Carboniferous Hydrocarbon Geology of the Southern North Sea and Surrounding Onshore Areas*. Occasional Publications, **7**. Yorkshire Geological Society, 135–146.

Moscariello, A. 2003. The Schooner Field, Blocks 44/26a, 43/30a, UK North Sea. *In*: Gluyas, J.G. & Hichens, H.M. (eds) *United Kingdom Oil and Gas Fields, Commemorative Millennium Volume*. Geological Society, London, Memoirs, **20**, 811–824, https://doi.org/10.1144/GSL.MEM.2003.020.01.68

Murchison, D. 2004. Aberrations in the coalification patterns of the offshore coalfields of Northumberland and Durham, United Kingdom. *International Journal of Coal Geology*, **58**, 133–146.

Murchison, D. & Pearson, J. 2000. The anomalous behaviour of properties of seams at the Plessey (M) horizon of the Northumberland and Durham coalfields. *Fuel*, **79**, 865–871.

National Coal Board 1960. *The Coalfields of Great Britain: Variation in Rank of Coal*. National Coal Board, Scientific Department, Coal Surveys, London.

Nesbit, R. & Overshott, K. 2010. Overcoming multiple uncertainties in a challenging gas development: Chiswick Field UK SNS. *In*: Vining, B.A. & Pickering, S.C. (eds) *Petroleum Geology: From Mature Basins to New Frontiers – Proceedings of the 7th Petroleum Geology Conference*. Geological Society, London, 315–323, https://doi.org/10.1144/0070315

O'Mara, P.T., Merryweather, M., Stockwell, M. & Bowler, M.M. 1999. The Trent Gas Field: correlation and reservoir quality within a complex Carboniferous stratigraphy. *In*: Fleet, A.J. & Boldy, S.A.R. (eds) *Petroleum Geology of Northwest Europe: Proceedings of the 5th Conference*. Geological Society, London, 809–821, https://doi.org/10.1144/0050809

O'Mara, P.T., Merryweather, M., Stockwell, M. & Bowler, M.M. 2003*a*. The Trent Gas Field, Block 43/24a, UK North Sea. *In*: Gluyas, J.G. & Hichens, H.M. (eds) *United Kingdom Oil and Gas Fields, Commemorative Millennium Volume*. Geological Society, London, Memoirs, **20**, 835–849, https://doi.org/10.1144/0050809

O'Mara, P.T., Merryweather, M. & Cooper, D. 2003*b*. The Tyne Gas Fields, Block 44/18a, UK North Sea. *In*: Gluyas, J.G. & Hichens, H.M. (eds) *United Kingdom Oil and Gas Fields, Commemorative Millennium Volume*. Geological Society, London, Memoirs, **20**, 851–860, https://doi.org/10.1144/GSL.MEM.2003.020.01.71

Oudmayer, B.C. & de Jager, J. 1993. Fault reactivation and oblique-slip in the Southern North Sea. *In*: Parker, J.R. (ed.) *Petroleum Geology of Northwest Europe: Proceedings of the 4th Conference*. Geological Society, London, 1281–1290, https://doi.org/10.1144/0041281

Parnell, J. 1988. Lacustrine petroleum source rocks in the Dinantian Oil Shale Group, Scotland: a review. *In*: Fleet, A.J., Kelts, K. & Talbot, M.R. (eds) *Lacustrine Petroleum Source Rocks*. Geological Society, London, Special Publications, **40**, 235–246, https://doi.org/10.1144/GSL.SP.1988.040.01.20

Parsons, T. & Trythall, R. 2015. New plays in a mature basin: a review of the Lower Carboniferous of the southern margin of the Mid-North Sea High [abstract]. Paper presented at 50 Years of Learning – a Platform for Present Value and Future Success, 8th Petroleum Geology of Northwest Europe Conference, London, 28–30 September 2015, Programme and Abstract Book, 46, www.petroleumgeologyconference.com

Partington, M.A., Copestake, P., Mitchener, B.C. & Underhill, J.R. 1993. Biostratigraphic calibration of genetic stratigraphic sequences in the Jurassic–lowermost

Cretaceous (Hettangian to Ryazanian) of the North Sea and adjacent areas. *In*: PARKER, J.R. (ed.) *Petroleum Geology of Northwest Europe: Proceedings of the 4th Conference*. Geological Society, London, 371–386, https://doi.org/10.1144/0040371

PATIJN, R.J.H. 1964. Die Enstehung von Erdgas infolge der Nachinkohlung im Nordosten der Niederlande. *Erdöl und Kohle: Erdgas, Petrochemie*, **17**, 2–9.

PEARCE, T.J., WRAY, D., RATCLIFFE, K., WRIGHT, D.K. & MOSCARIELLO, A. 2005*a*. Chemostratigraphy of the Upper Carboniferous Schooner Formation, southern North Sea. *In*: COLLINSON, J.D., EVANS, D.J., HOLLIDAY, D.W. & JONES, N.S. (eds) *Carboniferous Hydrocarbon Geology of the Southern North Sea and Surrounding Onshore Areas*. Occasional Publications, **7**. Yorkshire Geological Society, 147–164.

PEARCE, T.J., MCLEAN, D., WRAY, D., WRIGHT, D.K., JEANS, C.V. & MEARNS, E.W. 2005*b*. Stratigraphy of the Upper Carboniferous Schooner Formation, southern North Sea: chemostratigraphy, mineralogy, palynology and Sm–Nd isotope analysis. *In*: COLLINSON, J.D., EVANS, D.J., HOLLIDAY, D.W. & JONES, N.S. (eds) *Carboniferous Hydrocarbon Geology of the Southern North Sea and Surrounding Onshore Areas*. Occasional Publications, **7**. Yorkshire Geological Society, 165–182.

PEARSON, J. 1988. *Coalification studies in the Northumberland and Durham coalfields using vitrinite reflectances*. MSc dissertation, University of Newcastle.

PEARSON, M.J. & RUSSELL, M.A. 2000. Subsidence and erosion in the Pennine Carboniferous Basin, England: lithological and thermal constraints on maturity modelling. *Journal of the Geological Society, London*, **157**, 471–482, https://doi.org/10.1144/jgs.157.2.471

PEPPER, A.S. & CORVI, P.J. 1995. Simple kinetic models of petroleum formation. Part III: Modelling an open system. *Marine and Petroleum Geology*, **12**, 417–452.

PHARAOH, T.C., DUSAR, M. ET AL. 2010. Tectonic evolution. *In*: DOORNENBAL, J.C. & STEVENSON, A.G. (eds) *Petroleum Geological Atlas of the Southern Permian Basin Area*. EAGE Publications, Houten, 25–57.

PLETSCH, T., APPEL, J. ET AL. 2010. Petroleum generation and migration. *In*: DOORNENBAL, J.C. & STEVENSON, A.G. (eds) *Petroleum Geological Atlas of the Southern Permian Basin Area*. EAGE Publications, Houten, 225–253.

POWELL, T.G., DOUGLAS, A.G. & ALLAN, J. 1976. Variation in the type and distribution of organic matter in some Carboniferous sediments from northern England. *Chemical Geology*, **18**, 137–148.

QUIRK, D.G. 1997. Sequence stratigraphy of the Westphalian in the northern part of the Southern North Sea. *In*: ZIEGLER, K., TURNER, P. & DAINES, S.R. (eds) *Petroleum Geology of the Southern North Sea: Future Potential*. Geological Society London, Special Publications, **123**, 153–168, https://doi.org/10.1144/GSL.SP.1997.123.01.10

RIDD, M.F., WALKER, D.B. & JONES, J.M. 1970. A deep borehole at Harton on the margins of the Northumbrian Trough. *Proceedings of the Yorkshire Geological Society*, **38**, 75–103, https://doi.org/10.1144/pygs.38.1.75

RITCHIE, J.S. & PRATSIDES, P. 1993. The Caister Fields, Block 44/23a, UK North Sea. *In*: PARKER, J.R. (ed.) *Petroleum Geology of Northwest Europe: Proceedings of the 4th Conference*. Geological Society, London, 759–769, https://doi.org/10.1144/0040759

RODRIGUEZ, K., WRIGLEY, R., HODGSON, N. & NICHOLLS, H. 2014. Southern North Sea: unexplored multi-level exploration potential revealed. *First Break*, **32**, 107–113.

ROSCHER, M. & SCHNEIDER, J.W. 2006. Permo-Carboniferous climate: Early Pennsylvanian to Late Permian climate development of central Europe in a regional and global context. *In*: LUCAS, S.G., CASSINIS, G. & SCHNEIDER, J.W. (eds) *Non-Marine Permian Biostratigraphy and Biochronology*. Geological Society, London, Special Publications, **265**, 95–136, https://doi.org/10.1144/GSL.SP.2006.265.01.05

SCHEIDT, G. & LITTKE, R. 1994. Comparative organic petrology of interlayered sandstones, siltstones and coals in the Upper Carboniferous Ruhr Basin, northwest Germany, and their thermal history and methane generation. *Geologische Rundschau*, **78**, 375–390.

SCHROOT, B.M. & DE HAAN, H.B. 2003. An improved regional structural model of the Upper Carboniferous of the Cleaver Bank High based on 3D seismic interpretation. *In*: NIEUWLAND, D.A. (ed.) *New Insights into Structural Interpretation and Modelling*. Geological Society, London, Special Publications, **212**, 23–37, https://doi.org/10.1144/GSL.SP.2003.212.01.03

SCOTT, J. & COLTER, V.S. 1987. Geological aspects of current onshore exploration plays. *In*: BROOKS, J. & GLENNIE, K. (eds) *Petroleum Geology of North West Europe. Proceedings of the 3rd Conference on Petroleum Geology of North West Europe*. Graham & Trotman, London, 95–107.

SMITH, A.H.V. 1968. Seam profiles and seam characters. *In*: MURCHISON, D. & WESTOLL, T.S. (eds) *Coal and Coal-Bearing Strata*. Oliver & Boyd, Edinburgh, 31–40.

STERLING RESOURCES 2016. Recapitalization Overview and Corporate Update, www.sterling-resources.com/presentations.html

STEVENSON, R. 1999. Thoresby colliery outburst; a lesson learned. *International Mining and Minerals*, **August**, 211–222.

STEWART, S.A. & COWARD, M.P. 1995. Synthesis of salt tectonics in the southern North Sea, UK. *Marine and Petroleum Geology*, **12**, 457–475.

STONE, G. & MOSCARIELLO, A. 1999. Integrated modelling of the Southern North Sea Carboniferous barren red measures using production data, geochemistry, and pedofacies cyclicity. Paper presented at the SPE 1999 Offshore Europe Conference, Aberdeen, Scotland, 7–9 September 1999, SPE **56898**.

SULLIVAN, H.J. 1959. *The description and distribution of miospores and other microfloral remains in some sapropelic coals and their associated humic coals and carbonaceous shales*. PhD thesis, University of Sheffield.

TEICHMÜLLER, M., TEICHMÜLLER, R. & WEBER, K. 1979. Inkohlung und Illit-Kristallinität Vergleichende Untersuchen im Mesozoikum und Paläozoikum von Westfa len. *Fortschritte in der Geologie von Rheinland und Westfalen*, **27**, 201–276.

TRUEMAN, A. 1954. *The Coalfields of Great Britain*. Arnold, London.

VAN ADRICHEM BOOGAERT, H.A. & KOUWE, W.F.P. (Compilers) 1995. Stratigraphic Nomenclature of the Netherlands, revision and update by RGD and

NOGEPA, section C. *Mededelingen Rijks Geologische Dienst*, **50**.

van Buggenum, J.M. & den Hartog Jager, D.G. 2007. Silesian. *In*: Wong, Th.E., Batjes, D.A.J. & de Jager, J. (eds) *Geology of the Netherlands*. Royal Netherlands Academy of Arts and Sciences, Amsterdam, 43–62.

Vincent, C.J. 2015. *Maturity modelling of selected wells in the Central North Sea*. British Geological Survey Internal Report **CR/15/122**, http://nora.nerc.ac.uk/516764/

Ward, J., Chan, A. & Ramsay, B. 2003. The Hatfield Moors and Hatfield West Gas (Storage) Fields, South Yorkshire. *In*: Gluyas, J.G. & Hichens, H.M. (eds) *United Kingdom Oil and Gas Fields, Commemorative Millennium Volume*. Geological Society, London, Memoirs, **20**, 905–910, https://doi.org/10.1144/GSL.MEM.2003.020.01.76

Waters, C.N., Somerville, I.D. *et al.* 2011. *A Revised Correlation of Carboniferous Rocks in the British Isles*. Geological Society, London, Special Reports, **26**, https://doi.org/10.1144/SR26

Watts, A.B., McKerrow, W.S. & Fielding, E. 2000. Lithospheric flexure, uplift, and landscape evolution in south-central England. *Journal of the Geological Society, London*, **157**, 1169–1177, https://doi.org/10.1144/jgs.157.6.1169

White, N. & Lovell, B. 1997. Measuring the pulse of a plume with the sedimentary record. *Nature*, **387**, 888–891.

Whittaker, A., Holliday, D.W. & Penn, I.E. 1985. Geophysical Logs in British Stratigraphy. *Geological Society, London, Special Reports*, **18**.

Ziegler, P.A. 1990. *Geological Atlas of Western and Central Europe*. Shell Internationale Petroleum Maatschappij, The Hague.

Structural development of the Devono-Carboniferous plays of the UK North Sea

STAVROS ARSENIKOS[1]*, MARTYN QUINN[1], GEOFF KIMBELL[2], PAUL WILLIAMSON[2], TIM PHARAOH[2], GRAHAM LESLIE[1] & ALISON MONAGHAN[1]

[1]*British Geological Survey, Lyell Centre, Research Avenue South, Edinburgh EH14 4AP, UK*

[2]*British Geological Survey, Nicker Hill, Keyworth, Nottingham NG12 5GG, UK*

**Correspondence: sarsenikos@gmail.com*

Abstract: Decades of oil and gas exploration across the North Sea have led to a detailed understanding of its Cenozoic–Mesozoic structure. However, the deeper basin architecture of Paleozoic petroleum systems has been less well defined by seismic data. This regional structural overview of the Devono-Carboniferous petroleum systems incorporates interpretations from more than 85 000 line-kilometres of 2D seismic data and 50 3D seismic volumes, plus a gravity, density and magnetic study, from the Central Silverpit Basin to the East Orkney Basin. A complex picture of previously unmapped or poorly known basins emerges on an inherited basement fabric, with numerous granite-cored blocks. These basins are controlled by Devono-Carboniferous normal, strike-slip and reverse faults.

The main basins across Quadrants 29–44 trend NW–SE, influenced by the Tornquist trend inherited from the Caledonian basement. North of Quadrants 27 and 28, and the presumed Iapetus suture, the major depocentres are NE–SW (e.g. the Forth Approaches and Inner Moray Firth basins) to east–west (e.g. the Caithness Graben), and WNW–ESE trending (e.g. the East Orkney Basin), reflecting the basement structural inheritance. From seismic interpretation, there are indications of an older north–south fault trend in the Inner Moray Firth that is difficult to image, since it has been dissected by subsequent Permo-Carboniferous and Mesozoic faulting and rifting.

The Central and Northern North Sea (CNS and NNS, respectively) are key hydrocarbon provinces accounting for a large proportion of the UK's oil and gas production. Running from late 2014 to early 2016, and ahead of the release of UK Government seismic data and the 29th Offshore Licensing Round, the 21st Century Exploration Roadmap (21CXRM) Palaeozoic Project aimed to stimulate hydrocarbon exploration of the Paleozoic play across and around the Mid North Sea High (CNS study area, Quadrants 25–44: Fig. 1) to the Inner/Outer Moray Firth Basin and the East Shetland Platform (Orcadian study area, Quadrants 11–23: Fig. 1), focusing on Devonian and Carboniferous strata. This paper synthesizes the systematic regional study undertaken by an interpretation of 85 000 line- kilometres of the highest resolution seismic data (released and unreleased) across the study areas, integrated with gravity, magnetic, tectonics studies and onshore UK knowledge, to highlight key structural elements of the potential Upper Paleozoic petroleum systems.

More specifically, the Orcadian study area extends from the East Shetland Platform in the north (Quadrant 7) to Quadrants 14, 15 and 22–24 in the south. Farther south, the CNS study area includes the Forth Approaches Basin (Quadrants 26 and 27), across the Mid North Sea High and southwards to the northern margin of the Carboniferous–Permian gas basin of the Southern North Sea (SNS) (Quadrants 41–44: Cameron 1992, 1993; Cameron *et al.* 2005) (Fig. 1).

The Upper Paleozoic strata onshore UK have been heavily studied, reaching a broadly accepted understanding of the complex structural history of extension, transtension, transpression and inversion that these areas have undergone (Chadwick & Holliday 1991; Glennie & Underhill 1998; Underhill *et al.* 2008; Woodcock 2012; Woodcock & Strachan 2012). However, although the existence of potential offshore Devonian and Carboniferous petroleum systems across the study area has been previously documented (Evans *et al.* 2003; Hay *et al.* 2005; Doornenbal & Stevenson 2010; Milton-Worssell *et al.* 2010), regional seismic mapping and a structural overview have been lacking, resulting in a poorly understood structural setting for the Paleozoic basins.

Oil and gas production indicates that the majority of North Sea hydrocarbons are produced from Jurassic-sourced fields located in the heavily explored Cenozoic and Mesozoic successions of

From: Monaghan, A. A., Underhill, J. R., Hewett, A. J. & Marshall, J. E. A. (eds) 2018. *Paleozoic Plays of NW Europe*. Geological Society, London, Special Publications, **471**, 65–90.
First published online March 20, 2018, https://doi.org/10.1144/SP471.3

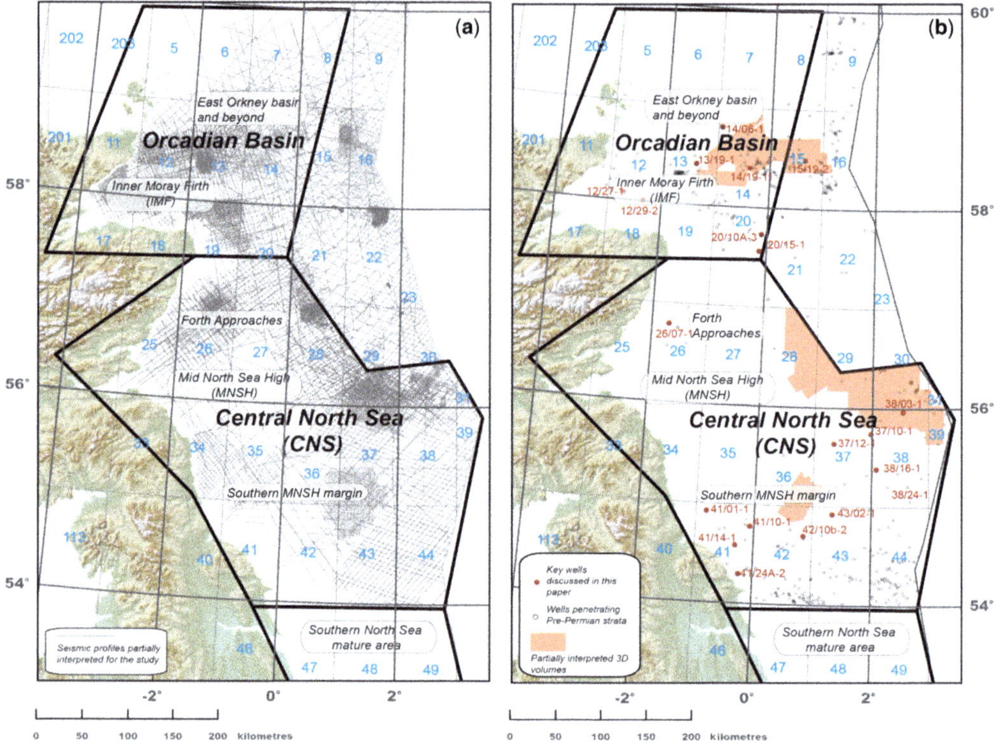

Fig. 1. (**a**) The 21CXRM Palaeozoic Project study areas. The CNS study area includes Quadrants 25–44 and the Orcadian Basin study area (Quadrants 7–22). Thin grey lines represent the 2D seismic profiles interpreted for the study. (**b**) Key wells discussed in this paper and wells penetrating pre-Permian strata (see also Kearsey *et al.* 2015; Whitbread & Kearsey 2016). Orange polygons are the 3D seismic volumes partially interpreted for the study. Includes mapping data licensed from the Ordnance Survey © Crown Copyright and/or database right 2018. License number 100021290 EU. Seismic line locations were facilitated by DECC/OGA.

the Central Graben and Moray Firth Basin (Abbotts 1991; Gluyas & Hichens 2003 and references therein). Across the 21CXRM Palaeozoic Project area, only a handful of fields (e.g. Buchan, Stirling, Claymore and Argyll/Ardmore) produce oil of assumed Jurassic source from a Devonian and Carboniferous reservoir (e.g. Edwards 1991; Robson 1991), whilst the Beatrice/Jacky and Breagh fields exemplify Devonian and Carboniferous sourcing, respectively (e.g. Stevens 1991; Symonds this volume, in press).

Tectonic setting

Episodic, plate-scale tectonism between Laurentia, Baltica, Gondwana and Avalonia was active during the Upper Paleozoic (Coward 1993; Domeier & Torsvik 2014). The tectonic framework of the study area is transitional between Iapetan and inherited Caledonian trends (NE–SW to ENE–WSW) in the north and Tornquist trends to the south (NW–SE) (British Geological Survey 1996; Pharaoh 1999). In Late Ordovician–Silurian times, Avalonia and Baltica were amalgamated by the closure of the Tornquist Sea (Pharaoh 1999; Torsvik & Rehnström 2003; Domeier & Torsvik 2014). By the Early Devonian, the Iapetus Ocean was closed, leading to a 'soft collision' between Laurentia and eastern Avalonia and to the infill of the sedimentary basins with large volumes of clastic sediments (Woodcock 2012). The location of the offshore Iapetus suture has been a subject of numerous studies (e.g. Klemperer & Matthews 1987; Soper *et al.* 1992). The exact location is still debated, but it is in the vicinity of the northern Mid North Sea High, with the Forth Approaches Basin lying to its north.

The most relevant tectonic models for the project area come from Coward (1993) and Maynard *et al.* (1997). Ziegler (1990) and Cocks & Torsvik (2006) have provided an overview of the wider palaeogeography of the region, local aspects of which have been updated through subsequent studies. According to Coward (1993), on the northern edge of the project area, during the Late Devonian–early Carboniferous, the NE expulsion of the Baltica microplate was accommodated by sinistral transtension in the vicinity of the Great Glen Fault, while the southern edge of the study area recorded dextral transtension as a combination of the Baltica microplate expulsion and the adjacent Variscan belt. Onshore southern UK, Late Devonian–early Carboniferous times were characterized by broadly north–south to NNW–SSE extension (Fraser & Gawthorpe 2003).

By late Carboniferous times, although Baltica moved back westwards, leading to stress reversal (Coward 1993), the regional transport direction (broadly NE–SW) would be expected to have remained the same as during the early Carboniferous (cf. De Paola *et al.* 2005). This change in the stress field would be expected to have reactivated suitably orientated structures as part of a regional transpressive regime.

Datasets and methodologies

Seismic data

The seismic dataset utilized in this study comprised released and unreleased 2D and 3D surveys provided to the British Geological Survey under contract from the Department of Energy and Climate Change/Oil and Gas Authority (DECC/OGA), covering the area from Quadrant 7 to Quadrant 44 (Fig. 1a). The 85 000 line-km of 2D data, including several regional-scale surveys, were the most important source of information for the study, due to their coverage and penetration of the Upper Paleozoic sequences between approximately 0.5 and 4 s two-way travel time (TWTT). The line spacing was irregular, from 2 to more than 10 km (Fig. 1a), but was considered to be adequate for regional structural insights across the majority of the study area. Data spacing across the Mid North Sea High (Quadrants 26–36) was between 5 and >30 km; this area was the focus of a gravity backstripping study to elucidate Carboniferous and Devonian basins, and of a subsequent UK Government seismic survey (released in 2016). Twenty-three 3D volumes were also consulted as source of information, and eight of them were partially interpreted, focusing on structurally complex areas. At the request of seismic data providers, these interpretations were resampled at 2D line spacing before inclusion in the 5 km-resolution TWTT and depth grids.

Well data

Of the thousands of exploration and production wells drilled across the CNS, only 550 have reached pre-Permian strata, most of them terminating after a few tens of metres in that sequence. Approximately 180 wells penetrating the pre-Permian (both Carboniferous and Devonian) were stratigraphically reinterpreted during the 21CXRM Palaeozoic Project (Fig. 2) (Kearsey *et al.* 2015, this volume, in press; Whitbread & Kearsey 2016) and, together with existing interpretations, these form the basis upon which the seismic and structural interpretations were made.

Calibration of well-to-seismic ties

Seismic calibration was achieved by the use of available time–depth pairs from downhole sonic logs and checkshots. Synthetic seismograms were produced for selected wells with a good penetration of Devonian and Carboniferous strata (e.g. 41/10-1 and 12/29-2). The comparison showed that there was a very good correspondence between the synthetic seismograms and the time–depth pairs.

Seismic interpretation/selected events

Ten seismic events were mapped through the Devonian, Carboniferous and Permian succession (Fig. 2). Events with the greatest acoustic impedance and greatest regional coverage were prioritized, as were events which represented either important intervals delineating the deep basins (e.g. Middle Devonian) or intervals crucial for a better understanding of the petroleum system (e.g. the Scremerston Formation and Middle Devonian source rocks). The study commenced with the interpretation of a regional grid of well-calibrated seismic profiles. Additional lines were interpreted as the characteristic reflectivity for events and packages was established across different surveys and sub-basins. The most challenging aspects of the interpretation were the loss of reflectivity at depth (i.e. loss of impedance contrast), the complicated diapiric geometries of the Zechstein affecting seismic imaging underneath (especially in the Forth Approaches Basin) and the limited penetration of the Paleozoic strata by wells.

Depth-conversion methodology

The very large area of the project area encompassed highly heterogeneous lithologies and time-equivalent depositional units with different burial and uplift histories, resulting in laterally and vertically varying interval velocities. Regional depth conversion was therefore a crucial, yet challenging, procedure, which took into account variations in the interval velocities both laterally and vertically.

Given the regional extent and variability, the layer-cake depth-conversion model was considered the most adequate approach. A 3D velocity model was constructed using 14 layers from the basement to the Cenozoic and was defined by the most significant variations in velocity.

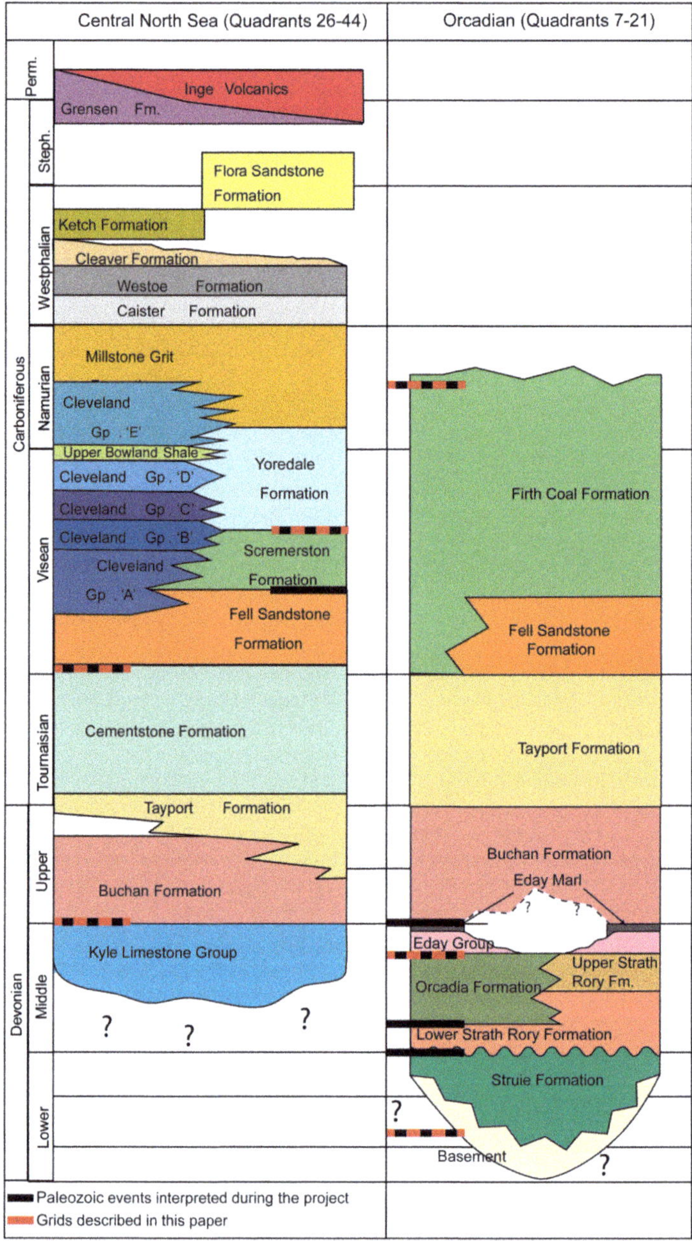

Fig. 2. Simplified stratigraphic chart of the CNS and Orcadian study areas. The ages and stratigraphic relationships are based on Kearsey *et al.* (2015) and Whitbread & Kearsey (2016). See also Kearsey *et al.* (this volume, in press). The thick black lines indicate the seismic events interpreted during the study (not all the events are shown in this paper). For the full grid dataset in TWTT and depth see Arsenikos *et al.* (2015, 2016).

Velocity data were collected from over 700 wells, with the most complete datasets of checkshots and/or velocity logs across Quadrants 11–44. Wells with anomalously high or low velocities were excluded.

In the northern part of the project area (Quadrants 11–22), the combination of the BGS well database with the interpreted horizons and faults allowed for the creation of a full 3D Structural Framework™ and a 3D Velocity Volume in Decision Space™.

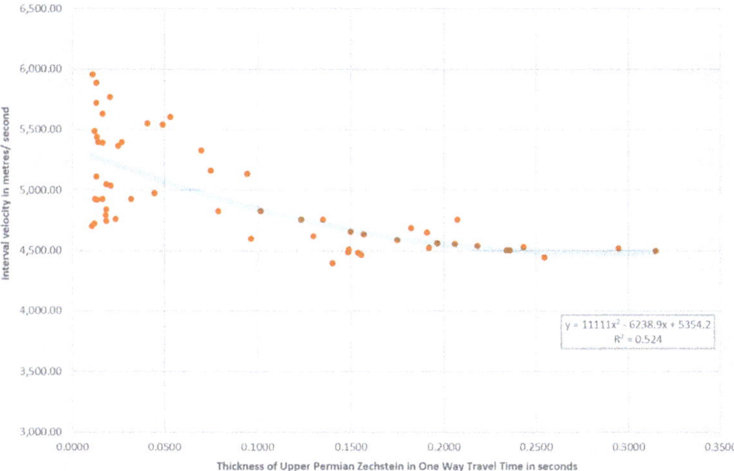

Fig. 3. Correlation between the Zechstein layer thickness (*x*-axis in seconds) and the interval velocity in the wells penetrating it (*y*-axis in m s^{-1}).

The depth-converted surfaces were quality-controlled against the drilled depths in the wells and corrected as necessary.

The upper Permian Zechstein Group needed to be treated differently between the southern and northern parts of the study. This is due to the presence of thick evaporite successions (halite, anhydrite and gypsum) and diapirs to the south (Quadrants 25–44), which gradually become more clastic-dominated to the north (Quadrants 11–22). The velocity of the Zechstein layer in The Netherlands is a function of its thickness rather than its burial depth (Velmod-1 and Velmod-2: Van Dalfsen *et al.* 2006). This applies also to the CNS study area, and the method followed respects the relationship between thickness and velocity. For thicknesses of more than 150 ms, a constant velocity of 4500 m s^{-1} OWT was used; while for thicknesses of less than 150 ms, a velocity function was applied based on the statistics from 65 wells in the CNS study area (Fig. 3).

Figure 4 shows an example of the very good correlation of the final model between interval velocity, structures and depth-converted surfaces.

Gravity modelling

Only an outline of the gravity-modelling methodology is provided below; for full details see Kimbell & Williamson (2015, 2016). Gravity data acquired by the British Geological Survey (BGS 2017*a*, *d*) were employed, and the analysis of the results was supplemented by comparison with magnetic data (BGS 2017*b*, *c*).

Downhole density logs from 146 wells in the CNS study area and 179 wells in the Orcadian study area were analysed. The logs were divided into units separated by the seismically defined boundaries that would be used in subsequent gravity modelling. The sampling in the CNS study area was relatively poorly distributed, so a predictive model was used to simulate the density of the post-Zechstein sequence in that area. The model employed compaction trends and burial anomalies based on the analysis of the available log data and integration with the results of previous studies. Compaction trends were derived from the shale and chalk models of Sclater & Christie (1980), and burial anomalies were based primarily on the results of Japsen (1998, 1999, 2000). The efficacy of the predictions was tested at the well sites and, as a result, overcompaction effects were incorporated in full, but thresholds were applied to the influence of undercompaction (overpressure) to avoid overcorrection. There is an inverse correlation between the average density of the Zechstein Group in the CNS study area and its thickness, which results from a greater proportion of low-density halite where the unit is thick, and higher-density dolomite and anhydrite where it is thin. This relationship was used to develop a density model for this sequence based on regression kriging, with well logs (where at least 60% of the sequence was sampled) as the primary control, and the relationship between thickness and density as the secondary drift.

Compaction trends and burial anomalies were used to estimate densities in the post-Chalk sequence in the Orcadian study area, but the older strata were relatively well sampled and their density was modelled by empirical Bayesian kriging. In both areas, densities of 2.75 and 1.03 Mg m^{-3} were assumed for basement and seawater, respectively.

The gravity modelling involved the removal ('stripping') of the effect of the shallower part of

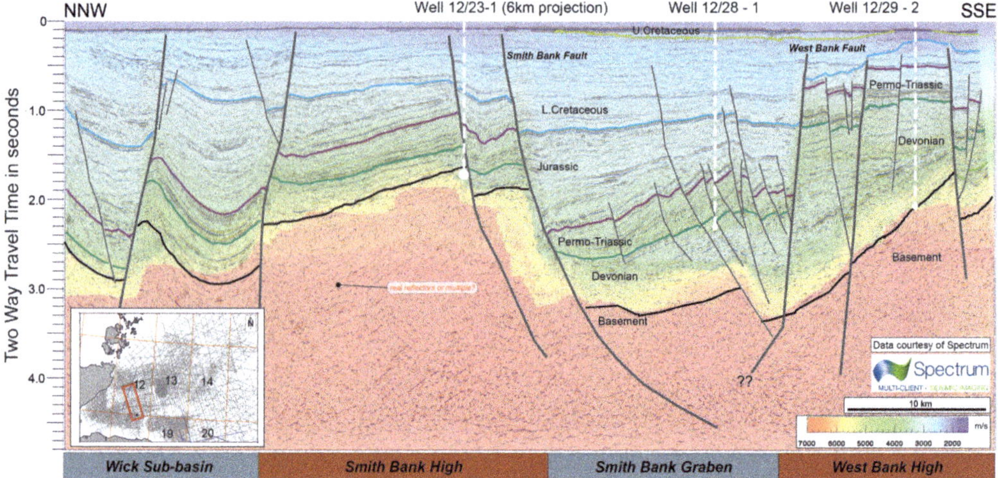

Fig. 4. Seismic profile (in the background) and velocity model (coloured, in the foreground) across the Inner Moray Firth Basin illustrating the very good correlation between the structures and the velocity model applied during depth conversion.

the sequence in order to isolate that of underlying structure. It was conducted using GM-SYS 3D routines within the Geosoft Oasis Montaj software package. In the CNS study area, the stripping extended to the base of the Zechstein Group; but in the Orcadian study area, it only extended to the top of the Zechstein because of limitations in the information available on the thickness of that unit. The structural inputs to the gravity stripping were depth-converted grids from the seismic interpretation. For the purposes of gravity modelling, the 5 km grids were resampled at 2.5 km node spacings to allow an improved resolution where the structure was smoothly varying, although this does not circumvent the resolution limits where the seismically defined structure contains short wavelengths. Both models included boundaries at the seabed, Top Chalk, Base Chalk and Top Zechstein; the CNS study area model also included the base Zechstein surface, and the Orcadian study area model included the base Cretaceous (Cimmerian Unconformity) and Top Triassic surfaces.

Stripped gravity fields were produced for both study areas, and these contained pronounced regional gradients relating to deep crustal structure (in particular, a reduction in the depth to the Moho and an increase in deep crustal density towards the central axis of the North Sea). These effects were removed in the form of a regional field which was constrained, in a generalized fashion, where there was sufficient control (good evidence of depth to basement in areas remote from major granite plutons) and allowed to vary smoothly in-between. Subtraction of the regional trend resulted in a residual stripped gravity field which was employed in further analysis. In the case of the Orcadian study area, this analysis was qualitative; but with the CNS study area, a further quantitative step was undertaken which involved inversion of the residual stripped anomalies in terms of a new depth interface. The density assumed for the unit between base Zechstein and the gravity inversion surface was based on a generalized model of the density of the pre-Zechstein Upper Paleozoic rocks, so the surface provides a simulation of the depth to top basement. This is not the case, however, where basement density contrasts and, in particular, low-density granite plutons affect the inversion. Although it is possible to excise the influence of granites from the results of gravity inversion (e.g. Milton-Worssell *et al.* 2010), in the present study we omitted this step in order to: (i) facilitate the integrated (seismic/gravity/magnetic) analysis of areas where granites and basins were in close proximity and the partitioning of their effects was difficult; and (ii) avoid prejudging the interpretation where intrusive bodies were identified with less confidence. The gravity inversion surface was converted into a horizon in TWTT and imported into the seismic interpretation environment to aid this integration.

The apparent thickness between the base Zechstein and the gravity inversion surface in the CNS study area is illustrated in Figure 5a, and the residual stripped gravity anomalies in the Orcadian study area are illustrated in Figure 5b.

Devonian and Carboniferous basin geometry and evolution

The mapping of the Paleozoic basins from the entire project area is described in two major study areas with eight subareas/basins. The CNS study area includes the North Dogger Basin, the Mid North Sea High, the Silverpit Basin, the offshore Northumberland Trough and the Forth Approaches Basin. The Orcadian study area includes the Inner Moray Firth Basin, the East Orkney Basin and the Outer Moray Firth. Figure 6 provides the regional synthesis of the individual mapped basins, their trends and their geographical relationship with onshore.

North Dogger Basin

The North Dogger Basin, initially described as a deep Carboniferous basin by Milton-Worssell *et al.* (2010), trends in a NW–SE direction, from the southern edge of Quadrant 29, across the NW corner of Quadrant 37 and into the central part of Quadrant 38 (Fig. 6). To the NNE, the basin margin is delineated by the Auk Ridge (Trewin & Bramwell 1991; Gatliff *et al.* 1994) and is well resolved in the gravity model (Fig. 5a: Tornquist trend). The basin is delineated by the Dogger Granite High to the SE. The Dogger Granite is clearly identifiable in the gravity-modelling results (and has a strong magnetic effect) with possible north–south and east–west extensions identifiable in both gravity and magnetic data (Fig. 5a) (Kimbell & Williamson 2015). The basin continues into the Dutch sector north of the Elbow Spit Platform (Wride 1995; ter Borgh *et al.* this volume, in press). The top of the Middle Devonian Kyle Group is the most prominent reflector in the Devonian–Carboniferous sequence, defining the geometry of the North Dogger Basin, as proven in wells 30/16-5, 30/24-3, 30/25a-2, 37/12-1 and 38/03-1 surrounding the basin (Fig. 7). Although the interpretation is based on the strong, characteristic Kyle Group reflector on the highs, it is possible that the Middle Devonian is represented by deeper-water facies into the basin. Lower and middle Carboniferous strata are proven in 11 wells and can be mapped across the basin. The limited number of well ties across Quadrants 29, 30, 37 and 38 remains the biggest constraint for a more detailed tectonostratigraphic model of the North Dogger Basin. The Scremerston Formation (lower Carboniferous) has been interpreted in an area more extensive than previously mapped. It has been interpreted based on its characteristic seismic signature across the basin (high-frequency, high-amplitude reflectors), in agreement with previous studies (Hay *et al.* 2005).

The North Dogger Basin is separated into two sub-basins by a prominent elevated block, the NW–SE-orientated North Dogger Horst (Fig. 7), which is constrained by well 37/10-1 proving uppermost Devonian and basal Carboniferous stratigraphy (the Tayport and Buchan formations: see Kearsey *et al.* 2015, this volume, in press). The North Dogger Horst is also identifiable as a high in the gravity model and as a magnetic anomaly (Fig. 5a).

The second sub-basin is located in Quadrant 29 (Figs 6 & 7) and is considered to be a fault-controlled Devono-Carboniferous depocentre (Milton-Worssell *et al.* 2010; this study). The varying levels of well control in the basin have an impact on the certainty of interpretation but seismic data clearly show deep reflectors which define a 1.5 s-thick TWTT basin under the Permian sequences. Devonian and Carboniferous sequences, which are eroded (or non-deposited) on top of the regionally extensive Dogger Granite High in Quadrants 37 and 38, reach a thickness of over 1 s TWTT around 50 km NNE of the high into the basin. The Top Kyle Group seismic pick plunges to depths of more than 7 km (3.5 s TWTT) in the centre of the North Dogger Basin (Quadrant 38: Fig. 8) until imaging is unclear. In the Auk-Flora Ridge area, reflectors of the well-calibrated Kyle Group (wells 30/16-5, 30/24-3 and 30/25a-2) are clearly visible on seismic data, offset by major normal faults trending broadly NE–SW (Fig. 7). The presence of Upper Devonian and Carboniferous strata in wells 37/10-1, 37/12-1, 38/16- 1, 38/18-1, 38/22-1, 38/24-1 and 39/07-1, combined with seismic data, indicate that the Upper Devonian and lower–middle Carboniferous intervals have infilled the available space of the North Dogger Basin. On seismic data, these are condensed sequences which appear to onlap onto the Dogger Granite High; and to the NE of the granite in Quadrant 38, they become thicker and deeper, infilling an underfilled basin (Fig. 8).

In summary, the North Dogger Basin is interpreted as having an overall (N)NW–(S)SE structural trend with more than 2.5 km (more than 1.2 s TWTT) of Upper Devonian–lower Carboniferous sediments.

Mid North Sea High: Dogger Granite High

The term Mid North Sea High, initially used in the description of the palaeogeographical division of the northern and southern Permian basins (Donato *et al.* 1983; Jenyon *et al.* 1984), is also used for the geographical area across Quadrants 27, 28, 35 and 36 (Fig. 6). Tectonic and summary maps are lacking detail (British Geological Survey (BGS) 1996; PESGB 2017), due to several kilometres' spacing of legacy seismic data (in places >20 km) and very limited well penetrations. As part of the CNS study area, seismic mapping across the deepest parts of the Mid North Sea High has added detail and defined a 'high' across Quadrants 26–28 and 35–36 that is less extensive than previously thought (Fig. 6: largely in the area of the offshore Southern Uplands), along with a series of highs and basins to its north, east and south margins.

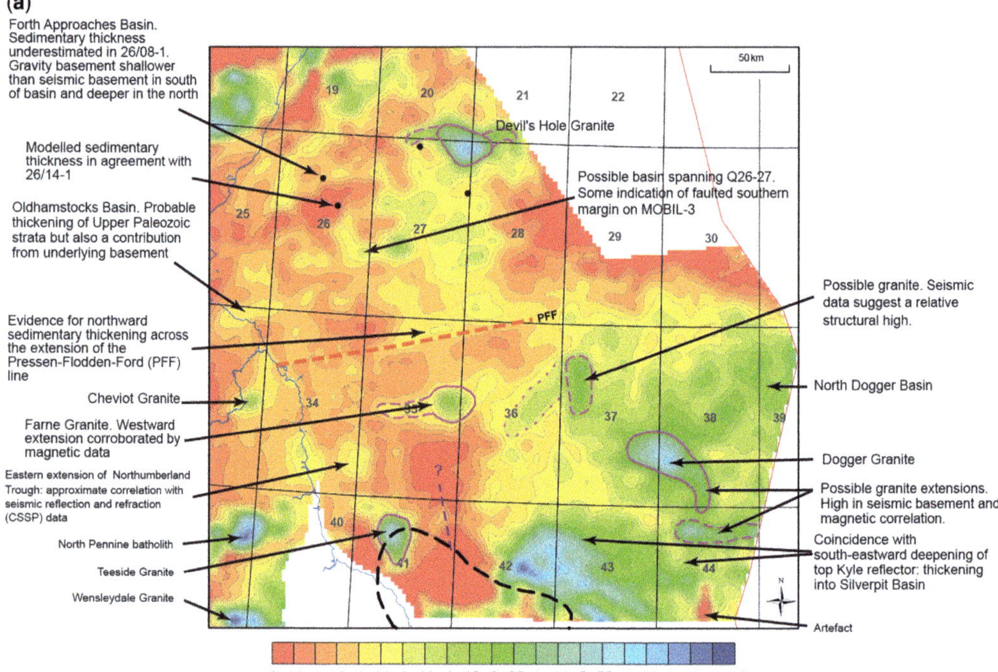

Fig. 5. (**a**) Apparent thickness between the base of the Zechstein and the gravity inversion surface across the CNS study area (after Kimbell & Williamson 2015).

The gravity model indicates structures with ENE–WSW trends crossing the Mid North Sea High (e.g. the offshore continuation of the Pressen-Flodden-Ford line, the Oldhamstocks Basin and a possible basin spanning the boundary of Quadrants 26 and 27: Fig. 5a). Magnetic features associated with Permo-Carboniferous dykes also follow this trend (Kimbell & Williamson 2015).

In Quadrants 26–27–28 and 34–35–36, regional mapping of three interpreted intervals (Top Cementstone Formation–early Carboniferous/Tournaisian, Top Fell Sandstone Formation–early/middle Carboniferous/Visean and Top Scremerston Formation–middle Carboniferous/Visean), constrained by wells to the north and to the south of the area, shows that the Mid North Sea High is a relatively flat, tilted terrace deepening eastwards with post?-Permian onlapping sequences. In more detail, this geometry is mapped across Quadrant 28 and northern Quadrant 36, and becomes truncated by the fault-controlled North Dogger Basin in Quadrant 29 and in the northern Quadrant 37. Moving west towards the UK coast (Quadrant 27 and the northern Quadrant 35), Paleozoic (i.e. Kyle Group or time-equivalent) seismic reflectors are interpreted as dipping up towards shallower depths in the subsurface. The northern margin of the Mid North Sea High is marked by several ENE-trending faults that downthrow northwards to the Forth Approaches Basin and by the Devil's Hole Horst block and granite.

At the boundary of Quadrants 36 and 37, the SE extremity of the Mid North Sea High is characterized by the Western Arcuate Fault described by Jenyon *et al.* (1984). It consists of a near-vertical NE–SW-trending fault mapped towards the NW–SE-trending faults bounding the North Dogger Basin (Figs 6 & 7). On the east side of Quadrant 36, a restricted Devono-Carboniferous (pull-apart?) basin is mapped. The gravity and magnetic modelling confirms the presence of this feature and the NE–SW trend identified as the Western Arcuate Fault (Fig. 5a). 3D seismic interpretation and coherence volumes over the northern North Dogger Basin (Quadrant 29) suggest an along-strike NE–SW-trending structure at approximately 3–4 s TWTT depth, offsetting the NW–SE-trending faults and with some evidence of pop-up flower structures (Fig. 7). This may represent an extension of the Western Arcuate Fault, or a similarly orientated system, and is consistent with the regional Devono-Carboniferous structural grain (e.g. Coward 1993; British Geological Survey (BGS) 1996; Coward *et al.* 2003; Fraser & Gawthorpe 2003; De Paola *et al.* 2005 and references therein). Further

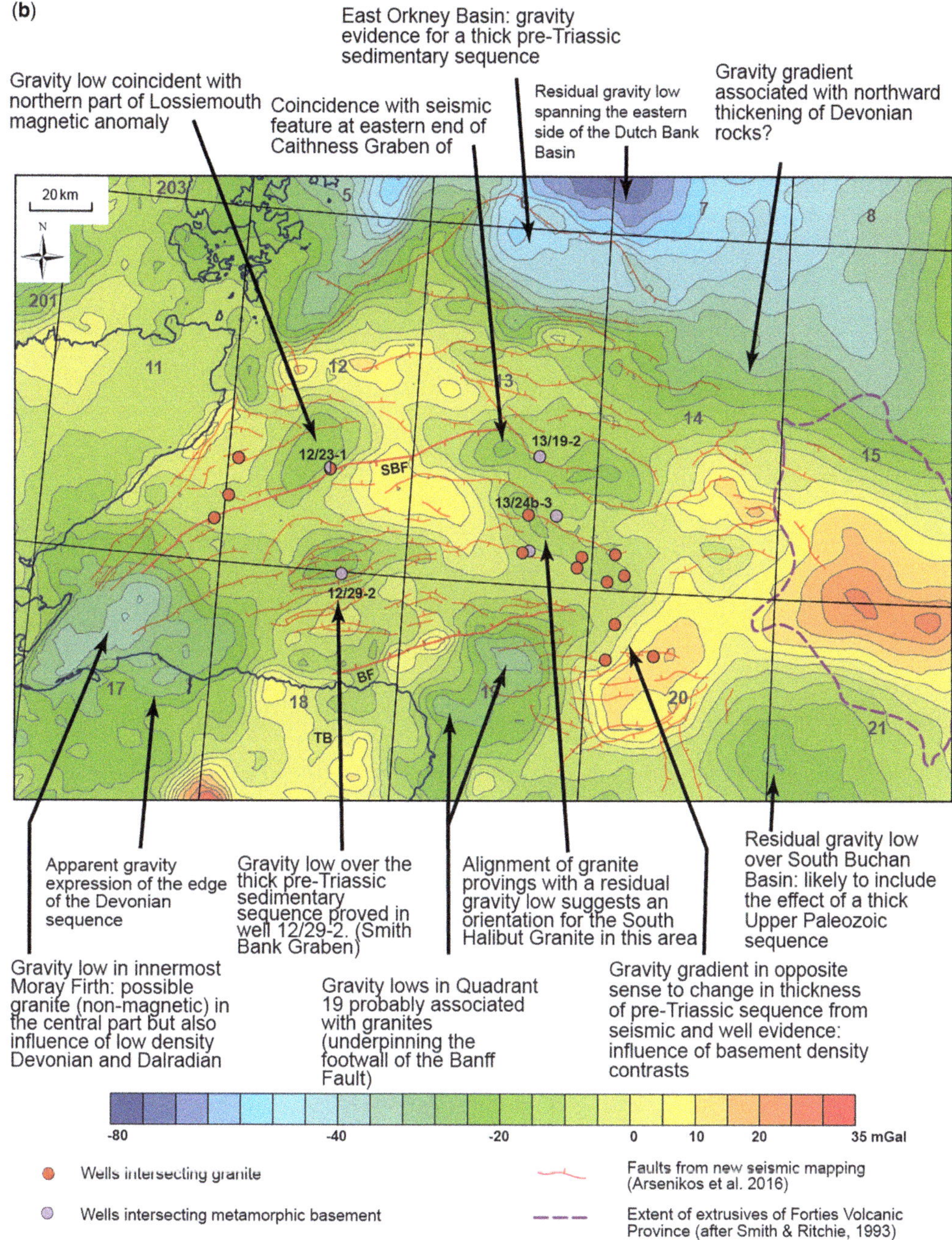

Fig. 5. (*Continued*) (**b**) Residual stripped (to Top Zechstein) gravity anomalies across the Orcadian study area (after Kimbell & Williamson 2016). For details and comments on the various basins see the text and the reports cited above. For the faults documented by the new seismic mapping, see Arsenikos *et al.* (2016) and for the extent of extrusives in the Forties Volcanic Province, see Smith & Ritchie (1993).

detailed structural mapping and analysis is required to deduce whether Devono-Carboniferous strain partitioning resulted in wrench- and extension-dominated domains across the Dogger Granite High and North Dogger Basin as a result of an oblique regional transport direction (NNW–SSE:

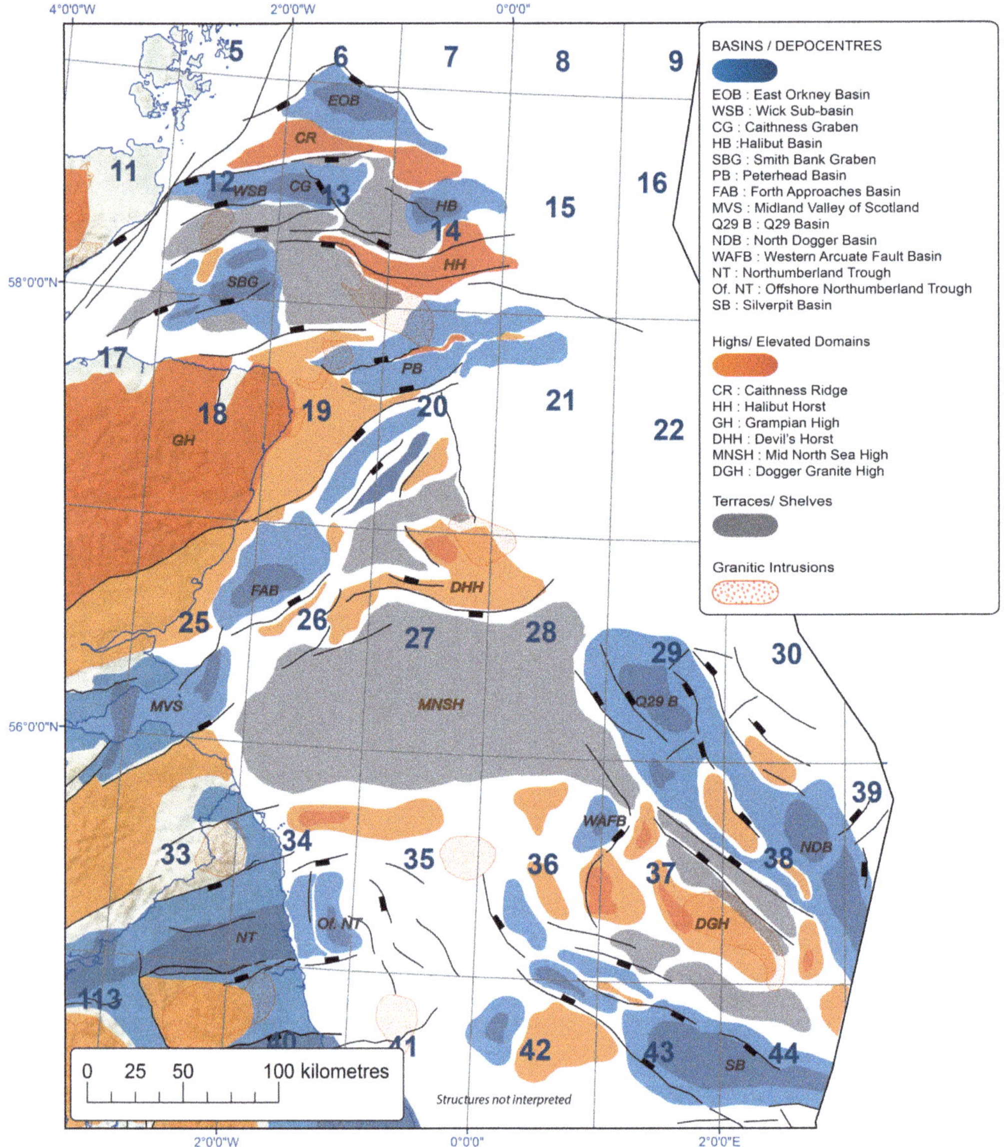

Fig. 6. Regional structural synthesis resulting from the mapping of the structures across the study areas. Illustrated here are the major basin-bounding faults. The Mid North Sea High area is significantly smaller than in previously published maps, and a series of basins and highs surround it to the north and to the SSE (e.g. Western Arcuate Fault Basin, Q29 basin).

Coward 1993; De Paola *et al.* 2005; Leslie *et al.* 2015) and so resolve the timing of the observed NW–SE and cross-cutting NE–SW ?oblique slip faults.

Silverpit Basin

Forming one of the major structures of the SNS Carboniferous gas basin, the Silverpit Basin is orientated NW–SE across Quadrants 43 and 44, extending northwards into Quadrant 36 (Bailey *et al.* 1993; Cameron 1993; Cameron *et al.* 2005) (Fig. 6). Since the margin of the basin is characterized by the presence of the Breagh Field (blocks 42/13a and 42/12a: Symonds this volume, in press), the rest of the basin margin is also a key zone for understanding the Visean–Namurian sedimentation and potentially prospective petroleum

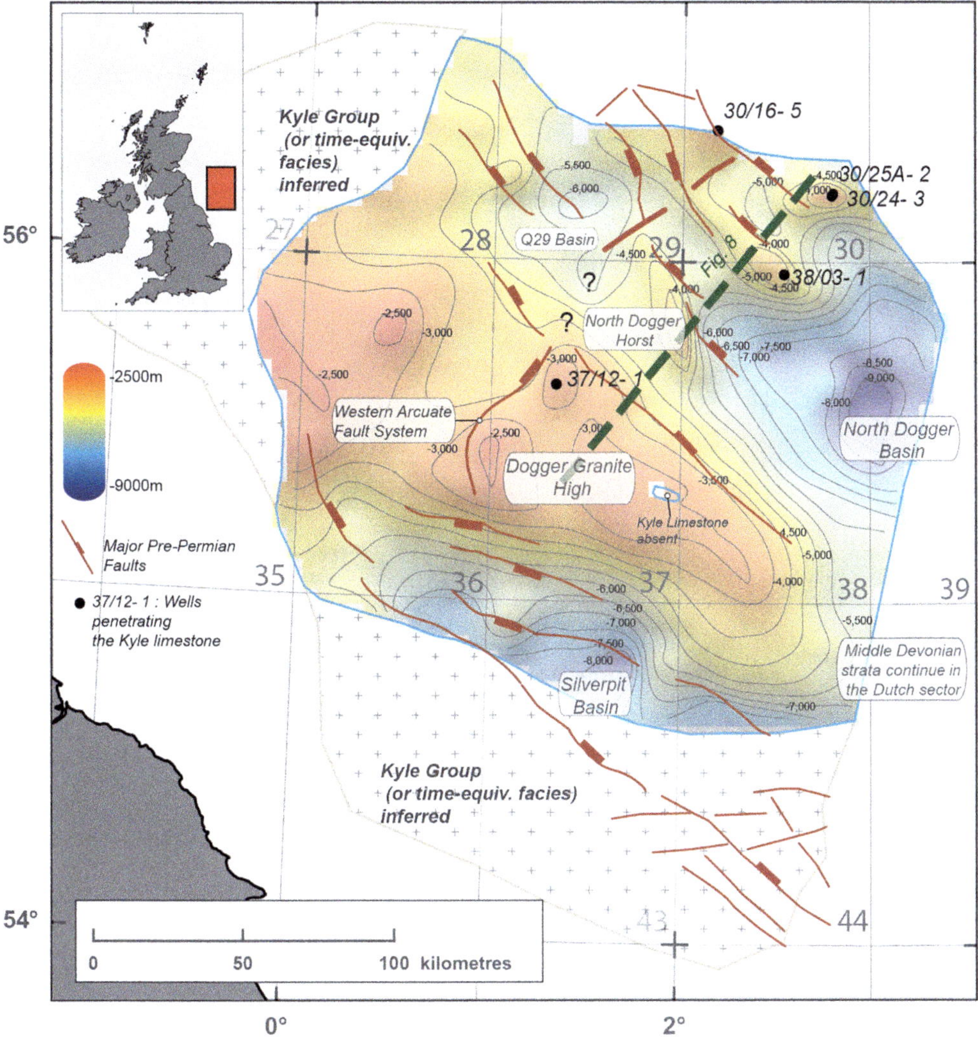

Fig. 7. Depth to the Middle Devonian Top Kyle Group (5 km resolution grid), and major faults in the North Dogger Basin and the Silverpit Basin. The most prominent feature is the deep Middle Devonian basin in Quadrant 38 (North Dogger Basin). As well as its extension further NE (the Q29 basin). The green line shows the approximate location of the seismic profile shown in Figure 8. The Western Arcuate Fault System is also illustrated. A discontinuity with a similar trend exists at depth in the Q29 basin but it is unclear whether it is the extension of the Western Arcuate Fault System or a separate feature.

systems (Monaghan *et al.* 2015). A well-calibrated seismic interpretation combined with the gravity model (Fig. 5a) provide a good outline of the basin margins. On the NE basin margin, Middle (?)/Upper Devonian and lower Carboniferous sequences onlap the SW flanks of the Dogger Granite High. Carboniferous strata tied to wells in Quadrants 42–43–44 constrain phases of Late Devonian–earliest Carboniferous normal faulting and mid-Carboniferous post-rift sedimentation, followed by Variscan inversion and the development of NW-trending gentle folds. Some NW–SE-trending faults were reactivated in post-Permian times, offsetting the Zechstein Group.

Offshore Northumberland Trough

The onshore Northumberland Trough (Fig. 6) is an ENE-trending Carboniferous basin that is bounded to the north and south by structural highs (the Cheviot and Alston blocks, respectively) which are underpinned by low-density granitic intrusions (Kimbell

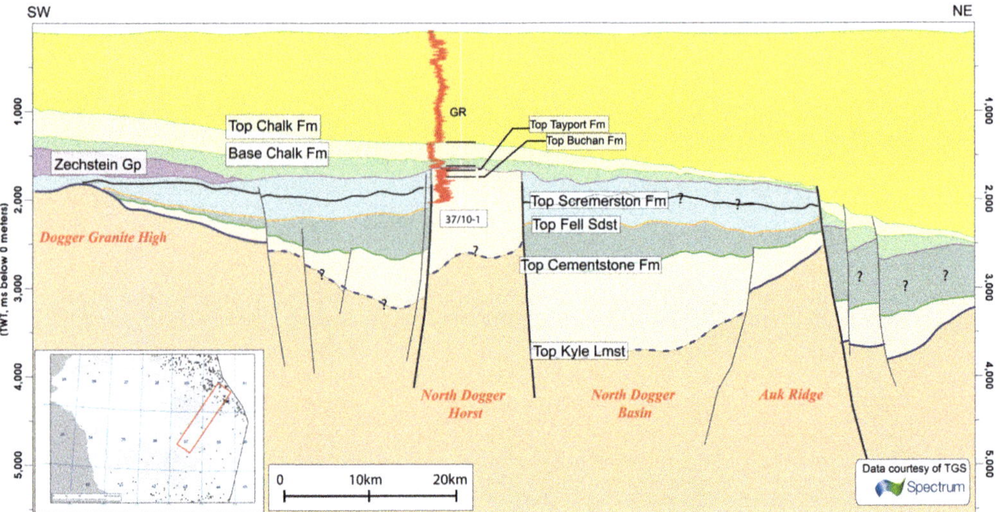

Fig. 8. Interpretation of a seismic section across the North Dogger Basin. The basin margins are the Dogger Granite High to the SW and the Auk-Flora Ridge to the NE. At the NE extremity of Quadrant 37, the basin is partitioned in two sub-basins separated by the North Dogger Horst; on the top of which well 37/10-1 penetrated Upper Devonian strata (the Tayport and Buchan formations). The Top Kyle Group reflector is strong and easily recognizable on the highs, and becomes gradually less evident at deeper levels. This could be related both to imaging issues and to a change of facies (becoming more distal basinwards).

et al. 1989; Chadwick & Holliday 1991; De Paola *et al.* 2007). Offshore, Carboniferous rocks subcrop the seabed adjacent to the Northumberland coast, and are succeeded by Permian and Triassic, Jurassic and Cretaceous successions eastwards. Seismic interpretation was poorly constrained in this area as the nearest deep offshore well (41/01-1) lies to the south of the area (Fig. 9). A thick Carboniferous succession comprising sandstone, coal, mudstone and limestone is expected to be present within the area by analogy with well 41/01-1 and the onshore succession within the Northumberland Trough. The near Top Cementstone Formation (lower Carboniferous) reflector was interpreted across southern Quadrants 34 and 35 (Fig. 9) in order to define faults, basins and highs, while, where present, Top and Base Chalk, and Top and Base Zechstein were interpreted to facilitate depth conversion. Well 41/01-1, 25 km south of the area, penetrates 159 m of Cementstone Formation sediments. Although there is no clear seismic reflector defining the top of this formation, a contrasting juxtaposition of the seismic packages above and below the boundary (higher amplitude, more continuous reflectors above with a more transparent seismic package below) enabled an interpretation and delineation of a primarily early Carboniferous basin infill into the area to be made (Fig. 10). Outcrop onshore and penetrations by offshore BGS shallow boreholes have allowed the offshore subcrop of Westphalian strata to be mapped but a lack of well ties, along with erosion due to Variscan inversion, resulted in discontinuous picks and precluded the interpretation of upper Carboniferous surfaces in the offshore Northumberland Trough.

Onshore, the major bounding faults to the south of the Northumberland Trough are the ENE-trending Stublick and Ninety-Fathom faults. They show evidence of syndepositional extensional faulting in the lower Carboniferous strata, with up to 4 km of Tournaisian and Visean sediments being deposited adjacent to the faults (Kimbell *et al.* 1989; De Paola *et al.* 2007). Offshore, the eastwards continuation of the Ninety-Fathom Fault can be mapped for approximately 30 km (Figs 6 & 9). The northern margin of the basin is characterized by an ENE-trending fault which can be mapped for approximately 20–25 km offshore in Quadrant 34, a continuation of the Hauxley Fault of Kimbell *et al.* (1989). Between the ENE-trending basin-bounding faults, a system of typically steep to vertical NNW- to NNE-trending faults with small throws of less than 150 m (*c.* 70 ms TWTT) and occasional reverse displacements that appear to be related to tight folds within the pre-Permian succession are interpreted from seismic data. The spacing and quality of the seismic dataset and the lack of well ties precludes a detailed model of the Carboniferous structural evolution in this area. Here the offshore extension of the Northumberland Trough,

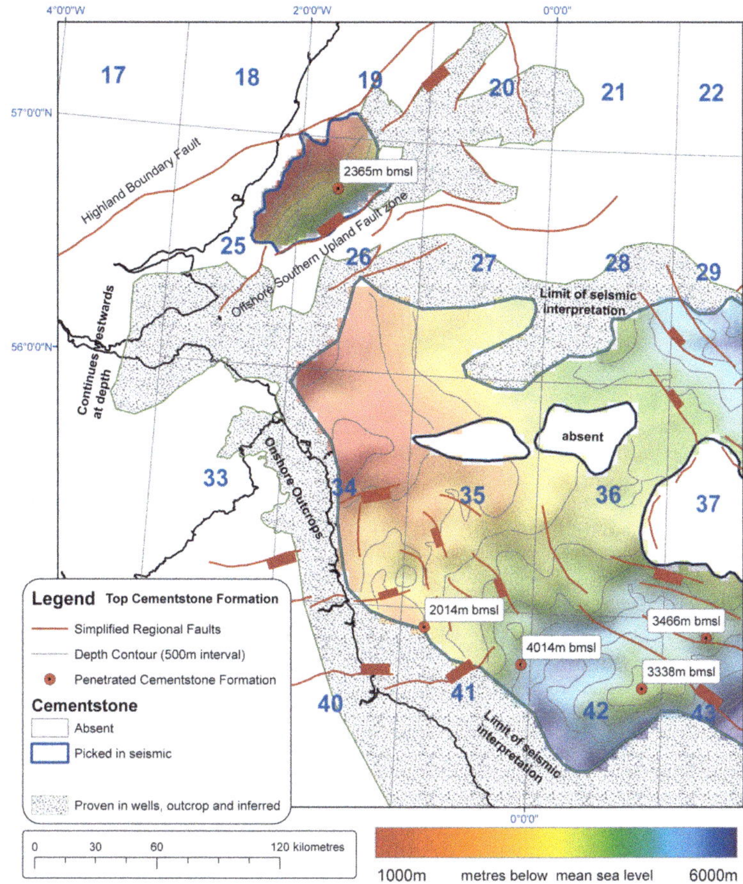

Fig. 9. Depth to the early Carboniferous (Tournaisian) Cementstone Formation (5 km resolution). The formation has been regionally mapped in the CNS study area, and is present in the offshore extension of the Northumberland Trough (Quadrants 34 and 35) and the Forth Approaches Basin (Quadrants 25 and 26). In the offshore Northumberland area, there are two major fault trends mapped: NE and NW, in agreement with De Paola *et al.* (2005) onshore. m bmsl, metres below mean sea level.

representative of the extension-dominated domain of De Paola *et al.* (2005), passes eastwards towards the NW–SE inherited Tornquist trend dominant within the mapped fault pattern in Quadrant 35 and 36 (Fig. 6).

Forth Approaches Basin

The Forth Approaches Basin is the eastern continuation offshore of the Paleozoic basins of the Midland Valley of Scotland (MVS). It is bounded to the north and to the south by the seawards extensions of the Highland Boundary and Southern Upland faults, respectively (Fig. 9). In this area, Early Devonian extension followed the Caledonian compressive regime and led to the establishment of small basins infilled with continental sediments (Marshall & Hewett 2003). There is no evidence of a Middle Devonian sedimentary succession in the Forth Approaches Basin area, which is interpreted as forming a relative high during this time. However, from latest Mid to Late Devonian times, fluvial coarse-grained clastic sediments spread south of the Highland Boundary Fault into the area. Onshore, and in the Firth of Forth (Quadrant 25), a series of north- to NNE-trending synsedimentary Carboniferous synclines, that are highly oblique to the regional faults, are interpreted to have developed in a predominantly dextral strike-slip regime during the mid to the late Carboniferous (Read *et al.* 2002; Ritchie *et al.* 2003; Underhill *et al.* 2008). Late Carboniferous tightening of the folds (Cartwright *et al.* 2001; Underhill *et al.*

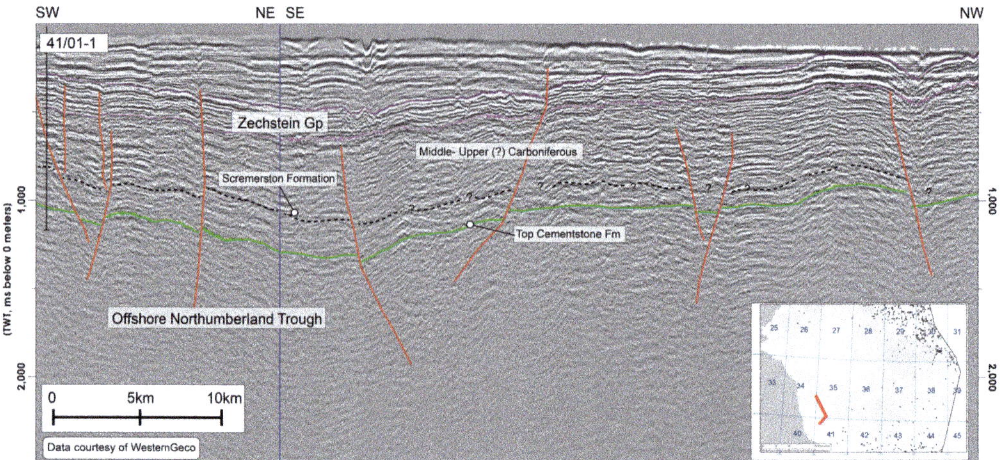

Fig. 10. Composite interpreted seismic section running SW–NE and SE–NW across the offshore Northumberland Trough. The Top Cementstone Formation (Tournaisian) is illustrated in green and the Top Scremerston (Visean) in dotted black. The faults (red lines) were active (or reactivated?) until upper Permian times.

2008) in the offshore may indicate Variscan, pre-Permian inversion.

Seismic ties from two key wells (26/07-1 and 26/08-1: Fig. 1b) within the SW part of the Forth Approaches Basin prove a thick coal-, mudstone- and sandstone-bearing Visean–Namurian succession (Firth Coal Formation = Scremerston Formation equivalent) within a half-graben geometry (Figs 6, 9 & 11). Southeasterly deepening is shown on the Top Cementstone Formation depth map (Fig. 9) against the fault system defining the northern edge of the Mid North Sea High. Significant SE thickening of Rotliegend sandstones is also observed, interpreted as synrift deposition during Early Permian extensional reactivation (Cartwright *et al.* 2001; Underhill *et al.* 2008). The southern boundary of the basin is represented by the offshore extension of a northwards-stepping en echelon system of faults that onshore include the Dunbar-Gifford and Lammermuir faults; the latter onshore faults form

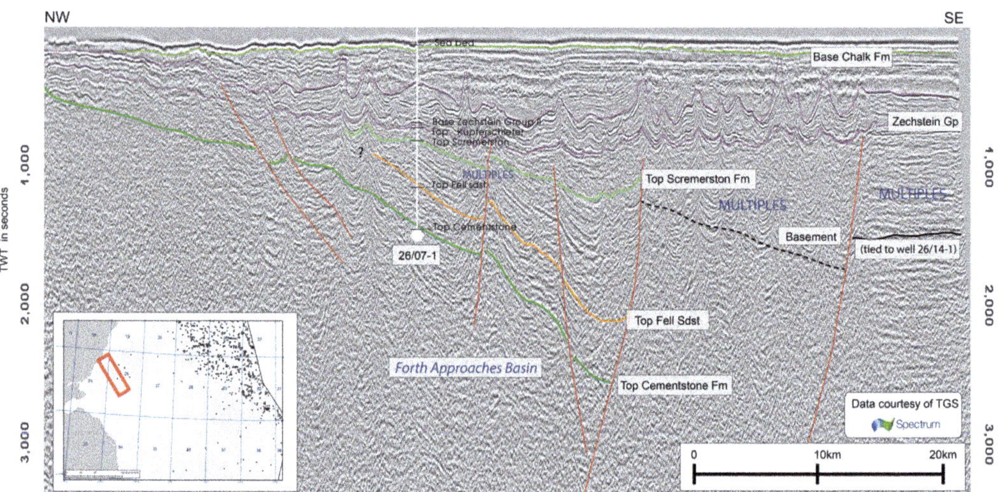

Fig. 11. Interpreted seismic section across the Forth Approaches Basin. The highly asymmetrical, half-graben geometry is controlled by the major fault to the SE. The Top Cementstone pick has been mapped across the area, whereas the Top Scremerston event, although present in well 26/07-1, proved challenging to map in detail. Both formations dip to greater depths to the SE and gradually shallow up to the NW.

part of the Southern Upland Fault system (British Geological Survey (BGS) 1996) (Figs 6 & 9). Offshore, faulting is interpreted to be offset northwards into Quadrant 25, adjacent to a NNE-trending syncline, before veering to a NE–SW trend and extending across Quadrants 26 and 20 (Fig. 9). Seismic interpretation between the Upper Devonian footwall succession proven in 26/12-1 (Fig. 9) and an interpreted Top Cementstone Formation succession in the hanging wall to the NW indicates a throw across the basin-bounding structures of over 3000 m (Figs 9 & 11).

Towards the NE extremity of the basin (Quadrants 19 and 20), a gravity low (Fig. 5a) has been mapped that could be interpreted as a thick Permo-Carboniferous basin similar to that present to the SW. However, seismic reflections beneath the Base Permian Unconformity at this location are lower amplitude and less continuous, and do not image any obvious thickening Carboniferous succession as observed to the SW; here, a relatively thin Carboniferous succession of approximately 400 m is interpreted and the gravity low may be in response to a thick Devonian succession.

The MVS contains oil and gas fields (Midlothian, D'Arcy-Cousland), and a proven working petroleum system (Underhill *et al.* 2008). The lower Carboniferous strata act simultaneously as the main source rock and a trap (Hallett *et al.* 1986; Underhill *et al.* 2008). Its offshore extension, represented by the Forth Approaches Basin, would need to constrain two main critical factors in order to prove that it can be a viable play: (a) the maturity; and (b) the volumes of the source rock. The main source rock, the Firth Coal Formation (Scremerston equivalent), is gas-prone but, according to the basin analysis model conducted by Vincent (2015), the gas window was not reached at the location of the modelled well 26/08-1. However, well reports document oil and gas shows from a Carboniferous source rock in the Forth Approaches (details in Vincent 2015).

Northern Outer Moray Firth

The Outer Moray Firth–Witch Ground Graben (Figs 5 & 6) area and the edge of the East Shetland Platform area are two other frontier areas where the Paleozoic strata could play a significant role in new plays.

The Witch Ground Graben is characterized by extensional faulting during the Jurassic and Cretaceous (Beach 1984). However, while the main focus of previous studies has been the Mesozoic structural evolution and petroleum potential of the graben (Beach 1984; Glennie & Underhill 1998; Jones *et al.* 1999), wells drilled into it (e.g. 14/19-12 and 15/19-2) prove that it also contains Carboniferous strata, which provide insights to the Paleozoic petroleum system and the deeper structural styles. The uppermost Devonian–lower Carboniferous intervals are represented by the Tayport Formation. The early Carboniferous (Tournaisian–Visean) is characterized by the Firth Coal Formation. These two formations can act both as potential source rocks (Firth Coal/coal-rich sequences) and reservoirs (parts of Tayport and Firth Coal/sand-rich facies).

As a regional observation, gravity and magnetic data indicate that the basement deepens towards the NE corner of the modelled area (Fig. 5b). In Quadrant 14, thick Middle Devonian sequences have been proven in wells (Marshall & Hewett 2003; Whitbread & Kearsey 2016) and are seismically interpreted as reaching over 700 m in thickness in depocentres such as the Halibut Basin. Moving to the east and basinwards of the Caithness Ridge and the West Fladen High, the major ENE-trending bounding faults offset the Devonian strata to depths of over 3 s TWTT. Patruno & Reid (2016, 2017) interpreted Devonian strata as present across Quadrant 14 and farther north towards Quadrants 7–9. The interpretation of Devonian strata in modern 3D volumes suggests that what has been considered as acoustic basement across some of the highs (such as the Halibut Horst) consists of tilted, deformed and truncated Devonian reflectors (Fig. 12). These deformed reflectors are penetrated by well 14/19-11, proving more than 700 m of Middle Devonian lacustrine sediments (Whitbread & Kearsey 2016). The top of the Devonian sequence is characterized by a large hiatus between Devonian and Cretaceous strata.

Inner Moray Firth Basin

Situated in Quadrants 11–13, the Inner Moray Firth Basin is characterized by the presence of Devonian lacustrine source rocks present within confined basins between fault-bounded highs. The area has been subjected to major tectonic episodes during the Devono-Carboniferous, Permo-Triassic, Jurassic–Early Cretaceous and Late Cretaceous–Cenozoic (Andrews *et al.* 1990). Regional Cenozoic erosion played a major role in the area, with estimates of approximately 1 km of sediments being removed across the entire basin and with more erosion to the west than to the east (Hillis *et al.* 1994).

The seismic profile in the background of Figure 4 is a representative example of the horst–graben geometry of the area and its complex tectonic history. At the SSE end of the profile, a thick Devonian sequence rests on top of the West Bank High (proven in well 12/29-2). The same sequence is interpreted on the hanging wall of the West Bank Fault in depths >3 s TWTT (approximately 4 km) in the Smith Bank Graben area (the top of the Middle Devonian

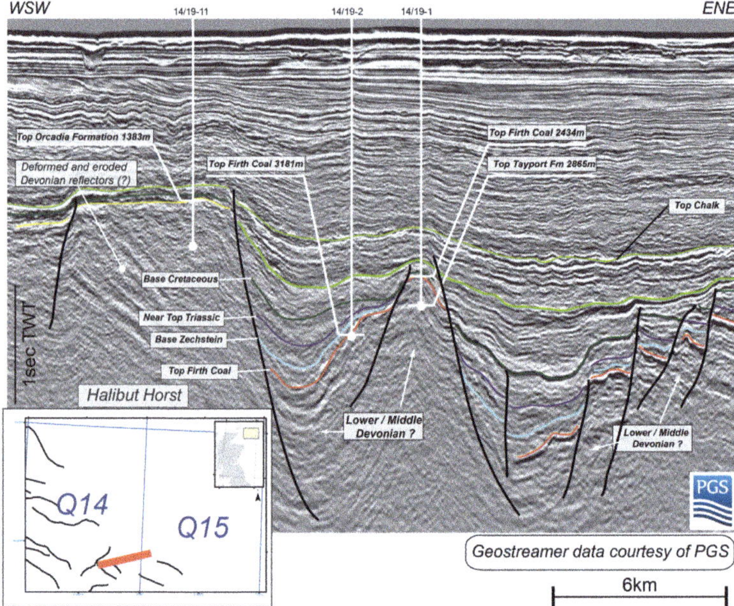

Fig. 12. Interpreted 3D seismic line across the Halibut Horst and the NW segment of the Witch Ground Graben. The Top Firth Coal Formation is penetrated on the highs by wells 14/19-1 and 14/19-2, and is interpreted as present in the hanging walls at depth. The Devonian strata are also present (proven in well 14/19-11), and they are interpreted as deformed and truncated on top of the Halibut Horst.

is penetrated in well 12/28-1). On top of the Smith Bank High, Devonian strata are present but thinner (proven in well 12/23-1), suggesting that the area was probably already an elevated (intra-basinal?) high at this time. In the NNW end of the profile, the Wick Sub-basin shows a very complex geometry and Paleozoic strata reaching depths of 3 s TWTT. The complex structures are related to the proximity of the basin to the major Great Glen Fault, its Paleozoic strike-slip activity (Roberts *et al.* 1990), and the subsequent Mesozoic normal faulting of the adjacent Wick and Helmsdale faults (Underhill 1991). Seismic evidence also confirms the presence of contractional features, as previously observed (Roberts *et al.* 1990; Underhill & Brodie 1993).

While development of the Devonian and Carboniferous basins is thought to have been controlled by strike-slip movement on the Great Glen and associated faults (Leslie *et al.* 2016 and references therein), interpretation of offshore seismic data and onshore field observations show that, during the Mesozoic, the development of the Inner Moray Firth Basin was the result of normal faulting in an extensional regime with a minimal strike-slip component (Underhill 1991; Thomson & Underhill 1993; Glennie & Underhill 1998). During the Mesozoic in the Inner Moray Firth Basin, the controlling fault was the Helmsdale Fault (situated west of the Great Glen Fault along the Scottish coast) while the Great Glen Fault played a minor strike-slip role (Andrews *et al.* 1990; Underhill 1991).

There are a significant number of publications on the Paleozoic intervals present in the Orcadian study area. However, the majority discuss the onshore stratigraphy, facies analysis and the depositional environments of the study area and the adjacent domains (e.g. Astin 1985; Duncan & Buxton 1995; Clarke & Parnell 1999; Marshall & Hewett 2003; Marshall *et al.* 2011).

Interpretation of the basement provided the first-order structure of the basins in Quadrants 12–20. The Top Basement reflector has been defined as the metamorphosed Lower Devonian or older Lower Paleozoic and Precambrian rocks or granite (e.g. wells 12/29-2 and 11/30a-10). The near Top Basement pick can be located above a more transparent and featureless seismic package (i.e. acoustic basement) immediately beneath Devonian, Carboniferous or younger successions. It may also be represented by an angular unconformity.

The Top Basement mapping depicts the remnants of the pre-Permian basin geometry (Fig. 13); however, this geometry has been overprinted by Mesozoic and Cenozoic events, and the present-day configuration is the combination of reactivated,

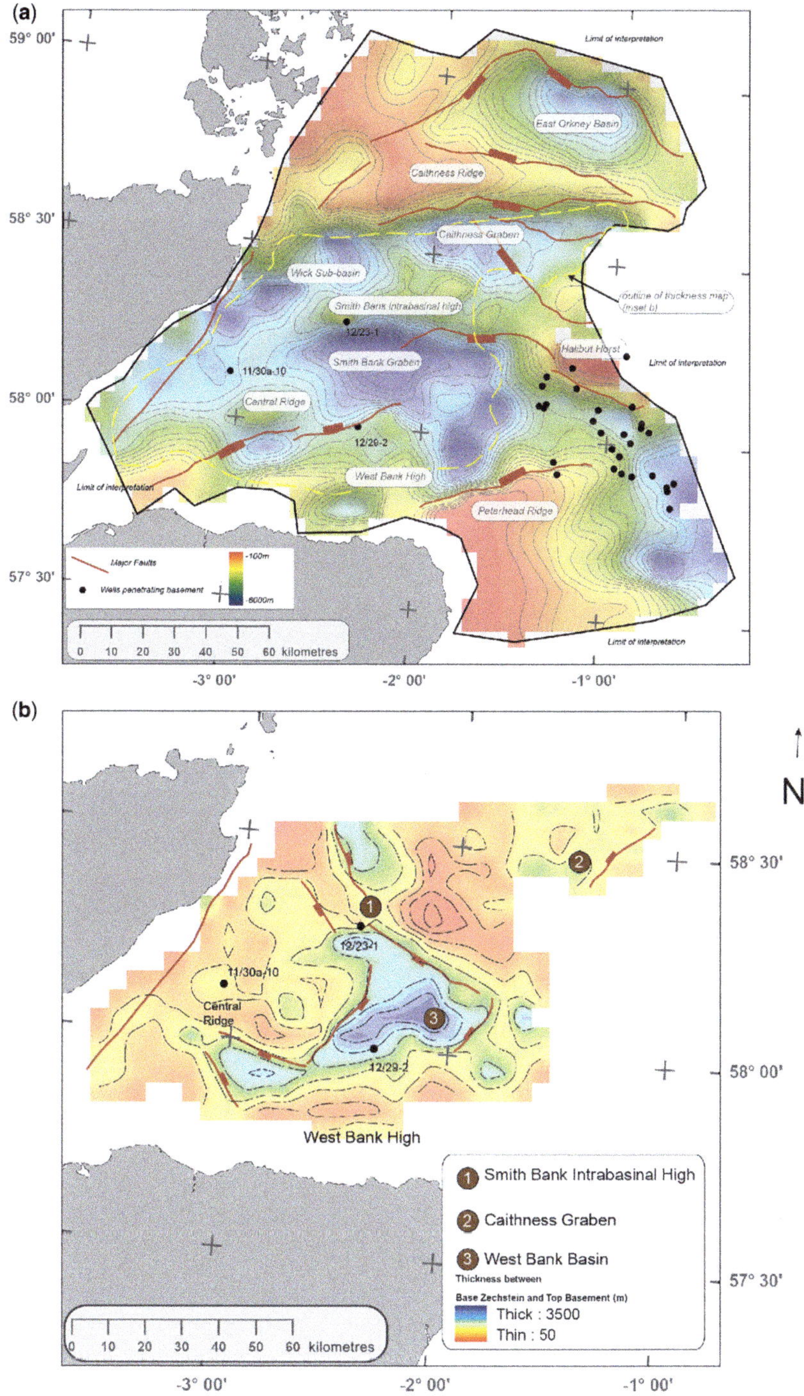

Fig. 13. (**a**) Depth to Top Basement pick across the Orcadian study area (5 km resolution). The regional trend is the well-known (E)NE–(W)SW direction in the Moray Firth area, resulting from Mesozoic faulting. (**b**) Thickness map between the Base of the Zechstein and the Top Basement (Base Devonian) pick. The most prominent depocentre is in Quadrant 12. It is possible in places to distinguish NNE–SSW and NNW–SSE trends that have been heavily overprinted by the NE–SW Permo-Carboniferous and reactivated Mesozoic trends.

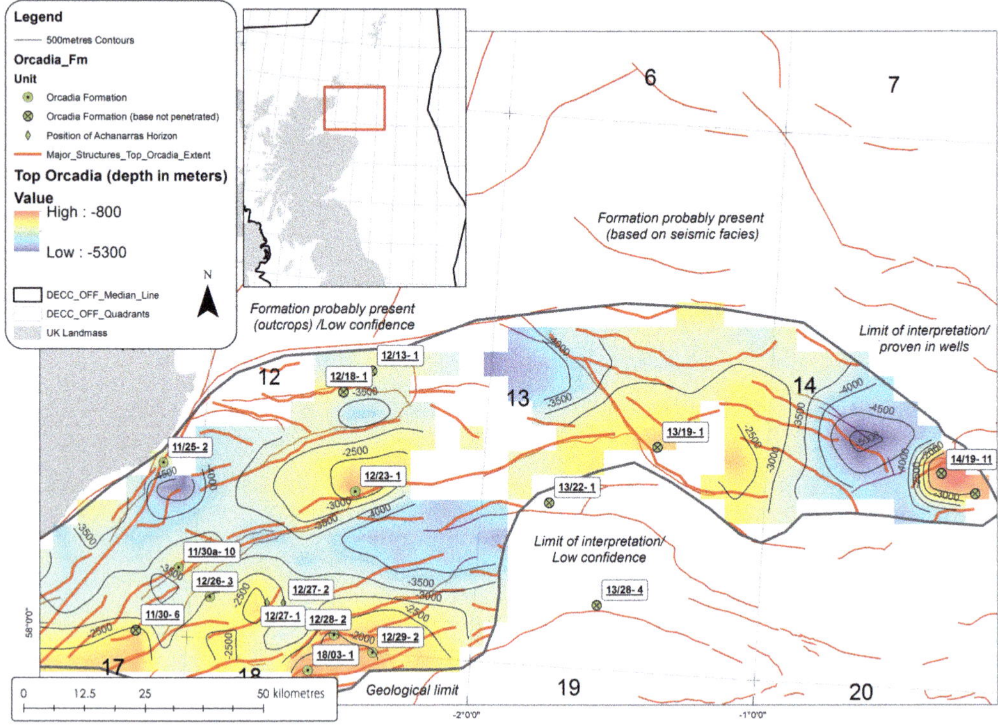

Fig. 14. Depth to the Mid-Devonian Top Orcadia Formation (5 km resolution) showing the regional extent of the current interpretation. The formation is probably also present in the East Orkney Basin (see the inset in Fig. 15).

inverted and eroded features across the Inner Moray Firth Basin.

The mapping has been aided by the use of gravity and magnetic modelling. Figure 5b shows the stripped gravity grid. It is important to highlight that there are multiple sources for a stripped gravity low over the Inner Moray Firth Basin. In addition to Devonian sedimentary rocks, contributions are likely from low-density Dalradian (Grampian Group) sediments and from granites, possibly including the source of the Lossiemouth magnetic anomaly (Dimitropoulos & Donato 1981; Pilkington *et al.* 1995). Stripped gravity lows are evident over the Smith Bank Graben and the eastern end of the Caithness Graben, and the Caithness High is characterized by a stripped gravity high and shallow magnetic sources (Kimbell & Williamson 2016).

The most extensive remnant of the Devonian depocentre is located across the south of Quadrant 12 in the Smith Bank Graben area (Figs 13 & 14).

South of the Caithness Ridge (Quadrant 13: Figs 6 & 13), a buried depocentre termed the Caithness Graben is infilled with Devonian, Permian and Early Mesozoic sediments that underpins the Halibut Platform (Fig. 15). The broadly ENE–WSW-trending depocentre is along strike from the Wick Sub-basin (Fig. 13).

The Top Basement mapping, and the subsequent Devonian thickness map in Figure 13b, show that apart from the dominant ENE–WSW trend of the basin, there are other less obvious trends. Interpreting in a regional tectonic model context, the en echelon configuration of the Central Ridge–West Bank High and Peterhead Ridge on the Top Basement and thickness maps (Fig. 13a, b) indicates that N (NW)–S(SE)- and NNE–SSW-trending discontinuities could be anticipated in the Moray Firth area, but they are less obvious due to Mesozoic structural overprinting. These structures may relate to Late Devonian–early Carboniferous intracontinental extensional stress. Such structures are similar to the faulting pattern associated to onshore Devonian outliers in the Moray–Buchan area (e.g. the Turriff outlier: Stephenson & Gould 1995; Trewin 2002) and also described in the Helmsdale region (Underhill & Brodie 1993; Leslie *et al.* 2016).

Apart from the Top Basement pick, the Top Orcadia Formation (Middle Devonian) is an indicative event of the Devonian basin configuration across the majority of the Orcadian study area. The interval

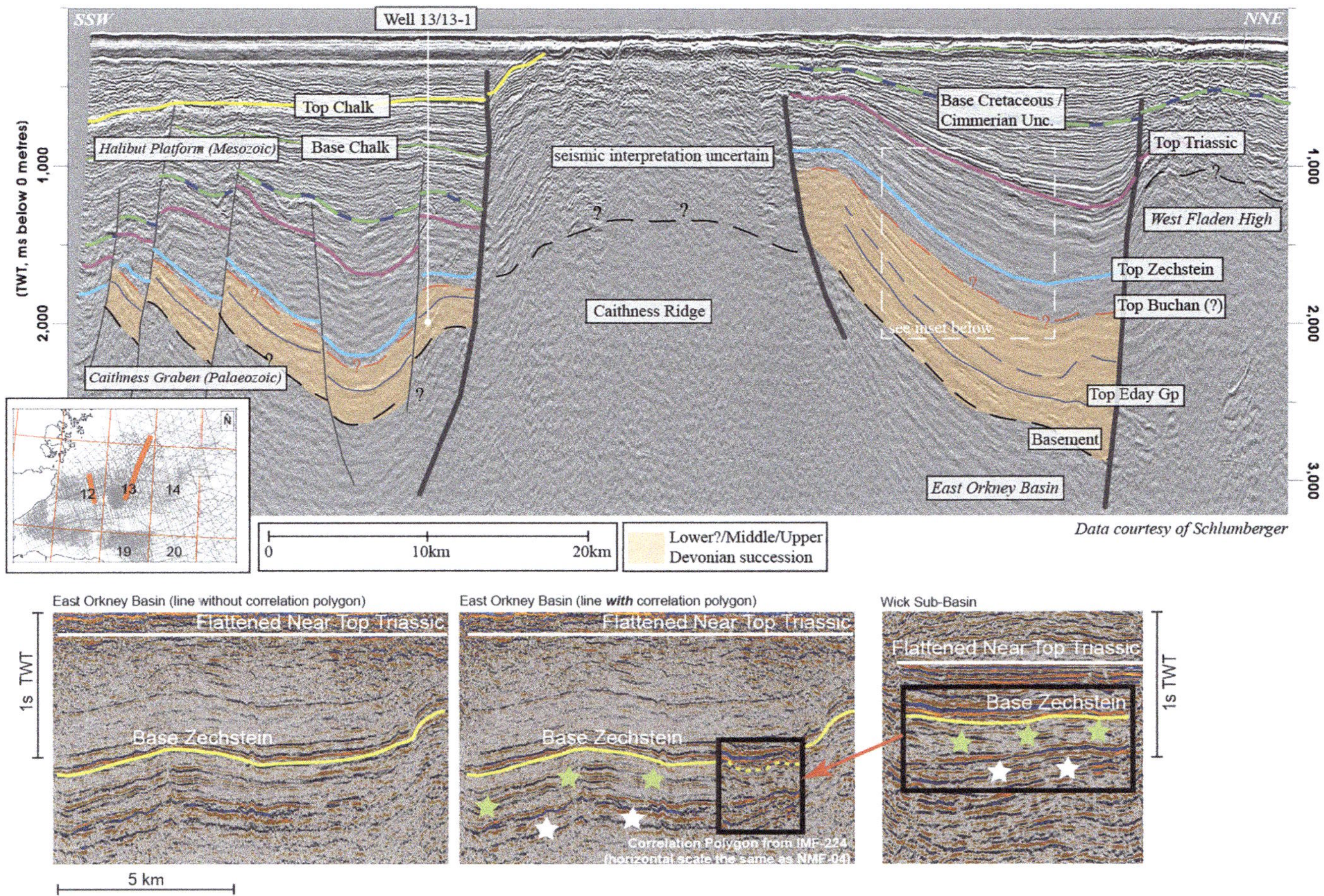

Fig. 15. Interpreted seismic section across the East Orkney Basin (NNE) and the Caithness Graben (SSW). The two depocentres are characterized by deeply buried Devonian strata. The inset illustrates a detail from the profile compared to a correlation polygon from the Wick Sub-basin area, a few kilometres SW of the Caithness Graben. It illustrates the almost identical seismic character between a seismic sample from the East Orkney Basin (far right square) and one from the Wick Sub-basin (left and middle rectangles). The stars indicate comparable stratified sequences, suggesting that buried Devonian sediments in the East Orkney Basin are similar to the ones in the Wick Sub-basin (i.e. Lower?–Middle Devonian lacustrine sediments proven in wells: Arsenikos *et al.* 2016; Whitbread & Kearsey 2016).

is regionally extensive, it reaches thicknesses of more than 750 m in Quadrant 12 and at least 700 m in well 14/19-11. Present-day depths are of the order of more than 4–4.5 km deep in the Smith Bank Graben (Fig. 14). Onshore, the formation is present as the equivalent Caithness Flagstone and Stromness Flagstone in Caithness and Orkney, with thicknesses of at least several hundreds of metres (900 m suggested by Astin 1990).

Finally, as part of the Inner Moray Firth Basin, the Caithness Graben remains an unexplored area, containing deeply buried pre-Permian strata underneath a thick Mesozoic pile. The presence (or absence) of the Devonian Struie Formation and Orcadia Formation source rocks will be primarily controlled by the extent of the intra-basinal Smith Bank High and its role during Devonian times. Poor seismic imaging related to the significant depths has reduced interpretation confidence but their presence in the Caithness Graben has been indirectly proven in wells on highs in the proximity, such as 12/18-1, 12/13-1, 13/19-1 and 13/22-1.

East Orkney Basin

Bounded to the north by the West Fladen High and to the south by the Caithness Ridge, the East Orkney Basin is an east–west to WNW–ESE fault-controlled half-graben located in Quadrants 6 and 13 (e.g. Andrews *et al.* 1990; Marshall *et al.* 1996) (Fig. 6).

No wells have been drilled in this basin. However, a conspicuous stripped gravity low over the area (Fig. 5b), abundant outcrops on the Orkney Islands and well penetrations further NE (8/04-1, 9/07-1 and 9/16-3), together with characteristic well-stratified reflectors correlated to probable Early–Mid-Devonian-aged sedimentary sequences proven farther south (inset in Fig. 15), strengthen the interpretation that thick fluvio-lacustrine sequences of the Eday Marl, Eday Flagstone and Orcadia formations are also present in the East Orkney Basin. Seismic data (Fig. 15) suggest that the Paleozoic strata in the basin are deeply buried, thus providing a potentially mature source rock in the area. Richardson *et al.* (2005) have reported a seep survey in the area near the basin and mapped oil seeps on the seabed. This observation, combined with the suggestion that the Jurassic source rock is immature to early mature in the East Orkney Basin area (Kubala *et al.* 2003), leads to the hypothesis that the oil seeps observed come from a different source rock. The lacustrine Orcadia Formation would be the best candidate for such a hypothesis and, even though no wells have proven it inside the East Orkney Basin, there are wells around the area and outcrops in the proximity which prove its presence (Marshall & Hewett 2003; Whitbread & Kearsey 2016).

Discussion

Regional context and basin geometries

The geometrical variability of the described deep basins is largely controlled by the inherited Caledonian structural grain (Iapetan v. Tornquist-related), which affects the accommodation space and the basin evolution. The regional tectonic framework is a broadly NW–SE-trending system south of the Iapetus suture zone (Mid North Sea High area) and a broadly (E)NE–(W)SW-trending system farther north due to Caledonian inheritance (Fig. 6).

Regional tectonic models (Coward 1993; Coward *et al.* 2003; Fossen 2010) suggest that during Late Devonian–early Carboniferous times, the lateral expulsion of Baltica relative to Laurentia and Avalonia would result in a NE–SW orientated stretch and regional transport direction across the CNS study area (e.g. Mid North Sea High), and strike-slip faulting along E(NE)–W(SW) trends (cf. De Paola *et al.* 2005). In the Orcadian study area, an E(SE)–W(NW)-directed stretching would have been anticipated in the Inner Moray Firth Basin related to the Great Glen–Helmsdale faults. By late Carboniferous times, although plate-scale motion of the Baltica microplate would have been reversed from a NE- to a west- or SW-directed motion, the overall regional transport directions across large-scale fault structures would have remained similar to that in Late Devonian–early Carboniferous times: that is, broadly aligned on a NE–SW axis (cf. De Paola *et al.* 2005). To the south of the CNS study area, inversion related to the Variscan Orogeny would have been recorded in the early Carboniferous and younger strata. The observations from the major basins of the study, such as the North Dogger Basin, the southern margin of the Mid North Sea High, the Silverpit Basin and the Inner Moray Firth Basin, are in agreement with these regional models superimposed upon the inherited structural framework. Observing the complexity of the basin geometries mapped across the study area, one can conclude that in order to better constrain the exact timing and direction of the faulting, it is essential to work in a local basin-by-basin basis, whilst keeping in mind the regional overview.

For these mapped Paleozoic basins, the source rock palaeogeography, burial and uplift history, erosion, and potential fault breach are all controlling factors of a functioning Paleozoic petroleum system. The combination of the proximity to the main kitchen areas with efficient migration routes and non-breached faults could potentially lead to prospects and successful plays.

Concerning the Mid North Sea High area, basin mapping suggests that the southern margin consists of a series of basins and blocks, and not a simple regional high as it is for the Permian and post-Permian succession.

The granitic intrusions are interpreted to have played a crucial role in strain partitioning, superimposed upon the Devono-Carboniferous structural trends. The Dogger Granite is a representative example. NE and SW of the granite margins, the fault trends are NW–SE and there is a Devono-Carboniferous sequence onlapping on the margins of the high. However, north of the Dogger Granite, the northeast-southwest trends are mapped and in places the faults are interpreted to follow the edge of igneous intrusive bodies (e.g. the Western Arcuate Fault).

Onshore, similar observations have led to the hypothesis of time-equivalent, spatially differentiated fault trends, such as the Northumberland Trough–Cheviot Pluton (De Paola *et al.* 2005).

Thickness maps derived from the depth-converted surfaces (see Arsenikos *et al.* 2015) show that the lower–mid-Carboniferous succession reaches thicknesses of up to 2 km in the North Dogger Basin, and 1.5 km in the offshore part of the Northumberland Trough (Quadrant 34) and Quadrant 42. These values are comparable to those in the literature for onshore Carboniferous basins (Fraser & Gawthorpe 1990, 2003; Waters & Davies 2006: thicknesses of the Carboniferous range from 1.5 to 3 km).

Implications for source rock extents

Seismic and well interpretation has constrained four major source rock intervals in the CNS and Orcadian study areas: the coal-bearing Scremerston Formation (Visean) in the CNS study area, the lacustrine source rocks of the Struie (Lower Devonian) and Orcadia (Middle Devonian) formations, and

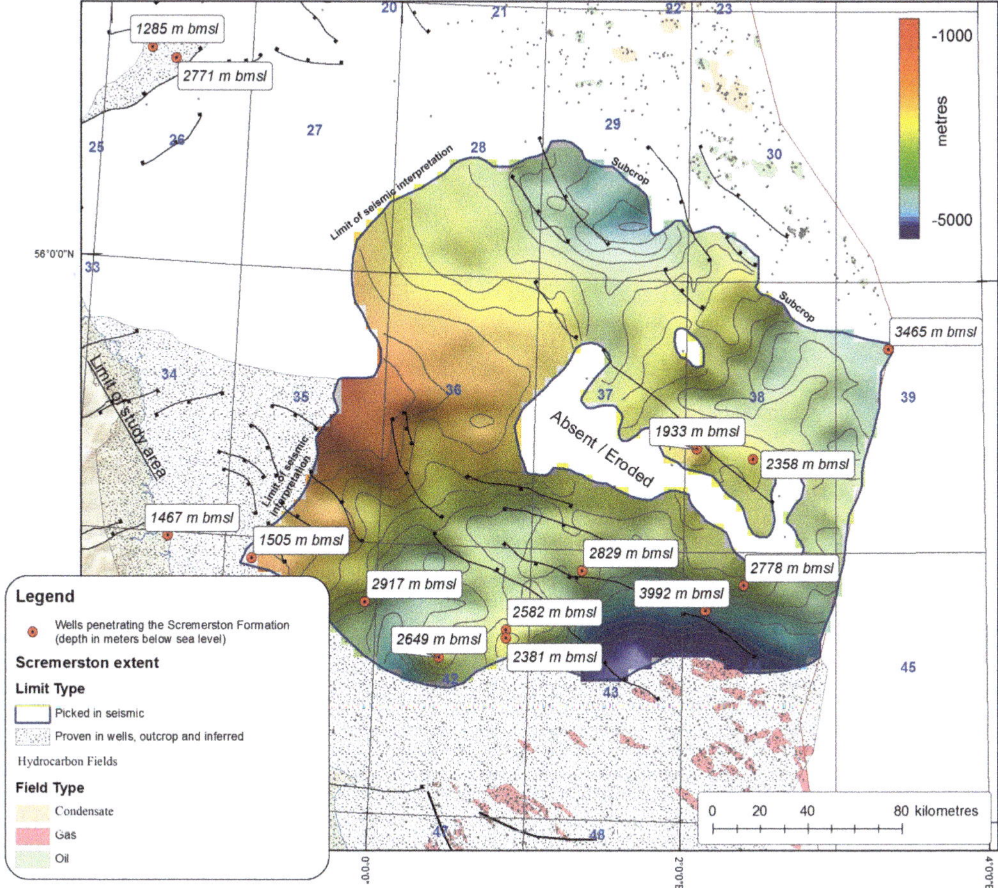

Fig. 16. Depth to Top Scremerston Formation in metres (5 km grid resolution). The formation is interpreted across the North Dogger Basin, the Q29 basin, on the southern part of the Mid North Sea High, in the Silverpit Basin and toward the offshore Northumberland Trough. Wells indicate the depth in metres below mean sea level as interpreted from Kearsey *et al.* (2015). m bmsl, metres below mean sea level.

the coal-bearing Firth Coal Formation (Visean–Namurian) in the Orcadian study area.

The Scremerston Formation (CNS study area) has been interpreted as present and mapped more extensively than previously recognized in reports (e.g. Hay *et al.* 2005) (Fig. 16). The Scremerston Formation has been penetrated both south of the Mid North Sea High (e.g. well 38/18-1) and north in the Forth Approaches Basin (e.g. 26/07-1), indicating its regional deposition. Based on its characteristic reflectivity (high-frequency/high-amplitude well-stratified reflectors), it was interpreted in the North Dogger Basin (at depths of the order of 3–3.5 km) and in the area adjacent to the southern margin of the Dogger Granite High and the Silverpit Basin (2.5–3 km depth). It is a good to excellent quality source rock; and basin modelling indicates that in southern Quadrants 41–43, it could be petroleum generating (Vincent 2015).

Farther north, in the Orcadian study area, the Firth Coal Formation has been proven in more than 15 wells across Quadrants 14 and 15, and in more than 10 wells in Quadrants 20 and 21 (Kearsey *et al.* 2015; Whitbread & Kearsey 2016). The potential coal-, mudstone- and oil-shale-bearing source rocks are commonly intercalated with potential reservoir sand bodies. These wells provided a good constraint and led to a confident interpretation of the Firth Coal Formation in the structurally complex depocentres, such as the westernmost end of the Witch Ground Graben, adjacent to the Halibut Horst (Figs 12 & 17). In this area, the formation has been interpreted as reaching depths of the order of 3.5–4 km.

The Orcadia Formation has been extensively mapped on seismic data for the first time in the Inner Moray Firth Basin and the Outer Moray Firth (Fig. 14), and it has also been interpreted in wells as far north as Quadrants 7, 8 and 9 (Patruno & Reid 2016, 2017). The interval is interpreted in the major Paleozoic depocentres in Quadrants 12–14, such as the Smith Bank Graben, the Caithness Graben, the Halibut Basin and the Witch Ground Graben. The formation is also proven in wells on some highs, such as the Halibut Horst (>700 m thick in well 14/19-11).

Seismic interpretation suggests that the distribution of the Struie Formation lacustrine source rocks is probably restricted in Quadrant 12, south of the Smith Bank intrabasinal high.

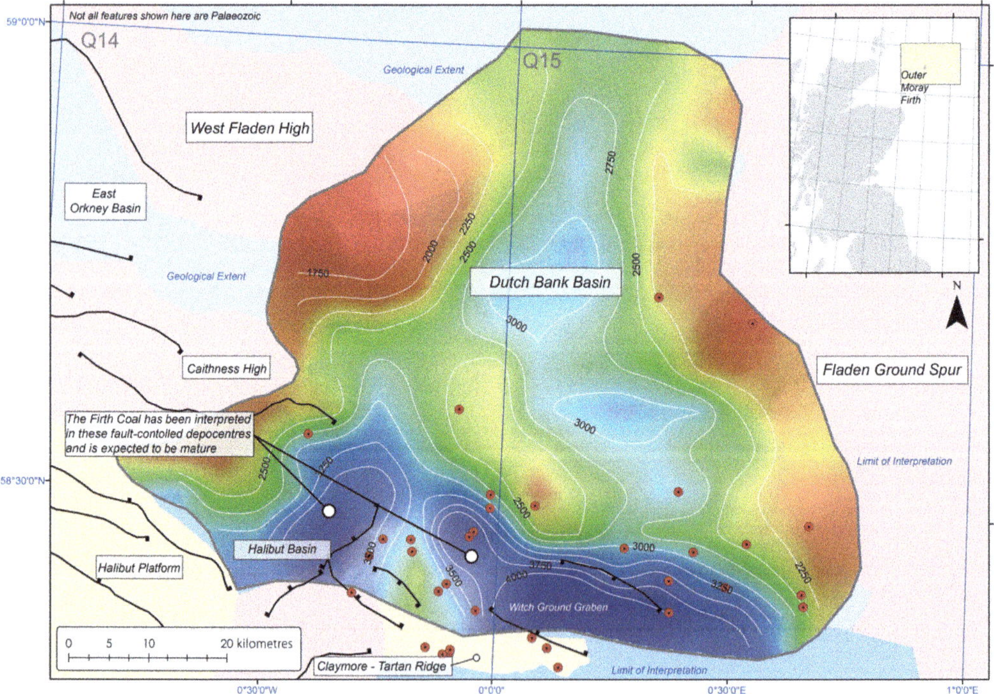

Fig. 17. Depth to Top Firth Coal Formation in metres (5 km grid resolution). The formation has been proven by wells (red dots) on elevated domains, and has been interpreted in deeper graben (e.g. Witch Ground Graben: see Fig. 12) and in a significant part of the Dutch Bank Basin. Depth-converted time values suggest that the Firth Coal Formation is present at depths of 3.5–4 km.

Conclusions

A series of Devonian and Carboniferous basins have been mapped from the margins of the East Shetland Platform, southwards to the northern margins of the Southern North Sea (SNS). The interpretation was based on 85 000 line-km of seismic data, tied to more than 180 wells and a regional gravity/magnetic study. An inherited Caledonian, Iapetan and Tornquist structural fabric emerges. The partitioned stress is related to transtensional and transpressional tectonic regimes, resulting in a variety of Devono-Carboniferous-age NW–SE- and NE–SW-orientated basins.

The granite-cored blocks have been long-lived highs, playing a significant role in the distribution of the basins and the extent of the sedimentary rocks they contain.

It is notable that, although some parts of the Mid North Sea High are underpinned by elevated domains and platforms, there are a series of potentially prospective Devono-Carboniferous basins over and around the 'high'. The deepest of these basins is the NW-trending North Dogger Basin across Quadrant 38.

Using the best released and unreleased seismic data, source rocks intervals such as the Scremerston, Firth Coal and Orcadia formations have been extensively mapped to depths of more than 4 km (Scremerston Formation).

Acknowledgements This paper derives from the 21st Century Exploration Roadmap (21CXRM) Palaeozoic Project from the British Geological Survey. The authors would like to readily acknowledge the assistance of several BGS colleagues and especially Ian Andrews for his review. Also, thank you to the anonymous reviewers for agreeing to review this paper and for their constructive comments. The authors would like to thank PGS, Spectrum, Schlumberger/Western Geco, CGGVeritas and TGS for giving permission for the reproduction of selected seismic lines and the release of the 5 km resolution grids. Richard Milton-Worssell is thanked for his contribution to the project and for requesting seismic data from companies as part of the BGS/OGA contract. This paper is published with the permission of the Executive Director, British Geological Survey (NERC).

Funding The project was funded by DECC/OGA, industry sponsors, BGS, and Oil and Gas UK.

References

Abbotts, I.L. (ed.) 1991. *United Kingdom Oil and Gas Fields: 25 Years Commemorative Volume*. Geological Society, London, Memoirs, **14**, https://doi.org/10.1144/GSL.MEM.1991.014.01.69

Andrews, I.J., Long, D., Richards, P.C., Thompson, A.R., Brown, S., Chesher, J.A. & McCormac, M. 1990. *The Geology of the Moray Firth*. United Kingdom Offshore Regional Report HMSO, London.

Arsenikos, S., Quinn, M.F., Pharaoh, T., Sankey, M. & Monaghan, A.A. 2015. *Seismic Interpretation and Generation of Key Depth Structure Surfaces within the Devonian and Carboniferous of the Central North Sea, Quadrants 25–44 Area*. British Geological Survey Commissioned Report CR/15/118, http://nora.nerc.ac.uk/516758/

Arsenikos, S., Quinn, M.F., Johnson, K., Sankey, M. & Monaghan, A. 2016. *Seismic Interpretation and Generation of Key Depth Structure Surfaces within the Carboniferous and Devonian of the Orcadian Study Area, Quadrants 7–9, 11–15 and 19-21*. British Geological Survey Commissioned Report CR/16/033N, http://nora.nerc.ac.uk/516771/

Astin, T.R. 1985. The palaeogeography of the Middle Devonian Lower Eday Sandstone, Orkney. *Scottish Journal of Geology*, **21**, 353–375, https://doi.org/10.1144/sjg21030353

Astin, T.R. 1990. The Devonian lacustrine sediments of Orkney, Scotland; implications for climate cyclicity, basin structure and maturation history. *Journal of the Geological Society, London*, **147**, 141–151, https://doi.org/10.1144/gsjgs.147.1.0141

Bailey, J.B., Arbin, P., Daffinoti, O., Gibson, P. & Ritchie, J.S. 1993. Permo-Carboniferous plays of the Silver Pit Basin. *In*: Parker, J.R. (ed.) *Petroleum Geology of Northwest Europe: Proceedings of the 4th Conference*. Geological Society, London, 707–715, https://doi.org/10.1144/0040707

Beach, A. 1984. Structural evolution of the Witch Ground Graben. *Journal of the Geological Society, London*, **141**, 621–628, https://doi.org/10.1144/gsjgs.141.4.0621

British Geological Survey (BGS) 1996. *Tectonic map of Britain, Ireland and Adjacent Areas: Sheet 1. 1:1.500 000*. British Geological Survey, Keyworth, Nottingham, UK.

BGS 2017*a*. *BGS Marine Gravity Survey*. British Geological Survey, Keyworth, Nottingham, UK.

BGS 2017*b*. *BGS Marine Magnetic Survey*. British Geological Survey, Keyworth, Nottingham, UK.

BGS 2017*c*. *GB Aeromagnetic Survey*. British Geological Survey, Keyworth, Nottingham, UK.

BGS 2017*d*. *GB Land Gravity Survey*. British Geological Survey, Keyworth, Nottingham, UK.

Cameron, D., Munns, J. & Stoker, S. 2005. Remaining hydrocarbon exploration potential of the Carboniferous fairway, UK southern North Sea. *In*: Collinson, J.D., Evans, D.J., Holliday, D.W. & Jones, N.S. (eds) *Carboniferous Hydrocarbon Geology: The Southern North Sea and Surrounding Onshore Areas*. Yorkshire Geological Society, Occasional Publications, **7**, 209–224.

Cameron, T.D.J. 1992. *The Geology of the Southern North Sea*. HMSO, London.

Cameron, T.D.J. 1993. Carboniferous and Devonian of the Southern North Sea. *In*: Knox, R.W.O. & Cordey, W.G. (eds) *Lithostratigraphic Nomenclature of the UK North Sea*. British Geological Survey, Keyworth, Nottingham, UK.

Cartwright, J., Stewart, S. & Clark, J. 2001. Salt dissolution and salt-related deformation of the forth approaches basin, UK North Sea. *Marine and Petroleum Geology*, **18**, 757–778, https://doi.org/10.1016/S0264-8172(01)00019-8

Chadwick, R.A. & Holliday, D.W. 1991. Deep crustal structure and Carboniferous basin development within the Iapetus convergence zone, northern England. *Journal of the Geological Society, London*, **148**, 41–53, https://doi.org/10.1144/gsjgs.148.1.0041

Clarke, P. & Parnell, J. 1999. Facies analysis of a back-tilted lacustrine basin in a strike-slip zone, Lower Devonian, Scotland. *Palaeogeography, Palaeoclimatology, Palaeoecology*, **151**, 167–190, https://doi.org/10.1016/S0031-0182(99)00020-6

Cocks, L.R.M. & Torsvik, T.H. 2006. European geography in a global context from the Vendian to the end of the Palaeozoic. *In*: Gee, D.G. & Stephenson, R.A. (eds) *European Lithosphere Dynamics*. Geological Society, London, Memoirs, **32**, 83–95, https://doi.org/10.1144/GSL.MEM.2006.032.01.05

Coward, M.P. 1993. The effect of Late Caledonian and Variscan continental escape tectonics on basement structure, Paleozoic basin kinematics and subsequent Mesozoic basin development in NW Europe. *In*: Parker, J.R. (ed.) *Petroleum Geology of Northwest Europe: Proceedings of the 4th Conference*. Geological Society, London, 1095–1108, https://doi.org/10.1144/0041095

Coward, M.P., Dewey, J.F., Hempton, M. & Holroyd, J. 2003. Tectonic evolution. *In*: Evans, D., Graham, C., Armour, A. & Bathurst, P. (eds) *The Millennium Atlas: Petroleum Geology of the Central and Northern North Sea*. Geological Society, London, 17–33.

De Paola, N., Holdsworth, R.E., McCaffrey, K.J.W. & Barchi, M.R. 2005. Partitioned transtension: an alternative to basin inversion models. *Journal of Structural Geology*, **27**, 607–625, https://doi.org/10.1016/j.jsg.2005.01.006

De Paola, N., Holdsworth, R.E., Collettini, C., McCaffrey, K.J.W. & Barchi, M.R. 2007. The structural evolution of dilational stepovers in regional transtensional zones. *In*: Cunningham, W.D. & Mann, P. (eds) *Tectonics of Strike-Slip Restraining and Releasing Bends*. Geological Society, London, Special Publications, **290**, 433–445, https://doi.org/10.1144/SP190.17

Dimitropoulos, K. & Donato, J.A. 1981. The Inner Moray Firth Central Ridge, a geophysical interpretation. *Scottish Journal of Geology*, **17**, 27–38, https://doi.org/10.1144/sjg17010027

Domeier, M. & Torsvik, T.H. 2014. Plate tectonics in the late Paleozoic. *Geoscience Frontiers*, **5**, 303–350, https://doi.org/10.1016/j.gsf.2014.01.002

Donato, J.A., Martindale, W. & Tully, M.C. 1983. Buried granites within the Mid North Sea High. *Journal of the Geological Society, London*, **140**, 825–837, https://doi.org/10.1144/gsjgs.140.5.0825

Doornenbal, H. & Stevenson, A. (eds). 2010. *Petroleum Geological Atlas of the Southern Permian Basin Area*. European Association of Geoscientists and Engineers (EAGE), Houten, The Netherlands.

Duncan, A.D. & Buxton, N.W.K. 1995. New evidence for evaporitic Middle Devonian lacustrine sediments with hydrocarbon source potential on the East Shetland Platform, North Sea. *Journal of the Geological Society, London*, **152**, 251–258, https://doi.org/10.1144/gsjgs.152.2.0251

Edwards, C.W. 1991. The Buchan Field, Blocks 20/5a and 21/1a, UK North Sea. *In*: Abbotts, I.L. (ed.) *United Kingdom Oil and Gas Fields: 25 Years Commemorative Volume*. Geological Society, London, Memoirs, **14**, 253–259, https://doi.org/10.1144/GSL.MEM.1991.014.01.31

Evans, D., Graham, C., Armour, A. & Bathurst, P. (eds). 2003. *The Millennium Atlas: Petroleum Geology of the Central and Northern North Sea*. Geological Society, London.

Fossen, H. 2010. Extensional tectonics in the North Atlantic Caledonides: a regional view. *In*: Law, R.D., Butler, R. W.H., Holdsworth, R.E., Krabbendam, M. & Strachan, R.A. (eds) *Continental Tectonics and Mountain Building: The Legacy of Peach and Horne*. Geological Society, London, Special Publications, **335**, 767–793, https://doi.org/10.1144/SP335.31

Fraser, A.J. & Gawthorpe, R.L. 1990. Tectono-stratigraphic development and hydrocarbon habitat of the Carboniferous in northern England. *In*: Hardman, R.F.P. & Brooks, J. (eds) *Tectonic Events Responsible for Britain's Oil and Gas Reserves*. Geological Society, London, Special Publications, **55**, 49–86, https://doi.org/10.1144/GSL.SP.1990.055.01.03

Fraser, A.J. & Gawthorpe, R.L. 2003. *An Atlas of Carboniferous Basin Evolution in Northern England*. Cambridge University Press, Cambridge, https://doi.org/10.1017/S0016756803228788

Gatliff, R.W., Richard, P.C. *et al.* 1994. *The Geology of the Central North Sea*. HMSO, London.

Glennie, K.W. & Underhill, J.R. 1998. Origin, development and evolution of structural styles. *In*: Glennie, K.W. (ed.) *Petroleum Geology of the North Sea: Basic Concepts and Recent Advances*. 4th edn. Blackwell Science, Oxford, 42–84, https://doi.org/10.1002/9781444313413.ch2

Gluyas, H.M. & Hichens, J.G. (eds). 2003. *United Kingdom Oil and Gas Fields Commemorative Millennium Volume*. Geological Society, London.

Hallett, D., Durant, G.P. & Farrow, G.E. 1986. Oil exploration and production in Scotland. *Scottish Journal of Geology*, **21**, 547–570, https://doi.org/10.1144/sjg21040547

Hay, S., Jones, C.M., Barker, F. & He, Z. 2005. *Exploration of Unproven Plays; Mid North Sea High*. Department of Energy and Climate Change (DECC), London.

Hillis, R.R., Thomson, K. & Underhill, J.R. 1994. Quantification of Tertiary erosion in the Inner Moray Firth using sonic velocity data from the Chalk and the Kimmeridge Clay. *Marine and Petroleum Geology*, **11**, 283–293, https://doi.org/10.1016/0264-8172(94)90050-7

Japsen, P. 1998. Regional velocity–depth anomalies, North Sea Chalk: a record of overpressure and Neogene uplift and erosion. *AAPG Bulletin*, **82**, 2031–2074, https://doi.org/10.1306/00AA7BDA-1730-11D7-8645000102C1865D

Japsen, P. 1999. Overpressured Cenozoic shale mapped from velocity anomalies relative to a baseline for marine shale, North Sea. *Petroleum Geoscience*, **5**, 321–336, https://doi.org/10.1144/petgeo.5.4.321

JAPSEN, P. 2000. Investigation of multi-phase erosion using reconstructed shale trends based on sonic data. Sole Pit axis, North Sea. *Global and Planetary Change*, **24**, 189–210, https://doi.org/10.1016/S0921-8181(00)00008-4

JENYON, M.K., CRESSWELL, P.M. & TAYLOR, J.C.M. 1984. Nature of the connection between the Northern and Southern Zechstein Basins across the Mid North Sea High. *Marine and Petroleum Geology*, **1**, 355–363, https://doi.org/10.1016/0264-8172(84)90136-3

JONES, G., RORISON, P., FROST, R.E., KNIPE, R. & COLLERAN, J. 1999. Tectono-stratigraphic development of the southern part of UKCS Quadrant 15 (eastern Witch Ground Graben): implications for the Mesozoic–Tertiary evolution of the Central North Sea Basin. *In*: FLEET, A.J. & BOLDY, S.A.R. (eds) *Petroleum Geology of Northwest Europe: Proceedings of the 5th Conference*. Geological Society, London , 133–152, https://doi.org/10.1144/0050133

KEARSEY, T., ELLEN, R., MILLWARD, D. & MONAGHAN, A.A. 2015. *Devonian and Carboniferous Stratigraphical Correlation and Interpretation in the Central North Sea, Quadrants 25–44*. British Geological Survey Commissioned Report CR/15/117N, http://nora.nerc.ac.uk/516755/

KEARSEY, T.I., MILLWARD, D., ELLEN, R., WHITBREAD, K. & MONAGHAN, A.A. In press. Revised stratigraphic framework of pre-Westphalian Carboniferous petroleum system elements from the Outer Moray Firth to the Silverpit Basin, North Sea, UK. *In*: MONAGHAN, A.A., UNDERHILL, J.R., MARSHALL, J.E.A. & HEWETT, A.J. (eds) *Paleozoic Plays of NW Europe*. Geological Society, London, Special Publications, **471**, https://doi.org/10.1144/SP471.11

KIMBELL, G.S. & WILLIAMSON, J.P. 2015. *A Gravity Interpretation of the Central North Sea*. British Geological Survey Commissioned Report CR/15/119N, http://nora.nerc.ac.uk/516759/

KIMBELL, G.S. & WILLIAMSON, J.P. 2016. *A Gravity Interpretation of the Orcadian Basin Area*. British Geological Survey Commissioned Report CR/16/034N, http://nora.nerc.ac.uk/516772/

KIMBELL, G.S., CHADWICK, R.A., HOLLIDAY, D.W. & WERNGREN, O.C. 1989. The structure and evolution of the Northumberland Trough from new seismic reflection data and its bearing on modes of continental extension. *Journal of the Geological Society, London*, **146**, 775–787, https://doi.org/10.1144/gsjgs.146.5.0775

KLEMPERER, S.L. & MATTHEWS, D.H. 1987. Iapetus suture located beneath the North Sea by BIRPS deep seismic reflection profiling. *Geology*, **15**, 195–198, https://doi.org/10.1130/0091-7613(1987)152.0.CO;2

KUBALA, M., BASTOW, M., THOMPSON, S., SCOTCHMAN, I. & OYGARD, K. 2003. Geothermal regime, petroleum generation and migration. *In*: EVANS, D., GRAHAM, C., ARMOUR, A. & BATHURST, P. (eds) *The Millennium Atlas: Petroleum Geology of the Central and Northern North Sea*. Geological Society, London, 289–315.

LESLIE, A.G., MILLWARD, D., PHARAOH, T.C., MONAGHAN, A., ARSENIKOS, S. & QUINN, M.F. 2015. *Tectonic Synthesis and Contextual Setting for the Central North Sea and Adjacent Onshore Areas, 21CXRM Palaeozoic Project*. British Geological Survey Commissioned Report CR15//125N, http://nora.nerc.ac.uk/516757/

LESLIE, A.G., MONAGHAN, A., ARSENIKOS, S. & QUINN, M.F. 2016. *Tectonic Synthesis and Contextual Setting for the Palaeozoic of the Moray Firth Region, Orcadian Basin*. British Geological Survey Commissioned Report CR16/039N, http://nora.nerc.ac.uk/516782/

MARSHALL, J.E.A. & HEWETT, A.J. 2003. Devonian. *In*: EVANS, D., GRAHAM, C., ARMOUR, A. & BATHURST, P. (eds) *The Millennium Atlas: Petroleum Geology of the Central and Northern North Sea*. Geological Society, London, 65–81.

MARSHALL, J.E.A., ROGERS, D.A. & WHITELEY, M.J. 1996. Devonian marine incursions into the Orcadian Basin, Scotland. *Journal of the Geological Society, London*, **153**, 451–466, https://doi.org/10.1144/gsjgs.153.3.0451

MARSHALL, J.E.A., BROWN, J.F. & ASTIN, T.R. 2011. Recognising the Taghanic Crisis in the Devonian terrestrial environment and its implications for understanding land–sea interactions. *Palaeogeography, Palaeoclimatology, Palaeoecology*, **304**, 165–183, https://doi.org/10.1016/j.palaeo.2010.10.016

MAYNARD, J.R., HOFMANN, W., DUNAY, R.E., BENTHAN, P.N., DEAN, K.P. & WATSON, I. 1997. The Carboniferous of Western Europe; the development of a petroleum system. *Petroleum Geoscience*, **3**, 97–115, https://doi.org/10.1144/petgeo.3.2.97

MILTON-WORSSELL, R., SMITH, K., MCGRANDLE, A., WATSON, J. & CAMERON, D. 2010. The search for a Carboniferous petroleum system beneath the Central North Sea. *In*: VINING, B.A. & PICKERING, S.C. (eds) *Petroleum Geology: From Mature Basins to New Frontiers – Proceedings of the 7th Petroleum Geology Conference*. Geological Society, London, 57–75, https://doi.org/10.1144/0070057

MONAGHAN, A., ARSENIKOS, S. ET AL. 2015. *Palaeozoic Petroleum Systems of the Central North Sea/Mid North Sea High*. British Geological Survey Commissioned Report CR/15/124N, http://nora.nerc.ac.uk/516766/

PATRUNO, S. & REID, W. 2016. New plays on the Greater East Shetland Platform (UKCS Quadrants 3, 8–9, 14–16) – part 1: Regional setting and a working petroleum system. *First Break*, **34**, 33–45.

PATRUNO, S. & REID, W. 2017. New plays on the Greater East Shetland Platform (UKCS Quadrants 3, 8-9, 14-16) – part 2: Newly reported Permo-Triassic intra-platform basins and their influence on the Devonian-Paleogene prospectivity of the area. *First Break*, **35**, 59–69.

PESGB 2017. *Structural Framework of the North Sea and Atlantic Margin*. Petroleum Exploration Society of Great Britain (PESGB), London.

PHARAOH, T.C. 1999. Palaeozoic terranes and their lithospheric boundaries within the Trans-European Suture Zone (TESZ): A review. *Tectonophysics*, **314**, 17–41, https://doi.org/10.1016/S0040 1951(99)00235 8

PILKINGTON, M., ABDOH, A. & COWAN, D.R. 1995. Pre-Mesozoic structure of the Inner Moray Firth Basin: constraints from gravity and magnetic data. *First Break*, **13**, 291–300.

READ, W., BROWNE, M.A.E., STEVENSON, D. & UPTON, B.G.J. 2002. Carboniferous. *In*: TREWIN, N.H. (ed.) *The Geology of Scotland*. 4th edn. Geological Society, London, 251–299, https://doi.org/10.1144/GOS4P.9

RICHARDSON, N.J., ALLEN, M.R. & UNDERHILL, J.R. 2005. Role of Cenozoic fault reactivation in controlling pre-rift plays, and the recognition of Zechstein Group evaporite–carbonate lateral facies transitions in the

East Orkney and Dutch Bank basins, East Shetland Platform, UK North Sea. *In*: Doré, A.G. & Vining, B. A. (eds) *Petroleum Geology: North-West Europe and Global Perspectives – Proceedings of the 6th Petroleum Geology Conference*. Geological Society, London, 337–348, https://doi.org/10.1144/0060337

Ritchie, J.D., Johnson, H., Browne, M.A.E. & Monaghan, A.A. 2003. Late Devonian–Carboniferous tectonic evolution within the Firth of Forth, Midland Valley; as revealed from 2D seismic reflection data. *Scottish Journal of Geology*, **39**, 121–134, https://doi.org/10.1144/sjg39020121

Roberts, A.M., Badley, M.E., Price, J.D. & Huck, I.W. 1990. The structural history of a transtensional basin: Inner Moray Firth, NE Scotland. *Journal of the Geological Society, London*, **147**, 87–103, https://doi.org/10.1144/gsjgs.147.1.0087

Robson, D. 1991. The Argyll, Duncan and Innes Fields, Block 30/24 and 30/25a, UK North Sea. *In*: Abbotts, I.L. (ed.) *United Kingdom Oil and Gas Fields: 25 Years Commemorative Volume*. Geological Society, London, Memoirs, **14**, 219–226, https://doi.org/10.1144/GSL.MEM.1991.014.01.27

Sclater, J.G. & Christie, P.A.F. 1980. Continental stretching: An explanation of the Post-Mid-Cretaceous subsidence of the central North Sea Basin. *Journal of Geophysical Research: Solid Earth*, **85**, 3711–3739, https://doi.org/10.1029/JB085iB07p03711

Smith, K. & Ritchie, J.D. 1993. Jurassic volcanic centres in the Central North Sea. *In*: Parker, J.R. (ed.) *Petroleum Geology of Northwest Europe: Proceedings of the 4th Conference*. Geological Society, London, 519–531, https://doi.org/10.1144/0040519

Soper, N.J., England, R.W., Snyder, D.B. & Ryan, P.D. 1992. The Iapetus suture zone in England, Scotland and eastern Ireland: a reconciliation of geological and deep seismic data. *Journal of the Geological Society, London*, **149**, 697–700, https://doi.org/10.1144/gsjgs.149.5.0697

Stephenson, D. & Gould, D.E. 1995. *The Grampian Highlands*. British Regional Geology Series. HMSO, London.

Stevens, V. 1991. The Beatrice Field, Block 11/30a, UK North Sea. *In*: Abbotts, I.L. (ed.) *United Kingdom Oil and Gas Fields: 25 Years Commemorative Volume*. Geological Society, London, Memoirs, **14**, 245–252, https://doi.org/10.1144/GSL.MEM.1991.014.01.30

Symonds, R. In press. The Breagh field. *In*: Monaghan, A. A., Underhill, J.R., Hewett, A.J. & Marshall, J.E.A. (eds) *Paleozoic Plays of NW Europe*. Geological Society, London, Special Publications, **471**.

ter Borgh, M.M., Jaarsma, B. & Rosendaal, E.A. In press. Structural development of the Mid North Sea area of the Dutch North Sea: Paleozoic–present. *In*: Monaghan, A. A., Underhill, J.R., Hewett, A.J. & Marshall, J.E.A. (eds) *Paleozoic Plays of NW Europe*. Geological Society, London, Special Publications, **471**, https://doi.org/10.1144/SP471.5

Thomson, K. & Underhill, J.R. 1993. Controls on the development and evolution of structural styles in the Inner Moray Firth Basin. *In*: Parker, J.R. (ed.) *Petroleum Geology of Northwest Europe: Proceedings of the 4th Conference*. Geological Society, London, 1167–1178, https://doi.org/10.1144/0041167

Torsvik, T.H. & Rehnström, E.F. 2003. The Tornquist Sea and Baltica–Avalonia docking. *Tectonophysics*, **362**, 67–82, https://doi.org/10.1016/S0040-1951(02)00631-5

Trewin, N.H. (ed.) 2002. *The Geology of Scotland*. Geological Society, London.

Trewin, N.H. & Bramwell, M.G. 1991. The Auk Field, Block 30/16, UK North Sea. *In*: Abbotts, I.L. (ed.) *United Kingdom Oil and Gas Fields: 25 Years Commemorative Volume*. Geological Society, London, Memoirs, **14**, 227–236, https://doi.org/10.1144/GSL.MEM.1991.014.01.28

Underhill, J.R. 1991. Implications of Mesozoic–Recent basin development in the western Inner Moray Firth, UK. *Marine and Petroleum Geology*, **8**, 359–369, https://doi.org/10.1016/0264-8172(91)90089-J

Underhill, J.R. & Brodie, J.A. 1993. Structural geology of Easter Ross, Scotland: implications for movement on the Great Glen fault zone. *Journal of the Geological Society, London*, **150**, 515–527, https://doi.org/10.1144/gsjgs.150.3.0515

Underhill, J.R., Monaghan, A.A. & Browne, M.A.E. 2008. Controls on structural styles, basin development and petroleum prospectivity in the Midland Valley of Scotland. *Marine and Petroleum Geology*, **25**, 1000–1022, https://doi.org/10.1016/j.marpetgeo.2007.12.002

Van Dalfsen, W., Doornenbal, J.C., Dortland, S. & & Gunnink, J.L. 2006. A comprehensive seismic velocity model for the Netherlands based on lithostratigraphic layers. *Geologie en Mijnbouw – Netherlands Journal of Geosciences*, **85**, 277–292.

Vincent, C.J. 2015. *Maturity Modelling of Selected Wells in the Central North Sea*. British Geological Survey Commissioned Report CR/15/122, http://nora.nerc.ac.uk/516764/

Waters, C.N. & Davies, S.J. 2006. Carboniferous: extensional basins, advancing deltas and coal swamps. *In*: Brenchley, P.J. & Rawson, P.F. (eds) *The Geology of England and Wales*. Geological Society, London, 173–223.

Whitbread, K. & Kearsey, T. 2016. *Devonian and Carboniferous Stratigraphical Correlation and Interpretation in the Orcadian Area, Central North Sea, Quadrants 7–22*. British Geological Survey Commissioned Report CR/16/032N, http://nora.nerc.ac.uk/516770/

Woodcock, N.H. 2012. Early Devonian sedimentary and magmatic interlude after Iapetus closure. *In*: Woodcock, N.H. & Strachan, R. (eds) *Geological History of Britain and Ireland*. 2nd edn. Wiley-Blackwell, Chichester, UK, 193–209, https://doi.org/10.1002/9781118274064.ch12

Woodcock, N.H. & Strachan, R. (eds). 2012. *Geological History of Britain and Ireland*. 2nd edn. Wiley-Blackwell, Chichester, UK, https://doi.org/10.1002/9781118274064

Wride, V.C. 1995. Structural features and structural styles from the Five Countries Area of the North Sea Central Graben. *First Break*, **13**, 395–407.

Ziegler, P.A. 1990. *Geological Atlas of Western and Central Europe*. Shell Internationale Petroleum Maatschappij, The Hague.

Revised stratigraphic framework of pre-Westphalian Carboniferous petroleum system elements from the Outer Moray Firth to the Silverpit Basin, North Sea, UK

T. I. KEARSEY*, D. MILLWARD, R. ELLEN, K. WHITBREAD & A. A. MONAGHAN

British Geological Survey, The Lyell Centre, Research Avenue South, Edinburgh EH14 4AP, UK

**Correspondence: timk1@bgs.ac.uk*

Abstract: Spatially and temporally variable Tournaisian to Namurian Carboniferous fluvial, fluvio-deltaic, platform carbonate and shale-dominated basin sedimentary successions up to 3.5 km thick are preserved in a complex series of basins from the Outer Moray Firth (Quadrant 14) to the Silverpit Basin (Quadrant 44). Differences in stratigraphic nomenclature in the areas surrounding the Mid North Sea High and onshore, combined with sparse biostratigraphic data, have hindered the systematic regional understanding of the timing and controls on stacked source and reservoir rock intervals. Over 125 well reinterpretations, tied to seismic interpretations, provide evidence of the inception and extent of a delta system. Regional time slices highlight a long-lived laterally equivalent basinal, mud-rich succession across Quadrants 41–44. They also show that the area from the Outer Moray Firth to the Silverpit Basin was part of the same sedimentary system up to at least Namurian times. All of this is placed within a simplified stratigraphic framework.

Supplementary material: Appendix A in the Supplementary Material contains the stratigraphic intervals interpreted on each well and highlights which intervals have biostratigraphic control. Supplemental Figures 1 and 2 are larger scale versions of Figures 6–8. The Supplementary Material is available at https://doi.org/10.6084/m9.figshare.c.4087046

The Westphalian gas-prone rocks of the Southern North Sea have been the main focus of Carboniferous hydrocarbon exploration and production in the North Sea (Cameron 1993*a*; Bruce & Stemmerik 2003; Underhill 2003; Kombrink *et al.* 2010). There has been a growing interest in the older Carboniferous strata, with production from the Yoredale Formation in the Breagh Field (Symonds *et al.* 2015), and increasing interest in Namurian and older deeper water systems (Rodriguez *et al.* 2014) and the offshore equivalents of the Bowland Shale Formation, highlighted for unconventional exploration (Andrews 2013). In response to the Wood Review (Wood 2014) and following consultation with industry, the British Geological Survey led the 21st Century Exploration Roadmap (21CXRM): Palaeozoic Project to provide regional interpretations of Devonian and Carboniferous petroleum systems from south of the Mid North Sea High to the Moray Firth and East Shetland Platform (Monaghan *et al.* 2017).

This paper reassesses the stratigraphy and facies architecture of the pre-Westphalian Carboniferous strata in the North Sea, integrating and summarizing results from the detailed lithostratigraphic descriptions contained in the 21CXRM project reports (Kearsey *et al.* 2015; Whitbread & Kearsey 2016). More than 550 wells drilled in the study area penetrate Paleozoic strata and, of these, 125 wells contain long sections through the Devonian or Carboniferous sequences. These 125 wells have been reinterpreted, incorporating biostratigraphic information, and integrated with seismic and gravity interpretations from Arsenikos *et al.* (2018). Utilizing sedimentological interpretations from well records and onshore datasets, the regional facies architecture of the North Sea sedimentary system from the early to mid-Carboniferous (Tournaisian–Namurian) from the Outer Moray Firth Basin to the Silverpit Basin is described in a series of time slices. The laterally and temporally variable sedimentary facies that developed in response to active tectonism, varying sediment supply and glacio-eustatic sea-level variations is rationalized within an integrated model of the Carboniferous system to aid exploration in these areas.

Background

The geological evolution of the Carboniferous succession beneath the North Sea has been interpreted

From: Monaghan, A. A., Underhill, J. R., Hewett, A. J. & Marshall, J. E. A. (eds) 2018. *Paleozoic Plays of NW Europe*. Geological Society, London, Special Publications, **471**, 91–113.
First published online May 3, 2018, https://doi.org/10.1144/SP471.11

in a number of key publications. Cameron (1993*b*) described the lithostratigraphy of the Devonian to Permian rocks to the north of the Mid North Sea High and Cameron (1993*a*), Bruce & Stemmerik (2003) and Kombrink *et al.* (2010) detailed the lithostratigraphy to the south. The scheme of Cameron (1993*a*) for the Pennsylvanian rocks of Quadrants 41–44 was subsequently modified by Besly (2005). South of the Mid North Sea High (Fig. 1), Collinson (2005) interpreted the depositional systems of the Mississippian and early Pennsylvanian age strata in Quadrants 41–43. To the north of the Mid North Sea High, strata of Carboniferous age are only preserved offshore in the east of the region in Quadrants 14, 15, 20, 21, 26, 27 and 29 (Fig. 1). Leeder & Boldy (1990) undertook a regional assessment of Carboniferous strata based on the analysis of wells in Quadrants 14 and 15. However, poor well coverage between Quadrants 14 and 15 and south of the Mid North Sea High has limited the degree to which it has been possible to correlate the Carboniferous of the North Sea as a whole. Together with a lack of published biostratigraphic studies, this has led to various stratigraphic schemes and has hindered regional interpretations.

By contrast, there are extensive studies of the onshore geology of the Devonian and Carboniferous of the Midland Valley of Scotland and northern England. Summaries of the lithostratigraphy, palaeogeography and tectonic evolution, for example, have been published by Chadwick *et al.* (1995), Browne *et al.* (1999), Read *et al.* (2002), Trewin & Thirlwall (2002), Waters & Davies (2006) and Stone *et al.* (2010). The Carboniferous stratigraphy detailed in these accounts has been summarized further, along with that offshore, by Waters *et al.* (2011). The revision of the onshore lithostratigraphy is based on the distribution of the predominant lithofacies associations throughout the sequence, thus unifying, as far as is practicable, the many previous regional schemes (Browne *et al.* 2003; Waters *et al.* 2007; Dean *et al.* 2011).

Our study retains the offshore scheme more familiar to the hydrocarbon industry, making comparison with the onshore equivalents where appropriate. We use a biostratigraphic time slice approach to outline the spatial extent of time-equivalent sedimentary facies interpretations from well records and onshore outcrops. Data integration reveals the large-scale sedimentary architecture and we attempt, at a high level, to unify the system as a whole.

Methods

Many of the 550 wells that have penetrated Devonian or Carboniferous rocks in Quadrants 14–44 contain only a short succession of these strata and their distribution is uneven (Fig. 2), with very few located outside Quadrants 14–15 (Outer Moray Firth) and Quadrants 41–44 (south of the Mid North Sea High). Of these wells, 125 had sections of Carboniferous strata >100 m or were long enough to identify stratigraphic units. Of those, 69 had biostratigraphic control, predominately in the form of unpublished palynology reports (Fig. 2; Appendix A).

This subset of 125 wells formed the basis for the reinterpretation of the sedimentary facies and stratigraphy in this study. A set of key onshore wells provided a link with the offshore stratigraphy. Three onshore wells in northern England were included: Seal Sands 1 (Johnson *et al.* 2011), Harton 1 (Ridd *et al.* 1970) and Kirby Misperton 1 (Andrews 2013). In Scotland, the Firth of Forth 1 and Milton of Balgonie 1 wells were considered (see Monaghan 2014).

For each well, composite logs, wireline geophysical logs (mainly gamma, caliper, sonic, neutron porosity and density), biostratigraphic and petrographic reports, and core photographs were examined. Full suites of wireline logs and biostratigraphy were unavailable for some wells. Previous interpretations of formation tops/bases in completion reports and in published papers were reassessed and modified with new biostratigraphic information or interpretations informed by regional study. Seismic interpretations (Arsenikos *et al.* 2016) were also used to aid well interpretation.

This study had access to all company biostratigraphic reports stored on the UK Oil and Gas Data (CDA) portal (www.ukoilandgasdata.com), previously unreleased British Geological Survey palynological reports and data donated by sponsors of the 21CXRM Palaeozoic Project. The CDA reports cited in this paper are released documents from www.ukoilandgasdata.com. The biostratigraphic reports vary in date of completion, the amount and quality of the data and the interpretation that they contain, resulting in uncertainties in the precise biostratigraphic age of parts of the succession. Where multiple interpretations exist, the most modern interpretation is taken. Ambiguities in the biostratigraphic ages were analysed against the sedimentology and seismic picks (see Arsenikos *et al.* 2018) and the most parsimonious solution was used. The miospore biozones for the Carboniferous have been revised several times. Waters *et al.* (2011) lists how the older nomenclature ('former index') correlates with the newer nomenclature ('index'). We differentiate between the former and current index where possible. A critical appraisal of the biostratigraphic data would probably refine the results presented in this paper further, but could not be completed within the timescale of the 21CXRM Project.

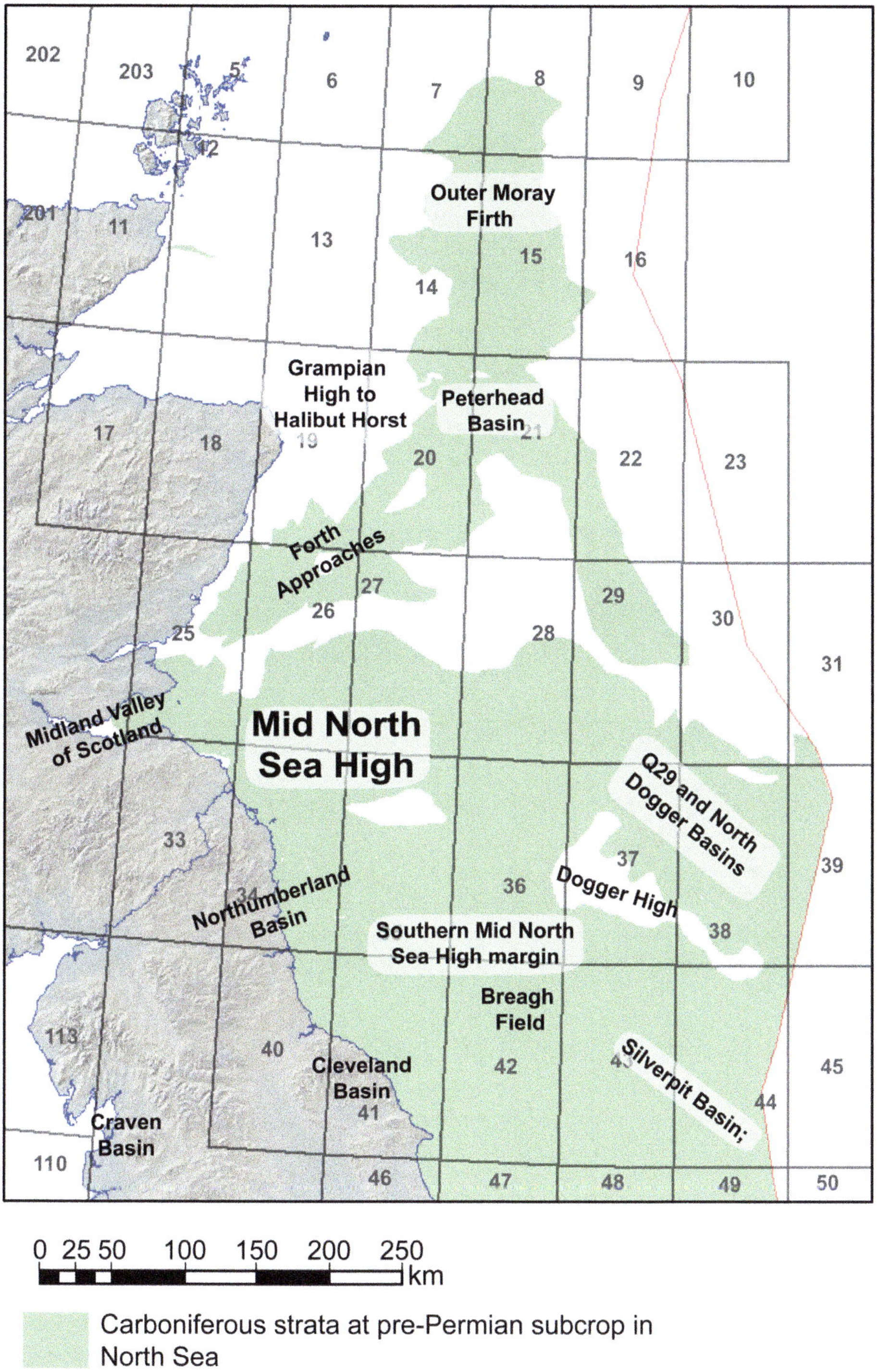

Fig. 1. Distribution of structural domains (including basins and highs) in the North Sea, as named in this study. The map also shows the distribution of Carboniferous strata at subcrop. Includes mapping data licensed from Ordnance Survey. ©Crown Copyright and/or database right 2017. Licence number 100021290 EU and NEXTMap Britain elevation data from Intermap Technologies.

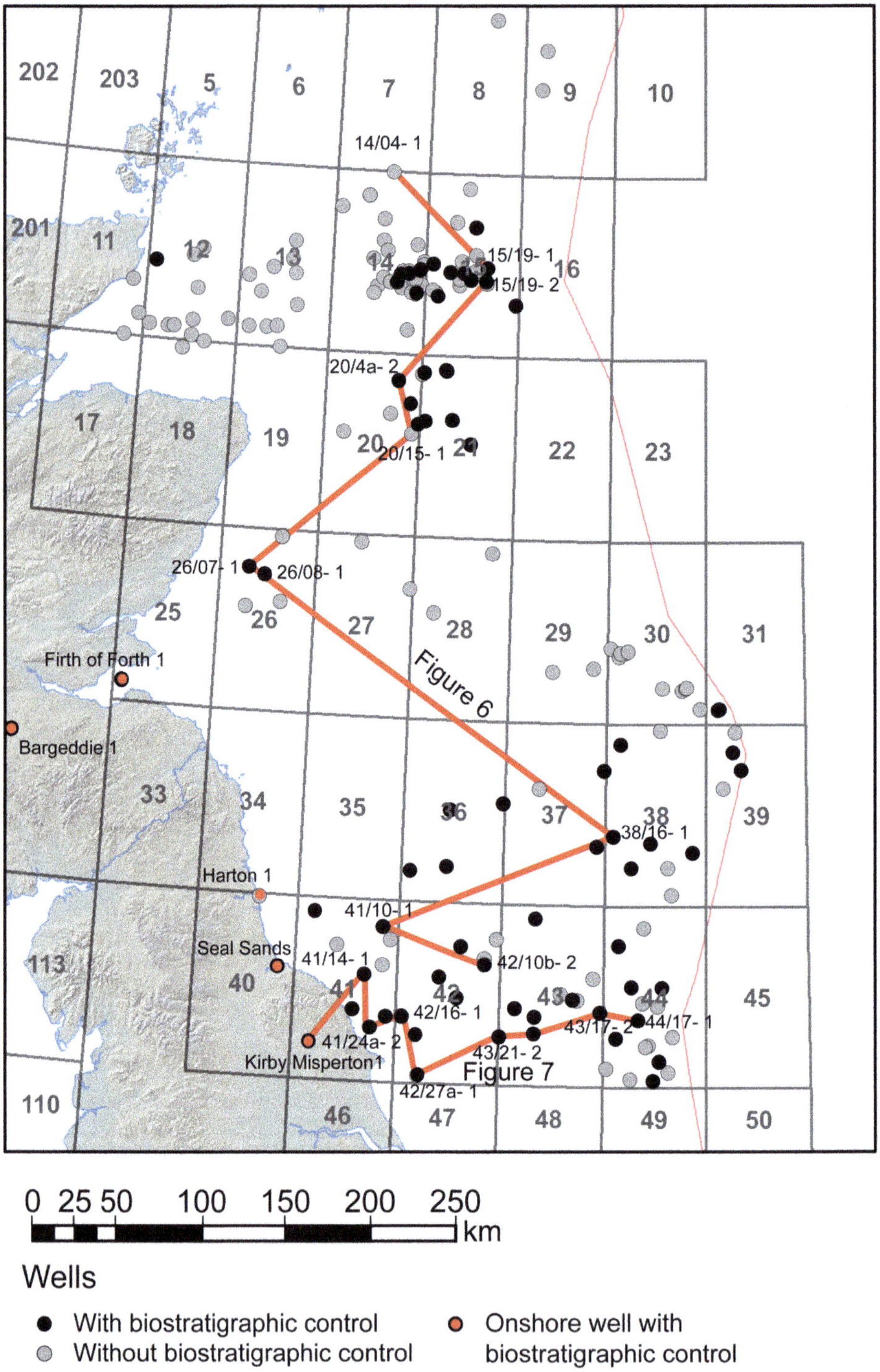

Fig. 2. Distribution of reinterpreted wells containing Carboniferous strata. Wells with biostratigraphic data are marked by black circles. The paths of the correlation panels in Figures 6 and 7 are shown. Includes mapping data licensed from Ordnance Survey. ©Crown Copyright and/or database right 2017. Licence number 100021290 EU and NEXTMap Britain elevation data from Intermap Technologies.

Stratigraphic framework

Eight Tournaisian (lowest Carboniferous) to Kinderscoutian (middle Namurian) lithostratigraphic units were identified in wells in the study area. In chronological order (from oldest to youngest), the Tayport, Fell Sandstone, Scremerston and Millstone Grit formations occur to the north and south of the Mid North Sea High (Figs 3 & 4), whereas the Cementstone, Firth Coal and Yoredale formations and the Cleveland Group are more locally restricted (Figs 3 & 4). Many of these units consist of different facies, but are time-equivalent (Fig. 5).

Tayport Formation

The Tayport Formation is proved in wells in Quadrants 12, 14, 15, 20 and 21 in the Outer Moray Firth and West Central Shelf (e.g. Figs 6 & 7). It is also interpreted in wells in Quadrants 36, 37, 38, 42 and 44 (Fig. 3) near the Dogger High.

Alternations of sandstone and mudstone characterize the Tayport Formation north of the Mid North Sea High (Cameron 1993*a*). The sandstone units are predominantly 15 m thick (Bruce & Stemmerik 2003). In Quadrants 14–15 the formation may also contain sporadic tuffaceous and rhyolitic beds, typically 1–5 m thick (Cameron 1993*a*); examples occur in well 15/25b-1a.

South of the Mid North Sea High, the Tayport Formation is known in well E06-01 in the Dutch sector, where it is described as ‘red to red-brown and grey claystone and siltstone, interspersed with well-developed sandstone beds’ (Van Adrichem Boogaert & Kouwe 1993–97*a*). The sandstone beds, 2–5 m thick, are interbedded with claystone and siltstone. The presence of thin limestone beds has also been inferred from the wireline logs.

In the Outer Moray Firth Basin, the Tayport Formation is thought to be least 500 m thick (Cameron 1993*b*) and spans the Devonian–Carboniferous boundary. It has a latest Famennian to Tournaisian age, proved palynologically (LN and CM biozones) in well 15/25b-1a (CDA, unknown date). South of Quadrants 14 and 15, the Tayport Formation is at least 350 m thick (42/10b-2, Fig. 6) and, where dated, is latest Devonian in age. The most southerly occurrence of the Tayport Formation, in well E06-01 in the Dutch sector (Van Adrichem Boogaert & Kouwe 1993–97*a*), is inferred to be late Famennian based on the occurrence of *Retispora lepidophyta* (Van Adrichem Boogaert & Kouwe 1993–1997*a*), which indicates the LL to LN biozones. The only occurrence of Tournaisian Tayport Formation south of Quadrants 14 and 15 is in well 37/10-1 (McLean 2013). The proximity of this well to the Dogger High may suggest that this occurrence represents a localized debris fan because the surrounding Tournaisian sedimentary rocks in wells such as 37/25-1 and 38/24-1 are of the Cementstone Formation (Fig. 5).

The sandstone units are fluvial channel fills, partly stacked to form thicker sandstone units in the northern area and representing a proximal fluvial channel system with associated overbank deposits; evidence of a marine influence is sparse (Leeder & Boldy 1990). In the south, the claystones and siltstones are floodplain deposits, their red coloration suggesting that they are well-drained palaeosols. There appears to have been no development of wetlands or mires in this succession (Van Adrichem Boogaert & Kouwe 1993–97*a*).

Cementstone Formation

The Cementstone Formation occurs in wells around the southern margin of the Mid North Sea High (Figs 3 & 6). It has been picked from seismic sections across the Mid North Sea High and in the Forth Approaches Basin (Fig. 3; Arsenikos *et al.* 2018). The Cementstone Formation is composed of red–brown and green–grey siltstone and thin, fine- to coarse-grained sandstone units up to 25 m thick, with dolostone and limestone beds between 1 and 4 m thick (Cameron 1993*b*). The gamma log signature varies abruptly, alternating between low and high gamma values; except for some sandstones, the bed thickness rarely exceeds 5 m, reflecting the thinly interbedded character. The thickness of the Cementstone Formation proved in wells is from 100 to 500 m.

The conformable top and basal contacts of the formation occur in well 42/10b-2 (Fig. 6). The formation is also present in the Dutch well E06-01 (Van Adrichem Boogaert & Kouwe 1993–97*b*). Its onshore correlatives are the Ballagan and Lyne formations of the Midland Valley, Tweed and Northumberland basins (Dean *et al.* 2011; Waters *et al.* 2011). The Ballagan Formation consists of fluvial sandstone, coastal plain siltstone interbedded with lacustrine siltstone, and dolostone (cementstone); the dolostone, typically in beds up to 1 m thick, was probably deposited in saline lagoons (Anderton 1985; Scott 1986; Andrews *et al.* 1991; Bennett *et al.* 2016; Kearsey *et al.* 2016). The Lyne Formation represents more sustained marginal marine intervals (Leeder 1974; Ward 1997) with marine limestones 1–10 m thick in cyclical sequences of sandstone, siltstone and mudstone.

The Cementstone Formation is dated palynologically as Tournaisian to early Visean in wells 38/22-1, 41/01-1 and 43/02-1 (Cater *et al.* 1967; Mahdi 1992; Fig. 5). At the base of well 26/07-1 in the Forth Approaches Basin, a limestone unit yielded spores from the Pu Biozone, suggesting that this well penetrated the very top of the

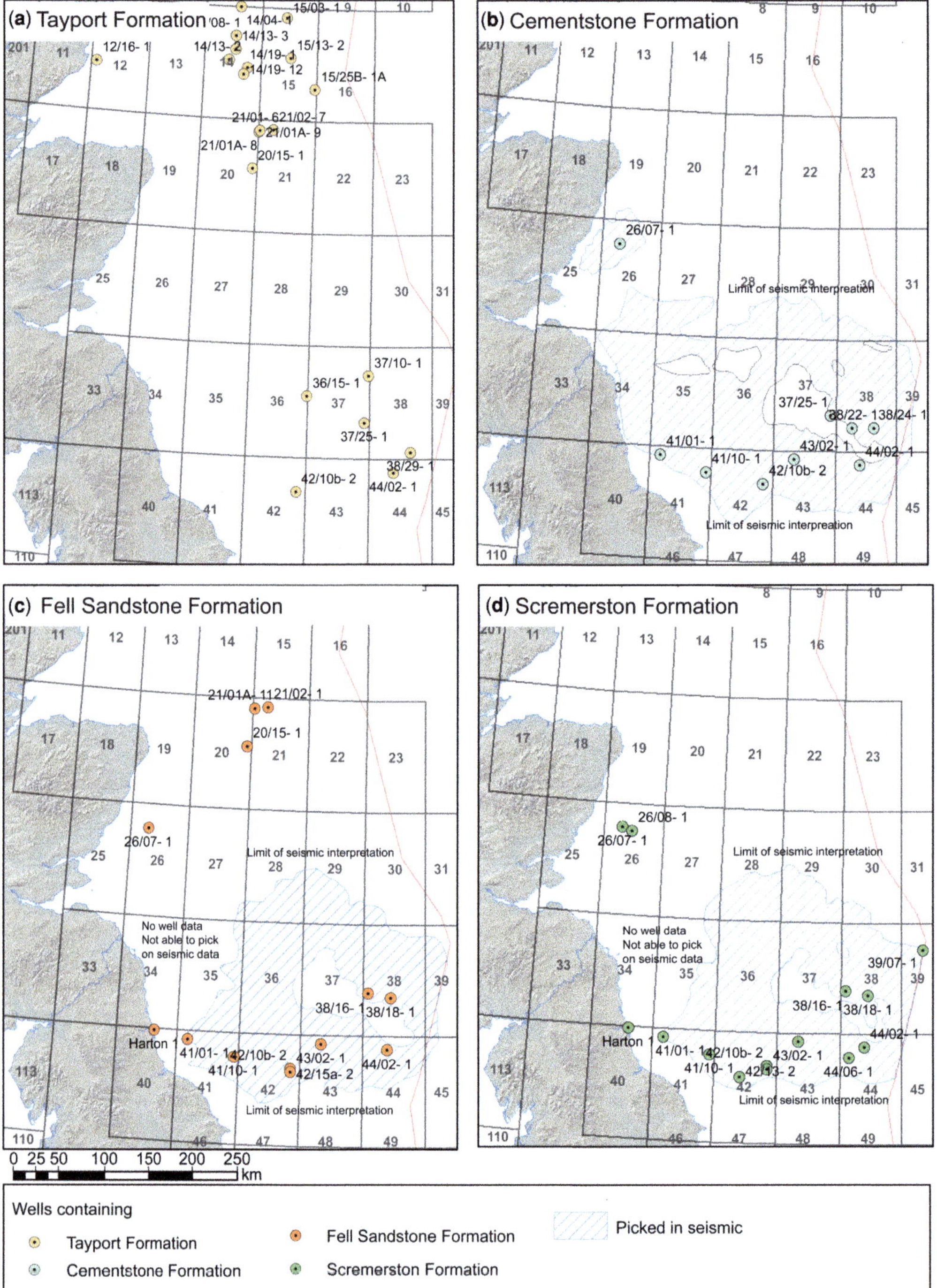

Fig. 3. Distribution of wells and extent of seismic picks of the (**a**) Tayport Formation (wells), (**b**) Cementstone Formation (wells), (**c**) Fell Sandstone Formation (wells and seismic picks) and (**d**) Scremerston Formation (wells and seismic picks). For information on the seismic picks, see Arsenikos *et al.* (2018). Includes mapping data licensed from Ordnance Survey. ©Crown Copyright and/or database right 2017. Licence number 100021290 EU and NEXTMap Britain elevation data from Intermap Technologies.

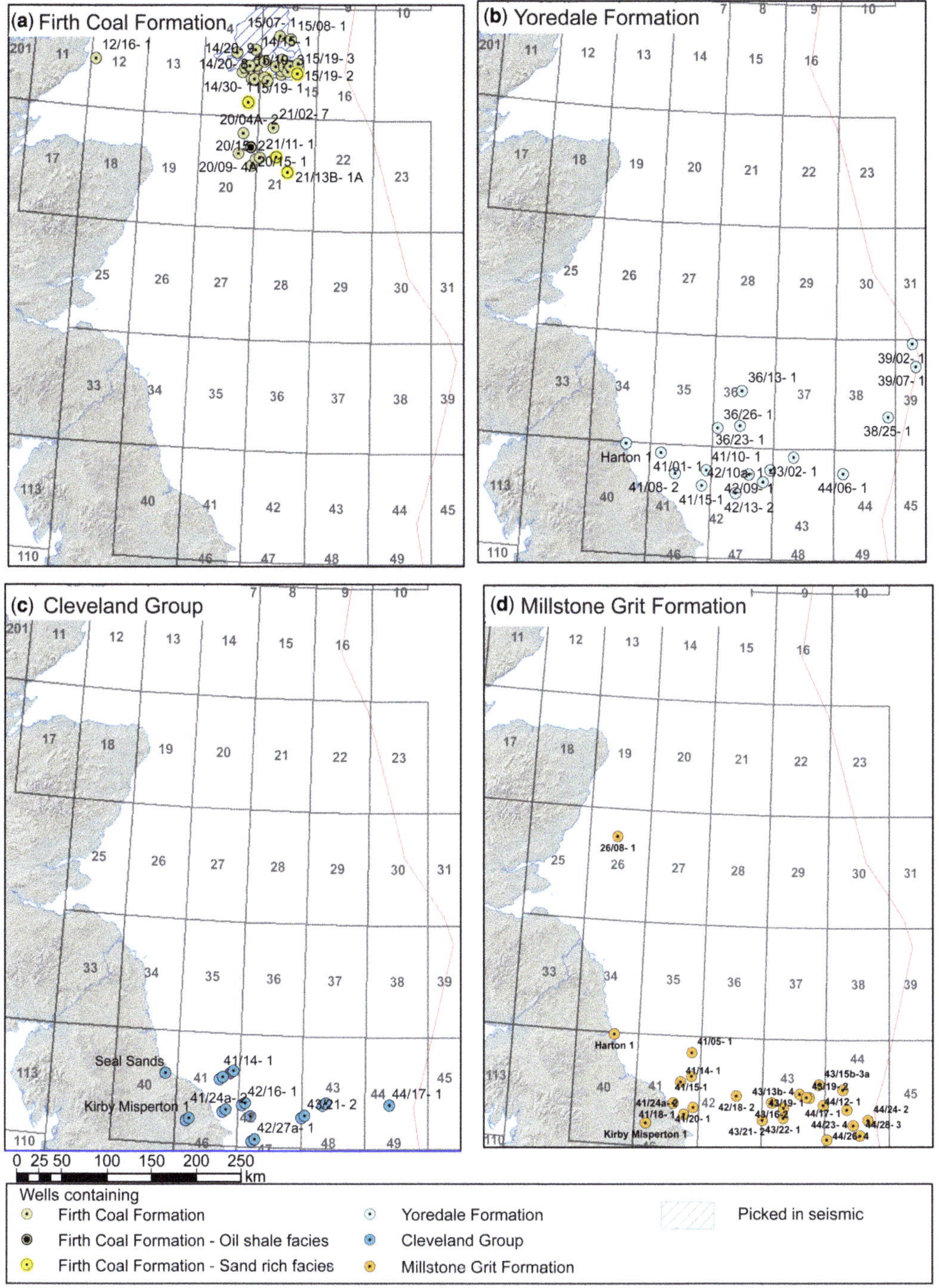

Fig. 4. Distribution of wells and extent of seismic picks of the (**a**) Firth Coal Formation (wells and seismic picks), (**b**) Yoredale Formation (wells), (**c**) Cleveland Group (wells) and (**d**) Millstone Grit Formation (wells). For information on the seismic extent information, see Arsenikos *et al.* (2018). Includes mapping data licensed from Ordnance Survey. ©Crown Copyright and/or database right 2017. Licence number 100021290 EU and NEXTMap Britain elevation data from Intermap Technologies.

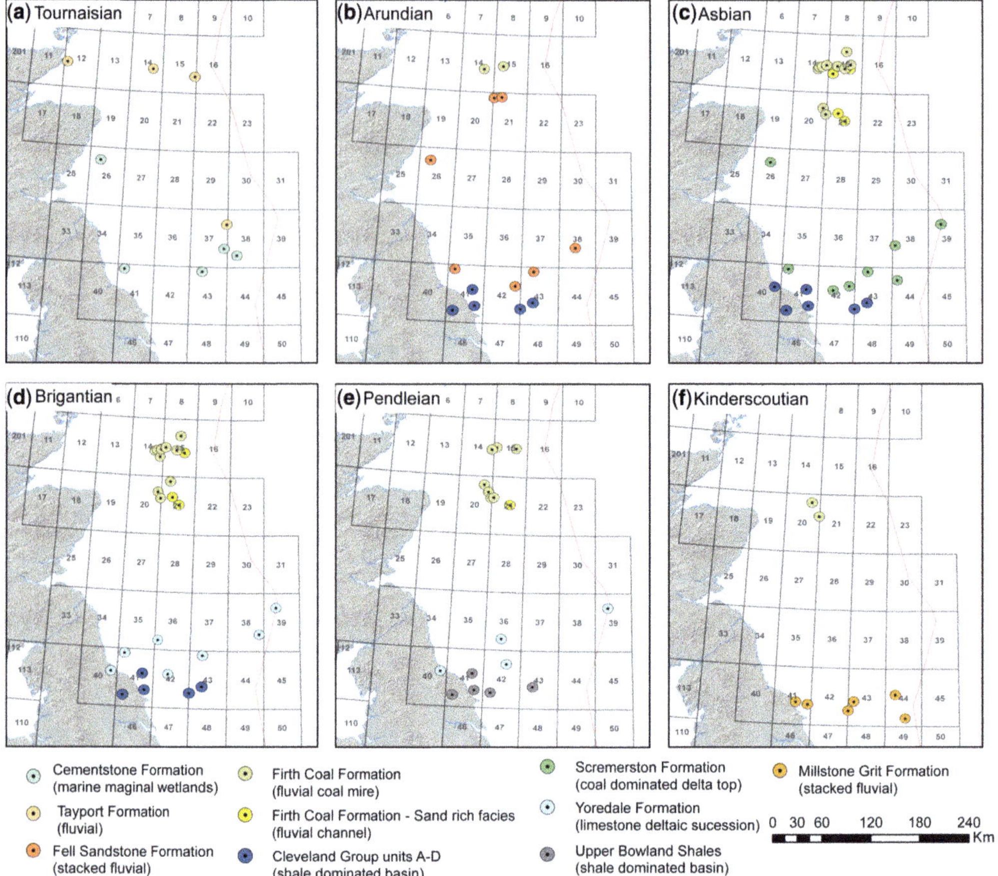

Fig. 5. Maps showing stratigraphic units proved in wells with biostratigraphic control in the (**a**) Tournaisian (earliest Carboniferous), (**b**) Arundian (early Visean), (**c**) Asbian (late Visean), (**d**) Brigantian (late Visean), (**e**) Pendleian (early Namurian) and (**f**) Kinderscoutian (mid-Namurian). Includes mapping data licensed from Ordnance Survey. ©Crown Copyright and/or database right 2017. Licence number 100021290 EU and NEXTMap Britain elevation data from Intermap Technologies.

Cementstone Formation. This mirrors the range found in the Midland Valley of Scotland and adjacent Tweed Basin, where the Ballagan Formation extends from the VI (Smithson *et al.* 2012) to the lower Pu Biozone (McAdam *et al.* 1985).

Fell Sandstone Formation

The Fell Sandstone Formation is found onshore in Northumberland (NE England) and Berwickshire (SE Scotland). There is no sedimentary equivalent in the Midland Valley of Scotland, where Chadian to Asbian volcanic activity gave rise to the extrusive igneous rocks of the Garleton Hills, Arthur's Seat and Clyde Plateau volcanic formations (Stephenson *et al.* 2004; Monaghan & Parrish 2006). The Fell Sandstone Formation occurs widely across the south of the Mid North Sea High (Cameron 1993*a*; Bruce & Stemmerik 2003; Collinson 2005), but prior to this study it had not been identified to the north. It is now interpreted in well 26/07-1 in the Forth Approaches Basin and in wells in Quadrants 20 and 21, east of the Grampian High and Halibut Horst (Figs 3 & 6).

The Fell Sandstone Formation has been picked in seismic surveys over the eastern side of the Mid North Sea High and in the North Dogger and Q29 basins (Fig. 3; Arsenikos *et al.* 2018) and is assumed to extend across the southern edge of the Mid North Sea High to the outcrops in Northumberland (Fig. 3). The formation is not identified north of Quadrants 21 and 22.

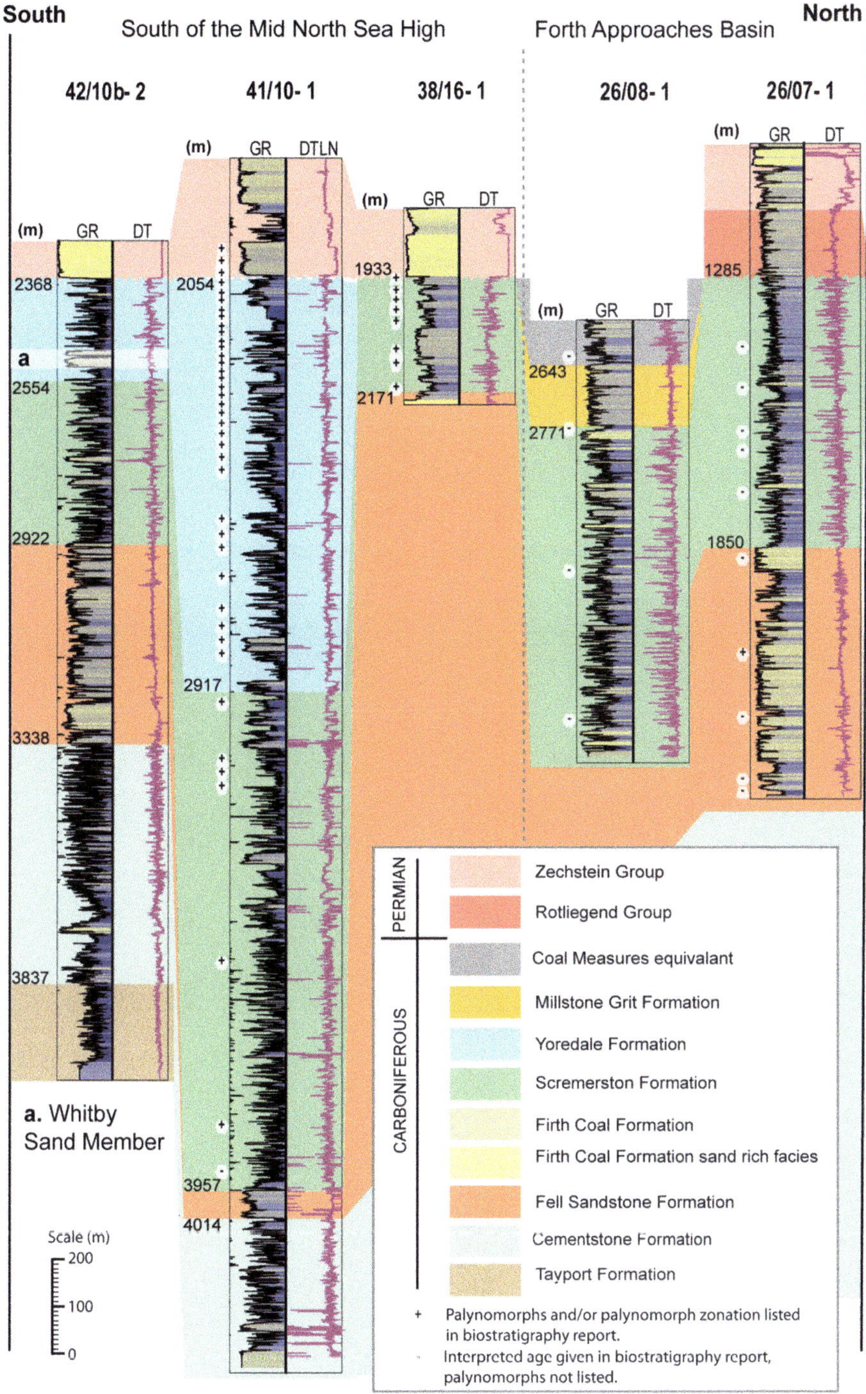

Fig. 6. Correlation panel of Carboniferous strata from south of the Mid North Sea High through the Forth Approaches. The position of the Whitby Sandstone Member is marked by the letter '**a**'. The gamma curve (GR) is scaled from 0 to 150 API and the sonic curve (DT) is scaled from 140 to 40 μs ft^{-1}. For biostratigraphic data, see Appendix A. A large-scale version of Figures 6 and 7 as a single panel can be found as Supplemental Figure 1 (some well curves supplied by CGG).

The Fell Sandstone Formation is typically characterized by massive sandstone units up to 50 m thick, interbedded with units of siltstone and mudstone up to 20 m thick. The gamma log signature is dominated by low gamma sandstones, which commonly have a more 'blocky' signature compared with the sandstones in the formations above and below (Figs 6 & 7). In some wells (e.g. 26/07-1), the succession consists of clean sandstone with only thin intercalations of siltstone and mudstone, but elsewhere (42/10b-2) the succession is more heterogeneous (Fig. 6).

The Fell Sandstone Formation ranges from 200 to 500 m thick in wells. Beyond well data coverage, seismic data (Arsenikos *et al.* 2018) show that it thickens into basins and has an average thickness of 500 m. The contrast between the seismic response from the sandstone facies of the Fell Sandstone Formation and the response from the coal facies of the overlying Scremerston Formation produces an event that can be picked in seismic data from the Forth Approaches Basin to Quadrants 41–44 (Arsenikos *et al.* 2015, 2018). This has allowed the offshore intersections of the Fell Sandstone to be identified with confidence in wells such as 38/16-1 (Fig. 6). The base of the Fell Sandstone Formation is taken at the abrupt downward change from thick sandstone units to the interbedded sandstone, siltstone and mudstone of the underlying Cementstone Formation, a transition marked by a distinct change in wireline log character (Fig. 6). The top of the formation is taken at the top of the uppermost thick sandstone unit below a coal-bearing succession of interbedded mudstone, siltstone and sandstone of the Scremerston Formation.

Palynological evidence from close to the top and base of the Fell Sandstone in wells 41/01-1, 43/02-1 and 42/15a-2 indicates that the Fell Sandstone Formation is Chadian (Pu Biozone) to earliest Asbian (TS Biozone) (Fig. 5). Onshore, in the Northumberland–Solway Basin, the base of the Fell Sandstone Formation is diachronous, becoming generally younger towards the SW (Stone *et al.* 2010). In the Forth Approaches Basin, a unit previously identified as the Tayport Formation in well 26/7-1 by Cameron (1993*a*) and Bruce & Stemmerik (2003) is reinterpreted in this study as the Fell Sandstone Formation, based on the mid-Holkerian (V3a) to Chadian (V1a) age (Welsh & Walton 1985) and the 'blocky' gamma signature.

The Fell Sandstone Formation is interpreted to be the deposit of a major sandy braided river system (Turner & Monro 1987) formed of stacked multi-storey channel fills and separated by mudstone intervals, some of which contain marine microfauna and are thought to represent maximum flooding surfaces (Turner *et al.* 1997; Collinson 2005).

Scremerston Formation

The Scremerston Formation is interpreted in wells and has been picked on seismic data from Quadrants 37, 38, 39, 40, 41, 42, 43 and 44 (Fig. 3d). It is likely to be continuous in extent from the Northumberland coast across the eastern side of the Mid North Sea High and along its southeastern margin (Cameron 1993*a*). It has also been identified in wells 26/07-1 and 26/08-1 in the Forth Approaches (Fig. 6). Previously classified in these wells as the Firth Coal Formation by Cameron (1993*b*) and Bruce & Stemmerik (2003), these rocks have a Holkerian to Asbian biostratigraphic age and a range of facies, within that of the Scremerston Formation, and are herein assigned to this formation. It has been picked in seismic data over the eastern side of the Mid North Sea High and in the Q29 and North Dogger basins (Fig. 3; Arsenikos *et al.* 2018). The Scremerston Formation broadly coincides with the onshore Tyne Limestone Formation, which includes the locally dominant Scremerston Coal Member (Dean *et al.* 2011; Waters *et al.* 2011).

The Scremerston Formation is characterized by coal-rich intervals within a mudstone-dominated sequence, with alternations of sandstone, siltstone and some thin limestone or dolostone beds (Cameron 1993*b*; Bruce & Stemmerik 2003). Upwards-coarsening parasequences are sporadically present within the upper parts of the formation, but it contains many more and thicker coals than the overlying Yoredale Formation. Coals are up to 3 m thick in exceptional cases, but are typically *c.* 1 m thick (Cameron 1993*b*). The sandstone units probably include both single and stacked channel fills. Homogeneous dark grey and black mudstones generate a uniformly high gamma log response (Fig. 6). In wells, the Scremerston Formation ranges in thickness from 90 to 1000 m, similar to that interpreted from seismic data (Arsenikos *et al.* 2018). This is significantly thicker than the onshore Scremerston Coal Member at Berwick-upon-Tweed, which is *c.* 300 m thick (Stone *et al.* 2010; Waters *et al.* 2011).

A late Holkerian to Asbian age for the Scremerston Formation is proved palynologically in well 41/01-1, but elsewhere south of the Mid North Sea High (e.g. wells 41/01-1, 43/02-1 and 44/06-1) it is solely Asbian in age. The two wells in the Forth Approaches have yielded palynomorphs of late Arundian and Holkerian to late Asbian age (McLean & Neves 1988).

The Scremerston Formation represents widespread delta plain and back-swamp mire environments (Bruce & Stemmerik 2003). In the Forth Approaches Basin it is thought to be dominantly terrestrial, whereas south in Quadrants 41–42 it is a deltaic, marine to terrestrial cyclic succession (Kearsey *et al.* 2015).

Firth Coal Formation

The Firth Coal Formation is found in Quadrants 14 and 15 (e.g. wells 14/04-1, 15/19-1 and 15/19-2), in the Witch Ground Graben and may extend northwards. The unit has also been interpreted in wells in the Outer Moray Firth Basin in Quadrants 20 and 21 (e.g. wells 20/04a-2 and 20/15-1, Fig. 7). It has been picked in seismic data in the Outer Moray Firth Basin (Fig. 4; Arsenikos *et al.* 2018).

The formation is lithologically similar to the Scremerston Formation, consisting of coal-rich intervals within a mudstone-dominated succession, with alternations of sandstone, siltstone and some thin limestone (Figs 6 & 7). A 6 m thick bed of oil shale has been identified in well 20/10a-3, highlighting the fact that organic-rich lacustrine units are present offshore. As well data are sparse in the Forth Approaches and towards the Outer Moray Firth, a more extensive oil shale source rock basin may exist there, but remains unproved. The Firth Coal Formation is 200 m thick on average. Although the top contact has been eroded from beneath the Permian or Cretaceous unconformities, the Firth Coal Formation is significantly thinner than the Scremerston Formation and was deposited over a longer time interval. It is retained as a separate unit in this northern area.

The oldest palynological date for the Firth Coal Formation is early Visean (Chadian, Pu Biozone) in well 15/17-1A (Bagnall *et al.* 1973) and in other wells in Quadrants 14–15 it spans the Chadian to Pendleian (Bagnall *et al.* 1973; Figs 6 & 7). The youngest biostratigraphic age is late Namurian (Kinderscoutian, KV Biozone) in well 20/15-2 (Harris *et al.* 1998) and the range in Quadrants 20 and 21 is Holkerian to Kinderscoutian. This would make some parts of the Firth Coal Formation equivalent in age to the Yoredale and Millstone Grit formations to the south.

A sand-rich facies is identified in the Firth Coal Formation in Quadrants 14, 15 and 21, e.g. wells 15/19-1 (Fig. 7), 15/19-3, 15/21a-7, 20/04a-2 and 21/13b-1a. Here, massive sandstone units up to 40 m thick are separated by siltstone and mudstone interbeds 10–20 m thick. By contrast, wells close by, such as 15/19-2 (Fig. 7), contain strata of equivalent age, but are represented by a mudstone-dominated, coal-rich succession. This suggests rapid lateral facies change and that the sand-rich facies represents channel infill. The mudstone-dominated facies of the Firth Coal Formation is found both below (20/04a-2, Fig. 7) and above (e.g. 15/19-3) the sand-rich facies. The sand-rich facies is similar in character to the Fell Sandstone Formation. However, seismic data do not show these to be laterally extensive units (Arsenikos *et al.* 2016). Also, in well 21/13b-1a, Asbian–Pendleian (NM-NC biozone) palynomorphs (Appendix A) have been recorded, which suggests that, in this well, it is equivalent in age to the Yoredale Formation south of the Mid North Sea High. However, it is possible that some examples of the sand-rich facies are of Fell Sandstone age and represent the feeder channels to the delta front.

Yoredale Formation

The Yoredale Formation occurs in wells across Quadrants 36, 38, 39, 40, 41, 42, 43 and 44. Its depositional extent is likely to have been continuous from northern England across the Mid North Sea High (Fig. 4). The Yoredale Formation is equivalent to onshore sedimentary rocks in northern England consisting of stacked sequences of 'Yoredale cycles' (Cameron 1993*a*). This includes the part of the Tyne Limestone Formation that lies above the Scremerston Coal Member, along with the Alston and Stainmore formations (Dean *et al.* 2011; Waters *et al.* 2011). Onshore, the Great Limestone is the thickest limestone of the Alston Formation at *c.* 20 m and is the uppermost bed of the Alston Formation. A comparably thick limestone is seen in some wells offshore (e.g. 44/06-1) and is equated with the Great Limestone, indicating that the equivalent of the Stainmore Formation is preserved, in part, to the south of the Mid North Sea High (Kearsey *et al.* 2015).

The offshore Yoredale Formation consists of a repeated cyclic succession beginning with a marine limestone, overlain by an upwards-coarsening succession of marine mudstone and sandstone, and capped by a seat earth palaeosol (a grey or whitish clay rock containing fossilized plant roots and occurring beneath coal seams) and coal. Some fining-upwards cycles are also present. The lower part of the Yoredale Formation locally contains thick multi-storey channel sandstones interpreted to be infilled palaeovalleys (Collinson 2005). Such distinctive sandstone units are encountered in wells 41/10-1, 42/10b-2 and 43/02-1, and are referred to as the Whitby Sandstone Member (Fig. 6; Maynard & Dunay 1999). This unit is distinguished on wireline logs by two thick units of sandstone separated by a mudstone interval. The geometry of these sandstone bodies is not known. They may represent a wide, single fluvial channel system, or separate, narrow channel fills. In many wells the Yoredale Formation is truncated by the unconformity beneath the base of the Permian strata, but, where not eroded, it is >800 m thick (Jones 2007).

Compared with the underlying Scremerston Formation, the Yoredale Formation contains thicker limestones and the common presence of coarsening-upwards parasequences (Stephenson *et al.* 2010). Although present at outcrop onshore, typically beneath the limestones, the thin coals within the

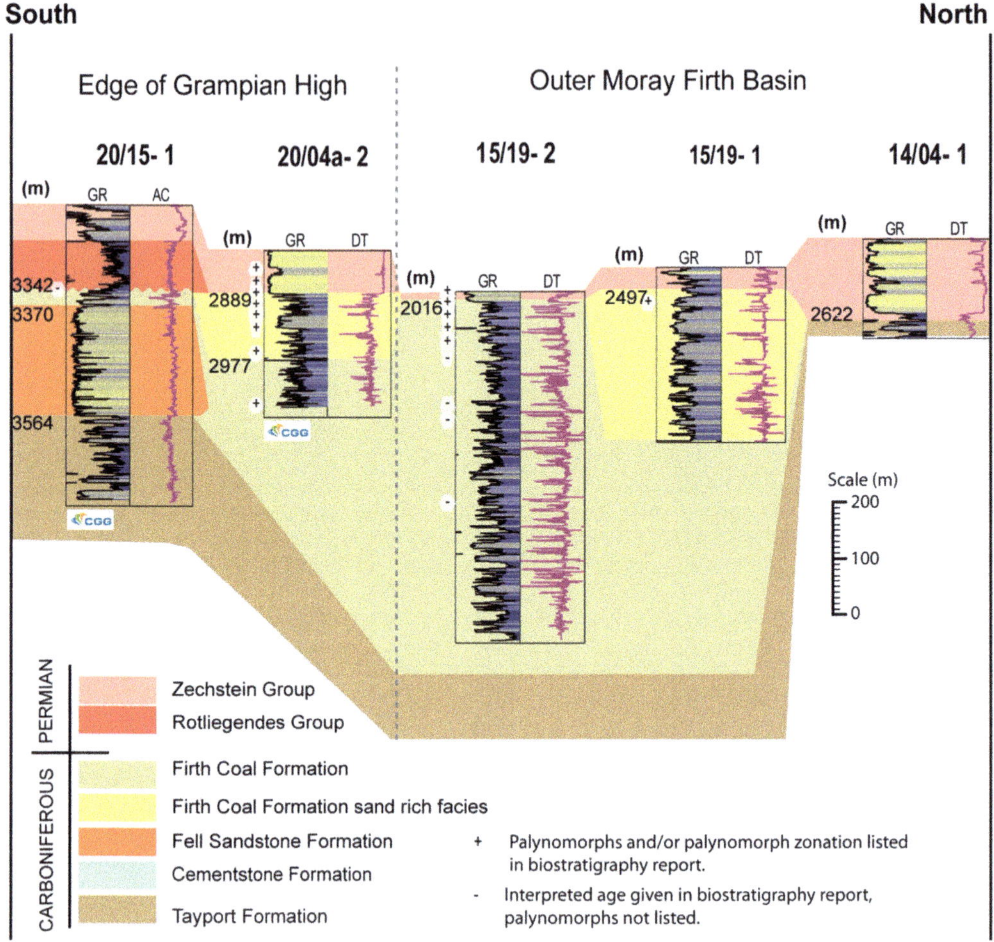

Fig. 7. Correlation panel of Grampian high and Outer Moray Firth Basins. The thickness of Carboniferous strata decreases markedly north of the Forth Approaches when compared to Figure 6. For biostratigraphic data see Appendix 1. A large scale version of Figures 6 and 7 as a single panel can be found in Supplemental Figures 1. (Well curve data for 20/04a-2 and 20/15-1 supplied by CGG for the 21CXRM Palaeozoic Project results upon which this paper is based).

Yoredale Formation are mostly below wireline logging resolution.

The Yoredale Formation is late Asbian and Brigantian to early Namurian in age, based on palynological data from wells 38/25-1, 39/07-1, 41/01-1, 42/09-1, 43/02-1 and 44/06-1.

Cleveland Group

The Visean and lower Namurian succession in the central and southern part of Quadrants 41–44 is characterized by apparently homogenous units of shale exceeding 100 m thick, interbedded with cycles of limestone, sandstone, mudstone and sporadic coals similar to those seen in the Scremerston and Yoredale formations to the north (Figs 4 & 8). A similar succession, *c.* 1400 m thick, occurs in the onshore well Kirby Misperton 1. The base of the succession is not proved and the Millstone Grit Formation overlies it. In the upper part, there is a high gamma unit of black shale, which Cameron (1993*a*) termed the Bowland Shales Formation. We refer to the high gamma unit as the Upper Bowland Shales to avoid direct comparison with the eponymous formation, which has a wider definition and age range across northern England (Aitkenhead *et al.* 1992).

Emphasizing the shale-rich character of this succession, Kearsey *et al.* (2015) assigned these rocks informally to the existing Cleveland Group. Johnson

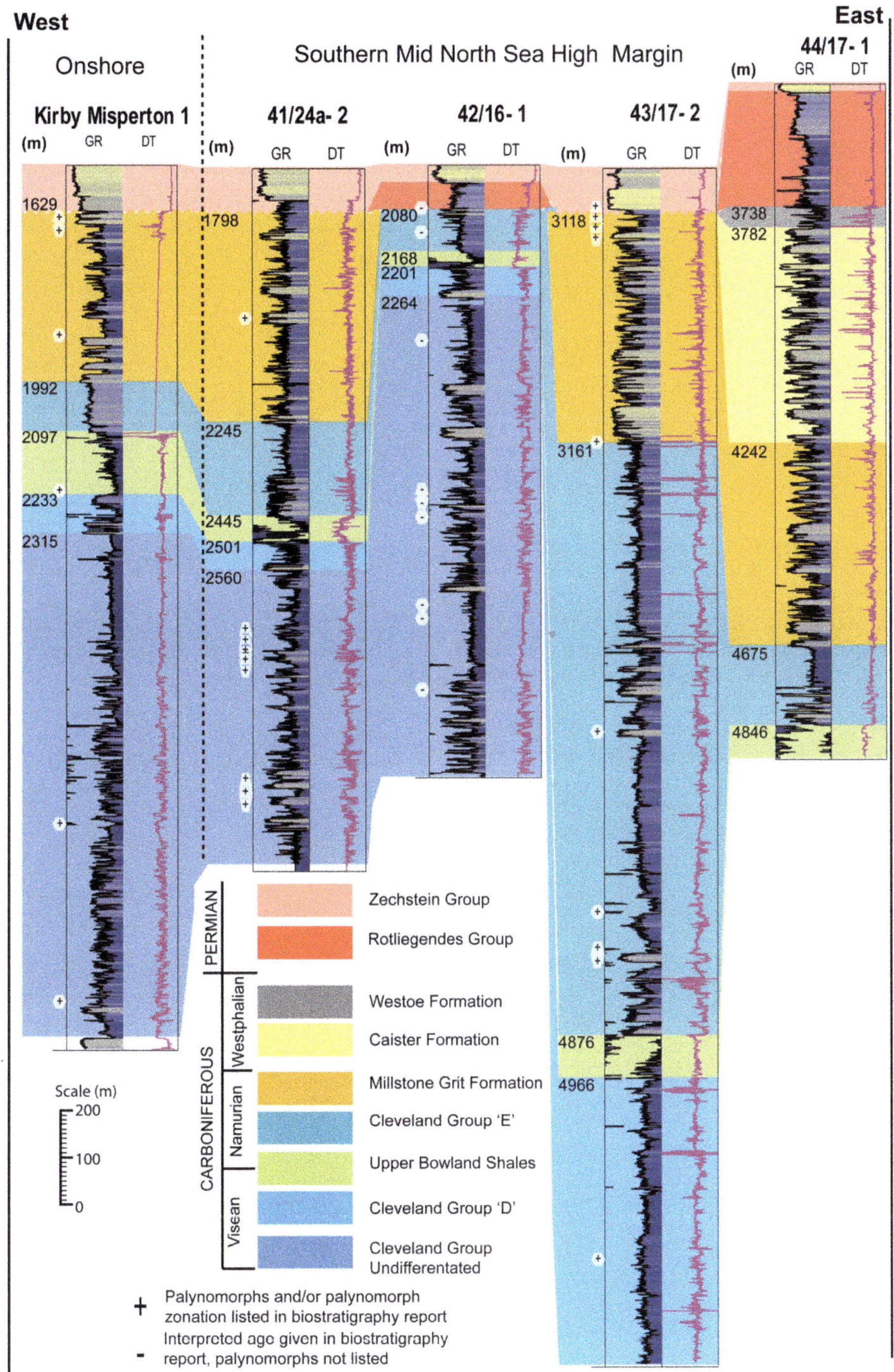

Fig. 8. Correlation panel of Carboniferous strata from west to east across Quadrants 41, 42 and 43 showing the distribution of the Cleveland Group and Upper Bowland Shales, representative of the 3 km thick basinal mudstone succession. The transition from the Cleveland Group E and the Millstone Grit Formation is diachronous and both were being deposited in different parts of the basin. For biostratigraphic data see, Appendix A. A large-scale version this figure can be found as Supplemental Figure 2.

et al. (2011) introduced this term for the thick, middle Visean, shale-rich succession proved in Seal Sands No. 1 well. However, significant intervals in that well consist of Yoredale facies, and parts are also coal-rich; as a result, this succession has only a superficial resemblance to that seen in Kirby Misperton and offshore. Although the term Cleveland Group may not be entirely appropriate given the small amount of data available, and its difference from the original 'type' section, we prefer to retain the informal term here, redefining it as in Kearsey *et al.* (2015).

Based on gross characteristics, Kearsey *et al.* (2015) divided the group in Kirby Misperton 1 into six informal units, consisting of units A–E and the Upper Bowland Shales, which lie below unit E. High gamma values characterizing the Upper Bowland Shales show this unit to be a consistent marker across all of the wells penetrating the group (Fig. 7). The Upper Bowland Shales has an average thickness of 60 m in wells and ranges in thickness from 17 to 136 m. As suggested by Kearsey *et al.* (2015), there is potential for the correlation of shale packages beneath the Upper Bowland Shales. For example, unit D shows consistent characteristics across many of the wells, although units A–C below this cannot be easily correlated between wells. Based on dip measurements, Collinson (2005) suggested that the interval identified by Kearsey *et al.* (2015) as Cleveland D in well 43/21-2 represents a slumped slope deposit. Coals are absent from the lower part of the group in Kirby Misperton 1, suggesting that this part represents a mudstone-dominated pro-delta and basinal succession. By contrast, the apparent intercalation of Yoredale-type facies with homogenous shales implies alternations between the delta top and slope-and-basin facies. In this respect, the northernmost well of this group, 41/14-1, and Seals Sands No. 1 more closely resemble the Yoredale Formation.

Although miospore assemblage data are few (Appendix A), a consistent pattern of ages emerges across the area. Samples from wells 41/14-1, 41/24a-2 and Kirby Misperton 1 indicate that the Cleveland Group units below the Upper Bowland Shales lie within the Arundian to Brigantian interval. The overlying Upper Bowland Shales contain upper Brigantian and Pendleian miospores. The highest unit (E) in the group in Quadrant 41 and Kirby Misperton 1 has yielded Pendleian to Arnsbergian ages, whereas in Quadrant 43 this range extends into the Alportian.

The succession above the Upper Bowland Shales, seen in wells 43/21-2, 43/17-2 and 44/17-1, may be the lateral equivalent of the Epen Formation in the Dutch sector (Van Adrichem Boogaert & Kouwe 1993–97*c*). This similarity is best seen in wells Nagele-1 and Luttelgeest-1 in the Netherlands. The Epen Formation also contains the Geverik Member, which is a high gamma shale similar to the Upper Bowland Shales (Van Adrichem Boogaert & Kouwe 1993–97*c*). However, the Geverik Member has been dated as Arnsbergian to Alportian, younger than the Upper Bowland Shales and so may not represent the same anoxic event. The high gamma values of the Upper Bowland Shales similarly occur within the Bowland–Hodder unit of the onshore Craven Basin, providing a crucial correlation with that basin through into the Irish Sea (Andrews 2013; Wakefield *et al.* 2016). This also shows that the Cleveland and Craven basins became linked during the deposition of the Upper Bowland Shales (Andrews 2013).

Millstone Grit Formation

The Millstone Grit Formation is widespread south of the Mid North Sea High to the London–Brabant Massif (Cameron 1993*a*). It is interpreted in wells in Quadrants 40, 41, 42, 43 and 44 (Figs 4 & 7; Cameron 1993*a*). A 126 m thick sandstone unit with a blocky gamma ray trace recorded in well 26/08-1 in the Forth Approaches Basin (Figs 6 & 7) is tentatively correlated with the Passage Formation in the Midland Valley. The Passage Formation is roughly equivalent to the Millstone Grit Formation (Cameron 1993*a*; Read *et al.* 2002; Waters *et al.* 2014).

The Millstone Grit Formation consists of cycles of sandstone, interbedded mudstone and siltstone, thin marine bands and sporadic thin coals (Collinson 1988; Cameron 1993*a*). The sandstone is typically coarse to very coarse grained, with some units containing common 'floating' quartz pebbles (Cameron 1993*a*). It lacks the upwards-coarsening cycles with limestone of the Yoredale Formation. It ranges from 10 to 1027 m thick across the study area (Fig. 8).

The offshore Millstone Grit Formation is Pendleian to Yeadonian in age. However, the base of the formation is markedly diachronous. Onshore, in the Craven Basin, the lowest formations in the Millstone Grit Group are Pendleian in age (Aitkenhead *et al.* 2002), whereas in Northumberland the lowest unit is Kinderscoutian (Waters *et al.* 2014), which is comparable with that recorded in biostratigraphic logs offshore (Appendix A).

Collinson (1988) interpreted the offshore Millstone Grit Formation as a succession of prograding fluvial and delta front successions accumulating in basinal areas.

Discussion

Previously, the Mid North Sea High has been used to separate Carboniferous stratigraphic nomenclature to the north and south of it (Cameron 1993*a*, *b*;

Bruce & Stemmerik 2003). However, Arsenikos *et al.* (2018) show that a series of basins underlay much of the area during the Carboniferous. It is also evident from wells in this study that the Tayport, Cementstone, Fell Sandstone, Scremerston and Millstone Grit formations are present to the north, south and over large areas of the eastern side of the Mid North Sea High (Fig. 9). Although these units thin across this area (Arsenikos *et al.* 2018), it is clear that the Mid North Sea High did not form a barrier to sedimentary systems during the Carboniferous.

This interpretation is supported by the identification, in the Forth Approaches Basin, of the Fell Sandstone and Scremerston formations in wells (26/07-1 and 26/08-1), suggesting that it was part of the same fluvio-deltaic system extending to Quadrants 41–44 (Fig. 6). To the south, the Cleveland Group has been interpreted in wells as far east as Quadrant 44, suggesting that this marine basin is extensive (Figs 7 & 8). Many of the formations are time-equivalent and suggest that the southern Central North Sea was dominated by pro-delta and basinal successions from the Visean to the Namurian, whereas fluvial facies are more common to the north.

Provenance of fluvial systems

Based on the regional tectonic history, it is suggested that the Carboniferous deltaic system is likely to have drained from the north within Laurussia in the early to mid-Carboniferous (Gilligan 1919; Cliff *et al.* 1991; Coward 1993; Morton *et al.* 2001). Establishing whether the fluvial systems are sourced locally from upland areas, such as the Grampian Highlands/Grampian High and Southern Uplands/offshore Mid North Sea High, or from much farther northwards within the more northern parts of the Caledonian orogeny, is crucial to understanding the size of this fluvial system. Leeder & Boldy (1990) suggested that the source of the Firth Coal Formation sediment was local, from the surrounding Grampian Highlands. However, zircon analysis has subsequently shown that the source region included a wide range of mid-Proterozoic and Archean rocks that are more likely to be from East Greenland, in addition to material derived locally (Morton *et al.* 2001). Morton *et al.* (2001) suggested that the zircon populations of the Firth Coal Formation are similar to those seen in the Namurian Ashover Grit and Rough Rock in the Pennine Basin, indicating a common provenance for both systems lying to the north of the modern North Sea. This finding supports the hypothesis presented here that the sand-rich facies of the Firth Coal Formation may represent feeder channel systems for the coeval Yoredale Formation and later Millstone Grit delta front system south of the Mid North Sea High. In addition, using a multiproxy provenance approach, Lancaster *et al.* (2017) suggested that the sand in the Millstone Grit Group in Yorkshire derived dominantly from the Greenland Caledonides, but with additional material from western Norway and NE Scotland.

The provenance of the Tournaisian and Visean sandstones is poorly constrained. Palaeocurrent analysis of the Fell Sandstone Formation in Northumberland indicates palaeoflow from the NE (Robson 1956; Turner *et al.* 1997). The presence of the Fell Sandstone Formation as far north as Quadrants 20 and 21, and its coexistence with the age-equivalent siltstone-dominated terrestrial succession and abundant coal mires of the Firth Coal Formation present to the north, suggest that the fluvial system had reached its apex point and become more channelized (Fig. 10b). This may indicate that, in the Arundian at least, the whole of the North Sea was part of the same fluvial system receiving sediment from the Caledonide mountains or from farther to the north.

Sedimentary basin evolution

High-resolution sequence stratigraphic examination of offshore wells was not attempted in this study because of the lack of correlative chronostratigraphic data and the regional scale. However, the onshore work of Church & Gawthorpe (1994), Hampson *et al.* (1996), Leeder & Stewart (1996), Davies (2008) and Stephenson *et al.* (2010) appears highly relevant to the Visean–Namurian delta plain to basinal setting. These studies used marine bands to identify transgressions and their erosion to identify the sequence boundaries; however, in our study, it was difficult to consistently pick out such marine bands and correlate them between wells. Sequence stratigraphy could be applied if the chronostratigraphic constraints were improved.

For the reasons given in the preceding paragraph, a sequence stratigraphic model for the Carboniferous has not been derived. Instead, the sedimentary facies interpretations are divided into selected time slices to create a regional palaeogeographical understanding of how the facies relate to each other and how the palaeogeography and sediment system evolved from the Tournaisian to Namurian (Fig. 10). The overall picture during the Mississippian and early Pennsylvanian is that there were two periods of voluminous sediment influx and large-scale progradation of the fluvial system in the Arundian and Kinderscoutian, possibly driven by uplift to the north (Coward 1993). Between these periods, the gross sediment influx reduced and eustatic/climatic controls dominated.

Tournaisian. In the Tournaisian, the fluvial Tayport Formation, which in the Devonian had been dominant south of the Mid North Sea High, became restricted to the Outer Moray Firth and some small

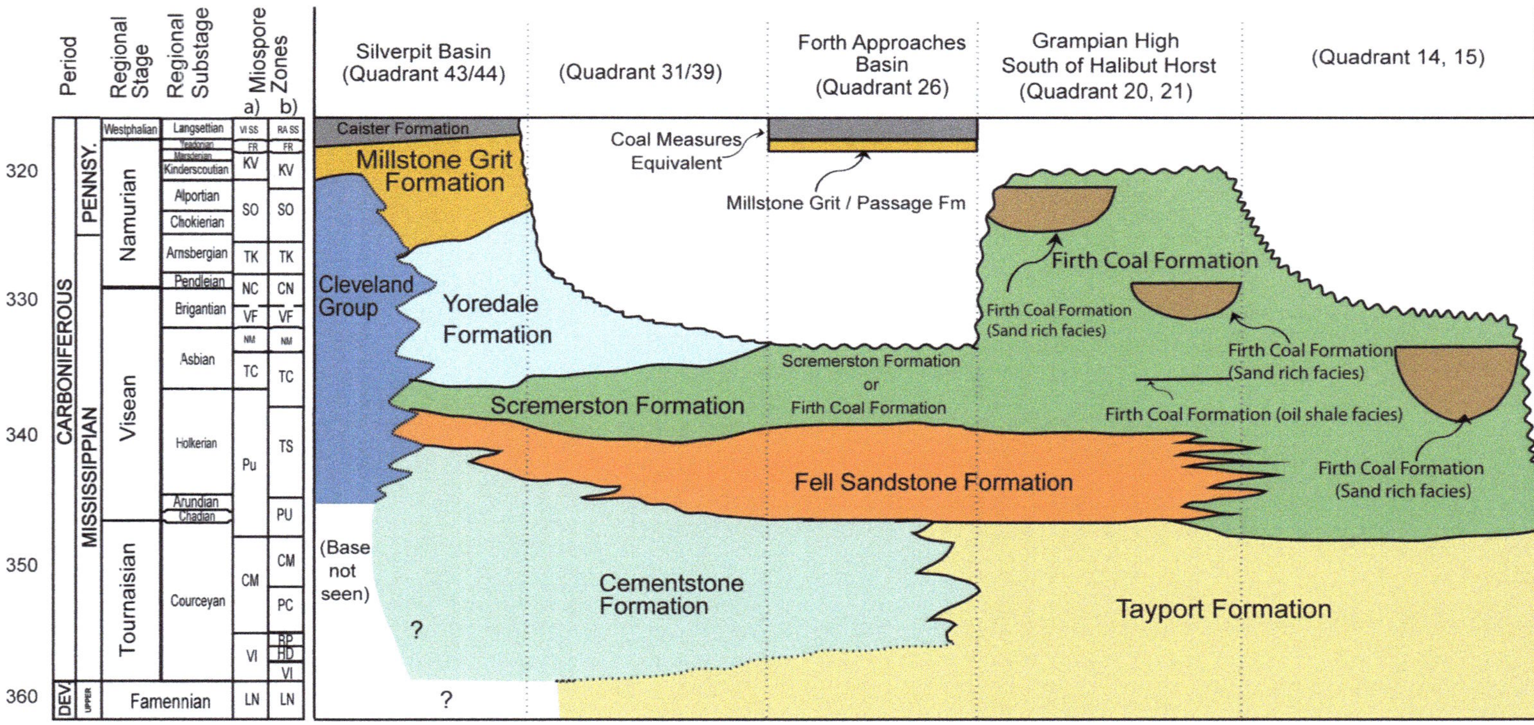

Fig. 9. Stratigraphic relationships of Upper Devonian to Carboniferous strata between the Outer Moray Firth Basin and the Silverpit Basin. Miospore zone (a) is the 'former index' and miospore zone (b) is the 'index' described by Waters *et al.* (2011). Correlation between miospore biozones and the global timescale after Davydov *et al.* (2004).

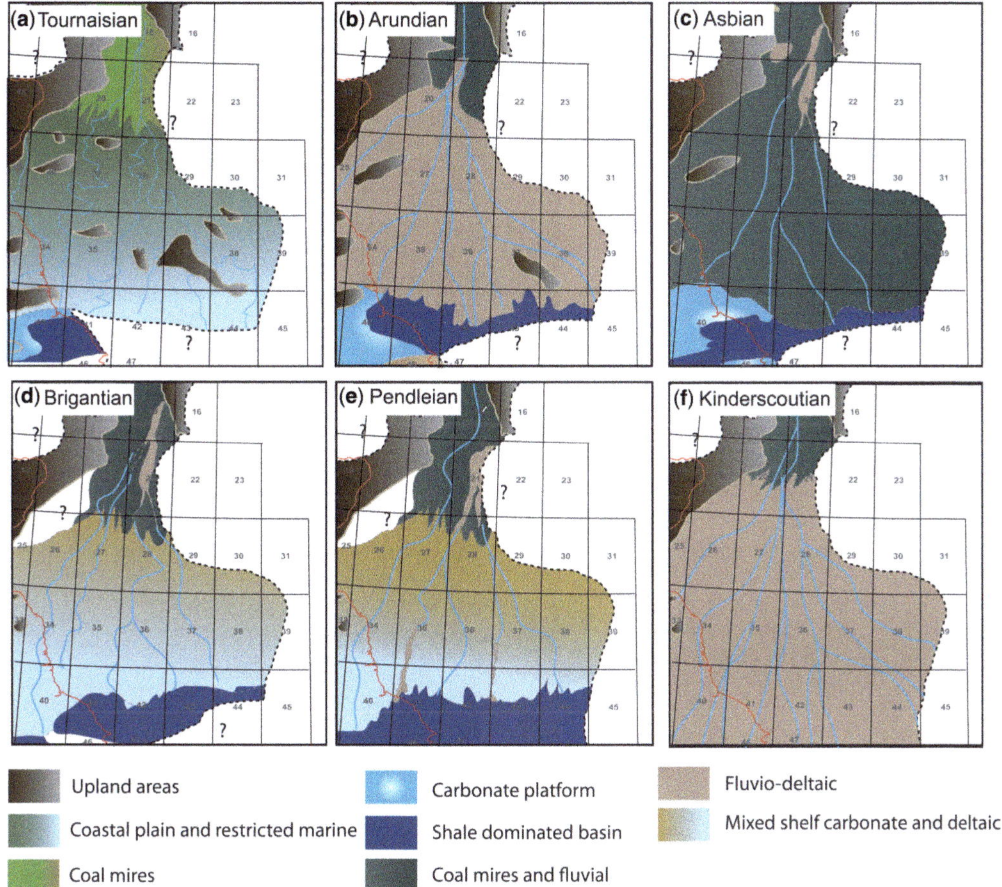

Fig. 10. Schematic evolution model of the facies distribution in the North Sea based on well data during the (**a**) Tournaisian (earliest Carboniferous), (**b**) Arundian (early Visean), (**c**) Asbian (late Visean), (**d**) Brigantian (late Visean), (**e**) Pendleian (early Namurian) and (**f**) Kinderscoutian (mid-Namurian). The dashed line at the eastern margin of each time slice reflects the extent of the data available for this study (within UK waters).

areas near to basin highs. To the south, the coastal plain and marginal marine Cementstone Formation is interpreted as far south as Quadrant 41. The transition from the Cementstone Formation to the Tayport Formation is not seen, but occurs somewhere between Quadrants 26 and 20–21 (Fig. 10a). However, an unconformity is identified in the underlying Upper Devonian in the area between the Grampian High and the Forth Approaches Basin (Arsenikos *et al.* 2016).

Studies of the Ballagan Formation onshore near Berwick-upon-Tweed have shown the presence of a dynamic coastal plain dominated by overbank flooding, lakes, sabkhas and abundant palaeosols, and subjected to short episodes of marine flooding (Bennett *et al.* 2016, 2017; Kearsey *et al.* 2016). Seismic mapping shows the Tournaisian succession to thin against older rocks, such as those of the granite-cored Dogger High, which was probably elevated above the coastal plain (Milton-Worssell *et al.* 2010; Arsenikos *et al.* 2018). On the south side of the Mid North Sea High, limestones, typically 1–4 m thick, but sporadically up to 10 m thick and containing a generally restricted marine invertebrate fauna, replace the ferroan dolostones that characterize the formation to the north. The limestones represent marine intervals and are characteristic of the Visean Lyne Formation in Northumberland (formerly the Lower Border Group *sensu* Day 1970).

Chadian: Holkerian. A major clastic system was established across the region by Arundian times (Fig. 10b). A braided fluvial system, probably originating in the Caledonide mountains or farther to the north, spread coarse-grained sand in stacked, multi-storey sheets southwards (Turner & Monro 1987; Morton *et al.* 2001). In north Northumberland the sand sheets probably fill an incised channel, but in the Northumberland Basin sandstone sheets in the upper part of the formation are intercalated with marine siltstones and the Fell Sandstone Formation progrades southwestwards (Turner *et al.* 1997).

The northernmost extent of the Fell Sandstone Formation is in Quadrants 20 and 21 (Fig. 5). To the north of this, in Quadrants 14 and 15, the Firth Coal Formation represents the alluvial plain developed upstream from the major braid plain, where coal mires, fluvial channel sandstones and minor lacustrine units dominated.

To the south, in the Cleveland Basin, the time-equivalent rocks are the interbedded sandstones and mudstones of the oldest known part of the Cleveland Group, dominated by pro-delta and basal successions. The sandstone units probably represent the distal facies of the Fell Sandstone.

Asbian. By the late Asbian, coal mires and fluvial facies had developed within the Carboniferous basins from the Outer Moray Firth Basin to the southern edge of the Mid North Sea High (Fig. 10c).

Over Quadrants 26–44, the Scremerston Formation represents a range of depositional environments, including fluvio-deltaic, lacustrine, wetland and marine-influenced bay deposits associated with a major deltaic system (Leeder *et al.* 1989; Leeder & Boldy 1990). Glacio-eustatic fluctuations in sea-level affected the deltas, with episodic rises in sea-level flooding the area and allowing carbonates to be deposited (e.g. Wright & Vanstone 2001). The marine fauna found in the carbonate rocks suggest that they were formed at similar shallow water depths, regardless of whether the carbonate formed on the block or basin (Brand 2011). However, there are significant changes in thickness in the clastic sediment between limestone units, across basin-bounding faults. This is thought to be controlled by differential subsidence and syn-depositional faulting in the Northumberland Basin (Leeder *et al.* 1989) and may also have continued into the North Sea. The limestone part of the cycle is generally short-lived in the Scremerston Formation and the limestones may be very thin or absent from some cycles in north Northumberland and the Forth Approaches Basin. By contrast, towards the south, the limestones are thicker and more consistent, suggesting that there was a greater effect from marine transgressions.

The Firth Coal Formation continued to be deposited to the north of the Forth Approaches Basin. Asbian strata record the first dated occurrence of the Firth Coal Formation sand-rich facies (Fig. 6), which is contemporaneous with the coal-rich intervals. The occurrence of sand-rich lithofacies in the Firth Coal Formation indicates that there may have been large-scale channel systems running north to south across the area. In the south, in Quadrants 41–43, the mudstone-dominated succession in the lower part of the Cleveland Group was deposited in the Cleveland Basin (Fig. 5).

Brigantian. At the onset of Brigantian times, Yoredale-type fluvio-deltaic clastic sediments and shallow carbonate seas overstepped the Asbian carbonate platforms of the Alston and Askrigg blocks (Fig. 10d). The Yoredale Formation was not deposited in the Forth Approaches Basin or was subsequently eroded, suggesting either that the basin was tectonically active at this time or that sediment had bypassed it (see well 26/08-1 in Figs 6 & 9).

The position of the transition to the mudstone-dominated pro-delta and basinal succession of the Cleveland Group had not grossly changed since the Arundian. However, by Brigantian times, units of sandstone with subordinate interbedded mudstone, siltstone and some limestone were deposited (see Cleveland Group; Fig. 8), suggesting that the delta had prograded into the Cleveland Basin.

Pendleian. The Yoredale fluvio-deltaic clastic sediments and shallow carbonate seas continued to be deposited through Pendleian and Arnsbergian times (Fig. 10e). Substantial areas of these rocks across the Mid North Sea High were eroded away in late Pennsylvanian and early Permian times. Fluvial fine- to medium-grained sandstones became progressively more significant within the succession over time. Incised fluvial channel systems are of the same order of magnitude as those seen in the Scremerston Formation (Jones 2007; Waters *et al.* 2014). In general, both the limestone and sandstone beds are typically below seismic resolution, except with the highest quality data.

Siltstone-dominated terrestrial successions with abundant coal mires and fluvial channels continued through this period in Quadrants 20, 21, 14 and 15, with deposition of the Firth Coal Formation (Fig. 7).

In the Cleveland Basin, a package of mudstones and silty mudstones with high gamma values (correlated to the Upper Bowland Shales, Fig. 8) has a late Brigantian to Pendleian age. This suggests that a link between the Cleveland and Craven basins was established for the first time in the Pendleian.

Arnsbergian: Kinderscoutian. The Arnsbergian sees the first occurrence of the Millstone Grit Formation in the North Sea. This is represented by sheets of deltaic coarse sand spreading from the NE due to renewed uplift in the Caledonide source region (Hallsworth & Chisholm 2008). The Millstone Grit Formation is inferred to have spread across much of the North Sea through the Arnsbergian to the Alportian, filling the Cleveland Basin in the study area by the Kinderscoutian (Collinson 2005) (Fig. 10f). The Millstone Grit Formation was eroded from the Mid North Sea High during late Pennsylvanian and Early Permian times (Hallsworth & Chisholm 2008).

North of the Forth Approaches in Quadrant 20, the long-established regime of coal mires and fluvial environments continued throughout this period (Fig. 10). After this time, either the younger Carboniferous sediments in the Outer Moray Firth Basin had been removed by erosion, or deposition in these basins ceased after the Kinderscoutian.

Sedimentological controls on the petroleum system

The Visean and Namurian basinal facies have been proved to be a working petroleum system onshore in the East Midlands and Cleveland basins (Pletsch *et al.* 2010). Our study has identified that the Cleveland Basin extends eastwards, at least as far as Quadrant 44, and that the organic-rich Upper Bowland Shales is present throughout (Monaghan *et al.* 2017). The lower–middle Carboniferous succession is >2 km thick in the south of Quadrants 41–44. The high gamma values for the Upper Bowland Shales offshore (Fig. 8) and source rock geochemistry data for the Cleveland Group suggest that these rocks have potential as source rocks, although they may be over-mature at deeper levels (Monaghan *et al.* 2017). Offshore, the Scremerston, Firth Coal and, to a lesser extent, Yoredale and Millstone Grit formations contain many coals, which could also act as source rocks, although this depends on their maturation history (Monaghan *et al.* 2017). Previously, shales and coals of the Scremerston Formation have been documented as a rich potential source rock and are dominantly gas-prone (Collinson *et al.* 1995). Mudstones in the Midland Valley equivalent of the Cementstone Formation, the Ballagan Formation, have low total organic carbon values and probably do not have significant source rock potential (Monaghan 2014).

The Fell Sandstone and Millstone Grit formations are dominated by fluvial sandstones up to 50 m thick, which, depending on their physical properties, could act as reservoirs within the Carboniferous succession (Monaghan *et al.* 2017). The Scremerston, Firth Coal and Yoredale formations contain channel or deltaic coarse sandstone bodies and the interbedded, extensive mudstones could provide intraformational seals. Multi-storey fluvial sandstone bodies with a ribbon-shaped geometry up to 30 m thick, up to several kilometres wide and tens of kilometres long are present in the Tyne Limestone Formation onshore (Jones 2007). The channel systems could provide pathways to allow hydrocarbons generated in the Cleveland Group to migrate up dip into the clastic succession (Monaghan *et al.* 2017).

North of the Forth Approaches Basin and in the Outer Moray Firth Basin, the identification of sand-rich facies within the Firth Coal Formation suggests that significant channels are present in this unit, which could act as both intraformational reservoirs and as pathways for hydrocarbons generated from the coals to migrate into younger strata.

Conclusions

Wells from the Outer Moray Firth Basin to the Silverpit Basin containing Carboniferous sedimentary rocks, just over half of which have available biostratigraphic reports, have been reinterpreted and integrated with seismic mapping. There is continuity between the sedimentary systems on both sides of the Mid North Sea High, suggesting that these systems were linked in the Carboniferous.

The extent and inception of a long-lived Carboniferous fluvio-deltaic system and laterally equivalent basinal, mud-rich successions have been documented using a regional well dataset interpreted in a series of time slices. Variations in the palaeogeography and facies architecture evolved in response to basin-forming tectonic events, sea-level changes and changes in sediment supply. Facies variations, which have resulted in complex nomenclatures, can be resolved into a stratigraphy utilizing newly released biostratigraphic data and a regional approach. Uncertainty has been reduced in the likely regional extent and character of coal, oil shale and marine mudstone-bearing source rock intervals (such as the Upper Bowland Shale, Scremerston and Firth Coal formations). Our understanding of the stratigraphic position and regional extent of potential reservoir intervals, in both stacked fluvial (Fell Sandstone and Millstone Grit formations) and channel (Yoredale, Scremerston and Firth Coal formations) systems, has been improved.

Acknowledgements This work was undertaken as a joint-industry project for the former Department of Energy and Climate Change (DECC) now Oil and Gas Authority (OGA), Oil and Gas UK, and 49 oil and gas company sponsors as part of the 21CXRM (21st Century Exploration Roadmap) Palaeozoic Project. Thanks to Richard Milton-Worsell for his enthusiastic guidance, and to the Technical Steering Committee of the project. We also gratefully acknowledge the provision of data CDA and UKOGL data libraries. We acknowledge LR Senergy for use of ODM under licence. CGG are thanked for the provision of LAS files for selected wireline logs. We thank Bernard Besly and Duncan McLean for their insightful reviews. This paper is published with the permission of the Executive Director, British Geological Survey (NERC).

Funding The project was funded by DECC/OGA, industry sponsors, BGS, and Oil and Gas UK. The work was completed independently by BGS with peer review by industry and OGA.

References

AITKENHEAD, N., BRIDGE, D.M., RILEY, N.J. & KIMBELL, S.F. 1992. *Geology of the Country Around Garstang*. British Geological Survey, Memoirs, Sheet 67 (England and Wales). HMSO, London.

AITKENHEAD, N., BARCLAY, W.J., BRANDON, A., CHADWICK, R.A., CHISHOLM, J.I., COOPER, A.H. & JOHNSON, E.W. 2002. *British Regional Geology: the Pennines and Adjacent Areas*. 4th edn. HMSO, London.

ANDERTON, R. 1985. Sedimentology of the Dinantian of Foulden, Berwickshire, Scotland. *Transactions of the Royal Society of Edinburgh, Earth Sciences*, **76**, 7–12.

ANDREWS, I.J. 2013. *The Carboniferous Bowland Shale Gas Study: Geology and Resource Estimation*. British Geological Survey, London.

ANDREWS, J.E., TURNER, M.S., NABI, G. & SPIRO, B. 1991. The anatomy of an early Dinantian terraced floodplain: palaeo-environment and early diagenesis. *Sedimentology*, **38**, 271–287.

ARSENIKOS, S., QUINN, M., JOHNSON, K., SANKEY, M. & MONAGHAN, A.A. 2015. *Seismic Interpretation and Generation of Key Depth Structure Surfaces within the Devonian and Carboniferous of the Central North Sea, Quadrants 25–44 Area*. British Geological Survey Commissioned Report **CR/15/118**, http://nora.nerc.ac.uk/516758/

ARSENIKOS, S., QUINN, M., JOHNSON, K., SANKEY, M. & MONAGHAN, A.A. 2016. *Seismic Interpretation and Generation of Key Depth Structure Surfaces within the Carboniferous and Devonian of the Orcadian Basin, Quadrants 7–9, 11–15 and 19–21 Area*. British Geological Survey Commissioned Report **CR/16/033**, http://nora.nerc.ac.uk/516771/

ARSENIKOS, S., QUINN, M., KIMBELL, G., WILLIAMSON, P., PHARAOH, T. & MONAGHAN, A. , 2018. Structural development of the Devono-Carboniferous plays of the UK North Sea. *In*: MONAGHAN, A.A., UNDERHILL, J.R., HEWETT, A.J. & MARSHALL, J.E.A. (eds) *Paleozoic Plays of NW Europe*. Geological Society, London, Special Publications, **471**, https://doi.org/10.1144/SP471.3

BAGNALL, J.M., CHURCH, J.W., HASKINS, C.W., KNOWLES, R.A. & NEVILLE, R.S.W. 1973. *The Micropalaeontology, Palynology and Stratigraphy of the Occidental 15/17-1a North Sea Well*. Robertson Research Report, downloaded from CDA, www.ukoilandgasdata.com [last accessed 15 January 2017].

BENNETT, C.E., KEARSEY, T.I., DAVIES, S.J., MILLWARD, D., CLACK, J.A., SMITHSON, T.R. & MARSHALL, J.E.A. 2016. Early Mississippian sandy siltstones preserve rare vertebrate fossils in seasonal flooding episodes. *Sedimentology*, **63**, 1677–1700.

BENNETT, C.E., HOWARD, A.S. ET AL. 2017. Ichnofauna record cryptic marine incursions onto a coastal floodplain at a key Lower Mississippian tetrapod site. *Palaeogeography, Palaeoclimatology, Palaeoecology*, **468**, 287–300.

BESLY, B. 2005. Late Carboniferous redbeds of the UK southern North Sea, viewed in a regional context. *In*: COLLINSON, J., EVANS, D., HOLLIDAY, D.W. & JONES, N. (eds) *Carboniferous Hydrocarbon Geology: the Southern North Sea and Surrounding Onshore Areas*. Yorkshire Geological Society, Occasional Publications, **7**, 225–226.

BRAND, P.J. 2011. The Serpukhovian and Bashkirian (Carboniferous, Namurian and basal Westphalian) faunas of northern England. *Proceedings of the Yorkshire Geological Society*, **58**, 143–165, https://doi.org/10.1144/pygs.58.3.283

BROWNE, M.A.E., DEAN, M.T., HALL, I.H.S., MCADAM, A.D., MUNRO, S.K. & CHISHOLM, J.I. 1999. *A Lithostratigraphic Framework for the Carboniferous Rocks of the Midland Valley of Scotland*. British Geological Survey Research Report **RR/99/07**.

BROWNE, M.A.E., MONAGHAN, A., RITCHIE, J.D., UNDERHILL, J.R., HOOPER, M.D., SMITH, R.A. & AKHURST, M.C. 2003. Revised tectono-stratigraphical framework for the Carboniferous of the Midland Valley of Scotland, UK. *In*: WONG, T.E. (ed.) *Proceedings of the XVth International Congress on Carboniferous and Permian Stratigraphy*. Utrecht, The Netherlands, Royal Netherlands Academy of Arts and Sciences, 10–16.

BRUCE, D. & STEMMERIK, L. 2003. Carboniferous. *In*: EVANS, D., GRAHAM, C., ARMOUR, A. & BATHURST, P. (eds) *The Millennium Atlas: Petroleum Geology of the Central and Northern North Sea*. Geological Society, London, 7–1, 7-11.

CAMERON, T. 1993*a*. Carboniferous and Devonian of the southern North Sea. *In*: KNOX, R. & CORDEY, W. (eds) *Lithostratigraphical Nomenclature of the UK North Sea*. British Geological Survey, Keyworth.

CAMERON, T. 1993*b*. Triassic, Permian and Pre-Permian (Central and Northern North Sea). *In*: KNOX, R. & CORDEY, W. (eds) *Lithostratigraphical Nomenclature of the UK North Sea*. British Geological Survey, Keyworth.

CATER, M.C., CHURCH, J.W., GREENWOOD, R.J. & HASKINS, C.W. 1967. *The Micropalaeontology and Stratigraphy of the Amoseas 38/22-1 North Sea Well*. Robertson Research, Report Confidential to DECC/OGA, www.ukoilandgasdata.com [last accessed 15 January 2017].

CHADWICK, R.A., HOLLIDAY, D.W., HOLLOWAY, S. & HULBERT, A.G. 1995. *The Structure and Evolution of the Northumberland–Solway Basin and Adjacent Areas*. British Geological Survey, HMSO, London, Subsurface Geology Memoirs, DY005.

CHURCH, K.D. & GAWTHORPE, R.L. 1994. High resolution sequence stratigraphy of the late Namurian in the Widmerpool Gulf (East Midlands, UK). *Marine and Petroleum Geology*, **11**, 528–544.

CLIFF, R.A., DREWERY, S.E. & LEEDER, M.R. 1991. Sourcelands for the Carboniferous Pennine river system: constraints from sedimentary evidence and U–Pb geochronology using zircon and monazite. *In*: MORTON, A.C., TODD, S.P. & HAUGHTON, P.D.W. (eds) *Developments in Sedimentary Provenance Studies*. Geological Society, London, Special Publications, **57**, 137–159, https://doi.org/10.1144/GSL.SP.1991.057.01.12

COLLINSON, J.D. 1988. Controls on Namurian sedimentation in the Central Province basins of northern England. *In*: BESLY, B.M. & KELLING, G. (eds) *Sedimentation in a*

Synorogenic Basin Complex: the Upper Carboniferous of Northwest Europe. Blackie, London, 86–101.

Collinson, J. 2005. Dinantian and Namurian depositional systems in the southern North Sea. *In*: Collinson, J., Evans, D., Holliday, D. & Jones, N. (eds) *Carboniferous Hydrocarbon Geology: the Southern North Sea and Surrounding Onshore Areas*. Yorkshire Geological Society, Occasional Publications, **7**, 35–56.

Collinson, J.D. & Jones, C. & IGI Ltd 1995. *Source Rock Potential of the Sub-Westphalian Carboniferous of the Southern North Sea*. Report and Appendix [PDF available from British Geological Survey].

Coward, M.P. 1993. The effect of Late Caledonian and Variscan continental escape tectonics on basement structure, Paleozoic basin kinematics and subsequent Mesozoic basin development in NW Europe. *In*: Parker, J.R. (ed.) *Petroleum Geology of Northwest Europe: Proceedings of the 4th Conference*. Geological Society, London, 1095–1108, https://doi.org/10.1144/0041095

Davies, S.J. 2008. The record of Carboniferous sea-level change in low-latitude sedimentary successions from Britain and Ireland during the onset of the late Paleozoic ice age. *In*: Fielding, C.R., Frank, T.D. & Isbell, J.L. (eds) *Resolving the Late Paleozoic Ice Age in Time and Space*. Geological Society of America, Special Papers, **441**, 187–204.

Davydov, V., Wardlaw, B.R. & Gradstein, F.M. 2004. The Carboniferous period. *In*: Gradstein, F.M., Ogg, J.G. & Smith, A.G. (eds) *A Geological Time Scale 2004*. Cambridge University Press, London, 603–651.

Day, J.B.W. 1970. *Geology of the Country Around Bewcastle*. Explanation of One-Inch Geological Sheet 12, New Series. British Geological Survey, Memoirs. HMSO, London.

Dean, M., Browne, M., Waters, C. & Powell, J. 2011. *A Lithostratigraphical Framework for the Carboniferous Successions of Northern Great Britain (Onshore)*. British Geological Survey Research Report **RR/10/07**.

Gilligan, A. 1919. The petrography of the Millstone Grit of Yorkshire. *Quarterly Journal of the Geological Society, London*, **75**, 251–294, https://doi.org/10.1144/GSL.JGS.1919.075.01-04.23

Hallsworth, C.R. & Chisholm, J.I. 2008. Provenance of late Carboniferous sandstones in the Pennine Basin (UK) from combined heavy mineral, garnet geochemistry and palaeocurrent studies. *Sedimentary Geology*, **203**, 196–212.

Hampson, G.J., Elliott, T. & Flint, S.S. 1996. Critical application of high resolution sequence stratigraphic concepts to the Rough Rock Group (Upper Carboniferous) of northern England. *In*: Howell, J.A. & Aitken, J.F. (eds) *High Resolution Sequence Stratigraphy: Innovations and Applications*. Geological Society, London, Special Publications, **104**, 221–246, https://doi.org/10.1144/GSL.SP.1996.104.01.14

Harris, C.R., Marshall, K.L., Matheson, F.E., McNaughan, S.Y. & Rich, B. 1998. *Well 20/15-2, British North Sea, Biostratigraphy of the Intervals*. Robertson Research, report downloaded from CDA, www.ukoilandgasdata.com [last accessed 15 January 2017].

Johnson, G.A.L., Somerville, I.D., Tucker, M.E. & Cozar, P. 2011. Carboniferous stratigraphy and context of the Seal Sands No. 1 Borehole, Teesmouth, NE England: the deepest onshore borehole in Great Britain. *Proceedings of the Yorkshire Geological Society*, **58**, 173–196, https://doi.org/10.1144/pygs.58.3.231

Jones, N.S. 2007. *The Scremerston Formation: Results of a Sedimentological Study of Onshore Outcrop Sections and Offshore Well 42/13-2*. British Geological Survey Commissioned Report **CR/07/101**, http://nora.nerc.ac.uk/512724/

Kearsey, T., Ellen, R., Millward, D. & Monaghan, A.A. 2015. *Devonian and Carboniferous Stratigraphical Correlation and Interpretation in the Central North Sea, Quadrants 25–44*. British Geological Survey Commissioned Report **CR/15/117**, http://nora.nerc.ac.uk/516755/

Kearsey, T.I., Bennett, C.E. *et al.* 2016. The terrestrial landscapes of tetrapod evolution in earliest Carboniferous seasonal wetlands of SE Scotland. *Palaeogeography, Palaeoclimatology, Palaeoecology*, **457**, 52–69.

Kombrink, H., Besly, B. *et al.* 2010. Carboniferous. *In*: Doornenbal, J.C. & Stevenson, A.G. (eds) *Petroleum Geological Atlas of the Southern Permian Basin Area*. EAGE, Houten, 81–99.

Lancaster, P.J., Daly, J.S., Storey, C.D. & Morton, A.C. 2017. Interrogating the provenance of large river systems: multi-proxy in situ analyses in the Millstone Grit, Yorkshire. *Journal of the Geological Society*, **174**, 75–87, https://doi.org/10.1144/jgs2016-069

Leeder, M.R. 1974. Lower Border Group (Tournaisian) fluvio-deltaic sedimentation and palaeogeography of the Northumberland Basin. *Proceedings of the Yorkshire Geological Society*, **40**, 129–180, https://doi.org/10.1144/pygs.40.2.129

Leeder, M.R. & Boldy, S.A.R. 1990. The Carboniferous of the Outer Moray Firth Basin, quadrants 14 and 15, Central North Sea. *Marine and Petroleum Geology*, **7**, 29–37.

Leeder, M.R. & Stewart, M.D. 1996. Fluvial incision and sequence stratigraphy: alluvial responses to relative sea-level fall and their detection in the geological record. *In*: Hesselbo, S.P. & Parkinson, D.N. (eds) *Sequence Stratigraphy in British Geology*. Geological Society, London, Special Publications, **103**, 25–39, https://doi.org/10.1144/GSL.SP.1996.103.01.03

Leeder, M.R., Fairhead, D., Lee, A., Stuart, G., Clemmey, H., Green, C. & Al-Haddeh, B. 1989. Sedimentary and tectonic evolution of the Northumberland Basin. *In*: Arthurton, R.S., Gutteridge, P. & Nolan, S.C. (eds) *The Role of Tectonics in Devonian and Carboniferous Sedimentation in the British Isles*. Occasional Publication No. 6. Yorkshire Geological Society, **47**.

Mahdi, S.A. 1992. *Shell/Esso 41/1-1 British North Sea Well Palynological Analysis of the Interval 2400′ – 7050′ (SWC)*. Simon-Robertson, Report Confidential to DECC/OGA, www.ukoilandgasdata.com [last accessed 15 January 2017].

Maynard, J.R. & Dunay, R.E. 1999. Reservoirs of the Dinantian (Lower Carboniferous) play of the southern North Sea. *In*: Fleet, A.J. & Boldy, S.A.R. (eds) *Petroleum Geology of Northwest Europe: Proceedings of the 5th Conference*. Geological Society, London, 729–745, https://doi.org/10.1144/0050729

McAdam, A.D., Tulloch, W., Elliott, R.W., Floyd, J.D., Wilson, R.B. & Graham, D.K. 1985. *Geology of the*

Haddington District. British Geological Survey, Memoirs, 1:50 000 Sheet 33W and Part of Sheet 41. HMSO, London.

McLean, D. 2013. *Palynostratigraphy of the Carboniferous Interval in Southern North Sea Well 37/10-1*. MB Stratigraphy, Report No. 83a, Report Confidential to DECC/OGA, www.ukoilandgasdata.com [last accessed 15 January 2017].

McLean, D. & Neves, R. 1988. *Palynostratigraphical Review of the Carboniferous Interval in Well 26/7-1*. Arco British, Report Confidential to DECC/OGA, www.ukoilandgasdata.com [last accessed 15 January 2017].

Milton-Worssell, R., Smith, K., McGrandle, A., Watson, J. & Cameron, D. 2010. The search for a Carboniferous petroleum system beneath the Central North Sea. *In*: Vining, B.A. & Pickering, S.C. (eds) *Petroleum Geology: From Mature Basins to New Frontiers – Proceedings of the 7th Petroleum Geology Conference*. Geological Society, London, 57–75, https://doi.org/10.1144/0070057

Monaghan, A.A. 2014. *The Carboniferous Shales of the Midland Valley of Scotland: Geology and Resource Estimation*. British Geological Survey, Keyworth, www.gov.uk/government/publications/bgs-midland-valley-of-scotland-shale-reports

Monaghan, A.A. & Parrish, R.R. 2006. Geochronology of Carboniferous–Permian volcanism in the Midland Valley of Scotland: implications for regional tectonomagmatic evolution and the numerical timescale. *Journal of the Geological Society of London*, **163**, 15–28, https://doi.org/10.1144/0016-764904-142

Monaghan, A., Arsenikos, S. *et al.* 2017. Carboniferous petroleum systems around the Mid North Sea High, UK. *Marine and Petroleum Geology*, **88**, 282–302.

Morton, A.C., Claoué-Long, J.C. & Hallsworth, C.R. 2001. Zircon age and heavy mineral constraints on provenance of North Sea Carboniferous sandstones. *Marine and Petroleum Geology*, **18**, 319–337.

Pletsch, T., Appel, J. *et al.* 2010. Petroleum generation and migration. *In*: Doornenbal, J.C. & Stevenson, A.G. (eds) *Petroleum Geological Atlas of the Southern Permian Basin Area*. EAGE, Houten, 225–253.

Read, W.A., Browne, M.A.E., Stephenson, D. & Upton, B.G.J. 2002. Carboniferous. *In*: Trewin, N.H. (ed.) *The Geology of Scotland*. 4th edn. Geological Society, London, 251–299, https://doi.org/10.1144/GOS4P.9

Ridd, M.F., Walker, D.B. & Jones, J.M. 1970. A deep borehole at Harton on the margin of the Northumbrian Trough. *Proceedings of the Yorkshire Geological Society*, **38**, 75–103, https://doi.org/10.1144/pygs.38.1.75

Robson, D.A. 1956. A sedimentary study of the Fell Sandstone of the Coquet Valley, Northumberland. *Quarterly Journal of the Geological Society*, **112**, 241–262, https://doi.org/10.1144/GSL.JGS.1956.112.01-04.12

Rodriguez, K., Wrigley, R., Hodgson, N. & Nicholls, H. 2014. Southern North Sea: unexplored multi-level exploration potential revealed. *First Break*, **32**, 107–113.

Scott, W.B. 1986. Nodular carbonates in the Lower Carboniferous Cementstone Group of the Tweed Embayment, Berwickshire: evidence for a former sulphate evaporite facies. *Scottish Journal of Geology*, **22**, 325–345, https://doi.org/10.1144/sjg22030325

Smithson, T.R., Wood, S.P., Marshall, J.E.A. & Clack, J.A. 2012. Earliest Carboniferous tetrapod and arthropod faunas from Scotland populate Romer's Gap. *Proceedings of the National Academy of Sciences of the USA*, **109**, 4532–4537.

Stephenson, M.H., Williams, M. *et al.* 2004. Palynomorph and ostracod biostratigraphy of the Ballagan Formation, Midland Valley of Scotland and elucidation of intra-Dinantian unconformities. *Proceedings of the Yorkshire Geological Society*, **55**, 131–143, https://doi.org/10.1144/pygs.55.2.131

Stephenson, M.H., Angiolini, L., Cózar, P., Jadoul, F., Leng, M.J., Millward, D. & Chenery, S. 2010. Northern England Serpukhovian (early Namurian) farfield responses to southern hemisphere glaciation. *Journal of the Geological Society*, **167**, 1171–1184, https://doi.org/10.1144/0016-76492010-048

Stone, P., Millward, D., Young, B., Merritt, J.W., Clarke, S.M., McCormac, M. & Lawrence, D.J.D. 2010. *British Regional Geology: Northern England*. 5th edn. British Geological Survey, Keyworth.

Symonds, R., Lippman, R., Mueller, B. & Kohok, A. 2015. Yoredale Sandstone architecture in the Breagh Field (UK SNS). Paper presented at the Geological Society Conference Sedimentology of Paralic Reservoirs: Recent Advances, London, 18–19 May 2015.

Trewin, N.H. & Thirlwall, M.F. 2002. Old Red Sandstone. *In*: Trewin, N.H. (ed.) *The Geology of Scotland*. Geological Society, London, 213–249, https://doi.org/10.1144/GOS4P.8

Turner, B.R. & Monro, M. 1987. Channel formation and migration by mass-flow processes in the Lower Carboniferous fluviatile Fell Sandstone Group, northeast England. *Sedimentology*, **34**, 1107–1122.

Turner, B.R., Dewey, C. & Fordham, C.E. 1997. Marine ostracods in the Lower Carboniferous fluviatile Fell Sandstone Group: evidence for base level change and marine flooding of the central graben, Northumberland Basin. *Proceedings of the Yorkshire Geological Society*, **51**, 297–306, https://doi.org/10.1144/pygs.51.4.297

Underhill, J.R. 2003. The tectonic and stratigraphic framework of the United Kingdom's oil and gas fields. *In*: Gluyas, J.G. & Hichens, H.M. (eds) *United Kingdom Oil and Gas Fields, Commemorative Millennium Volume*. Geological Society, London, Memoirs, **20**, 17–59, https://doi.org/10.1144/GSL.MEM.2003.020.01.04

Van Adrichem Boogaert, H.A. & Kouwe, W.F.P. 1993–1997a. Tayport Formation ORTP. Stratigraphic Nomenclature of the Netherlands, www.dinoloket.nl/tayport-formation-ortp [last accessed 15 January 2017].

Van Adrichem Boogaert, H.A. & Kouwe, W.F.P. 1993–1997b. Cementstone Formation ORTP. Stratigraphic Nomenclature of the Netherlands, www.dinoloket.nl/node/5203 [last accessed 15 January 2017].

Van Adrichem Boogaert, H.A. & Kouwe, W.F.P. 1993–1997c. Epen Formation DCGE. Stratigraphic Nomenclature of the Netherlands, www.dinoloket.nl/epen-formation-dcge [last accessed 15 January 2017].

Wakefield, O., Waters, C.N. & Smith, N.J.P. 2016. *Carboniferous Stratigraphical Correlation and Interpretation in the Irish Sea*. British Geological Survey Commissioned Report **CR/16/040**, http://nora.nerc.ac.uk/516783/

Ward, J. 1997. Early Dinantian evaporites of the Easton-1 well, Solway basin, onshore, Cumbria, England. *In*: Meadows, N.S., Trueblood, S.R., Hardman, M. & Cowan, G. (eds) *Petroleum Geology of the Irish Sea and Adjacent Areas*. Geological Society, London, Special Publications, **124**, 277–296, https://doi.org/10.1144/GSL.SP.1997.124.01.17

Waters, C.N. & Davies, S.J. 2006. Carboniferous: extensional basins, advancing deltas and coal swamps. *In*: Brenchley, P.J. & Rawson, P.F. (eds) *The Geology of England and Wales*. Geological Society, London, 173–223, https://doi.org/10.1144/GOEWP.9

Waters, C.N., Browne, M., Dean, M. & Powell, J. 2007. *Lithostratigraphical Framework for Carboniferous Successions of Great Britain (Onshore)*. British Geological Survey, Research Report **RR/07/001**.

Waters, C.N., Somerville, I.D. *et al.* (eds) 2011. *A Revised Correlation of Carboniferous Rocks in the British Isles*. Geological Society, London, Special Report, **26**, 82–88.

Waters, C.N., Millward, D. & Thomas, C.W. 2014. The Millstone Grit Group (Pennsylvanian) of the Northumberland-Solway Basin and Alston Block of northern England. *Proceedings of the Yorkshire Geological Society*, **60**, 29–51, https://doi.org/10.1144/pygs2014-341

Welsh, A. & Walton, W. 1985. *Palaeontological Summary for Well 26/7-1*. Report downloaded from CDA, www.ukoilandgasdata.com [last accessed 15 January 2017].

Whitbread, K. & Kearsey, T. 2016. *Devonian and Carboniferous Stratigraphical Correlation and Interpretation in the Orcadian Area, Central North Sea, Quadrants 7–22*. British Geological Survey Commissioned Report **CR/16/032**, http://nora.nerc.ac.uk/516770/

Wood, I. 2014. *UKCS Maximising Recovery Review: Final Report*. Report for Department of Energy & Climate Change Department of Energy & Climate Change, London, https://assets.publishing.service.gov.uk/government/uploads/system/uploads/attachment_data/file/471452/UKCS_Maximising_Recovery_Review_FINAL_72pp_locked.pdf [last accessed March 2018].

Wright, V.P. & Vanstone, S.D. 2001. Onset of Late Palaeozoic glacio-eustasy and the evolving climates of low latitude areas: a synthesis of current understanding. *Journal of the Geological Society*, **158**, 579–582, https://doi.org/10.1144/jgs.158.4.579

Structural development of the northern Dutch offshore: Paleozoic to present

M. M. TER BORGH*, B. JAARSMA & E. A. ROSENDAAL

EBN BV, Daalsesingel 1, 3511 SV, Utrecht, The Netherlands

M.t.B., 0000-0002-1698-2411

**Correspondence: marten.borgh-ter@ebn.nl*

Abstract: We used new high-quality 2D and 3D seismic data covering the northern Dutch offshore to develop a structural framework for Paleozoic–recent times. Early Carboniferous extension was accommodated predominantly along WNW–ESE-trending faults, and was characterized by an alternation of highs and lows; in the northern Dutch offshore, the principal high coincides with the present-day Elbow Spit Platform. An Early Carboniferous low, the North Elbow Basin, is present north of this high in the A and B quadrants. Lower Carboniferous deposits have been preserved here and on the Elbow Spit Platform. Late Carboniferous–Early Rotliegend deformation is accommodated primarily along normal faults with a NE–SW trend. These faults commonly show no significant offset of Upper Rotliegend and younger units. Development of the Dutch Central Graben and Step Graben occurred during the Triassic–Early Cretaceous, primarily along north–south-trending faults, but reactivation of pre-existing faults as oblique-slip faults occurred as well. Associated with these north–south-trending faults was another, not previously described, family of WSW–ENE-trending dextral strike-slip faults which are proposed to represent transfer faults that accommodated strain partitioning.

The North Sea is a mature petroleum province where exploration has historically focused on two main fairways: the Southern North Sea area (Fig. 1), where production is mainly from Permian, Carboniferous and Triassic reservoirs; and the Northern and Central North Sea (Fig. 1), where most production is from Mesozoic to Paleogene deposits that are associated with the Central Graben system (e.g. Evans *et al.* 2003; de Jager & Geluk 2007; Doornenbal & Stevenson 2010).

Until recently, the hydrocarbon potential of the area situated in-between these two fairways, the Mid North Sea area (Fig. 1), was generally considered to be low. One reason was the supposed absence of reservoir rocks; the reservoir presence of Permian Rotliegend rocks was suspected but unproven this far north (e.g. Geluk 2007). Another reason was the perceived absence of a mature source rock, as the Upper Carboniferous, from which most gas in the Southern North Sea is commonly thought to be sourced, is mostly absent (e.g. van Buggenum & den Hartog Jager 2007).

However, recent discoveries suggest that alternative reservoir and source rocks are present. The Lower Carboniferous is an example. It was not ruled out that this stratigraphic interval had reservoir potential, but its economic potential was generally considered unproven (e.g. de Jager & Geluk 2007). This changed following the discovery of the 20 Bcm (billion cubic metres) Lower Carboniferous Breagh Field in the UK offshore (McPhee *et al.* 2008) (Fig. 1). Further assessment of the potential of the Lower Carboniferous requires a good understanding of the structural development of the area, as deposition during the Early Carboniferous was strongly fault controlled. Interestingly, not much is known yet about the structural development of the Mid North Sea area during this period.

Another recent game-changing discovery north of the main Southern North Sea fairway is the UK Cygnus Field (Taggart & Catto 2015) (Fig. 1), proving that relatively coarse Rotliegend deposits can be found this far north. Moreover, Carboniferous deposits were encountered that showed better reservoir quality than expected. Apparently, a local high provided a sediment source. Predicting if similar coarse siliciclastic sediments are present in the Dutch sector requires a better understanding of the structural evolution during this period.

A question closely related to this issue is the onset of deformation in the Dutch Central Graben (DCG) and Step Graben (SG) (Fig. 1). The main phase of activity of these structures was during the Triassic and Jurassic, but activity as early as the Devonian has been suggested (e.g. Ziegler 1990). Understanding when these graben were active is a key constraint for depositional models, for modelling source-rock maturation and for predicting reservoir quality. It is also relevant for the Triassic and Jurassic plays in the Mid North Sea area; reservoir-quality rocks have been encountered for both intervals (e.g. de Jager & Geluk 2007; Kortekaas *et al.* 2016). To understand why, and to further predict reservoir presence, a better understanding of the structural

From: Monaghan, A. A., Underhill, J. R., Hewett, A. J. & Marshall, J. E. A. (eds) 2018. *Paleozoic Plays of NW Europe*. Geological Society, London, Special Publications, **471**, 115–131.
First published online March 20, 2018, https://doi.org/10.1144/SP471.4

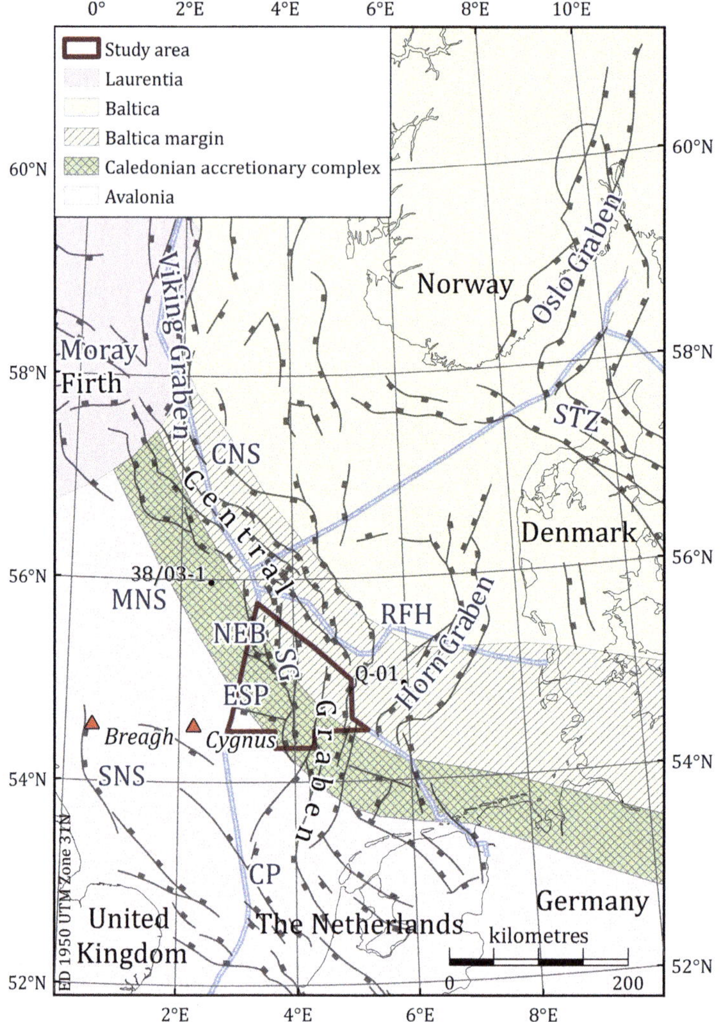

Fig. 1. Regional overview map showing the plate tectonic setting (after Smit *et al.* 2016) and large-scale structures (Ziegler 1990; Wride 1995; McCann *et al.* 2006; de Jager 2007; and this study). CNS, Central North Sea; CP, Cleaver Bank Platform; ESP, Elbow Spit Platform; MNS, Mid North Sea area; NEB, North Elbow Basin; SG, Step Graben; SNS, Southern North Sea; STZ, Sorgenfrei-Tornquist Zone; RFH, Ringkøbing-Fyn High.

evolution is required. This would also aid in establishing migration routes from the Jurassic Posidonia Formation source rock, which is known to be present and mature in the DCG (de Jager & Geluk 2007), to overlying (Jurassic, Chalk) and juxtaposed deposits. Finally, a better understanding of the structural evolution of the area will help in identifying trapping geometries, fault seal and reservoir compartmentalization.

In summary, an improved structural framework for the Mid North Sea area will be of significant benefit for assessing the economic potential of the area. This study will focus on the Dutch part of the Mid North Sea area: the northern part of the Dutch offshore.

Geological outline

The study area is located on the boundary between different types of basement (Fig. 1). The basement below most of the Dutch offshore is commonly considered to be part of the Avalonia microcontinent, which is presently wedged between two larger palaeocontinents: Baltica to the north and Gondwana to the south (e.g. Pharaoh 1999). Avalonia diverged from Gondwana during rifting of probable Early Ordovician age, forming the Rheic Ocean. Avalonia subsequently moved north and was accreted to the Baltica and Laurentia continents (Fig. 1) in a three-way collision during the Caledonian Orogeny in Late Ordovician–earliest Silurian times (Cocks *et al.* 1997). The study area was proposed to coincide with a separate crustal unit consisting of a collapsed Caledonian accretionary complex located between Baltica and Avalonia (Fig. 1) (Smit *et al.* 2016).

During the Devonian, closure of the Rheic Ocean was initiated, causing Gondwana to follow Avalonia (Ziegler 1990). During convergence, significant extension occurred in the Southern North Sea area and a basin formed. The mechanism responsible for this extension is subject to debate: either back-arc extension (e.g. Leeder 1988), escape tectonics (e.g. Maynard *et al.* 1997) or a combination of both (e.g. Coward 1993). Either way, the evolution of the basin was strongly linked to plate margin processes; convergence occurred during the Late Devonian–Carboniferous, resulting in the Variscan Orogeny and the formation of the Pangaea supercontinent (Fig. 2). During convergence and collision, compressional and extensional pulses alternated, which are reflected in the subsidence history of the Southern North Sea area (Ziegler 1990). Subsidence in the basin was predominantly fault-controlled during the Devonian and Visean (Early Carboniferous), with thermal subsidence gaining importance from the Namurian onwards. During this period, subsidence was larger than sedimentation in large parts of the basin, resulting in significant water depths. As the collision progressed, compressional forces finally overcame extensional forces and inversion of the study area occurred during the Late Carboniferous (Fig. 2) (Ziegler 1990). Sediment influx from the rising Variscan mountains increased during this period and the basin was filled by the end of the Carboniferous (e.g. van Buggenum & den Hartog Jager 2007). The inversion intensity decreased towards the north; in the Southern North Sea area, north of the Variscan thrust front, the inversion resulted in the reactivation of Late Devonian–Early Carboniferous normal faults as reverse faults, and long-wavelength folding (e.g. Schroot & de Haan 2003).

Inversion had largely ceased during the Carboniferous–Permian transition. What followed was a period with extensive magmatism and crustal thinning (van Wees *et al.* 2000; Neumann *et al.* 2004). South of the study area, the magmatism was associated with strike-slip deformation along NW-trending faults (Ziegler 1990). North of the study area, rifting led to the formation of the Oslo Graben (e.g. Neumann *et al.* 2004). During the Autunian (Early Permian), the magmatism caused thermal uplift of the Southern North Sea area and the Dutch–German–

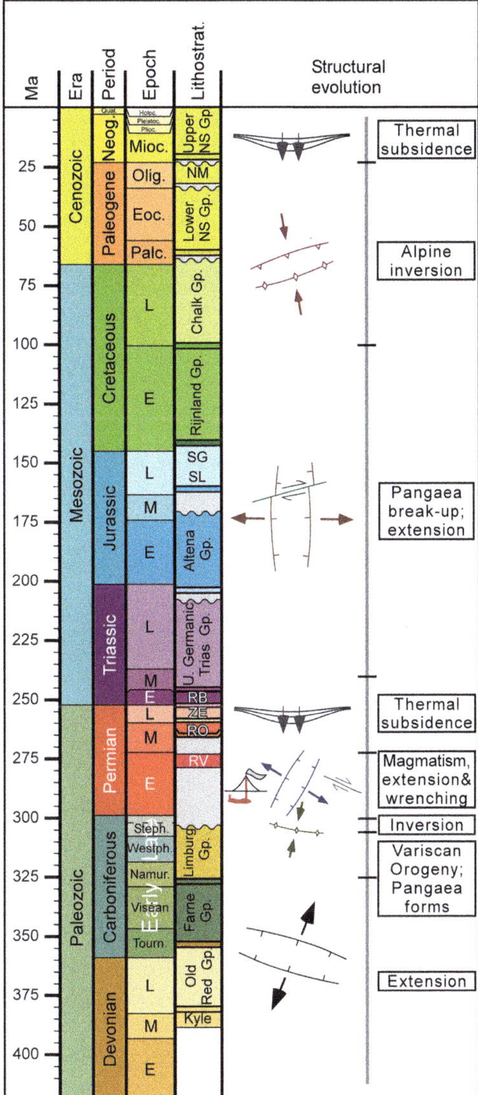

Fig. 2. Stratigraphic/event chart. NM, Middle North Sea Group; SG, Scruff Group; SL, Schieland Group; RB, Lower Germanic Trias Group; ZE, Zechstein Group; RO, Upper Rotliegend Group; RV, Lower Rotliegend Group.

Polish lowlands (Ziegler 1990; van Wees *et al.* 2000), resulting in widespread erosion and the development of the large regional Base Permian Unconformity (BPU). As a result of thermal subsidence, widespread sedimentation started again with the deposition of the Upper Rotliegend Group in the middle Permian (Fig. 2). Early Permian rocks have a limited areal extent and the early Permian is, thus, poorly preserved in the sedimentary record; in most of the

Southern North Sea area, there is a 40–60 myr hiatus between Carboniferous and Permian rocks (Fig. 2) (Geluk 2007). The remainder of the Permian is commonly considered a period of relative tectonic quiescence, with long-wavelength thermal subsidence accentuated by minor tectonic pulses (Geluk 2007). During the Late Permian, restricted marine conditions were established, resulting in the widespread deposition of Zechstein evaporites and carbonates (Fig. 2).

During the Triassic, Pangaea started breaking up (Fig. 2), which led to the formation of the Central Graben system in the North Sea area. The fault system essentially consists of three rift arms (Fig. 1): the Central Graben in the south, the Viking Graben in the north and the Moray Firth Basin in the west. Extension in the Central Graben was orientated east–west to NE–SW (Zanella & Coward 2003), and faults in the DCG and SG have a north–south trend (Fig. 1). Extension reached the Southern North Sea in the Middle Triassic (de Jager 2007). During the Middle Jurassic, uplift centred at the triple junction of the three rift segments led to the formation of the Central North Sea Rift Dome (Ziegler 1990; Underhill & Partington 1993). As a result, widespread erosion occurred which extended into the study area, and sediments were preserved in the graben only. In the Late Jurassic, the region started subsiding again and sedimentation outside the graben resumed. Late in the Early Cretaceous, continental break-up of Pangaea was reached in the North Atlantic. As a result, extension in the Central Graben system died out, leaving the system as a failed rift (Ziegler 1990). From the Late Cretaceous onwards, subsidence resulted predominantly from thermal subsidence. The development of the Alpine orogenic system led to a number of inversion events during the Late Cretaceous and Paleogene (de Jager 2007). The Neogene and Quaternary were periods of relative tectonic quiescence characterized by thermal subsidence (de Jager 2007).

The easternmost part of the study area is dominated by the Triassic–Jurassic DCG (Figs 3 & 4). West of the graben, we find from north to south (Fig. 1): (1) the SG, a structure that formed during the same period as the DCG, but where significantly less extension was accommodated; (2) the Elbow Spit Platform (ESP), a structurally high area where Lower Carboniferous and Devonian deposits are present at relatively shallow depths; and (3) a relative low situated south of the ESP. The topmost part of the ESP is the Elbow Spit High; here, Upper Cretaceous deposits can be found directly on top of the Devonian. The term Elbow Spit High was previously used for the entire ESP (Kombrink *et al.* 2012). The ESP is commonly considered to be an extension of the Mid North Sea High which formed a high from Permian times onwards, and extends into the UK, German, Danish and Norwegian sectors.

Methods

The dataset used to study fault trends and fault activity consists of 2D and 3D seismic data from a variety of sources. The southern part of the area is covered by the DEF survey, a 7950 km^2 3D multiclient survey (Fig. 4, inset) that was acquired in 2011 and 2012. The streamer length for this survey was 7.5 km, allowing for good imaging below the Upper Permian Zechstein salt. The northern part of the study area is partly covered by 3D surveys acquired in the period 1994–2006 that have since been made public. Imaging of the pre-salt units is commonly fair to poor on these surveys. 2D seismic data were used where 3D seismic data were unavailable or where the imaging of the 2D seismic data was better. The entire study area is covered by the North Sea Renaissance (NSR) 2D multiclient survey, which consists of a grid of regional seismic lines shot in NW–SE and NE–SW directions, with a spacing between lines of 5.3 km (Fig. 4, inset). The survey was acquired between 2005 and 2010, and the Dutch part of the survey has since become public. This survey was also acquired with long streamers and imaging of the pre-salt is good.

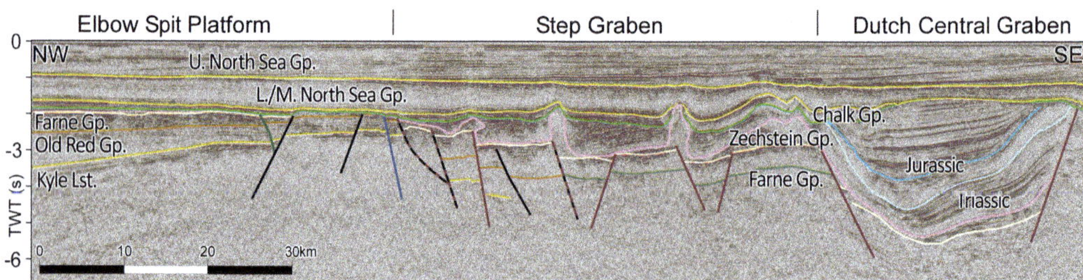

Fig. 3. Seismic section showing the three main structural elements in the study area: Elbow Spit Platform, Step Graben and Dutch Central Graben. The legend to the horizons is given in Figure 2; the location is shown in Figure 4. Public seismic line NSR 1061.

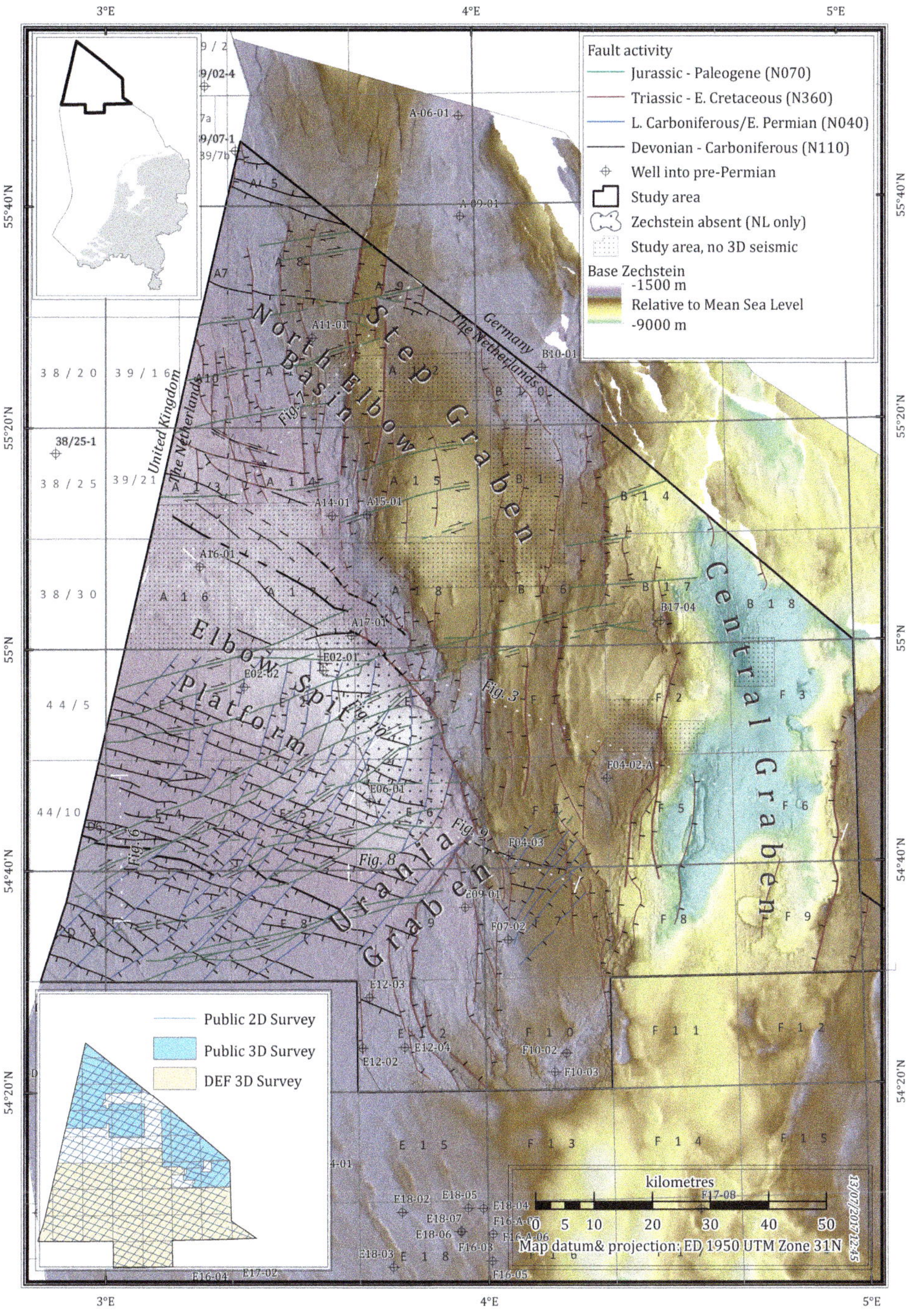

Fig. 4. Structural framework for the northern Dutch offshore. Faults are shown at the Base Permian level or, in case of older fault activity, at the topmost affected horizon. Along with the faults, the depth to the base of the Zechstein Group is shown; where the Zechstein is absent, the base of the first younger unit is shown. The Base Zechstein surface outside the study area is after Kombrink *et al.* (2012) and Arfai *et al.* (2014). Faults are coloured according to their activity; reactivated faults are shown with dashed lines. Apparent fault density of the N110° and N040° faults is high on the ESP where seismic imaging is good, and decreases in the DCG and SG. This is thought to be related to imaging, not an actual decrease in fault density.

For horizon interpretation, well data from over 60 wells were used. The wells were tied to the seismic data using check shots. All public seismic surveys and well data were requested from NLOG (http://www.nlog.nl/), the Dutch repository for deep subsurface data. Seismic horizons were interpreted in the time domain and converted to depth using a layer-cake velocity model. Between the base of the Chalk Group and the Base Permian Unconformity (BPU), the Velmod-2 velocity model (van Dalfsen *et al.* 2006, 2007) was used. For the Lower–Upper North Sea groups and the Chalk Group pseudo-seismic picks of 82 wells within the study area were used to construct a V_0 map, which was possible because the top and base of the Chalk Group are distinct seismic reflections in the study area. *K*-constants from Velmod-2 were used. From the BPU downwards, insufficient data points were available in the study area to construct a V_0 map; all wells with check shot or VSP data in the study were investigated and it was concluded that this interval is best modelled using a constant interval velocity of 4200 m s^{-1}.

Fault trends were mapped using coherency cubes in Petrel; structural smoothing was applied to the 3D seismic cubes first, and the variance was calculated subsequently. Mapping of fault trends was done on time slices, and on levels at fixed times below key horizons; these horizons, most prominently the BPU, were interpreted first, and coherency values were subsequently extracted from the coherency cube at this horizon and at fixed time below it. Based on the resulting time slices and horizon extractions, fault trends were mapped, focusing on faults that caused offset at the BPU level. In an iterative process, interpretations were checked and updated using further seismic interpretation. Fault offset and the timing of fault activity were constrained using the interpreted horizons.

Results

The coherency cubes reveal a complex fault pattern (Fig. 5); at least four fault trends can be mapped (Fig. 4): normal faults with a N110°-(WNW–ESE) trend (displayed in black in all figures), normal faults with a N040°-(NE–SW) trend (displayed in blue), normal faults with a N360°-(north–south) trend (brown) and strike-slip faults with a N070°-(WSW–ENE) trend (green).

N110° faults

Faults with a WNW–ESE trend are well represented on the ESP in units subcropping the BPU (Fig. 4; displayed in black). These faults limit the platform on its southern side (Fig. 6) and the offset along the faults is commonly normal, varying in magnitude from the resolution of the seismic data (approximately 30 m) to hundreds of metres. Recent long-offset 2D seismic data show for the first time that these faults also limit the ESP at its northern boundary (Fig. 7); Upper Devonian and Lower Carboniferous units thicken and have been downfaulted. Most faults offset the Lower Carboniferous and older units only; the base of the Permian is usually not affected. Where the BPU is affected, the offsets are smaller than the offset within the Carboniferous.

When the offset within the Carboniferous is of the order of hundreds of metres, the thickening of the Lower Carboniferous is accompanied by a

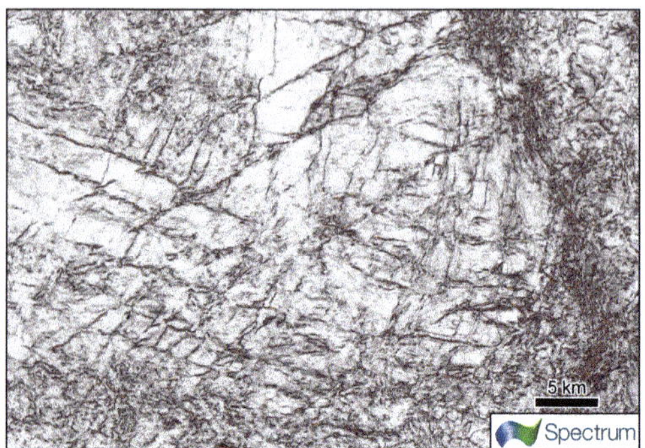

Fig. 5. Coherency cube extraction at 200 ms TWT below the Base Permian Unconformity, showing the main fault trends observable on the Elbow Spit Platform. DEF 3D seismic data courtesy of Spectrum ASA.

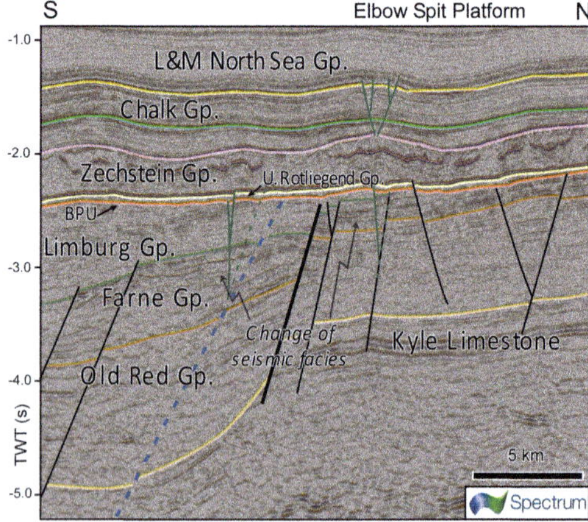

Fig. 6. Late Devonian–Early Carboniferous faulting on the southern flank of the Elbow Spit Platform. The legend to the horizons is given in Figure 2; the approximate location is shown in Figure 4. DEF 3D seismic data courtesy of Spectrum ASA.

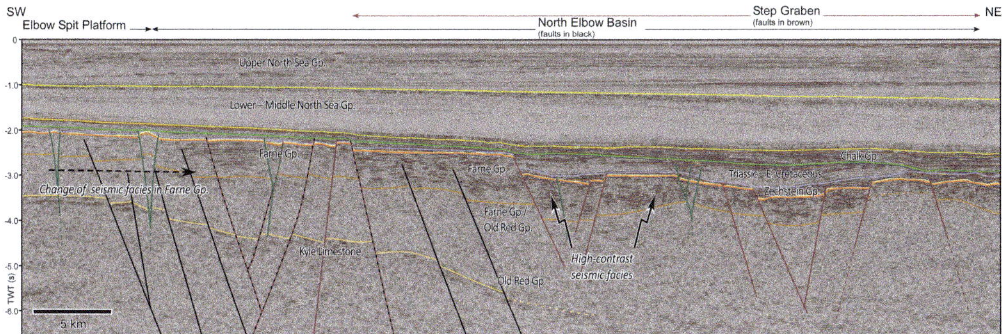

Fig. 7. Seismic section across the North Elbow Basin, a major low that lies hidden within the Step Graben. The Lower Carboniferous Farne Group has a high-contrast seismic facies that is likely to have been caused by the presence of coals. The legend to the horizons is given in Figure 2; the location is shown in Figure 4. Public seismic line NSR32294.

change in seismic facies. Continuous high-frequency parallel reflections are observed above the footwall, while above the hanging wall the configuration of the reflections changes to low-frequency divergent discontinuous (Fig. 6). This is evident in Upper Devonian, Tournaisian, Visean and Namurian units. Altogether, these observations show that the ESP represented a horst during the Late Devonian–Early Carboniferous, with lows on its northern and southern flanks.

Faults with this trend are also present on the ESP, but here offsets are smaller than on the northern and southern flanks. The faults are observed in all regions when the Carboniferous is sufficiently well imaged, which is the case on the ESP and its southern and northern flanks, and in the part of the SG directly east of the present-day ESP (Fig. 4). As a result of the large burial depth of the Carboniferous in the DCG and the remainder of the SG, it was not possible to constrain the fault trends in the Carboniferous here. This should not be taken as proof of absence, however.

N040° faults

A second prominent fault trend consists of NE–SW-trending faults (Fig. 4; displayed in blue). Like the N110° trend, these faults show a normal offset of pre-Permian units, and in most cases no offset at the Base Permian level. The magnitude of offset varies greatly; on top of the ESP, for instance in blocks E01 and E02, offset is commonly small, just above the seismic resolution (*c.* 30 m). However, close to the eastern flank of the ESP, in block E06, a fault is present with a vertical offset in the Carboniferous of approximately 0.5 s two-way travel time (TWT) (Fig. 8), which, assuming an interval velocity for the Carboniferous of 4.2 km s^{-1}, converts to a throw of about 1 km. No significant offset of the Upper Rotliegend Group is observed, but the overlying Rotliegend deposits show a gentle fold above the crest of the footwall. Further east, faults with similar trends and offsets are observed in the area that is now the SG (Figs 4 & 9). Again, offset of the Upper Rotliegend Group is small, or absent. Faults from this family offset the Visean Elleboog and Yoredale formations (Farne Group: see Fig. 2), which can be readily recognized on seismic data because of their distinct high-continuity, medium-contrast seismic facies, showing that activity post-dated the Visean. In blocks D06–E06, the Westphalian appears and can be observed to thicken across these faults.

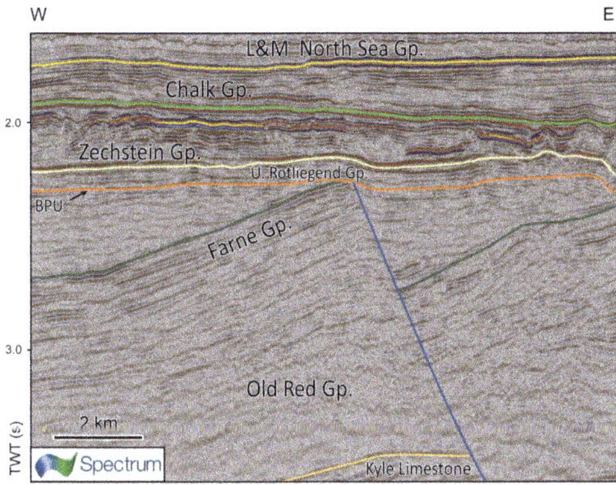

Fig. 8. Section across block E6 showing a N040°-trending fault that has an offset of approximately 0.5 s TWT. The interval velocity within the Carboniferous is approximately 4 km s^{-1} (*c.* 1 km). Activity of the fault post-dates the deposition of the Farne Group and predated the deposition of the Permian Upper Rotliegend Group. The legend to the horizons is given in Figure 2; the approximate location is shown in Figure 4. DEF 3D seismic data courtesy of Spectrum ASA.

In some cases, these faults clearly offset the N110° faults, which is commonly the case when the offset along the N040° fault is significant, of the order of hundreds of metres. When offset is at or just above the seismic resolution, no offset of the N110° faults is evident. Although atypical, faults with a N040° trend offset the Base Permian in a few cases: for instance, in block E09. In that particular case, the post-Carboniferous deformation is associated with the north–south-trending faults that form the SG and DCG, which is be described below.

Similar to the N110° trend, these faults can be recognized whenever the Carboniferous is sufficiently well imaged, showing that they constitute a regionally pervasive fabric. This is the case for the parts of the ESP and SG covered by the DEF survey (Fig. 4). In the DCG, the Carboniferous is buried to depths not imaged on the seismic data. In the A quadrant, only vintage 3D data are available and imaging of the Carboniferous is poor. As a consequence, it was not possible to confirm the absence or presence of faults with a N040° trend. Fault throws are limited on the ESP, and are of the order of hundreds of metres to 1 km on the eastern flank of the ESP and the region east of it (Fig. 9).

In summary, these faults were active post-dating the deposition of Visean units (Early Carboniferous) and predating the deposition of the Upper Rotliegend Group.

N360° faults

This fault trend accommodated the opening of the DCG and SG, but did not leave a significant imprint on the ESP (Fig. 4; displayed in brown). The faults trend north–south, and the offsets are commonly of the order of hundreds of metres up to kilometres. The faults that bound the DCG have a significantly larger offset than any other faults in the area: up to about 3500 m normal offset (Fig. 4). Unlike the previously described fault trends, these faults offset the BPU and affect units as young as the Early Cretaceous. West of the DCG, the SG consists of a number of smaller parallel graben, leading to an alternation of highs and lows. The throw along these faults is significant (up to *c.* 900 m), but is smaller than in the DCG (Fig. 4).

On the northern and eastern flanks of the ESP, north–south-trending faults can be observed to merge with faults with other trends: on the eastern flank, with faults with a N040° trend; and on the northern flank, with faults with a N110° trend (Fig. 4). Where this occurs, these faults offset the Permian, which is atypical for faults with these N110° and N040° trends, suggesting that in these cases these faults have been reactivated when the north–south-trending faults became active (Fig. 9).

Features associated with salt tectonics, such as salt walls and diapirs, commonly align with these faults (Fig. 3) and where this is the case, imaging of the subsalt units, including the fault itself, is poor.

N070° faults

On time slices, a fourth fault trend is visible that offsets all previously described fault trends. The trend of these faults is generally N070°, although the faults undulate locally (Fig. 4; displayed in green). Where

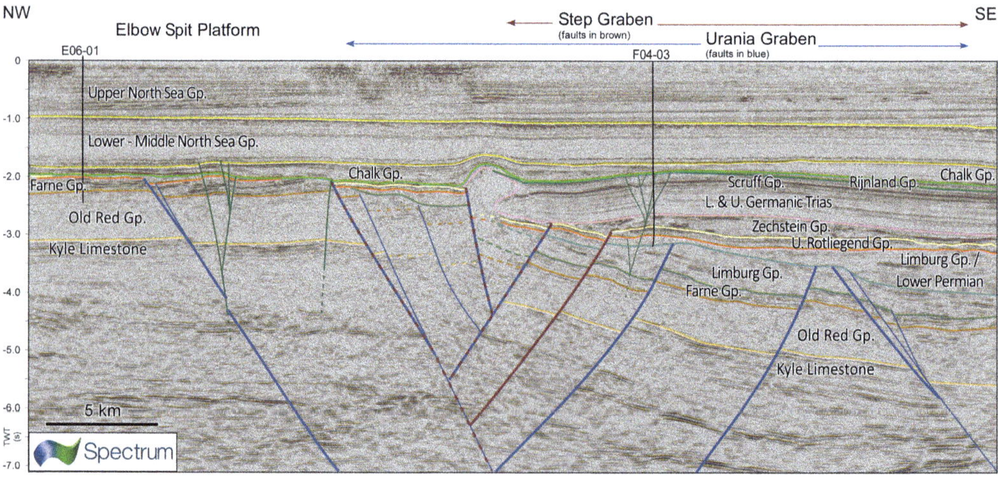

Fig. 9. Seismic section from the Elbow Spit Platform (left) to the Step Graben (right). The section reveals that the Urania Graben is an older structure, present below the Step Graben. Faults are coloured according to the period in which they were active: the legend is shown in Figure 4. The legend to the horizons is given in Figure 2; the approximate location is shown in Figure 4. DEF 3D seismic data courtesy of Spectrum ASA.

visible, the offsets consist of dextral strike-slip, with offsets of up to 1 km (e.g. Fig. 4, blocks A9 and A11). The faults are especially profound on the ESP, where they can be mapped over distances of over 90 km (Fig. 4). The faults affect units as young as the Late Cretaceous and possibly the Paleogene, but offsets are largest in units older than the Late Cretaceous.

The structural styles of these faults vary systematically; where the faults are observed to undulate in map view, a cross-section of the fault shows pop-up and pop-down geometries (Fig. 10); within these structures vertical offsets of up to hundreds of metres in magnitude are observed. The boundaries of the pop-ups often have trends that coincide with one of the previously described older fault trends (Fig. 4), suggesting that these faults have been reactivated. In-between pop-ups and pop-downs, the vertical offsets along these faults are small and, as a result, the faults are hard to identify on vertical sections. The faults are often readily visible on time slices and horizon extractions from the coherency cubes (Fig. 5).

The Zechstein salt appears to have a strong effect on the structural styles of these faults: where salt is present, the effect on units overlying the Zechstein is mild at most, while offsets at the Base Zechstein level can be significant. The system appears to have been decoupled in these cases, with the Zechstein salt acting as the décollement. When the salt is absent or thin, younger units are affected as well.

These faults are also observed within the SG and Central Graben, but locating them is more difficult than on the ESP; units above thick Zechstein salt are commonly not affected, meaning that the faults are limited to units below the salt, the base of which is at a significantly greater depth than on the ESP, resulting in poorer imaging of these units. Above salt windows, units younger than the Permian are affected by these faults as well.

Other fault trends

A number of subordinate fault trends have been identified in the study area. In block E12, faults with a NW–SE trend are observed. This fault trend is much more prominent further south: for instance, on the Cleaver Bank Platform and on the Groningen Platform, but seems not to be present north of E12. Above the Zechstein salt, a large number of faults associated with salt movement are present; most prominently, these consist of collapse graben on top of, and of radial faults surrounding, salt features.

Structural framework

Late Devonian–Early Carboniferous extension

Late Devonian–Early Carboniferous extension in the area was accommodated along N110°-trending faults. The newly available seismic data show that a low, the North Elbow Basin, was present north of the Elbow Spit Platform (ESP) (Fig. 1), which is in line with the identification of the North Dogger Basin in the adjacent UK sector (Milton-Worssell *et al.* 2010; Arsenikos *et al.* this volume, in press). The implications of the presence of such a low are major; first, it shows that the Mid North Sea High was not a high

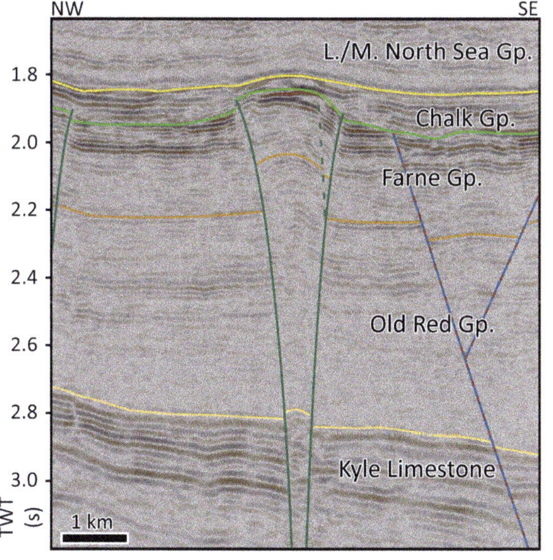

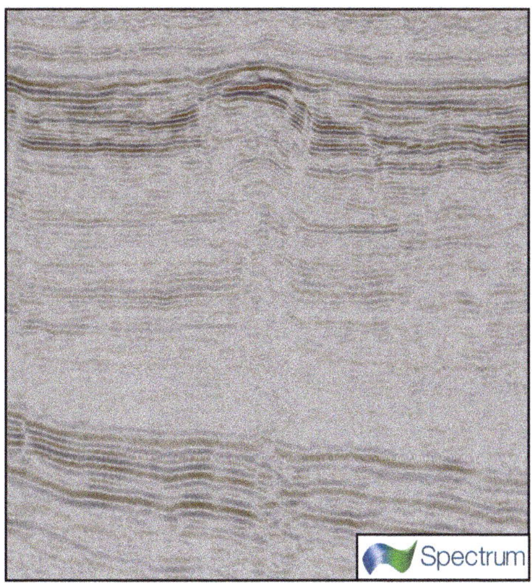

Fig. 10. Pop-up structure in the E2 and E3 blocks, which offsets units as young as the Late Cretaceous. The legend to the horizons is given in Figure 2. DEF 3D seismic data courtesy of Spectrum ASA.

during the Carboniferous but that the area consisted of an alternation of highs and lows. The Mid North Sea High area only became a long-lived high later on. It also leads to a reconsideration of the hydrocarbon potential of the newly identified low and surrounding areas (ter Borgh *et al.* this volume, in press).

Structurally, the picture that emerges is one of granite-cored highs, in line with a concept that was inferred previously in the UK onshore, and was extended to the UK offshore and northern Dutch offshore (Donato *et al.* 1983; Besly 1998). In summary, the granites strengthen the crust locally, and deformation is accommodated in the surrounding, weaker, regions, but not the high itself. On the ESP, biotite monzogranite has been encountered in well A17-01, which was dated at 410 ± 9 Ma (T. Pharaoh pers. comm.; cf. Geluk *et al.* 2007). It should be noted that only 31 m of strongly altered rocks were encountered in the lowermost part of the well; whether they represent the top part of a batholith or core complex, or merely a granitic intrusion, remains uncertain.

The presence of the North Elbow Basin is also reflected in the sedimentary record; the Upper Devonian and Lower Carboniferous are significantly thicker north (Fig. 7) and south (Fig. 6) of the ESP. A condensed Lower Carboniferous section on the ESP is in line with what was observed for the UK onshore Alston block (Kimbell *et al.* 1989), where Visean deposits were found to increase in thickness from around 400 m above the block to 4000 m in the adjacent low. In that case, deposits on the high were limited to the Asbian and Brigantian (the top part of the Visean), with older parts of the Visean being present only in the adjacent low. Extension first created accommodation space in the hanging wall. Subsidence of the footwall was less profound, but sufficient during the later phases of extension to accommodate sedimentation.

The situation on the ESP is largely similar, but not completely. In contrast to the Alston block, where sedimentation occurred only during the latest part of the Visean (Kimbell *et al.* 1989), all Visean substages are represented on the ESP, as shown by wells E02-01 and E06-01. A significant part of the thickening that is evident in the North Elbow Basin is seen to occur in the deposits underneath the reflective package. The top of the reflective package is shown by the wells to be of Visean age, implying that the transparent seismic facies below represents the Tournaisian and/or Upper Devonian and possibly the lower part of the Visean. We therefore propose that the North Elbow Basin was in an 'overfilled' state, meaning that on average sediment influx into the area exceeded the amount of new accommodation space created by subsidence. This allowed part of the sediment that entered the area from the north (Collinson 2005) to bypass the low, and be deposited on the ESP. The accommodation space on the platform was limited as well, and part of the sediment therefore bypassed the platform and was deposited in the low south of the platform. The sediment influx there was not high enough to balance the subsidence caused by faulting, leaving the basin south of the ESP in a 'starved' state.

A Devonian–Early Carboniferous proto-Central Graben?

The new, long-offset DEF seismic survey permits an improved reconstruction of fault activity prior to the Zechstein. The seismic data show that a high-contrast continuous Yoredale-type seismic facies also occurs east of the ESP, in the present-day SG area (Fig. 9: blocks F01, F04, F07 and E09). Although the exact depositional environment cannot be constrained based on the seismic facies alone, the facies is similar to the one observed on top of the ESP (Fig. 7) (ter Borgh *et al.* this volume, in press), where well control is available, and distinct from the inferred basinal section south of the ESP. This suggests strongly that in the Early Carboniferous the ESP extended further east than it does at present and that the SG had not formed. Whether shallow-marine to fluvial conditions prevailed as far east as the present-day DCG cannot be derived from the seismic data at present, as the quality of the imaging below the thick infill of the DCG does not permit this.

Although the imaging of the pre-Zechstein is poor below the DCG, the new findings from this study can be used to test the assumptions that were used in the past to propose activity of the DCG during the Devonian and Carboniferous (e.g. Ziegler 1990; de Jager 2007). Precursors to the DCG (the 'proto-Central Graben') have been proposed to have existed as early as the Devonian, based on the presence of the Middle Devonian seaway, as proposed by Ziegler (1990). The constraints that were available when this seaway was proposed were limited; marine Middle Devonian limestones were drilled in a number of UK wells, including well 38/03-1 (Fig. 1), while the Orcadian basin and Caledonian uplands to the NE and north were known to be continental. The palaeogeographical implication of these two observations was that a connection to the marine realm had to be found towards the south and/or east (Belgium, Germany). Continental Devonian deposits had been encountered in Dutch well A17-01 (Fig. 4) and German well Q-01 (Fig. 1), with the latter well containing (possibly reworked) marine fossils (Best *et al.* 1983). The seaway was therefore proposed to have been located in-between these wells, roughly corresponding to part of the later SG and DCG. This interpretation was perfectly plausible with the constraints available then.

The results from the present study provide additional constraints and lead to a different interpretation; a seismic facies indicative of Devonian ('Kyle') limestones is observed below the ESP, in the area that was interpreted by Ziegler (1990) to be the continental area west of the seaway during the Middle Devonian. The Kyle seismic facies can be observed throughout the parts of the study area where the Devonian is sufficiently well imaged (Figs 3, 6, 7 & 9); Middle Devonian marine deposition affected a much larger part of the study area. Reflection seismic data show that the Kyle seismic facies is, indeed, absent near well A17-01, but appears a few kilometres SE of the well and thickens from thereon. The absence of limestones at the location of this well can be attributed to the fact that the well was drilled at the top of the Elbow Spit High which had already apparently formed a high during the Devonian.

As a consequence of the new findings, it is no longer necessary to interpret a narrow Middle Devonian seaway between wells Q-01 and A17-01; much larger parts of the study area were flooded. Without this seaway, there is no need, or proof, for a Devonian Central Graben either. In summary, significant extension occurred during the Devonian–Early Carboniferous, accommodated along N110°-trending faults.

A Late Carboniferous–Early Permian proto-Central Graben?

Now that it has been shown that it was unlikely that a proto-Central Graben was present during the Devonian–Early Carboniferous, the possibility of the presence of such a feature during the Late Carboniferous–Early Permian will be discussed. During this period, faults with a N040° trend were active in the area (Figs 8 & 9); activity post-dated the deposition of the Visean–early Namurian Farne Group and predated the deposition of the Upper Rotliegend Group. This fault trend is observed throughout the study area, in places where the imaging of the pre-Rotliegend is sufficient. It has been described in other parts of the Dutch subsurface: for instance, in the Cleaver Bank Platform area, where the vertical offset has been described as 'minimal' (Oudmayer & de Jager 1993). This is also the case for most of the ESP, but elsewhere intra-Carboniferous offsets can be large, even in areas that are commonly considered to be part of a high; an example is the fault shown in Figure 8 that is located on the eastern flank of the ESP in block E06 which has a fault throw of around 1 km. Another example is found in the prerift sequence of the SG, where faults with similar large throws offset the Farne Group (Fig. 9); the combined throw of the Carboniferous units is a few kilometres. The fact that the faults significantly offset the Carboniferous while offsets at the base of the Permian are commonly insignificant (in this case, most probably the base of the middle Permian-age Upper Rotliegend Group: Fig. 2) shows that significant deformation must have occurred during the Late Carboniferous–middle Permian (Fig. 9).

Effects of Carboniferous–Permian deformation on deposition. Late Carboniferous–Early Permian deformation helps to explain the preservation of progressively younger units below the BPU from west to east; on the ESP, Lower Carboniferous units subcrop the BPU (Fig. 3). Upper Carboniferous deposits, sometimes as young as Westphalian D and, possibly, even Stephanian, were encountered below the BPU east of the ESP in blocks B17, F04, F07, F10, and, possibly, E09 and E12. So far, subcrop maps commonly use north–south-trending faults to explain the presence of these relatively young Carboniferous deposits (van Adrichem Boogaert & Kouwe 1993–1997; Besly 1998; Mijnlieff 2003; van Buggenum & den Hartog Jager 2007; Kombrink *et al.* 2010). The newly identified N040°-trending faults should be expected to also affect the subcrop pattern (Fig. 9).

A related issue is the variation in the thickness of the Rotliegend, which is thicker in the SG area than above the ESP (e.g. Geluk 2007). This thickening is sometimes cited as proof that activity along the SG and Central Graben started during the Permian. Figure 9 shows that there is a direct relationship between the thickness of the Westphalian and the Rotliegend: where the Westphalian is thick, so is the Rotliegend. We propose that three processes may have contributed to this change in thickness; first, the fact that extension was relatively large in the area where Westphalian deposits were preserved may have resulted in stronger post-rift thermal subsidence. Secondly, a thicker syndepositional package results in higher subsequent compaction, providing additional accommodation space. This may cause differential compaction between areas located over hanging walls and areas located above footwalls, resulting in gentle folds such as the one described in block E06. Thirdly, erosion may have had a stronger effect on the shale-rich deposits of the Westphalian Step Graben Formation than on the Lower Carboniferous deposits, which contain more competent lithologies, such as limestones and sandstones, analogous to the model proposed by Mijnlieff & Geluk (2011).

In summary, significant extension occurred in an area that is now part of the southern SG, but extension was not accommodated along the north–south-trending faults that characterize the Mesozoic SG and DCG but by N040°-trending faults. Referring to the structures formed by the N040° faults as a proto-Central Graben or proto-Step Graben can therefore be confusing, and we propose to call this structure the Urania Graben system instead. The implications for understanding deposition during the Late Carboniferous and Permian are great; faults with the offsets observed could have affected deposition significantly, and any exploration focusing on deposits of this age should take this into account.

Regional tectonic setting. The finding that significant graben, such as the Urania Graben, formed during the Late Carboniferous–Early Permian is somewhat surprising as this period is generally associated with wrench tectonics, not regional extension (Ziegler 1990; Geluk 2007). Efforts have been made to find strike-slip faults of this age in the study area, but none were found. The study area is not the first region where extension was observed during the Late Carboniferous–Early Permian, however. Near well 39/2- 4 (Fig. 4), just 12 km NW of the study area, Late Carboniferous–Early Permian extensional faulting was observed (Heeremans *et al.* 2004). In the German and Danish offshore, the southern segment of the Horn Graben also shows a NE–SW orientation and is thought to have been active during the Late Carboniferous–Early Permian, although stratigraphic control is poor here (Best *et al.* 1983; Vejbæk 1990; Abramovitz *et al.* 1998; Abramovitz & Thybo 1999). Significant extension also occurred further north, in the Oslo Graben area (Neumann *et al.* 2004). In addition to the structures mentioned, other examples exist in the vicinity, supporting the finding that extension affected a wide region; further south, on the Cleaver Bank Platform, normal displacement along NE–SW-trending faults was also observed, along with strike-slip displacement (Schroot & de Haan 2003). Strain partitioning may have caused the alternation between extension and wrenching; some areas, such as the study area, the Horn Graben and the Oslo Graben, could have been dominated by extension while other zones were dominated by wrenching. The fact that the underlying lithosphere is highly heterogeneous (Fig. 1) may have contributed to strain partitioning.

Late Carboniferous–Early Permian extension is generally considered to be genetically related to magmatism, as igneous rocks and normal faults formed during this period are often observed together (Heeremans *et al.* 2004; Neumann *et al.* 2004). This is also the case in the study area, where Lower Rotliegend igneous rocks are present (e.g. Geluk 2007; de Bruin *et al.* 2015). The causal relationship between the two phenomena is subject to debate: upwelling of hot mantle material may have resulted in magmatism and active rifting above the resulting thermal anomaly or, alternatively, extension resulting from far-field stresses may have caused the lithosphere to heat and melt (Neumann *et al.* 2004).

No major extension had been observed during the Late Carboniferous–Early Permian in the study area before. We explain this from the later Mesozoic rifting, which buried the Urania Graben below Triassic–Jurassic deposits, to depths at which seismic imaging was poor on vintage seismic data. The new seismic surveys used in this study provided the required imaging.

Formation of the Central Graben and the Step Graben (Mesozoic)

The present study improves the mapping of fault geometries of Mesozoic structures. An example is the SG in the A quadrant. On overview maps, the SG is commonly represented by a single NNW-trending fault (e.g. Kombrink *et al.* 2012). The findings from this study underline that the SG is made up of multiple horsts and graben, with a north–south trend (Fig. 4). The transition from the SG to the ESP is marked by a clear-cut change in trends; the north–south faults merge with faults with a N110° trend. We interpret this to reflect a reactivation of these older Carboniferous faults as oblique-slip faults. The boundary coincides roughly with the northern boundary fault that separated the ESP from the North Elbow Basin during the Early Carboniferous, and we propose that this was a rheological boundary; the ESP, which had already withstood Devonian–Early Carboniferous extension, resisted deformation again. The region north of it that had already suffered significant extension during the Early Carboniferous, was again extended during the Triassic–Jurassic.

Reactivation of the other Carboniferous fault trend (N040°) also occurred: for example, on what is now the eastern flank of the ESP, in block E09. Here, the base of the Upper Rotliegend Group was offset by a fault with an orientation parallel to the Urania Graben (Fig. 4).

Although salt tectonics are beyond the scope of this paper, we note that there is a clear relationship between the location of salt walls, pillows and diapirs, and basement faults; the former commonly occur above the latter (Fig. 3).

Late Jurassic–Early Cretaceous strain partitioning

Prominent strike-slip faults were found with a N070° trend. The faults have rather large horizontal offsets, of the order of hundreds of metres to over 1 km where the sense of shear is dextral (Figs 4 & 5). Individual fault zones can be continued over as much as 90 km. On the ESP, where the quality of the seismic imaging is high, the faults are regularly spaced, commonly around 5 km, although outliers exist. In some cases, the faults anastomose and form a fault zone. In other cases, the spacing can be as high as 15 km. The faults cause significant offset up to the Early Cretaceous, and minor offsets of units as young as the Paleogene; the former is interpreted to represent the main phase of activity of these faults, and the latter to be a reactivation during Late Cretaceous–Paleogene inversion.

Deformation along these faults was strongly influenced by the presence of Zechstein salt, which, where present, acted as a décollement, meaning that the units above and below the salt were deformed in a different manner. Above the salt, deformation is commonly mild because the salt caused deformation to be distributed over a wide area and, as a result, commonly no faults with significant offsets are visible on seismic data. Below the salt, however, deformation is accommodated by a smaller number of faults, causing clear offset along these faults. Where the salt is absent or thin, units younger than the Zechstein are also deformed. In these cases, flower structures, forming both pop-ups and pop-downs, are commonly present, as faults extended up to the surface. Pop-ups and pop-downs often result from the local reactivation of older fault trends; reactivation of the N110° trend predominantly produces pop-ups, while reactivation of the N040° trend produces pop-downs.

The fact that decoupling in many cases limits offsets to pre-Zechstein units can easily confuse seismic interpreters. As a result, activity along the N070°-trending faults can be erroneously interpreted to have occurred during the Permian: for instance, by linking them to Early Permian wrenching. Activity during this period cannot be ruled out, but supporting evidence is lacking.

Faults with this trend went unrecognized previously in the northern Dutch offshore as a result of three factors: first, strike-slip faults are generally harder to spot on seismic data than normal or reverse faults because vertical offsets occur only locally. Secondly, the fact that deformation during this time was strongly affected by decoupling along the Zechstein salt means that, in many cases, deformation is visible only below the Zechstein salt. Thirdly, the imaging of subsalt units was poor on vintage seismic data. Long-offset seismic surveys, such as NSR and DEF, provide the required imaging below the salt. A combination of the preceding factors mean that identifying such faults in studies carried out at the scale of individual hydrocarbon fields or blocks is particularly challenging.

We propose that these strike-slip faults formed as a result of strain partitioning. North of the study area, in the Danish Central Graben, extension was primarily accommodated along north–south-trending faults up to the Kimmeridgian (Late Jurassic), and along NW–SE-trending faults during the Kimmeridgian, Volgian (Late Jurassic) and Early Cretaceous (e.g. Møller & Rasmussen 2003). In the DCG, NNW–SSE (*c.* N155°)-trending Jurassic faults are rare. Similar net strain to the Danish Central Graben could have occurred in the DCG, however, if strain was partitioned with a roughly east–west extension along N360°-trending faults and dextral slip along N070°-trending faults. Transfer faults have previously been proposed as being associated with extension along the Central Graben in the Central North Sea (Fraser *et al.* 2003; Zanella *et al.* 2003) and in the German and Danish parts of the Central Graben (Cartwright 1989; Sundsbø & Magsen 1993; Wride 1995: the so-called ‘transverse zones’).

The recognition of this fault trend could be relevant for estimating the reservoir potential of the Jurassic in the study area; deposition at this time was strongly affected by fault activity (Barnes & Hodgkinson 1993; Bartholomew *et al.* 1993; Sears *et al.* 1993; Wride 1995; Fraser *et al.* 2003; Wonham *et al.* 2014); strike-slip faulting may have contributed to localized uplift and erosion, both as a result of forming pop-ups and by activating salt movement, and may thus have controlled sediment pathways between the highs and the graben.

Inversion events

Compared to the Cleaver Bank Platform area located south of the study area, inversion associated with both the Variscan Orogeny and the Late Cretaceous–Paleogene inversion events is less profound. The effects of the Variscan inversion cannot be identified unequivocally; a gentle folding of the Lower Carboniferous on the ESP could be related to the inversion, but can also be explained by Early Carboniferous extension.

Late Cretaceous and Paleogene inversion in the study area caused activation of salt movement, long-wavelength regional folding and local reactivation of normal faults (van Wijhe 1987; de Jager 2003). A possible example of fault reactivation is the eastern bounding fault of the DCG, where uplift of the Base Chalk within the DCG relative to the area east of the DCG suggests that this fault was reactivated as a reverse fault (van der Molen *et al.* 2005). Unfortunately, imaging of the fault itself is poor as it is overlain by a salt wall, making it difficult to establish what part of the offset of the Base of the Chalk is caused directly by fault movement and what part by salt movement.

Some indications of reactivation can also be observed on the N070°-trending strike-slip faults, but the offset is much smaller than during the Jurassic and Early Cretaceous. It is therefore hard to constrain whether the faults were reactivated at all and, if so, whether they were reactivated as thrusts or as strike-slip faults. The latter seems most likely, considering the nearly vertical dip of the faults. In this case, sinistral displacement would be expected based on the regional stress field. No sinistral offset was observed on the N070° faults in the study area, but this is to be expected; pre-Late Cretaceous offset along these faults is relatively large (hundreds of metres) and, as long as the offset due to Late Cretaceous–Paleogene inversion is smaller, the net offset will still be dextral.

A notable difference between the study area and the rest of the Dutch subsurface is the paucity of NW–SE-trending faults. These faults are thought to have been active during Variscan inversion and have been reactivated many times since, most recently during the Late Cretaceous–Paleogene inversion (e.g. Frikken 1999). In the study area, they have not been observed north of block E12, which coincides with a major tectonic boundary (Fig. 4): to the north, the lithosphere is thought to consist of remnants of the Caledonian accretionary complex, while the remainder of the Dutch subsurface forms part of the Avalonia microplate (Smit *et al.* 2016) (Fig. 1). We suggest that the contrast in structural style results from this contrast in lithospheric structure.

Hydrocarbon potential

The new structural framework sheds new light on the hydrocarbon potential of the study area. An important factor that has so far discouraged exploration in the northern Dutch offshore is the perceived absence of source rock. Significant amounts of Visean Scremerston (Lower Carboniferous) coals have been described in the Lower Carboniferous in two nearby wells: 39/07- 1 and A09-01 (Fig. 4) (ter Borgh *et al.* this volume, in press). The age-equivalent interval has not been penetrated by wells in the Dutch A quadrant, and from vintage seismic data it was uncertain whether these deposits were present here. The identification of the North Elbow Basin makes it plausible that the coals have been preserved here (ter Borgh *et al.* this volume, in press).

The structural framework can also be used for assessing reservoir development, source-rock maturation, trap formation and hydrocarbon preservation. An example of the relevance of the N360° trend for trap formation is shown in Figure 4; in the A quadrant, the alternation of horsts and graben creates potential closures. The likely seal for such closures would be the Permian Silverpit Formation and Zechstein salt. Faults with a N040° trend are known to cause reservoir compartmentalization in the Cleaver Bank Platform area (Oudmayer & de Jager 1993; van Hulten 2010) and may also seal on geological timescales. The N110° and N040° trends have the potential for creating intra-Carboniferous closures. Faults with a N070° trend may have controlled deposition during the Jurassic and Early Cretaceous, and pop-ups formed along this trend may form traps when a seal is present above.

Conclusions

Recently acquired 3D and 2D seismic data allow for a much more detailed reconstruction of fault patterns and fault activity than was possible previously. Extension during the Late Devonian–Early Carboniferous was accommodated along N110°-trending normal faults. The new findings show that an alternation of highs and lows was present in the study area, similar in setting to the UK onshore. The Elbow Spit Platform (ESP) was already a high during this period, and earlier findings that it was bound on its southern side by a major normal fault are confirmed. In contrast to previous findings, however, a significant low was found to be present north of the present-day ESP: the North Elbow Basin, which we propose is the continuation of the North Dogger Basin in the adjacent UK sector. This has major implications for source-rock and reservoir development, and shows that part of the present-day Mid North Sea High was, in fact, a basin during the Late Devonian–Early Carboniferous.

In most of the A blocks, the North Elbow Basin underlies the SG, which is a younger geological feature. In the easternmost part of the A quadrant, it lies hidden within what is commonly considered to be the extension of the Mid North Sea High. This shows that structural elements as they are present now are, in many cases, not representative of the Paleozoic structuration.

The SG and DCG have previously been proposed to have originated in the Devonian or Early Carboniferous. No indication for activity of these north–south-trending structures during this period was found. Nevertheless, significant extension did occur during the Late Carboniferous and/or Early Permian. The extension was accommodated by the Urania Graben system, which consists of faults with a NE–SW trend, at a 40° angle to the main trend of the DCG and SG. The offset and timing of activity of these faults is consistent with Late Carboniferous–Early Permian normal faulting described elsewhere in Europe.

Although activity of the Urania Graben seems to have predated the deposition of the Upper Rotliegend Group, it is likely that residual topography, post-rift subsidence and compaction still affected deposition and accommodation space. The ESP and Elbow Spit High may have formed a local sediment source, and clastic sediments with reservoir potential similar to the ones found in the Cygnus Field may well be present in the Rotliegend on and off the flanks of the Elbow Spit Platform.

The SG and DCG formed during the Triassic, Jurassic and Early Cretaceous. The present study has significantly improved the understanding of the geometries of these features. Improved imaging of the pre-salt shows that salt diapirs are often located above normal faults. Extension was associated with N070°-trending transfer faults that are clearly visible below Zechstein salt and in places where the Zechstein salt is absent. Further research on the interaction between these faults and the depositional systems above the salt is recommended.

The structural framework described in this study marks a significant improvement in the understanding of the geological development of the Mid North Sea area from the Devonian to the present. The framework provides a firm basis for assessing the hydrocarbon potential of the area.

Acknowledgements Reviewers Harmen Mijnlieff and Stavros Arsenikos, and editor Tony Hewett, are thanked for their constructive comments. This study benefited from assistance and comments from numerous EBN colleagues but we want to thank Walter Eikelenboom in particular. MtB thanks Jeroen Smit for discussions. We thank Spectrum ASA for allowing us to show data from the DEF survey.

References

Abramovitz, T. & Thybo, H. 1999. Pre-Zechstein structures around the MONA LISA deep seismic lines in the southern Horn Graben area. *Bulletin of the Geological Society of Denmark*, **45**, 99–116.

Abramovitz, T., Thybo, H. & MONA LISA Working Group. 1998. Seismic structure across the Caledonian Deformation Front along MONA LISA profile 1 in the southeastern North Sea. *Tectonophysics*, **28**, 153–176, https://doi.org/10.1016/S0040-1951(97)00290-4

Arfai, J., Jähne, F., Lutz, R., Franke, D., Gaedicke, C. & Kley, J. 2014. Late Palaeozoic to Early Cenozoic geological evolution of the northwestern German North Sea (Entenschnabel): New results and insights. *Netherlands Journal of Geosciences*, **93**, 147–174, https://doi.org/10.1017/njg.2014.22

Arsenikos, S., Quinn, M., Kimbell, G., Williamson, P., Pharaoh, T., Leslie, G. & Monaghan, A. In press. Structural development of the Devono-Carboniferous plays of the UK North Sea. *In*: Monaghan, A.A., Underhill, J.R., Hewett, A.J. & Marshall, J.E.A. (eds) *Paleozoic Plays of NW Europe*. Geological Society, London, Special Publications, **471**, https://doi.org/10.1144/SP471.3

Barnes, K.R. & Hodgkinson, R.J. 1993. Rectilinear structuration of the kingdom of Norway – the neglected NW–SE lineament. Conference abstract presented at the EAGE 55th Meeting and Technical Exhibition, 9 June 1993, Stavanger, Norway, http://doi.org/10.3997/2214-4609.201411468

Bartholomew, I.D., Peters, J.M. & Powell, C.M. 1993. Regional structural evolution of the North Sea: oblique slip and the reactivation of basement lineaments. *In*: Parker, J.R. (ed.) *Petroleum Geology of Northwest Europe: Proceedings of the 4th Conference*. Geological Society, London, 1109–1122, https://doi.org/10.1144/0041109

Besly, B.M. 1998. Carboniferous. *In*: Glennie, K.W. (ed.) *Petroleum Geology of the North Sea: Basic Concepts and Recent Advances*. 4th edn. Blackwell Science, Oxford, 104–136.

Best, G., Kockel, F. & Schöneich, H. 1983. Geological history of the Southern Horn Graben. *In*: Kaasschieter, J.P.H. & Reijers, T.J.A. (eds) *Petroleum Geology of the Southeastern North Sea and the Adjacent Onshore Areas. Geologie en Mijnbouw*, **62**, 25–33.

Cartwright, J.A. 1989. The kinematics of inversion in the Danish Central Graben. *In*: Cooper, M.A. & Williams, G.D. (eds) *Inversion Tectonics*. Geological Society, London, Special Publications, **44**, 153–175, https://doi.org/10.1144/GSL.SP.1989.044.01.10

Cocks, L.R.M., McKerrow, W.S. & Van Staal, C.R. 1997. The margins of Avalonia. *Geological Magazine*, **134** , 627–636.

Collinson, J.D. 2005. Dinantian and Namurian depositional systems in the southern North Sea. *In*: Collinson, J.D., Evans, D.J., Holliday, D.W. & Jones, N.S. (eds) *Carboniferous Hydrocarbon Geology: The Southern North Sea and Surrounding Onshore Areas*. Yorkshire Geological Society, Occasional Publications, **7**, 35–56.

Coward, M.P. 1993. The effect of Late Caledonian and Variscan continental escape tectonics on basement structure, Paleozoic basin kinematics and subsequent Mesozoic basin development in NW Europe. *In*: Parker, J.R. (ed.) *Petroleum Geology of Northwest Europe: Proceedings of the 4th Conference*. Geological Society, London, 1095–1108, https://doi.org/10.1144/0041095

de Bruin, G., Bouroullec, R. et al. 2015. *New Petroleum Plays in the Dutch Northern Offshore*. TNO Report TNO 2015 R10920.

de Jager, J. 2003. Inverted basins in the Netherlands, similarities and differences. *Netherlands Journal of Geosciences*, **82**, 355–366, https://doi.org/10.1017/S0016774600020175

de Jager, J. 2007. Geological development. *In*: Wong, Th. E., Batjes, D.A.J. & de Jager, J. (eds) *Geology of the Netherlands*. Royal Netherlands Academy of Arts and Sciences, Amsterdam, 5–26.

de Jager, J. & Geluk, M.C. 2007. Petroleum geology. *In*: Wong, Th.E., Batjes, D.A.J. & de Jager, J. (eds) *Geology of the Netherlands*. Royal Netherlands Academy of Arts and Sciences, Amsterdam, 241–264.

Donato, J.A., Martindale, W. & Tully, M.C. 1983. Buried granites within the Mid North Sea High. *Journal of the Geological Society, London*, **140**, 825–837, https://doi.org/10.1144/gsjgs.140.5.0825

Doornenbal, J.C. & Stevenson, A.G. (eds). 2010. *Petroleum Geological Atlas of the Southern Permian Basin Area*. European Association of Geoscientists and Engineers (EAGE), Houten, The Netherlands.

Evans, D., Graham, C., Armour, A. & Bathurst, P. (eds). 2003. *The Millennium Atlas: Petroleum Geology of the Central and Northern North Sea*. Geological Society, London.

Fraser, S., Robinson, A., Johnson, H., Underhill, J. & Kadolsky, D. 2003. Upper Jurassic. *In*: Evans, D.,

Graham, C., Armour, A. & Bathurst, P. (eds) *The Millennium Atlas: Petroleum Geology of the Central and Northern North Sea*. Geological Society, London, 157–189.

Frikken, H. 1999. *Reservoir-geological aspects of productivity and connectivity of gasfields in the Netherlands. PhD thesis*, TU Delft, Delft, The Netherlands.

Geluk, M.C. 2007. Permian. *In*: Wong, Th.E., Batjes, D.A.J. & de Jager, J. (eds) *Geology of the Netherlands*. Royal Netherlands Academy of Arts and Sciences, Amsterdam, 63–83.

Geluk, M.C., Dusar, M. & de Vos, W. 2007. Pre-Silesian. *In*: Wong, Th.E., Batjes, D.A.J. & de Jager, J. (eds) *Geology of the Netherlands*. Royal Netherlands Academy of Arts and Sciences, Amsterdam, 27–42.

Heeremans, M., Timmerman, M.J., Kirstein, L.A. & Faleide, J.I. 2004. New constraints on the timing of late Carboniferous–early Permian volcanism in the central North Sea. *In*: Wilson, M., Neumann, E.-R., Davies, G.R., Timmerman, M.J., Heeremans, M. & Larsen, B.T. (eds) *Permo-Carboniferous Magmatism and Rifting in Europe*. Geological Society, London, Special Publications, **223**, 177–194, https://doi.org/10.1144/GSL.SP.2004.223.01.08

Kimbell, G.S., Chadwick, R.A., Holliday, D.W. & Werngren, O.C. 1989. The structure and evolution of the Northumberland Trough from new seismic reflection data and its bearing on modes of continental extension. *Journal of the Geological Society, London*, **146**, 775–787, https://doi.org/10.1144/gsjgs.146.5.0775

Kombrink, H., Besly, B. *et al.* 2010. Carboniferous. *In*: Doornenbal, J.C. & Stevenson, A.G. (eds) *Petroleum Geological Atlas of the Southern Permian Basin Area*. European Association of Geoscientists and Engineers (EAGE), Houten, The Netherlands, 81–99.

Kombrink, H., Doornenbal, J.C., Duin, E.J.T., Den Dulk, M., van Gessel, S.F., ten Veen, J.H. & Witmans, N. 2012. New insights into the geological structure of the Netherlands; results of a detailed mapping project. *Netherlands Journal of Geosciences*, **91**, 419–446, https://doi.org/10.1017/S0016774600000329

Kortekaas, M., Böker, U., van der Kooij, C., Jaarsma, B. & Rosendaal, E. 2016. The Triassic Main Buntsandstein play – new prospectivity in the Dutch northern offshore. Conference abstract presented at the Mesozoic Resource Potential in the Southern Permian Basin, 7–9 September 2016, London, UK.

Leeder, M.R. 1988. Recent developments in Carboniferous geology: a critical review with implications for the British Isles and N.W. Europe. *Proceedings of the Geologists' Association*, **99**, 73–100.

Maynard, J.R., Hofmann, W., Dunay, R.E., Bentham, P.N., Dean, K.P. & Watson, I. 1997. The Carboniferous of Western Europe: the development of a petroleum system. *Petroleum Geoscience*, **3**, 97–115, https://doi.org/10.1144/petgeo.3.2.97

McCann, T., Pascal, C. *et al.* 2006. Post-Variscan (end Carboniferous–Early Permian) basin evolution in Western and Central Europe. *In*: Gee, D.G. & Stephenson, R.A. (eds) *European Lithosphere Dynamics*. Geological Society, London, Memoirs, **32**, 355–388. http://doi.org/10.1144/GSL.MEM.2006.032.01.22

McPhee, C.A., Judt, M.R., McRae, D. & Rapach, J.M. 2008. Maximising gas well potential in the Breagh field by mitigating formation damage. Paper SPE 115690 presented at the 2008 SPE Asia Pacific Oil & Gas Conference and Exhibition, 20–22 October 2008, Perth, Australia, http://doi.org/10.2118/115690-MS

Mijnlieff, H.F. 2003. Top pre-Permian distribution map and some thematic regional geological maps of the Netherlands. Poster presentation at the 15th International Congress on Carboniferous and Permian (ICCP), 10–16 August 2003, Utrecht, The Netherlands, retrieved from http://www.nlog.nl/posters/

Mijnlieff, H.F. & Geluk, M. 2011. Palaeotopography-governed sediment distribution – A new predictive model for the Permian Upper Rotliegend in the Dutch sector of the Southern Permian Basin. *In*: Grötsch, J. & Gaupp, R. (eds) *The Permian Rotliegend of the Netherlands*. SEPM, Special Publications, **98**, 147–159.

Milton-Worssell, R., Smith, K. & McGrandle, A. 2010. The search for a Carboniferous petroleum system beneath the Central North Sea. *In*: Vining, B.A. & Pickering, S.C. (eds) *Petroleum Geology: From Mature Basins to New Frontiers – Proceedings of the 7th Petroleum Geology Conference*. Geological Society, London, 57–75, https://doi.org/10.1144/0070057

Møller, J.J. & Rasmussen, E.S. 2003. Middle Jurassic–Early Cretaceous rifting of the Danish Central Graben. *Geological Survey of Denmark and Greenland Bulletin*, **1**, 247–264.

Neumann, E.-R., Wilson, M., Heeremans, M., Spencer, A. E., Obst, K., Timmerman, M.J. & Kirstein, L. 2004. Carboniferous–Permian rifting and magmatism in southern Scandinavia, the North Sea and northern Germany: a review. *In*: Wilson, M., Neumann, E.-R., Davies, G.R., Timmerman, M.J., Heeremans, M. & Larsen, B.T. (eds) *Permo-Carboniferous Magmatism and Rifting in Europe*. Geological Society, London, Special Publications, **223**, 11–40, http://doi.org/10.1144/GSL.SP.2004.223.01.02

Oudmayer, B.C. & de Jager, J. 1993. Fault reactivation and oblique-slip in the Southern North Sea. *In*: Parker, J.R. (ed.) *Petroleum Geology of Northwest Europe: Proceedings of the 4th Conference*. Geological Society, London, 1281–1290, https://doi.org/10.1144/0041281

Pharaoh, T. 1999. Palaeozoic terranes and their lithospheric boundaries within the Trans-European Suture Zone (TESZ): a review. *Tectonophysics*, **314**, 17–41, https://doi.org/10.1016/S0040-1951(99)00235-8

Schroot, B.M. & de Haan, H.B. 2003. An improved regional structural model of the Upper Carboniferous of the Cleaver Bank High based on 3D seismic interpretation. *In*: Nieuwland, D.A. (ed.) *New Insights into Structural Interpretation and Modelling*. Geological Society, London, Special Publications, **212**, 23–37, https://doi.org/10.1144/GSL.SP.2003.212.01.03

Sears, R.A., Harbury, A.R., Protoy, A.J.G. & Stewart, D.J. 1993. Structural styles from the Central Graben in the UK and Norway. *In*: Parker, J.R. (ed.) *Petroleum Geology of Northwest Europe: Proceedings of the 4th Conference*. Geological Society, London, 1231–1243, https://doi.org/10.1144/0041231

Smit, J., van Wees, J.-D. & Cloetingh, S. 2016. The Thor suture zone: From subduction to intraplate basin setting. *Geology*, **44**, 707–710, https://doi.org/10.1130/G37958.1

SUNDSBØ, G.O. & MAGSEN, J.B. 1993. Structural styles in the Danish Central Graben. *In*: PARKER, J.R. (ed.) *Petroleum Geology of Northwest Europe: Proceedings of the 4th Conference*. Geological Society, London, 1255–1267, https://doi.org/10.1144/0041255

TAGGART, S. & CATTO, R. 2015. New opportunities on the northern feather-edge play of the Silverpit Basin: Sedimentology and facies distribution of the Permian Lower Leman and Carboniferous Ketch Formations, Cygnus field, Southern North Sea. Conference abstract presented at the 8th Petroleum Geology of Northwest Europe Conference, 28–30 September 2015, London.

TER BORGH, M.M., EIKELENBOOM, W. & JAARSMA, B. In press. Hydrocarbon potential of the Visean and Namurian in the northern Dutch offshore. *In*: MONAGHAN, A.A., UNDERHILL, J.R., HEWETT, A.J. & MARSHALL, J.E.A. (eds) *Paleozoic Plays of NW Europe*. Geological Society, London, Special Publications, **471**, https://doi.org/10.1144/SP471.5

UNDERHILL, J.R. & PARTINGTON, M.A. 1993. Jurassic thermal doming and deflation in the North Sea: implications of the sequence stratigraphic evidence. *In*: PARKER, J.R. (ed.) *Petroleum Geology of Northwest Europe: Proceedings of the 4th Conference*. Geological Society, London, 337–345, https://doi.org/10.1144/0040337

VAN ADRICHEM BOOGAERT, H.A. & KOUWE, W.F.P. 1993–1997. Stratigraphic nomenclature of the Netherlands, revision and update by RGD and NOGEPA. *Mededelingen Rijks Geologische Dienst*, **50**.

VAN BUGGENUM, J.M. & DEN HARTOG JAGER, D.G. 2007. Silesian. *In*: WONG, TH.E., BATJES, D.A.J. & DE JAGER, J. (eds) *Geology of the Netherlands*. Royal Netherlands Academy of Arts and Sciences, Amsterdam, 43–62.

VAN DALFSEN, W., DOORNENBAL, J.C., DORTLAND, S. & GUNNINK, J.L. 2006. A comprehensive seismic velocity model for the Netherlands based on lithostratigraphic layers. *Netherlands Journal of Geosciences*, **85**, 277–292, https://doi.org/10.1017/S0016774600023076

VAN DALFSEN, W., VAN GESSEL, S.F. & DOORNENBAL, J.C. 2007. *Velmod-2*. TNO Report TNO 2007-U-R1272C, retrieved from http://www.nlog.nl/

VAN DER MOLEN, A.S., DUDOK VAN HEEL, H.W. & WONG, T.E. 2005. The influence of tectonic regime on chalk deposition: examples of the sedimentary development and 3D-seismic stratigraphy of the Chalk Group in the Netherlands offshore area. *Basin Research*, **17**, 63–81, https://doi.org/10.1111/j.1365-2117.2005.00261.x

VAN HULTEN, F.F.N. 2010. Geological factors affecting compartmentalization of Rotliegend gas fields in the Netherlands. *In*: JOLLEY, S.J., FISHER, Q.J., AINSWORTH, R.B., VROLIJK, P.J. & DELISLE, S. (eds) *Reservoir Compartmentalization*. Geological Society, London, Special Publications, **347**, 301–315, http://doi.org/10.1144/SP347.17

VAN WEES, J.-D., STEPHENSON, R.A. *ET AL.* 2000. On the origin of the Southern Permian Basin, Central Europe. *Marine and Petroleum Geology*, **17**, 43–59, https://doi.org/10.1016/S0264-8172(99)00052-5

VAN WIJHE, D.H. 1987. Structural evolution of inverted basins in the Dutch offshore. *Tectonophysics*, **137**, 171–219, https://doi.org/10.1016/0040-1951(87)90 320-9

VEJBÆK, O.A. 1990. The Horn Graben, and its relationship to the Oslo Graben and the Danish Basin. *Tectonophysics*, **178**, 29–49, https://doi.org/10.1016/0040-1951(90)90458-K

WONHAM, J.P., RODWELL, I., LEIN-MATHISEN, T. & THOMAS, M. 2014. Tectonic control on sedimentation, erosion and redeposition of Upper Jurassic sandstones, Central Graben, North Sea. *In*: MARTINIUS, A.W., RAVNÅS, R., HOWELL, J.A., STEEL, R.J. & WONHAM, J.P. (eds) *From Depositional Systems to Sedimentary Successions on the Norwegian Continental Margin*. International Association of Sedimentologists, Special Publications, **46**, 473–512, https://doi.org/10.1002/9781118920435.ch17

WRIDE, V.C. 1995. Structural features and structural styles from the Five Countries Area of the North Sea Central Graben. *First Break*, **13**, 395–407, https://doi.org/10.3997/1365-2397.1995020

ZANELLA, E. & COWARD, M.P. 2003. Structural framework. *In*: EVANS, D., GRAHAM, C., ARMOUR, A. & BATHURST, P. (eds) *The Millennium Atlas: Petroleum Geology of the Central and Northern North Sea*, Geological Society, London, 45–59.

ZANELLA, E., COWARD, M.P. & MCGRANDLE, A. 2003. Crustal structure. *In*: EVANS, D., GRAHAM, C., ARMOUR, A. & BATHURST, P. (eds) *The Millennium Atlas: Petroleum Geology of the Central and Northern North Sea*. Geological Society, London, 35–42.

ZIEGLER, P.A. 1990. *Geological Atlas of Western and Central Europe*, 2nd revised edn. Shell Internationale Petroleum Maatschappij, The Hague.

Hydrocarbon potential of the Visean and Namurian in the northern Dutch offshore

M. M. TER BORGH*, W. EIKELENBOOM & B. JAARSMA
EBN BV, Daalsesingel 1, 3511 SV, Utrecht, The Netherlands
M.t.B, 0000-0002-1698-2411
**Correspondence: marten.borgh-ter@ebn.nl*

Abstract: Following the play-opening successes of the Breagh and Pegasus gas fields, we evaluated the potential of the Visean and Namurian (Carboniferous) petroleum plays in the northern Dutch offshore. This evaluation incorporated seismic and well data from the Dutch, British and German North Sea sectors. The abundance and thickness of reservoir-quality Visean–Namurian sandstones was found to increase from Breagh towards the NE. Visean–Namurian coals and shales are considered promising source rocks to charge these reservoirs in the Dutch Central Graben (DCG) and Step Graben (SG). The presence of a mature Paleozoic source rock in the SG and DCG is supported by hydrocarbon shows and vitrinite reflectance data. In the southern E and F blocks, charge may also occur laterally from Upper Carboniferous Westphalian coals. A regional post-well analysis showed that the Visean and Namurian plays are virtually untested in the Dutch northern offshore. Two tests were positive but had high N_2 contents, one was negative, while 10 wells were drilled off-structure and are therefore considered invalid tests of this play. Hence, it is concluded that the Visean and Namurian in the northern Dutch offshore have significant hydrocarbon potential.

Production of hydrocarbons from Carboniferous reservoirs in the Southern North Sea comes primarily from the Westphalian. The discovery and development of gas fields with reservoirs in the pre-Westphalian, such as the Visean Breagh Field (Fig. 1) (McPhee *et al.* 2008) and the Namurian Pegasus Field, triggered new interest in the hydrocarbon potential of the Visean and Namurian in the Southern North Sea and Mid North Sea High area. In this study, a regional analysis of the hydrocarbon potential of this stratigraphic interval is presented focusing on the Dutch offshore, while incorporating data from the UK and German offshore.

To assess the viability of the Visean and Namurian plays, well results were analysed initially. Next, well and seismic data were used to constrain depositional trends and palaeogeography, and, based on this analysis, the distribution of potential reservoirs was established. To further quantify the reservoir potential, core-plug measurements are used. To establish the chances of hydrocarbon charge and migration, shows of hydrocarbons in existing wells are summarized, and the presence and maturity of source rocks is estimated from seismic and well data. A recently developed structural framework for the area (ter Borgh *et al.* this volume, in press) is used to further assess reservoir development, source-rock presence and maturation, and trap formation.

Geological setting

During the Early Carboniferous, the study area was located on the southern margin of the continent Laurussia, which had formed during the Ordovician–Silurian collision of three older continents during the Caledonian Orogeny: Avalonia, Baltica and Laurentia (e.g. Ziegler 1990; Smit *et al.* 2016). During the Early Carboniferous, Laurussia was converging with Gondwana, located to the south of Laurussia, which resulted in significant extension on Laurussia's southern margin due to either back-arc extension (e.g. Leeder 1988), continental escape (e.g. Maynard *et al.* 1997) or a combination of both (e.g. Coward 1993). This caused a roughly east–west-orientated basin to form between present-day Ireland and Poland.

Sediment influx into the basin occurred from the Caledonides in the north (e.g. Collinson 2005). In the northern part of the basin, fluvial and paralic conditions prevailed, while greater water depths developed in the central part of the basin. The study area is located at the boundary between these two domains (Smit *et al.* 2016). The London–Brabant Massif forms the southern margin of the deep part of the basin (Ziegler 1990).

Extension was distributed over a wide area, and resulted in a large-scale alternation of highs and lows, with widths of the order of tens of kilometres each. The highs in many cases have a core that

From: Monaghan, A. A., Underhill, J. R., Hewett, A. J. & Marshall, J. E. A. (eds) 2018. *Paleozoic Plays of NW Europe*. Geological Society, London, Special Publications, **471**, 133–153.
First published online March 20, 2018, https://doi.org/10.1144/SP471.5

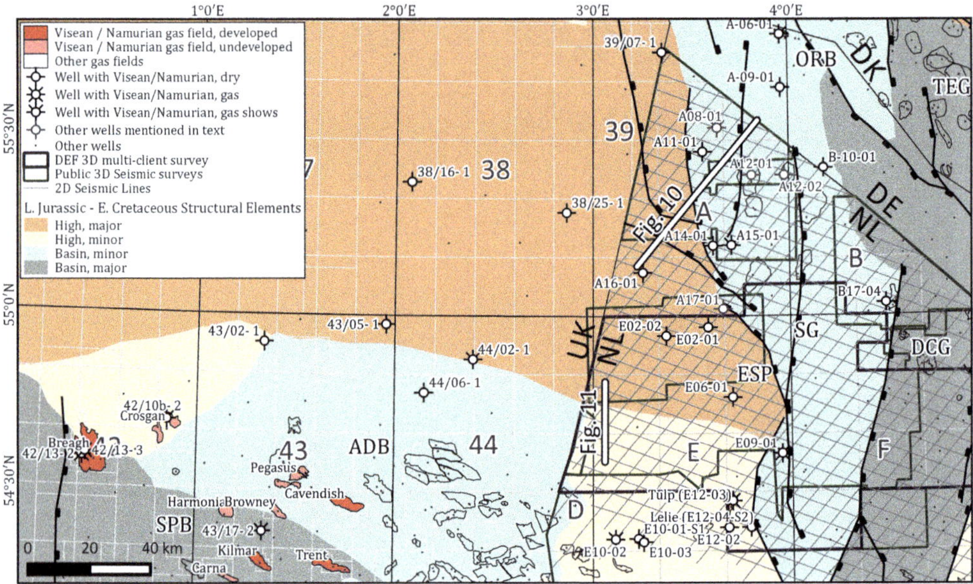

Fig. 1. Overview of wells and gas fields discussed in this paper. All fields are shown where the gas column extends into Visean or Namurian reservoirs. In fields with Namurian reservoirs, the gas column commonly extends into the Westphalian. Late Jurassic–Early Cretaceous structural elements (modified after Doornenbal & Stevenson 2010) are shown. ADB, Anglo-Dutch Basin; DCG, Dutch Central Graben; ESP, Elbow Spit Platform; ORB, Outer Rough Basin; SG, Step Graben; SPB, Silver Pit Basin; TEG, Tail End Graben.

consists of a Caledonian granite (Donato *et al.* 1983). The Elbow Spit Platform (ESP), located in the northern Dutch offshore (Figs 1 & 2), is an example of such a high. At its northern side, it is bounded by the North Elbow Basin, where subsidence rates were higher than on the ESP, but not high enough for significant water depths to develop as sedimentation rates exceeded subsidence (Figs 2 & 3). As a result, sedimentation extended onto the ESP. Sedimentation also occurred south of the ESP, but sediment influx during the Visean was insufficient to fill this part of the basin completely (ter Borgh *et al.* this volume, in press) and significant water depths developed there, except on intra-basinal highs where carbonate platforms formed (Kombrink *et al.* 2010*b*; van Hulten 2012).

Continental collision of Laurussia and Gondwana finally occurred during the late Visean–Westphalian, leading to the Variscan Orogeny. Extension ceased and inversion of the basin occurred, causing folding and erosion of the basin fill (Ziegler 1990). Late Carboniferous–Early Permian volcanism was accompanied by additional uplift and erosion. Inversion and thermal uplift resulted in the formation of the Base Permian Unconformity (BPU). During the Permian, sedimentation resumed following a 40–60 myr hiatus with the deposition of Upper Rotliegend reservoirs, and Silverpit and Zechstein Formation seals (Geluk 2007). The subcrop at the BPU in the study area varies from Lower Carboniferous to Lower Permian.

During the Mesozoic, the Central Graben system formed, leading to the formation of the Dutch Central Graben (DCG) and the Step Graben (SG) in the study area (Fig. 1). Extension along these structures has been proposed to have started as early as the Devonian, but recent data suggest otherwise (ter Borgh *et al.* this volume, in press). Extension led to regional subsidence and progressive burial of the Carboniferous, but subsidence was interrupted during the Middle Jurassic, when uplift centred at the triple junction of the three rift segments of the Central Graben system led to the formation of the Central North Sea Rift Dome (Ziegler 1990; Underhill & Partington 1993). Extension and subsidence subsequently resumed. Extension ceased in the Early Cretaceous, but post-rift subsidence allowed for the deposition of up to 2 km of Cenozoic deposits.

As a result of the post-Carboniferous development of the area, the present-day burial depth of Carboniferous deposits varies from about 2.2 km on the crest of the ESP to over 9 km in the DCG. As a result, reservoir quality may be expected to range from very good to very poor, and source-rock maturity from immature to over mature.

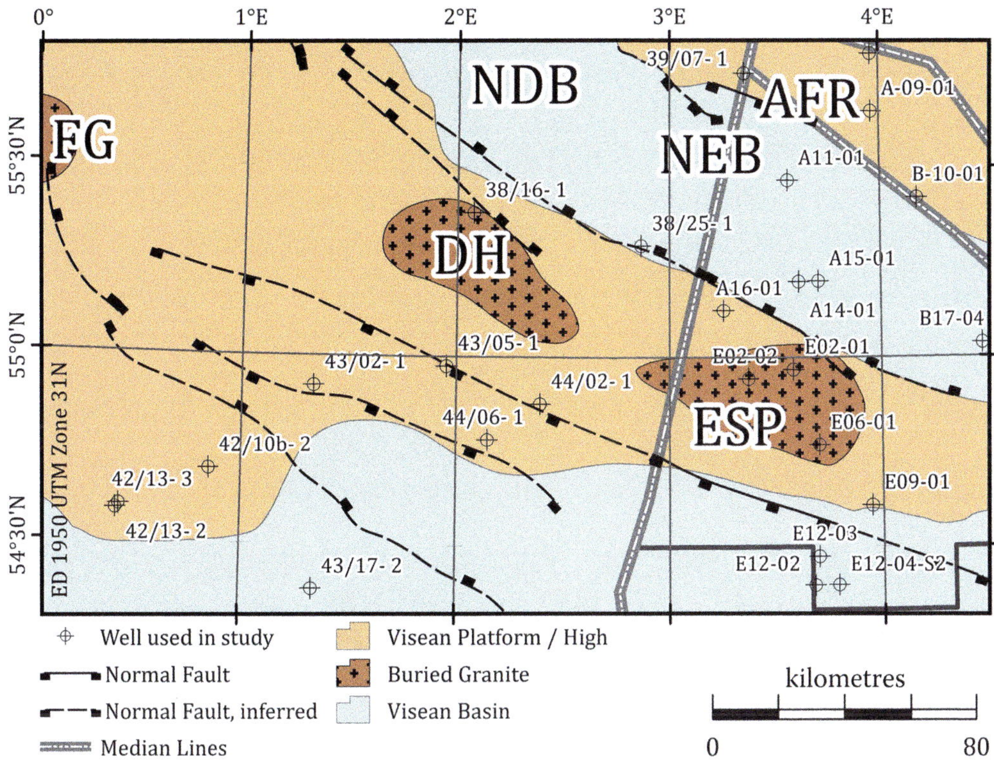

Fig. 2. Structural elements during the Visean. UK sector after Milton-Worssell *et al.* (2010), Arsenikos *et al.* (2015) and Kearsey *et al.* (2015). AFR, Auk-Flora Ridge; DH, Dogger High; ESP, Elbow Spit Platform; FG, Farne Granite; NDB, North Dogger Basin; NEB, North Elbow Basin.

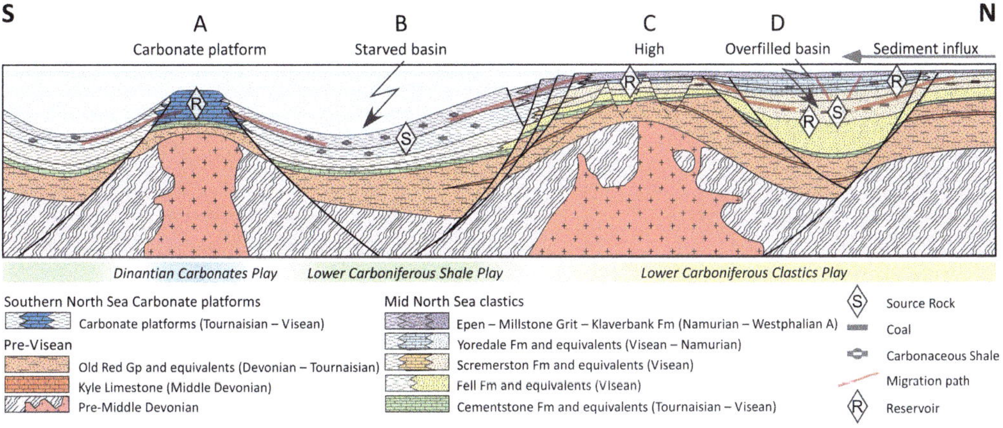

Fig. 3. Diagram illustrating the structural geology and play elements of the Visean and Namurian in the Mid North Sea area. The Elbow Spit Platform is an example of a high (C), while the North Elbow Low is an example of an overfilled basin (D). South of the study area, carbonate platforms are present on intra-basinal highs such as the Groningen High (A) (Kombrink *et al.* 2010*b*; van Hulten 2012). Between the highs, basins were present where subsidence exceeded sedimentation (B).

Stratigraphical framework

This publication focuses on the two Carboniferous stages predating the Westphalian: the Visean and the Namurian (Fig. 4). The stratigraphic subdivision of the Carboniferous commonly used in Western Europe is distinct from the international system (Kombrink *et al.* 2010*a*) (Fig. 4). The subdivision of the Carboniferous in The Netherlands has historically been based on lithostratigraphy, not chronostratigraphy (Kombrink *et al.* 2010*a*). This has a significant impact as many formation boundaries

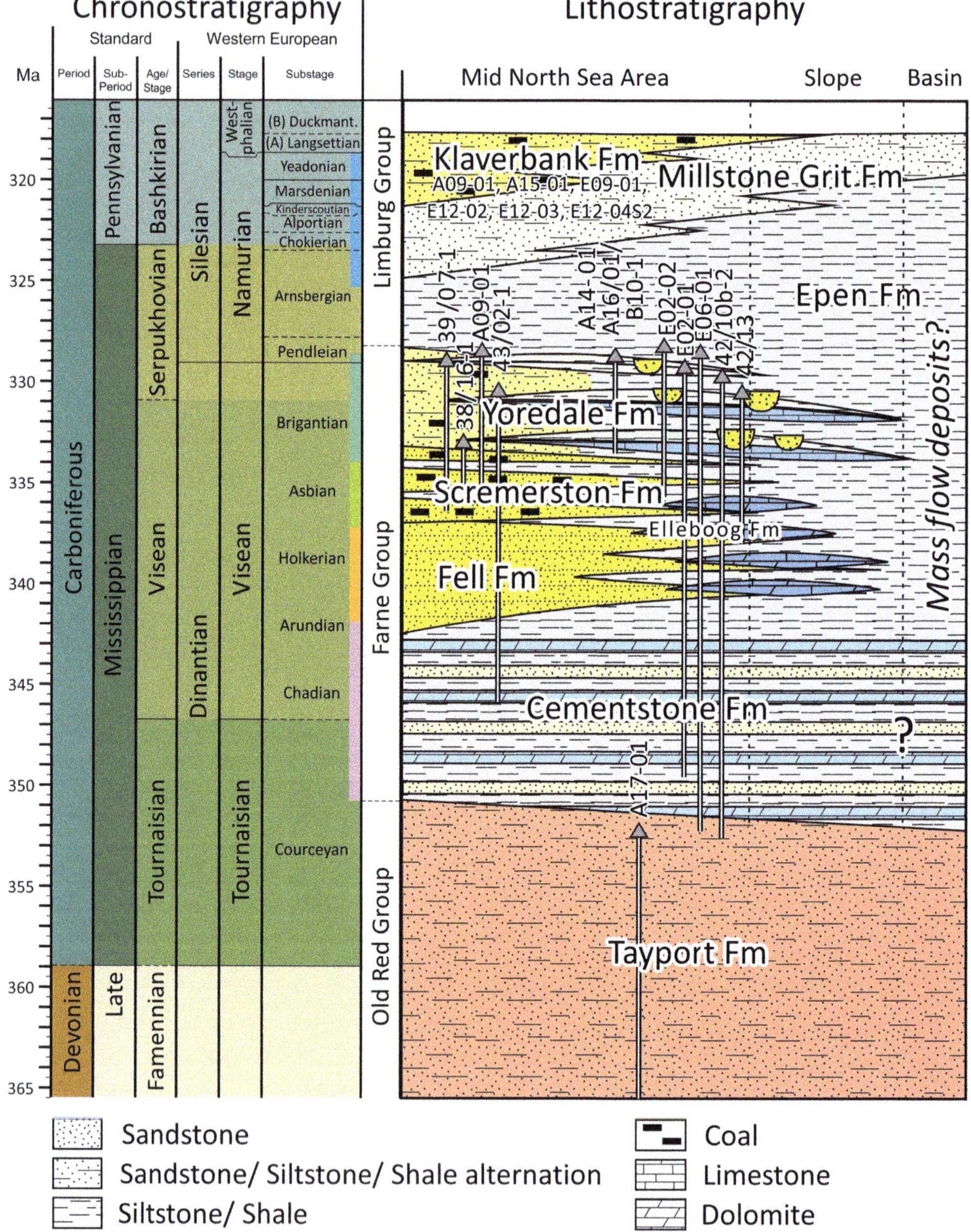

Fig. 4. Stratigraphic chart for the Visean and Namurian of the Mid North Sea area. Well penetrations are shown. Colours in the substage column refer to the maps in Figure 5.

within the Carboniferous are diachronous; the base of the Yoredale Formation, for instance, is known in northern England to range from the early Brigantian to the Pendleian (Waters *et al.* 2007). To predict where reservoir rocks and source rocks were deposited, it is necessary to reconstruct the depositional system and, hence, to know which sections are age equivalent. Where available, biostratigraphic constraints have therefore been used to correlate between wells, using the results from the study by Schroot *et al.* (2006) as a starting point. The resolution of biostratigraphy during this time interval is limited, however, and significant uncertainties remain.

Data and methods

Seismic and well data

2D and 3D seismic data from various vintages were used. The DEF survey, a 7950 km^2 3D multiclient survey, covers the Dutch D quadrant and the southern E and F quadrants (Fig. 1). The northern E and F, and the A and B quadrants are partly covered by publicly available 3D surveys acquired in the period between 1994 and 2006. 2D data were used in places where no 3D data were available: the entire study area is covered by the North Sea Renaissance (NSR) 2D multiclient survey, which consists of a grid of regional seismic lines shot in the NW–SE and NE–SW directions, with a spacing between lines of 5.3 km. The survey was acquired between 2005 and 2010, and the Dutch part of the survey has since become public. Seismic interpretation was carried out in the time domain and the non-SEGY seismic convention was used.

For the Dutch part of the area, data from over 60 wells acquired from NLOG (http://www.nlog.nl/) were used for seismic interpretation, 14 of which encountered the Visean and/or Namurian (Fig. 1). In the UK sector, data from eight public wells that encountered these stages were used (Fig. 1). Data availability from the German sector was limited to a well that was formerly located in the Dutch sector (B10-01) and a well published in the Southern Permian Basin atlas (A-09-01: Kombrink *et al.* 2010*a*). The wells were tied to seismic data using checkshots.

Analysis of well results

To assess the hydrocarbon potential of the Visean and Namurian, the results from existing wells in the northern Dutch offshore were evaluated with the aim of establishing whether a particular well can be considered a valid test of the Visean and Namurian plays, and, if so, whether the result was positive or negative. This evaluation consisted of four steps:

(1) The presently available seismic data were used to establish whether a valid trap is present at the well location. If it is clear from the seismic data that no trap is present, the well is considered an invalid test. This commonly occurs when a well targeted a closure at another level; if the well targeted a closure at the Chalk level, for instance, no structure needs to be present within the Visean. Another common reason for an invalid test due to the absence of a valid trap is that old 2D data were used.
(2) Establish the presence of reservoir from well data. This was considered positive if enough reservoir rock is present within the stratigraphic interval so that if gas bearing, the well would have been a technical success.
(3) Evaluate the presence of a seal.
(4) Finally, the presence of hydrocarbons ('charge') was evaluated. If enough hydrocarbons are present to consider the well a technical success, the well was considered a valid positive test.

If the test is considered valid (see step 1) and either reservoir, seal or charge is absent or ineffective, the well was considered a valid negative test.

Lithological classification

To assess the source-rock and reservoir potential of the Visean and Namurian, a lithological classification was performed for all selected wells based on wireline logs, including the gamma-ray and sonic logs. Cut-off values to constrain lithological classes were defined empirically by comparing log response, mud logs, and previous interpretations by Collinson Jones Consulting (1997) and Schroot *et al.* (2006). An overview of the classes and cut-off values used is given in Table 1. The lithological classification was checked against cuttings descriptions if available.

Porosity and permeability measurements

To assess the reservoir potential, an analysis of porosity and permeability data was carried out. For Dutch wells, core-plug measurements were acquired from reports and data available on http://www.nlog.nl/. For the UK wells, data from publicly available well reports were used. To each measurement, the result from the lithological classification was added in order to establish the reservoir properties for each lithological class.

Results

Well results

As the number of wells penetrating the Visean and Namurian in the northern Dutch offshore is low, only two play types are distinguished. In principle,

Table 1. *Cut-off values used in lithological classification*

	Gamma ray (GR, API)		Sonic (DT, μs/ft)	
	>	≤	>	≤
Carbonate		60		57–68*
Sandstone		60	57–68*	
Coal			100	
Shaly sandstone	60	100	68	
Marl	60	100		57–68*
Shale	100		68	
Igneous rock	From mud log/operator composite log			

To take into account the effects of depth of burial on the seismic velocity, the values marked with an asterisk were made depth dependent. At depths lower than 3000 m, a value of 68 μs/ft was used. In wells 42/13- 2 and 42/10b- 2, a value of 60 μs/ft was used below 3000 m measured depth (MD), and a value of 57 μs/ft below 3250 m MD. In wells 39/07- 1, A-09-01 and B10-01, a value of 65 μs/ft was used below 3000 m MD.

it would be possible to make a more detailed classification but this would not yield statistically meaningful results:

- The Visean clastics play: reservoirs in the Elleboog Formation or Yoredale Formation, sourced from Visean coals, Visean–Namurian basinal shales, Westphalian coals or Zechstein Group source rocks, and trapped below the Permian Silverpit Formation or Zechstein Group deposits.
- The Namurian clastics play: source and seal as above, but with a Namurian (post-Yoredale) reservoir.

Visean play. Five out of six wells penetrating the Visean in the study area are considered invalid tests of the Visean play as they did not drill valid structures, results from the sixth well are inconclusive. This is not unexpected, as none of these wells targeted the Visean. In addition to this, most wells were drilled in a period when 3D seismic data were not widely available; the most recent well that encountered the Visean was drilled in 1990 (E02-02), the other wells were drilled between 1972 and 1983. In most cases the presently available seismic data show that no structure is present at the well location, whereas vintage 2D seismic data may have suggested that a closure was present.

All wells contained at least some layers with reservoir potential (Table 2). Rocks with sealing potential were present in most cases. In four wells, sealing was provided by relatively low-risk Silverpit shales. In one case, an intra-Carboniferous seal may have been present but as a trap was absent in this case, it is unclear whether the seal was effective. In well E02-01, the Carboniferous was overlain by Chalk deposits. The Chalk is considered a riskier seal than the Silverpit and no detailed analysis was carried out to assess its sealing potential here. Minor or doubtful shows were found in two wells. The absence of shows in the other wells should not be used to conclude that charge was absent; as traps were absent, no trapping of hydrocarbons would be expected to occur in the first place.

Namurian play. Eight wells in the study area encountered the Namurian (Table 3). Similar to the findings for the Visean, six of these wells are considered invalid tests as presently available seismic data show that the wells did not drill a valid structure. In two cases, accumulations were found: the Tulp and Lelie fields (Fig. 1), which are likely to have been charged from downthrown Westphalian Coal Measures and possibly the Namurian. As all play elements worked and gas was tested at economic flow rates, these two wells are considered technical successes and valid tests. It should be noted, however, that significant amounts of N_2 were encountered in both wells, rendering these discoveries uneconomic.

Further north, where no Westphalian Coal Measures are expected, well A15-01 (Fig. 1) drilled only a thin section of Namurian rocks but the overlying Zechstein carbonates were found to be gas bearing, with oil shows. The interval was tested and a flare was lit, but production rates were low. The test was adversely affected by a poor cement job, which makes it likely that communication existed between the tested carbonates and previously tested

Table 2. *Well results for the Visean.*

Well	Trap	Reservoir	Seal	Charge	Conclusion
A14-01	Absent (2D)	Present	Doubtful (thin intra-formation shale)	Gas shows	Invalid test
A16-01	Probable (2D)	Present	Present, thin	No shows	Negative/ Invalid test
B10-01	Absent (2D)	Present	Present	No shows	Invalid test
E02-01	Absent (3D)	Present	Doubtful (Chalk)	Doubtful shows	Invalid test
E02-02	Absent (3D)	Present	Present, thin	No shows	Invalid test
E06-01	Doubtful (3D)	Present	Present	No shows	Invalid test

Table 3. *Well results for the Namurian.*

Well	Trap	Reservoir	Seal	Charge	Conclusion
A11-01	Absent (3D)	Present	Present	Weak gas shows	Invalid test
A15-01	Absent (2D)	Inconclusive	Lower Rotliegend volc.	Gas in Zechstein (16% N_2)	Invalid test
B17-04	Absent (2D)	Tight (high depth; 4600 m)	Present	Mature source rock in well	Invalid test
E06-01	Doubtful (3D)	Only 17 m, possibly part of Yoredale.	Present	No shows	Invalid test?
E09-01	Inconclusive	Inconclusive	Present	Present, 85% N_2	Invalid test
E12-02	Absent (3D)	Probable	Present	Gas shows	Invalid test
E12-03	Present	Present	Present	Present, 33% N_2	Positive
E12-04-S2	Present	Present	Present	Present, 65% N_2	Positive

water-bearing intervals. Nitrogen was present, but not in problematic quantities (16%). Although it remains unclear what source the gas was derived from, the test proves that a mature Paleozoic source rock is present. Candidates are the Visean and Namurian (Elleboog, Klaverbank and Yoredale formations), the Zechstein itself, and the Devonian.

Reservoir: porosity, permeability and net-to-gross ratios

Visean. To constrain the distribution of reservoir rocks, the results from the lithological classification were summed for each stratigraphic interval, and the results were displayed in map view (Fig. 5). The results show in broad terms that most formations become more sand-rich towards the north. The characteristics of core plugs taken from intervals classified as sandstone or shaly sandstone are shown in Figure 6, confirming the presence of reservoir quality sandstones within the Visean and Namurian; based on this result, intervals classified as sandstone are considered net reservoir. Intervals classified as shaly sandstone can be considered upside. Based on the data, no clear lower limit for the reservoir potential can be given; the deepest well with significant amounts of data, E12-04-S2, shows decent porosities (average 12.7%, $\sigma = 3.0$) at 3.8 km. The reservoir potential of the Visean is probably higher than the core-plug measurements suggest; channel-fill sandstones are commonly considered the best Visean reservoirs and, although indications for such sandstones are present in all wells penetrating the Farne Group in the northern Dutch offshore, they have only been cored in well E02-01. Porosities as high as 25% (average 12.9%, $\sigma = 4.1$) and permeabilities as high as 269 mD (average 42.7 mD) were measured here (Fig. 6).

Net-to-gross ratios vary per interval (Fig. 5). In general, proximal deposits have greater amounts of sand, meaning that net-to-gross ratios increase towards the north/NE. However, reservoir rocks may occur in all parts of the system. Figure 5 shows a map with lithological characteristics for the first 100 m below the BPU, giving an indication of net-to-gross ratios for closures at the Base Permian level.

Namurian. Sandstone intervals with a sharp base and top, and little internal variation on log data ('blocky' sandstones) are found throughout the Namurian Klaverbank Formation in the study area. These sandstones resemble producing Westphalian Klaverbank Formation reservoirs from the D and southern E quadrants (Fig. 4); the depositional environments are similar but, as progradation progressed from north to south, they differ in age. The net-to-gross ratio of this stratigraphic interval is generally high (*c.* 45%: Fig. 5a); not just in the E12 quadrant but also further north.

Visean–Namurian source rocks

In the British and German offshore. Visean and Namurian coals are predominantly found in the Scremerston Formation. The Dutch equivalent, the Elleboog Formation, has been encountered in three wells in The Netherlands: E02-01, E02-02 and E06-01. Coal streaks recognizable on logs have been found in well E02-02, and thin coal seams below the resolution of log data have been identified in cores and mud logs from wells E06-01 and E02-01. Although the number of wells is limited, the coal content of the Scremerston Formation appears to increase northwards (Fig. 5c): well 39/07- 1, located less than 1 km across the median line from the A quadrant, contains 23 m of coal (Figs 7 & 8), but this should be considered a minimum thickness as the base of the Scremerston Formation was not reached in this well. German well A-09-01, located 13 km across the border, captures a complete Scremerston section (Kombrink *et al.* 2010*a*) (Figs 7 & 9); proportionally, it has a similar amount of coal (14%) but,

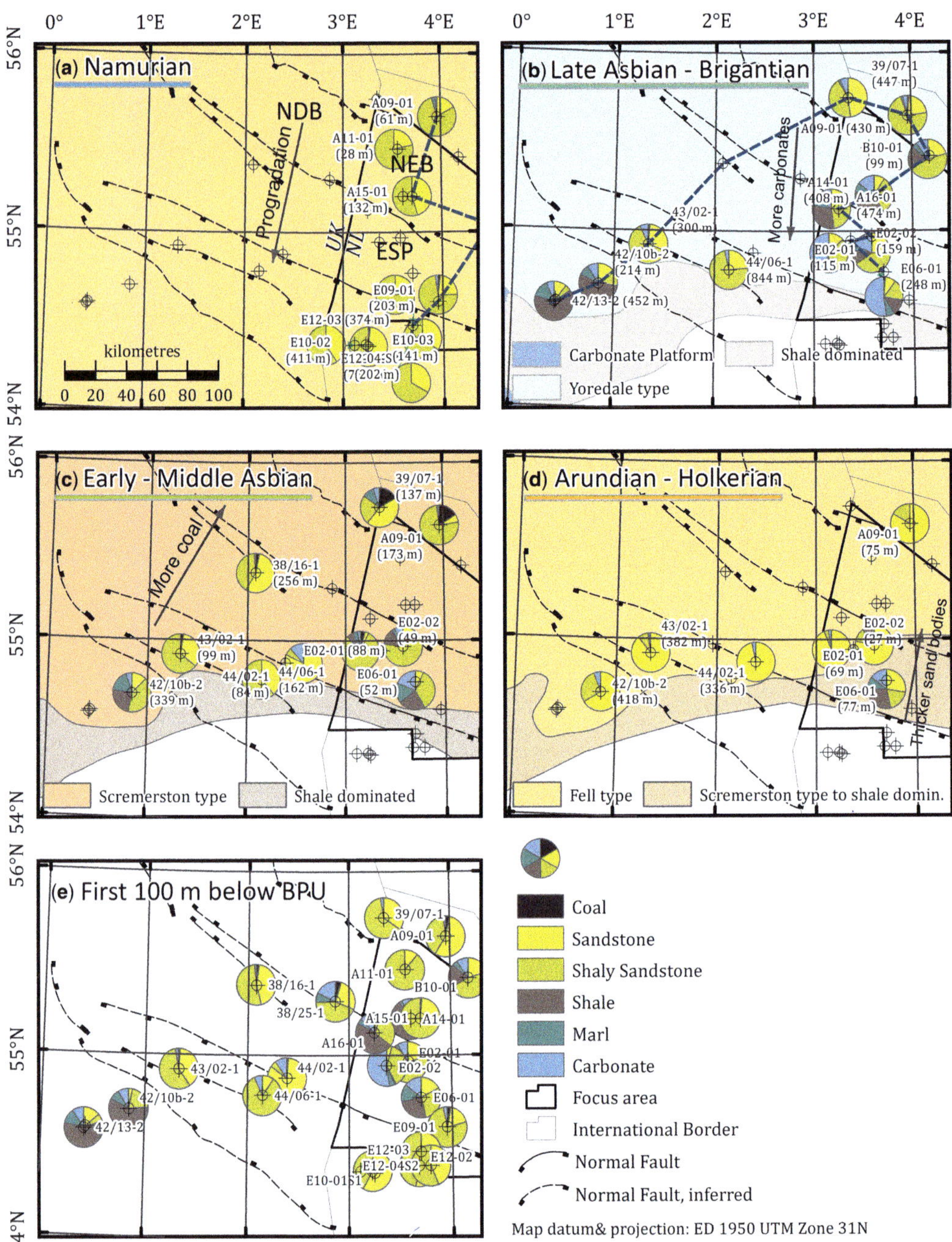

Fig. 5. (**a**)–(**d**) Palaeogeographical charts and lithological statistics, see also Figure 4. (**e**) Lithological statistics for the first 100 m below the Base Permian Unconformity (BPU), showing that the chances of encountering sandstone in the Visean or Namurian below the Silverpit Formation and Zechstein Group seals are good. Note that not all wells represent full penetrations of a given interval; see Figures 7–9. UK palaeogeography after Kearsey *et al.* (2015; this volume, in review); UK structures after Arsenikos *et al.* (2015). ESP, Elbow Spit Platform; NDB, North Dogger Basin; NEB, North Elbow Basin.

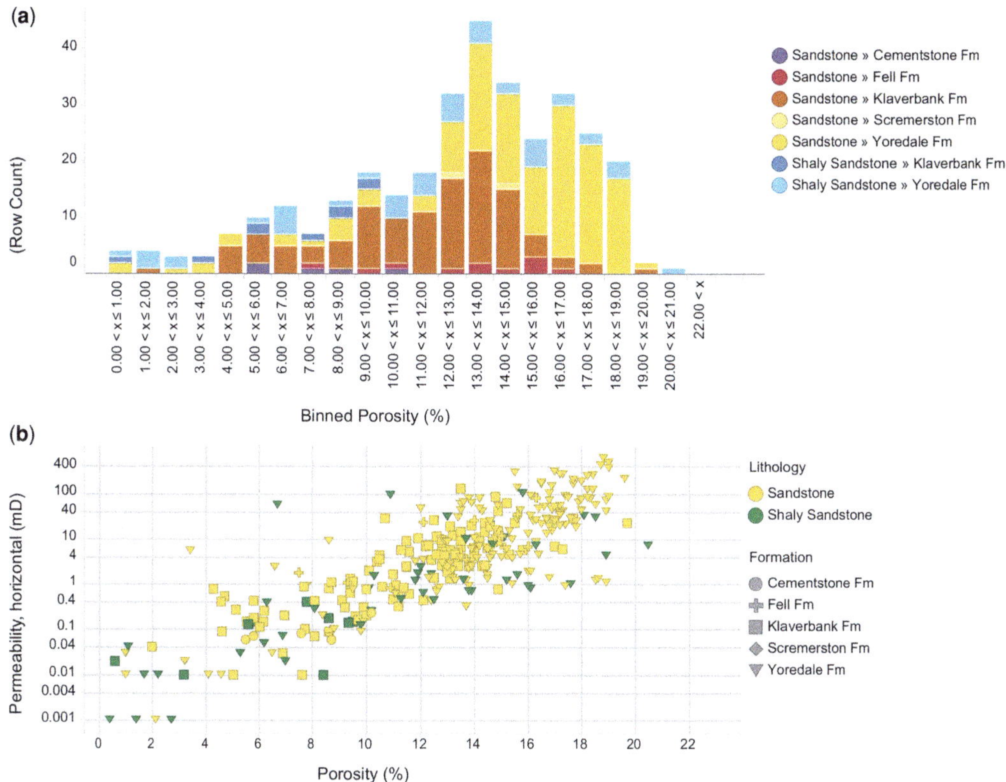

Fig. 6. Porosity and permeability measurements from Visean and Namurian intervals classified as sandstone or shaly sandstone. $N = 328$. Data are from wells A14-01, A16-01, B10-01, E02-01, E06-01, E12-02, E12-03, E12-04-S2, 43/02- 1 and 42/13- 2. Note that measurements below the detection limit are plotted at the detection limit. (**a**) porosity distribution; (**b**) porosity–permeability plot.

because the formation is completely penetrated, the total amount of coal encountered is greater (30 m).

In well 39/07- 1, the presence of coals is accompanied by a typically high-contrast seismic facies. This facies can be mapped a small distance into the Dutch offshore, but cannot be continued across a WNW–ESE-trending normal fault south of the well location that downthrows the Visean (Fig. 2). It may be the case that deposition and preservation of the coals is fault-controlled, but the downthrown fault block does not appear to be imaged on the seismic data covering this area. Further west, indications for the presence of coals are given by NSR line 32294, which has a better imaging quality at these depths (Fig. 10); on the ESP, the seismic facies shows only moderate contrast, but further north and up to the German border the contrast increases (Fig. 10). In the continuation of the line, 13 km further to the NE, the coal-bearing well A-09-01 is located (Fig. 1). Altogether, these observations are compatible with the presence of coals in the region north of the ESP, although the higher contrast may also result from lithologies other than coal.

Coals have also been encountered in the Yoredale Formation: 4 m in 39/07- 1, 3 m in well A-09-01, 4 m in well A16-01, 0.3 m in A14-01 and 2.7 m in E06-01. The same is true for the Namurian: wells 39/07- 1, B17-04, E09-01, E10-03, E12-02, E12-03 and E12-04-S2 all contain between 1 and 3 m of coal. The estimates given are conservative as they only take into account coal seams that are thick enough to be distinguishable on the sonic log.

Maturity measurements

The available well data are of limited use for assessing maturity as it primarily covers the present-day highs (Fig. 1), and not the lows where source rocks have been buried to greater depths and hydrocarbon kitchens are expected. The base of the Carboniferous is located at approximately 2000 m true vertical depth (TVD) in well E02-01 and at 2200 m in well E06-01, while the Carboniferous is locally buried up to depths of more than 5500 m in the SG, and to even greater depths in the DCG.

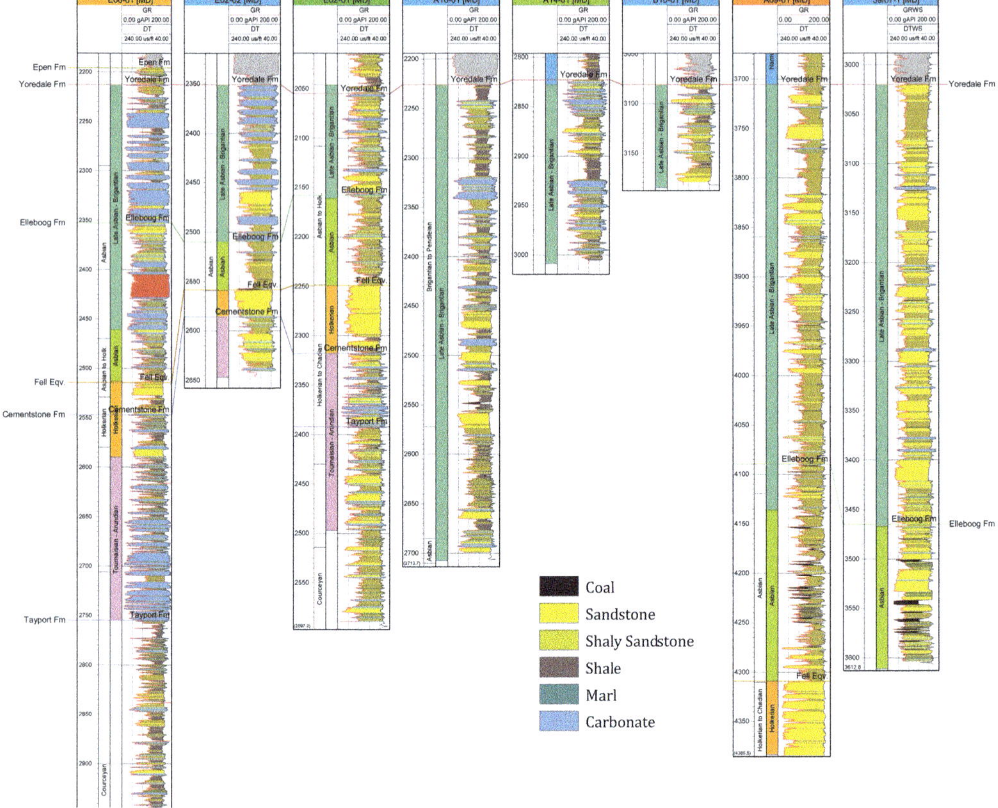

Fig. 7. Well correlation panel, Visean, Dutch offshore. The location is shown in Figure 5. Red denotes igneous rocks. Two columns with biostratigraphy are given: available data are shown in the first column, the resulting interpretation (which is also used in Fig. 5) is in the second column. Measured depth is used as the vertical scale. All wells are vertical, meaning that the difference between true vertical depth and measured depth is negligible.

The only well that can be considered representative of the present-day lows (i.e. the SG and DCG) in terms of maturity is well B17-04, where mature coals were found. Palynological dating suggests a Namurian (Pendleian–Arnsbergian) age (Cutler & Darlington 1990), and vitrinite reflectance data show that the coals have reached the peak gas window (1.4%Ro).

Maturity data from the present-day highs show that source-rock-bearing intervals are presently in the late oil–early gas window. In well A14-01, a value of *c.* 0.74%Ro was measured at around 2850 m and well 39/07- 1 shows a value of *c.* 0.8%Ro at 3500 m. In well 38/16- 1, the vitrinite reflectance increases from 0.75 to 1.00%Ro over an interval of 200 m. A similar gradient is apparent in well 41/24a- 2. The measurements from well E06-01 do not display any relationship with depth; vitrinite reflectances vary around 1.0%Ro. This may result from the (local) occurrence of igneous intrusions near this well, which could have led to a local increase in temperature.

Geological evolution and palaeogeography

During the Early Carboniferous, the northern Dutch offshore experienced siliciclastic influx from the north. The region was subject to lithosphere-scale extension, leading to an alternation of highs and lows (Fig. 3). A major facies boundary traversed the E and F quadrants, roughly trending WNW–ESE (Figs 2 & 5). To the north of the boundary, paralic conditions prevailed and to the south of it predominantly deeper-water conditions.

The facies boundary was structurally controlled and coincided with a major normal fault. North of the fault, a long-lived high was present: the ESP (Figs 2 & 5). The facies boundary represents the area where sediment influx and subsidence were

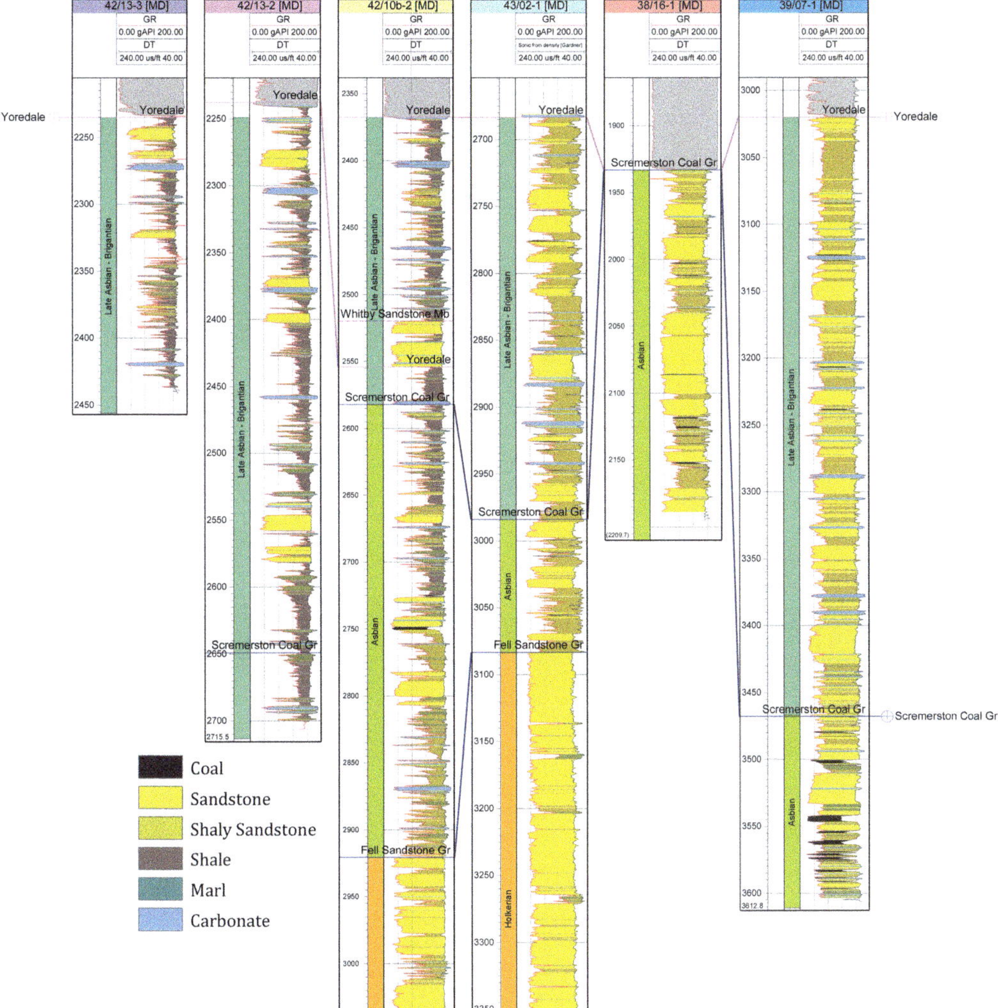

Fig. 8. Well correlation panel, Visean, UK offshore. The location is shown in Figure 5.

roughly in balance. North of the boundary, sediment influx was higher than subsidence, leading to overfilled basins; south of the boundary, subsidence was higher than sediment influx, leading to starved basins (Fig. 3). Seismic data, however, indicate that the Lower Carboniferous and Devonian also thicken north of the ESP, and this is interpreted to result from the presence of the North Elbow Low (Figs 3 & 10).

Although the boundary between deep water and shallow water is thought to have existed throughout the Visean (Fig. 5), fluctuations of relative sea level caused the overfilled area to drown episodically, establishing shallow-water marine conditions, followed by an infilling of the accommodation space by fluvial processes. On the ESP, this led to a cyclic alternation of shallow-marine carbonates and fluvio-deltaic clastic sediments, including back-swamp deposits and coals: the so-called Yoredale cycles (e.g. Waters *et al.* 2007). The Yoredale Formation is best known for this type of cyclicity but similar cycles can be observed in the older formations. These short timescale sea-level fluctuations were accompanied by steady regional tectonic subsidence, allowing for the preservation of sediments.

Towards the north, depositional environments were progressively more proximal, as can be inferred from a decrease in the amount of carbonate, and an increase in clastics and coals (Fig. 5). Carbonate beds are present at least as far north as well

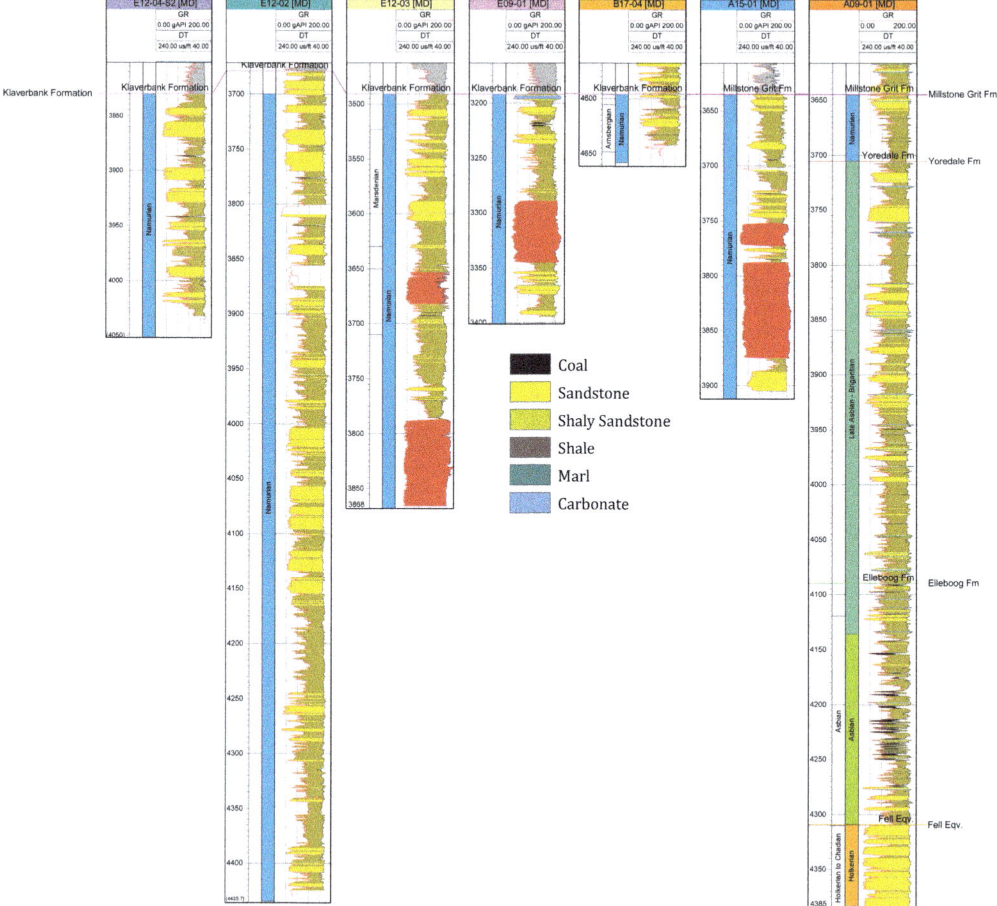

Fig. 9. Well correlation panel, Namurian, Dutch offshore. The location is shown in Figure 5.

39/07- 1, however, suggesting that the entire study area experienced episodic drowning. An alternative explanation for the presence of these beds that cannot be ruled out at present is that they formed in a lacustrine setting.

Courceyan–Chadian: Tayport and Cementstone Formation

Courceyan–Chadian (Tournaisian–lowermost Visean: Fig. 4) deposits have been encountered in the Dutch northern offshore in wells E06-01, E02-01 and, possibly, in well E02-02, for which no biostratigraphic constraints are available. The oldest deposits found are fluvial and playa lake deposits from the Tayport Formation (Old Red Group); the majority of the Old Red Group, including the lower part of this formation, is Devonian in age, but sedimentation continued into the Tournaisian (Early Carboniferous) (Cameron 1993; van Adrichem Boogaert & Kouwe 1993–97) (Fig. 4).

The Tayport Formation is covered conformably by the sediments of the Cementstone Formation, which is characterized by metre-scale alternations of carbonates and fine clastic sediments, representative of deposition in an alluvial-plain to marginal-marine flat environment that was affected by periodic desiccation and fluctuations in salinity, in a semi-arid climate (e.g. Waters *et al.* 2007). In the UK onshore, the occurrence of cementstone facies is commonly limited to troughs associated with (half-) graben; on the highs, the laterally equivalent cornstone facies is developed (Waters *et al.* 2007). As the Tayport Formation can be considered to have been developed in the cornstone facies (Waters *et al.* 2007), it cannot be ruled out that the Tayport Formation and Cementstone Formation are laterally equivalent with the boundary between the two formations being diachronous.

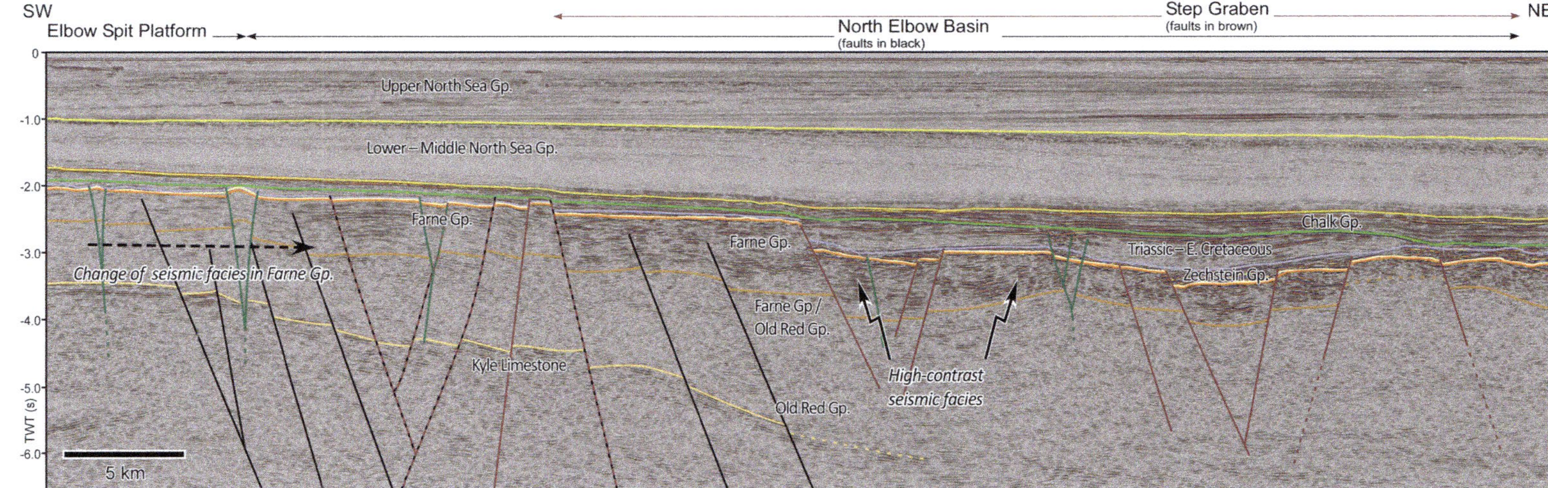

Fig. 10. Seismic section across the North Elbow Basin. The Visean Elleboog Formation (see Fig. 4) has a high-contrast seismic facies that is likely to have been caused by the presence of coals. The location is shown in Figure 1. Public seismic line NSR32294. TWT, two-way travel time.

Owing to the limited well penetrations in the Dutch offshore, little is known about the direction of sediment transport but it seems likely that the sediment source area was located in the north, as has been inferred for the rest of the Visean. It is likely that local palaeocurrents were affected by tectonics, aligning to the axes of (half)graben. As only very limited well data are available for this stratigraphic interval in the Dutch sector (Fig. 7), little is known about the basin structure during this period. Local highs may have existed on which no deposition occurred or, in the case of sufficient sediment influx and accommodation space, fluvial and/or playa lake deposits may have been deposited. A transition to deeper-marine conditions further south and a gradual transition to more proximal facies to the north may be inferred.

Arundian–Holkerian: lower part of the Elleboog Formation

During the Arundian–Holkerian, deposition occurred in a fluvial to paralic setting; rivers deposited thick stacked channel fills, whilst shales and carbonates were being deposited in areas experiencing a marine influence (Cameron 1993; van Adrichem Boogaert & Kouwe 1993–97).

We interpret the thick sand bodies at the base of the Elleboog Formation in wells E02-01 and E02-02 as representing a more distal equivalent of the more massive sandstones of the Fell Formation of the UK offshore (Cameron 1993). The north–south proximal–distal trend is further illustrated by sediments encountered in well E06-01, which were deposited close to the basin edge (Fig. 5d), where carbonate beds appear, alternated with shales and blocky sandstones. This alternation is interpreted to be similar to the cyclicity observed in the Asbian–Pendleian Yoredale Formation. It is hard to place this well in the standard lithostratigraphic nomenclature (van Adrichem Boogaert & Kouwe 1993–97); in The Netherlands, it is commonly placed in the Elleboog Formation, but the presence of carbonate beds is at odds with the definition of this formation. This is due to its position close to the platform edge; the current lithostratigraphic framework works better for deposits in a more proximal setting.

Asbian: upper part of the Elleboog Formation

Asbian deposits on the ESP are generally characterized by an alternation of sandstone, shale and carbonate, interpreted to have been deposited in a paralic setting, and more shale and coal-rich deposits in back-swamp areas. They form the upper part of the Elleboog Formation which, as mentioned earlier, is the Dutch equivalent of the UK Scremerston Formation. Similar to the preceding period, the well-derived facies show a transition from proximal to distal towards the south (Cameron 1993; van Adrichem Boogaert & Kouwe 1993–97). On seismic data, this alternation of deposits produces a distinct high-frequency, high-continuity seismic facies, which is continuous over the ESP. A change in seismic facies to lower frequencies and lower continuity occurs across the major normal fault that bounds the platform on its southern side; this is interpreted to represent the transition to basinal conditions (Fig. 11). This change in seismic facies was mapped as a palaeogeographical boundary (Fig. 5c). Deposits from this interval show a gradual northwards change to more proximal conditions, reflected in an increasing sand and coal content, and a decreasing carbonate and shale content.

Late Asbian–Pendleian: Yoredale Formation

Towards the end of the Asbian, a gradual relative sea-level rise occurred, as can be inferred from an upwards increase in the incidence and thickness of carbonate beds (Figs 7 & 8). This trend continues into the Brigantian. The Yoredale Formation consists of a cyclic alternation of carbonates and clastics, with rare coal seams (Fig. 4). It was deposited in a paralic setting and the cyclicity resulted from sea-level fluctuations. A typical Yoredale cycle starts with a sea-level rise establishing marine conditions, resulting in the deposition of carbonates and clays. Progradation subsequently results in a coarsening-upwards clastic succession that may end with a coal layer or palaeosol (Waters *et al.* 2007). Similar to the situation during the Asbian, the deposits are sandier in the north; the Brigantian in well E06-01

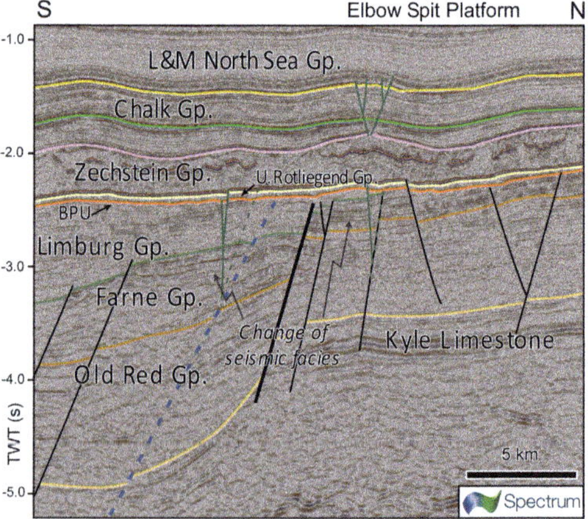

Fig. 11. Late Devonian–Early Carboniferous faulting on the southern flank of the Elbow Spit Platform. The approximate location is shown in Figure 1. DEF 3D seismic data courtesy of Spectrum ASA.

in the SE consists almost entirely of carbonate, in well A16-01 of roughly equal amounts of carbonate, clay and sand, whilst well 39/07- 1 in the NE consists mostly of sand and clay with minor carbonate beds (Figs 5b & 7). The youngest Yoredale deposits encountered in wells in the Dutch northern offshore that have been dated with reasonable certainty are Brigantian in age. Deposits from the Pendleian, the earliest substage of the Namurian (Fig. 4), may have been drilled locally (for instance, in well A16-01), meaning that Yoredale-type deposition possibly continued into the early Namurian.

Namurian

At the start of the Namurian, the Visean bathymetry essentially still existed. It is commonly viewed that extension had ceased, although indications exist that significant extension must still have occurred (Kombrink *et al.* 2008). During the Namurian, a turbidite-fronted delta system prograded southwards, the sediment source area was located towards the north (Collinson 2005). South of the delta, shales and turbidites were deposited. Whilst no penetrations of this interval exist in the northern Dutch offshore (Fig. 4), their presence can be inferred from correlations to the UK onshore and well 43/17-2 (Fig. 2) (Collinson Jones Consulting 1995; Collinson 2005). The foresets of the prograding delta were characterized by Millstone Grit and Klaverbank deposition (Fig. 4). Since the system was prograding, the formation boundaries are diachronous and expected to be older in the north than in the south. This progradation continued during the Westphalian; Namurian Klaverbank deltaic deposits, for instance, closely resemble Westphalian A Klaverbank deltaic deposits, but these facies were located further north during the Namurian. Similar to the rest of the Carboniferous, significant higher-order cyclicity occurred during this period. Once a basin (or part of a basin) had been filled, however, the basin never reverted back to fully basinal conditions (Collinson 2005).

Discussion

Reservoir potential

Visean. Two end-member models exist for predicting reservoir presence in the Visean: one model predicts reservoir presence based on fault control (Turner *et al.* 1993); the other focuses on the transgressive infill of palaeovalleys that incised during a sea-level drop (Maynard & Dunay 1999). Both models are probably valid end members.

The incised valley model is highly relevant, in particular for the area close to the basin edge, as is the case for the Breagh Field example. A prediction of reservoir presence here requires an understanding of the location of the valleys, which is challenging owing to their limited thickness and the limited well control. Somewhat surprisingly, the Breagh Field is located in a zone with a relatively large proportion of shales (Fig. 5: well 42/13- 2).

The ESP (Fig. 2) was affected by episodic drowning and progradation, expressed in the cyclic alternation of sandstones, shales, coals and carbonates. This is well documented for the Yoredale Formation, but the results from this study show that this cyclicity is also present in the underlying Elleboog Formation. The well data suggest that the proportion of clastics in both formations increases northwards. North of the ESP, Late Devonian–Early Carboniferous faulting led to the formation of the North Elbow Basin (Fig. 2) (ter Borgh *et al.* this volume, in press). The basin trends WNW–ESE, measures about 60 km from SSE to NNW and can be continued across the median line (Fig. 2) (Milton-Worssell *et al.* 2010; Arsenikos *et al.* 2018). More accommodation space was available in the basin during the Visean; the exact impact on reservoir potential is unclear and requires further study, but the overall trend of an increasing clastic sediment content towards the north is supported by wells north of this basin.

The results from this study and the structural framework developed for the study area (ter Borgh *et al.* this volume, in press) demonstrate the activity of syndepositional WNW–ESE-trending faults during the Late Devonian–Early Carboniferous, with the southern boundary fault of the ESP as the most prominent example. South of it, basinal conditions prevailed; north of it, paralic conditions. Considering the significant offset at the fault and the rapid change of seismic facies across it, mass-flow processes should be expected, and, although undrilled, reservoir-bearing turbidites may be present south of the fault (Fig. 11).

Namurian. The Namurian part of the Klaverbank Formation shows high net-to-gross deltaic sandstones across the study area, in line with some of the most prospective reservoir intervals of the Westphalian (Figs 5 & 9). The Tulp and Lelie fields (Fig. 1) confirm the reservoir potential of this interval.

Hydrocarbon charge from the Visean and Namurian

An important factor that has so far discouraged exploration in the northern Dutch offshore is the perceived absence of source rock outside the DCG. As of yet, producing gas fields in this area are limited to Mesozoic fields charged from the Early Jurassic Posidonia Shale Formation, and Neogene shallow gas fields. Whether shallow gas originates from thermogenic or biogenic sources is subject to debate, although a substantial biogenic contribution seems likely (Schroot & Schüttenhelm 2003; van den Boogaard & Hoetz 2014).

There is a significant sampling bias of potential source-rock intervals towards samples from the present-day structural highs and towards samples of coals. The bias towards the highs results from the fact that the wells have been primarily drilled here. The bias to coals results from the often-made assumption that all gas from the Carboniferous is sourced from coals, meaning that sampling and observations focused on these coals. Shales and dispersed organic (plant) material may also have significant potential (e.g. Besly 2016), however, but have historically been sampled less frequently. Coals are present not only in the Westphalian, but also within the Visean Elleboog Formation (the upper part of which is the equivalent of the UK Scremerston Formation), the Yoredale Formation and the Namurian Klaverbank Formation (Fig. 4).

The sampling bias towards samples from the present-day highs also affects maturity data – the available maturity data come mostly from the highs, and maturity in the lows should be expected to be higher. The source rocks in these lows are expected to have reached maximum burial during the Neogene. The Lower Carboniferous on the present-day highs is also expected to be at or close to maximum burial; Visean and Namurian deposits may have been exhumed during the formation of the BPU by about 1250 m (van Buggenum & den Hartog Jager 2007), but have since been buried again to depths greater than 1900 m. It is therefore encouraging that the measurements from wells E02-01 and E06-01, taken on the Elbow Spit High, are already in the late oil–early gas window. Wells A14-01 and A16-01 only penetrate part of the Namurian; this interval is in the oil window and immature for gas. The Visean has not been penetrated in these wells. The only sample representative for the present-day lows, from well B17-04, show that the Carboniferous has entered the main gas window here.

Well and seismic data show that it is likely that significant amounts of Visean Scremerston coals are present north of the ESP. This probably results from a facies shift; the wells in the E quadrant were located close to the shoreline, where limited amounts of coals have been preserved; wells 39/07- 1 and A-09-01 that contain significant amounts of coal are located in a more proximal setting (Fig. 5c). The trend of increasing coal content towards the north has been recognized previously in the UK offshore (Leeder 1988; Cameron 1993). The deposits are presently buried in the SG area to depths of up to at least 5500 m, with most of the deposits at depths of between 3000 and 5000 m, meaning that if coals are, indeed, present they are most likely to be gas mature.

Organic material is dispersed throughout Visean and Namurian units, and the source-rock potential of these deposits can be assessed using samples from the UK onshore, and from the offshore well 43/17-2, that drilled the Namurian in a basinal section. The so-called marine bands within the basinal sequence show an average total organic content (TOC) of 4.1 ± 0.78%, and the non-marine bands 1.89 ± 0.6%. Within the delta, these values are 3.86 ± 2.47 and 1.6 ± 0.76%, respectively (Collinson Jones Consulting 1995). Similar deposits are to be expected in the southernmost part of the E and F quadrants.

Basin modelling in the German offshore directly north of the study area (Arfai & Lutz 2018) shows that Visean–Namurian source rocks first become mature in the Late Carboniferous, and that peak generation and expulsion in the SG area occurs prior to Late Cretaceous inversion and continues to the present day. Visean–Namurian source rocks are interpreted to have charged the A6-A gas field (Fig. 1, well A-06-01). In the Central Graben, generation occurred until the Late Jurassic, at which point the source rocks were buried to depths where they became overmature (Arfai & Lutz 2018).

Hydrocarbon charge from other source rocks

In the F quadrant and the eastern and southern parts of the E quadrant, Westphalian Coal Measures – the predominant source rock in the Dutch onshore and offshore – are mature present day. A lateral charge of Westphalian gas into Visean and Namurian reservoirs is possible on the southern flank of the ESP, as the large-scale structure permits updip migration from downthrown kitchens (Fig. 3).

Charge was confirmed in blocks E12 and E09, but these discoveries were uneconomic as a result of the high nitrogen content of the gas (Table 3). Possible sources of the nitrogen include very early mature or overmature organic sources and inorganic sources. In a regional evaluation, Verweij *et al.* (2017) proposed that, in the case of the E12 and E09 blocks, the nitrogen originated from coals and shales that were rapidly heated by magmatic intrusions. Intrusions were identified in two of the three wells with high nitrogen contents (Table 3 & Fig. 9). Vitrinite reflectance data show that coals directly adjacent to the intrusion are overmature (>5%Ro) (Philippe *et al.* 1992). Sills are often readily observed on seismic data in this area, allowing for an assessment of the economic risks associated with high nitrogen contents. The only other gas sample from a Paleozoic reservoir in the northern Dutch offshore showed acceptable nitrogen values (Table 3, A15-01: 16%).

Considering the significant normal faulting in the DCG and SG, lateral charge from stratigraphically younger units is possible. Within the DCG, the organic-rich Jurassic Posidonia Shale Formation is preserved; indeed, a number of oil discoveries in the DCG show that the formation is oil mature and locally gas mature. Basinal facies within the Zechstein also have source-rock potential, and the

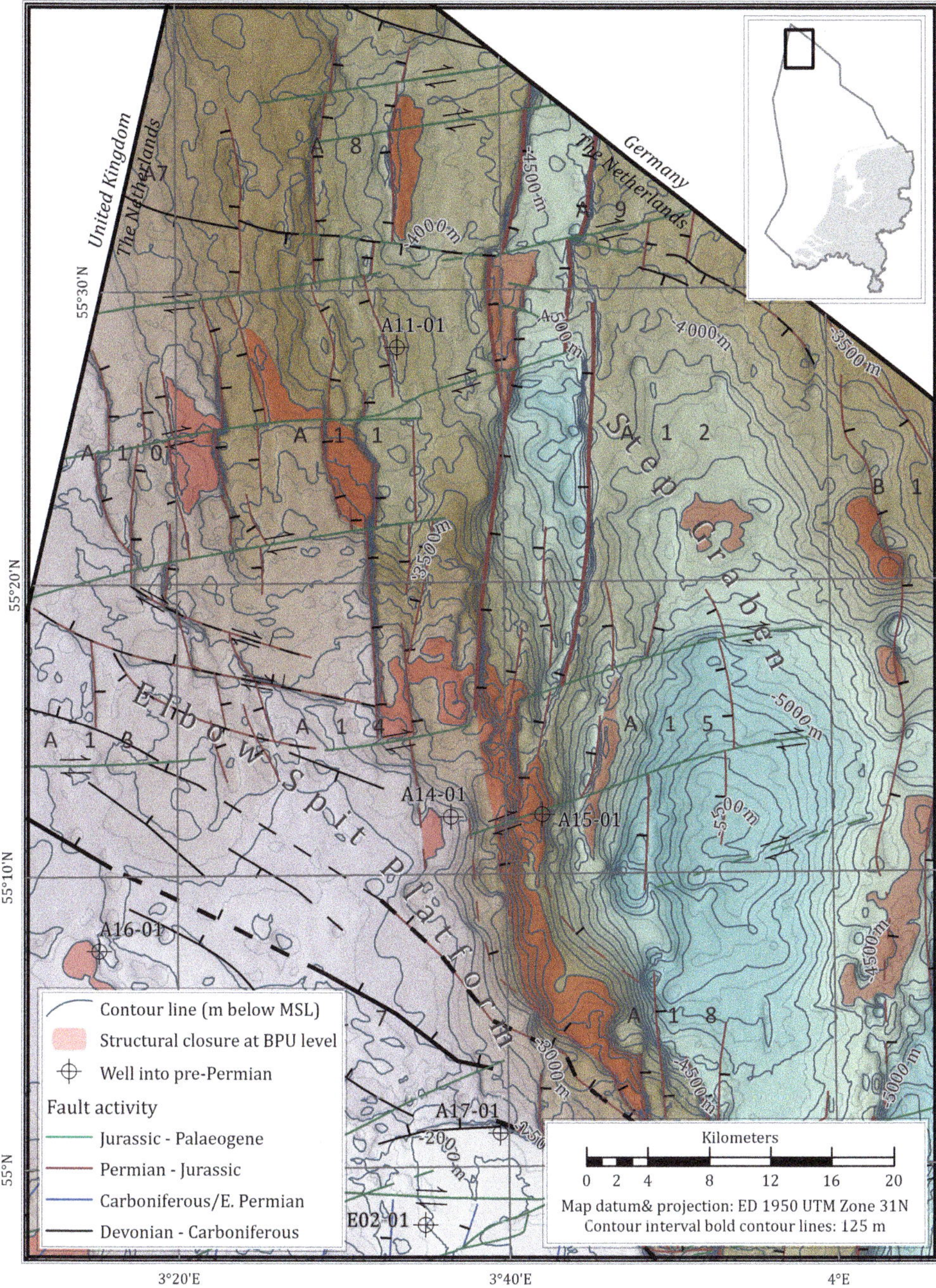

Fig. 12. Structures at the BPU level in the A quadrant, illustrating the types of structures that may form traps for hydrocarbons. The figure should not be regarded a detailed assessment of the prospectivity of the area as it is based on a regional seismic interpretation and time–depth conversion, and only structures with a height greater than the contour interval (62.5 m) are indicated. MSL, mean sea level; see Figure 1 for other abbreviations.

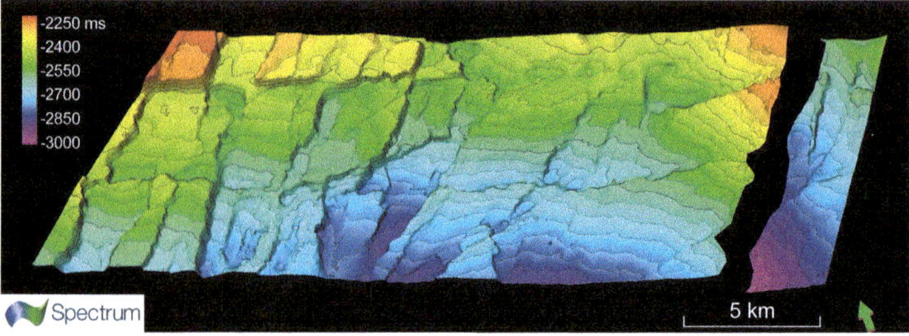

Fig. 13. Top Yoredale two-way travel time (TWT) map (in ms) illustrating structures at the Yoredale Formation level in the E blocks. Mapping is on the 3D DEF survey, courtesy of Spectrum ASA.

Kupferschiefer at the base of the Zechstein may form an auxiliary source rock too. Although often overlooked, significant shows occur in Paleozoic and Mesozoic units in a number of wells in the A quadrant, providing proof that a mature source rock is present in this area as well. The number of wells that drilled into Paleozoic strata in this area is limited, but most of the wells that are available have significant shows; well A15-01 tested gas from tight Zechstein carbonates and had oil shows in the Zechstein and Triassic. Well A08-01 and A12-01 had gas shows throughout the Mesozoic and down to the Zechstein, and well A12-02 had oil shows in the Chalk Group.

Trap formation, top seal and fault seal

The structural geology of the study area is discussed in detail in ter Borgh *et al.* (this volume, in press). Based on that study, trap formation is discussed. An example of the relevance of the SG trend for trap formation is shown in Figure 12; in the A quadrant, the alternation of horsts and graben creates potential closures. The likely seal for such closures would be provided by the Rotliegend Silverpit Formation and the Zechstein salt, the latter of which can, in some cases, be observed to drape the horsts. The main risks for these traps is the presence of an effective seal, as the Zechstein salt thins out towards the west, and the Silverpit Formation may, in fact, contain sandstone streaks locally. In the case that these are of sufficient reservoir quality they may be the target, but otherwise they form a waste zone. Another reservoir and potential target that may be present below the Silverpit and Zechstein are the Lower Rotliegend volcanics (de Bruin *et al.* 2015), although the flow capacity of this unit has not been demonstrated.

Faults with a N040° trend are known to cause compartmentalization in the Cleaverbank High area (Oudmayer & de Jager 1993; van Hulten 2010), and it is not unlikely that they would have the same effect in the study area. On the Cleaverbank High, vertical offsets are generally in the order of tens of metres. This is also true for part of the study area, but large offsets (hundreds of metres to 1 km) occur as well, with the Urania Graben located on the eastern flank of the ESP as the most prominent example (ter Borgh *et al.* this volume, in press). Establishing under what conditions these faults act as a seal, and whether an increase in offset improves or degrades the sealing and compartmentalization potential of the faults, could help to de-risk the sealing potential of this fault trend.

The N110° and N040° trends have the potential for creating intra-Carboniferous closures; for instance, in horsts or footwalls created by the N110° trend (Fig. 13). The hydrocarbon potential of these structures depends on the presence of either the Silverpit Formation or intra-Carboniferous seals as a top seal, combined with side seals along or across the fault. The charge would have to be provided from downthrown Upper Carboniferous sources. The charge and sealing risks for these structures are considerable. This may be offset by their relatively large volumes.

Strike-slip faults with a N070° trend were recently identified on the ESP (ter Borgh *et al.* this volume, in press). As they have not been identified in public literature before, no studies are presently available to assess their sealing potential. Where pop-up features are overlain by a seal, traps may be present.

Conclusions

Visean and Namurian deposits in the northern Dutch offshore are found to offer a significant hydrocarbon potential. The post-well analysis shows that out of the 14 wells that encountered the Visean and Namurian in the northern Dutch offshore, only three can be considered a valid test: two were positive and one negative. The other wells were invalid tests of the play because they did not test a valid structure. This has two main causes: first, all but two wells were targeting a structure at another stratigraphic level, whereas no structure was present in the Visean or Namurian; and, secondly, often structures were mapped using 2D seismic data only, whereas presently available 3D data show that no structure is present at these well locations. Structures do exist, however: in the Step Graben (SG) area Triassic–Early Cretaceous extension caused structures to form, with closures in footwalls. Top and, in some cases, side seals would have to be provided primarily by the Permian Silverpit and Zechstein Group deposits. Carboniferous–Early Permian extension caused structures to form on the Elbow Spit Platform (ESP); traps at these levels would require intra-Carboniferous seals.

Core-plug data confirm that sandstones of Visean and Namurian age with sufficient reservoir quality are present across the area. Chances of encountering reservoir rocks increase towards the north; the Breagh Field, although a commercial success, is located on the southern edge of the area where Visean–Namurian fluvial and shallow-marine reservoir rocks are expected.

A number of potential source rocks have been identified: the presence of coal in nearby wells and the palaeogeographical reconstruction suggest that significant amounts of Visean Scremerston coals may have been preserved in the A and B quadrants. Additional coals were found in the Yoredale Formation and the Namurian, and TOC measurements show that dispersed organic material is present within the Visean and Namurian. In the southern E and F quadrants, lateral charge may have come from downthrown Westphalian deposits, and from basinal Visean and Namurian shales that have not been drilled but whose presence may be inferred from the palaeogeographical reconstruction. Additional charge may have come from downthrown basinal Zechstein deposits and by long-distance migration from the Lower Jurassic Posidonia Shale Formation in the Dutch Central Graben (DCG). Hydrocarbon shows that support this do occur. Hydrocarbon generation and expulsion is expected to have occurred north, east and south of the ESP, and updip migration onto the platform is considered possible.

Acknowledgements We thank reviewers Andrea James and Malcolm Gall, and editor Alison Monaghan, for their constructive comments. Permission from Spectrum ASA to show data from the DEF survey is much appreciated.

References

Arfai, J. & Lutz, R. 2018. 3D basin and petroleum system modelling of the NW German North Sea (Entenschnabel). *In*: Bowman, M. & Levell, B. (eds) *Petroleum Geology of NW Europe: 50 Years of Learning – Proceedings of the 8th Petroleum Geology Conference*. Geological Society, London, 67–86, https://doi.org/10.1144/PGC8.35

Arsenikos, S., Quinn, M.F., Pharaoh, T., Sankey, M. & Monaghan, A. 2015. *Seismic Interpretation and Generation of Key Depth Structure Surfaces within the Devonian and Carboniferous of the Central North Sea, Quadrants 25–44 Area*. British Geological Survey Commissioned Report CR/15/118.

Arsenikos, S., Quinn, M., Kimbell, G., Williamson, P., Pharaoh, T., Leslie, G. & Monaghan, A. 2018. Structural development of the Devono-Carboniferous plays of the UK North Sea. *In*: Monaghan, A.A., Underhill, J.R., Hewett, A.J. & Marshall, J.E.A. (eds) *Paleozoic Plays of NW Europe*. Geological Society, London, Special Publications, **471**. First published online 19 February, 2018, https://doi.org/10.1144/SP471.3

Besly, B. 2016. Carboniferous exploration in the Central and Southern North Sea: a 30 year personal retrospective. Abstract presented at the Palaeozoic Plays of NW Europe Conference, 26–27 May 2016, London.

Cameron, T.D.J. 1993. 5. Carboniferous and Devonian of the Southern North Sea. *In*: Knox, R.W.O'B. & Cordey, W.G. (eds). *Lithostratigraphic Nomenclature of the UK North Sea*. British Geological Survey, Keyworth, Nottingham, UK. Released proprietary report.

Collinson, J.D. 2005. Dinantian and Namurian depositional systems in the southern North Sea. *In*: Collinson, J.D., Evans, D.J., Holliday, D.W. & Jones, N.S. (eds) *Carboniferous hydrocarbon Geology: The Southern North Sea and Surrounding Onshore Areas*. Yorkshire Geological Society, Occasional Publications, **7**, 35–56.

Collinson Jones Consulting 1995. *Source Rock Potential of the Sub-Westphalian Carboniferous of the Southern North Sea*. Released proprietary report. Collinson Jones Consulting, Shrewsbury, UK.

Collinson Jones Consulting 1997. *Reservoir Distribution and Controls in the Dinantian and Lower Namurian of Quadrants 41, 42 and 43, Southern North Sea, Volumes 1–3*. Released proprietary report. Collinson Jones Consulting, Shrewsbury, UK.

Coward, M.P. 1993. The effect of Late Caledonian and Variscan continental escape tectonics on basement structure, Paleozoic basin kinematics and subsequent Mesozoic basin development in NW Europe. *In*: Parker, J.R. (ed.) *Petroleum Geology of Northwest Europe: Proceedings of the 4th Conference*. Geological Society, London, 1095–1108, https://doi.org/10.1144/0041095

Cutler, I. & Darlington, C. 1990. *Geochemical Evaluation of the section 600' to 15251' in the Arco B/17A-4 well, Dutch North Sea*. Robertson Group plc report for Arco Netherlands Inc. Retrieved from http://www.nlog.nl/

de Bruin, G., Bouroullec, R. *et al.* 2015. *New Petroleum plays in the Dutch Northern Offshore*. TNO Report TNO 2015 R10920.

Donato, J.A., Martindale, W. & Tully, M.C. 1983. Buried granites within the Mid North Sea High. *Journal of the Geological Society, London*, **140**, 825–837, https://doi.org/10.1144/gsjgs.140.5.0825

DOORNENBAL, J.C. & STEVENSON, A.G. (eds) 2010. *Petroleum Geological Atlas of the Southern Permian Basin Area*. European Association of Geoscientists and Engineers (EAGE), Houten, The Netherlands.

GELUK, M.C. 2007. Permian. *In*: WONG, TH.E., BATJES, D.A.J. & DE JAGER, J. (eds) *Geology of the Netherlands*. Royal Netherlands Academy of Arts and Sciences, Amsterdam, 63–83.

KEARSEY, T., ELLEN, R., MILLWARD, D. & MONAGHAN, A.A. 2015. *Devonian and Carboniferous stratigraphical correlation and interpretation in the Central North Sea, Quadrants 25–44*. British Geological Survey Commissioned Report CR/15/117.

KEARSEY, T.I., MILLWARD, D., ELLEN, R., WHITBREAD, K. & MONAGHAN, A.A. In press. Unified stratigraphy of Carboniferous petroleum system elements from the Outer Moray Firth to the Silverpit Basin, North Sea, UK. *In*: MONAGHAN, A.A., UNDERHILL, J.R., HEWETT, A.J. & MARSHALL, J.E.A. (eds) *Paleozoic Plays of NW Europe*. Geological Society, London, Special Publications, **471**, https://doi.org/10.1144/SP471.11

KOMBRINK, H., LEEVER, K.A., VAN WEES, J.D., VAN BERGEN, F., DAVID, P. & WONG, T.E. 2008. Late Carboniferous foreland basin formation and Early Carboniferous stretching in Northwestern Europe – Inferences from quantitative subsidence analyses in the Netherlands. *Basin Research*, **20**, 377–395, https://doi.org/10.1111/j.1365-2117.2008.00353.x.

KOMBRINK, H., BESLY, B. *ET AL.* 2010*a*. Carboniferous. *In*: DOORNENBAL, J.C. & STEVENSON, A.G. (eds) *Petroleum Geological Atlas of the Southern Permian Basin Area*. European Association of Geoscientists and Engineers (EAGE), Houten, The Netherlands, 81–99.

KOMBRINK, H., VAN LOCHEM, H. & VAN DER ZWAN, K.J. 2010*b*. Seismic interpretation of Dinantian carbonate platforms in the Netherlands; implications for the palaeogeographical and structural development of the Northwest European Carboniferous Basin. *Journal of the Geological Society, London*, **167**, 99–108, https://doi.org/10.1144/0016-76492008-149

LEEDER, M.R. 1988. Recent developments in Carboniferous geology: a critical review with implications for the British Isles and N.W. Europe. *Proceedings of the Geologists' Association*, **99**, 73–100.

MAYNARD, J.R. & DUNAY, R.E. 1999. Reservoirs of the Dinantian (Lower Carboniferous) play of the Southern North Sea. *In*: FLEET, A.J. & BOLDY, S.A.R. (eds) *Petroleum Geology of Northwest Europe: Proceedings of the 5th Conference*. Geological Society, London, 729–745, https://doi.org/10.1144/0050729

MAYNARD, J.R., HOFMANN, W., DUNAY, R.E., BENTHAM, P.N., DEAN, K.P. & WATSON, I. 1997. The Carboniferous of Western Europe: the development of a petroleum system. *Petroleum Geoscience*, **3**, 97–115, https://doi.org/10.1144/petgeo.3.2.97

MCPHEE, C.A., JUDT, M.R., MCRAE, D. & RAPACH, J.M. 2008. Maximising gas well potential in the Breagh Field by mitigating formation damage. Paper SPE 115690 presented at the 2008 SPE Asia Pacific Oil & Gas Conference and Exhibition, 20–22 October 2008, Perth, Australia, http://doi.org/10.2118/115690-MS

MILTON-WORSSELL, R., SMITH, K. & MCGRANDLE, A. 2010. The search for a Carboniferous petroleum system beneath the Central North Sea. *In*: VINING, B.A. & PICKERING, S.C. (eds) *Petroleum Geology: From Mature Basins to New Frontiers – Proceedings of the 7th Petroleum Geology Conference*. Geological Society, London, 57–75, http://doi.org/10.1144/0070057

OUDMAYER, B.C. & DE JAGER, J. 1993. Fault reactivation and oblique-slip in the Southern North Sea. *In*: PARKER, J.R. (ed.) *Petroleum Geology of Northwest Europe: Proceedings of the 4th Conference*. Geological Society, London, 1281–1290, https://doi.org/10.1144/0041281

PHILIPPE, B., NICOLAS, G. & PENIGUEL, G. 1992. *Geochemical Evaluation of the Carboniferous Section (3492–3865 m TD) in the E/12-3 Well (The Netherlands)* Elf-Aquitaine internal report. Retrieved from http://www.nlog.nl/

SCHROOT, B.M. & SCHÜTTENHELM, R.T.E. 2003. Expressions of shallow gas in the Netherlands North Sea. *Netherlands Journal of Geosciences*, **82**, 91–105, https://doi.org/10.1017/S0016774600022812

SCHROOT, B.M., VAN BERGEN, F., ABBINK, O.A., DAVID, P., VAN EIJS, R. & VELD, H. 2006. *Hydrocarbon Potential of the Pre-Westphalian in the Netherlands On- and Offshore – Report of the PETROPLAY project*. TNO, Utrecht, The Netherlands, retrieved from http://www.nlog.nl/

SMIT, J., VAN WEES, J.-D. & CLOETINGH, S. 2016. The Thor suture zone: From subduction to intraplate basin setting. *Geology*, **44**, 707–710, https://doi.org/10.1130/G37958.1

TER BORGH, M.M., JAARSMA, B. & ROSENDAAL, E.A. In press. Structural development of the northern Dutch offshore: Paleozoic to present. *In*: MONAGHAN, A.A., UNDERHILL, J.R., HEWETT, A.J. & MARSHALL, J.E.A. (eds) *Paleozoic Plays of NW Europe*. Geological Society, London, Special Publications, **471**, https://doi.org/10.1144/SP471.4

TURNER, B.R., YOUNGER, P.L. & FORDHAM, C.E. 1993. Fell Sandstone lithostratigraphy south-west of Berwick-upon-Tweed: implications for the regional development of the Fell Sandstone. *Proceedings of the Yorkshire Geological Society*, **49**, 269–281, http://doi.org/10.1144/pygs.49.4.269

UNDERHILL, J.R. & PARTINGTON, M.A. 1993. Jurassic thermal doming and deflation in the North Sea: implications of the sequence stratigraphic evidence. *In*: PARKER, J.R. (ed.) *Petroleum Geology of Northwest Europe: Proceedings of the 4th Conference*. Geological Society, London, 337–345, https://doi.org/10.1144/0040337

VAN ADRICHEM BOOGAERT, H.A. & KOUWE, W.F.P. 1993–97. Stratigraphic nomenclature of the Netherlands, revision and update by RGD and NOGEPA. *Mededelingen Rijks Geologische Dienst*, **50**.

VAN BUGGENUM, J.M. & DEN HARTOG JAGER, D.G. 2007. Silesian. *In*: WONG, TH.E., BATJES, D.A.J. & DE JAGER, J. (eds) *Geology of the Netherlands*. Royal Netherlands Academy of Arts and Sciences, Amsterdam, 43–62.

VAN DEN BOOGAARD, M. & HOETZ, G. 2014. Derisking shallow gas as exploration target by seismic characterisation. Extended abstract presented at the EAGE Shallow Anomalies Workshop, 23–26 November 2014, Malta, http://doi.org/10.3997/2214-4609.20147437

VAN HULTEN, F.F.N. 2010. Geological factors affecting compartmentalization of Rotliegend gas fields in the Netherlands. *In*: JOLLEY, S.J., FISHER, Q.J., AINSWORTH,

R.B., Vrolijk, P.J. & Delisle, S. (eds) *Reservoir Compartmentalization*. Geological Society, London, Special Publications, **347**, 301–315, http://doi.org/10.1144/SP347.17

van Hulten, F.F.N. 2012. Devono-Carboniferous carbonate platform systems of the Netherlands. *Geologica Belgica*, **15**, 284–296.

Verweij, H., Nelskamp, S., Goldberg, T. & Hettelaar, J. 2017. Nitrogen in Dutch natural gas accumulations revisited. Conference Abstract presented at the AAPG SEG International Conference & Exhibition 2017, 15–18 October 2017, London.

Waters, C.N., Browne, M.A.E., Dean, M.T. & Powell, J.H. 2007. *Lithostratigraphical Framework for Carboniferous Successions of Great Britain (Onshore)*. British Geological Survey Research Report RR/07/01.

Ziegler, P.A. 1990. *Geological Atlas of Western and Central Europe*, 2nd revised edn. Shell Internationale Petroleum Maatschappij, The Hague.

The role of palaeorelief in the control of Permian facies distribution over the Mid North Sea High, UK Continental Shelf

PHILIP MULHOLLAND[1]*, PAOLO ESESTIME[2], KARYNA RODRIGUEZ[2] & PHILLIP JOHN HARGREAVES[2]

[1]*Mulholland Geoscience, Weybridge, Surrey, UK*

[2]*Spectrum Geo Ltd, Dukes Court, Duke Street, Woking, Surrey GU21 5BH, UK*

P.M., 0000-0002-6539-3306

**Correspondence: philip.mulholland@uclmail.net*

Abstract: The Mid North Sea High (MNSH) consists of a ridge of Paleozoic strata located in the centre of the North Sea between 55° N and 56° N. In 2010, interest in the Permian of the MNSH was revived by the discovery in Quadrant 44 of the Cygnus gas field. This study focuses on the Zechstein carbonates of the MNSH and uses play concepts that draw an analogy with the Zechstein oil and gas fields discovered in Denmark and Poland, and also with the Wissey gas field in Quadrant 53.

New 2D seismic establishes the presence of a significant Zechstein reef that blocks the southern entrance of the Jenyon Gap onto the MNSH. Seismic data show that the reef developed in stages and its presence can be inferred from the occurrence of isolated lagoons within the centre of the build-up. The barrier reef's existence explains both the presence of the rare hygroscopic mineral tachyhydrite in the centre of the MNSH and also the observed isopach difference in the Zechstein cycles over the MNSH, as the barrier restricted marine-water ingress onto the MNSH and allowed the creation of a 'crystal-starved basin', the evaporitic equivalent of a sediment-starved basin.

Early hydrocarbon exploration in the North Sea (Kent 1967) revealed that a west–east partial barrier located across quadrants 35–39 between 55° N and 56° N of the UK Continental Shelf (UKCS), and named as the Mid North Sea High by Rhys (1974) and Kent (1975), divides the North Sea Basin. Throughout the Permian, the Mid North Sea High (MNSH) formed a palaeo-massif separating the Northern and Southern Permian basins of western Europe (Fig. 1). Hydrocarbon exploration of the Auk Field in Block 30/16 on the western flank of the Central Graben has established that at this location the buried core of the MNSH comprises folded pre-Devonian basement rocks (Trewin *et al.* 2003, fig. 3). The MNSH can therefore be considered as a continuation of the Silurian and Ordovician outcrop of the Southern Uplands of Scotland, extending east beneath the North Sea and linking farther east, beyond the Central Graben, with the similarly formed and buried Ringkøbing-Fyn High in Denmark (Cartwright 1990).

In 2010, interest in the Permian of the MNSH was revived by the discovery in Quadrant 44 of the Cygnus gas field partly reservoired in Lower Permian distal fluvial sands that were transported south by desert wadis draining from the MNSH (Catto *et al.* 2017). This study focuses on the Zechstein carbonates of the MNSH, and uses play concepts that draw an analogy with the Zechstein oil and gas fields discovered onshore to the east in Denmark (Maver 1995), Poland (Dyjaczynski *et al.* 2001) and also with the Wissey gas field in Quadrant 53 of the Southern Permian Basin (Duguid & Underhill 2010). We have combined the results of a reanalysis of the Permian strata of 41 MNSH exploration wells drilled over a 45 year-period between 1965 and 2010 with two 2D long-offset seismic exploration datasets recently acquired during 2013 and 2015 to assess the validity of the Permian hydrocarbon play concepts used during prior exploration drilling, and to propose new interpretation strategies to achieve exploration success.

The new seismic data establishes the presence of a significant Zechstein reef complex in the south of Quadrant 37 that blocks the southern entrance of the Jenyon Gap onto the MNSH. This barrier reef complex is a new play concept for the MNSH, the presence of the early formed strait-blocking barrier reef restricted marine water ingress onto the MNSH from the Anglo-Dutch Basin to the south. Seismic data show that the reef developed in stages and its presence can be inferred from the occurrence of isolated lagoons within the centre of the build-up. These lagoons are observed on seismic to be filled

From: MONAGHAN, A. A., UNDERHILL, J. R., HEWETT, A. J. & MARSHALL, J. E. A. (eds) 2018. *Paleozoic Plays of NW Europe*. Geological Society, London, Special Publications, **471**, 155–175.
First published online May 3, 2018, https://doi.org/10.1144/SP471.8

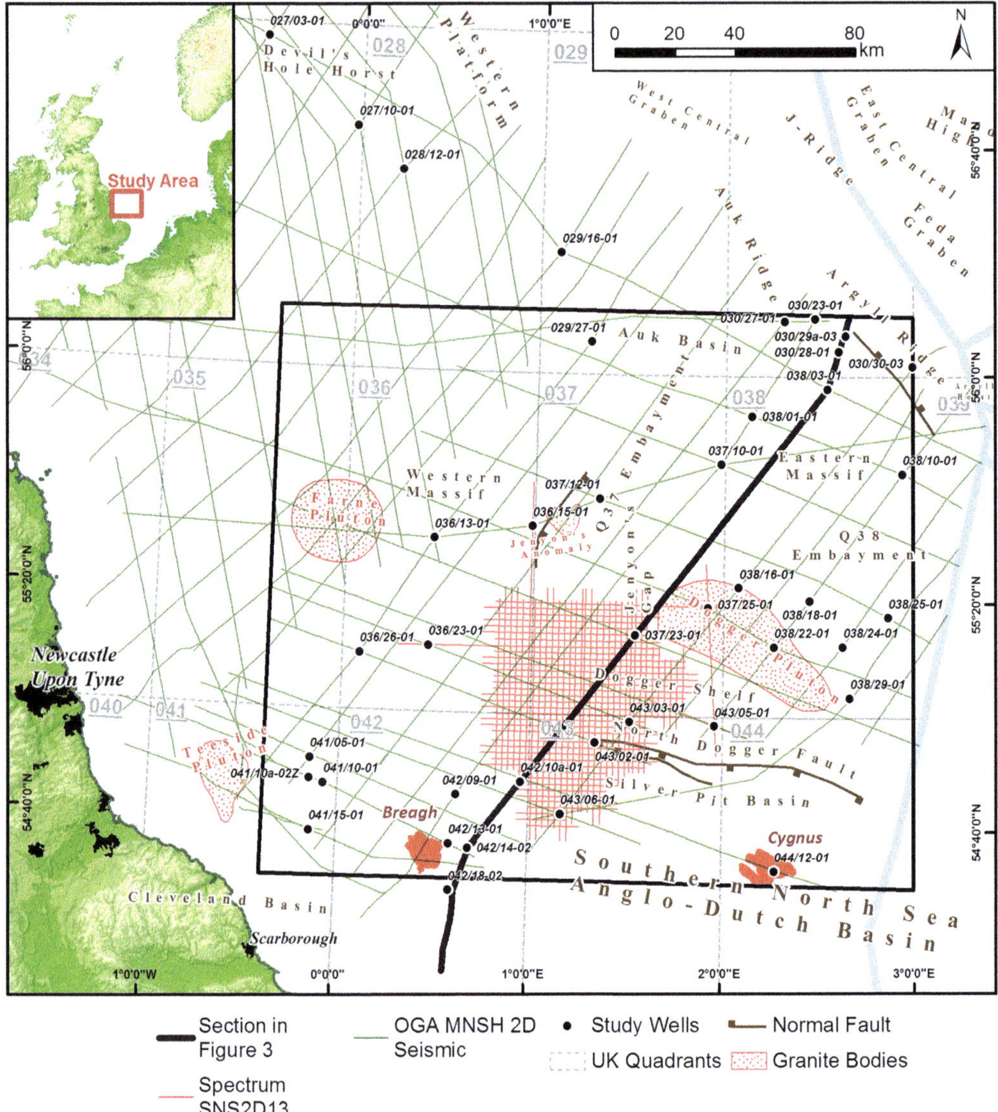

Fig. 1. Location of the study area, the distribution of Paleozoic structural elements, and the offshore wells and 2D seismic data used in this analysis of the MNSH.

with Zechstein salt that was then mobilized during the earliest Triassic by overburden sediment loading.

In the south of Quadrant 37, we therefore expect to find a play fairway of porous reservoir carbonates sealed by anhydrites and charged with either early oil that is now bypassed or filled by a later gas flush, and that these hydrocarbons have migrated north onto the MNSH from the Carboniferous source-rock kitchen of the Anglo-Dutch Basin. The barrier reef's existence explains both the presence of the mineral tachyhydrite (a hydrous chloride of calcium and magnesium) in the centre of the MNSH reported by Taylor (1993) and also explains the observed isopach difference in the Zechstein cycles over the MNSH, as the barrier restricted marine water ingress onto the MNSH and allowed the creation of a 'crystal starved basin', the evaporitic equivalent of a sediment-starved basin.

Geological and geophysical data

The 41 exploration wells (Oil & Gas Authority 2015) used in this analysis were drilled on and around the

MNSH over the 45 year-interval between 1965 and 2010, and 30 of these wells form the bulk of this study (Table 1). They were used to perform a geological reanalysis of the proven Permian stratigraphy on the MNSH, in the offshore UKCS sector of the North Sea (Fig. 2). These well data were studied alongside the lines from two 2D long-offset seismic exploration datasets, recently acquired during 2013 and 2015, to assess the validity of the Permian and older strata in this lightly explored sector of the North Sea.

The 2013 Southern North Sea (SNS) 2D seismic data were acquired as a speculative survey targeting the Rotliegend and Carboniferous plays of the Cygnus and Breagh gas fields on the southern margin of the MNSH. This survey is located in the southern boundary of Quadrant 37 with Quadrant 43 and consists of 4017 km of 2D seismic lines in a north–south- and east–west-orientated grid typically spaced 2 km apart and covering a core area of 4400 km^2. The survey was shot and processed by Spectrum Geo Ltd in 2013 (Table 2), and was designed with tie lines to nine exploration wells to aid seismic interpretation and geological analysis (Fig. 1).

The 2015 MNSH 2D seismic survey was commissioned by the Oil and Gas Authority (OGA), and was shot and processed by Western Geco Ltd (Table 3). The survey was designed with a NW–SE- and NE–SW-orientated grid of long 2D lines variously spaced between 10 and 25 km apart, and chosen to link the historical exploration wells in this poorly surveyed region of the UK North Sea. This study uses a subset of these OGA lines that extend across the 40 000 km^2 area of interest, centred on UK quadrants 36–38 of the Central North Sea (Fig. 1).

Revised lithostratigraphic nomenclature

The thick succession of Upper Permian carbonate evaporitic sediments found offshore in the UK sector of the southern Zechstein Basin have been described and studied by various authors (Taylor 1998 and key references therein). The stratigraphic nomenclature used on borehole composite logs to describe the Carboniferous, Permian and Triassic formations has changed as knowledge of the North Sea Basin subsurface has developed through time, following intensive hydrocarbon exploration of this prolific gas basin.

In this paper, we have adopted the stratigraphic nomenclature for the Upper Permian (Zechstein) used by Tucker (1991). For a comparison of the historical English and German stratigraphic nomenclature used previously, see the composite log for well 42/10a-1 (Oil & Gas Authority 2015) and also Peryt *et al.* (2010). For the Rotliegend, we have followed current practice used offshore for the Southern Permian Basin hydrocarbon exploration composite logs (e.g. 44/12-1, Oil & Gas Authority 2015) but, to take account of the predominance of desert pavement and wadi deposits observed on the MNSH, we propose that the arenaceous wadi deposits in this area be called the Cygnus Sandstone Formation rather than the Lower Leman Sandstone (Table 4).

Geological history and structural evolution

The structural evolution of the MNSH commenced between 490 and 390 Ma, during the Ordovician and Silurian–earliest Devonian interval, as part of the subduction and closure of the Iapetus Ocean (Klemperer & Matthews 1987), and the subsequent continental collision between the palaeo-North American (Laurentian) and palaeo-European (Avalonian) continents during the Caledonian Orogeny (Chadwick & Holliday 1991). Following the orogeny, a system of extensional basins formed, associated with the development of the Rheic Ocean to the south, and the older Paleozoic core of the MNSH was buried with Devonian and early Carboniferous sediments (Leeder & Hardman 1990). Later compressional movements that occurred during the late Carboniferous and early Permian, and which were brought about by the assembly of Pangaea and the end of the Paleozoic Era, are associated with both the Variscan Orogeny (van Wees *et al.* 2000) and the closure of the Rheic Ocean (Nance *et al.* 2010). This second orogeny created the series of basins and highs that constitute the present structural morphology of the MNSH, and determined the pattern and distribution of Permian sediments on and around the massif (Fig. 2).

The structural elements of the MNSH can be identified on seismic data as two distinct classes of seismic facies that can be used as a guide to interpretation (Fig. 3). Basins that preserve the reflective folded and fractured deltaic clastics with coals and earlier marine carbonates of the Lower Carboniferous (e.g. well 38/18-1) are located between, and downfaulted against, the generally less folded Devonian strata that form the rigid sandstone blocks of the structural highs (e.g. well 38/03-1). These Devonian highs are locally underpinned by low-density granitic intrusions, which range from Ordovician to early Devonian in age (Chadwick & Holliday 1991) and have an associated gravity low anomaly (Fig. 1). Three Bouguer anomalies, which were first identified by marine gravity surveys over the MNSH (Donato *et al.* 1983) and named as the Dogger (1228 km^2), Farne (658 km^2) and Teesside (275 km^2) plutons by Leeder & Hardman (1990, fig. 7), are comparable to Bouguer anomalies and associated granite plutons seen onshore to the west.

Table 1a. *Meta-data for 30 selected MNSH**

UKCS well name	Year	Operator	Licence	Intent	Crustal element	Prospect	Final plugged and abandoned (P&A) status	Spud date
38/29-1	1965	American Overseas Petroleum Ltd	P023	Exploration	Dogger Shelf	Wildcat	P&A Dry	26 December 1964
43/03-1	1966	Rycade Oil Corporation	P048	Exploration	Anglo-Dutch Basin	Wildcat	Minor gas shows in Bunter Shale	18 June 1966
27/03-1	1967	Amoco (UK) Petroleum Ltd	P019	Exploration	Devil's Hole Horst	Wildcat	P&A Dry	1 July 1967
36/13-1	1967	Arpet Petroleum Ltd	P035	Exploration	Western Massif	Wildcat	P&A Dry	28 May 1967
36/15-1	1967	Arpet Petroleum Ltd	P035	Exploration	Western Massif	Wildcat	Dead Oil	31 July 1967
38/16-1	1967	Amoco (UK) Petroleum Ltd	P018	Exploration	Dogger Pluton	Wildcat	P&A Dry	10 September 1967
38/18-1	1967	Arpet Petroleum Ltd	P036	Exploration	Dogger Shelf	Wildcat	P&A Dry	15 June 1967
38/22-1	1967	American Overseas Petroleum Ltd	P023	Exploration	Dogger Pluton	Wildcat	P&A Dry	11 October 1967
36/23-1	1969	Home Oil (UK) Ltd	P041	Exploration	Anglo-Dutch Basin	Wildcat	P&A Dry	6 April 1969
37/10-1	1969	Amoco (UK) Petroleum Ltd	P018	Exploration	Eastern Massif	Wildcat	P&A Dry	14 June 1969
38/25-1	1969	Burmah Oil Exploration Co Ltd	P005	Exploration	Dogger Shelf	Wildcat	P&A Dry	21 May 1969
27/10-1	1970	Amoco (UK) Petroleum Ltd	P065	Exploration	Devil's Hole Horst	Wildcat	P&A Dry	9 June 1970
36/26-1	1970	Placid Oil Co (UK)	P046	Exploration	Western Massif	Wildcat	P&A Dry	16 May 1970
37/23-1	1970	Texaco Production Services Ltd	P034	Exploration	Jenyon's Embayment	Wildcat	P&A Dry	8 July 1970
28/12-1	1971	BP Petroleum Development Ltd	P059	Exploration	Devil's Hole Horst	Wildcat	P&A Dry	23 August 1971
29/16-1	1973	Ashland Exploration Company	P189	Exploration	West Central Shelf	Wildcat	P&A Dry	1 October 1973
30/23-1	1973	Arpet Petroleum Ltd	P188	Exploration	Auk/Argyll High	Wildcat	P&A Dry	11 February 1973
38/01-1	1974	Texas Gas Exploration (UK) Corporation	P238	Exploration	Eastern Massif	Wildcat	P&A Dry	2 November 1974
38/03-1	1975	Mobil North Sea Ltd	P217	Exploration	Duncan Embayment	Wildcat	P&A Dry	11 August 1975
42/10a-1	1982	Total Oil Marine Plc	P034	Exploration	Anglo-Dutch Basin	Wildcat	Minor gas shows in Zechstein	31 May 1982
38/24-1	1983	Amoco (UK) Exploration Co	P443	Exploration	Dogger Shelf	Wildcat	P&A Dry	26 August 1983
37/12-1	1985	Murphy Petroleum Ltd	P438	Exploration	Jenyon's Embayment	Wildcat	P&A Dry	18 October 1985
29/27-1	1987	Amerada Hess Ltd	P509	Exploration	MNSH	Wildcat	P&A Dry	9 September 1987
44/12-1	1988	Marathon Oil UK Ltd	P514	Exploration	Cleaver Bank High	Wildcat	Gas shows in Leman	24 July 1988
41/15-1	1991	Conoco (UK) Ltd	P604	Exploration	Cleveland Basin	Wildcat	Gas shows in Z2 Haupt Dolomite	3 April 1991
30/29a-3	1996	Agip UK Ltd	P079	Exploration	Duncan Embayment	Wildcat	P&A Dry	27 April 1996
38/10-1	2002	Murphy Petroleum Ltd	P948	Exploration	Eastern Massif	Contessa	P&A Dry	6 July 2002
41/05-1	2004	Walter UK Exploration & Production Ltd	P1129	Exploration	Cleveland Basin	Wildcat	P&A Dry	29 September 2004
37/25-1	2009	Shell UK Ltd	P1259	Exploration	Dogger Pluton	Corbenic	P&A Dry	7 February 2009
42/14-2	2010	RWE Dea UK SNS Ltd	P1327	Exploration	Breagh Swell	Macanta	Minor gas shows in Zechstein	20 June 2010

Table 1b. *Meta-data for 30 selected MNSH** (*Continued*)

UKCS well name	Date the total depth (TD) was reached	Datum elevation (ft)	Water depth (ft)	(TD) True vertical depth subsea (ft)	Formation at total depth (TD)	Primary target	Main play	Key wells
38/29-1	16 March 1965	83	85	9221	Devonian	Permian	Groningen Sandstone	First well in North Sea. Zechstein cores (BGS core photographs)
43/03-1	22 July 1966	94	88	10266	Carboniferous	Zechstein	Platten Dolomite	
27/03-1	8 September 1967	84	246	6442	Devonian	Zechstein	Carbonates	Proved Devonian Kyle Limestone. Zechstein cores (BGS core photographs)
36/13-1	12 June 1967	80	255	4423	Carboniferous	Zechstein	Platten Dolomite	Type well: Zechstein Carbonate Bank and Lower Permian caliche. Regional geochemistry report
36/15-1	20 August 1967	89	254	5812	Devonian	Zechstein	Carbonates	Geochemistry report: Zechstein dead-oil staining
38/16-1	9 October 1967	94	133	7123	Carboniferous	Zechstein	Carbonates	Zechstein reservoir. Zechstein cores (BGS core photographs)
38/18-1	29 July 1967	83	131	8004	Carboniferous	Carboniferous Visean-Tournasian	Red Beds	Type well: Quadrant 38 Embayment. Oil-prone Carboniferous source rocks. Dead oil in Zechstein Dolomites
38/22-1	3 September 1967	101	116	7289	Carboniferous	Zechstein	Carbonates	Palynology report. Porous Zechstein carbonates. Zechstein cores (BGS core photographs)
36/23-1	24 April 1969	87	275	5863	Carboniferous	Zechstein	Platten Dolomite	
37/10-1	6 July 1969	78	272	8030	Carboniferous	Carboniferous	Sands	Type well: karstification. Geochemistry report (36/15-1)
38/25-1	17 June 1969	99	100	7367	Devonian	Zechstein	Carbonates	
27/10-1	26 June 1970	94	262	4829	Devonian	Zechstein	Carbonates	Barrier reef on the northern margin of the MNSH. TD apatite fission-track analysis (AFTA) age: 570–408 Ma
36/26-1	2 June 1970	88	238	4842	Carboniferous	Zechstein	Carbonates	Zechstein algal pellets
37/23-1	27 July 1970	102	125	8222	Carboniferous	Zechstein	Magnesian Limestone	Type well: Quadrant 37 channel. Geochemstry report (36/15-1)
28/12-1	29 September 1971	32	285	7373	Devonian	Lower Tertiary	Sandstone	Coastal sabkha on the northern margin of the MNSH

(*Continued*)

Table 1b. *Meta-data for 30 selected MNSH** *(Continued)*

UKCS well name	Date the total depth (TD) was reached	Datum elevation (ft)	Water depth (ft)	(TD) True vertical depth subsea (ft)	Formation at total depth (TD)	Primary target	Main play	Key wells
29/16-1	29 November 1973	85	292	10615	Rotliegend	Paleocene Sands & Fractured Chalk	Sandstone	Type well: Lower Permian Auk Sandstone
30/23-1	12 March 1973	64	248	9436	Devonian	Zechstein	Carbonates	Poor oil shows in Zechstein carbonates. Rotliegendes cores (BGS core photographs)
38/01-1	21 November 1974	64	279	6400	Carboniferous	Zechstein	Carbonates	
38/03-1	4 October 1975	85	249	12375	Devonian	Zechstein	Carbonates and sandstones	Devonian Kyle Limestone core (BGS core photographs)
42/10a-1	14 July 1982	79	200	9742	Carboniferous	Triassic	Bunter	Zechstein lithostratigraphy gives UK and German nomenclaure (Comp Log records TWTT in seconds)
38/24-1	27 September 1983	110	105	9000	Carboniferous	Zechstein	Carbonates	Zechstein cores (BGS core photographs) Westphalian Barren Measures at TD
37/12-1	13 December 1985	84	265	9198	Carboniferous	Devonian	Limestone	Key MNSH exploration review
29/27-1	21 October 1987	90	274	9413	Rotliegend	Zechstein	Reefs	Zechstein tachyhydrite salt mounds not reefs. Zechstein cores (BGS core photographs)
44/12-1	17 October 1988	121	66	15177	Carboniferous	Carboniferous	Westphalian A and B	Type well: Zechstein salina and Lower Permian Basin
41/15-1	1 June 1991	80	221	11220	Carboniferous	Carboniferous	Visean	Zechstein cores (BGS core photographs)
30/29a-3	14 May 1996	**85**	**259**	**8719**	Rotliegend	Jurassic	Sandstone	
38/10-1	27 July 2002	**85**	**217**	**8445**	Devonian	Upper Jurassic	Sandstones	Type well: Lower Permian Inge Volcanics
41/05-1	16 October 2004	83	274	6197	Carboniferous	Zechstein	ZS3 Haupt Dolomite	
37/25-1	22 March 2009	**154**	**130**	**7585**	Silurian	Devonian	Kyle Limestone	Type well: BPU pavement. Located on top of the Dogger Pluton
42/14-2	31 July 2010	110	218	7946	Carboniferous	Carboniferous	Sandstones	Comprehensive Carboniferous nomeclature

*The data for these wells were released by the UK Oil and Gas Authority under a UK Open Government Licence permit for public sector information and hosted on the CDA UK Oil and Gas Data website as a Public Access Data Package (OGA 2015 Seismic Programme for the Mid North Sea High: Deviation Surveys, Joined Logs, Checkshots/VSP, Reports, Log Images (89 wells)). See https://www.nationalarchives.gov.uk/doc/open-government-licence/version/3/.

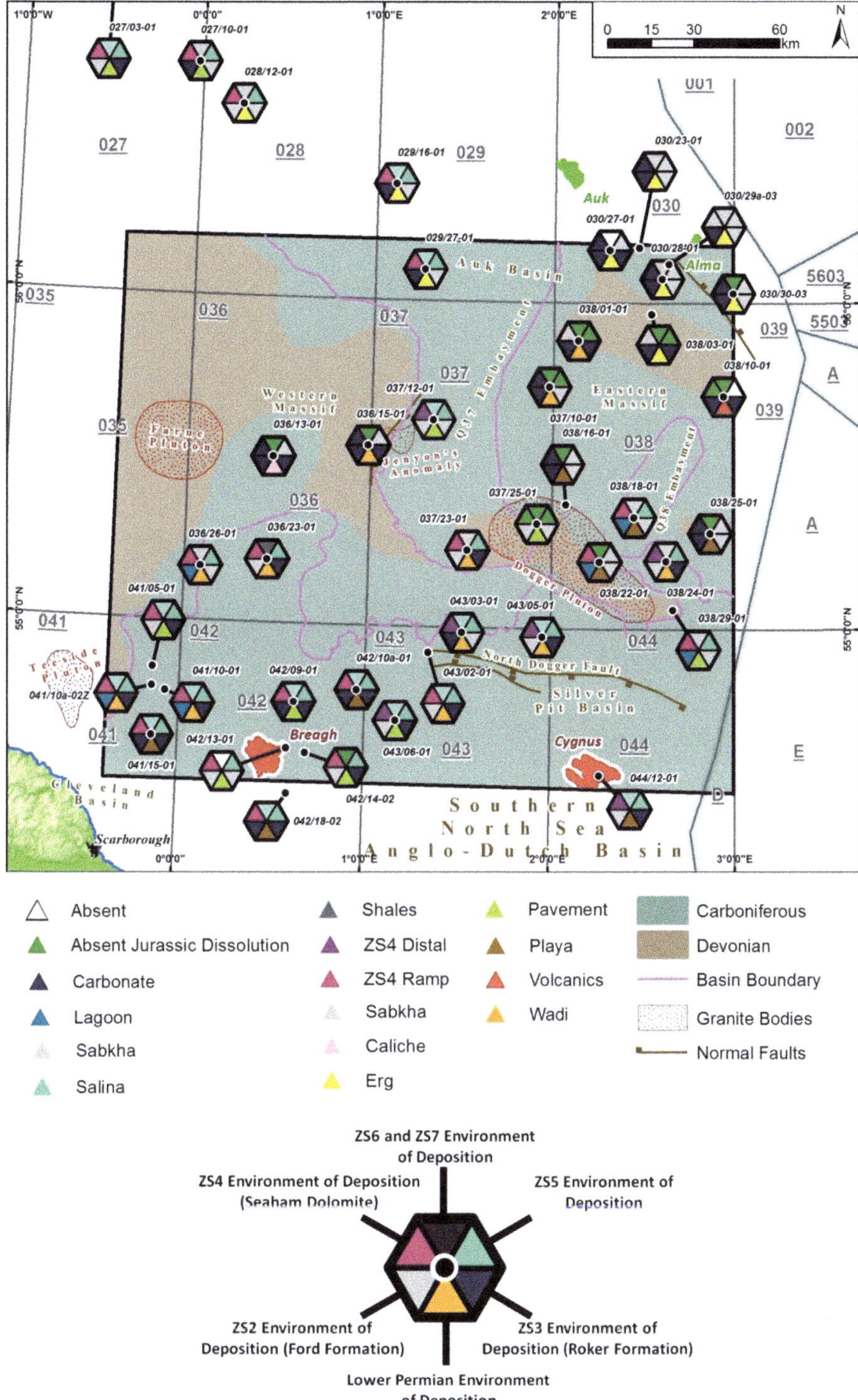

Fig. 2. The distribution of Permian facies recorded in wells located in and around the study area showing the relationship between the inferred environments of deposition, the formation age of the base Permian subcrop, and the boundary between the paleo-highs and adjacent depositional lowland basins.

Table 2. *Spectrum Geo Ltd – 2013 Southern North Sea Gas Basin 2D Seismic Survey SNS2D-13 seismic data acquisition and processing parameters*

Spectrum Geo Ltd: Southern Gas Basin SNS2D-13 acquisition parameters		Spectrum Geo Ltd: SNS2D-13 Processing sequence December 2013 (Final)	
		Processing: 6.25 m Common depth point (CDP) interval	Common offset domain Fourier-X filter
Year of acquisition:	2013	Reformat, output to internal processing format	First pass Kirchhoff pre-stack time migration (PSTM)
Acquisition contractor:	Seabird Exploration	Zero-phase designature and debubble, using provided far-field signature	Migration velocity analysis (1 km)
Acquisition vessel:	M/V *Harrier Explorer*	Resample to 4 ms sample rate with application of zero-phase anti-alias filter	Second-pass Kirchhoff pre-stack time migration (PSTM) using 1 km velocities
Shot-point interval:	25 m	SOD delay: −40 ms shift	Noise attenuation in CDP domain using FK: noise modelled in FK space and subtracted from input data
Source:	4210 cu. in. Bolt airguns/2000 psi	Low-cut filter (3/18 Hz/db Oct.)	High-density residual velocity analysis (fourth order – VELTUNE)
Source depth:	7 m	Amplitude recovery: linear T 1.8 gain correction	Inner and outer trace muting
Streamer length:	8000 m	Swell noise attenuation: two passes in shot domain and CDP domain	Stack: full-offset plus angle stacks
Group interval:	12.5 m	Tau-P linear filtering: noise removal in shot domain	Gun and cable static
Number of channels:	640	Tau-P Gap Deconvolution: two passes in shot domain and receiver domain	Post-stack noise attenuation
Streamer depth:	9 m	Multiple attenuation: surface-related multiple elimination (SRME)	Time- and space-varied bandpass filtering
Sample rate:	2 ms	Initial velocity analysis (2 km)	Expanding window automatic gain control (AGC)
Record length:	10.0 s	Multiple attenuation using high-resolution radon demultiple	SEGY output

In northern England, the Cheviot Granite (61 km^2), with an estimated K/Ar age of 380 Ma (Mitchell 1972), is a late Devonian pluton that crops out in Northumberland (Jhingran & Tomkeieff 1942) and is associated with a Bouguer gravity anomaly (714 km^2) (Chadwick & Holliday 1991, fig. 1). The Cheviot gravity anomaly is similar in size to the Farne gravity anomaly seen offshore in Block 35/15 of the MNSH (Fig. 1). To the south of the Cheviot Block, across the Carboniferous basin of the Northumberland Trough, lies the Alston Block. This structurally high area of Cumbria and County Durham, with its outcrop of Lower Carboniferous limestones, contains a major Bouguer anomaly (1535 km^2) that was identified by gravity surveys and is associated with the pluton of the Weardale Granite. This batholith was proved by the 1961 Rookhope Borehole that found an altered granite erosion surface directly beneath Carboniferous Visean Yoredale-type facies at a subsurface depth of 1281 ft (390.5 m) (Dunham *et al.* 1965).

A deep reflection seismic line across the Weardale Granite exhibits lower-crustal reflectivity at 8 s two-way travel time (TWT) (Kimbell *et al.* 2010, fig. 7), similar reflections at 8 s TWT are also seen on OGA seismic lines across the Farne, Teeside and Dogger anomalies offshore. These reflections are estimated to occur at a depth of *c.* 20 km and are believed to be cumulate rocks: layers of dense crystal accumulations formed below deep-seated Devonian granites on the aggrading floor of the original magma chamber (cf. Collins *et al.* 2006). The Dogger Bank Bouguer anomaly centred in blocks 37/25 and 38/21 on the southern

Table 3. *UK Oil and Gas Authority-2015 Mid North Sea High 2D Seismic Survey OGA 2015 seismic data acquisition and processing parameters**

Western Geco Ltd: Mid North Sea High 2015 Oil & Gas Authority (OGA) Survey		
Survey summary		**Processing**
Type:	2D long offset	EDGF noise attenuation and deghosting
Broadband:	Yes	Demultiple (GSMP) modelling (velans)
Broadband type:	single streamer deghosting	Demultiple DWD modelling and subtraction
Size:	10 849 km (89 lines)	Demultiple GSMP modelling
Acquisition year:	2015	Demultiple GSMP model subtraction
Completion of processing:	2016	2 km velocity analysis
Record length:	8 s	PSTM 500 m Velan generation
Sample interval:	2 ms	Second-pass layer stripped GSMP and model subtraction
Vessels:	WesternGeco *Tasman* and *Regent*	DAR, Tau-P decon and statistical debubble
Acquisition parameters		Line merging, tidal statics
Streamer length:	8000 m	Radon demultiple 500 ms cut after enhanced demultiple flow
Streamer Type:	Q Marine (solid)	PRIMAL Demultiple
Streamer:	Slant 8–35 m depth	KPreSTM migration
Source volume:	Bolt 5085 cu in,	Post-migration radon
Source pressure:	2000 psi	SCVA: Continuous RNMO – moveout correction
Source depth:	Delta three-source, 9–6–9 m array	Post migration CMP gather PRIMAL/FK
Source signature:	Calibrated marine source	Final PSTM stacking
Shot interval:	18.75 m	

*The data were released by the UK Oil and Gas Authority under a UK Open Government Licence permit for public sector information and hosted on the CDA UK Oil and Gas Data website as a Public Access Data Package (OGA 2015 Seismic Programme for the Mid North Sea High).

margin of the MNSH (Fig. 1) and first identified by Donato *et al.* (1983) is therefore comparable both in size and also seismic expression to the proven North Pennine batholith onshore.

In 2009, hydrocarbon exploration well 37/25-1 (Oil & Gas Authority 2015) was drilled offshore on a structural high by Shell UK Ltd, as the operator on behalf of ExxonMobil, and provides a valid test of the Dogger Bank gravity anomaly. The well proved a thin succession of Zechstein anhydrites and dolomites (226 ft (69 m)) and also demonstrated the absence of Permian clastic sediment at this location. Below the Base Permian Unconformity (BPU), the well found only relatively thin (833 ft (254 m)) Old Red Sandstone (?Upper Devonian) sediments overlying quartzite basement. The results of well 37/25-1 show that the bright dome-shaped reflector, the original Middle Devonian Kyle Limestone target for this well, may instead be the seismic expression of the BPU and its dome shape is associated with the form of the underlying granite batholith of the Dogger Pluton (Fig. 3).

Palaeoenvironmental and sedimentological history

In the late Carboniferous and early Permian, the MNSH experienced a phase of basaltic extrusive volcanism associated with the initiation of the continental fragmentation of Pangea, the development of the proto-Central Graben in the North Sea (Ziegler *et al.* 1979), and the creation of a north-south-trending half-graben in Quadrant 37 that links the Northern and Southern Permian basins (Jenyon *et al.* 1984). In the north of Quadrant 38, well 38/10-1 encountered 164 ft (50 m) of igneous volcanics which were assigned to the Rotliegend by comparison with offset wells 39/1-1 and 39/2-1 (38/10-1 composite log). No specific dating was carried out for well 38/10-1, but the Quadrant 39 volcanics have been dated as Permian and assigned to the Inge Volcanics Formation (Heeremans *et al.* 2004).

The Lower Permian is usually recorded as being absent in historical wells drilled over the bulk of the MNSH with Zechstein carbonates and basal anhydrites overlying, in a major hiatus, desert reddened Carboniferous and Devonian strata. During the early Permian, the MNSH was located in the centre of the supercontinent of Pangea (Scotese & Langford 1995), and the desert surface of the MNSH massif was exposed to deep subsurface oxidation, subaerial erosion, fluvial transport and local deposition of clastic wadi and playa sediments.

Lower Permian sediments in the Silver Pit Basin (Bailey *et al.* 1993; Taylor 1998) on the southern margin of the MNSH show an aridization trend with fining-upwards sand to shale sediment patterns

Table 4. *Stratigraphy of the Permian on the MNSH and in the Anglo-Dutch Basin, and their relationship to the onshore provinces of the UK and north Germany/The Netherlands (adapted from Tucker (1991, table 1)*

Zechstein sequences	Durham Province	(Tucker 1991, table 1)	Yorkshire Province	MNSH North (27/10-1 RRI Oilfield Report # 378)	MNSH South (38/18-1 composite log)	MNSH South: (38/18-1 Revised)	Anglo-Dutch Basin (42/10a-1 composite log (Anglo))	Anglo-Dutch Basin (42/10a-1 composite log (German))	North Germany and The Netherlands	Zechstein cycles
ZS7	Roxby Formation		Roxby Formation Littlebeck Anhydrite Sleights Siltstone Sneaton Halite	Absent		Littlebeck Anhydrite Sleights Siltstone Absent	Top Anhydrite Upper Halite	Grezanhydrit Aller Halite	Ohre Anhydrit Untere Ohre Ton Alle Salze	EZ5
ZS6	Sherburn Anhydrite Upgang Formation Rotten/Carnallitic Marl			Upper Anhydrite Dolomite Absent	Upper Evaporite Group	Sherburn Anhydrite Absent Absent	Upgang Formation Carnallitic Marl	Pegmatitanhydrit Roter Saltzon	Pegmatit Anhydrit thin carbonate Roter Salzton	EZ4
ZS5	Boulby Halite Billingham Anhydrite			Middle Salt Billingham Main Anhydrite		Boulby Halite Billingham Main Anhydrite	Boulby Halite Billingham Main Anhydrite	Leine Halite Hauptanhydrit	Leine Salze Hauptanhydrit	EZ3
ZS4	Seaham Formation	Illitic Shale	Brotherton Formation	Upper Magnesian Limestone	Upper Magnesian Limestone	Brotherton Limestone	Upper Magnesian Limestone	Plattendolomit	Plattendolomit	
						Deck Anhydrite	Grauer Saltzon & Deckanhydrit	Grauer Saltzon & Deckanhydrit	Grauer Saltzon	
		Fordon Evaporites		Lower Evaporite Group	Lower Evaporite Group	Fordon Evaporites Basal Anhydrite	Fordon Evaporites Basal Anhydrite	Stassfurt Halite Basalanhydrit	Stassfurt Evaps Basalanhydrit	EZ2
ZS3	Roker Formation Hartlepool Anhydrite		Kirkham Abbey Formation Hayton Anhydrite	Roker Dolomite Hartlepool Dolomite		Kirkham Abbey Dolomite	Kirkham Abbey Formation (Middle Magnesian Limestone)	Hauptdolomit	Hauptdolomit	
ZS2	Ford Formation	Cadeby Formation	Sprotborough Member	Middle Magnesian Limestone	Lower Magnesian Limestone	Hayton Anhydrite	Hayton Anhydrite Equivalent	Werradolomit	Werradolomit	EZ1
ZS1	Raisby Formation	Marl Slate	Wetherby Member	Lower Magnesian Limestone		Cadeby Dolomite Marl Slate	Lower Magnesian Limestone Marl Slate	Zechsteinkalk Kupferschiefer	Zechsteinkalk Kupferschiefer	
Lower Permian	Yellow Sands and breccias			Auk Sandstone	Absent	Cygnus Sandstone	Silverpit Shale	Barren Red Measures	Rotliegendes	

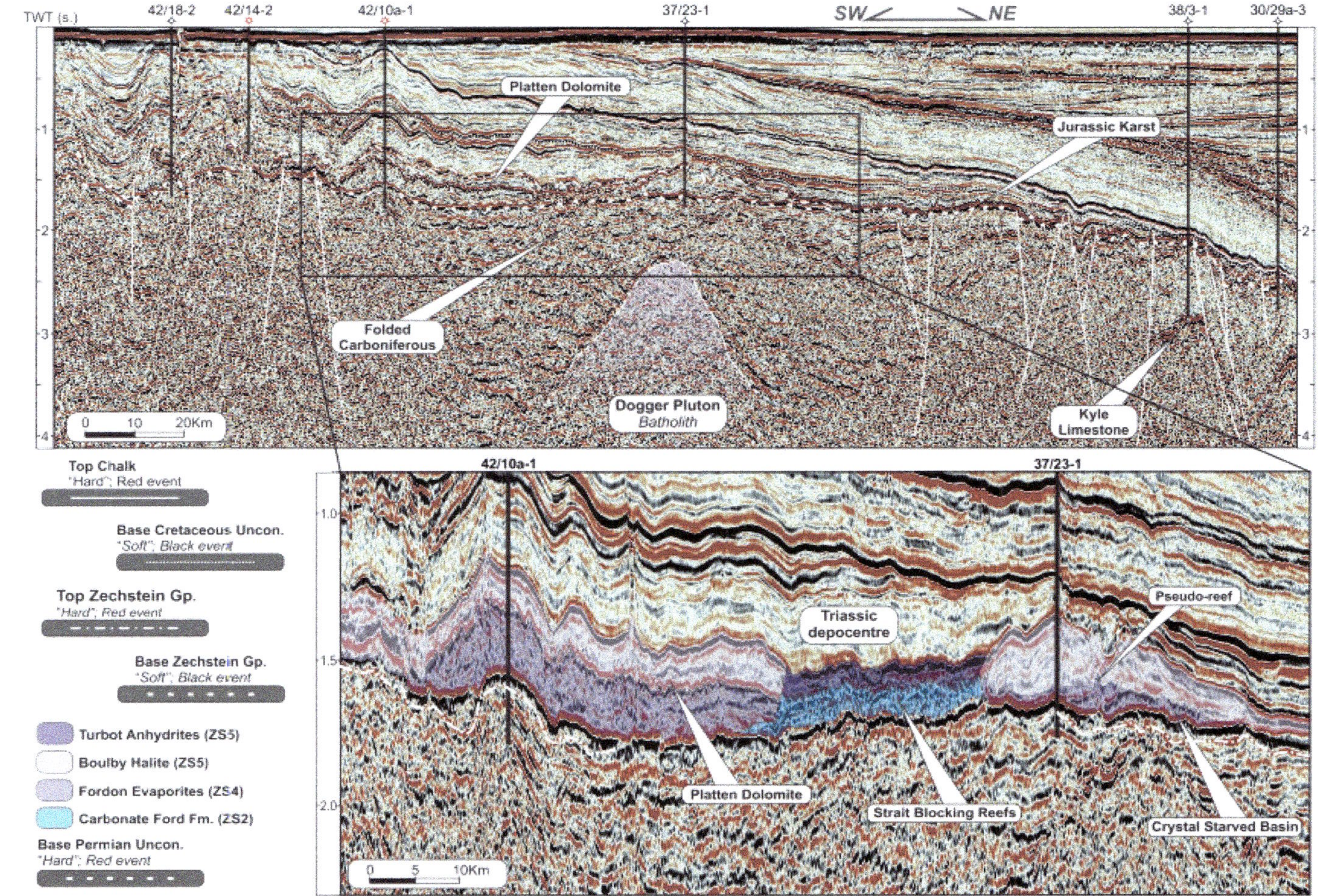

Fig. 3. Seismic cross-section over the MNSH from the Anglo-Dutch Basin to the Argyll Ridge; showing the uplift effect of the Dogger Pluton on the Permian paleo-relief, the impact of this on the location of the ZS2 strait-blocking reefs, the formation of the ZS4 crystal-starved back basin with associated pseudo-reef, and the Triassic sediment-loading-induced structural inversion.

characteristic of humid and desiccation cycles associated with the climatic cycles of the Gondwana ice cap (Roscher & Schneider 2006). Reanalysis of exploration wells drilled in UKCS quadrants 36, 37 and 38 suggests that a thin deposit of Lower Permian clastics can exist over those parts of the MNSH with low palaeorelief. Well data also show that the BPU can form an indurated desert pavement, locally associated with a palaeosoil caliche that is overlain by a thin desert regolith (e.g. well 36/13-1 at 4170 ft measured depth (MD)).

The Lower Leman Sandstone and the overlying Silver Pit Shale formations proved in well 44/12-1 are the distal lacustrine delta and playa equivalents for the network of proximal stream wadi sediments that are now believed to locally occur on the MNSH (e.g. well 37/23-1). In palaeolows, where the thickness of these Lower Permian clastic sediments exceeds wavelet resolution, it is possible to identify them on seismic; here the BPU reflector can be observed to have a hard seismic response that identifies it as the palaeo-land surface of the early Permian desert.

The Late Permian Zechstein Sea

At the close of the Middle Permian Kazanian (Maokovian) stage, the marine waters of the Boreal Sea, to the north of Pangea, connected into the isolated intercontinental Permian desert basin during an episode of catastrophic marine flooding, thereby forming the epicontinental Zechstein Sea (Paul 1995). Consequently, the depositional environment over the MNSH abruptly switched from subaerial desert clastics to episodic Zechstein marine carbonates and evaporites (Tucker 1991). The process of Zechstein marine flooding was so rapid that the relief of the former early Permian land surface was preserved in the bathymetry of the Zechstein Sea. Over the local region of the MNSH, the deepest Zechstein waters were located over the sites of the former early Permian wadis and lowland playa lakes, whereas the shallowest marine waters were located over the sites of the drowned basement mesas (Fig. 4a).

The Zechstein Super Group (Bailey *et al.* 1993) is a cyclic carbonate–evaporite succession deposited across the late Permian basin of Western Europe. Tucker (1991) divided the Zechstein into seven depositional sequences which show a progressive change from the earliest ZS1 unit of the Raisby Formation (formerly the Lower Magnesian Limestone of the Durham Province), a marginal-ramp to basin-wide marine carbonate, indicating that the first flooding occurred with cold well-oxygenated marine waters (Peryt *et al.* 2012). The second ZS2 unit, the Ford Formation (formerly the Middle Magnesian Limestone), involved reefal developments at the basin margins, with attendant build-ups and a partial blocking of connecting marine straits by barrier reefs observed on 2D seismic, such as occurred at the southern end of the Jenyon Gap, the channel in Quadrant 37 (Fig. 4b) that links the Northern and Southern Permian basins across the MNSH (Jenyon *et al.* 1984).

The presence of the strait-blocking ZS2 barrier reefs suggests that the marine connection of the Southern Permian Basin to the Boreal Sea was progressively restricted at critical narrows along its length, and, consequently, the Zechstein Sea of the Southern Permian Basin entered a phase of sea-level drawdown and basin-margin anhydrite precipitation (Hartlepool–Werra Anhydrite). Global sea-level glacioeustatic variations allowed for further flooding episodes; but with these subsequent cycles, the lack of vertical accommodation space in the poorly subsiding Zechstein Basin meant that shallow-water oolite carbonate banks and distal carbonate sediment ramps, rather than reefs, developed during the ZS3 Roker Dolomite and ZS4 Seaham Formation highstands (Tucker 1991, fig. 6). The oolitic carbonates of the ZS4 Seaham Formation were the last episode of major carbonate deposition seen on the MNSH (e.g. well 36/13-1); all subsequent sequences involved highstand shallow-water gypsum sabkhas (e.g. the Turbot Bank Anhydrite Formation in 30/29a-3) over the former carbonate platforms and lowstand halite deposits (ZS5 Boulby Halite to ZS7 Sneaton Halite) in the surrounding basinal salinas (e.g. well 37/12-1).

Throughout the Zechstein, because the MNSH remained a positive region separating the Northern and Southern Permian Basins, the accommodation spaces over this area were limited and carbonate reef development at the adjacent basin margins quickly filled to the global base level of Panthalassa, the Permian world ocean. This limitation of upwards growth potential meant that while the earliest ZS1 and ZS2 reefal carbonates were able to locally fill the marine accommodation space on the margins of the MNSH, aggrading carbonate growth was precluded during later ZS3 and ZS4 cycles, and was replaced by prograding shallow-water oolitic carbonate banks and adjacent ramp slopes (Fig. 4c). These carbonate-sediment factories fed oolitic sands, sourced from the agitated shallow coastal waters sited over the older carbonate reefs on the core of the MNSH (e.g. well 36/13-1), into the surrounding basins. Consequently, extensive submarine-fan deposition of ramp-derived oolitic grainstones occurred in the distal deep waters on the margins of the MNSH during these marine highstands (e.g. well 38/16-1 Seaham Dolomite Core #1; BGS, Offshore Hydrocarbon Wells: http://www.bgs.ac.uk/data/offshore Wells/).

During global lowstands, the Zechstein Sea was isolated from ocean-water recharge and extensive

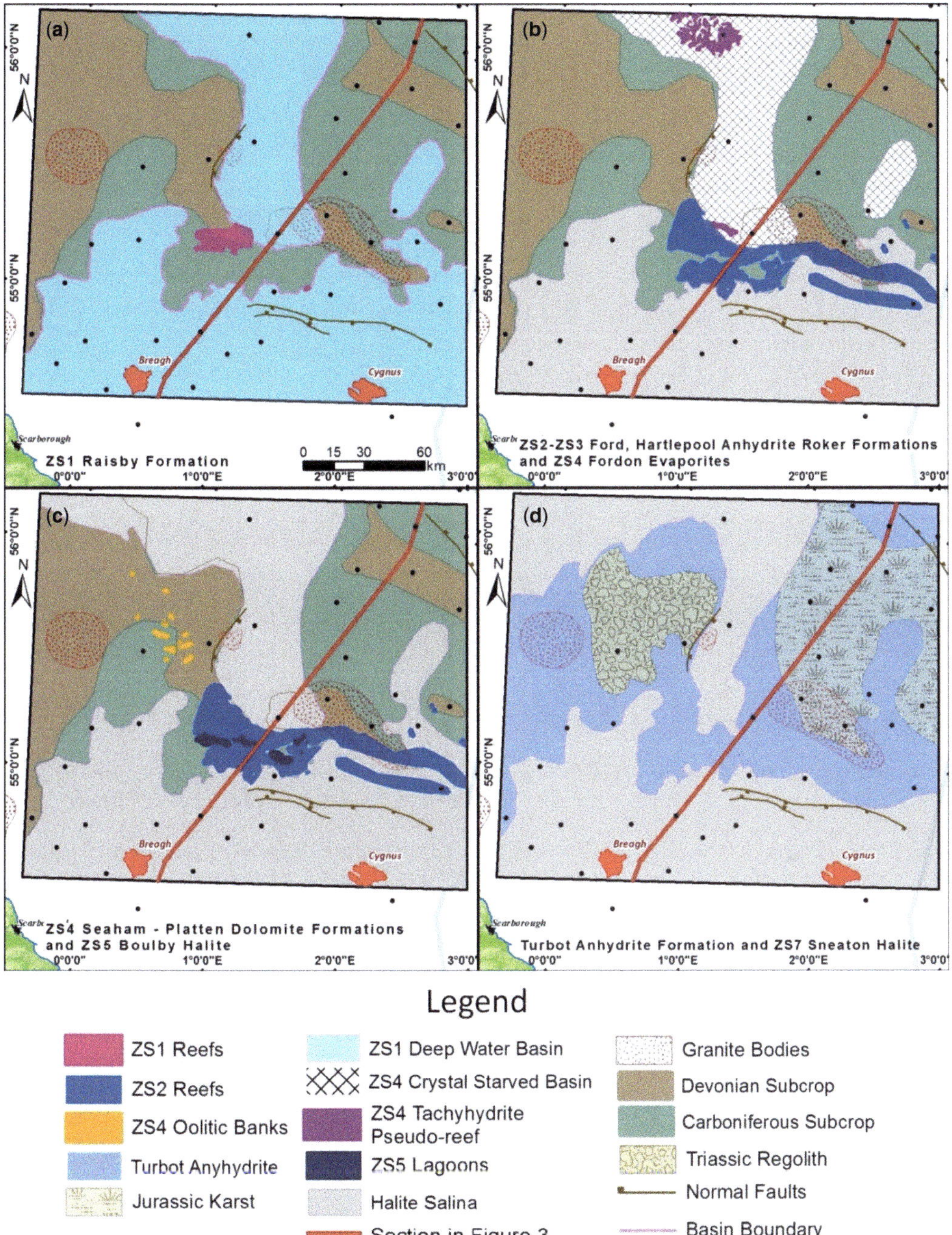

Fig. 4. The Zechstein environments of deposition for the study area showing the progressive changes from: (**a**) the earliest ZS1 open-marine carbonates; through (**b**) the ZS2–ZS3 carbonate reef build-ups; the subsequent ZS4 desiccation salina and crystal-starved basins; (**c**) the change to shallow-water ZS4 prograding oolitic carbonates followed by the ZS5 Boulby Halite; and (**d**) the final phases of proximal Turbot anhydrites on top of the former carbonate platforms, with ZS7 Sneaton halites in the surrounding distal basins.

salinas formed in the Cleveland and Sole Pit Basins on the southern margin of the MNSH (e.g. well 44/12-1). The first episode of catastrophic basin-wide desiccation commenced with the formation of the lowstand basin-fill halite of the ZS4 Fordon Evaporites (Tucker 1991, fig. 6) when halite preferentially filled the lows of the marine palaeodeeps. Over the bulk of the MNSH, this marine drawdown created subaerial conditions and exposed the strait-blocking reefs at the southern end of the Quadrant 37 and Quadrant 38 embayments. As a consequence of this, both of these local basins on the MNSH experienced a unique environment of extreme desiccation in which the hygroscopic mineral tachyhydrite ($CaMg_2Cl_6 \cdot 12H_2O$) was precipitated forming pseudo-reefs, as observed on 2D seismic and reported in well 29/27-1 by Taylor (1993, fig. 7). Our assessment of the age of these thin halites in the Quadrant 37 Jenyon Embayment and the Quadrant 38 Embayment as belonging to the ZS4 Fordon Evaporites supports the work of Taylor (1993).

In well 29/27-1, drilled in 1987 in the North Permian Basin by Amerada Hess Ltd targeting Zechstein 1 and 2 reef mounds, the operator suggested that the upper unit of thick halite (composite log for well 29/27-1) was part of the Z2 Stassfurt Evaporites (=ZS4 Fordon Evaporites) overlying Z2 Haupt Dolomite and Z1 Werra Anhydrite containing thin tachyhydrite. However, in a reassessment, Taylor (1993) concluded that the thick upper halite belonged instead to the ZS5 Boulby Halite.

Taylor (1993) established that the thin lower unit of tachyhydrite salts in well 29/27-1 are equivalent to the Fordon Evaporites of the East Yorkshire Province (Smith *et al.* 1974, table III), but here was deposited in an extreme sabkha environment that the seismic data show was separated from the main salina of the Anglo-Dutch Basin by the strait-blocking ZS2 Ford Formation reefs located in the south of Quadrant 37 (Fig. 4b). The magnesium and calcium chloride brines that form tachyhydrite are produced by a geochemical process of chromatographic cation separation that filters seawater passing through the phreatic zone of surrounding subaerially exposed carbonate reefs. The deliquescent tachyhydrite salts are then crystallized in groundwater-fed ponds located in isolated perched back basins, such as the Jenyon Embayment on the crest of the MNSH (see also Wardlaw 1972 for more details of this geochemical process).

In essence, the ZS4 tachyhydrite salts are deposited in a 'crystal-starved basin'. This observation explains not only the extreme thickness of later ZS5 Boulby Halite in the Jenyon Embayment on the MNSH, but also the low location of the distal ZS4 Seaham Formation carbonates that were deposited into the accommodation space left unused during the crystallization of the earlier thin Fordon Evaporites (e.g. well 37/23-1). The identification on seismic at the southern end of the Jenyon Gap of a new pseudo-reef (Fig. 3), which is located along strike from well 37/23-1 where potassium salts are recorded in the Fordon Evaporites, means that we prefer a primary Zechstein environment of deposition as the mode of formation for the plastically mobile tachyhydrite, rather than secondary hydrothermal volcanic fluids as suggested by Taylor (1993).

Hydrocarbon exploration review

Exploration History

The first deep exploration well drilled offshore in the North Sea was UK wildcat well 38/29-1 (Oil & Gas Authority 2015) that was spudded on the Dogger Bank in December 1964 by American Overseas Petroleum exploring the Lower Permian Groningen gas play on the MNSH (Table 1). Well 38/29-1 found the expected Quaternary, Tertiary, Cretaceous, Triassic and Upper Permian sequences, but the Jurassic and Carboniferous were absent, and the well was completed in Upper Devonian (Upper Famennian) dark red sandstones, which were cored 66 ft (20 m) below the base of the Permian.

The well found 1927 ft (587 m) of Zechstein anhydrite, halite and dolomite, but only the 171 ft (52 m) of oolitic dolomite with fossil debris were named on the composite log and ascribed to the Middle Magnesian Limestone by the operator and therefore equivalent to the ZS2 Ford Dolomite of the onshore Durham Province. Below these dolomites, the well found 512 ft (156 m) of unnamed anhydrite which is dolomitic towards the base. This sequence is considered here to be Hayton Anhydrite passing down into a thin (<1 m) distal basin sequence of ZS2 and ZS1 dolomite that rests directly on the Devonian. Neither the ZS1 Marl Slate (Kupferschiefer) nor any Lower Permian Yellow Sands, both seen onshore in the Durham Province, were identified in this well.

The key to understanding the depositional history of the Zechstein carbonates, anhydrites and halites on the MNSH is to identify correctly the age of the thick succession of halite over quadrants 28 and 29 to the north and its extension south into the Jenyon Embayment in the northern part of Quadrant 37 (Jenyon *et al.* 1984).

Exploration well 29/16-1 (Oil & Gas Authority 2015) drilled in 1973 by Ashland Exploration Company had as its primary objective the Lower Permian Rotliegend (Auk) sandstone. This well proved a 4760 ft succession of Zechstein anhydrite, halite, dolomite and polyhalite with an 8 ft-thick unit at the base, originally reported as black soft lignitic shale but really a ZS1 Kupferschiefer equivalent unit. The well found three main halite intervals; the top unit is 285 ft thick and is separated at its base by 47 ft of anhydrite and 18 ft of non-calcareous clay from the second middle unit of halite. This clay is inferred to be the Carnallitic Marl, the unit that marks the onset of the ZS6 cycle of marine flooding and salina development.

The middle unit consists of 3860 ft of halite above a 70 ft basal anhydrite and a 75 ft unit of dolomite identified as being the ZS4 Platten Dolomite at the base of a cycle of marine flooding and salina development. The third and lowest evaporitic unit found in well 29/16-1 consists of 195 ft of carnallite and polyhalite with a 25 ft unit of anhydrite at the base. This anhydrite sequence is proposed as being the equivalent to the ZS2 cycle of the Southern Permian Basin. The presence of carnallite, a deliquescent potassium magnesium chloride salt, suggests that the environment of deposition was a shallow-water sabkha located at a considerable distance from open-marine conditions in a crystal-starved basin (Fig. 4b).

Below the ZS2 sabkha, a 90 ft unit of dolomite was proved in this well, and this unit is inferred to have been deposited under open-marine pelagic conditions typical of the ZS1 pattern of marine flooding and carbonate deposition, with the 9 ft (3 m) Marl Slate diagnostic of the initial catastrophic basin-filling marine flooding at its base.

Play concepts

The Zechstein carbonate play was targeted by 22 wells drilled between 1966 and 2007 on and around the MNSH but without success. The pre-eminent target for this historical exploration was the Middle Magnesian Limestones (e.g. American Overseas Petroleum Ltd 1967 well 38/22-1, Core #1; BGS, Offshore Hydrocarbon Wells) in a structurally high location. The last well to target Zechstein was the Lundin well 41/10a-2Z drilled in 2007 targeting both the ZS4 Platten Dolomite and the ZS3 Haupt Dolomite but without success. The reasons for the failure of the ZS3 Roker/Haupt Dolomite play on the MNSH compared to its success to the south in the Southern Permian Basin are not clear; particularly as wells 36/15-1 and 38/18-1 (Table 1) in the centre of the MNSH recorded trace dead-oil staining in Zechstein carbonates, suggesting oil migration from a mature source kitchen in these localities.

In this analysis, we have separated the Zechstein carbonate play into two distinct types based on the environment of deposition and so recognize, in total, four hydrocarbon play concepts that can be applied to the exploration of the Permian over the MNSH:

- The Lower Permian fluvial clastic 'wadi play' (Fig. 2). This play was proved in 2010 by the successful appraisal drilling of the Cygnus gas field, located in blocks 44/11 and 44/12, on the southern margin of the MNSH. Well 44/11a-4 drilled in 2010 by GDF Suez showed that the Cygnus Field included the productive Lower Permian fluvial Cygnus Sandstone as a valid hydrocarbon reservoir (Catto *et al.* 2017).
- The ZS2 Carbonate Reef Play (Fig. 4b). The success of this play requires careful targeting of build-ups that can both preserve porosity and acquire hydrocarbon charge by fluid migration from an adjacent mature kitchen. This play has been proved along strike onshore in Denmark on the southern margin of the Ringkøbing-Fyn High (Maver 1995), and also farther east in Poland where the Zechstein reefs of the Wolsztyn Ridge region of western Poland form a productive hydrocarbon province (Dyjaczynski *et al.* 2001).
- Over the major part of the MNSH, the shallow waters of the Zechstein Sea were not an optimum location for bioherm growth, and gypsum sabkha was the dominate environment of deposition. However, along its southern margin and, in particular, at the southern end of the Quadrant 37 channel, the Dogger Pluton and the uplift against it of compressional folded Carboniferous provided a shallow-water substrate in an optimum location for reef development (Fig. 3). Consequently, the tidal water interchange between the northern and southern basins through this channel permitted the growth of a strait-blocking ZS2 barrier reef that forms the best untested target for hydrocarbon exploration on the MNSH.
- The ZS4 Carbonate Bank Play (Fig. 4c). This play was targeted in 1967 by Arpet Petroleum Ltd with two wells both drilled in 1967. Well 36/13-1 was spudded first, although the well failed to find hydrocarbons; the ZS4 Seaham Formation oolitic grainstone reservoir target contained very good visible porosity (36/13-1 composite log) and, although the adjacent well 36/15-1 was also dry, this well found dead-oil traces at two levels in both the ZS4 Seaham Dolomite and also the ZS2 Ford Dolomite (36/15-1 composite log). Coincident with the MNSH exploration, this play was also explored in an analogous structural setting on the northern margin of the London–Brabant Platform. The Wissey Field of Quadrant 53, located on the Winterton High at the southern margin of the Southern Permian Basin, was first discovered in 1967 with well 53/04-1. The field was subsequently developed in 2008 by Tullow Oil with well 53/04d-11, and produced gas from a fractured ZS4 Platten Dolomite reservoir deposited at the shelf break of the carbonate ramp (Duguid & Underhill 2010, fig. 3).
- The Zechstein Karst Play (Fig. 4d). This play occurs in the northern part of quadrants 37 and

38, and also locally in the eastern part of Quadrant 36, over those parts of the MNSH that were subaerially exposed and weathered in the early–mid Jurassic, prior to the paralic sedimentation of Upper Jurassic coals and coastal marine sands. The play was tested by well 37/10-1, which found vuggy porosity in Zechstein dolomites, but no hydrocarbons.

Failure analysis

Zechstein halite-cored swells that form Triassic structural highs can be observed on 2D seismic over parts of the MNSH (Fig. 3). These anticlines were targeted by those early hydrocarbon exploration wells that had Zechstein reefs as their primary objective. Wells drilled on these structures (e.g. well 36/26-1) typically found that the presence of a Triassic anticline did not guarantee the presence of an underlying Zechstein reef. Instead, if the well proved a reduced stratigraphic thickness of Bunter shales, then the anticline below contained a thick sequence of halite that had been mobilized by overburden sediment loading, during the early Triassic, from an underlying Zechstein evaporitic basin.

Key geological observations

Lower Permian Rotliegend

The Rotliegend forms an extensive sequence of aeolian, distal fluvial, lacustrine playa and evaporitic deposits in the two Permian basins that flank the MNSH. In the Northern Permian Basin and on the northern margins of the MNSH is the Auk Sandstone, an aeolian sequence that was deposited in an extensive desert erg and forms the proven reservoir of the Auk Field in UKCS Block 30/16 (Trewin & Bramwell 1991). Similarly in the Southern Permian Basin, the Leman Sandstone, a separate aeolian sequence that was also deposited in an extensive desert erg, forms the main gas reservoir in the Anglo-Dutch Basin (Hillier & Williams 1991).

North of the Leman Sandstone fairway on the southern margin of the MNSH is the Silver Pit Formation. This sequence of shales and evaporites was deposited in an extensive playa lake, and forms a competent top seal to older Lower Permian and Upper Carboniferous reservoirs (Bailey *et al.* 1993). The Lower Leman Sandstone (below the Silver Pit Shale) is an early Permian fluvial sandstone derived from wadis draining south from the MNSH (e.g. well 44/12-1). This sandstone forms the main reservoir of the Cygnus gas field and is herein informally renamed as the Cygnus Sandstone. This suggested change in name is to clarify the fact that the environment of deposition of this sandstone is fluvial, that it is similar in age and environment of deposition to the Lower Leman Sandstone of the Cleeton Field (Warren & Smalley 1994), and is older than the aeolian Leman Sandstone *sensu stricto* of the Leman Field (Hillier & Williams 1991).

Upper Permian Zechstein

The Zechstein Supergroup forms a regional sequence of marine carbonates, evaporitic sulphates and extreme desiccation halites deposited across the major Permian basins of the North Sea, the intervening palaeohighs of the MNSH, the Ringkøbing-Fyn High, and also across the basinal surrounds of the Durham and Yorkshire provinces and elsewhere to the east in The Netherlands, Germany, Denmark and Poland (Peryt *et al.* 2012).

Following its initial catastrophic flooding formation, the overall salinity of the Zechstein Sea increased during the Wuchiapingian and Changhsingian of the late Permian. Cyclical marine-water refreshment occurred, but the reducing accommodation spaces with time limited the potential environmental recovery of the Zechstein Sea and progressively led to the formation of a hypersaline basin by the end of the last marine cycle.

The extensive regional occurrence of the Zechstein deposits means that onshore analogues to the MNSH can be identified and compared with the proven borehole and seismic data obtained over this offshore province. In Poland, a zone of Zechstein reefs are observed on the margins of palaeohighs and generally located 2 km offshore from the palaeo-coastline (Paul 1995, fig. 9). In Denmark along strike from the MNSH, Zechstein reefs have been proved along the southern margin of the Ringkøbing-Fyn High (Maver 1995).

Geological studies have established that there is a three-stage generation of Zechstein carbonates which could form potential reservoirs. These stages are as follows:

- In ZS1, small local bioherms were formed in proximal shallow coastal waters.
- In ZS2, extensive bioherm build-ups developed, often co-located with the earlier ZS1 bioherms, and these structures filled the accommodation space up to the Zechstein Sea surface (Dyjaczynski *et al.* 2001).
- During ZS3 and ZS4 sequences, vertical carbonate growth was restricted due to the lack of subsidence and so, during the marine highstand phase, extensive deposits of oolite were formed over the existing ZS2 reef flats, and the carbonate platforms prograded basinwards. Precipitated carbonate sand grains were exported downslope from these shallow-marine banks to form an extensive sedimentary apron of Platten Dolomite *sensu*

stricto in the distal deep-water basin, on top of the halites of the former lowstand salina.

- During Z5–Z7, no carbonate deposition is observed in the MNSH region, as elsewhere in the Southern Permian Basin, and only sabkha anhydrite and salina halite, along with fine clastics, were deposited. By this time, the Zechstein Sea had become hypersaline, was no longer biologically active and so was not depositing organic carbonates.

Triassic overburden loading

At the commencement of the Triassic, the environment of deposition across the MNSH changed from that of a marine evaporitic basin to that of a subaerial clastic desert. With the ingress of Triassic deposits, the overburden sediment loading of the Bacton shales induced salt pillow mobilization in the underlying Zechstein marine halite, thereby creating a process of loading-induced structural inversion (Fig. 3).

At the southern margin of the MNSH, the Triassic overburden load was focused over, and supported by, the rigid structure of the underlying Zechstein carbonate build-ups, producing a 'room and pillar' effect. The Triassic sediment load was carried by the incompressible rigid strength of the Zechstein carbonate reefs, the 'pillars', whereas the mobile plastic halite, deposited in the adjacent Zechstein salinas and isolated saline lagoons, ballooned upwards, forming the 'rooms' (Fig. 3). The penecontemporaneous salt mobilization accentuated the Triassic accommodation space in the troughs directly over the carbonate reefs, creating local Triassic depocentres and this inversion also accounts for the observed thinning of the overburden shales over the salt-cored structural arches.

Consequently, there is a direct relationship between the thickness of the Lower Triassic shales and the nature of the underlying Zechstein strata. The thickest Triassic shales are directly located over the thinner rigid Zechstein carbonate build-ups, whereas the thinnest Triassic shales are located over the thicker Zechstein salt pillows. This inverse structure creates a Zechstein carbonate exploration strategy best summed up by this sentence: Target the thickest Triassic deposits, particularly those where the sediments also form anticlines, following local uplift post-Triassic.

Post-Triassic erosion and karstification

The post-Permian geology can be divided into three regions roughly coincident with the three quadrants of the MNSH. In the west, Quadrant 36 has the thinnest Tertiary cover and, below the Cretaceous Chalk, thin Lower Cretaceous and Upper Jurassic shales; it retains a thin cover of Lower Triassic shales, as exemplified by well 36/13-1. The seismic data establish the presence of thicker Triassic sediments over the bulk of Quadrant 36, away from the eastern margin fault, that precludes any possibility of salt dissolution or karstification of the underlying Permian succession. The observed absence of the Zechstein salt on the high itself is due to the original distribution of the evaporites being located on the flanks of this carbonate-dominated palaeostructure. The existence of the Farne Pluton in Quadrant 35 to the west also provided a control on the Lower Permian palaeorelief and on the patterns of Zechstein halite deposition (Fig. 4).

In the centre of the MNSH in Quadrant 37, the presence of a salt-filled channel connecting the Northern and Southern Permian basins was described by Jenyon *et al.* (1984). This structural low formed a connecting channel between these major basins following the flooding of the Zechstein Sea. Consequently, this palaeolow was filled with Zechstein salt formed during the episodes of desiccation; salt that still remains there today beneath a protecting cover of Triassic shale. There is, therefore, no potential for karst development along this channel axis.

To the east, the presence of karstified Zechstein carbonates on the MNSH was established during the exploration of both the Auk Field (Trewin & Bramwell 1991) and the Argyll Field (Robson 1991). Jenyon *et al.* (1984) established that to the east of the Quadrant 37 salt-filled channel, the Zechstein salt is absent over Quadrant 38 due to post- depositional dissolution (Fig. 3). Their surmise that this dissolution was associated with the Base Cretaceous Unconformity (BCU) and the removal of the Triassic cover can now, instead, be ascribed to an early Jurassic uplift and erosion in light of later well evidence that shows the presence of late Jurassic shales in the east of Quadrant 37 (e.g. well 37/25-1), and that on seismic evidence these late Jurassic strata extend east across Quadrant 38 (Fig. 3).

Conclusions

The future

The MNSH was the target of hydrocarbon exploration in the earliest phase of activity on the UKCS. This structural province remains a poorly tested region, with oil and gas potential in the Permian that warrants further exploration in UK quadrants 36–38.

Exploration potential

Four Permian plays have been identified as having the potential to host hydrocarbons on the MNSH, a Lower Permian clastic reservoir play: the Cygnus Sandstone wadi play and three Zechstein carbonate reservoir plays – the ZS2 reef build-up play, the ZS4 oolitic bank play and the Zechstein karst play. The optimum locations for these plays are as follows (Fig. 2):

- Quadrants 37 and 38 embayments – the Cygnus Sandstone wadi gas play.
- Quadrants 37 south and 38 south – the ZS2 reef build-up oil and gas play on the margin of the MNSH.
- Quadrant 36 Western Massif – the ZS4 oolitic bank oil play.
- Quadrants 37 and 38 Eastern Massif – the karst oil play.

The Lower Permian Cygnus Sandstone wadi play. This play was proved in the Cygnus Field by the discovery of gas in Lower Permian sands below the Silver Pit Shale (Catto *et al.* 2017). The wadi system draining the MNSH is the proximal source of the Cygnus sands. The potential of this play on the MNSH can be realized by targeting the wadi system in the palaeolows, such as the Jenyon Embayment in Quadrant 37 and the Quadrant 38 Embayment (Fig. 1).

Well data for well 37/23-1, drilled in a sediment-filled wadi, shows that the overlying Zechstein anhydrite has an increased thickness co-located with the wadi. The stratigraphic nature of this play requires the use of 3D seismic to explore fully its potential.

The ZS2 carbonate reef play. The presence of ZS2 Ford Formation barrier reefs on the Dogger Shelf explains the following features of the regional geology of the area.

These newly identified Dogger Shelf barrier reefs account for the blocking of the marine connection between the Southern and Northern Permian basins across the MNSH during the precipitation of the lowstand basin-floor halite of the Fordon Evaporites in the salina of the Anglo-Dutch Basin.

The distinct difference in halite fill, both in age of formation, mineral types and thicknesses, between the northern and southern basins can be explained if the presence of similar barrier reefs is invoked in the other marine straits that divide the MNSH (Smith & Taylor 1989), and also in the structural gaps that separate the MNSH from the Ringkøbing-Fyn High to the east.

We suggest that the Tethys Ocean to the east was the source of the ZS2 sequence marine waters, bringing warm higher-salinity and more dense seawater with their attendant Tethyian open ocean biota (nautiloids as reported by Peryt *et al.* (2012, section 4.1.4) into the Zechstein Sea, and that this process occurred gradually with the Tethys-derived waters being initially stratified below the less saline and, therefore, less dense waters derived from the Boreal Sea during ZS1.

We predict that on the northern margin of the MNSH, a zone of ZS2 barrier reefs can be expected that separates the Northern Permian Basin from the shallow waters of the Auk Basin in Quadrant 28 (Fig. 1). For example, well 27/10-1, located on the Devil's Hole Horst, proved 165 ft (50.3 m) of carbonate (ZS1 and ZS2), whereas well 28/12-1, located to the south, proved only 32 ft (9.8 m) of basal dolomite followed by 60 ft (18.3 m) of Hartlepool Anhydrite, a sequence pattern and isopach relationship indicative of a coastal sabkha environment of deposition that confirms the separation of the proximal MNSH from the distal Northern Permian Basin.

We suggest that the optimum location for ZS2 reef development is on the southern margin of the MNSH adjacent to the deep-water environment of the Anglo-Dutch Basin, but north of the North Dogger Fault in a play fairway location similar to that reported for Denmark by Maver (1995) and also seen in the offshore Dutch sector of the MNSH by Jaarsma *et al.* (2014) and Tolsma (2015). In the south of Quadrant 37, we expect to find porous reservoir carbonates sealed by anhydrites and charged with either early oil that is now bypassed or filled by a later gas flush, and that these hydrocarbons have migrated north onto the MNSH from the Carboniferous source-rock kitchen of the Anglo-Dutch Basin.

We believe that the failure to find successful Zechstein carbonate prospects in the initial exploration phase of the MNSH derives from the assumption that valid Zechstein carbonate traps are co-located below Triassic structural highs, when in fact the converse relationship is more likely to occur. The following observation of Maver (1995, p. 44) accounts for the failure of well 43/03-1 drilled in 1966 by Rycade Oil Corporation: 'Exploration wells that target the porous Ca-2 [i.e. ZS3] carbonate interval at the platform edge, have to be located carefully to avoid encountering a thick Na-2 halite layer instead'.

The primary objective of well 43/03-1 was the Zechstein Platten Dolomite and the target was at a Top Zechstein structural anticline. The well failed because the Triassic structural high was Zechstein salt cored and, although the Platten Dolomite was found, the reservoir was tight as this formation was deposited in a distal deep-water environment prior to structural inversion. The former ZS2 proximal reef flat, the sediment source of the oolitic carbonate sands, remains as a valid untested target 5 km to the NE of this dry well (Fig. 3).

The ZS4 oolitic bank play. The lack of vertical accommodation space at the end of the ZS2 sequence led to the development of lateral prograding carbonate ramps on the flanks of the MNSH in a depositional setting comparable to that of the Wissey Field on the Winterton High in Quadrant 53 (Duguid & Underhill 2010).

The following geological features are required to achieve play fairway success:

- The proximity of a prospect to the former reef flats, such as those of the Dogger Shelf, which created the sediment source for the oolitic grainstone deposits of the ZS4 Platten Dolomite in the Anglo-Dutch Basin to the south and also the Jenyon Embayment to the north in Quadrant 37.
- The progradation of the ZS4 carbonate ramp off the high that allowed for the development, behind the ramp, of a coastal microbial mat sabkha containing gypsum. During subsequent marine regression, the depocentre for this coastal sabkha then moved basinwards over the earlier formed oolites.
- This overlying gypsum subsequently underwent diagenesis to anhydrite, during subaerial exposure, and created a competent top seal during the peak regression of the desiccation lowstand. The optimum preservation of porosity seen in the oolites of well 36/13-1 on the Western Massif (Fig. 2) can therefore be explained because the diagenesis to anhydrite of the overlying gypsum sabkha produced an early formed seal that precluded groundwater invasion into the oolites during subsequent marine transgressions. See also the well 38/16-1 ZS4 Seaham Dolomite with medium–high porosity (Core #1, 5629–5636 ft; BGS, Offshore Hydrocarbon Wells). The presence of an overlying cap of Triassic shales in Quadrant 36 also means that here these oolitic grainstone reservoirs have preserved porosity as they did not experience any significant diagenesis by meteoric groundwater invasion during the Jurassic, a reservoir-occluding process which did occur to the east in Quadrant 38.

The Karst Play. Oil-charged Zechstein carbonate reservoirs with vuggy porosity associated with karstification occur in both the Auk Field (Trewin & Bramwell 1991) and the Argyll Field, where an organic-rich dolomitic interval, the Sapropelic Dolomite, separates the upper and lower parts of the Zechstein succession. The dolomite shows evidence of extreme karstification, with common vugs and collapse breccias (Robson 1991).

The karst play has not been successfully explored over the MNSH and the potential remains unrealized. The optimum location for this play is in the northern parts of quadrants 37 and 38, adjacent to the uplift associated with the Auk–Argyll Ridge on the western margin of the Central Graben, with additional potential over the eastern part of Quadrant 36 where the protecting overburden of the Triassic shales is thin or absent (Fig. 4d). The presence of vuggy porosity in Zechstein carbonates below Jurassic shales in well 37/10-1 is evidence of karstification. The failure of this well to find hydrocarbons is believed to be due to poor targeting that missed the crest of the trap.

Lessons learned

The key strategy in any interpretation of new data is to work from the known to the unknown. For the Southern Permian Basin, the Zechstein strata of the Anglo-Dutch Basin form a distinctive seismic pattern that permits the ZS4 Platten Dolomite reflector to be identified away from well control. Starting with this known seismic facies, the identification of this seismic pattern in small isolated areas of the Dogger Shelf, at the southern end of Quadrant 37 (Fig. 3), required explanation.

The density of the SNS 2D lattice of seismic lines (Fig. 1) allowed the hypothesis that these isolated areas are ponded lagoons to be tested. The former presence of individual ponds requires the existence of surrounding barriers, and the ZS2 Ford Formation carbonate reefs provided a credible explanation for the nature of this barrier. Although these barrier reefs are not easily seen on the seismic data, their inferred presence allows for the creation of a predictive geological model that can be used to guide seismic interpretation.

Acknowledgement We would like to thank the management of Spectrum Geo Ltd for permission to use the SNS 2D survey that forms the basis of this study and for permitting access to, and use of, their computer facilities during the analysis of these data.

References

Bailey, J.B., Arbin, P., Daffinoti, O., Gibson, P. & Ritchie, J.S. 1993. Permo-Carboniferous plays of the Silver Pit Basin. *In*: Parker, J.R. (ed.) *Petroleum Geology of Northwest Europe: Proceedings of the 4th Conference.* Geological Society, London, 707–715, https://doi.org/10.1144/0040707

Cartwright, J. 1990. *The Structural Evolution of the Ringkøbing-Fyn High. Tectonic Evolution of the North Sea Rifts.* Clarendon Press, Oxford.

Catto, R., Taggart, S. & Poole, G. 2017. Petroleum geology of the Cygnus gas field, UK North Sea: from discovery to development. *In*: Bowman, M. & Levell, B. (eds) *Petroleum Geology of NW Europe: 50 Years of Learning – Proceedings of the 8th Petroleum Geology Conference.* Geological Society, London. First published online March 23, 2017, https://doi.org/10.1144/PGC8.39

Chadwick, R.A. & Holliday, D.W. 1991. Deep crustal structure and Carboniferous basin development within the Iapetus convergence zone, northern England. *Journal of the Geological Society, London*, **148**, 41–53, https://doi.org/10.1144/gsjgs.148.1.0041

Collins, W.J., Wiebe, R.A., Healy, B. & Richards, S.W. 2006. Replenishment, crystal accumulation and floor aggradation in the megacrystic Kameruka Suite, Australia. *Journal of Petrology*, **47**, 2073–2104.

Donato, J.A., Martindale, W. & Tully, M.C. 1983. Buried granites within the mid North Sea High. *Journal of the Geological Society, London*, **140**, 825–837, https://doi.org/10.1144/gsjgs.140.5.0825

Duguid, C. & Underhill, J.R. 2010. Geological controls on Upper Permian Plattendolomit Formation reservoir prospectivity, Wissey Field, UK Southern North Sea. *Petroleum Geoscience*, **16**, 331–348, https://doi.org/10.1144/1354-0793/10-021

Dunham, K.C., Dunham, A.C., Hodge, B.L. & Johnson, G.A.L. 1965. Granite beneath Viséan sediments with mineralization at Rookhope, northern Pennines. *Quarterly Journal of the Geological Society*, **121**, 383–414, https://doi.org/10.1144/gsjgs.121.1.0383

Dyjaczynski, M.G., Mamczur, S. & Peryt, T.M. 2001. Reefs in the basinal facies of the Zechstein Limestone (Upper Permian) of Western Poland: a new gas play. *Journal of Petroleum Geology*, **24**, 265–285.

Heeremans, M., Timmerman, M.J., Kirstein, L.A. & Faleide, J.I. 2004. New constraints on the timing of late Carboniferous–early Permian volcanism in the central North Sea. *In*: Wilson, W., Neumann, E.R., Davies, G.R., Timmerman, M.J., Heeremans, M. & Larsen, B.T. (eds) *Permo-Carboniferous Magmatism and Rifting in Europe*. Geological Society, London, Special Publications, **223**, 177–193, https://doi.org/10.1144/GSL.SP.2004.223.01.08

Hillier, A.P. & Williams, B.P.J. 1991. The Leman Field, Blocks 49/26, 49/27, 49/28, 53/1, 53/2, UK North Sea. *In*: Abbotts, I.L. (ed.) *United Kingdom Oil and Gas Fields: 25 Years Commemorative Volume*. Geological Society, London, Memoirs, **14**, 451–458, https://doi.org/10.1144/GSL.MEM.1991.014.01.56

Jaarsma, B., Schneider, J., Tolsma, S.P. & Lutgert, J.E. 2014. A new chance for the Zechstein-2 carbonate play in the Mid North Sea High area? Extended abstract presented at the *76th EAGE Conference and Exhibition 2014, Extended Abstracts*, 16–19 June 2014, Amsterdam.

Jenyon, M.K., Cresswell, P.M. & Taylor, J.C.M. 1984. Nature of the connection between the northern and southern Zechstein Basins across the Mid North Sea High. *Marine and Petroleum Geology*, **1**, 355–363.

Jhingran, A.G. & Tomkeieff, S.I. 1942. The Cheviot granite. *Quarterly Journal of the Geological Society*, **98**, 241–259, https://doi.org/10.1144/GSL.JGS.1942.098.01-04.14

Kent, P.E. 1967. Outline geology of the southern North Sea Basin. *Proceedings of the Yorkshire Geological Society*, **36**, 1–22, https://doi.org/10.1144/pygs.36.1.1

Kent, P.E. 1975. The tectonic development of Great Britain and the surrounding seas. *In*: Woodland, A.W. (ed.) *Petroleum and the Continental Shelf of North-West Europe; Volume 1. Geology*. Applied Science, London, 3–28.

Kimbell, G.S., Young, B., Millward, D. & Crowley, Q.G. 2010. The North Pennine batholith (Weardale Granite) of northern England: new data on its age and form. *Proceedings of the Yorkshire Geological Society*, **58**, 107–128, https://doi.org/10.1144/pygs.58.1.273

Klemperer, S.L. & Matthews, D.H. 1987. Iapetus suture located beneath the North Sea by BIRPS deep seismic reflection profiling. *Geology*, **15**, 195–198.

Leeder, M.R. & Hardman, M. 1990. Carboniferous geology of the Southern North Sea Basin and controls on hydrocarbon prospectivity. *In*: Hardman, R.F.P. & Brooks, J. (eds) *Tectonic Events Responsible for Britain's Oil and Gas Reserves*. Geological Society, London, Special Publications, **55**, 87–105, https://doi.org/10.1144/GSL.SP.1990.055.01.04

Maver, K.G. 1995. Seismically detected porous Zechstein carbonates in Southern Jutland, Denmark. *Bulletin of the Geological Society of Denmark*, **42**, 34–46.

Mitchell, J.G. 1972. Potassium–argon ages from the Cheviot Hills, northern England. *Geological Magazine*, **109**, 421–426.

Nance, R.D., Gutiérrez-Alonso, G. *et al.* 2010. Evolution of the Rheic ocean. *Gondwana Research*, **17**, 194–222.

Oil and Gas Authority 2015. *Mid North Sea High, MNSH Area Well Data: Deviation Surveys, Joined Logs, Checkshots/VSP, Reports, Log Images*. CDA UK Oil and Gas Data website, Public Access Data Package, OGA 2015 Seismic Programme (89 wells), https://www.ukoilandgasdata.com

Paul, J. 1995. Stromatolite reefs of the upper Permian Zechstein basin (Central Europe). *Facies*, **32**, 28–31.

Peryt, T.M., Geluk, M.C., Mathiesen, A., Paul, J. & Smith, K. 2010. Zechstein. *In*: Doornenbal, H. & Stevenson, A. (eds) *Petroleum Geological Atlas of the Southern Permian Basin Area*. European Assocation of Geoscientists & Engineers (EAGE), Houten, Netherlands, 123–147.

Peryt, T.M., Raczyński, P., Peryt, D. & Chłódek, K. 2012. Upper Permian reef complex in the basinal facies of the Zechstein Limestone (Ca1), western Poland. *Geological Journal*, **47**, 537–552.

Rhys, G.H. 1974. *(Compiler) A Proposed Standard Lithostratigraphic Nomenclature for the Southern North Sea and an Outline Structural Nomenclature for the Whole of the (UK) North Sea*. Institute of Geological Sciences, London.

Robson, D. 1991. The Argyll, Duncan and Innes Fields, Block 30/24 and 30/25a, UK North Sea. *In*: Abbotts, I.L. (ed.) *United Kingdom Oil and Gas Fields: 25 Years Commemorative Volume*. Geological Society, London, Memoirs, **14**, 219–226, https://doi.org/10.1144/GSL.MEM.1991.014.01.27

Roscher, M. & Schneider, J.W. 2006. Permo-Carboniferous climate: Early Pennsylvanian to Late Permian climate development of central Europe in a regional and global context. *In*: Lucas, S.G., Cassinis, G. & Schneider, J.W. (eds). *Non-Marine Permian Biostratigraphy and Biochronology*. Geological Society, London, Special Publications, **265**, 95–136, https://doi.org/10.1144/GSL.SP.2006.265.01.05

Scotese, C.R. & Langford, R.P. 1995. Pangea and the paleogeography of the Permian. *In*: Scholle, P.A., Peryt, T.M. & Ulmer-Scholle, D.S. (eds) *The Permian of Northern Pangea*. Springer, Berlin, 3–19.

Smith, D.B. & Taylor, J.C.M. 1989. A 'North-west Passage' to the southern Zechstein Basin of the UK North Sea. *Proceedings of the Yorkshire Geological Society*, **47**, 313–320, https://doi.org/10.1144/pygs.47.4.313

Smith, D.B., Brunstrom, R.G.W., Manning, P.I., Simpson, S. & Shotton, F.W. 1974. *A Correlation of Permian Rocks in the British Isles*. Geological Society, London, Special Report, 5.

Taylor, J.C.M. 1993. Pseudo-reefs beneath Zechstein salt on the northern flank of the Mid North Sea High. *In*: Parker, J.R. (ed.) *Petroleum Geology of Northwest Europe: Proceedings of the 4th Conference*. Geological Society, London, 749–757, https://doi.org/10.1144/0040749

Taylor, J.C.M. 1998. Upper Permian–Zechstein. *In*: Glennie, K.W. (ed.) *Petroleum Geology of the North Sea: Basic Concepts and Recent Advances*. 4th edn. Blackwell Science, Oxford, 174–211.

Tolsma, S.P. 2015. *Seismic characterization of the Zechstein carbonates in the Dutch northern offshore. Master's thesis*, Utrecht University, Utrecht, The Netherlands (Open Access version available via Utrecht University Repository).

Trewin, N.H. & Bramwell, M.G. 1991. The Auk Field, Block 30/16, UK North Sea. *In*: Abbotts, I.L. (ed.) *United Kingdom Oil and Gas Fields: 25 Years Commemorative Volume*. Geological Society, London, Memoirs, **14**, 227–236, https://doi.org/10.1144/GSL.MEM.1991.014.01.28

Trewin, N.H., Fryberger, S.G. & Kreutz, H. 2003. The Auk Field, Block 30/16, UK North Sea. *In*: Gluyas, J.G. & Hichens, H.M. (eds) *United Kingdom Oil and Gas Fields, Commemorative Millennium Volume*. Geological Society, London, Memoirs, **20**, 485–496, https://doi.org/10.1144/GSL.MEM.2003.020.01.39

Tucker, M.E. 1991. Sequence stratigraphy of carbonate-evaporite basins: models and application to the Upper Permian (Zechstein) of northeast England and adjoining North Sea. *Journal of the Geological Society, London*, **148**, 1019–1036, https://doi.org/10.1144/gsjgs.148.6.1019

van Wees, J.D., Stephenson, R.A. *et al.* 2000. On the origin of the southern Permian Basin, Central Europe. *Marine and Petroleum Geology*, **17**, 43–59.

Wardlaw, N.C. 1972. Unusual marine evaporites with salts of calcium and magnesium chloride in Cretaceous basins of Sergipe, Brazil. *Economic Geology*, **67**, 156–168.

Warren, E.A. & Smalley, P.C. 1994. Part 1: Compendium of North Sea oil and gas fields. *In*: Warren, E.A. & Smalley, P.C. (eds) *North Sea Formation Waters Atlas*. Geological Society, London, Memoirs, **15**, 3–77, https://doi.org/10.1144/GSL.MEM.1994.015.01.02

Ziegler, A.M., Scotese, C.R., McKerrow, W.S., Johnson, M.E. & Bambach, R.K. 1979. Paleozoic paleogeography. *Annual Review of Earth and Planetary Sciences*, **7**, 473–502.

Polyphase tectonic inversion and its role in controlling hydrocarbon prospectivity in the Greater East Shetland Platform and Mid North Sea High, UK

STEFANO PATRUNO[1]*, WILLIAM REID[1], CHRISTIAN BERNDT[2] & LAURENT FEUILLEAUBOIS[3]

[1]*PGS, 4 The Heights, Brooklands, Weybridge, Surrey KT13 0NY, UK*

[2]*GEOMAR Helmholtz Centre for Ocean Research Kiel, Wischhofstrasse 1-3, D-24148 Kiel, Germany*

[3]*PGS, Lilleakerveien 4, 0283 Oslo, Norway*

S.P., 0000-0002-6375-995X

**Correspondence: stefano.patruno@gmail.com*

Abstract: Thick Paleozoic successions are buried under the Greater East Shetland Platform (ESP) and Mid North Sea High (MNSH), two large underexplored platform regions flanking the structural depocentres of the North Sea. Here, newly acquired broadband seismic data are interpreted to provide a novel assessment of the regional tectonostratigraphic evolution and its influence on hydrocarbon prospectivity. Numerous working reservoir units are present over these two frontier areas, together with large Paleozoic traps. Hydrocarbon charge occurs either via a likely maximum 30–40 km lateral migration from the Jurassic/Carboniferous basinal source kitchens or, possibly, via vertical/lateral migration from deeper Devono-Carboniferous source intervals. The two regions underwent a largely similar evolution, consisting of at least eight successive switch-overs between regional compression/uplift and extension/subsidence in the last 420 myr. However, on the Greater MNSH, the lack of significant Permo-Triassic rifting probably resulted in too little subsidence for the lower Carboniferous interval to reach sufficient burial depth for gas maturation. Seep and fluid escape data suggest a working 'deep' source in the Greater ESP. Here, the presence of localized Permo-Triassic intra-platform grabens and half-grabens provided sufficient subsidence for the oil-prone middle Devonian unit to eventually enter the oil maturation window and faults provide easy conduits for the upwards migration of oil.

The present day structural configuration of the Central and Northern North Sea is that of a failed trilete Late Jurassic rift basin (Fig. 1; Pegrum & Spencer 1990; Ziegler 1992; Coward *et al.* 2003; Fraser *et al.* 2003; Johnson *et al.* 2005). The three main graben axes are the Viking Graben to the north, the Central Graben to the south and the Moray Firth and Witch Ground Graben to the west (Fig. 1). Conversely, the Southern North Sea is centred on a broad Permian structural low (the 'Southern Permian Basin'), which in the UK sector is represented mostly by the Anglo-Dutch Basin (Fig. 1; Glennie & Underhill 1998; Pharaoh *et al.* 2010).

In the UK Continental Shelf (UKCS), a broad Mesozoic platform area is situated to the west of the Viking Graben and Central Graben depocentres. The Moray Firth and Witch Ground Graben Mesozoic depocentres subdivide this wide platform into two regions, known, respectively, as the Greater East Shetland Platform (ESP) to the north and the Greater Mid North Sea High (MNSH) to the south (Fig. 1). These two 'platform' areas correspond to two broad Variscan highs, which were largely unaffected by the Late Jurassic extension (Glennie & Underhill 1998). The Greater ESP covers a total area of 62 000 km^2 from the Shetland and Orkney coastlines to the master bounding faults of the Viking Graben and Witch Ground Graben; the Greater MNSH region spans 84 000 km^2 from the present British coastline to the edges of the Central Graben and the Anglo-Dutch Basin (Fig. 1b). Several subregions and intra-platform basins can be defined within each of these two broad platform areas (Fig. 1a). The regional chronostratigraphic framework for the ESP and adjacent basinal depocentres is shown in Figure 2 and that for the MNSH is shown in Figure 3 (see Table 1 for key to the minor stratigraphic units in Figs 2 & 3).

Most North Sea hydrocarbon discoveries are located within, or in close proximity to, the mature source kitchens corresponding to three Late Jurassic graben axes in the Northern and Central North Sea and to the Paleozoic depocentres of the 'Southern Permian Basin' in the Southern North Sea ('Anglo-Dutch Basin' in Fig. 1). By contrast, the Greater ESP and MNSH have been underexplored, with only ten

From: Monaghan, A. A., Underhill, J. R., Hewett, A. J. & Marshall, J. E. A. (eds) 2018. *Paleozoic Plays of NW Europe*. Geological Society, London, Special Publications, **471**, 177–235.
First published online May 4, 2018, https://doi.org/10.1144/SP471.9

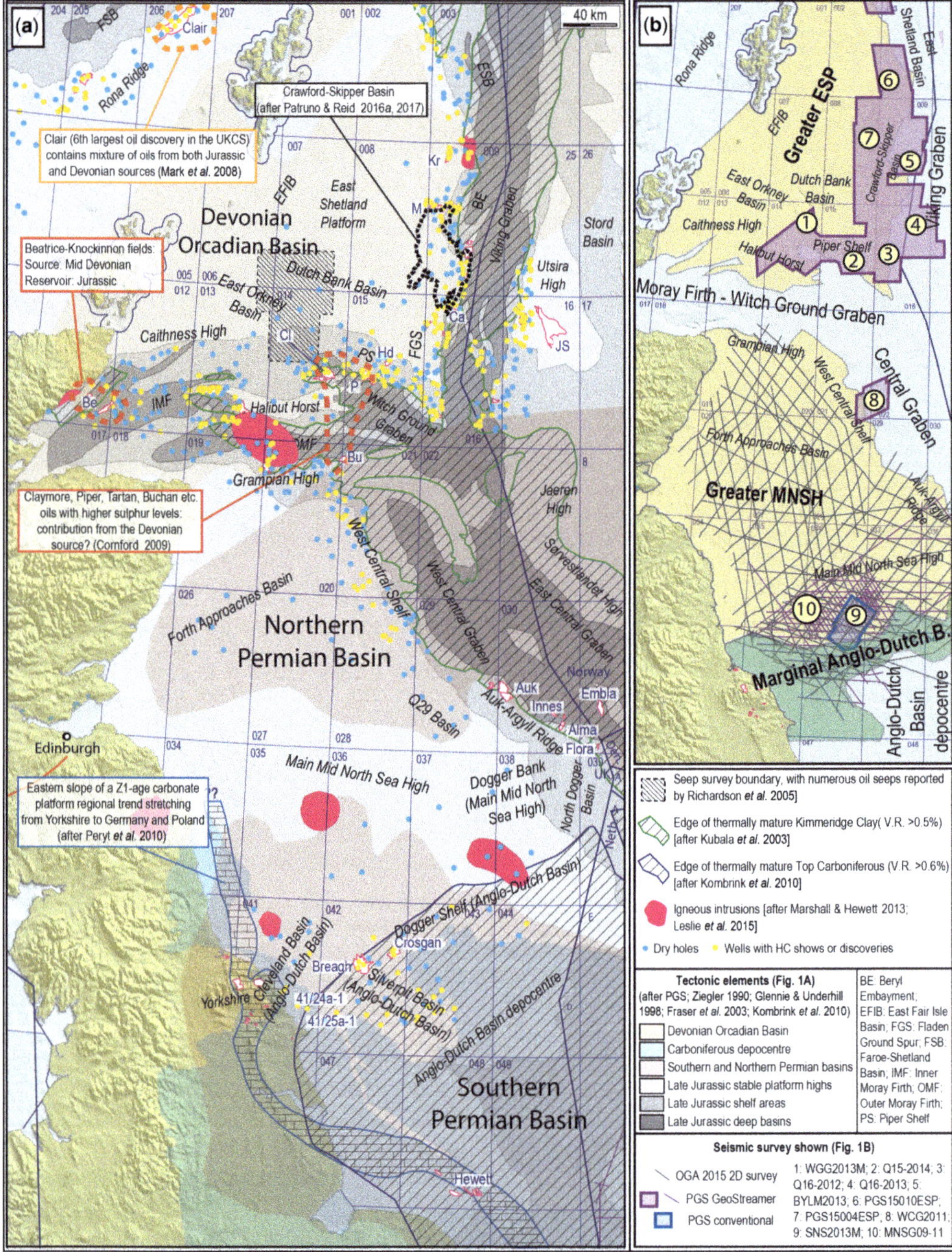

Fig. 1. (**a**) Location map of the study area highlighting the data used in this study and the main basin areas. The outlines of the Orcadian Basin and the Northern and Southern Permian basins are after Ziegler (1990) and Glennie & Underhill (1998). (**b**) Outline of the regional study area and overview of PGS data used in this study. The Greater East Shetland Platform and Greater Mid North Sea High are shown as yellow polygons and the marginal Anglo-Dutch Basin (corresponding to the Dogger Shelf, Cleveland Basin and Silverpit Basin of Fig. 1a) is shown as a green polygon. Be, Beatrice; Bu, Buchan; Ca, Cairngorm; Cl, Claymore; ESP, East Shetland Platform; Hd, Hood; JS, Johan Sverdrup; Kr, Kraken; M, Mariner; MNSH, Mid North Sea High; P, Piper.

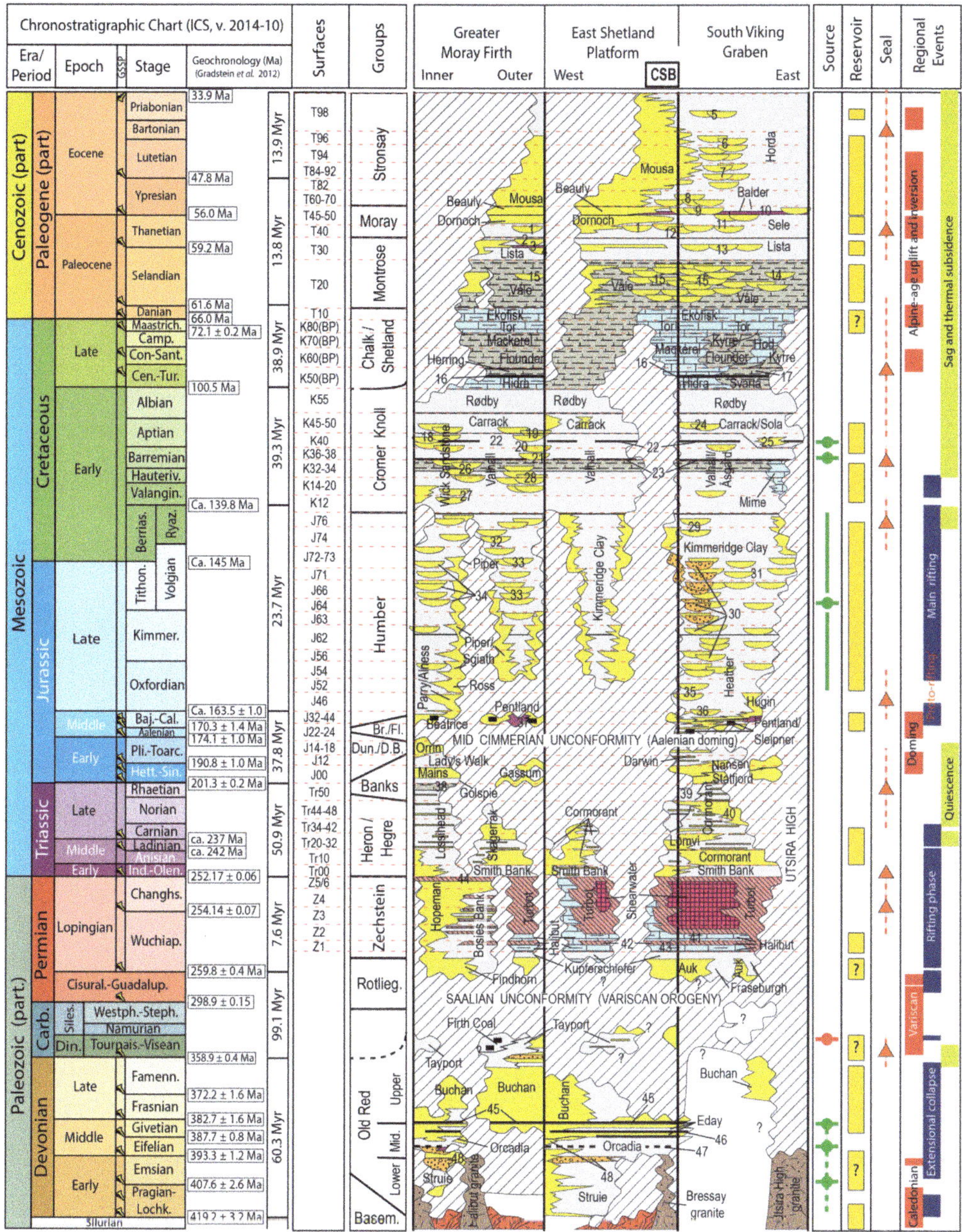

Fig. 2. Regional chronostratigraphic framework for the East Shetland Platform and adjacent basinal depocentres (Outer Moray Firth–Witch Ground Graben and South Viking Graben). CSB, Crawford–Skipper Basin; Br./Fl., Brent Group/Fladen Group; Dun./D.B., Dunlin Group/Dunrobin Bay Group; Rotlieg., Rotliegend Group; see Table 1 for key to the minor stratigraphic units, here numbered from 1 to 48. For the key to the lithological symbols, see Figure 3. Main source for the stratigraphy shown in this figure is Patruno & Reid (2016*c*) and references cited therein.

and one well per 1000 km^2, respectively (Fig. 1). The main reason for this historical lack of interest in platform areas is the perceived absence of mature source rocks and effective sealing, as detailed in Patruno & Reid (2016*a*, 2017). Furthermore, existing Paleogene discoveries in the Greater ESP are mostly characterized by heavy oils (e.g. 11–19° API in Mariner and Bressay), which has acted as an additional historical

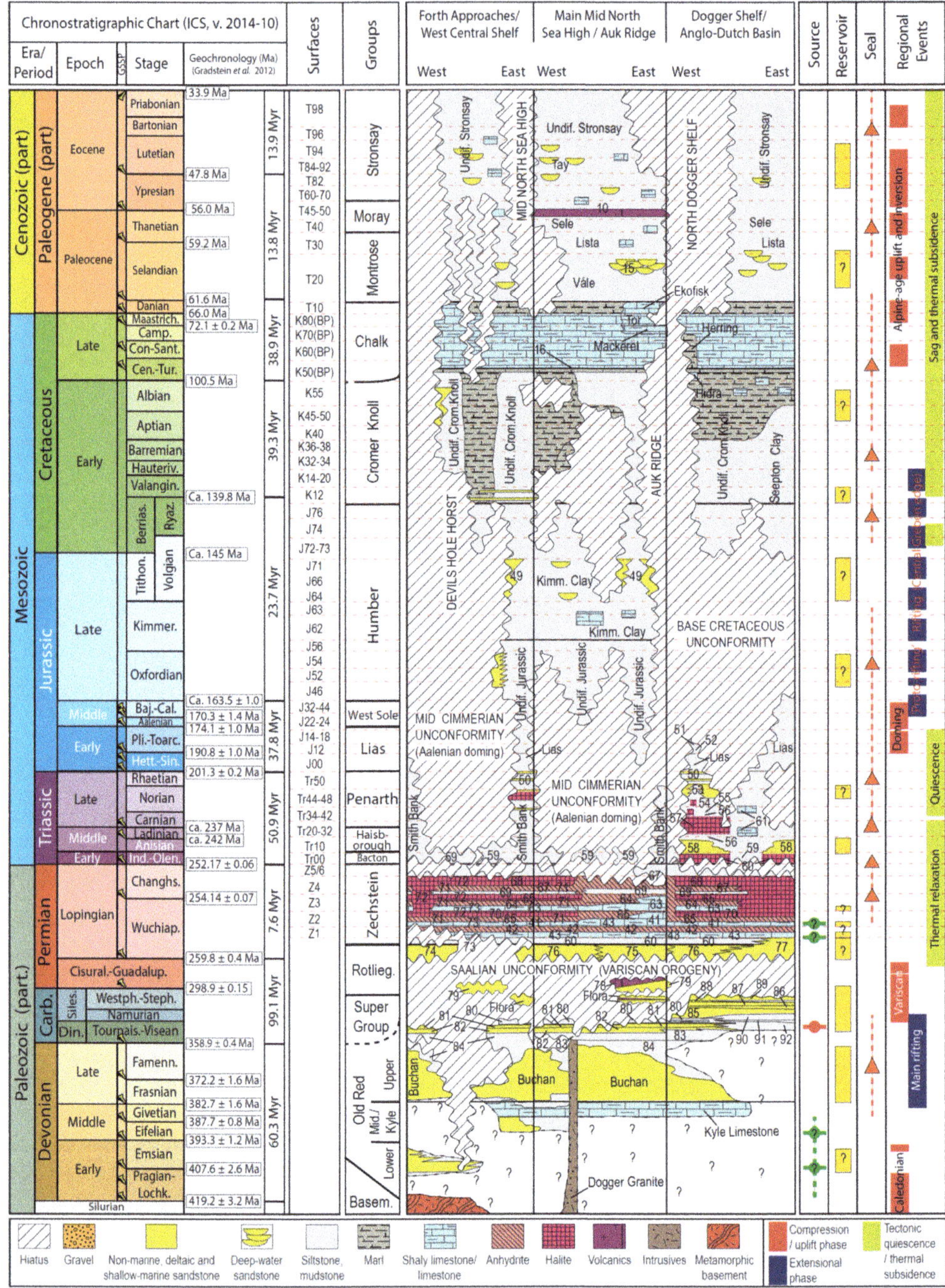

Fig. 3. Regional chronostratigraphic framework for the Greater Mid North Sea High and adjacent basinal depocentres (Forth Approaches, Main Mid North Sea High, Marginal Anglo-Dutch Basin, including Dogger Shelf). Rotlieg., Rotliegend Group; see Table 1 for key to the minor stratigraphic units, here numbered up to number 92.

deterrent to further exploration. Modern technologies, however, allow for economic production of hydrocarbons from large-sized heavy oil discoveries, as demonstrated, for example, by the recent development of the Mariner Field (250 MMbbl oil recoverable reserves) by Statoil.

Renewed attention to frontier platform plays started in 2010, when the Johan Sverdrup oilfield was discovered on the Utsira High, a platform on the eastern flank of the South Viking Graben (Fig. 1). This large structural–stratigraphic trap is characterized by relatively thin, but laterally extensive, reservoir intervals of upper Jurassic, middle Jurassic and upper Permian age (IHS Edin 2017). With expected resources of 1900–3000 million barrels of oil equivalent (MMboe), this field will be one of the most important industrial projects of the next few decades and, at peak production, will constitute 25% of all Norwegian petroleum production during that time (Statoil 2017). This super-giant field is the latest of several oil and gas discoveries that straddle the platform margins, with reservoir ages ranging from Paleozoic fractured basement (e.g. Cairngorm on the edge of the ESP), to Devonian (e.g. Clair, West of Shetlands) to Devonian–Permian (e.g. Auk and Argyll on the western flank of the Central North Sea) to Paleogene (e.g. Mariner, Bressay, Bentley, Kraken) (Figs 4 & 5). Most of these fields are characterized by more than 100 MMboe ultimate recoverable reserves and Clair, in particular, contains *c.* 1100 MMboe (see Patruno & Reid 2016*a* for a full discussion of the existing discoveries).

The discovery of the Johan Sverdrup Field was followed by numerous initiatives, driven by both oil companies and the government sector (e.g. the UK 21st Century Road Map) aimed at revisiting our current understanding of the North Sea platform exploration potential. Within this context, PGS has acquired in excess of 17 500 km^2 of recent (2011–15) three-dimensional (3D) towed dual-sensor broadband seismic data on the outer ESP (UKCS Quadrants 3, 9, 14–16; see Fig. 1). Furthermore, the UK Oil & Gas Authority (OGA) in 2015 commissioned the acquisition of a new regional broadband two-dimensional (2D) seismic programme over the Greater MNSH region with a total length of 10 849 line-km (Fig. 1). The following year, the OGA funded the acquisition of a new regional 2D broadband seismic programme over the inner part of the ESP.

In the context of the mature exploration stage in the UK North Sea, the Greater ESP and MNSH regions are probably the only areas where potentially large structural closures have still not been drilled, particularly in the deeper sub-Jurassic intervals (Reid & Patruno 2015; Patruno & Reid 2016*a*, *b*; Patruno 2017). Recent interpretation work on the 3D dual-sensor broadband surveys of the Greater ESP has highlighted deeper Paleozoic plays, including possible working source intervals and intra-platform basins (Patruno & Reid 2016*a*, 2017; Patruno 2017). On the southern edges of the Greater MNSH, frontier Zechstein (upper Permian) carbonate plays have also been analysed in three dimensions (Patruno *et al.* 2018). As proven by the existing fields, Paleozoic intervals could host sizeable hydrocarbon accumulations, usually in large structural closures, and should not be disregarded for future exploration (Patruno & Reid 2016*a*, *b*).

The ability to view deeper stratigraphy with more clarity is a key requisite to comprehending the deeper geological framework of the platforms (including the interactions between Caledonian shear zones, Devonian extensional structures and more recent tectonics) and to exploiting their full exploration potential (e.g. Patruno & Reid 2016*a*; 2017; Phillips *et al.* 2016; Fazlikhani *et al.* 2017; Patruno & Lampart 2018). Deeply buried Permo-Triassic rift systems have been known to exist within the platform area on the opposite margin of the Viking Graben (Sneider *et al.* 1995; Færseth 1996; Whipp *et al.* 2014; Patruno & Lampart 2018) and shallower Triassic and Jurassic age basins have been described in the western part of the ESP (e.g. the Unst Basin, East Orkney Basin and Dutch Bank Basin; Ziegler 1992; Coward *et al.* 2003; Johnson *et al.* 2005; Richardson *et al.* 2005) and of the East Shetland Basin (e.g. Claringbould *et al.* 2017). As a result of the limitations of legacy seismic data, however, most of the ESP and MNSH area has, for a long time, been conceived as a broad, flat high (Hospers & Ediriweera 1991; Platt 1995; Zanella & Coward 2003) with a shallow acoustic basement and few visible sub-Cretaceous structures (e.g. Skipper 1977; Ziegler 1992; Platt 1995; Sneider *et al.* 1995; Platt & Cartwright 1998; Zanella & Coward 2003; Johnson *et al.* 2005). This misconception is reflected by the tabular nature of the traditionally deepest regionally mappable horizons: the Base Cretaceous Unconformity (BCU) on the Greater ESP (Fig. 4) and the Base Zechstein Unconformity (BZU) on the Greater MNSH (Fig. 5). Clearly mappable Paleozoic reflectors, however, are now visible on the platform area, highlighting large undrilled fault blocks and anticlines, as well as newly described intra-platform basins (Fig. 6 and Patruno 2017; Patruno & Lampart 2018; Patruno & Reid 2016*a*, *b*, 2017; Reid & Patruno 2015).

The present work combines existing well data with all the available regional seismic data covering the Greater ESP and the Greater MNSH, including the latest OGA acquisition. The aim is to re-examine the tectonostratigraphic evolution of these two large regions, highlighting similarities and subtle differences between them, with a specific focus on the previously poorly imaged Paleozoic section. Particular attention will be paid to the relationships between hydrocarbon exploration plays and the sub-Cretaceous structural–stratigraphic framework.

Data and methods

In this study, well data were used to guide and constrain the seismic interpretation, the stratigraphic

Table 1. *List of minor ranking stratigraphic units (usually members) in the East Shetland Platform, South Viking Graben, Greater Moray Firth, Forth Approaches/West Central Shelf, Main Mid North Sea High/Auk Ridge and Dogger Shelf/Cleveland Basin/Anglo-Dutch Basin regions*

Numbers (in Figs 2 & 3)	Unit	Parent unit
1	Forties Sandstone Member	Sele Formation
2	Balmoral Sandstone (Mey) Member	Lista Formation
3	Balmoral Tuffite or Glamis (Mey) Member	Lista Formation
4	Andrew Sandstone (Mey) Member	Lista Formation
5	Brioc Sandstone Member	Horda Formation
6	Grid Sandstone Member	Horda Formation
7	Caran Equivalent Sandstone Member	Horda Formation
8	Frigg/Skroo Sandstone Member	Horda Formation
9	Odin (Beauly) Sandstone Member	Balder Formation
10	Balder Tuff Member and overlying Balder Claystone Member	Balder Formation
11	Flugga/Hermod Sandstone Member	Sele Formation
12	Teal/Skadan Sandstone Member	Sele Formation
13	Heimdal Sandstone Member	Lista Formation
14	Ty Sandstone Member	Våle Formation
15	Maureen Sandstone Member	Våle Formation
16	Black Band Bed Member (= Plenus Marl Member)	Herring Formation
17	Blodøcks and Tryggvason Formation	Chalk/Shetland Group
18	Captain Sandstone Member	Wick Sandstone Formation
19	Britannia Sandstone Member	Carrack Formation
20	Sloop Sandstone Member	Valhall Formation
21	Yawl Sandstone Member	Valhall Formation
22	Fischschiefer Bed Member	Valhall Formation
23	Munk Marl Bed Member	Valhall Formation
24	Skiff Sandstone Member	Carrack Formation
25	Ran Sandstone Member	Carrack/Sola Formation
26	Coracle Sandstone Member	Wick Sandstone Formation
27	Punt Sandstone Member	Wick Sandstone Formation
28	Scapa Sandstone Member	Valhall Formation
29	Birch Sandstone Member	Kimmeridge Clay Formation
30	Brae Formation	Humber Group
31	Miller Sandstone Member	Brae Formation
32	Dirk Sandstone Member	Kimmeridge Clay Formation
33	Claymore/Galley Sandstone Member	Kimmeridge Clay Formation
34	Ettrick and Burns Sandstone members	Kimmeridge Clay Formation
35	Bruce Sandstone Member	Heather Formation
36	Ling Sandstone Member	Heather Formation
37	Rattray Volcanic Member	Pentland Formation
38	Stotfield Calcrete Member	Lossihead Formation
39	Harris Member	Cormorant Formation
40	Lewis Sandstone Member	Cormorant Formation
41	Argyll Limestone Member	Halibut Formation
42	Iris Anhydrite Member	Halibut Formation
43	Innes Limestone Member	Halibut Formation
44	Morag Anhydrite Member	Turbot Formation
45	Eday Marl	Eday Formation
46	John O'Groats Fish Bed	Eday Formation
47	Achanarras Fish Bed	Orcadia Formation
48	Strath Rory Formation	Old Red Group
49	Fulmar Sandstone Formation	Humber Group
50	Winterton Formation	Penart Group
51	Offa Formation	Lias Group
52	Penda Formation	Lias Group
53	Rhaetian Sandstone Member	Penart Group
54	Triton Formation	Penart Group
55	Dudgeon Formation	Haisborough Group

(Continued)

Table 1. (*Continued*)

Numbers (in Figs 2 & 3)	Unit	Parent unit
56	Dowsing Formation	Haisborough Group
57	Rot Halite Member	Dowsing Formation
58	Bunter Sandstone Formation	Bacton Group
59	Bunter Shale Formation	Bacton Group
60	Brockelschiefer Member	Bunter Shale Formation
61	Principle Oolite Zone	Bunter Shale Formation
62	Kupferschiefer Formation	Zechstein Group
63	Plattendolomit Formation	Zechstein Group
64	Hauptanhydrit Formation	Zechstein Group
65	Basalanhydrit Formation	Zechstein Group
66	Leine Halite Formation	Zechstein Group
67	Pegmatitanhydrit Formation	Zechstein Group
68	Aller Halite Formation	Zechstein Group
69	Roter Salzton Formation	Zechstein Group
70	Stassfurt Halite Formation	Zechstein Group
71	Turbot Formation	Zechstein Group
72	Shearwater Formation	Zechstein Group
73	Halibut Formation	Zechstein Group
74	Fraserburgh Formation	Rotliegend Group
75	Auk Formation	Rotliegend Group
76	Leman Sandstone Formation	Rotliegend Group
77	Silverpit Formation	Rotliegend Group
78	Inge/Karl Volcanic Formation	Rotliegend Group
79	Grensen Formation	Rotliegend Group
80	Yoredale Formation	Farne Group
81	Scremerston Formation	Farne Group
82	Fell Sandstone Formation	Border Group
83	Cementstone Formation	Farne Group
84	Tayport Formation	Upper Old Red Group
85	Millstone Grit Formation	Geul Subgroup
86	Caister Coal Formation	Conybeare Group
87	Westoe Coal Formation	Conybeare Group
88	Cleaver Formation	Coal Measures Group
89	Ketch Formation	Conybeare Group
90	Cleveland Group E	Cleveland Group
91	Upper Bowland Shale	Craven Group
92	Cleveland Group A–D	Cleveland Group

The numbers in the first column of the table refer to those in the stratigraphic columns of Figures 2 and 3.

succession and the tectonostratigraphic events of the offshore Greater ESP, Greater MNSH and marginal Anglo-Dutch Basin. The Greater ESP corresponds to the main East Shetland Platform and adjoining intra-platform basins and marginal structural terraces (according to the definition of Patruno & Reid 2016*a*); the Greater MNSH is the wide Variscan and Mesozoic high delimited to the north by the Grampian High, to the east by the West Central Shelf and the Auk–Argyll Ridge and to the south by the Dogger Shelf; the marginal Anglo-Dutch Basin corresponds to the Dogger Shelf, Cleveland Basin and Silverpit Basin (i.e. parts of UKCS Quadrants 36–38 and 41–44) (see Fig. 1b for definition of these areas).

The well analysis consisted of the following sequential steps: (1) analysis of the stratigraphic succession recorded by all vertical exploration wells that penetrated Triassic or deeper strata (205 wells on the Greater ESP and 214 on the Greater MNSH) (Figs 1, 4, 5, 7 & 8); (2) the extraction of decompacted post-erosional sedimentation rate statistics from the two sets of wells and of simple presence/absence data for key stratigraphic intervals (Fig. 8a, b); (3) on the Greater ESP, one-dimensional (1D) basin modelling was performed at key well locations (with well information complemented by seismic interpretation for the stratigraphic horizons below the well total depth (TD)), applying the methodology of Allen & Allen (2005) to reconstruct the burial history and, ultimately, the tectonic subsidence rate through time (Fig. 8d).

The following limitations of the well database have to be taken into account when considering the well-based results shown in Figure 8: (1) as most

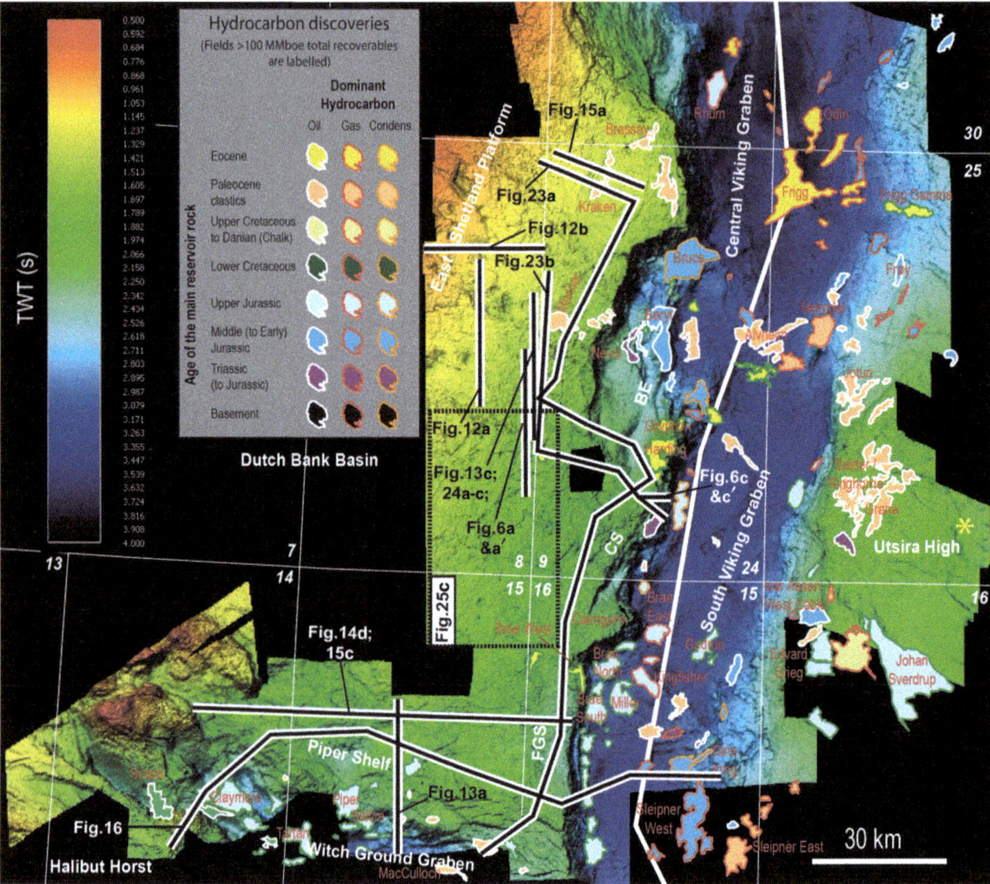

Fig. 4. Base Cretaceous Unconformity two-way travel time–structure map showing the line locations for the seismic cross-sections in this work. Hydrocarbon discoveries by main reservoir age and dominant hydrocarbon type are also shown. BE, Beryl Embayment; CS, Crawford Spur; FGS, Fladen Ground Spur.

of the wells were drilled in proximity of the basin-bounding faults (Fig. 1), the stratigraphic trends of the platform margins are over-represented in the statistical analyses; (2) the three wells selected for 1D basin modelling (Fig. 8d) are the only ones that encountered at least a part of all of the key chronostratigraphic units; as a consequence, the amount of Variscan uplift at these three well locations is probably an under-representation of the overall values on the Greater ESP; and (3) the definition of Paleozoic age boundaries from exploration wells is rather uncertain, particularly the difference between Triassic, Rotliegend and Devonian continental clastic strata based on lithostratigraphy and/or well logs alone, in the absence of more easily identifiable Zechstein (carbonate- and evaporite-rich) and Carboniferous (coal-rich) intervals. This latest issue is made even worse by the usual absence of palynomorphs within Devonian and Rotliegend successions (Marshall & Hewett 2003). As discussed later in the paper, we have employed the following criteria to accept or amend the original operator-defined Paleozoic boundaries: (1) utilization of biostratigraphic ages, wherever available; (2) presence of easily identifiable Zechstein and Carboniferous intervals which, in most cases, allows an informal sub-division of the more monotonous Triassic, Carboniferous and Devonian continental 'red beds'; (3) the presence of angular unconformities in the seismic data has usually been associated to known major erosional truncations (e.g. the Variscan Unconformity is usually placed at the base of the Permian section); and (4) informal lithostratigraphic well-log indicators help to differentiate Devonian and Rotliegend strata when these were in direct contact with each other (as detailed in the Devonian and Permian sections of this paper).

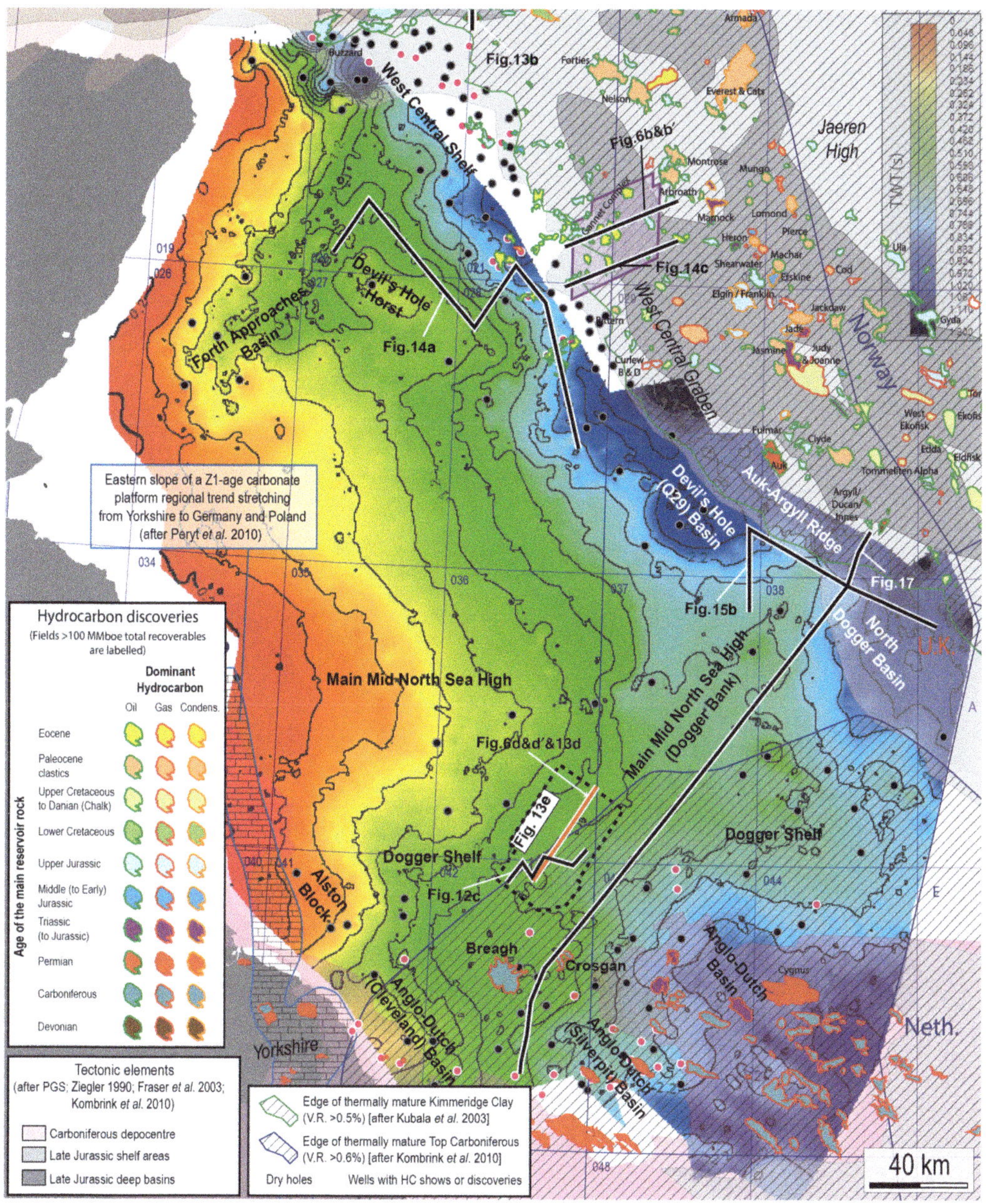

Fig. 5. Base Zechstein Unconformity two-way travel time–structure map showing the line locations for seismic cross-sections in this work. Hydrocarbon discoveries by main reservoir age and dominant hydrocarbon type are also shown. The edges of the 3D SN2013M survey are outlined with the dashed black line (cf. Figs 1b, 7 & 13e).

The available seismic dataset (Fig. 1) consisted of: (1) various 3D towed dual-sensor broadband surveys (GeoStreamer) acquired by PGS in 2011–15 on the outer ESP (part of UKCS Quadrants 3, 8–9, 14–16, on a total area spanning 17 500 km^2); (2) a PGS 3D survey (SNS2013M) acquired in 2013 at the southern margin of the Greater MNSH (part of UKCS Quadrants 36–37 and 42, on a total area spanning 1520 km^2), (3) a PGS regional 2D towed dual-sensor broadband survey (MNSG2009-2011) acquired in 2009–11 over the central and southern portion of the Greater MNSH region,

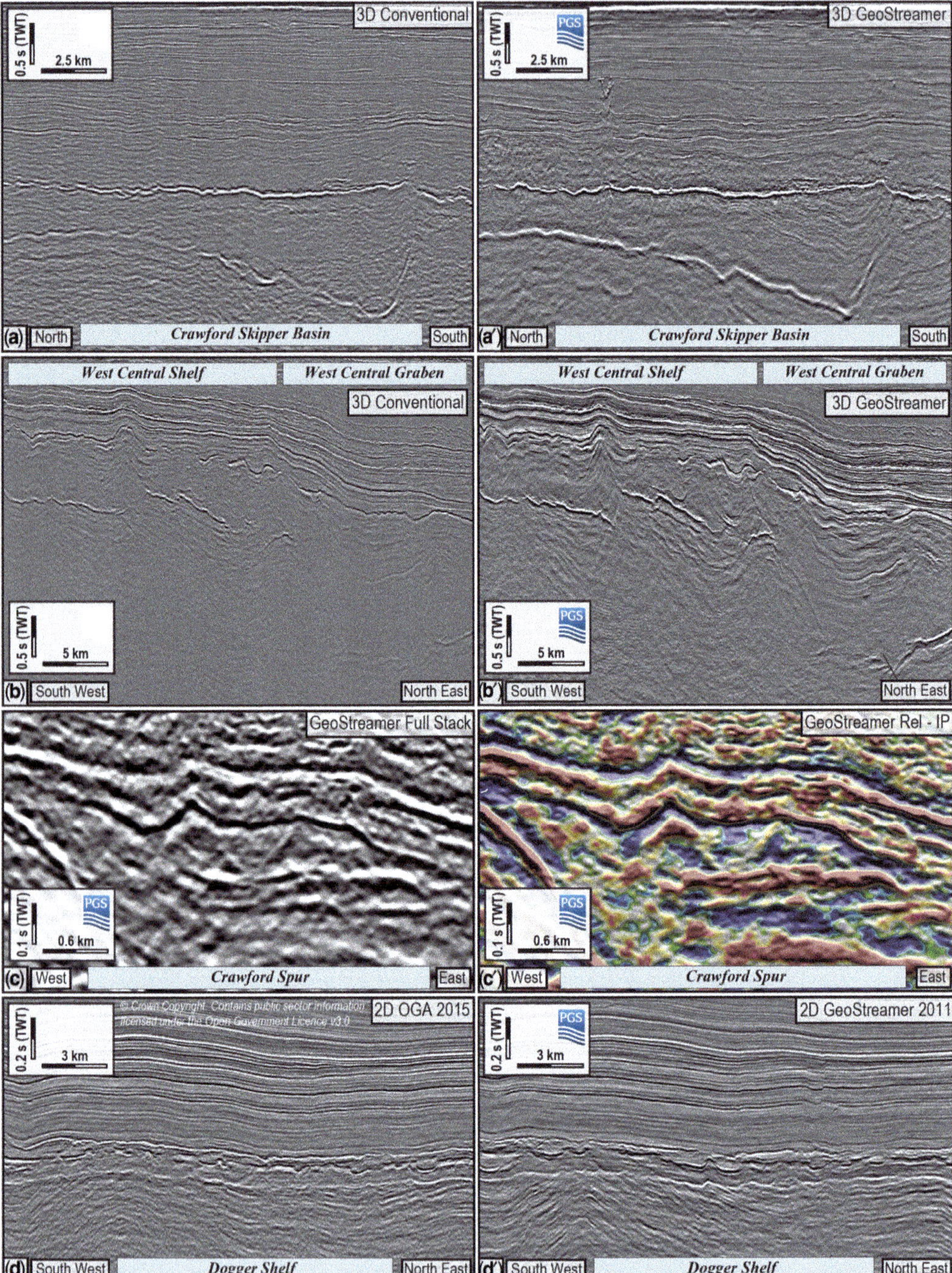

Fig. 6. Comparisons between (**a**, **b**) pre-existing 3D conventional and (**a′**, **b′**) newly acquired 3D broadband seismic in the (**a**, **a′**) East Shetland Platform and (**b**, **b′**) West Central Shelf areas. Comparison between (**c**) full-stack and (**c′**) pre-stack relative acoustic impedance imaging over the Crawford Field (East Shetland Platform), with the impedance section showing increased structural interpretability, with well-imaged minor Triassic tilted fault blocks. Comparison between (**d**) a UK Oil & Gas Authority 2D seismic line and (**d′**) a PGS 2D broadband line over the Dogger Shelf area (Greater Mid North Sea High).

Fig. 7. Regional Paleozoic tectonic elements and Base Permian subcrop map, redrafted after Marshall & Hewett (2003), Bruce & Stemmerik (2003), Glennie *et al.* (2003), Arsenikos *et al.* (2015) and Leslie *et al.* (2015). The edges of the GeoStreamer 3d surveys on the Greater ESP are outlined in purple (cf. Fig. 1b); the edges of the 3D SNS2013M survey are outlined in blue (cf. Figs 1b, 7 & 13e).

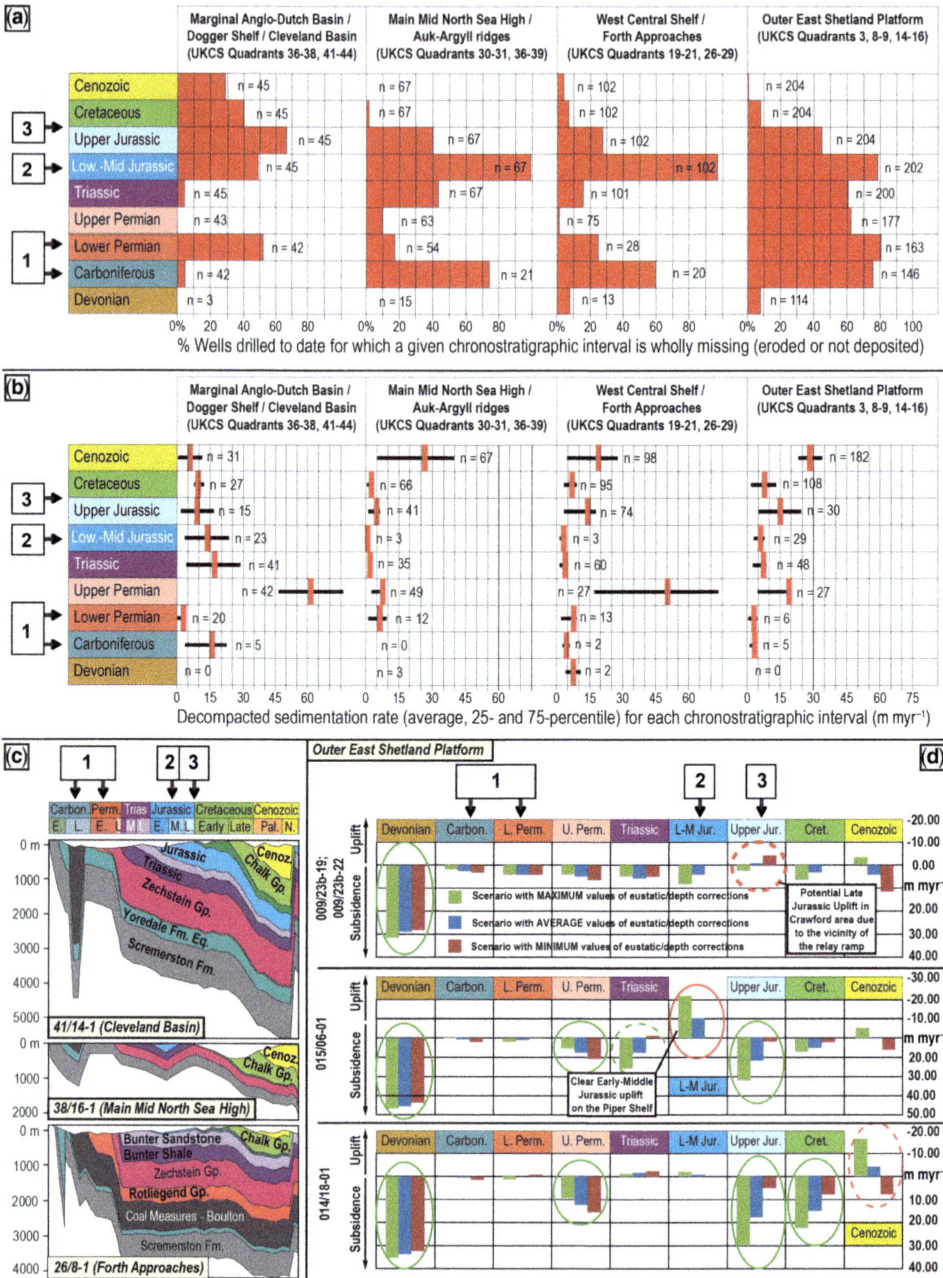

Fig. 8. (**a**) Percentage and total number (*n*) of wells for which a given chronostratigraphic interval is completely missing (not deposited or eroded). (**b**) Decompacted post-erosional sediment accumulation rates (average and 25–75 percentile range) for each chronostratigraphic interval. (**c**) Burial history plots for key wells over the Greater Mid North Sea High and marginal Anglo-Dutch Basin areas (redrafted after Hay *et al.* 2005; Vincent 2015). (**d**) Tectonic sedimentation rates extracted from the back-stripping workflow outlined by Allen & Allen (2005) for three key wells in the Greater East Shetland Platform area. In parts (a–d), annotations (1), (2) and (3) reflect, respectively, the Variscan (Saalian) Unconformity (Late Carboniferous to Early Permian), the Mid Cimmerian Unconformity (Middle Jurassic) and the Base Cretaceous Unconformity. Only the marginal Anglo-Dutch Basin (corresponding with parts of UKCS Quadrants 37–38, 41–42 and the northern 43–44, see Fig. 1) are included in this analysis (i.e. all the wells highlighted in Figs 5 & 20a).

Table 2. *List of interpreted seismic horizons in the study area*

	Horizons	Observations
1	Seabed	Primary (interpreted everywhere)
2	Near-Top Paleocene	Primary (interpreted everywhere)
3	Near-Base Tertiary = Top Chalk Group	Primary (interpreted everywhere)
4	Base Cretaceous Unconformity	Primary (interpreted everywhere)
5	Base of the preserved Jurassic = Mid-Cimmerian Unconformity	Primary (interpreted everywhere)
6	Top Permian = Top Zechstein Group	Primary (interpreted everywhere)
7	Top Rotliegend Group	Secondary horizon (only interpreted in the Crawford–Skipper area)
8	Top Carboniferous	Secondary horizon (only interpreted in the Quadrant 15 area)
9	Intra-Carboniferous	Secondary horizon (only interpreted in the Main Mid North Sea High)
10	Top Devonian = Top Old Red Sandstones	Primary (interpreted everywhere)
11	Middle Devonian dolomite/source rock	Secondary horizon (only interpreted in the Quadrant 8 area)
12	Top crystalline basement	Only interpreted where visible

with a total line-km of 5755 km; and (4) a regional 2015 2D broadband survey acquired by OGA on the Greater MNSH region consisting of 10 849 km.

A total of eight primary horizons and four secondary horizons were interpreted (Table 2). Seismic interpretation was carried out on 2D lines and 3D in-lines and cross-lines at every 0.8 km of horizontal distance; subsequently, whenever possible, the horizons were autotracked across the entire survey area (e.g. Fig. 4).

Numerous possible fluid escape features were recognized in the Greater ESP 3D surveys. To better characterize their map-view distribution and geometry, a series of coherency and RMS attributes were extracted.

Visible fault segments were interpreted in the 3D seismic datasets where possible and characterized statistically in terms of type (normal, inverted, reverse), the main age of activity, strike trends and maximum plan view length, which is an established proxy for total fault displacement (e.g. Cowie & Scholz 1992; Kim & Sanderson 2005; Figs 9–11). The timing of fault activity for each fault segment was estimated by noting the age of associated synrift growth packages whenever present. In the absence of synrift wedges, faults were more tentatively dated by tracing their dip extent and tipping points. For some faults, however, the dating was inconclusive and these were not utilized for the statistical analysis.

Paleozoic structural framework

The regional Paleozoic structural evolution of northern Britain and the eastern UKCS was dominated by intermittent activity along a few major lineaments, interpreted as possible inverted Caledonian faults (Ziegler 1988, 1990; Glennie & Underhill 1998; Coward *et al.* 1989; 2003): the Great Glen Fault, Wick Fault, Walls Boundary Fault, Halibut Horst Fault, Highland Boundary Fault, Southern Uplands Fault and the Devil's Hole Horst structure (Fig. 7). These faults represent the boundaries of three main Paleozoic basins, surrounded by positive areas of older deformed rocks (Figs 1 & 7): the Orcadian Basin to the north and the east–west-trending Southern and Northern Permian basins to the south (Glennie & Underhill 1998). The Orcadian Basin consists of the West Shetland Platform, ESP and proto-Moray Firth regions, whereas the entirety of the Greater MNSH (including the Forth Approaches Basin) constitutes only a part of the continental-scale Southern and Northern Permian basins (Figs 1 & 7).

The NE-trending lineament formed by the Highland Boundary Fault and the related Utsira High fault, in particular, represents a key regional boundary between the Orcadian Basin (to the north of the lineament) and the Northern and Southern Permian basins (to the south) (Fig. 7; Ziegler 1990; Arsenikos *et al.* 2015). The parallel lineament situated just further south (Southern Uplands Fault) corresponds to the Iapetus Suture, which formed from the subduction of the Iapetus Ocean between Laurentia and Baltica during the Caledonian Orogeny (Glennie & Underhill 1998; Coward *et al.* 2003; Stone 2012). The Northern and Southern Permian basins are separated by the Mid North Sea–Ringkøbing-Fyn system of highs (Glennie *et al.* 2003). The Southern Permian Basin continues southwards into the Southern North Sea and onshore the Netherlands, Germany, Poland, Latvia and Lithuania (Peryt *et al.* 2010; Patruno *et al.* 2018).

Permian deposits are consistently preserved over the whole Northern and Southern Permian basins, except for a narrow strip adjacent to the British

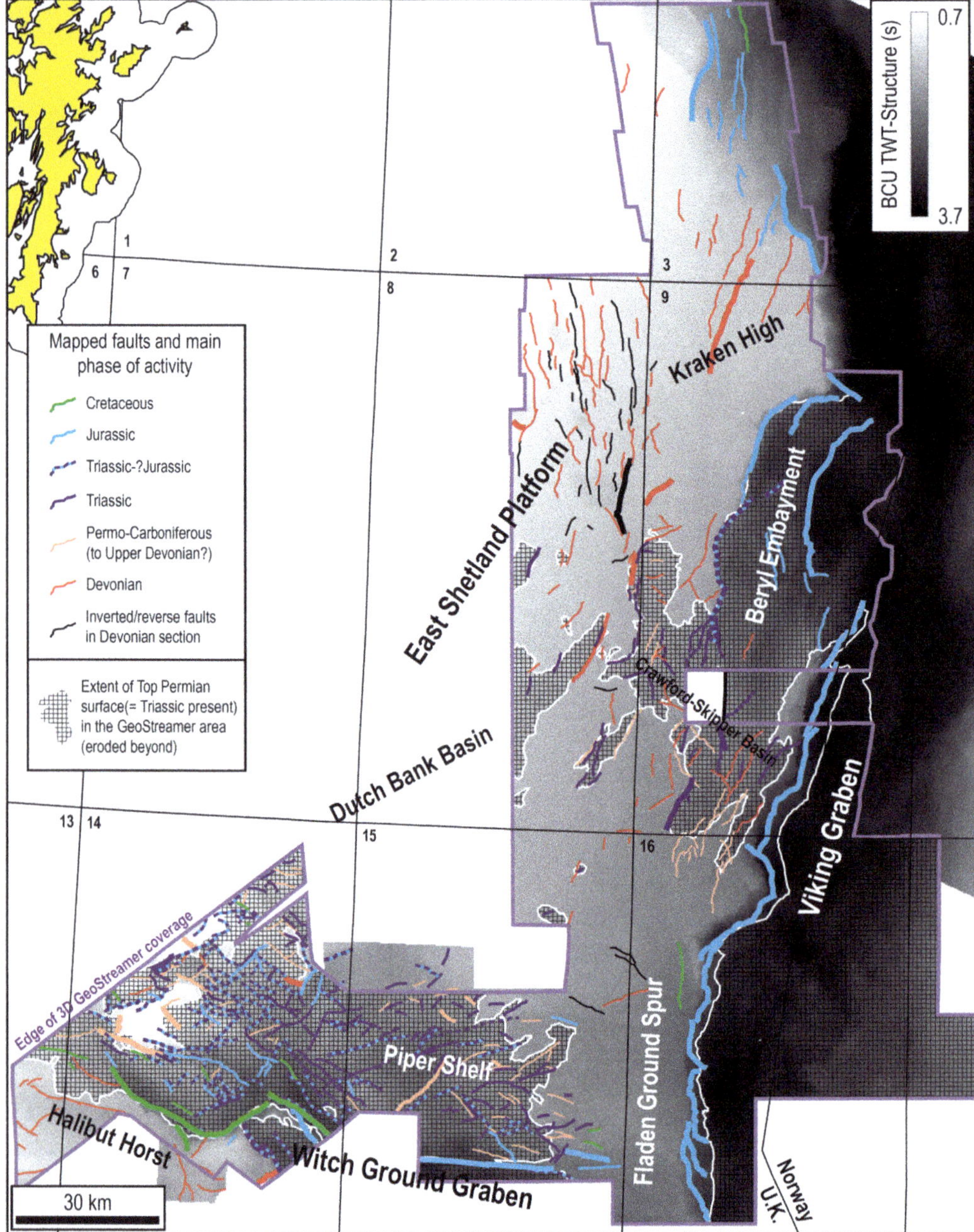

Fig. 9. Mapped intra-platform fault traces over the 3D surveys in the Greater East Shetland Platform region, representing the intersections between the fault surfaces and a Base Cretaceous Unconformity +0.7 s two-way travel time horizon (modified after Patruno 2017). The fault segments have been coloured by the main phase of activity. See Figures 1 and 4 for locations.

coastline, where they were removed by Neogene erosion (Fig. 7). In particular, an extensive thick Zechstein cover, with abundant mobile halite, is present (Fig. 7; Johnson *et al.* 1994; Kearsey *et al.* 2015).

The bulk of the Paleozoic succession on the Greater MNSH is represented by lower Carboniferous sediments that directly subcrop the Base Permian Unconformity (BPU), which in this area is

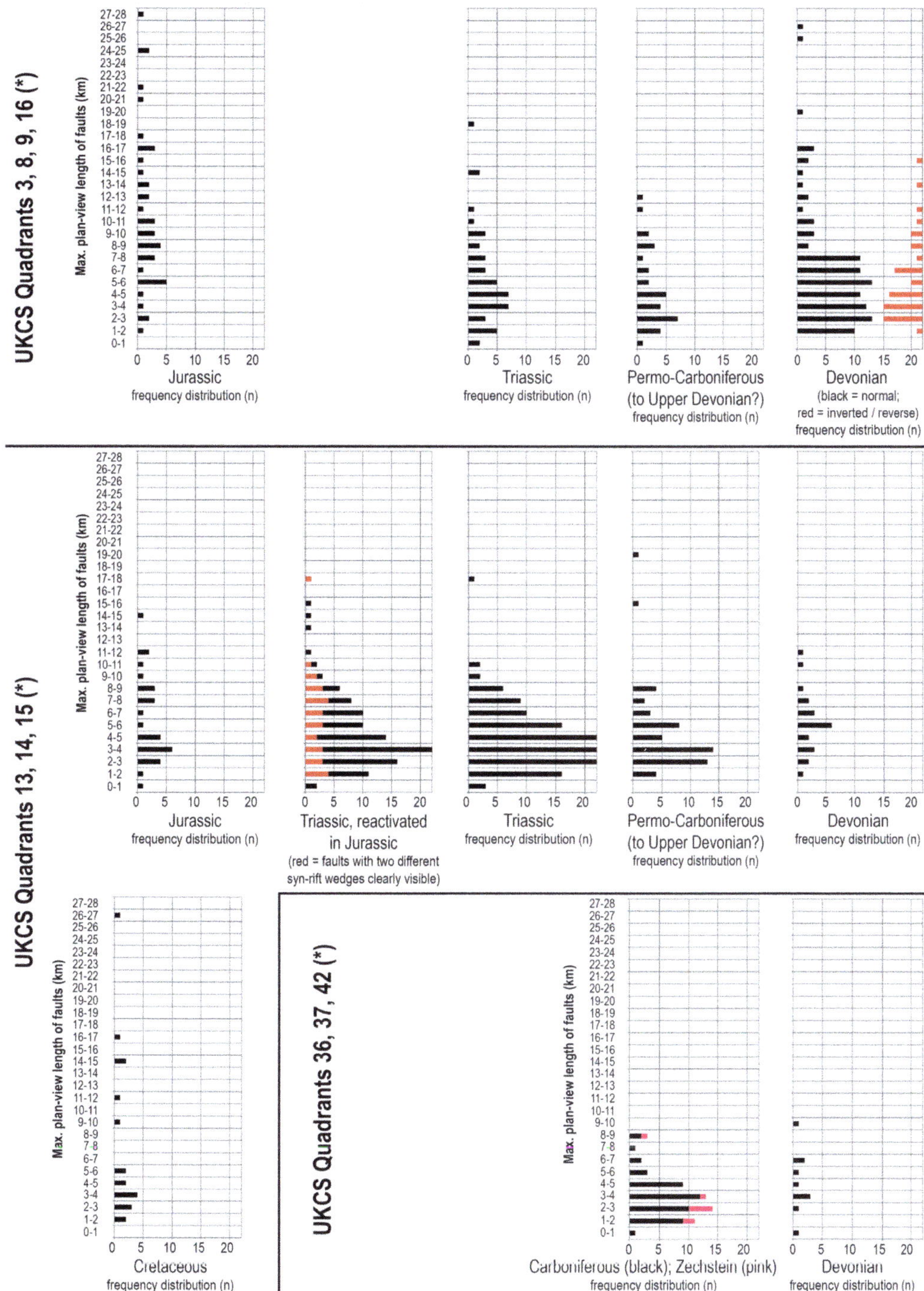

Fig. 10. Statistics of the mapped faults in the portions of the Viking Graben section of the Greater East Shetland Platform (Quadrants 3, 8, 9, 16), the Witch Ground Graben section (Quadrants 13, 14, 15) and the Dogger Shelf (Greater Mid North Sea High) (Quadrants 36, 37, 42) covered by 3D seismic (cf. Figs 1, 4 & 5). The frequency distribution of the maximum plan view length of fault segments is shown here. (*) Only the portion of UKCS Quadrants 36, 37 and 42 covered by PGS Survey SNS2013M (cf. Figs 1b & 5) is included in this fault analysis.

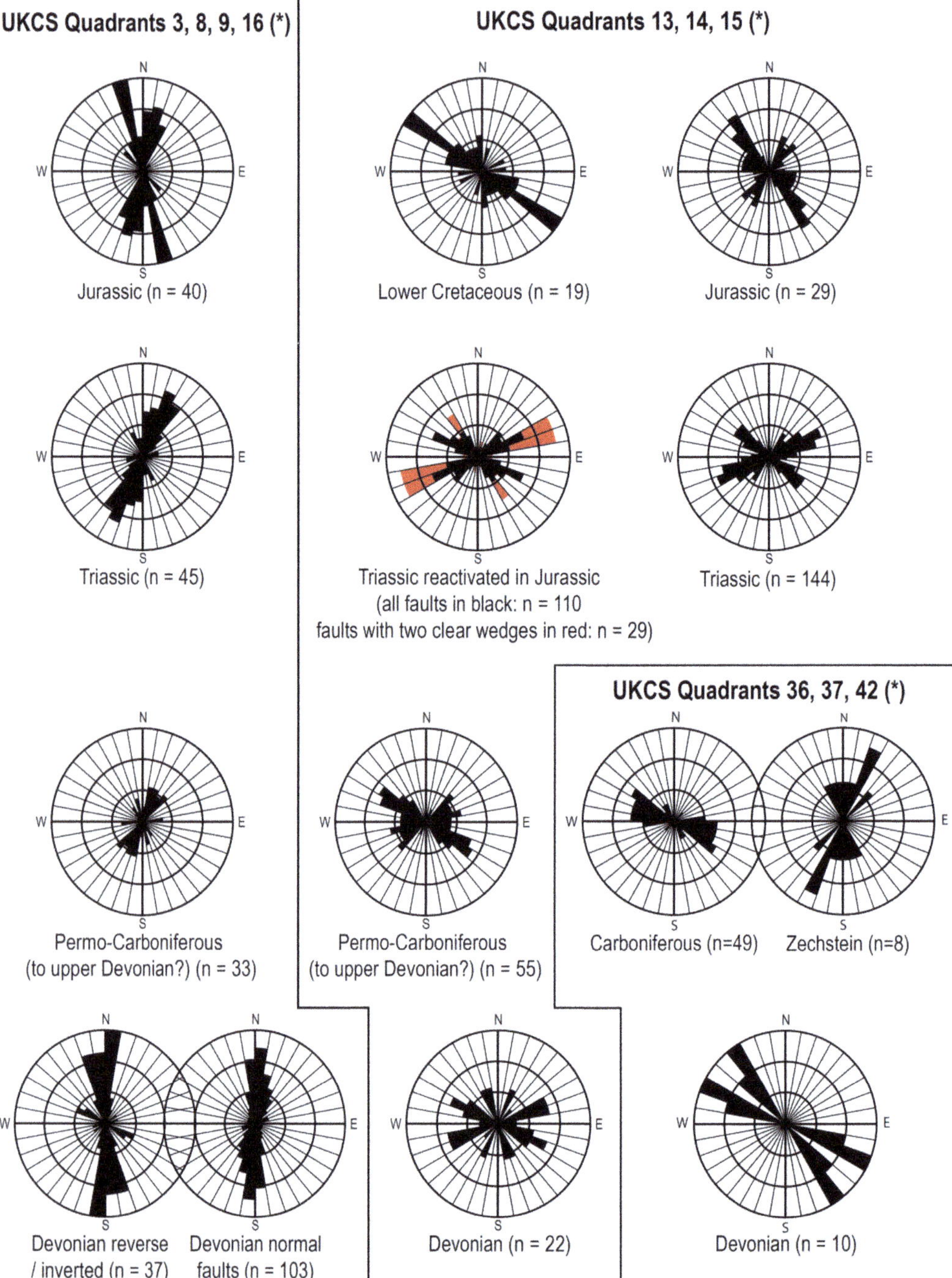

Fig. 11. Statistics of the mapped faults in the portions of the Viking Graben section of the Greater East Shetland Platform (Quadrants 3, 8, 9, 16), the Witch Ground Graben section (Quadrants 13, 14, 15) and the Dogger Shelf (Greater Mid North Sea High) (Quadrants 36, 37, 42) covered by 3D seismic (cf. Figs 1, 4 & 5). Rose diagrams highlighting main strike trends for the fault segments characterized by different main activity age are shown here (cf. Fig. 9). (*) Only the portion of UKCS Quadrants 36, 37 and 42 covered by PGS Survey SNS2013M (cf. Figs 1b & 5) is included in this fault analysis.

represented by the base of the Zechstein Group (Figs 3 & 7). On the structurally highest areas of the Greater MNSH, the Permian is directly overlying Devonian sandstones or older basement rocks (Cameron *et al.* 1992; Cameron 1993; Hay *et al.* 2005; Kombrink *et al.* 2010; Arsenikos *et al.* 2015). Three such areas can be distinguished (Fig. 7): (1) Devil's Hole Horst and an east-trending 40 km wide strip between the Southern Uplands Fault and this horst; (2) the whole eastern platform margin area, from the West Central Shelf to the Auk–Argyll Ridge; and (3) the Dogger Bank area (Quadrants 37–38). Between these structural highs, the remainder of the Greater MNSH is subdivided into lower Carboniferous intra-platform basins (Fig. 7): the Forth Approaches, the Main MNSH and the North Dogger and Q29 basins (Fig. 7; Arsenikos *et al.* 2015; Leslie *et al.* 2015).

Unlike the Southern and Northern Permian basins region, the entire Permian succession is absent across *c.* 60% of the Orcadian Basin (Figs 4 & 7; Bruce & Stemmerik 2003; Glennie *et al.* 2003). The Piper Shelf–Witch Ground Graben area (Quadrants 14–15) is the only region to the north of the Iapetus Suture where wells have consistently encountered Carboniferous rocks (Fig. 7; Leeder & Boldy 1990; Patruno & Reid 2017). Throughout the rest of the Orcadian Basin, either Devonian sediments or crystalline rocks subcrop the Permian or younger strata (Figs 4 & 7). A thick Devonian succession is usually present (*c.* 1–5 km of preserved thickness), with several Devonian depocentres bounded by syndepositional faults (Fig. 7; Marshall & Hewett 2003).

Tectonostratigraphic architecture

Below and in Table 3, the tectonostratigraphic architecture of the study areas has been constrained by integrating well data with seismic interpretation of faults and horizons.

Well data trends

The stratigraphic information derived from well penetrations reveals a broadly consistent tectonostratigraphic development of both the Greater ESP and MNSH areas, albeit with some significant differences (Fig. 8). In both areas, the Devonian interval is generally present everywhere as it has been encountered by most wells with a bottom-hole age older than Carboniferous (Fig. 8a). On the Greater ESP, the Devonian is characterized by very high average tectonic subsidence rates (30–50 m myr^{-1}) (Fig. 8d), which lead to total Devonian thicknesses of 1000–5000 m extending across much of the area (Glennie & Underhill 1998; Patruno & Reid 2016*a*, 2017). Over the Greater MNSH, it is difficult to characterize the total Devonian thickness and subsidence rates as a result of both incomplete well penetrations and challenging seismic imaging.

The Carboniferous interval tends to be present in virtually all wells drilled in the marginal Anglo-Dutch Basin within our study area, but it is completely absent in *c.* 60–75% of the wells drilled in the Greater ESP and MNSH areas (Fig. 8a). This is reflected by relatively high subsidence rates recorded in the marginal Anglo-Dutch Basin and by much lower (<5 m myr^{-1}) rates recorded by the wells that penetrated Carboniferous deposits in both the Greater MNSH and ESP areas (Figs 8b, d). Burial history modelling performed on pseudo-wells at the basin margins (Hay *et al.* 2005; Vincent 2015), however, reveal that the overall low Carboniferous subsidence rate on the Greater MNSH is due to rapid Early Carboniferous subsidence followed by abrupt uplift in the Late Carboniferous (Fig. 8c). In the marginal Anglo-Dutch Basin and the Forth Approaches area, a less significant late Carboniferous uplift leads to the preservation of the main source interval for the Southern North Sea (upper Carboniferous Coal Measures) (Figs 3, 7 & 8c), although in the Forth Approaches this interval is largely devoid of coal and the source rock potential in this area is restricted to Visean shale and coal intervals (Monaghan *et al.* 2015; Kearsey *et al.* 2015; Vincent 2015).

The overall lower–'middle' Permian trend is similar to the Carboniferous, with <10 m myr^{-1} of subsidence recorded by the wells that penetrate this whole interval (Fig. 8b, d). However, unlike the abrupt Carboniferous burial and uplift trends, the pre-Zechstein Permian interval is characterized by low overall sedimentation rates (Fig 8c, d). This interval is completely missing in *c.* 50 and 80% of the wells in the marginal Anglo-Dutch Basin and the Greater ESP, respectively. However, the lower–'middle' Permian Rotliegend Group has been encountered by most of the wells on the eastern and northern margins of the Greater MNSH (Forth Approaches, West Central Shelf and Auk–Argyll Ridge) (Fig 8a, c).

The total accumulation rates show a significant peak during the deposition of the upper Permian Zechstein interval in both the marginal Anglo-Dutch Basin and the West Central Shelf/Forth Approaches areas (15–75 m myr^{-1}), but only a minor increase in the Main MNSH (<10 m myr^{-1}) and the Greater ESP (5–20 m myr^{-1}) areas (Fig. 8b–d). Within the Greater ESP, a more sustained tectonic subsidence seems to have been experienced in the Piper Shelf areas Quadrants 14–15 (Fig. 8d). As a consequence, more wells with a bottom-hole age greater than or equal to the Carboniferous encountered the Zechstein interval than the underlying Carboniferous–'-middle' Permian units, particularly on the marginal Anglo-Dutch Basin and West Central Shelf/Forth Approaches areas, but also on the Main MNSH

Table 3. *Comparative table showing the main tectonic elements by age of activity, across the Greater East Shetland Platform (ESP) and Greater Mid North Sea High (MNSH)*

Period	Greater East Shetland Platform				Greater Mid North Sea High			Comments
	Distribution	Tectonics	Fault orientation, UKCS Quadrants 3, 8, 9 and 16	Fault orientaion, UKCS Quadrants 13, 14 and 15	Distribution	Tectonics	Fault orientation, UKCS Quadrants 36, 37 and 42	
Devonian	Everywhere	Extension	Normal 350–010° N	Multiple: (1) 020–030° N, (2) 060–080° N, (3) 280–300° N and (4) 330–350° N	Everywhere	Extension	Multiple: (1) 280–300° N, (2) 310–330° N	ESP: Mid-Devonian is lacustrine with limited marine facies MNSH: Mid-Devonian is dominated by a marine carbonate (Kyle Limestone)
Lower Carboniferous	Sporadic	Extension?	Faults difficult to differentiate from the Permian ones	Faults difficult to differentiate from the Permian ones	Everywhere	Extension	270–310° N	ESP: lower Carboniferous four times thinner than the same units of the MNSH and is restricted to Permo-Carboniferous mini basins MNSH: lower Carboniferous regionally distributed and affected by folding and faulting
Upper Carboniferous	Absent	Uplift/ compression			Absent	Uplift/ compression?		ESP: Absent MNSH: Absent
Lower Permian	Sporadic	Extension	020–050° N	Multiple: (1) 040–080° N, (2) 290–310° N	Absent/spo radic	Interval missing		ESP: lower–middle Permian Rotliegend is restricted to mini basins (i.e. active intra-platform grabens and half-grabens) MNSH: Rotliegend Group is generally less widespread and is absent north of Quadrant 40 with Zechstein (upper Permian) narrow directly lying on Carboniferous or older units
Upper Permian	Sporadic	Extension?	020–050° N	Multiple: (1) 040–080° N, (2) 290–310° N	Everywhere	Extension?/ quiescence	330–030° N	ESP: Zechstein is present within mini basins (i.e. active intra-platform grabens and half-grabens) and shows a condensed succession with diapirs being small to absent MNSH: Zechstein is thick and generally lying unconformably on Carboniferous/Devonian; salt facies are thick and have large diapiric structures

Triassic	Sporadic	Extension	000–040° N	Multiple: (1) 050–080° N, (2) 300–320° N	Everywhere	Quiescence	No fault observed	ESP: Triassic is largely absent and confined to mini basins MNSH: Triassic displays two relationships with the underlying Zechstein: (1) Zechstein halokinesis has occurred post-Triassic and (2) Triassic was deposited syn- or post-Zechstein mobilization
Lower–middle Jurassic	Absent	Uplift			Absent	Uplift		ESP: Absent MNSH: Absent, but preserved at MNSH Southern North Sea transition
Upper Jurassic	Sporadic	Extension/ quiescence	Multiple: (1) 340–350° N, (2) 000–020° N	Multiple: (1) 020–050° N, (2) 310–330° N; additionally, many Triassic faults were reactivated in the Jurassic, mostly trending N60–80° N and, secondarily, N290–310° N and N320–330° N	Sporadic	Extension/ quiescence	No fault observed	ESP: present in localized veneers and absent in structurally elevated areas MNSH: present in localized veneers and absent in structurally elevated areas
Lower Cretaceous	Absent/ condensed	Extension in the Witch Ground Graben in the Lower Cretaceous; Quiescence thereafter and elsewhere	No fault observed	280–320°N	Everywhere	Quiescence	No fault observed	ESP: thickness varies, missing from <5% of wells MNSH: thickness varies, missing from <5% of wells
Cenozoic	Everywhere	Quiscence/ inversion	No fault observed	No fault observed	Everywhere	Quiescence/ inversion	No fault observed	ESP: thick interval, 2 s TWT, forms key seal and reservoir interval; due to seawards tilting the succession thickens seawards and undergoes extensive erosion landward MNSH: thick interval 2 s TWT forms key seal and reservoir interval; due to seawards tilting the succession thickens seawards and undergoes extensive erosion landward

(see Figs 5 & 7), where only 10% of the wells did not encounter any upper Permian strata (Fig. 8a). Similarly, but less markedly, on the Greater ESP, *c.* 62% of the wells did not encounter upper Permian units, while the lower–'middle' Permian was not encountered by *c.* 80% of the wells.

Over the Greater ESP and MNSH areas, the whole Triassic interval is characterized by an overall lower subsidence rate than the upper Permian (Fig. 8b–d). There are areas, however, where the subsidence values, particularly of the lower Triassic units, are the same as those observed in the upper Permian (e.g. the Forth Approaches and parts of the Greater ESP, Figs 8c, d). The Triassic unit is widespread on the Greater MNSH and in the marginal Anglo-Dutch Basin, but is not as prevalent as the upper Permian interval, particularly on the Auk–Argyll Ridge (Fig. 8a). However, the Triassic interval has been encountered by a slightly higher proportion of wells than the upper Permian on the Greater ESP, where it is present in 'only' 40% of the wells (Fig. 8a).

The entire lower–middle Jurassic interval is missing in almost all Greater MNSH wells and in nearly 80% of the Greater ESP wells (Fig. 8a). This interval is, however, present in *c.* 50% of the marginal Anglo-Dutch Basin wells, with a continuous Jurassic section in the Cleveland Basin at the transition with the Greater MNSH (Fig. 8a, b). This is the result of an overall tectonic uplift, particularly in the proximity of the Central and South Viking Graben (e.g. wells 38/16–1 and 15/6–1 in Fig. 8c, d).

As a result of both eustatic rise and renewed rifting activity, the upper Jurassic is characterized by a new peak in depositional rates, particularly in regions adjacent to the Jurassic trilete North Sea graben systems, including over the Greater ESP (tectonic subsidence rates up to 20–30 m myr^{-1}), West Central Shelf and Auk–Argyll Ridges (Fig. 8b, d). As a consequence, unlike the lower–middle Jurassic, the upper Jurassic unit has been encountered in the majority of wells on the Greater ESP and MNSH (Fig. 8a). The opposite trend, however, has been noticed in the marginal Anglo-Dutch Basin, with a minor net uplift (Fig. 8a–c).

Cretaceous units are present in nearly all the Greater ESP and MNSH wells, but are missing in *c.* 40% of the marginal Anglo-Dutch Basin wells (Fig. 8a, b). Overall subsidence rates and tectonic subsidence rates are moderate (Fig. 8b–d), with the exception of the Witch Ground Graben margins, where a pronounced tectonic subsidence rate has been defined by well data (e.g. Fig. 8b and well 14/18–1 in Fig. 8d).

The Cenozoic interval is present in the vast majority of wells in the Greater MNSH and the ESP. However, closer to the British coastline the Cenozoic sediments tend to be absent (e.g. the Forth Approaches, Cleveland Basin) (Fig. 8a, c). Both on the Greater ESP and the MNSH, overall active subsidence is recorded on the eastern platform margin and uplift in the proximity of the British coastline, particularly (Fig. 8b–d). As a consequence, areas hosting both wells drilled in the proximity of the present day coastline and wells drilled much further seawards reveal a large spread in post-erosional sediment accumulation rates (5–40 m myr^{-1}) (e.g. the MNSH, see Fig. 8b).

Seismic interpretation: fault analysis

A total of 633 and 67 fault segments have been identified, respectively, on the Greater ESP 3D and the MNSH SNS2013M surveys (Figs 9–11; Patruno 2017; Patruno *et al.* 2018).

The Greater ESP area can be subdivided into two regions based on the diametrically different trends of fault orientation: a Viking Graben section (i.e. portions of Quadrants 3, 8–9 and 16) and a Witch Ground Graben section (i.e. portions of Quadrants 13, 14 and 15) (Fig. 9).

In terms of the number of measured fault segments, the Witch Ground Graben section of the Greater ESP is dominated by Permo-Carboniferous and Triassic extensional faults. The Viking Graben section of the Greater ESP hosts predominantly Devonian and, secondarily, Permo-Triassic fault segments. In the SNS2013M survey (parts of Quadrants 36–37 and 42 on the Greater MNSH, Fig. 1), mostly Carboniferous (to Early Permian?) extensional fault segments have been mapped (Figs 9 & 10).

On the Witch Ground Graben section of the Greater ESP, the longest fault segments (in plan view) are: (1) a lower Cretaceous Witch Ground Graben border fault (26–27 km) and (2) a few Permo-Carboniferous to Triassic–Jurassic faults that are to 17–20 km long (Fig. 10). In the Viking Graben section, the longest measured segments are upper Jurassic Viking Graben border faults and Devonian master faults (as shown in Fig. 14a), each with lengths of 24–28 km (Figs 9, 10). Triassic, Permo-Carboniferous and Devonian fault segments display a unimodal length distribution, with a mode of 2.5–6.0 km on the Greater ESP and 2–4 km on the Greater MNSH (Fig. 10). In comparison, Jurassic and Cretaceous faults on the Greater ESP display a plurimodal plan view length distribution with discrete occurrence peaks at 2–4, 5–9 and 14–17 km (Fig. 10).

The Viking Graben and Witch Ground Graben sections of the Greater ESP are characterized by different structural grains: a roughly north–south trend for the Viking Graben section and an ENE and ESE conjugate set for the Witch Ground Graben section (Figs 9–11). These overall trends are directly reflected by the orientation of the master faults bounding the main Jurassic–Cretaceous graben: NNE and SSE trends along the Viking Graben edge and an overall WNW trend along the Witch Ground Graben (Figs 9 & 11). Permo-Carboniferous and Triassic fault trends on the Viking Graben section display a noticeable clockwise rotation (0–50° N) when compared with the predominant north–south overall trends typical of Jurassic and Devonian faults (Fig. 11). In the Witch Ground Graben section, the conjugate ENE and ESE trends are clearly visible in pre-Jurassic faults (Fig. 11). However, a single NW–SE dominant trend has been documented for 'Late Jurassic only' and Early Cretaceous faults. As a consequence, the conjugate ENE set became statistically less significant in the Late Mesozoic and a main WNW trend was established for the major faults bounding the Witch Ground Graben (Figs 9 & 11).

Three fault sets were mapped in the SNS2013M survey (Quadrants 36–37 and 42 on the Greater MNSH, see Fig. 1b), two of which predate the deposition of the Zechstein (Fig. 13d). In particular, the faults tipping out just above the Kyle Limestone horizon (as shown in Fig. 13d) may reflect Late Devonian extension, whereas those offsetting the Carboniferous and terminating near the BZU (as shown in Fig. 13d) probably formed during Early Carboniferous extension (cf. the basin-wide Dinantian rifting event described by Kombrink *et al.* 2010). The dominant WNW to NW strike trend of the two Devonian–Carboniferous fault sets in the SNS2013M area is consistent with one of the two time-equivalent conjugate sets in the Witch Ground Graben section of the Greater ESP, as well as with the dominant Carboniferous structural grain of the main Southern North Sea to the SE (Fig. 11; Leeder & Hardman 1990; Geluk 2005; Pharaoh *et al.* 2010; Kombrink *et al.* 2010). This strike is markedly different from the NNE–SSW strike characterizing the Zechstein faults in the 3D survey area, suggesting a change in extension direction between the Devonian–Carboniferous and Permian extensional events (Fig. 11). In addition, the modal plan view length of the Carboniferous–Devonian faults is *c.* 1 km longer than that of the Permian faults (Fig. 10). This observation, together with the lower number of mapped Permian faults and previous regional-wide observations (e.g. Ziegler 1989; Geluk 1999), indicate that the Carboniferous extension over the SNS2013M area had been a statistically more significant event than the Permian one.

Seismic interpretation: stratigraphy and tectonics

Compressional folds and reverse faults (Fig. 12), extensional fault blocks (Fig. 13), Zechstein-related halokinetic structures (Fig. 14) and various types of inverted structures (Figs 15–17) have all been identified in the Greater MNSH and the ESP areas. Observations drawn from seismic interpretation of the studied seismic dataset are outlined in the following sections, from the oldest to the youngest stratigraphic level. A series of decompacted cartoons highlighting the structural evolution along two representative regional sections is shown in Fig. 18 and a schematic play concept for the Greater East Shetland Platform and Mid North Sea High regions is shown in Fig. 19.

Devonian units. In the Orcadian Basin, which includes the ESP, the Devonian Old Red Group consists of an upper and lower sandstone-prone continental unit, known as Buchan and Struie formations, respectively. The sandstone-prone upper and lower Devonian units are separated by a Middle Devonian interval that is instead rich in lacustrine claystones, source rock fish beds, coals, as well as possible marine anhydrites (e.g. well 9/16–3, see Duncan & Buxton 1995; Patruno & Reid 2016*a*, *b*, *c*, 2017) and dolomites (e.g. well 8/4-1). The Middle Devonian, in turn, is subdivided into a Givetian age Eday Group and the Eifelian age Orcadia and Strath Rory formations (Marshall & Hewett 2003) (Fig. 2).

The Givetian Eday Group contains both sandstone- and mudstone-prone units, including possible source rocks (Marshall & Hewett 2003). In particular, the distinctive Eday Marl Formation usually represents the uppermost sub-unit of the Eday Group. The Eday Marl is composed of siltstones with a significant content of carbonate and anhydrite, and is considered the time-equivalent of the Kyle Limestone, a fully marine carbonate-rich unit of the Central North Sea region (Cameron 1993; Esso 2009; Marshall & Hewett 2003). A clear intra-Eday Marl anhydrite marker was penetrated by well 9/16-3 in the Greater ESP (Marshall & Hewett 2003).

The Eifelian Orcadia Formation is a lacustrine mudstone-prone deposit – including distinctive organic-rich markers such as the Achanarras Fish Bed, which might represent an important and overlooked regional source rock – and is the source for the Inner Moray Firth Beatrice Field (Duncan & Buxton 1995; Marshall & Hewett 2003; Patruno & Reid 2016*a*, *b*, *c*). In the Greater ESP, the Orcadia mudstones are usually characterized by a higher than normal carbonate content, including several interbedded dolomitic layers (e.g. well 8/4-1). The Achanarras Fish Bed source rock has been penetrated by well 9/16-3 in the Greater ESP study area (Duncan & Buxton 1995).

The Devonian Old Red Sandstone Group is usually thicker than 0.5 s two-way time (TWT) (e.g. Fig. 15). No well samples the entire thick Devonian succession, but hundreds of wells penetrate the top of

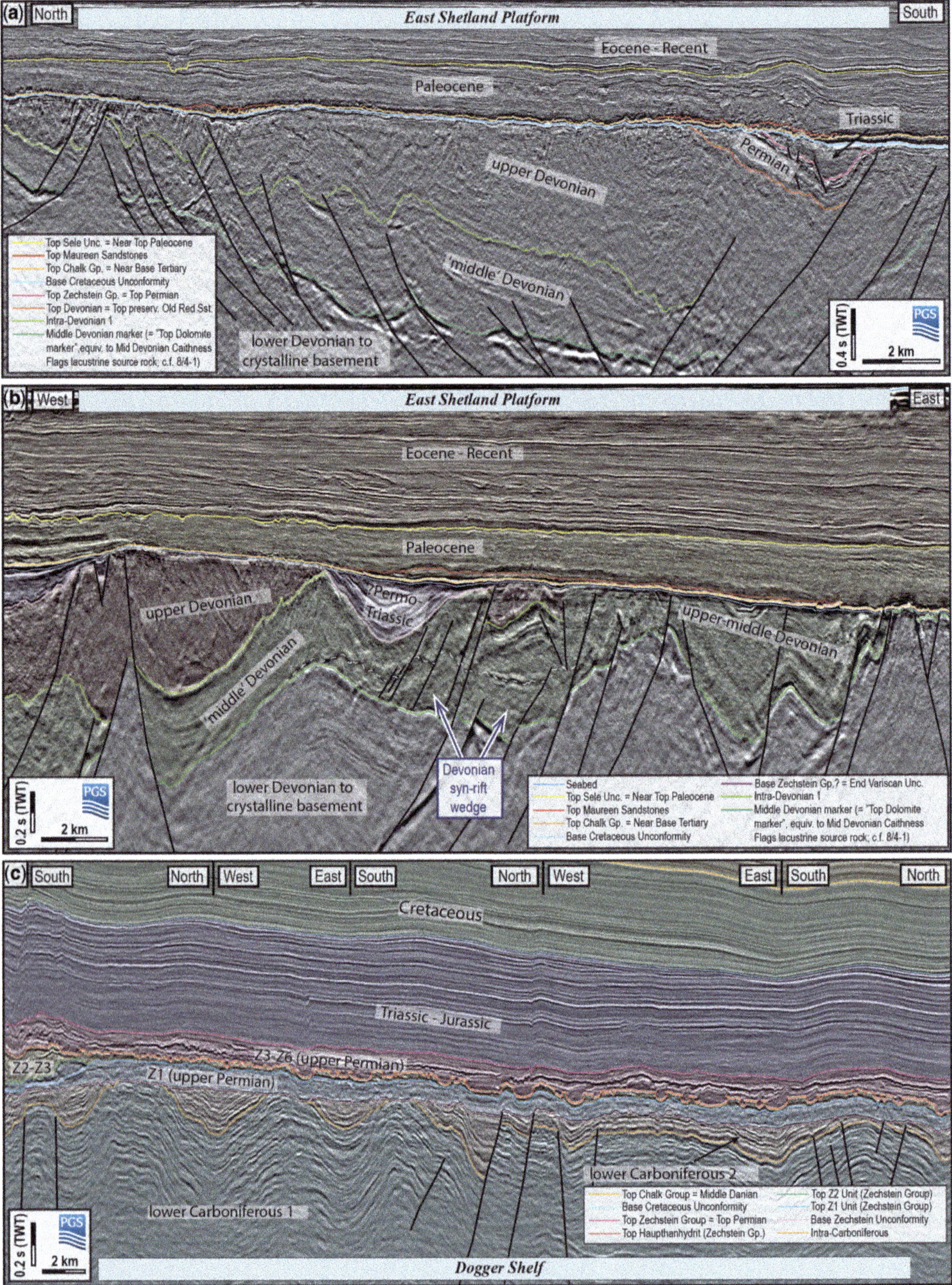

Fig. 12. Interpreted seismic cross-sections showing compressional folds and faults of Middle Devonian to Variscan age observed throughout the (**a, b**) Greater East Shetland Platform and (**c**) Mid North Sea High, and buried beneath the (**a, b**) Base Cretaceous Unconformity and (**c**) Base Zechstein Unconformity = Base Permian Unconformity. See Figures 4 and 5 for line locations.

the Old Red Sandstone, particularly on the Greater ESP (Figs 8a, 16a). As a consequence, the 'Top basement' horizon is often subjective (Figs 12, 13, 15–17). Two units are delimited by the 'Top Devonian' and 'middle Devonian marker', which we call 'upper' and 'middle Devonian' in the following.

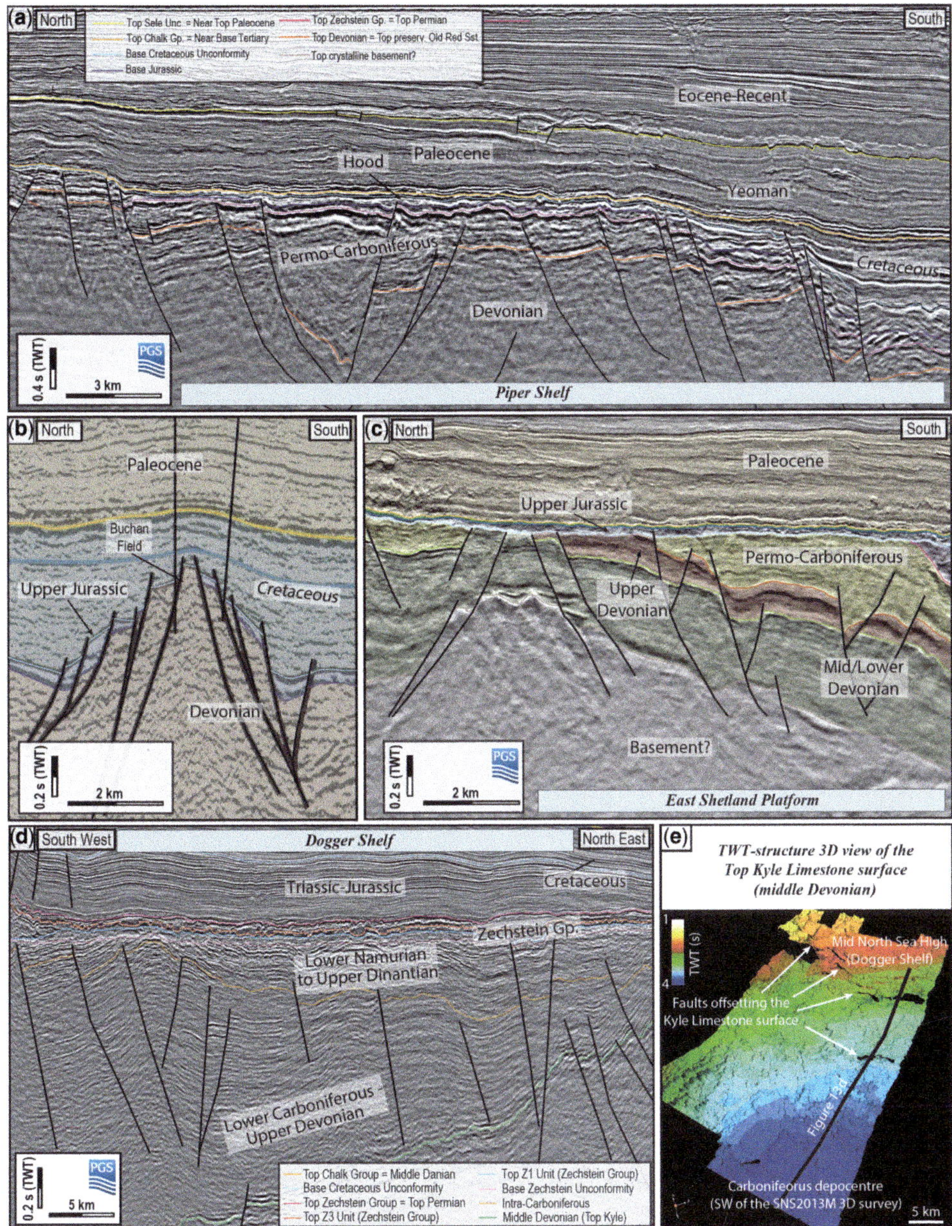

Fig. 13. Interpreted seismic cross-sections and 3D views showing extensional fault blocks observed throughout the (**a–c**) Greater East Shetland Platform and (**d, e**) Mid North Sea High. These are buried, respectively, beneath the Base Cretaceous Unconformity (**a–c**) and Base Zechstein Unconformity (**d, e**). In particular, parts (b) and (c), respectively, show the large Devonian Buchan Field (200 MMboe) in the Greater Moray Firth area (redrafted after Edwards 1991) and a similar undrilled horst block mapped over the Greater East Shetland Platform area, shown at the same scale. See Figures 4 and 5 for line locations. Figure 13e shows a 3D view of the interpreted Top Kyle Limestone in the SNS2013M survey in the southern Mid North Sea High (see Figs 1b and 4 for location).

The 'upper Devonian' is characterized by a relatively opaque low-amplitude seismic facies, which is suggestive of relatively uniform lithology (Figs 12a, 13a, 15a, b, 16b & 17). The 'middle Devonian' marker, where present, is a high-amplitude continuous 'hard-kick' event and the Middle Devonian

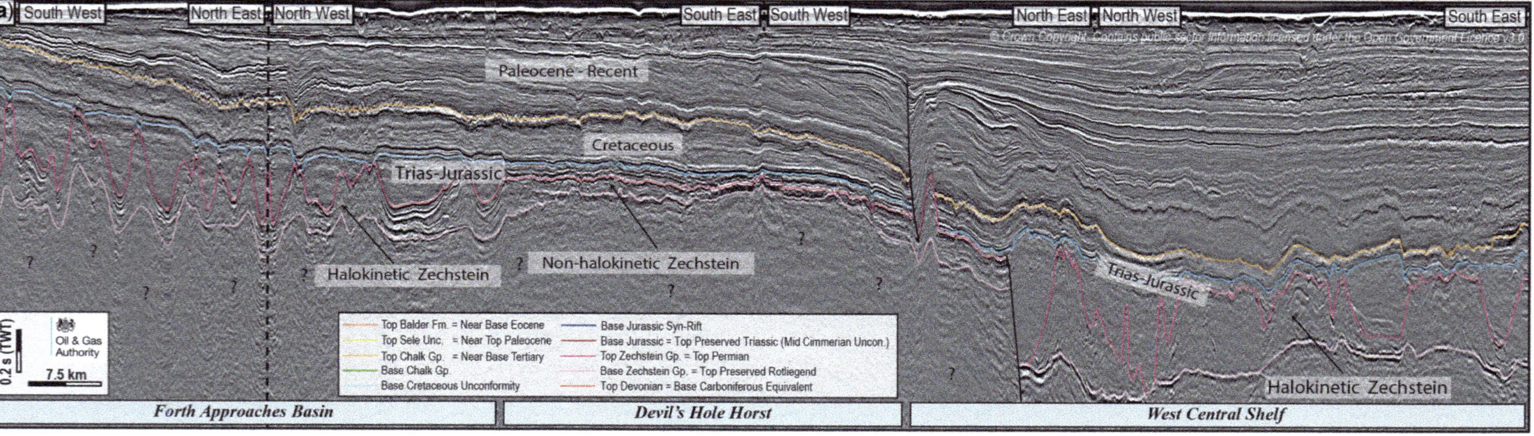

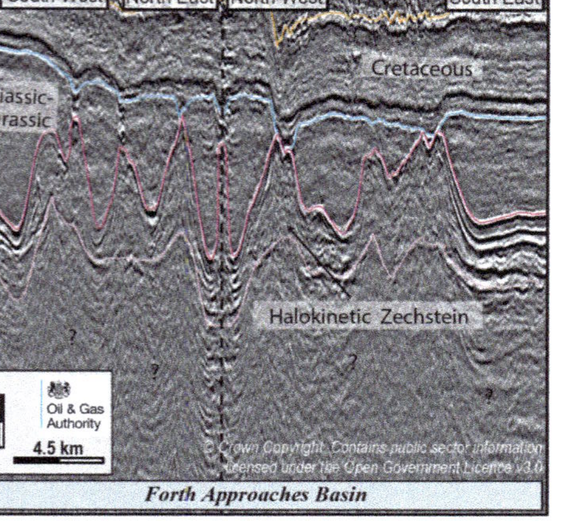

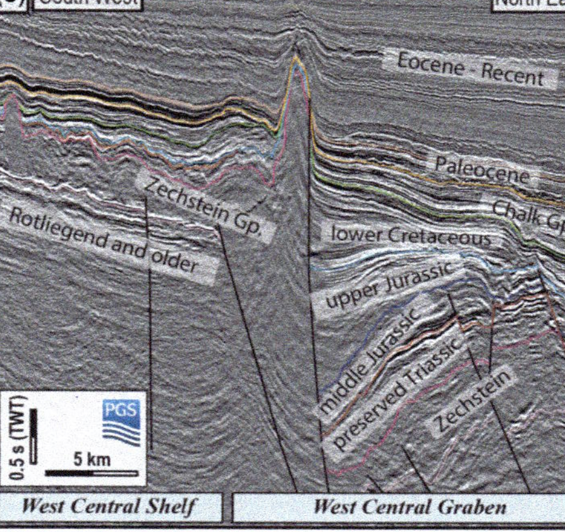

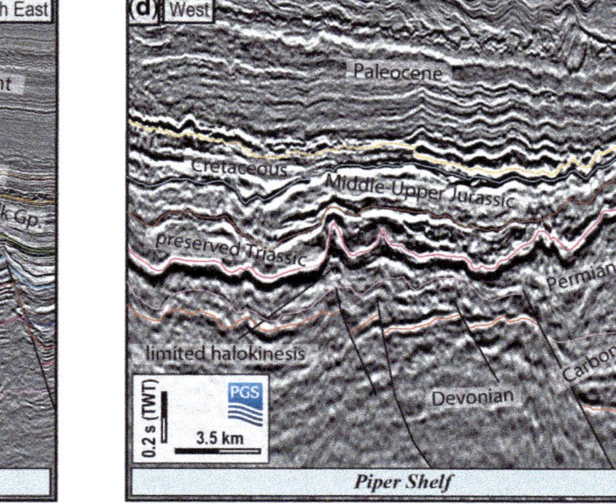

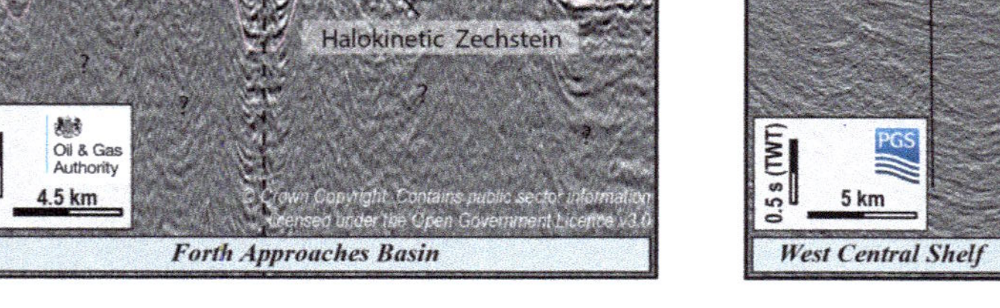

Fig. 14. Interpreted seismic cross-sections showing halokinetic structures suggesting different types and ages of Zechstein deformation over the (**a–c**) Greater Mid North Sea High and (**d**) East Shetland Platform areas. Part (b) is an enlargement of part of (a) (Forth Approaches Basin). See Figures 4, 5 for line locations.

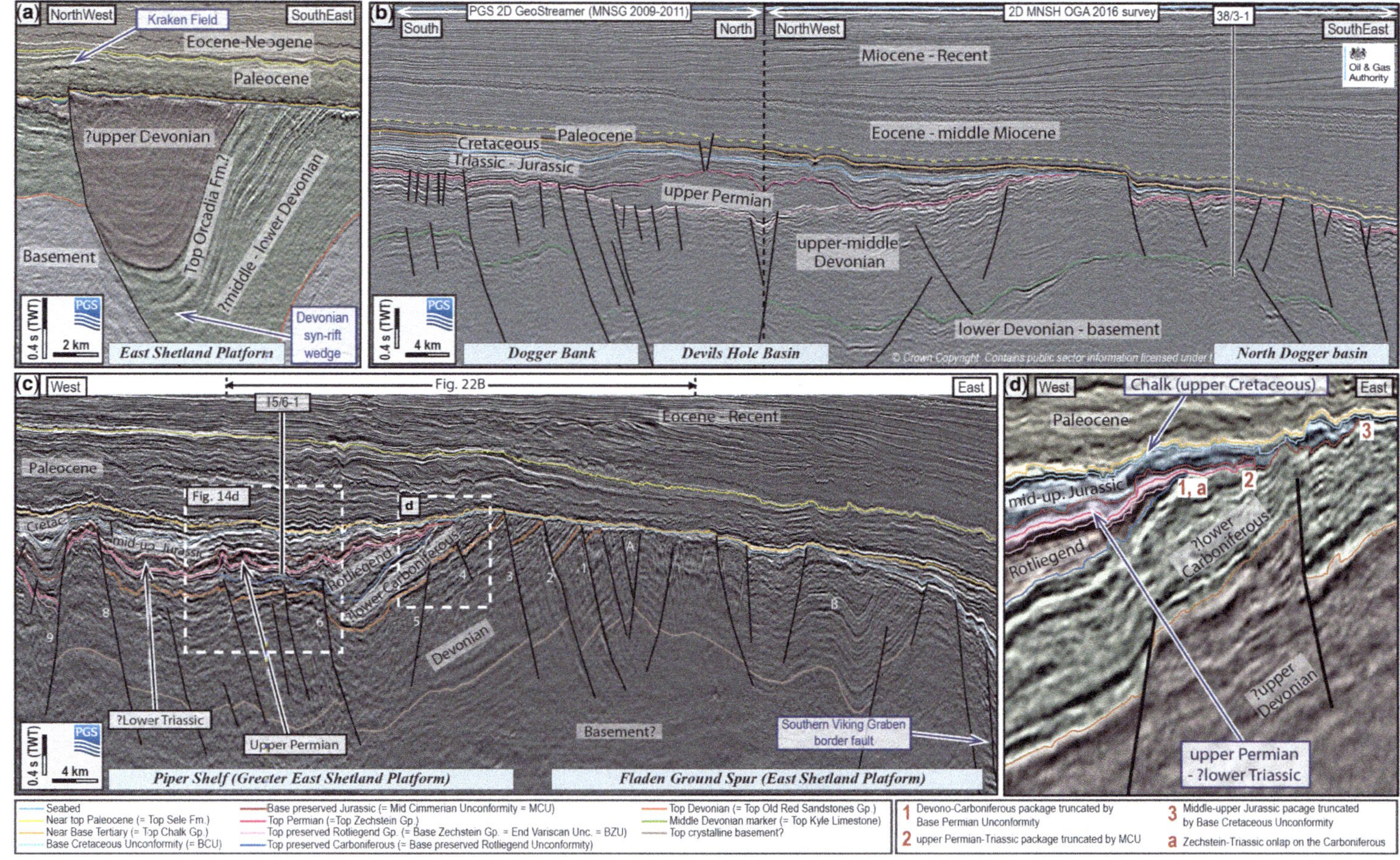

Fig. 15. Interpreted seismic cross-sections showing variable interactions between Paleozoic compressional folds and faults and pre-Cretaceous extensional fault blocks over the Greater East Shetland Platform (**a**, **c**, **d**) and (**b**) Mid North Sea High areas, with possible structural inversion episodes. See Figures 4 & 5 for line locations. The location of Figure 14d is highlighted in part (c). Part (c) after Patruno & Reid (2018) and the annotations for the faults (1–9) and folds (A, B) in this figure are the same as those utilized in that same publication and in Figure 22b. See Patruno & Reid (2018) for a further discussion on the timing and kinematics of these faults and folds. MCU, Mid Cimmerian Unconformity = Base preserved Jurassic Unconformity (cf. Figs 2–3).

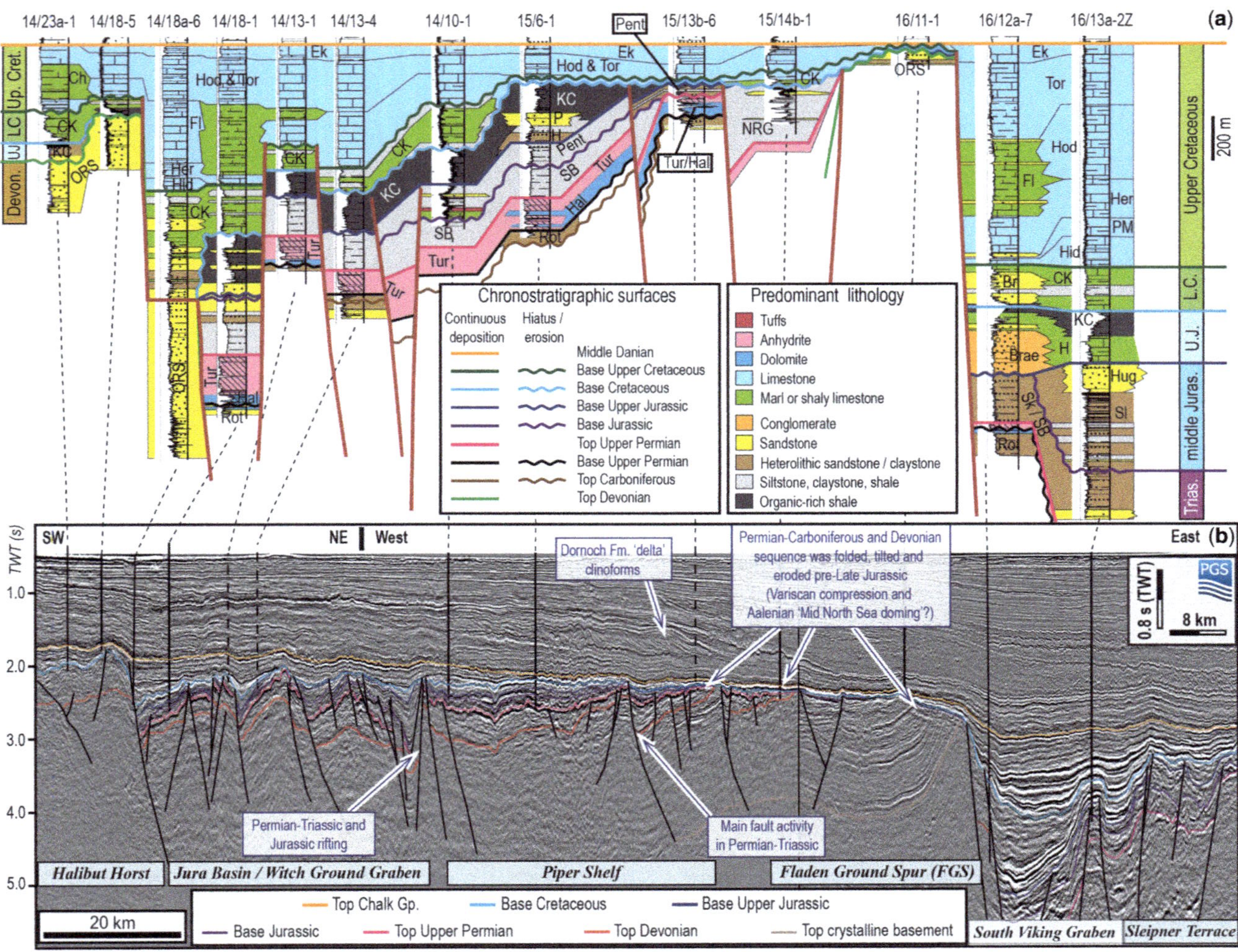
(a)
14/23a-1
14/18-5
14/18a-6
14/18-1
14/13-1
14/13-4
14/10-1
15/6-1
Pent
15/13b-6
15/14b-1
16/11-1
16/12a-7
16/13a-2Z
Tur/Hal
200 m
Devon.
UJ
LC
Up. Cret.
Upper Cretaceous
L.C.
U.J.
middle Juras.
Trias.
Chronostratigraphic surfaces
Continuous deposition
Hiatus / erosion
Middle Danian
Base Upper Cretaceous
Base Cretaceous
Base Upper Jurassic
Base Jurassic
Top Upper Permian
Base Upper Permian
Top Carboniferous
Top Devonian
Predominant lithology
Tuffs
Anhydrite
Dolomite
Limestone
Marl or shaly limestone
Conglomerate
Sandstone
Heterolithic sandstone / claystone
Siltstone, claystone, shale
Organic-rich shale
(b)
SW
NE
West
East
TWT (s)
1.0
2.0
3.0
4.0
5.0
Dornoch Fm. 'delta' clinoforms
Permian-Carboniferous and Devonian sequence was folded, tilted and eroded pre-Late Jurassic (Variscan compression and Aalenian 'Mid North Sea doming'?)
PGS
0.8 s (TWT)
8 km
Permian-Triassic and Jurassic rifting
Main fault activity in Permian-Triassic
Halibut Horst
Jura Basin / Witch Ground Graben
Piper Shelf
Fladen Ground Spur (FGS)
South Viking Graben
Sleipner Terrace
20 km
Top Chalk Gp.
Base Cretaceous
Base Upper Jurassic
Base Jurassic
Top Upper Permian
Top Devonian
Top crystalline basement

package is characterized by an overall higher amplitude and more continuous seismic facies than the overlying and underlying upper Devonian and lower Devonian intervals (Figs 12a, b, 13c, 15a, b). Given the regional stratigraphy, these two intervals might be referred to, respectively, as the Buchan Formation and the underlying Eifelian–Givetian succession. The high-amplitude 'middle Devonian' marker can be associated with the top of the coal-, shale- and carbonate-enriched middle Devonian interval.

The lower Devonian interval often has a similar seismic character to the crystalline basement. As a consequence, the deepest high-amplitude reflector at times corresponds to the 'middle Devonian' marker instead of Top Basement (Figs 12a, b & 15b). These two horizons can only be discriminated based on nearby mid-Devonian well penetrations (e.g. well 8/4-1; Fig. 12a, b).

There are also exceptions, with areas where the Devonian package cannot be subdivided into upper, middle and lower Devonian (e.g. Quadrants 15 and 16, or in southern Quadrant 8). This is caused by the absence of an interpretable high-amplitude 'middle Devonian marker'. As a consequence, the entire Devonian package consists of well-bedded continuous seismic character (see Figs 15c & 16b) or, conversely, it is opaque, discontinuous and resembles an acoustic basement (Figs 13a & 16b).

On much of the Greater MNSH, the only sub-Carboniferous interpretable horizon is the middle Devonian Top Kyle Limestone (Figs 13d, 15b & 17). If the upper Givetian to lower Frasnian Kyle Limestone Unit is absent, the whole sub-Carboniferous interval is characterized by opaque and discontinuous seismic facies. No 'top crystalline basement' horizon can be distinguished on the Greater MNSH, and the presence of a lower Devonian sedimentary interval is uncertain in the absence of well penetrations (Figs 3, 13b, 14a, b & 17).

The Devonian interval is variously deformed by compressional folds, reverse faults and thrust-related folds that are typical of external parts of compressional belts (e.g. Calamita *et al.* 2007) (e.g. Figs 12a, b), as well as extensional fault blocks (Figs 13a–c) and inverted extensional/compressional or possibly oblique-slip motifs (e.g. Figs 15b, c & 16b). The northern portion of Quadrant 8 on the Greater ESP, in particular, is characterized by likely flower-structure motifs within the well-imaged lower–middle Devonian interval, where several north-striking faults show either normal or reverse net-throw, with occasional association to fault-related folds or signs of inversion (Figs 9 & 11). These structures deform the lower–middle Devonian package, but do not seem to affect the upper Devonian and younger intervals.

Carboniferous units. Over the study areas, the Carboniferous, where present, is represented by heterolithic sandstone, claystone and coal-rich successions (Figs 2 & 3). The upper Carboniferous is usually missing (Figs 2 & 3).

On the Greater MNSH, the lower Carboniferous interval is visible in some places and is usually affected by folding and extensional faulting (Figs 7, 12c, 13d & 17). On 2D seismic, the Carboniferous imaging is challenging, particularly where it is buried under a thick Zechstein succession (Fig. 14a, b). The lower Carboniferous of the MNSH is a key interval both in terms of source and reservoir potential (Monaghan *et al.* 2017). In particular, the deltaic Yoredale Group is the youngest Carboniferous unit preserved beneath the Base Permian Unconformity (Cameron 1993) and may contain >60 m net sandstone with porosities of 6–13% (e.g. well 41/10-1). Gas shows were also encountered in well 42/9-1 within Yoredale sandstones, which appears to have been drilled in a syncline flanked by two undrilled anticlines (see Fig. 3c of Patruno *et al.* 2018).

On the Greater ESP, the only area where wells have consistently encountered Carboniferous strata is the Piper Shelf (Fig. 7). In this area, the Carboniferous reflectors form a parallel and continuous well-bedded seismic character. There is no angular unconformity with the underlying Devonian section (Figs 15c, d).

Permian units. Permian units are present throughout the Greater MNSH and over about half of the Greater ESP (Figs 20–22). They consist of the lower–'middle' Permian Rotliegend and upper Permian Zechstein groups. These units are directly underlain

Fig. 16. (**a**) Representative east–west-oriented regional well correlation panel and (**b**) seismic cross-section through UK Quadrants 14–16 on the Greater East Shetland Platform (Piper Shelf and Fladen Ground Spur). Modified after Patruno 2017. See Figure 4 for line location. ORS, Old Red Sandstone; Rot, Rotliegend Group; Hal, Halibut Formation (Zechstein Group); Tur, Turbot Formation (Zechstein Group); NRG, New Red Group; SB, Smith Bank Formation; Sk, Skagerrak Formation; Pent, Pentland Formation; Sl, Sleipner Formation; Hug, Hugin Formation; H, Heather Formation; P, Piper Formation; KC, Kimmeridge Clay Formation; Brae, Brae Formation; CK, Cromer Knoll Group; Br, Britannia Member; Hid, Hidra Formation (Chalk Group); PM, Plenus Marl Formation (Chalk Group); Her, Herring Formation (Chalk Group); Fl, Flounder Formation (Chalk Group); Hod, Hod Formation (Chalk Group); To, Tor Formation (Chalk Group); Ek, Ekofisk Formation (Chalk Group).

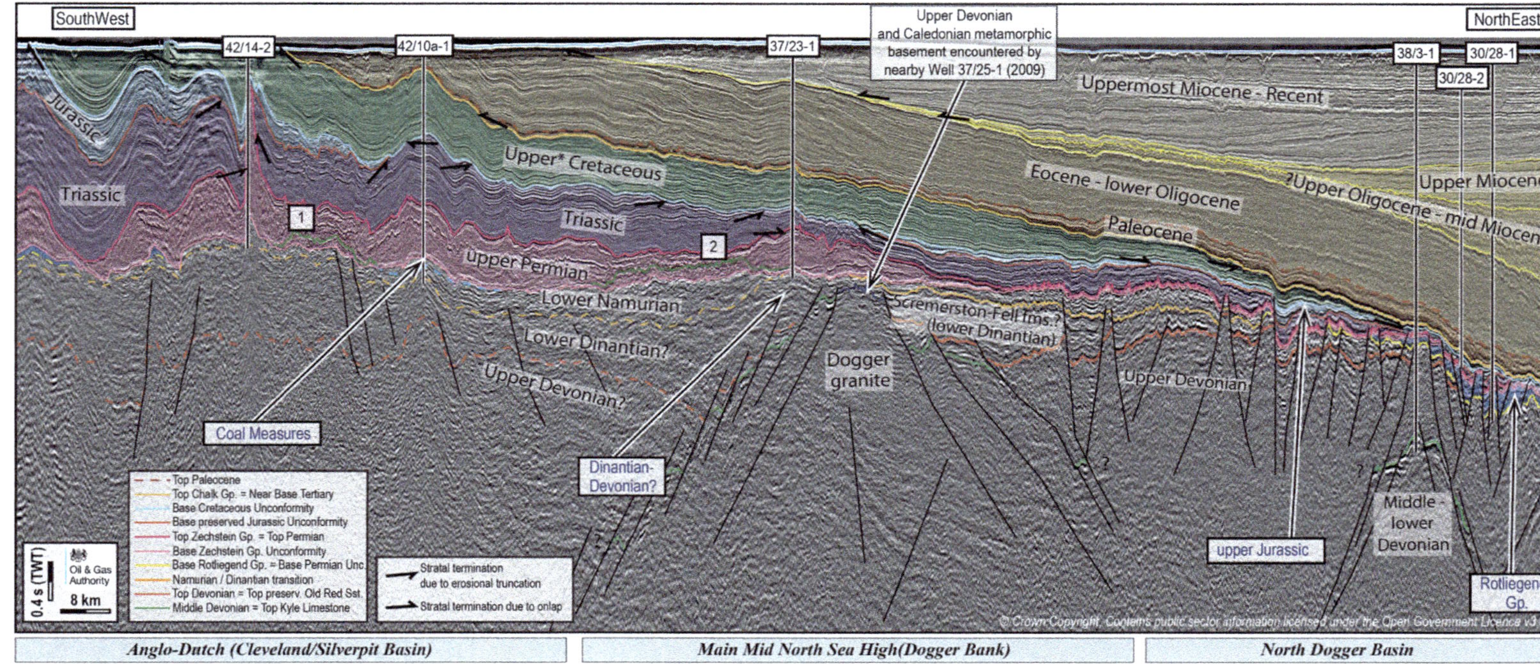

Fig. 17. Representative seismic cross-section highlighting the structural elements of UKCS Quadrants 42, 37, 38 (Greater Mid North Sea High and marginal Anglo-Dutch Basin). (1) and (2) indicate the two intra-Zechstein carbonate–sulphate complexes described by Patruno *et al.* (2018): (1) a small build-up that forms a part of the Crosgan gas fields; and (2) a larger undrilled complex situated between Quadrants 36, 37 and 42 (see Fig. 21). See Figure 5 for location. *, the Upper Cretaceous package locally comprises a thin veneer of Middle Cretaceous sediment, particularly towards the western depocentres of the line.

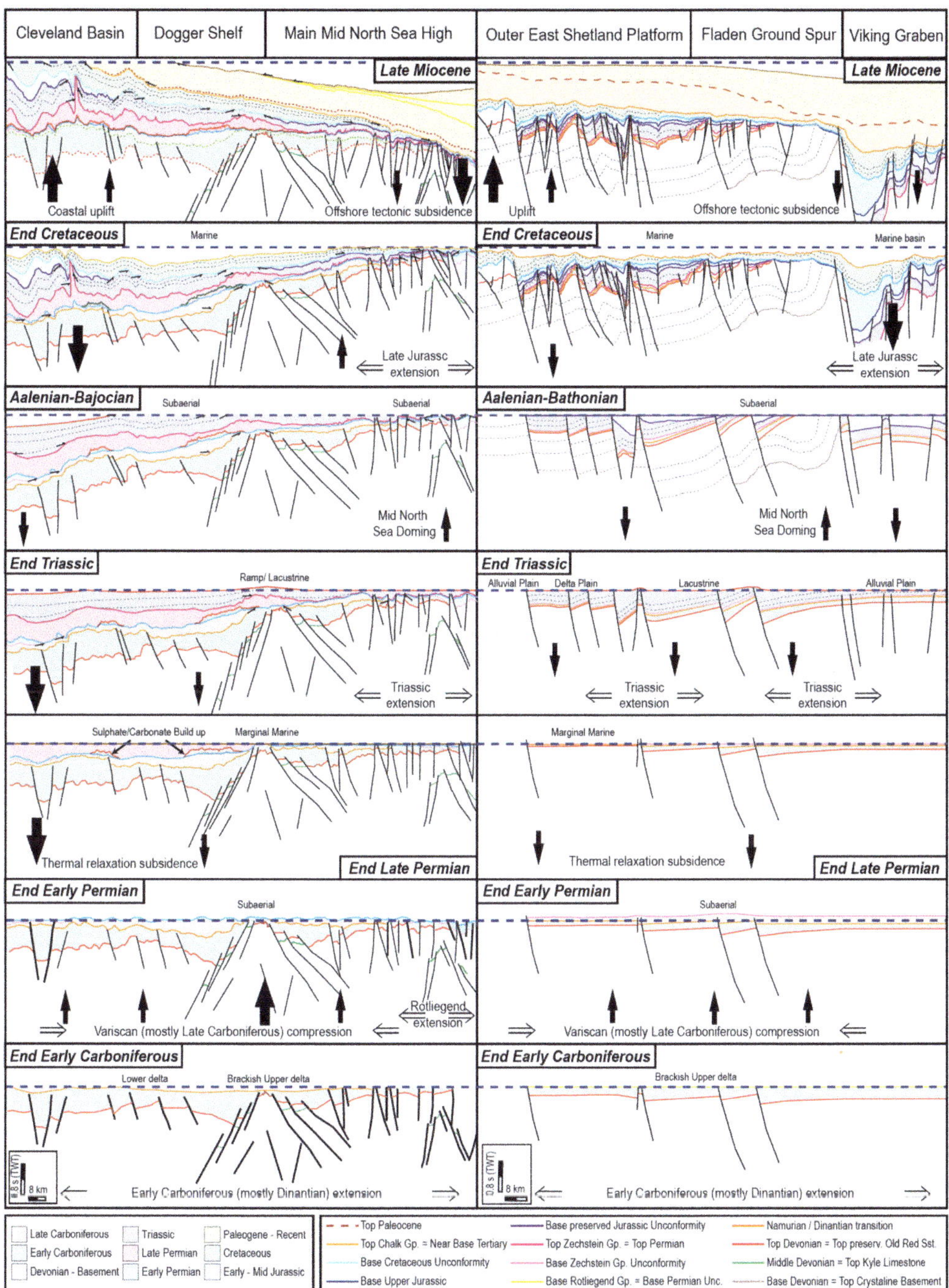

Fig. 18. A series of decompacted cartoons highlighting the structural evolution along two representative regional sections: the seismic cross-section shown in Figure 16 for the Greater East Shetland Platform (modified after Patruno 2017) and that shown in Figure 17 for the Greater Mid North Sea High. These evolutionary cartoons were reconstructed based on: stratal geometries (seismic interpretation); inferred depositional thickness trends (seismic interpretation and wells); back-stripping analysis of selected wells (Fig. 8); facies variations in the wells (cf. Figs. 8 & 16a); and background knowledge of regional tectonic events. The horizontal dashed blue lines in all figure parts represent the inferred erosional base level at each time.

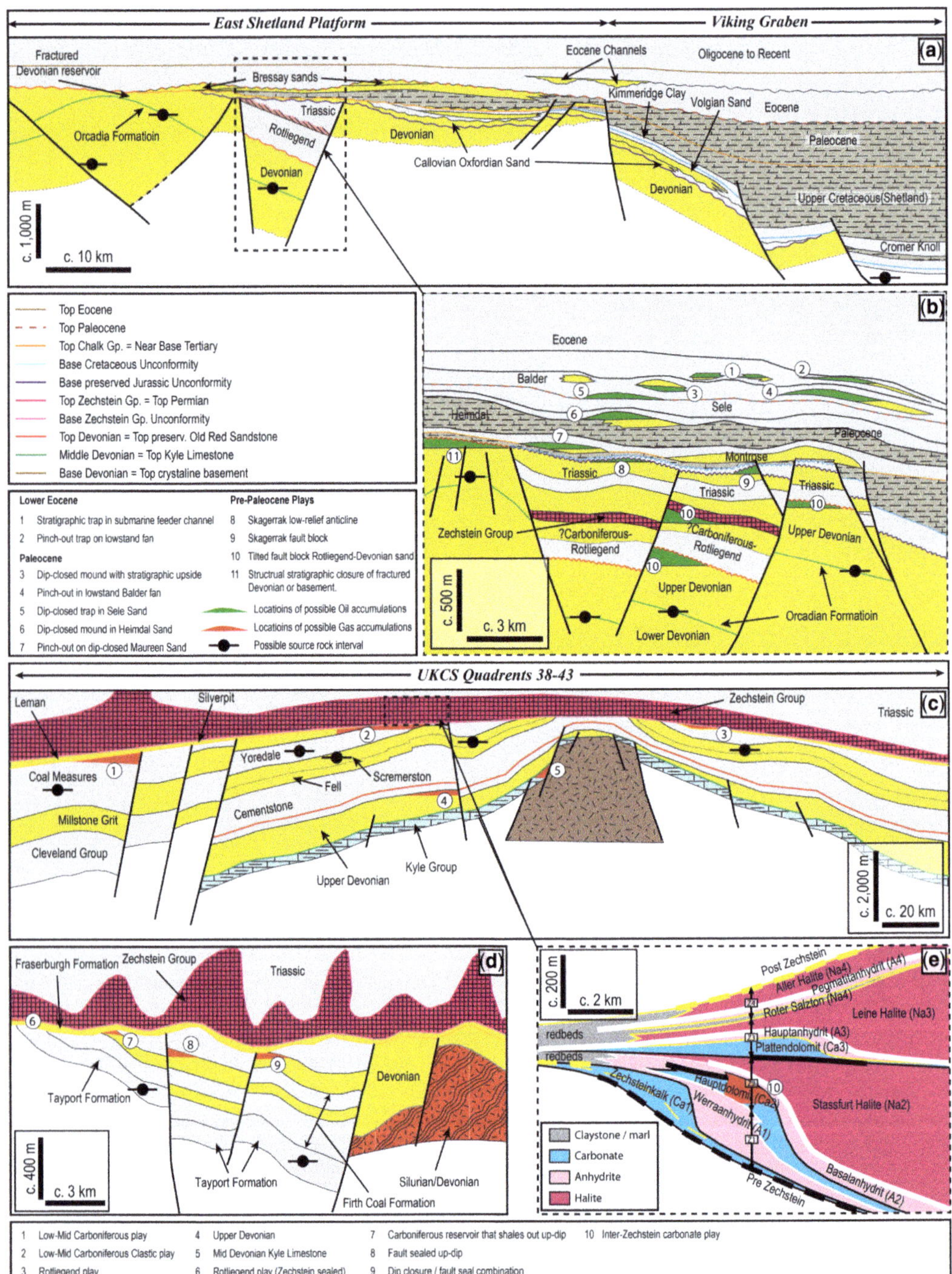

Fig. 19. Schematic play concept for the Greater East Shetland Platform and Mid North Sea High regions. (**a**) Idealized East Shetland Platform area (modified after United Kingdom Discovery Digest, 2001). (**b**) Idealized Crawford–Skipper Basin area on the Greater East Shetland Platform (modified after United Kingdom Discovery Digest, 1999). (**c**) Regional schematic diagram through the Dogger Shelf, Main Mid North Sea High and Dogger Basin areas (modified after Monaghan *et al.* 2015 and Arsenikos *et al.* 2015). (**d**) Idealized Forth Approaches area (modified after Monaghan *et al.* 2015 and Arsenikos *et al.* 2015); (**e**) Intra-Zechstein play schematic on the Main Mid North Sea High and Dogger Shelf area (after Patruno *et al.* 2018).

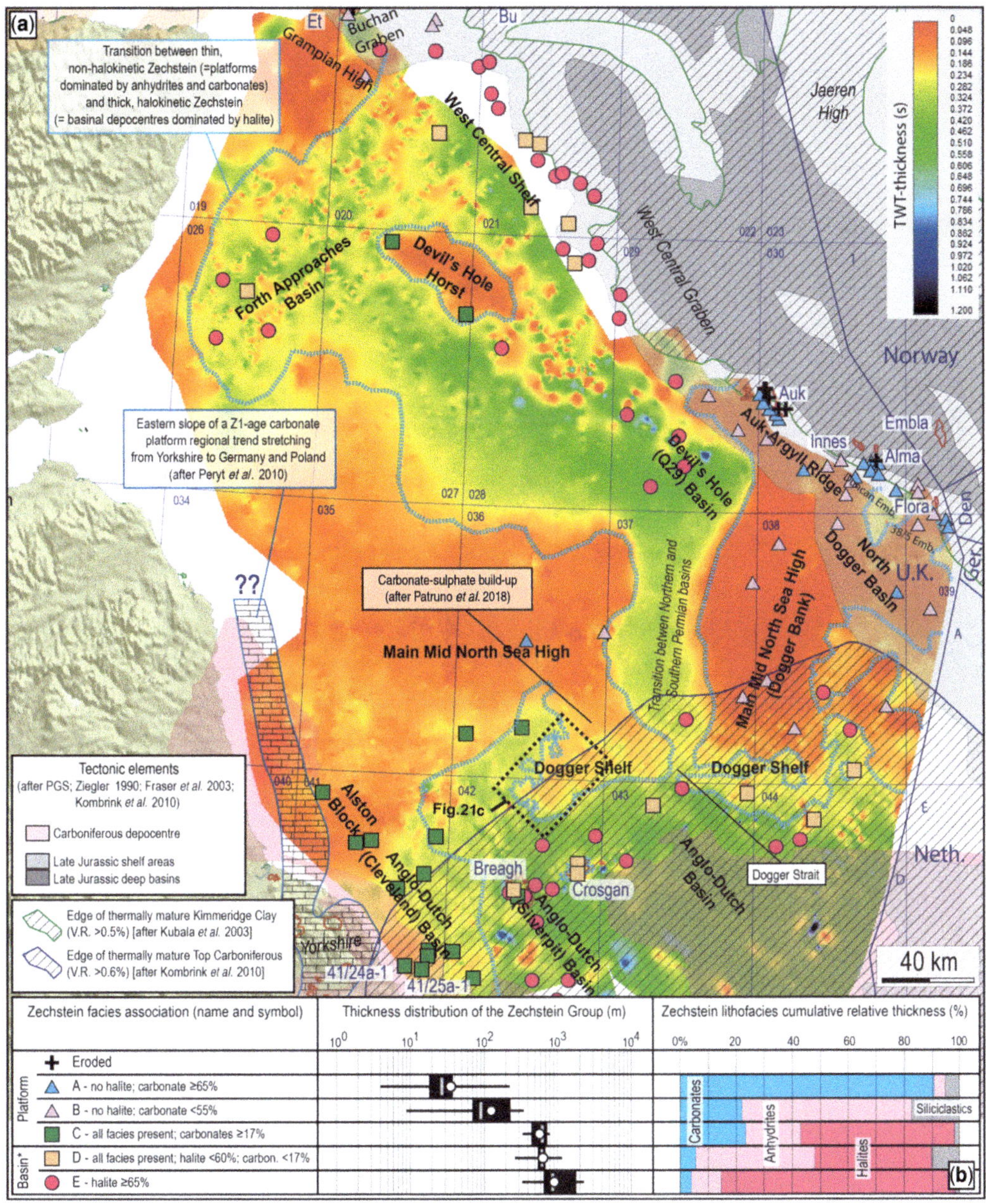

Fig. 20. (**a**) Zechstein two-way travel time thickness map on the Greater Mid North Sea High and Anglo-Dutch Basin area. (**b**) Well-based thickness and lithofacies statistics for the Zechstein Group in the Greater Mid North Sea High and marginal Anglo-Dutch Basin region (i.e. based on data from all the wells shown in part (a)). The transition between a thin, weakly deformed Zechstein and a thick, diapiric Zechstein closely mirrors the edge of the Zechstein platform facies (dominated by anhydrites and carbonates). The transition zone between the Northern Permian Basin and the Southern Permian Basin is also highlighted. Bu, Buchan Field; Et, Etive Field.

by a prominent Base Permian erosional truncation surface, reflecting widespread Late Carboniferous to Early Permian uplift in the foreland of the Variscan Orogeny. This surface is known as Base Permian Unconformity or Saalian Unconformity (Figs 2 & 3).

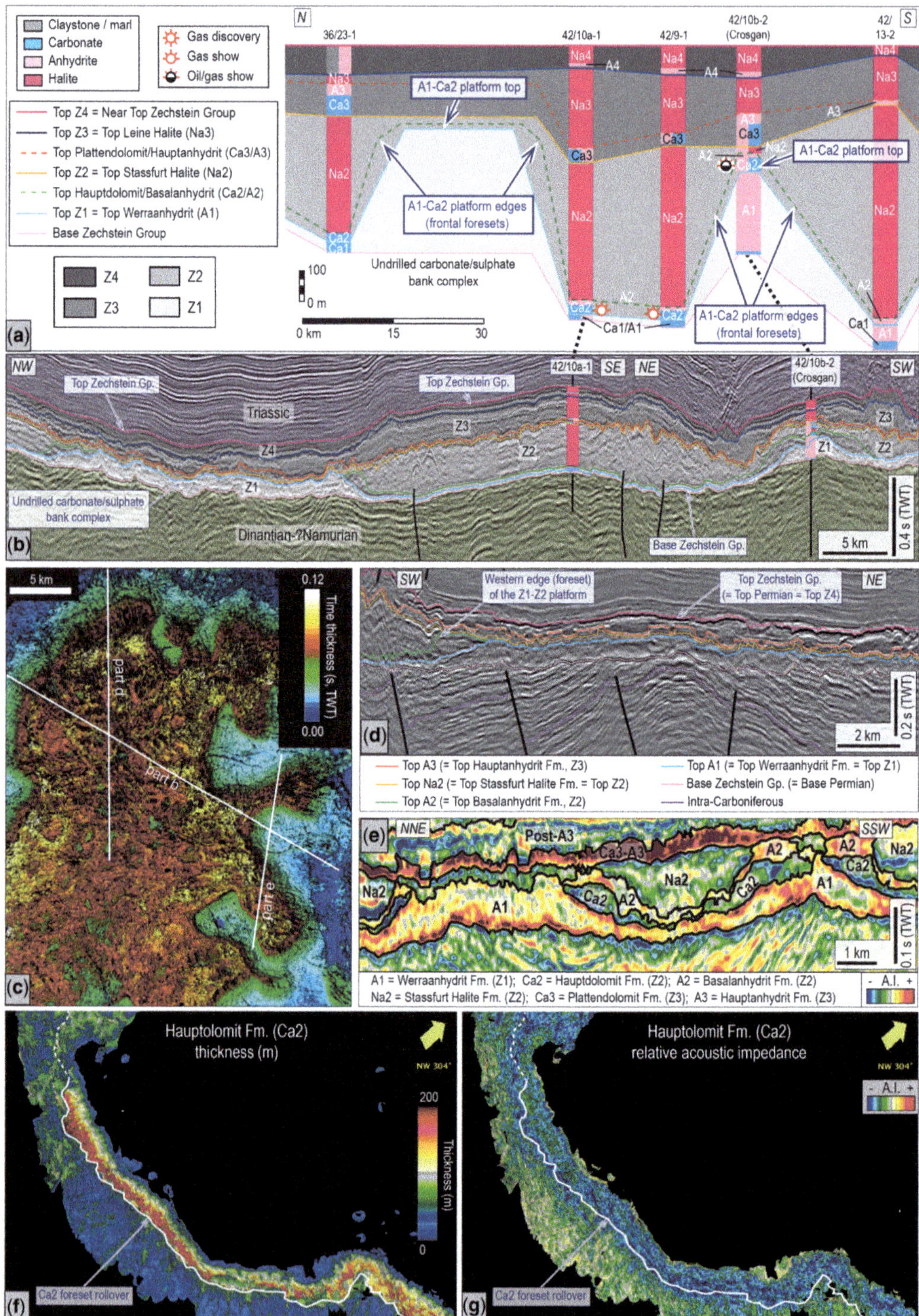

Fig. 21. Interpretation of the intra-Zechstein carbonate–sulphate complex in the southern Greater Mid North Sea High (after Patruno *et al.* 2018; see Fig. 20 for location map). (**a**) Well correlation panel. (**b**, **d**) Interpreted full-stack seismic lines showing the Crosgan Discovery and the undrilled complex in Quadrants 36, 37 and 42. (**c**) 3D time thickness map of the Z1 unit. (**e**) Pre-stack relative acoustic impedance showing intra-Zechstein lithology

The Top Permian horizon is a widespread high-amplitude 'hard-kick' marker on seismic data, which corresponds to the top of the carbonate- and evaporite-rich Zechstein unit. When compared with the thinner (generally <200 m) anhydrite- and carbonate-dominated Zechstein successions developed over the Greater ESP and on parts of the Greater MNSH, the Zechstein Group is significantly thicker (*c.* 500–2000 m) in those areas of the Greater MNSH where halite-dominated halokinetic basinal facies were developed (e.g. the marginal Anglo-Dutch Basin, Forth Approaches, West Central Shelf; see Figs 7, 8b & 20).

In cases when the Zechstein Group directly overlies the Saalian Unconformity, wherever thicker Zechstein units are developed, the Base Zechstein horizon is well imaged and appears at the base of Zechstein halokinetic structures as a near-planar unconformity that truncates the underlying Carboniferous and Devonian folds and fault blocks (Figs 14a–c, 15b, 17, 21b–e). Even though the Zechstein is significantly thinner on the Greater ESP, a similar relationship between its base and the underlying tilted and faulted Devono-Carboniferous succession is, at times, highlighted by subtle stratal architecture (Fig. 15d).

Some Zechstein diapirs are prominent structures up to 1 s TWT high, at times intimately associated with the development of later faults (Figs 14c & 17). Only sufficiently thick and halite-rich units can develop halokinesis (Waltham 1997) and, as a consequence, Zechstein platform facies are rarely diapiric, with one exception seen on the Greater ESP (Piper Shelf), where two minor isolated diapirs on an otherwise non-halokinetic and thin Zechstein succession can be observed (Fig. 14d).

The lower–'middle' Permian Rotliegend Group is less widespread, as the Zechstein is often lying directly on Carboniferous or older units, particularly on the Main MNSH. On the Greater ESP, the Rotliegend unit has been mapped in parts of the Piper Shelf and the Crawford–Skipper Basin (*sensu* Patruno & Reid 2017). The Rotliegend mapping in the Greater ESP has been based on several well penetrations, encountering usually sandstone-rich intervals (often associated with minor thin anhydritic layer intercalations) underlying the carbonates and anhydrites of the Zechstein Group (see well map in fig. 6 of Patruno & Reid 2017). The Rotliegend top is conformably overlain by the Zechstein Group, while the Carboniferous–Devonian units are unconformably overlain by the Rotliegend Base (e.g. Fig. 22b). The angular unconformity between Rotliegend (Permian) and Devonian to lower Carboniferous units probably corresponds to the Variscan-related Saalian Unconformity (Figs 2 & 3). In the absence of well-imaged angular unconformities in seismic and/or of an intervening coal-rich ?Carboniferous interval between the sandstone-prone Devonian and Rotliegend units, the stratigraphic distinction of Permian and Devonian intervals on a lithostratigraphic basis become more challenging. In addition, Rotliegend–Devonian continental rocks are usually barren of diagnostic biostratigraphic markers (Marshall & Hewett 2003). Informal diagnostic criteria to help distinguish Permian and Devonian 'red sandstones' in direct contact with each other (e.g. well 7/16–1) in the Greater ESP area were proposed by PGS (2017) based on established lithostratigraphic features from a number of wells (e.g. UK 9/16–2) and include: (1) an abrupt downward decrease in effective porosity logs (i.e. increase in density) across the Permian–Devonian boundary; (2) the usual absence of interpretable mudstone-rich intervals within the blockier Rotliegend logs, whereas these are usually present in the Devonian Old Red Sandstones (including the Buchan Formation); and (3) the usual absence of anhydrite layers intercalated within the upper Devonian Buchan Formation; see also the discussion on the definition of the Devonian–Permian boundary in Marshall & Hewett (2003).

Rotliegend seismic units are characterized by opaque seismic facies and form wedges between tilted Devono-Carboniferous structures and the overlying Base Zechstein (Figs 13c & 15d). Some of these wedges correspond to partly inverted synrift growth packages (e.g. Fault 6 in Fig. 16c). The time thickness map of the Piper Shelf (Fig. 22b) illustrates three Permo-Carboniferous fault-bounded depocentres that are largely infilled by Rotliegend sediments. On the Greater MNSH, the Rotliegend

Fig. 21. (*Continued*) discrimination. (**f**) Thickness map (in metres) of the Z2 carbonates (Hauptdolomit Formation or Ca2). (**g**) Minimum relative acoustic impedance map extracted at the top of the Ca2 unit. The lowest relative acoustic impedence response is associated with thicker packages. A new large sulphate–carbonate platform complex similar to the existing complex penetrated by the Crosgan well is interpreted based on the well data in part (a) integrated by full-stack imaging using well to seismic ties (parts (b)–(d)). Pre-stack analysis supported by a rock physics study based on the surrounding wells helped to discriminate the lithologies in 3D, where anhydrites were characterized by the highest acoustic impedance followed by carbonates and halites (part (e)) (see Patruno *et al.* 2018 for details). The intra-carbonate relative acoustic impedance variations agrees with the overall geological model, suggesting better porosity for the foreset carbonates (high hydrodynamic energy) than the thin and tight bottomset carbonates (low hydrodynamic energy). Figs c–g are located within the 3D SNS2013M dataset (cf. Figs 1b, 7 & 13e for location).

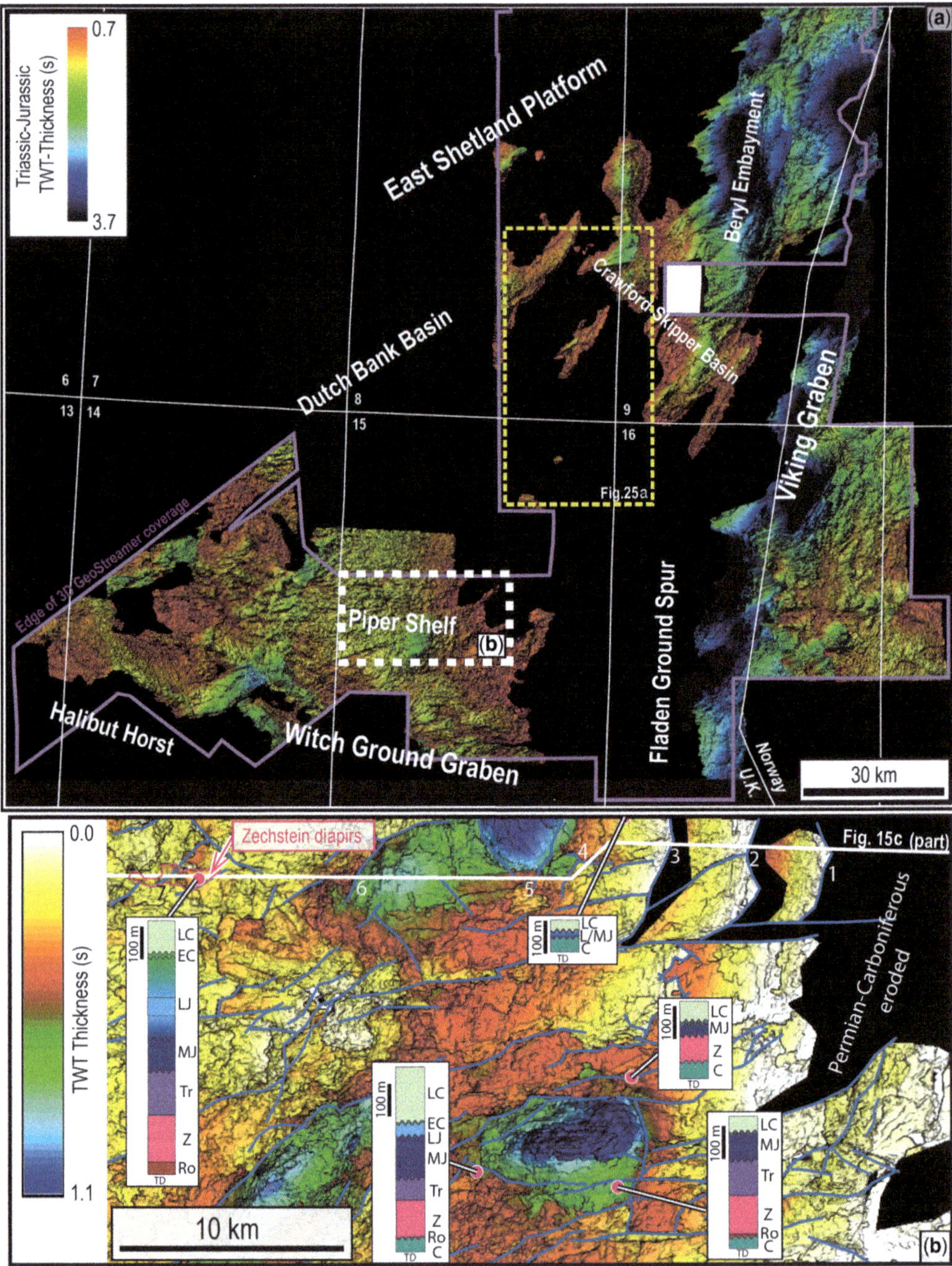

Fig. 22. Location of intra-platform basins within the East Shetland Platform. (**a**) Triassic–Jurassic two-way travel time thickness map, highlighting areas with preserved stratigraphy between the Variscan and the Base Cretaceous unconformities, most notably the Crawford–Skipper Basin *sensu* Patruno & Reid 2017 and the Piper Shelf. (**b**) Permian–Carboniferous two-way travel time thickness map over part of the Piper Shelf (Quadrant 15) identifying intra-platform Permo-Carboniferous depocentres (blue) (see Fig. 22a for map location). The annotations for the faults (1–9) in Fig. 22b are the same as those utilized in Patruno & Reid (2018) and in Figure 15c. See Patruno & Reid (2018) for a further discussion on timing and kinematics of these faults and fold.

units have only been interpreted on the eastern platform margins, where they are associated with extensional fault block systems (Fig. 17).

Triassic units. The Triassic interval tends to conformably overlie non-halokinetic or mildly halokinetic Zechstein successions (Figs 12c, 13c, d, 14d, 15b–d, 16 & 17). On the Greater MNSH, where the Triassic is associated with halokinetic Zechstein, two types of relationships with the underlying salt have been identified: (1) the Triassic shows a nearly constant thickness, lies conformably on the salt and is subject to the same amount of deformation as the salt (Figs 12c, 13d & 17); or (2) the Triassic shows significant thickness changes, is onlapping onto the salt walls and is mostly confined to the rim synclines (Figs 14a, b & 15b; the salt pillow penetrated by well 37/23-1 in Fig. 17). In the former case, halokinesis has occurred post-Triassic, whereas in the latter example it is syn- or pre-Triassic deposition.

The preserved Triassic seismic stratigraphic units generally display strong variations in time thickness. On the Greater MNSH, this is mainly controlled by underlying salt movements (Figs 14a–c, 15b & 17) and by successive erosional events towards the eastern platform margins (Figs 3, 13d & 17). On the Greater ESP, Triassic units are largely absent and confined to syntectonic intra-platform extensional fault-related blocks and mini-basins (Patruno & Reid 2017; Figs 2, 13a, 16, 22a, 23 & 24) and progressively onlap onto a deformed Paleozoic substrate. Here, Triassic reflectors are associated with syndepositional extensional faults and often display synrift wedging and possible Jurassic age reactivation (Fig 9; Fault 9 in Fig. 15c; Figs 23 & 24).

Triassic facies are fairly variable, with dominant mudstone-prone lacustrine units in the main depocentres to the west (e.g. the lower Triassic Smith Bank Formation in the Piper Shelf and Forth Approaches, Figs 2 & 3) and fluvial sandstone-prone facies towards the east on the Greater ESP (Cormorant Formation, Fig. 2) and towards the southern Greater MNSH (Bunter Sandstones, Fig. 3). On seismic, the preserved Triassic units appear dominated by low-reflectivity parallel-bedded seismic facies on the Greater ESP and the Northern Permian Basin (Figs 12a, 14, 23 & 24) and by higher reflectivity, continuous seismic facies on the Main MNSH and Dogger Shelf areas (Figs 12c & 17).

Jurassic units. Over the whole of the Greater ESP and MNSH areas, lower–middle Jurassic strata were either not deposited or have been eroded by the Mid-Cimmerian event (Figs 2, 3 & 8a). The lower Jurassic units, in particular, are usually regarded as having been initially deposited and then removed by Mid-Cimmerian Aalenian–Toarcian erosion (Steel 1993). Lower–middle Jurassic units are better preserved at the transition between the Greater MNSH and the Southern North Sea areas (e.g. the Cleveland Basin and Inner Moray Firth, Fig. 8b), where they are deformed by Zechstein halokinesis and eventually onlap northwards onto a tilted and eroded Triassic substrate (e.g. the southern half of the line in Fig. 17). In this area, lower–middle Jurassic reflectors form high-amplitude, continuous and well-bedded seismic facies (Fig. 17).

Conversely, Bathonian to upper Jurassic sediments deposited after the Mid-Cimmerian event have been penetrated by the majority of wells in both the Greater ESP and MNSH areas (Fig. 8a), but in some instances only form thin localized veneers and are still missing over the most structurally elevated areas (e.g. much of the Fladen Ground Spur) (Figs 2, 12a, b, 15 & 16). In some cases, particularly in areas close to the Southern Viking Graben and Central Graben, a subtle onlap of the Bathonian to upper Jurassic reflectors is noticeable on a tilted pre-mid-Cimmerian substrate, with eventual truncation against the structurally most elevated areas (Fig. 15d). On the platform margins, proximal to the Late Jurassic master faults that bound the main graben, the upper Jurassic packages are in some places subject to minor syndepositional faulting and display subtle synrift wedging (Figs 9, 15b & 16c, northeastern half of the seismic line in Fig. 17 & 23a). These minor extensional structures tend to be oriented sub-parallel to the closest graben-bounding master faults (Fig. 9).

Potential Jurassic reservoir sandstones could be present, particularly on the Greater ESP (Fig. 2). Therefore the ability to map the shape of these thin preserved Jurassic veneers on the platform is key to the definition of prospects similar to those successful on the Utsira High (e.g. Johan Sverdrup Field).

Cretaceous units. Cretaceous seismic stratigraphic units correspond to well-bedded and continuous seismic facies (Figs 12c, 13b, 14, 16 & 17). Similar to the upper Jurassic, the lower Cretaceous claystones and marlstones of the Cromer Knoll Group are largely absent or very thin on the structurally highest areas of the Greater ESP (e.g. Kraken High, Fladen Ground Spur, Halibut Horst) and on a large part of the Greater MNSH (Figs 2, 3, 16 & 17). The upper Cretaceous limestones and marlstones of the Chalk and Shetland groups are significantly more widespread (Figs 2 & 3) and vary in thickness from a very thin veneer, sealing tilted and eroded Devonian successions on the structurally highest areas (Figs 13a, 14d, 15, 16, 17, 23 & 24) up to 0.2–0.4 s TWT on intra-platform relative depocentres (Figs 13a, b, 13d, 14, 15b, c, 16 & 17). Nevertheless, <5% of the wells on the Greater ESP and MNSH did not encounter any Cretaceous sediment at all (Fig. 8a). In particular, a large number of

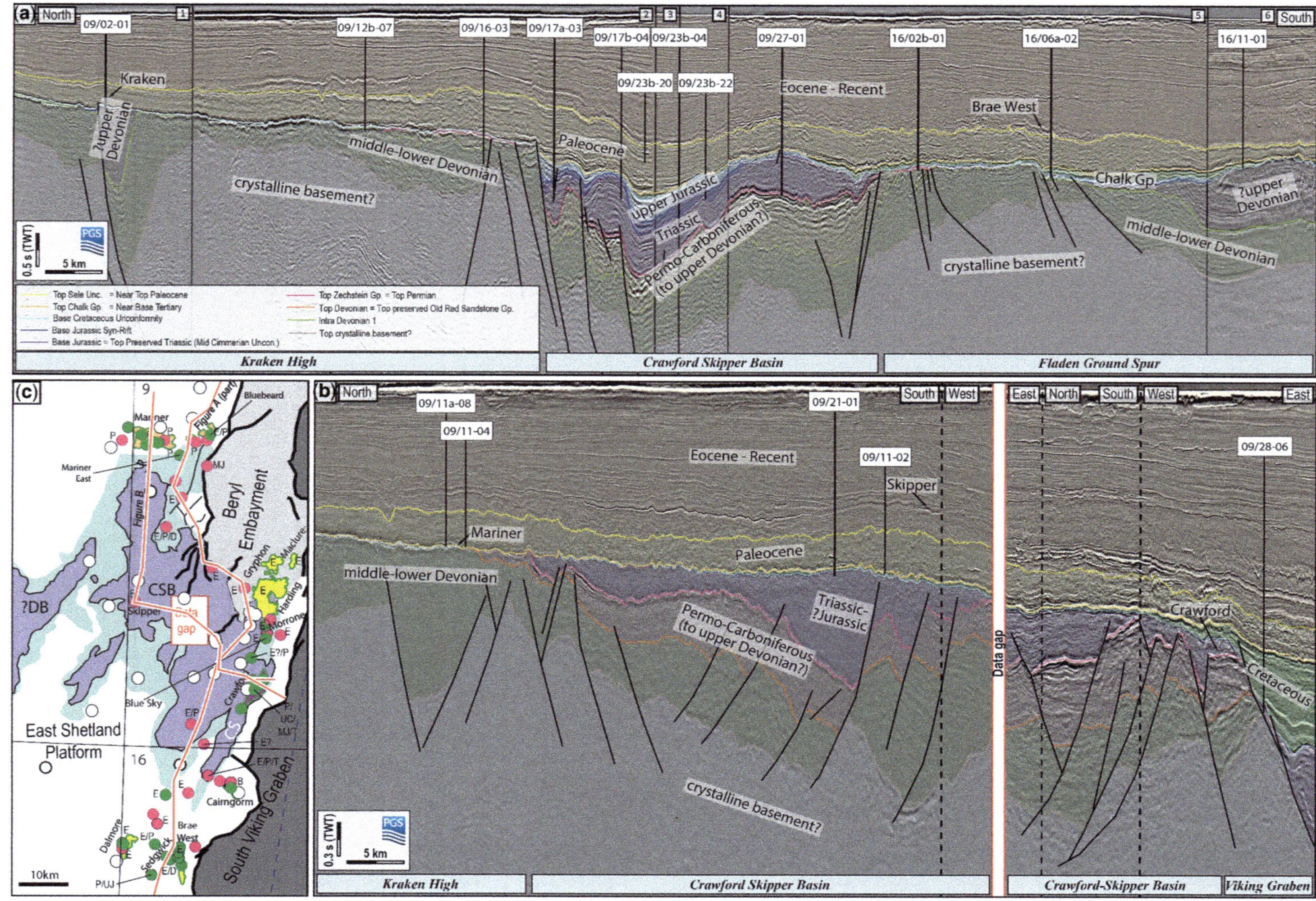

(a)
North
South
09/02-01
09/12b-07
09/16-03
09/17a-03
09/17b-04
09/23b-04
09/23b-20
09/23b-22
09/27-01
16/02b-01
16/06a-02
16/11-01
Kraken
?upper Devonian
middle-lower Devonian
crystalline basement?
Paleocene
Eocene - Recent
upper Jurassic
Triassic
Permo-Carboniferous (to upper Devonian?)
Brae West
Chalk Gp.
0.5 s (TWT)
5 km
PGS
Top Sele Unc. = Near Top Paleocene
Top Chalk Gp. = Near Base Tertiary
Base Cretaceous Unconformity
Base Jurassic Syn-Rift
Base Jurassic = Top Preserved Triassic (Mid Cimmerian Uncon.)
Top Zechstein Gp. = Top Permian
Top Devonian = Top preserved Old Red Sandstone Gp.
Intra Devonian 1
Top crystalline basement?
Kraken High
Crawford Skipper Basin
Fladen Ground Spur
(c)
East Shetland Platform
Beryl Embayment
South Viking Graben
?DB
CSB
Mariner
Mariner East
Skipper
Blue Sky
Brae West
Cairngorm
Sedgwick
Dalmore
Bluebeard
Harding
Gryphon
Maclure
Morrone
Figure A (part)
Figure B
Data gap
9
16
10km
(b)
North
South
West
East
09/11a-08
09/11-04
09/21-01
09/11-02
09/28-06
Mariner
Skipper
Crawford
Cretaceous
Paleocene
Eocene - Recent
Triassic-?Jurassic
Permo-Carboniferous (to upper Devonian?)
middle-lower Devonian
crystalline basement?
Data gap
0.3 s (TWT)
5 km
PGS
Kraken High
Crawford Skipper Basin
Crawford-Skipper Basin
Viking Graben

wells did not encounter Cretaceous strata in the marginal Anglo-Dutch Basin, chiefly in the western sub-basins (e.g. the Cleveland Basin and Alston Block).

Only a few faults are seen to propagate into the Cretaceous deposits. Cretaceous seismic reflectors indicate that the Cretaceous sediments are passively infilling the tilted and deformed palaeobathymetic relief created by pre-Cretaceous tectonic events (Figs 12c, 13a, b, d & 14). As a result, they progressively onlap onto the slopes of the most persistent structural highs (e.g. the Fladen Ground Spur, Kraken High, Halibut Horst, Dogger Bank) and usually end up covering them with a thin veneer of upper Chalk deposits (Figs 15b–d, 16 & 17). The only observable exception is the main structural depocentre of the northwestern Witch Ground Graben in Quadrant 14. This is bounded by master faults of Early Cretaceous timing and is associated with well-imaged lower Cretaceous synrift wedges (Fig. 9).

Cenozoic units. Paleocene to Recent reflectors form relatively high-amplitude continuous seismic facies. These correspond to a largely mudstone-prone succession, with interspersed deep marine and shelfal Paleogene working reservoir units on the Greater ESP (Figs 2 & 3; Patruno & Reid 2016*a*). In particular, the uppermost Paleocene to lowermost Eocene units were deposited as seawards-prograding mixed mudstone–sandstone-prone shelf-scale clinothems of the Dornoch Formation (Figs 2, 15c & 16b).

The Cenozoic seismic stratigraphic units display a wide variation in preserved total thickness, up to 2 s (TWT) over the eastern margins of the Greater MNSH and the ESP (Figs 12b, 15b, c, 16b, 14a & 17) to a progressively reduced thickness towards the west, where the strata are subject to regional westwards tilting and erosional truncation against an unconformity parallel to the present day seafloor (Figs 14a & 17). Well and seismic data indicate that, in the proximity of the Holocene British coastline, much of the Meso-Cenozoic strata were tilted and eroded (e.g. the Cleveland Basin, Alston Block, Forth Approaches Basin) (Figs 8a & 17). These tilting and erosional trends of units continue northwards in the undrilled parts of the western Greater MNSH and the ESP (Glennie & Underhill 1998; Fyfe *et al.* 2003, PGS 2017).

Similar to the upper Cretaceous, the Cenozoic succession is largely unaffected by faulting and only locally interrupted by a few minor Alpine-age compression and inversion episodes (e.g. Fig. 15a; Patruno & Reid 2016*a*, 2017).

Interpretation of tectonostratigraphic data

As explained below and summarized in Table 3, both the regional literature and the evidence given in the preceding section indicate active polyphase tectonics in the North Sea, with multiple well-documented tectonic inversions (also see Glennie & Underhill 1998; Scisciani *et al.* in press). The Greater MNH and ESP, in particular, provide compelling seismic stratigraphic evidence for at least eight documented switch-overs from compression to extension (and vice versa) in the last 420 myr:

(1) from Caledonian compression to ?Early Devonian extensional collapse;
(2) from ?Early Devonian extensional collapse to a transient Middle Devonian compressional/transtensional inversion;
(3) from the Middle Devonian compression back to ?Late Devonian to Early Carboniferous extension;
(4) from Early Carboniferous extension to the Late Carboniferous Variscan Orogeny;
(5) from the Late Carboniferous Variscan Orogeny to renewed Permo-Triassic extension;
(6) from Permo-Triassic extension to Aalenian thermal doming;
(7) from Aalenian thermal doming to Late Jurassic rifting;
(8) from Late Jurassic rifting to Cretaceous passive subsidence, locally interrupted by intraplate Alpine inversions, and regionally interrupted in proximity of the present-day British

Fig. 23. Seismic sections and map showing the areal distribution and stratal geometries of the Permian–Triassic Crawford–Skipper Basin (Greater East Shetland Platform). The vertical lines highlighted by numerical annotations (3, 4, 5, 6) in part (a) highlight 3D survey boundaries. The numbers refer to the PGS surveys mentioned in Fig. 1b. The Base Cretaceous Unconformity (BCU) subcrop map in the eastern part of the Greater East Shetland Platform shown in part (c) uses the following abbreviations: CS, Crawford Spur; CSB, Crawford–Skipper Basin (*sensu* Patruno & Reid 2017); ?DB, possible eastwards extension of the Dutch Bank Basin; FGS, Fladen Ground Spur. The colours of the fill and outline of the hydrocarbon discovery polygons reflect the age of the main reservoir interval and the type of hydrocarbon encountered, respectively; see Figures 4 and 5 for key. This map also shows the location and results of all the wells drilled on the Greater East Shetland Platform area. The wells symbols are as follows: white circles, dry wells; purple circles, hydrocarbon show wells; green circles, oil discovery wells. Letters annotated next to each show/discovery well highlights the age of the main successful reservoir interval: E, Eocene; P, Paleocene; UC, upper Cretaceous; LC, lower Cretaceous; UJ, upper Jurassic; MJ, middle Jurassic; T, Triassic; D, Devonian; B, crystalline basement.

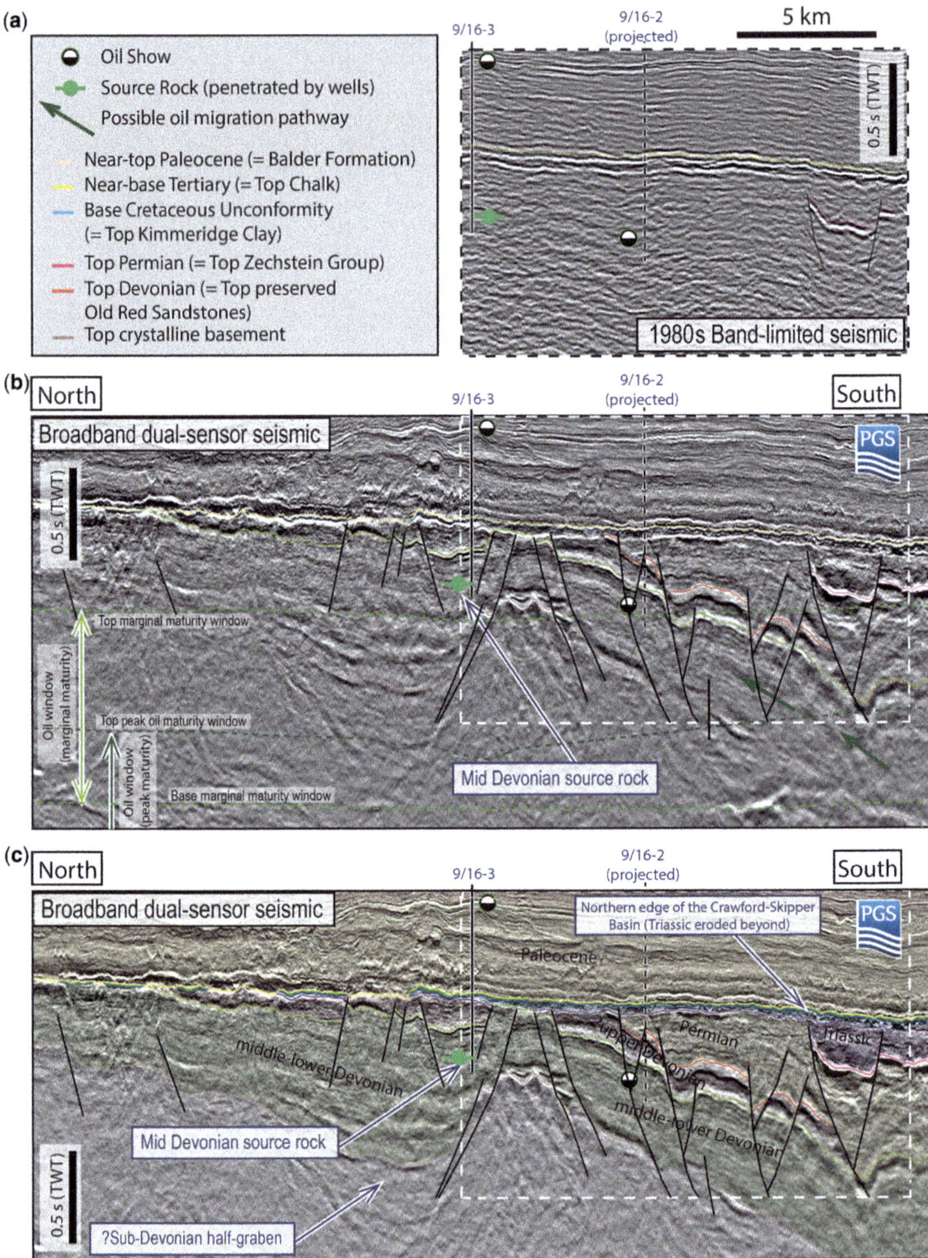

Fig. 24. (**a**) A conventional seismic line (1986 acquisition) compared with (**b**, **c**) a GeoStreamer line (MC3D-BYLM2013 survey) (modified after Patruno & Reid 2017). The eastern half of the GeoStreamer line (highlighted by a white dashed polygon) follows the same north–south transect as the conventional line. The main horizons and the computed depths of the present day oil window are shown. See Figures 4 and 9 for line location. The 1986 line does not show interpretable reflectors below the Top Permian; however, two wells penetrated thick Devonian strata containing oil shows (well 9/16-2) and a middle Devonian lacustrine source rock (well 9/16-3, see Duncan and Buxton 1995). These well results from the 1980s to 1990s are better understood thanks to the 2013 GeoStreamer seismic (b, c). These new data clearly show intra-Devonian reflectors and fault blocks, including an undrilled horst between the two wells. The 1980s band-limited data show the quality of seismic data at the time the wells 09/16-03 and 09/16-02 were drilled. Note the uplift in imaging of the Devonian, particularly the mid-Devonian source rock interval, which could explain the result of the oil show and source rock encountered in the wells.

coastlines by progressive seawards tilting of the platforms induced by the Paleogene Icelandic thermal plume.

These tectonostratigraphic events are discussed in greater detail in the following sections. Local variation between and within the Greater MNSH and ESP areas are also discussed as these might have played a significant part in the local hydrocarbon prospectivity.

Devonian to Early Carboniferous: pronounced regional subsidence and possible inversion

During the Devonian, the Orcadian Basin was subject to a large amount of tectonic subsidence due to the post-orogenic collapse of the Caledonian Orogen, leading to the deposition of a thick (1–5 km) succession (Fig. 8*a*, *d*; McClay *et al.* 1986; Coward *et al.* 1989; Seranne 1992; Ziegler 1992; Marshall & Hewett 2003; Wilson *et al.* 2010). An understanding of the Devonian on the Greater MNSH is difficult due to a lack of seismic imaging below the Kyle Limestone.

The thick Old Red Sandstone sediments accumulated in an overall continental setting on the Laurussia mega-continent as a result of the Caledonian Orogeny (Glennie & Underhill 1998). Fluvial/aeolian sandstones and lacustrine fish beds are working reservoirs and source rocks for several North Sea oil and gas fields (Marshall & Hewett 2003; Patruno & Reid 2016*a*). In particular, the upper Devonian Buchan Formation Equivalent is a proven reservoir for large fields (e.g. the Clair and Buchan fields), whereas oil sourced from middle Devonian lacustrine shales (e.g. the Orcadia Formation and Eday Group) have been associated with various fields in the Moray Firth and West of Shetlands (Figs 1–3) (Marshall & Hewett 2003; Patruno & Reid 2016*a*, *b*). Partial marine ingressions have been recorded by the deposition of middle Devonian carbonates on the Dogger Bank and Auk–Argyll Ridge area on the Main MNSH (Kyle Limestone) and parts of the ESP (e.g. dolomites in well 8/4-1 and anhydrites in 9/16-3). Several researchers have speculated that this seaway probably followed the present day north-trending Central Graben lineament as the northern arm of the Rheic Ocean, which existed until its Variscan closure (Ziegler 1990; Marshall *et al.* 1996; Glennie & Underhill 1998).

On the Greater ESP, numerous Devonian age extensional, compressional and possibly oblique shear and inverted faults with a north–south orientation have been mapped in the northern part of Quadrant 8, as well as between Quadrants 3 and 9 and under the Crawford–Skipper Basin (Figs 9–11).

Active extension is suggested by lower Devonian synrift wedging (e.g. Figs 12b, 15a, 23b, 24b, c). Some deep extensional faults on the Greater ESP seem restricted to sub-Devonian intra-basement reflectors (e.g. below the Kraken area) and are possibly associated with enough crustal stretching to generate seafloor spreading and oceanic crust development (e.g. the Scottish Highland Border Complex) (Glennie & Underhill 1998).

Compressional folds, thrusts and inverted faults (Figs 12a, b & 15c) deform the Devonian succession. Eastwards-dipping Caledonian thrust planes were identified north of the Scottish mainland, extending to depths of 30 km (Brewer & Smythe 1984; Snyder *et al.* 1997). In most cases, these fault types in the Greater ESP study area penetrate the lower–middle Devonian package, but do not seem to affect the upper Devonian interval (Fig. 12a–c). A similar middle Devonian unconformity is well known onshore (e.g. the Midland Valley; Mendum & Noble 2010) and is consistent with either a transient middle Devonian compressional/transpressive event (cf. Coward 1990; Coward *et al.* 2003) or with blind thrusting associated with Variscan age compression.

The lower Carboniferous interval has been identified on much of the Greater MNSH and on some of the intra-platform depocentres of the Greater ESP, although question marks remain in some cases due to the poor and partial biostratigraphic definition of the Paleozoic periods. In both areas, the lower Carboniferous is thought to consist of coal-rich clastic successions, which conformably overlie the upper Devonian strata (Figs 12c, 13d, 15c, d & 17). The difference in facies between the upper Devonian and the Carboniferous reflects a transition from arid subtropical conditions to humid equatorial conditions (Habicht 1979).

The preserved lower Carboniferous sediments over the Greater MNSH is up to four times thicker than the time-equivalent preserved units on the Greater ESP (cf. Figs 12c & 13d; Fig. 15c, d). Active extension has been documented in the Greater MNSH by Dinantian synrift wedging (e.g. Figs 13d, 17 & 18), which had largely abated in Namurian–Westphalian times (Glennie & Underhill 1998; Fig. 18). A conjugate NE- and NW-striking structural trend is typical of the mapped Permo-Carboniferous faults on both the MNSH and the Piper Shelf, still visible in Permo-Triassic fault trends in the latter (Figs 9 & 11).

The conjugate sets of faults identified by our work are consistent with a largely transtensional regional tectonics. The traditional interpretations see these same conjugate NE- and NW-striking trends developed during Caledonian time in the northern England Carboniferous Basin, the Midland Valley and the Southern North Sea, and consistently reactivated in the Carboniferous in an extensional and, in the Late Carboniferous, a compressional sense (Glennie & Underhill 1998; Geluk 2005). However, more recent analyses have cast doubts on the occurrence of 'real' Variscan inversion. For example, De Paola *et al.* (2005) reinterpreted the 'Variscan inversion' structures from the onshore Carboniferous Northumberland Basin as formed because of horizontal shortening components of deformation during progressive and partitioned transtension, with pre-existing basement structures playing a key part in fault localization. NNE-striking Carboniferous folds in the onshore Midland Valley have also been more recently reinterpreted as growth folds formed under a long-lived dextral strike-slip regime, which were further tightened during the Late Carboniferous (Variscan) deformation, with possibilities of Variscan mature source kitchens in local synclines (Underhill *et al.* 2008).

Variscan compression and inversion

An event entailing compression, uplift and erosion of Devono-Carboniferous strata is indicated by: (1) the absence of preserved upper Carboniferous deposits (Fig. 7); (2) erosional truncation on top of the pre-Variscan succession (Figs 12c, 15b, 15d, 17 & 21d); and (3) compressional folds, faults and inverted faults, within the preserved pre-Variscan succession on both the Greater ESP (Figs 12a, b, 15c, d) and the Greater MNSH (Figs 12c, 13d, 15b, 17, 18, 21b & 21d).

This net-erosional event can be followed throughout central and northern Europe and was directly caused by the Variscan Orogeny, which marked continental collision between Gondwana and Laurussia, the closure of the Rheic Ocean and the creation of the Pangaea supercontinent (Coward *et al.* 1989; Corfield *et al.* 1996; Glennie & Underhill 1998; Zanella & Coward 2003). Following this event, the British Isles and the North Sea have largely lain in an intra-plate setting (Glennie & Underhill 1998). The majority of the North Sea and central–northern Britain represented the foreland of the main Variscan fold–thrust belt (Glennie & Underhill 1998) and was only marginally affected by it. Nevertheless, compressional deformation is documented throughout the Variscan foreland, from England to the Midland Valley of Scotland and the Moray Firth (Fraser & Gawthorpe 1990; Thomson & Underhill 1993; Corfield *et al.* 1996). In the Greater MNSH, Variscan uplift was particularly pronounced on the Dogger Bank (Hay *et al.* 2005). The pronounced Middle Carboniferous to Middle Permian regional erosional truncation associated with Variscan compression and uplift events is often referred to as the Saalian Unconformity (Cameron 1993; Kombrink *et al.* 2010; Peryt *et al.* 2010; Figs 2 & 3).

In the Greater MNSH, in particular, the angular unconformity at the base of the Zechstein Group corresponds to the Variscan uplift-related Saalian Unconformity. Near the deeper, central part of the Southern Permian Basin, this unconformity becomes less prominent and the uppermost Carboniferous strata are usually preserved (Kombrink *et al.* 2010). As a consequence, it can be inferred that Variscan uplift was greater in the MNSH compared with the Southern Permian Basin depocentres (e.g. the marginal Anglo-Dutch Basin). Alternatively, if the Variscan uplift was regionally uniform, then greater magnitudes of erosion along the Saalian Unconformity indicate that the MNSH was a pre-Variscan structural high. This latter interpretation is supported by our fault analysis, which indicates that sub-Zechstein faults are small in terms of displacement and length (Fig. 10). This suggests that, even during Carboniferous and Devonian time, the MNSH region underwent limited fault-driven subsidence compared with the main Anglo-Dutch Basin depocentre. Fault-bound anticlines (e.g. Fig. 3a) probably formed in response to Middle Carboniferous to Middle Permian Variscan shortening and inversion (Cameron 1993; Kombrink *et al.* 2010) prior to extensive peneplanation and the successive formation of the capping Base Zechstein sediments. The exploration significance of Carboniferous anticlines is highlighted by neighbouring discoveries in Breagh and Crosgan, within lower Carboniferous Yoredale and Scremerston sandstone reservoirs (Figs 1, 5; Patruno *et al.* 2018).

Permian–Triassic: renewed sediment accumulation

With the exception of localized fault-bounded Rotliegend–Triassic grabens and half-grabens on parts of the Greater ESP, a large proportion of the Greater MNSH and ESP areas remained a stable Variscan subaerial high for much of Permian times and sedimentation did not begin again until the Late Permian onset of the Zechstein evaporitic cycles (*sensu* Richter-Bernburg 1955*a*, *b*). The Late Permian initial marine ingression was caused by active Arctic rifting combined with a glacio-eustatic sea-level rise, leading to the development of a seaway linking the Arctic Ocean and the Permian basins of northern and central Europe (Ziegler 1988, 1990; Torsvik *et al.* 2002; Coward *et al.* 2003). Following this initial flooding, the rate of water supply during the evaporitic cycles did not vary tectonically, but fluctuated in concert with the last Gondwana ice-cap glaciation cycles (Peryt *et al.* 2010).

With the exception of a few minor tilted fault blocks on the eastern margins of the Greater MNSH (Figs 17 & 18), very little syndepositional Permo-Triassic faulting took place in this whole region and Permo-Triassic reflectors can be followed uninterrupted by faulting over large areas (e.g. Figs 12c, 13d, 14a, b, 17 & 18; Patruno *et al.* 2018).

The observed minor faults on the eastern platform margins, coupled with the high Zechstein subsidence rates and the large spread in the sedimentation rates, suggest active, albeit minor, rifting in those areas (Fig. 8b). For the rest of the Greater MNSH and the Southern North Sea, the general consensus is that Permo-Triassic subsidence was driven mostly by regional thermal relaxation following the lower Rotliegend mafic and intermediate igneous episodes, with crustal extension playing only a minor part (Fig. 18; Lorenz & Nicholls 1984; Glennie & Underhill 1998; Glennie *et al.* 2003; Van Wees *et al.* 2000; Scheck *et al.* 2003; Ziegler *et al.* 2004, 2006). The spatial subsidence distribution varied uniformly and gradually over large regions, increasing progressively away from the Main MNSH. This area, underlain by granites, remained a stable positive area when the Southern and Northern Permian basins began to subside on either side of it (Glennie & Underhill 1998; Pharaoh *et al.* 2010; Figs 7 & 18).

As a consequence, although the Rotliegend strata are confined to the southern and eastern periphery of the Greater MNSH region, the Zechstein sediments blanket the entire area (Figs 7, 8a & 20). The Zechstein Group accumulated as a more or less planar-parallel unit over a deformed Devono-Carboniferous substrate, progressively onlapping the most persistent structural highs (e.g. Dogger Bank) (Figs 12c, 13d, 17 & 18; Taylor 1998; Pharaoh *et al.* 2010).

According to the well-established depositional model for the Zechstein (Fig. 19e; Taylor 1980, 1998; Tucker 1991), the distribution of Zechstein thickness and facies reflects the palaeobathymetry at the onset of the deposition of the unit. Within the Zechstein Group on the Greater MNSH, it is, in fact, possible to distinguish expanded successions composed of halite-dominated halokinetic basinal facies and condensed successions consisting of anhydrite- and carbonate-dominated non-halokinetic platform facies (Figs 14, 15b, 17, 20 & 21a–e). Areas with an initially deeper depositional surface were characterized by a greater overall accommodation (which accounts for the overall thicker Zechstein succession) and could host halite deposition during the latest drawdown stages of each evaporitic cycle. By contrast, areas with a shallower depositional surface were characterized by the growth of neritic carbonate and selenitic gypsum crystals during the initial phases of each evaporative drawdown cycle and were eventually exposed subaerially in the later stages when halite was being deposited in the 'basins'. The resulting internal Zechstein stratigraphic architecture consists of a series of basin-fill halite units and prograding carbonate- and sulphate-rich marginal sedimentary wedges ('clinothems') (Fig. 19d; Tucker 1991; Patruno *et al.* 2015*a*, *b*, *c*, 2018).

In the southern part of Quadrant 37, a narrow seaway existed between the Zechstein epicontinental platform and the Dogger Bank platform (Fig. 20; also see Hay *et al.* 2005). On the Greater MNSH area, at times when the Zechstein platforms were exposed subaerialy, this was the only direct connection between the Northern and Southern Permian basins, which represent areas covered by mainly Zechstein basinal facies, stretching north and south of this sea strait, which we hereby refer to as 'Dogger Strait' (Figs 1 & 20).

The Triassic stratal architecture in the Greater MNSH suggests a simple continuation of thermal subsidence, with the marginal Anglo-Dutch Basin area (as defined in Fig. 1) subsiding faster than the Greater MNSH (Fig. 18). Triassic reflectors are largely conformable over the Zechstein evaporites (Figs 12c, 13d, 17 & 18), except in cases of early halokinesis (Figs 14a, b & 15b, 37/23-1 salt pillow in Fig. 17). The Triassic interval has a TWT thickness of up to 0.5 s on the Greater MNSH, but significantly thicker (up to 1.0 s) in the marginal Anglo-Dutch Basin (Figs 17 & 18). Only on the eastern margin of the Greater MNSH (Dogger Shelf, Auk–Argyll Ridge, West Central Shelf) are numerous, albeit minor, syndepositional extensional faults visible on seismic (Figs 17 & 18). The Greater ESP region, by contrast, saw the creation of Permo-Triassic accommodation space via fault-driven tectonic subsidence focused in specific tectonic depocentres.

Permo-Triassic units are not present over at least half of the Greater ESP (Figs 7, 8a, 18, 22a & 23), and this interval is present in *c.* 40% of the wells (Fig. 7, 8a, 9 & 22a). On the Greater ESP, the Zechstein is almost exclusively represented by relatively thin (<200 m) platform facies with no widespread halokinesis (Fig. 14d). However, when present, the Permo-Triassic interval as a whole entails a stratigraphic succession that is both thicker and more complete (e.g. the Rotliegend is also present) than that on the Greater MNSH (Fig. 15c, d, 16, 18, 22–24). Unlike the Greater MNSH, on the Greater ESP these expanded and more stratigraphically complete Permo-Triassic successions are confined to fault-driven syndepositional intra-platform graben and half-graben, such as the Piper Shelf and the Crawford–Skipper Basin (Figs 18, 22–24; Patruno & Reid 2017; Patruno & Lampart 2018). Triassic, Zechstein, Rotliegend and possibly Carboniferous strata are all present within these fault-bounded depocentres and the Zechstein unit, albeit not halokinetic, is thicker than elsewhere on the Greater ESP (e.g. 474 m thick in well 9/28a-7) (Figs 13a–c, 14d, 15c, 22b, 23 & 24). As a consequence, nearly all the preserved Carboniferous–Mesozoic sedimentary volumes of the Greater ESP are concentrated within these structures (Figs 11b, 22 & 23; Leeder & Boldy 1990; Bruce & Stemmerik 2003; Patruno &

Reid 2017). The rest of the platform remained as an emergent Variscan high until at least the Late Jurassic, but more commonly until the latest Cretaceous (Figs 13c, 15c, d, 16 & 18). For example, the Fladen Ground Spur had been a high that sourced clastic deposition to the Viking Graben until the Middle Kimmeridgian (Eudoxus flooding), when shallow marine conditions affected the area for the first time (Freer *et al.* 1996; also see discussion in Turner *et al.* 2018).

The majority of the extensional faults mapped on the Greater ESP are Triassic in age (Figs 9–11). Triassic reflectors often display synrift wedging, and maximum Triassic–Jurassic TWT thickness is often >0.8 seconds (Figs 16b, 18, 22a & 23; Patruno & Reid 2017). This is consistent with supra-regional trends, as rifting intensified in the Greenland–Scandinavia region during the Early Triassic and propagated into the North Atlantic domain as well as into the North Sea, where major graben were activated along the Horda Platform and the Viking and Central graben (Ziegler 1988, 1990; Roberts *et al.* 1995; Færseth 1996; Bell *et al.* 2014). A particular dense and diffuse network of Permo-Triassic faults has been mapped in the Crawford–Skipper Basin and Piper Shelf areas, with conjugate sets of faults in each area oriented parallel to the direction of the future Late Jurassic major graben (Figs 9 & 11; Patruno & Reid 2017, 2018). The same Permo-Triassic NW–NE conjugate trends mapped throughout the southern Greater ESP and the MNSH (Figs 9 & 11) can be followed regionally over much of the Variscan foreland area and are associated with the post-orogenetic transtensional reactivation of older faults with similar orientation (Chadwick & Evans 1995; Glennie & Underhill 1998). On the Piper Shelf, for example, Triassic faults are often seen to have been reactivated or are oriented parallel to older, Rotliegend age intra-platform graben (Figs 9 & 22b, Fault 6 in Fig. 15c). Regional evidence points to Triassic rifting that is mostly confined to the lower Triassic units and is followed by a phase of post-rift sag in Late Triassic and Early Jurassic times (Glennie & Underhill 1998).

Similar Early Permian and Permo-Triassic post-orogenic rifting events, often related to the inversion of old Caledonide lines of weakness (Glennie & Underhill 1998), have been known to occur throughout the North Sea and northern Europe (Ziegler 1988, 1990, 1992; Glennie 1995; Brekke 2000; Stemmerik *et al.* 2000; Goldsmith *et al.* 2003; Wilson *et al.* 2010), including on the ESP itself (e.g. Unst, East Orkney and Dutch Bank basins, UKCS Quadrant 7; Ziegler 1992; Coward *et al.* 2003; Johnson *et al.* 2005; Richardson *et al.* 2005; PGS 2017) and on the platform on the opposite margin of the Viking Graben (Horda Platform; see Sneider *et al.* 1995; Færseth 1996; Whipp *et al.* 2014).

Middle Jurassic Mid North Sea doming event

Following a possible Early Jurassic phase of widespread post-rift sediment deposition (e.g. Steel 1993), the Toarcian–Aalenian was a time of widespread net uplift and erosion, particularly on the central–eastern portion of the platforms, where no lower–middle Jurassic and upper Triassic sediments were preserved. This regional uplift was driven by the Aalenian Mid North Sea doming, which reflects the development of a diffuse and transient mantle plume head around the North Sea 'triple junction' and is associated with the extrusion of the Rattray Series and the Forties Igneous Province (Figs 2 & 3; Ziegler 1992; Underhill & Partington 1993, 1994; Davies *et al.* 1999; Coward *et al.* 2003; Hendrie *et al.* 1993; Husmo *et al.* 2003; Zanella & Coward 2003; Johnson *et al.* 2005; Graversen 2006).

Further away from the area affected by this supra-regional doming, lower–middle Jurassic strata are preserved in the marginal Anglo-Dutch Basin and the northern Greater ESP. These were deposited under a period of relatively uniform regional thermal subsidence (Fig. 18). Middle Jurassic sandstone-prone reservoirs on the northern ESP include well-known reservoir intervals, such as the Brent and Emerald sandstones (Patruno & Reid 2016*a*, *b*, *c*).

In the Greater ESP region, the Aalenian doming was quickly followed by Bajocian–Callovian proto-rifting, with the onset of fault movement for some of the Late Jurassic faults (Fig. 18). On the main Bajocian–Callovian depocentres (e.g. the Piper Shelf and South Viking Graben), this proto-rifting lead to a direct erosional contact between the lower Triassic Smith Bank Formation and the late Middle Jurassic Pentland/Sleipner/Hugin formations (Figs 2 & 16a). These Middle Jurassic units are sporadically encountered in the Central North Sea, where they represent the oldest preserved reservoirs following the Aalenian doming (Davies *et al.* 1999; Patruno & Reid 2016*a*, *b*, *c*).

Away from these Middle Jurassic depocentres, the hiatus associated with the unconformity that separates these units, the Mid-Cimmerian Unconformity (Fig. 2), encompasses a progressively wider geological time. Over continuously emergent Variscan highs on the eastern margins of both the Greater ESP and MNSH regions (e.g. parts of the Fladen Ground Spur and Dogger Bank), the Mid-Cimmerian Unconformity merges with the Variscan-related unconformity, leading to a direct contact between a thin Chalk veneer and a deformed and truncated thick Devonian unit (Figs 15c & 16; Patruno & Reid 2016*a*, *b*, *c*, 2017). Some of these Variscan highs might have been rejuvenated by Middle Jurassic doming-related uplift, as indicated by tilting and deformation affecting both Devonian and Permo-Triassic units on the Fladen Ground Spur (Fig. 18).

The oldest sub-horizontal unit that post-dates the tilting and uplifting of this area is a thin upper Jurassic or Cretaceous veneer (Figs 16 & 18).

Late Jurassic rifting

The Late Jurassic rifting event shaped the present day structural grain of the Central and Northern North Sea and led to the eventual development of a trilete failed rift system: the master faults that bound the main graben developed, and the Greater ESP and MNSH platform regions became separated (Figs 9 & 18) (Pegrum & Spencer 1990; Ziegler 1990; Ziegler 1992; Rattey & Hayward 1993; Balson *et al.* 2002; Coward *et al.* 2003; Fraser *et al.* 2003). 3D fault mapping reveals that the basin-bounding faults consist of discrete segments, variously linking and overlapping to form relay ramps or transfer faults (Fig. 9; Glennie & Underhill 1998).

Several important synrift source and reservoir intervals were deposited at the edges of the main basinal graben of the Central and Northern North Sea (e.g. Fig. 2; Turner *et al.* 1984, 2018; Færseth & Pederstad 1988; Cherry 1993; Partington *et al.* 1993; Harker & Rieuf 1996; Fletcher 2003; Fraser *et al.* 2003; Patruno & Reid 2016*a*, *b*, *c*). The Kimmeridge Clay unit, in particular, is the key regional source rock (e.g. Pegrum & Spencer 1990; Balson *et al.* 2002; Gautier 2005).

In contrast with the basinal graben, the Greater ESP and MNSH areas were tectonically less active at this time. On the platform margin areas, however, a few relatively minor faults propagated during the Late Jurassic (for example, in the Piper Shelf area) and the regional subsidence distribution suggests minor rifting (Figs 8b, 9, 13a, 15b, 16–18). These trends support that, unlike during the more diffuse Permo-Triassic rifting event, the majority of the strain in the Late Jurassic was focused along the main basinal graben and the master faults bounding them. In the Greater MNSH area, relatively minor rifting took place on the eastern platform margins (e.g. the Auk–Argyll Ridge, West Central Shelf; Figs 17 & 18). The rest of the area was probably emergent during the Late Jurassic to Early Cretaceous (Late Cimmerian inversion), leading to a 'Late Cimmerian' hiatus (Figs 8a, 12c, 13d & 18).

The pluri-modal distribution of the maximum fault lengths of Late Jurassic and Cretaceous faults on the Greater ESP (Fig. 10) is evidence of sequential fault reactivation, growth and linkage (Cowie & Scholz 1992; Cowie & Roberts 2001). This interpretation, together with the observation of a substantially uniform strike trend for faults active from the Devonian to the Cretaceous (Figs 9 & 11), supports the idea that structural lineaments in the North Sea reutilized and reactivated inherent zones of weakness (e.g. basement and Caledonide lineaments) (e.g. Johnson & Dingwall 1981; Bartholomew *et al.* 1993; Glennie & Underhill 1998; Whipp *et al.* 2014). This resulted in a substantially uniform main structural grain, at least from the Caledonian Orogeny onwards.

Cretaceous–Paleogene thermal subsidence

During the Early Cretaceous, the only remaining active rifting was localized on the northwestern Witch Ground Graben, where active master faults formed (Fig. 9), leading to the deposition of synrift reservoir sandstones (Jeremiah 2000). Throughout the rest of the North Sea rift system, rifting abated and began focusing on the evolving Norwegian–Greenland Sea rift (Ziegler 1988; Stampfli & Borel 2002, 2004; Schmid *et al.* 2008).

In the Late Cretaceous and Paleogene, regional post-rift thermal subsidence ensued and, as a consequence, the Chalk and the Paleogene units are usually unfaulted. Their deposition on the Greater MNSH and ESP was widespread, including on the most structurally elevated areas. These were progressively onlapped by a thin veneer of Chalk (e.g. the Fladen Ground Spur and Dogger Bank; Figs 15b, c, d, 16–18) or Paleocene deposits (e.g. Kraken High and Quadrant 8; Figs 12a, b & 15a).

Thermal subsidence patterns were locally interrupted by relatively minor compressional events, leading to the partial local inversion of extensional structures (e.g. the Devonian master fault in Fig. 15a). This is consistent with the intra-plate propagation of Alpine compressional stresses that also affected areas elsewhere in Britain (e.g. Alberts & Underhill 1991; Butler 1998). The most significant of these events is the Aptian-aged Austrian orogenic phase (Glennie & Underhill 1998), possibly the main reason why lower Cretaceous sediments are thin or absent (Fig. 7). This is the age of the earliest indications of the relative switch of plate motion in the Tethys region, from passive margin to Eo-Alpine Orogeny (e.g. Doglioni & Bosellini 1987; Sibuet *et al.* 2004; Patruno *et al.* 2015*d*; Unida & Patruno 2016).

Zechstein halokinesis on the Greater MNSH occurred mostly post-Chalk deposition, with the whole Triassic–Paleogene succession equally deformed (Figs 17 & 18). There are, however, areas where early, Triassic age, halokinesis is prevalent (Figs 14 & 15b, salt pillow under 37/23–1 in Fig. 17).

Eocene–Neogene coastal uplift and supra-regional seawards tilting of the platform areas

A significant uplift of the British land mass took place during Eocene–Recent times, with subsequent large-scale seawards tilting of the platforms (Fig. 18). Increased Cenozoic subsidence and sediment accumulation towards the central–eastern portions of the Greater MNSH and ESP meant that a thick (>1.5 s TWT) mudstone-dominated Cenozoic succession accumulated on the areas, forming an effective topseal unit (Figs 15–18 & 23). By contrast, net uplift and erosion means that Permo-Carboniferous and Devonian units are subcropping the seafloor over the exhumed inner parts of the Greater MNSH and ESP (Figs 7 & 18; Underhill 1991; Hillis *et al.* 1994; PGS 2017).

On the outer part of the Greater ESP, the deposition of laterally extensive, basinwards-prograding, shelf-edge-scale clinoforms (Dornoch Formation, Fig. 2) and the delivery of large volumes of Paleogene clastics were triggered by the rejuvenation of the Orkney–Shetland hinterland via thermal uplift, with the subsequent basinwards tilt of the whole ESP area and enhanced erosion to the west and sediment supply to the east (Pegrum & Spencer 1990; Mudge & Copestake 1992; Jones *et al.* 2003).

The coastal uplift and large-scale seawards tilting of the Greater MNSH platform is mostly Neogene in age and has presumably been still active in the Quaternary as the present day seafloor acts as a planar truncation surface (Figs 17 & 18). The platform rotation was followed by the progradation of large lower Neogene shelfal clinoforms (Figs 15b, 17 & 18). The Neogene uplifting of the British Isles was a result of the igneous underplating linked to the Paleogene development of the Iceland hotspot and North Atlantic rifting (Glennie & Underhill 1998).

Play analysis and relationships with deep-seated structures

Reservoir rocks

On the Greater ESP, several reservoir units have been described by, among others, Marshall & Hewett (2003), Glennie *et al.* (2003), Ahmadi *et al.* (2003), Gray (2014), Patruno & Reid (2016*a*, *b*, c) and references cited therein. These can be summarized briefly as follows (Figs 2, 19a, b; Table 4): (1) upper Paleocene to lower Eocene shelfal clinoform sandstones and related channel sandstones in mainly stratigraphic traps (e.g. the Bressay and Bentley heavy oil discoveries; Figs 2, 4, 19a, b); (2) lower and middle Paleocene deep marine sandstones in mainly stratigraphic to mixed traps (e.g. the Yeoman, Kraken and Mariner heavy oilfields; Figs 2, 4, 13a; 15a & 19a, b); (3) upper Jurassic sandstones in thin veneers forming structural or pinchout traps (e.g. the Hood oil discovery on the ESP and the Johan Sverdrup Field on the Utsira High; Figs 2, 4, 13a & 19a); (4) Permo-Triassic sandstones and Zechstein carbonates in structural traps preserved in local intra-platform depocentres (e.g. the Crawford oil discovery and the Zechstein–Carboniferous reservoir components in large oilfields such as the Claymore and Johan Sverdrup fields (Figs 2, 4, 19a, b & 23b; Whitehead & Pinnock 1991); (5) Devonian fractured Old Red Sandstones in large structural closures (e.g. the Clair and Buchan oilfields; Figs 2, 4, 13b, c & 19a); and (6) Devonian–Rotliegend sandstones or fractured and weathered crystalline basement in structural traps along platform margins (e.g. the Cairngorm oil discovery on the ESP and the Whirlwind and Lancaster oil discoveries on the West Shetland Platform margins; Figs 2, 4 & 19a).

The ultimate recoverable reserve volumes of the largest fields with fractured Devonian or basement reservoirs demonstrate the potential reservoir quality of this interval: Clair (1100 MMboe); Buchan (220 MMboe); Lancaster (207 MMboe); Whirlwind (190 MMboe); Stirling (5 MMboe); and Cairngorm (30 MMboe) (Figs 1, 4 & 7; Patruno & Reid 2016*a*).

In the marginal Anglo-Dutch Basin, the main working reservoirs are lower–'middle' Permian Rotliegend sandstones (e.g. Silverpit and Leman sandstones) sitting on the upper Carboniferous Coal Measures source rock and sealed by the Zechstein salt (Figs 3 & 19c). As a result of the extensive Variscan uplift of the Greater MNSH, both the upper Carboniferous Coal Measures and the Rotliegend sandstones are missing, with the latter appearing again only west of the Dogger Granite (Figs 3, 7, 17, 19c; Hay *et al.* 2005; Kombrink *et al.* 2010; Arsenikos *et al.* 2015; Monaghan *et al.* 2015). As a result of the absence of the main graben reservoir and source intervals, all the current hydrocarbon discoveries straddle the eastern and southern edges of the region: on the Auk Ridge (Mesozoic, Zechstein and Devonian reservoirs; Patruno & Reid 2016*a*, *b*, *c*, 2017), West Central Shelf (Paleogene reservoirs) and at the transition with the marginal Anglo-Dutch Basin (Dinantian and Zechstein reservoirs for the Breagh and Crosgan fields; Figs 1, 3, 5, 17 & 19c; Patruno *et al.* 2018; Patruno & Reid 2016*a*, *b*, 2017).

Proven and potential reservoirs present on the Greater MNSH region are different from those observed in the marginal Anglo-Dutch Basin and include continental sandstones belonging to the lower Namurian Yoredale Formation (e.g. the Crosgan gas discovery, Fig. 21b), the upper Visean (Dinantian) Scremerston Formation (e.g. the Breagh and Crosgan gas discoveries), the lower Visean (Dinantian) Fell Sandstone Formation and the upper Devonian Old Red Sandstones Group (e.g. part of the reservoir for the Embla and Alma/Argyll oil and gas fields) (Cameron *et al.* 1992; Cameron 1993; Monaghan *et al.* 2017; Fig. 3). In addition, the lower–middle Triassic Bunter Sandstone unit is another proven reservoir in the Southern North Sea, but is again largely confined to the periphery of the Greater MNSH, (Figs 3 & 5; Hay *et al.*

Table 4. *List of petroleum elements for the Greater East Shetland Platform and Greater Mid North Sea High, the highlighted boxes indicate the positive presence of that element within the period defined**

Period	Greater East Shetland Platform				Greater Mid North Sea High			
	Present day maturity of source rock	Source rock presence	Reservoir	Seal	Present day maturity of source rock	Source rock presence	Reservoir	Seal
Devonian	Mostly mature	✓	✓	✓ (intraformational)	?	?	?	✓
Lower Carboniferous	Mostly immature	✓	✓?	✗	(intraformational)	✓	✓	✓?
Upper Carboniferous		✗	✗	✗		✗	✗	✗
Lower–‘middle’ Permian		✗	✓?	✗		✗	✓?	✗
Upper Permian		✗	✓?	✓	?	✓?	✓	✓
Triassic		✗	✓	✓?		✗	✓	✓ (intraformational)
Lower–middle Jurassic		✗	✗	✗		✗	✗	✗
Upper Jurassic	Immature	✓	✓?	✓		✗	✓?	✓
Cretaceous		✗	✓?	✓		✗	✗	✓
Cenozoic		✗	✓?	✓		✗	✓?	✓

✓?, not present everywhere; ✓, present; ✗, not present.
* The lower Carboniferous becomes mostly mature in the Anglo-Dutch Basin.

2005), whereas upper Jurassic and Paleogene sandstones have been successful on the West Central Shelf (e.g. the Gannet Complex, Bittern and Curlew oil and gas fields) (Figs 3 & 5; Hay *et al.* 2005). The more speculative middle Devonian Kyle Limestone Group has long been hoped (e.g. Esso 2009) to form carbonate build-ups on top of pre-existing highs (Figs 3, 5 & 19c). Z1-, Z2- and Z3-age carbonate units of the lower Zechstein Group are a proven working reservoir for several fields in both the Northern Permian Basin (e.g. the Johan Sverdrup, Claymore, Auk and Alma/Argyll oil and gas fields) and the Southern Permian Basin (e.g. the Hewett, Crosgan and several large fields onshore Netherlands, Germany and Poland; see Patruno *et al.* 2018) (Figs 2, 3, 5, 19c, e & 21).

The carbonate–sulphate Zechstein platform complex characterized in 3D by Patruno *et al.* (2018) is a peninsula projecting out of the main body of the epicontinental Zechstein platform (Figs 20 & 21). This complex consists of a series of prograding–aggrading clinothems of lower Zechstein age (Z1–Z3), with some composed of anhydrite (high impedance) and others of carbonate (lower impedance) (Fig. 21e, g). The Z2 carbonates form the most prospective clinothems, both regionally and locally (Karnin *et al.* 1996; Peryt *et al.* 2010; Patruno *et al.* 2018). Although gas shows have been identified in 21 nearby wells that penetrated the tight Z2 bottomset carbonates, the thickest and most permeable part of these clinothems (i.e. the foresets) has never been drilled in this area. This interpretation is consistent with the typical stratigraphic architecture of a prograding intra-Zechstein sulphate–carbonate build-up (cf. Taylor 1980, 1998; Tucker 1991; Patruno *et al.* 2018).

Source rocks

The highest exploration risk for the Greater ESP and MNSH areas is related to the hydrocarbon charge. The traditional source rocks (upper Jurassic in the Central and Northern North Sea and Carboniferous Westphalian in the Southern North Sea) are usually absent and, where present, they are thin and immature for hydrocarbon generation over large parts of the Greater MNSH and ESP areas (Hay *et al.* 2005; Kombrink *et al.* 2010; Patruno & Reid 2016*a*, 2017; Patruno *et al.* 2018; Fig. 19; Table 4). Alternative lower Carboniferous source rocks are very likely present at least in parts of the Greater MNSH and ESP (e.g. the Witch Ground Graben and Piper Shelf areas), although their maturation is doubtful (Monaghan *et al.* 2015, 2016; Table 4). In particular, the lower Carboniferous is currently mature for gas maturation between the southern edge of the Greater MNSH and the marginal Anglo-Dutch Bank Basin (Monaghan *et al.* 2015, 2017; Vincent 2015). In the Witch Ground Graben and Piper Shelf areas (southern Greater ESP), the lower Carboniferous is currently within marginal oil maturation on highs, with potential for gas maturation into the basins (Monaghan *et al.* 2016; Vane *et al.* 2016).

A successful lateral migration route of up to 30–40 km from the source kitchens has been demonstrated by hydrocarbon discoveries on both the Greater ESP/Utsira High (e.g. the Mariner, Kraken, Brae West, Hood, Johan Sverdrup oilfields; Figs 1 & 4) and, possibly, the margins of the Greater MNSH (e.g. the Breagh and Crosgan gas discoveries at the transition with the marginal Anglo-Dutch Basin and several Permian to Eocene oil and gas discoveries at the transition with the Central Graben) (Figs 1–5 & 19c). In addition, recent modelling points out that the gas from Breagh and Crosgan could feasibly have been sourced at least partly from vertical migration from the lower Carboniferous source rock (Monaghan *et al.* 2015, 2017; Vincent 2015).

To date, no oil seep nor direct hydrocarbon indicator (DHI) has been observed anywhere on the Greater MNSH (as defined in Fig. 1b) further than 40 km away from the closest basinal source kitchen, and no hydrocarbon show has been recorded by any of the few wells drilled in this area (Figs 1 & 4). Even though the coal-rich Yoredale and Scremerston lower Carboniferous source rock intervals represent the potential for shorter lateral migration, or even for vertical charge, of the Greater MNSH, these are out of the present day gas window and are immature over the entire area (Hay *et al.* 2005; Kombrink *et al.* 2010; Figs 3 & 19c). By contrast, just to the south of the Greater MNSH, there is evidence of a mature Visean–Namurian source rock in the marginal Anglo-Dutch Basin area, with several shows and gas discoveries (Breagh and Crosgan) in this area (Figs 1 & 5; Monaghan *et al.* 2015, 2017; Vincent 2015).

Compared with the Greater MNSH, oil seep data in parts of the Greater ESP (Quadrants 6, 7 and 14), suggest a working source and viable migration pathways at the very centre of the ESP, up to 80 km away from the closest Upper Jurassic Kimmeridge Clay source kitchen (Table 4; Fig. 1; Richardson *et al.* 2005). Possible hydrocarbon escape features observed on seismic data on the Greater ESP will be discussed later in this paper. All this evidence suggests that additional charge on the Greater ESP may, in fact, be provided both within and beyond the 30–40 km virtual platform 'margin' by vertical/lateral migration from working Middle Devonian organic-rich lacustrine source intervals (the Orcadia and Eday Flagstone formations; Fig. 2; Marshall & Hewett 2003), which have been penetrated by wells in the study area (e.g. 9/16-3 in Fig. 24; Duncan & Buxton 1995). In the Orcadian Basin, the Devonian is the known source for the Jurassic

Beatrice Field (Inner Moray Firth); this interval might also partly source the oil of the giant Clair oilfield (Rona Ridge) and of other large fields in the Moray Firth (e.g. the Claymore, Piper, Tartan and Buchan fileds) (Fig. 1; Marshall 1998; Marshall & Hewett 2003; Mark *et al.* 2008; Cornford 2009; Monaghan *et al.* 2016; Patruno & Reid 2016*a*, *b*).

There is the additional risk that the tectonostratigraphic history of the Greater ESP and MNSH regions switched off the Paleozoic source rock maturation, either during the Variscan uplift (e.g. Duncan & Buxton 1995) or by Neogene coastal uplift (Glennie & Underhill 1998). 1D basin modelling performed by Patruno & Reid (2016*a*; 2017) suggests that the Middle Devonian source interval might currently be mature for oil generation over a part of the Greater ESP, particularly where it is buried at greater depths beneath the infill of Permo-Triassic intra-platform extensional mini-basins.

Seal rocks

Within the Greater MNSH and ESP areas, the mudstone- and marlstone-prone upper Cretaceous to Cenozoic succession forms the overall topseal for the underlying Paleozoic and Mesozoic traps (Figs 2 & 3). An additional seal for Devonian–Rotliegend and older reservoirs is present where a thick Zechstein–Triassic succession was deposited within intra-platform fault-bound basins, particularly in association with halite-rich 'basinal' Zechstein facies and/or mudstone-rich lower Triassic Smith Bank Formation (Figs 2 & 3; Patruno & Reid 2016*a*, 2017; Table 4).

The Eocene to Recent large-scale tilting of the platforms resulted in enhanced Cenozoic thickness on the seaward (i.e. eastward) margins of the Greater MNSH and ESP areas (up to 2 s TWT) and therefore probably provided sufficient seal integrity (Figs 14–18; Patruno & Reid 2016*a*). At the same time, towards the west, the Cretaceous–Cenozoic units progressively thin until becoming absent (Figs 16–18). As a consequence, seal integrity and seal breaching risks become progressively higher westwards over the Greater ESP and MNSH areas.

Seismic evidence for recent fluid migration on the Greater ESP

Distribution of vertical seismic amplitude anomalies in the seismic data

Numerous vertical seismic amplitude anomalies occur within the Cenozoic succession of the Greater ESP (Fig. 25), whereas on the Greater MNSH only a few gas cloud anomalies at the eastern periphery of the region (Quadrant 29) have thus far been documented (Hay *et al.* 2005). The Greater ESP anomalies are clustered in three main areas: (1) on the Piper Shelf in UK Quadrant 15 (Q15–2014 survey as in Fig. 1; see fig. 2 of Patruno & Reid 2016*b*); (2) in Quadrants 2, 3 and 9 (PGS15010ESP survey as in Fig. 1), just above existing hydrocarbon discoveries (particularly the Bentley and Bressay heavy oilfields, with the lower Eocene to uppermost Paleocene Dornoch Formation reservoir; see Fig. 25a) and in close association with similar undrilled closures to the north, and with wells with oil and gas shows (e.g. well 3/16-1); and (3) in Quadrants 8, 9 and 15 (PGS15004ESP survey as in Fig. 1) in close association with the southeastern edge of the Crawford–Skipper Basin and other minor Permo-Triassic fault-bounded depocentres (Fig. 25b, d). For example, seven main pipe clusters can be distinguished in the area to the west and southwest of the Crawford–Skipper Basin: one consisting of five anomalies in the south, three areas with three anomalies each in the east and three areas with one anomaly each in the north. There are four single anomalies in the central and western parts of the study area (Fig. 25c).

The seismic character of the anomalies varies over the Greater ESP area. The structures associated with the uppermost Paleocene to lower Eocene Bentley and Bressay discoveries (Fig. 25a) reach from the Base Paleocene to about 0.2 s TWT below the seafloor. These are characterized by upwards curvature of the surrounding strata and high amplitudes above the near-Base Miocene. They are much less pronounced in the Paleocene and lower Eocene succession, where only slight distortion of the seismic image and an offset in the Base Paleocene reflector can be observed.

The vertical seismic anomalies in the areas at the edge of the Crawford–Skipper Basin (Fig. 25b–d) do not show strong upwards curvature of the surrounding strata and they extend *c.* 0.1 s TWT above the near-Base Miocene. Instead, they can be easily traced all the way down to the Base Paleocene reflector and at the truncational edge of the Triassic sedimentary fill of the fault-bounded Crawford–Skipper Basin. Figure 25b shows a typical vertical seismic amplitude anomaly in the area at the edge of the Crawford–Skipper Basin (PGS15004ESP survey). This is underlying a strong normal polarity seismic event at *c.* 300 ms TWT below the seafloor and diminishes downward into the Eocene succession. It is characterized by mostly lower seismic amplitudes and slight upwards curvature of the surrounding strata.

Interpretation of the vertical seismic amplitude anomalies

Vertical seismic amplitude anomalies are frequently caused by fluid migration through sedimentary rocks

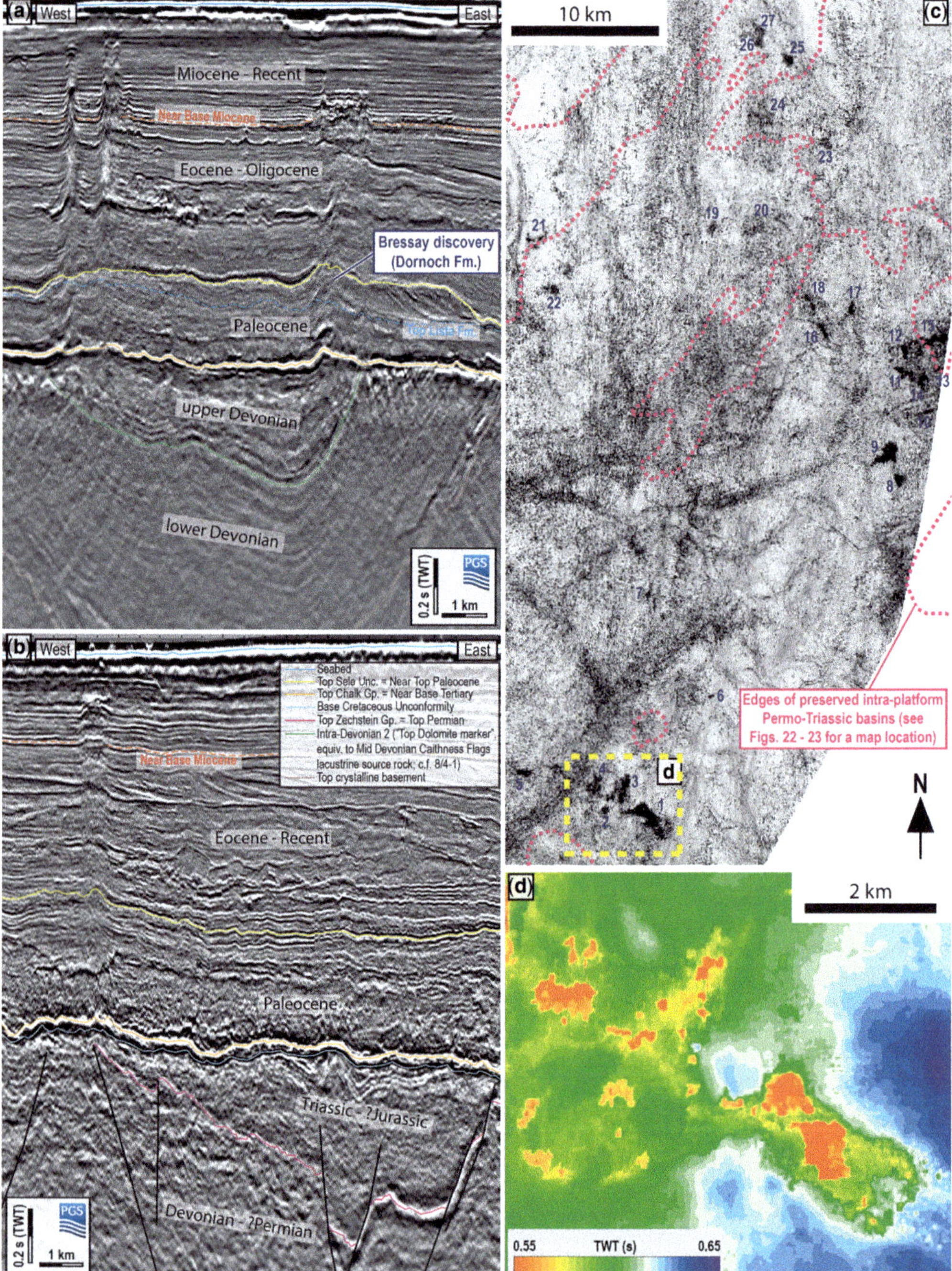

Fig. 25. Vertical amplitude anomalies (pipes) highlighting possible fluid escape features in the Greater East Shetland Platform area (**a**) over the Bressay Paleogene discovery and (**b**) at the edge of the preserved Triassic sedimentary wedges of the Crawford–Skipper Basin. (**c**) An RMS map of a coherency volume over part of Quadrant 8 showing 27 different pipes, mostly developed close to the edge of the Permo-Triassic intra-platform basins (see Figures 4 and 22a for location map). (**d**) Two-way travel time–structure map of the near-Top Miocene horizon showing the detailed geometry of several pipes in an area at the southern edge of part (c) (see Fig. 25c for location).

(Berndt *et al.* 2003; Cartwright *et al.* 2007; Løseth *et al.* 2011) and have been documented by their association with gas-seeping pock-marks and seismic evidence for free gas in the subsurface (Plaza-Faverola *et al.* 2010). The migrating fluids may be aqueous or contain free gas, resulting in different seismic characters. Commonly high-amplitude reflections within the anomalous zones are interpreted to result from free gas (Taylor *et al.* 2000), whereas the absence or dimming of the surrounding seismic reflectors within the anomalous zone is interpreted to be the result of the migration of aqueous fluids and the structural deformation that they are causing to the sediments (Berndt 2005). Upwards curvature of the sedimentary strata has been shown to result from real structural deformation (Plaza-Faverola *et al.* 2011), showing that the migration of fluids is able to perturb the rocks beyond hydrofracturing, which is in agreement with field observations in Utah and New Mexico (Karstens 2015). Based on outcrop-analogues and the careful interpretation of seismic data, it is possible to distinguish at least two major classes of fluid flow anomalies: (1) structures commonly called pipes, which have sharp vertical edges; and (2) high-amplitude reflectors and chimneys, which are characterized by chaotic internal seismic facies and irregular outlines in seismic sections. It is likely that the pipe structures result from fast deformation of the subsurface due to hydrofracturing, whereas chimneys result from the diffusive migration of gas into the overlying sediments when gas overcomes the capillary entry pressure of clay units (Arntsen *et al.* 2007; Karstens & Berndt 2015).

The vertical seismic anomalies linked to the Bentley and Bressay discoveries in the northern PGS15010ESP survey (Fig. 25a) are typical pipe structures with up-bend, high-amplitude reflections and sharp vertical edges. They were probably caused by hydrofracturing and the high amplitudes indicate that not only aqueous fluids, but also significant (>4%) amounts of free hydrocarbon gas, must have entered these structures because there is no other realistic source for free gas, such as active volcanism, in the study area. The fact that the pipe structures in Figure 25a, and particularly the westernmost structure, reach all the way to the Base Tertiary Unconformity, and the fact that this and the easternmost structures coincide with a step in the Base Paleocene reflector, suggest that the pipes extend all the way down through the Tertiary section and that the migrating fluids originate from the underlying Paleozoic strata. It is interesting to note that the pipe structures in Figure 25a seem to originate precisely from the two areas of subcrop and erosion of the interpreted Middle Devonian interval (potentially containing an oil-prone source rock) by the Base Tertiary Unconformity.

The seismic anomalies at the edge of the Crawford–Skipper Basin in the PGS15004ESP survey (Fig. 25b) are similar, extending from the base of the Paleocene through near-Base Miocene where they terminate in high-amplitude reflections, suggesting that gas accumulates here. Similar to the pipes around the Bressay discovery (i.e. Fig. 25a), its base almost coincides with the edge of the Triassic infill of the intra-platform Permo-Carboniferous Crawford–Skipper Basin. The presence of outwardly tilted Paleozoic monoclinal strata at the edge of the Crawford–Skipper Basin might represent an easy conduit to allow lateral migration of the hydrocarbon from a deep Devonian source kitchen towards and beyond the edge of the Permo-Triassic basin.

This explanation is compatible, for example, with the interpretation put forward by Patruno & Reid (2017) for the oil shows in the Devonian succession penetrated by well 9/16-2 (Fig. 24). Around the well 9/16-2 area, the simultaneous presence of a source kitchen for the Devonian source rock under the Crawford–Skipper Basin (with the middle Devonian source rock penetrated by the nearby well 9/16-3 and within the present day oil window under the Triassic infill) and of easy migration pathways due to the monoclinal dip of the Paleozoic reflectors might explain the presence of oil shows in the Devonian package penetrated by well 9/16-2 in the vicinity of an as-yet undrilled horst block (Fig. 24).

The stratal geometries discussed here suggest that the vertical hydrocarbon escape features in Figures 25a, b might be generated by vertical hydrocarbon migration within the Greater ESP area itself from deeply buried Devonian source rocks. There is no conclusive evidence to discount an origin from the lateral migration of hydrocarbons from the basinal Jurassic source kitchens; for example, via partial breaching of the Paleogene Bressay trap.

Relationships between hydrocarbon prospectivity and intra-platform Permo-Triassic basins

The key differences in the evolution of the Greater MNSH compared with that of the Greater ESP is the amount and type of Permo-Triassic tectonic subsidence and sedimentary fill. This can be summed up as a difference between thermal relaxation basins and diffuse rift basins and mini-basins (Fig. 18). In the case of the thermal relaxation basins on the Greater MNSH, syndepositional faulting is rare, the Permo-Triassic subsidence is driven by regionally consistent post-igneous thermal relaxation, and facies and thickness differences are controlled by the inherited morphology and palaeotopography of the depositional substrate, as well as by the amount of thermal cooling for each region (Figs 18 & 26;

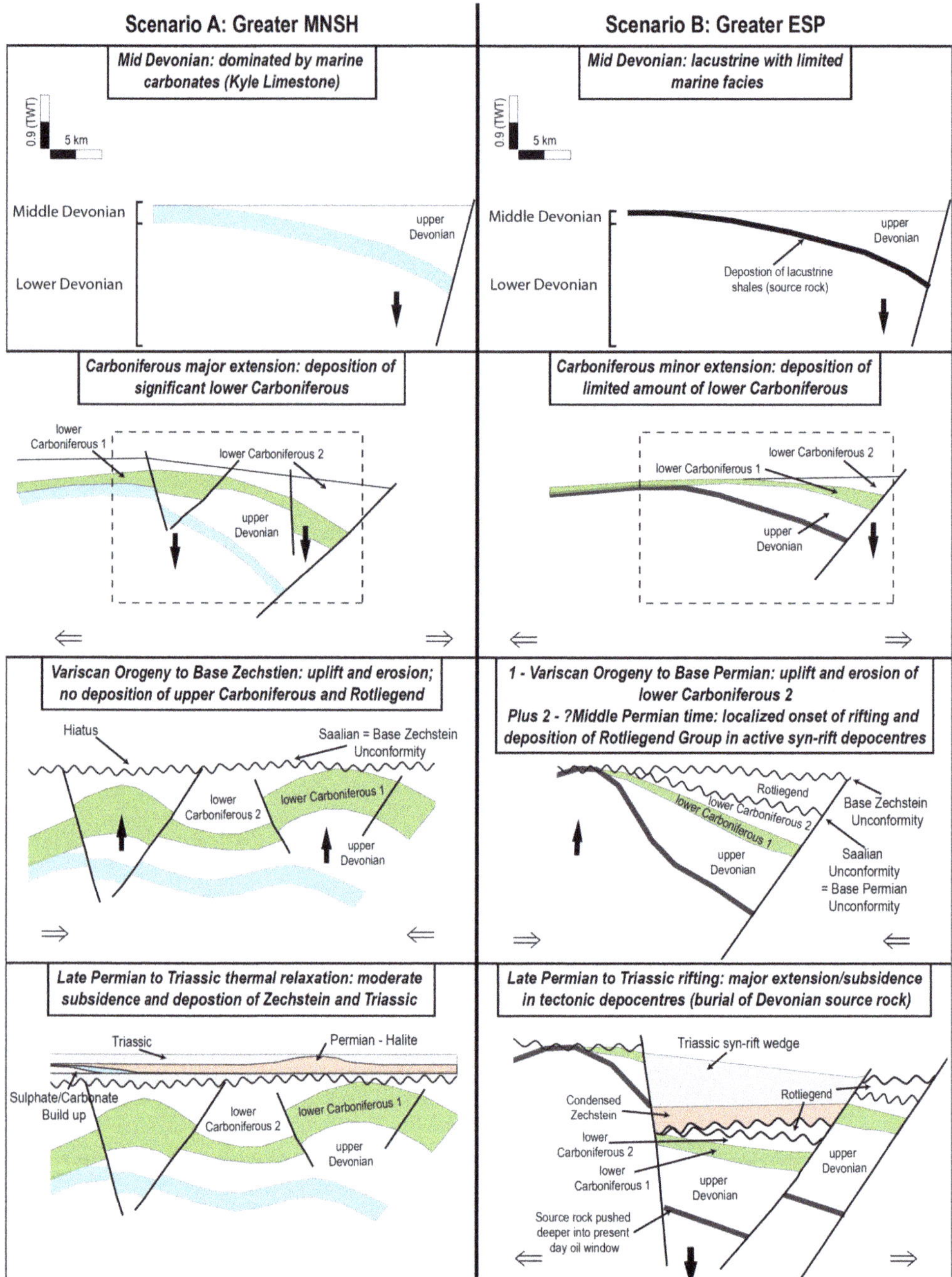

Fig. 26. Idealized sketch showing the relationships between the development of distinct Permo-Triassic intra-platform rift basins (Greater East Shetland Platform) or thermal relaxation basins (Greater Mid North Sea High) and the overall petroleum system.

Lorenz & Nicholls 1984; Van Wees *et al.* 2000; Ziegler *et al.* 2004, 2006). In the Permo-Triassic rift basins on the Greater ESP, syndepositional faulting is common and controls the presence, thickness and facies of Permo-Triassic sediments. Tectonic subsidence is fault-driven and focused into fault-bounded depocentres (Figs 18 & 26). The different Permo-Triassic architecture of the Greater ESP and

MNSH areas (Fig. 18) is likely to have played a part in their overall hydrocarbon potential (Fig. 26) (cf. Patruno & Lampart 2018). Two key differences between thermal relaxation and rift basins are likely to have affected the overall petroleum systems: (1) the maximum tectonic subsidence and Permo-Triassic thickness; and (2) the presence of faults.

The greater subsidence associated with the intra-platform fault-bounded depocentres of the Greater ESP allowed the deposition and preservation of Permo-Triassic, Jurassic and possibly Carboniferous reservoirs, as well as providing further potential seal intervals for deep closures (e.g. Devonian fault blocks) (Figs 19b & 26; Patruno & Reid 2017; Patruno & Lampart 2018). In the main fault-bounded depocentres on the Greater ESP, the Triassic–Jurassic time thickness is >0.8 s TWT (Figs 22a & 23) and the Permo-Carboniferous thickness is up to 1.1 s TWT (Figs 22b & 23). The Devono-Carboniferous is therefore buried deeper here than in the rest of the Greater ESP and this additional burial is often sufficient to push the Mid-Devonian interval into the present day oil maturation window (Figs 24 & 25b; Patruno & Reid 2017). Within this context, it is interesting to note that the oil seeps at the very centre of the Greater ESP (Quadrants 7, 14; Fig. 7) are situated in the area of the intra-platform Permo-Triassic Dutch Bank Basin (Richardson *et al.* 2005) and that several fluid escape pipes were observed in close proximity to the intra-platform Crawford–Skipper Basin (Fig. 25b–d).

By contrast, the maximum time thickness of the preserved Permian and Triassic–Jurassic units over the Greater MNSH area is *c.* 0.7 s TWT each (Figs 12c, 14a–c, 15b & 17). As a consequence, the burial of the lower Carboniferous source rocks is not normally sufficient to place them in the present day gas window. The lack of Permian age rifting means that most of the Greater MNSH remained largely above the basal erosional level during the pre-Zechstein Permian time, leading to the absence of the Rotliegend Group, which is the main reservoir interval in the Southern North Sea (Fig. 18). This unit has instead been penetrated, together with occasional Carboniferous strata in the depocentres of the intra-platform fault-bounded basins and mini-basins on the Greater ESP (e.g. Figs 7, 23 & 24; Patruno & Reid 2017and the well-base subcrop map shown in Fig. 6 of the latter paper).

Another important consequence associated with Permo-Triassic intra-platform graben and half-graben is the presence of associated deep-seated faults penetrating the Meso-Cenozoic package stemming from the deep Devonian units (e.g. Figs 15a, c, d, 16, 23 & 24). This structural configuration provides pathways for the vertical migration of hydrocarbons from the Devonian source kitchen towards the post-middle Devonian (e.g. Buchan Formation and Rotliegend sandstones) and the Meso-Cenozoic reservoir units and traps (e.g. Figs 23 & 24). Similarly, the presence of outwardly tilting Paleozoic monoclinal strata at the edge of the Crawford–Skipper Basin might represent another easy conduit to allow the lateral migration of hydrocarbon from the deep source kitchen towards and beyond the edge of the Permo-Triassic basin (Fig. 24) (Patruno & Reid 2017; Patruno & Lampart 2018). In the Greater MNSH, the scarcity of faults breaching the seal between the source interval (lower Carboniferous) and the post-Zechstein reservoir units means that the only reservoirs that might be charged are sub-Zechstein together with, perhaps, the lower Zechstein carbonates (Figs 12c, 13d, 14a–c, 15b, 17 & 21).

It is important to consider the role of deeper heterogeneity on the timing, locus and magnitude of post-Caledonian tectonic subsidence. The Greater ESP largely lies north of the Iapetus Suture, whereas the MNSH lies within the Midland Valley, Southern Uplands accretionary complex and Carboniferous block-faulted terranes (Coward *et al.* 2003). The Greater MNSH, due to its increased vicinity to the Variscan fold belt, has been feasibly subject to a significantly greater impact of Variscan uplift, which resulted in the absence of the Rotliegend reservoir interval and perhaps in the less active Permo-Triassic extension.

Summary and conclusions

The Greater ESP and MNSH are two underexplored large platform regions flanking the main structural depocentres of the North Sea mature hydrocarbon province. Since 2011, these frontier regions of the UKCS have been extensively covered by 2D and 3D conventional and broadband seismic data. Here, seismic interpretation of these data has been complemented by well analysis to shine new light on the complex tectonostratigraphic evolution of these regions and how this potentially controls hydrocarbon prospectivity.

The interpreted structures, together with the documented borehole, seismic stratigraphic and stratal trends, point to a largely consistent evolution of the Greater ESP and MNSH, intermittently influenced by tectonic inversion. The regional geological history of the study areas in the last 420 myr is a succession of eight (or more) regional tectonic inversions: (1) from Caledonian compression to ?Early Devonian extensional collapse; (2) from ?Early Devonian extensional collapse to a transient Middle Devonian compressional/transtensional inversion; (3) from the Middle Devonian compression back to ?Late Devonian to Early Carboniferous extension; (4) from Early Carboniferous extension to the Late

Carboniferous Variscan Orogeny; (5) from the Late Carboniferous Variscan Orogeny to renewed Permo-Triassic extension; (6) from Permo-Triassic extension to Aalenian thermal doming; (7) from Aalenian thermal doming to Late Jurassic rifting; and (8) from Late Jurassic rifting to Cretaceous passive subsidence, locally interrupted by intra-plate Alpine inversions and regionally interrupted by progressive seawards tilting of the platforms induced by the Paleogene Icelandic thermal plume. Regional variations of these polyphase tectonic cycles through the Greater MNSH and ESP areas have also been identified, however. These variations are likely to have played a fundamental part in the local hydrocarbon prospectivity.

The events that played a key part in the hydrocarbon plays of the two regions include: (1) on the Greater ESP, high Devonian subsidence due to the extensional collapse of the Caledonian Orogen, leading to the deposition of a thick continental clastic interval (Old Red Sandstones) with reservoir and source potential; (2) on the Greater MNSH, Early Carboniferous extension, leading to the deposition of a continental clastic interval with reservoir and source potential; (3) the Variscan Orogeny, leading to Late Carboniferous compression and uplift, which were likely more intense and long-lasting in the Greater MNSH; (4) Permo-Triassic rifting (on the Greater ESP) and thermal relaxation (on the Greater MNSH), leading to the deposition of clastic and evaporitic/carbonatic units, with reservoir and seal potential; (5) the Middle Jurassic Mid North Sea doming event, leading to the westwards tilting of the platforms and erosion of their eastern margin; (6) Late Jurassic extension; (7) Cretaceous–Paleogene thermal subsidence, leading to the deposition of a mudstone-rich marine succession with several significant sandstone reservoir intervals interspersed in it (particularly on the Greater ESP); and (8) Eocene–Neogene progressive coastal uplift and seawards tilting of the platforms, leading to enhanced subsidence on their eastern margins and to a deep erosional unconformity westwards, closer to the present-day British coastlines.

Several working reservoir units are present over large parts of the two study areas and structural traps have been imaged in the Paleozoic section by the new seismic data. Hydrocarbon charge could potentially come via lateral migration from the main basinal source kitchens to the east and south, but this migration may be unlikely beyond 30–40 km from the Viking/Witch Ground Graben Basin source kitchen margins. On the Greater ESP, the presence of seeps and fluid escape pipes up to 80 km away from the Viking/Central Graben Basin source kitchens suggests the possible existence of a deeper, most likely middle Devonian, lacustrine source (e.g. Orcadia Formation). This has been known throughout the Orcadian Basin as a proven or potential source for the hydrocarbons of large existing fields (e.g. the Beatrice, Clair and Claymore fields). Conversely, in the Greater MNSH, no show, seep or fluid escape pipe has been recorded in wells or seismic further than 40 km from the closest source kitchens (chiefly the Anglo-Dutch Basin and the Central Graben). Despite several lower Carboniferous source intervals being present, these appear to be largely immature away from the marginal Anglo-Dutch Basin. Furthermore, only non-source rock middle Devonian facies have thus far been penetrated by wells on the Greater MNSH (e.g. Kyle Limestone unit).

The different evolution of the two study areas during Permo-Triassic times is likely to have played a major part in controlling the maturation of the deeper source rocks and the presence of key Paleozoic–Jurassic reservoir units. On the Greater MNSH, the absence of significant Permo-Triassic (or successive) rifting events meant there was insufficient tectonic subsidence to bury the gas-prone lower Carboniferous to sufficient depths to reach the gas maturation window. On the Greater ESP, the presence of localized and diffuse Permo-Triassic intra-platform grabens and half-grabens provided sufficient subsidence to locally bury the oil-prone middle Devonian interval up to, and beyond, the oil maturation window, with faults providing conduits for oil migration towards shallower reservoirs. In this context, it is interesting to note that the oil seeps at the very centre of the Greater ESP (Quadrants 7 and 14) are situated in the area of the fault-bounded Permo-Triassic Dutch Bank Basin and that several fluid escape pipes were observed in close proximity to the intra-platform Permo-Triassic Crawford–Skipper Basin.

This paper shows that, although the Greater ESP and MNSH areas have generally similar tectono-stratigraphic histories, significant differences in the Middle Devonian and Permo-Triassic sections have resulted, in different hydrocarbon exploration potentials.

Acknowledgements The authors gratefully acknowledge PGS for providing the seismic data for this study. We are also grateful to PGS Reservoir, Henk Kombrink, Ian Sharp and an anonymous BGS reviewer for reviewing a previous draft of this paper, providing constructive and insightful feedback.

References

AHMADI, Z.M., SAWYERS, M., KENYON-ROBERTS, S., STANWORTH, C.W., KUGLER, K.A., KRISTENSEN, J. & FUGELLI, E.M. 2003. Paleocene. *In*: EVANS, D., GRAHAM, C., ARMOUR, A. & BATHURST, P. (eds) *The Millennium*

Atlas: Petroleum Geology of the Central and Northern North Sea. Geological Society, London, 235–259.

Alberts, M.A. & Underhill, J.R. 1991. The effect of Tertiary structuration on Permian gas prospectivity, Cleaver Bank area, southern North Sea, UK. *In*: Spencer, A.M. (ed.) *Generation, Accumulation, and Production of Europe's Hydrocarbons*. Special Publications of the European Association of Geologists, **1**, 161–173.

Allen, P.H. & Allen, J.R. 2005. Subsidence and thermal history. *In*: *Basin Analysis, Principles and Applications*, 2nd edn. Blackwell, Oxford, 349–395.

Arntsen, B., Wensaas, L., Løseth, H. & Hermanrud, C. 2007. Seismic modeling of gas chimneys. *Geophysics*, **72**, SM251–SM259.

Arsenikos, S., Quinn, M.F., Pharaoh, T., Sankey, M. & Monaghan, A. 2015. *Seismic Interpretation and Generation of Key Depth Structure Surfaces within the Devonian and Carboniferous of the Central North Sea, Quadrants 25-44 Area, 21CXRM Palaeozoic Project*. British Geological Survey, Energy and Marine Geoscience Programme Commissioned Report **CR/15/118**. British Geological Survey, Keyworth, http://nora.nerc.ac.uk/516758/

Balson, P., Butcher, A., Holmes, R., Johnson, H., Lewis, M. & Musson, R. 2002. *North Sea Geology. Strategic Environmental Assessment, SEA2&3*. Technical Report **008 Rev1**. British Geological Survey, Keyworth.

Bartholomew, I.D., Peters, J.M. & Powell, C.M. 1993. Regional structural evolution of the North Sea: oblique slip and the reactivation of basement lineaments. *In*: Parker, J.R. (ed.) *Petroleum Geology of Northwest Europe: Proceedings of the 4th Conference*. Geological Society, London, 1109–1122, https://doi.org/10.1144/0041109

Bell, R.E., Jackson, C.A.L., Whipp, P.S. & Clements, B. 2014. Strain migration during multiphase extension: observations from the northern North Sea. *Tectonics*, **33**, 1936–1963.

Berndt, C. 2005. Focused fluid flow in continental margins. *Philosophical Transactions of the Royal Society of London*, **363**, 2837–2854, https://doi.org/10.1098/rsta.2005.1662

Berndt, C., Bünz, S. & Mienert, J. 2003. Polygonal fault systems on the mid-Norwegian margin: a long term source for fluid flow. *In*: Van Rensbergen, P., Hillis, R.R., Maltman, A.J. & Morley, C.K. (eds) *Subsurface Sediment Mobilization*. Geological Society, London, Special Publications, **216**, 283–290, https://doi.org/10.1144/GSL.SP.2003.216.01.18

Brekke, H. 2000. The tectonic evolution of the Norwegian Sea continental margin with emphasis on the Vøring and Møre Basins. *In*: Nottvedt, A. (ed.) *Dynamics of the Norwegian Margin*. Geological Society, London, Special Publications, **167**, 327–378, https://doi.org/10.1144/GSL.SP.2000.167.01.13

Brewer, J.A. & Smythe, D.K. 1984. MOIST and the continuity of crustal reflector geometry along the Caledonian-Appalachian orogen. *Journal of the Geological Society, London*, **141**, 105–120, https://doi.org/10.1144/gsjgs.141.1.0105

Bruce, D.R.S. & Stemmerik, L. 2003. Carboniferous. *In*: Evans, D., Graham, C., Armour, A. & Bathurst, P. (eds) *The Millennium Atlas: Petroleum Geology of the Central and Northern North Sea*. Geological Society, London, 83–89.

Butler, M. 1998. The geological history of the Wessex basin: a review of new information from oil exploration. *In*: Underhill, J.R. (ed.) *Development and Evolution of the Wessex Basin*. Geological Society, London, Special Publications, **133**, 67–86.

Calamita, F., Patruno, S., Pomposo, G. & Tavarnelli, E. 2007. Geometria e cinematica delle anticlinali dell'Appenino centrale esterno: il ruolo delle faglie dirette giurassiche [Geometry and kinematics of the thrust-related anticlines from the central outer Apennines: the role of the Jurassic normal faults]. *Rendiconti Società Geologica Italiana, Nuova Serie*, **4**, 167–192.

Cameron, T.D.J. 1993. 5. Carboniferous and Devonian of the Southern North Sea. *In*: Knox, R.W.O'B. & Cordey, W.G. (eds) *Lithostrigraphic Nomenclature of the UK North Sea*. British Geological Survey (on behalf of the UK Offshore Operators Association), Keyworth.

Cameron, T.D.J., Crosby, A., Balson, P.S., Jefferey, D.H., Lott, G.K., Bulat, J. & Harrison, D.J. 1992. *The Geology of the Southern North Sea*. British Geological Survey, UK Offshore Regional Report. HMSO, London.

Cartwright, J., Huuse, M. & Aplin, A. 2007. Seal bypass systems. *American Association of Petroleum Geologists Bulletin*, **91**, 1141–1166.

Chadwick, R.A. & Evans, D.J. 1995. The timing and direction of Permo-Triassic extension in southern Britain. *In*: Boldy, S.A.R. (ed.) *Permian and Triassic Rifting in Northwest Europe*. Geological Society, London, Special Publications, **91**, 161–192, https://doi.org/10.1144/GSL.SP.1995.091.01.09

Cherry, S.T.J. 1993. The interaction of structure and sedimentary process controlling deposition of the Upper Jurassic Brae Formation Conglomerate, Block 16/17, North Sea. *In*: Parker, J.R. (ed.) *Petroleum Geology of Northwest Europe: Proceedings of the 4th Conference*. Geological Society, London, 387–400, https://doi.org/10.1144/0040387

Claringbould, J.S., Bell, R.E., *et al.* 2017. Pre-existing normal faults have limited control on the rift geometry of the northern North Sea. *Earth and Planetary Science Letters*, **475**, 190–206.

Corfield, S.M., Gawthorpe, R.L., Gage, M., Fraser, A.J. & Besley, A. 1996. Inversion tectonics of the Variscan foreland of the British Isles. *Journal of the Geological Society, London*, **153**, 17–32, https://doi.org/10.1144/gsjgs.153.1.0017

Cornford, C. 2009. Source rocks and hydrocarbons of the North Sea. *In*: Glennie, K.W. (ed.) *Petroleum Geology of the North Sea – Basic Concepts and Recent Advances*. 4th edn. Blackwell, Oxford, 376–462.

Coward, M.P. 1990. The Precambrian, Caledonian and Variscan framework to NW Europe. *In*: Brooks, J. & Hardman, R.F.P. (eds) *Tectonic Events Responsible for Britain's Oil and Gas Reserves*. Geological Society, London, Special Publications, **55**, 1–34, https://doi.org/10.1144/GSL.SP.1990.055.01.01

Coward, M.P., Enfield, M.A. & Fischer, M.W. 1989. Devonian basins of northern Scotland: extension and inversion related to Late Caledonian–Variscan tectonics. *In*: Cooper, M.A. & Williams, G.D. (eds) *Inversion Tectonics*. Geological Society, London, Special

Publications, **44**, 275–308, https://doi.org/10.1144/GSL.SP.1989.044.01.16

Coward, M.P., Dewey, J.F., Hempton, M. & Holroyd, J. 2003. Tectonic evolution. *In*: Evans, D., Graham, C., Armour, A. & Bathurst, P. (eds) *The Millennium Atlas: Petroleum Geology of the Central and Northern North Sea*. Geological Society, London, 17–33.

Cowie, P.A. & Scholz, C.H. 1992. Displacement–length scaling relationship for faults: data synthesis and discussion. *Journal of Structural Geology*, **14**, 1149–1156.

Cowie, P.A. & Roberts, G.P. 2001. Constraining slip rates and spacings for active normal faults. *Journal of Stratigraphic Geology*, **23**, 1901–1915.

Davies, R.J., O'Donnell, D., Bentham, P.N., Gibson, J.P.C., Curry, M.R., Dunay, R.E. & Maynard, J.R. 1999. The origin and genesis of major Jurassic unconformities within the triple junction area of the North Sea, UK. *In*: Fleet, A.J. & Boldy, S.A.R. (eds) *Petroleum Geology of Northwestern Europe: Proceedings of the 5th Conference*. Geological Society, London, 117–131, https://doi.org/10.1144/0050117

De Paola, N., Holdsworth, R.E., McCaffrey, K.J. & Barchi, M.R. 2005. Partitioned transtension: an alternative to basin inversion models. *Journal of Structural Geology*, **27**, 607–625.

Doglioni, C. & Bosellini, A. 1987. Eoalpine and mesoalpine tectonics in the Southern Alps. *Geologische Rundschau*, **76**, 735–754.

Duncan, W.I. & Buxton, W.K. 1995. New evidence for evaporitic Middle Devonian lacustrine sediments with hydrocarbon source potential on the East Shetland Platform, North Sea. *Journal of the Geological Society, London*, **152**, 251–258, https://doi.org/10.1144/gsjgs.152.2.0251

Edwards, C.W. 1991. The Buchan Field, Blocks 20/5a and 21/1a, UK North Sea. *In*: Abbotts, I.L. (ed.) *United Kingdom Oil and Gas Fields, 25 Years Commemorative Volume*, Geological Society, London, Memoirs, **14**, 253–259, https://doi.org/10.1144/GSL.MEM.1991.014.01.31

Esso 2009. *Relinquishment Report for Licence P1259*. Esso Exploration and Production UK, London.

Fazlikhani, H., Fossen, H., Gawthorpe, R.L., Faleide, J.I. & Bell, R.E. 2017. Basement structure and its influence on the structural configuration of the northern North Sea rift. *Tectonics*, **36**, https://doi.org/10.1002/2017TC004514

Færseth, R.B. 1996. Interaction of Permo-Triassic and Jurassic extensional fault-blocks during the development of the northern North Sea. *Journal of the Geological Society, London*, **153**, 931–944, https://doi.org/10.1144/gsjgs.153.6.0931

Færseth, R.B. & Pederstad, K. 1988. Regional sedimentology and petroleum geology of marine, late Bathonian–Valanginian sandstone in the North Sea. *Marine and Petroleum Geology*, **5**, 17–33.

Fletcher, K.J. 2003. The Central Brae Field, Blocks 16/07a, 16/07b, UK North Sea. *In*: Gluyas, J.G. & Hichens, H.M. (eds) *United Kingdom Oil and Gas Fields Commemorative Millennium Volume*. Geological Society, London, Memoirs, **20**, 183–190.

Fraser, A.J. & Gawthorpe, R.L. 1990. Tectono-stratigraphic development and hydrocarbon habitat of the Carboniferous in Northern England. *In*: Hardman, R.P.F. & Brooks, J. (eds) *Tectonic Events Responsible for Britain's Oil and Gas Reserves*. Geological Society, London, Special Publications, **55**, 49–86, https://doi.org/10.1144/GSL.SP.1990.055.01.03

Fraser, S., Robinson, A., Johnson, H., Underhill, A. & Kadolsky, D. 2003. Upper Jurassic. *In*: Evans, D., Graham, C., Armour, A. & Bathurst, P. (eds) *The Millennium Atlas: Petroleum Geology of the Central and Northern North Sea*. Geological Society, London, 157–189.

Freer, G., Hurst, A. & Middleton, P. 1996. Upper Jurassic sandstone reservoir quality and distribution on the Fladen Ground Spur. *In*: Hurst, A., Johnson, H.D., Burley, S.D., Canham, A.C. & MacKertich, D.S. (eds) *Geology of the Humber Group: Central Graben and Moray Firth, UKCS*. Geological Society, London, Special Publications, **114**, 235–249, https://doi.org/10.1144/GSL.SP.1996.114.01.11

Fyfe, J.A., Gregersen, U. *et al.* 2003. Oligocene to Holocene. *In*: Evans, D., Graham, C., Armour, A. & Bathurst, P. (eds) *The Millenium Atlas: Petroleum Geology of the Central and Northern North Sea*. Geological Society, London, 279–287.

Gautier, D.L. 2005. *Kimmeridgian Shales Total Petroleum System of the North Sea Graben Province*. US Geological Survey, Bulletin, **2204-C**.

Geluk, M.C. 1999. Late Permian (Zechstein) rifting in the Netherlands: models and implications for petroleum geology. *Petroleum Geoscience*, **5**, 189–199, https://doi.org/10.1144/petgeo.5.2.189

Geluk, M.C. 2005. *Stratigraphy and tectonics of Permo-Triassic basins in the Netherland and surrounding areas*. PhD thesis, Utrecht University.

Glennie, K.W. 1995. Permian and Triassic rifting in northwest Europe. *In*: Boldy, S.A.R. (ed.) *Permian and Triassic Rifting in Northwest Europe*. Geological Society, London, Special Publications, **91**, 1–5, https://doi.org/10.1144/GSL.SP.1995.091.01.01

Glennie, K.W. & Underhill, J.R. 1998. Origin, development and evolution of structural styles. *In*: Glennie, K.W. (ed.) *Petroleum Geology of the North Sea: Basic Concepts and Recent Advances*. 4th edn., Blackwell Science, Oxford, 42–84.

Glennie, K.W., Higham, J. & Stemmerik, L. 2003. Permian. *In*: Evans, D., Graham, C., Armour, A. & Bathurst, P. (eds) *The Millennium Atlas: Petroleum Geology of the Central and Northern North Sea*. Geological Society, London, 91–103.

Goldsmith, P.J., Hudson, G. & Van Veen, P. 2003. Triassic. *In*: Evans, D., Graham, C., Armour, A. & Bathurst, P. (eds) *The Millennium Atlas: Petroleum Geology of the Central and Northern North Sea*. Geological Society, London, 105–127.

Gradstein, F.M., Ogg, J.G., Schmitz, M. & Ogg, G. (eds) 2012. *The Geologic Time Scale 2012*. Elsevier.

Graversen, O. 2006. The Jurassic–Cretaceous North Sea Rift Dome and associated basin evolution. Search and Discovery Article #30040.

Gray, J. 2014. *Stratigraphic Plays of the UKCS*. UK Department of Energy & Climate Change, Online Report, www.og.decc.gov.uk/UKpromote/posters/Strat_Plays_of_UKCS.pdf [last accessed 21 September 2017].

Habicht, J.K.A. 1979. *Palaeoclimate, Palaeomagnetism and Continental Drift*. American Association of Petroleum Geologists, Studies in Geology, **9**, Tulsa, USA.

HARKER, S.D. & RIEUF, M. 1996. Genetic stratigraphy and sandstone distribution of the Moray Firth Humber Group (Upper Jurassic). *In*: HURST, A., JOHNSON, H.D., BURLEY, S.D., CANHAM, A.C. & MACKERTICH, D.S. (eds) *Geology of the Humber Group: Central Graben and Moray Firth, UKCS*. Geological Society, London, Special Publications, **114**, 109–130, https://doi.org/10.1144/GSL.SP.1996.114.01.05

HAY, S., JONES, C.M., BARKER, F. & HE, Z. 2005. *Exploration of Unproven Plays: Mid North Sea High*. PGL Report for EUPP Mid North Sea High Consortium. PGL, Aberdeen.

HENDRIE, D.B., KUSZNIR, N.J. & HUNTER, R.H. 1993. Jurassic extension estimates for the North Sea 'triple junction' from flexural backstripping: implications for decompression melting models. *Earth and Planetary Science Letters*, **116**, 113–127.

HILLIS, R.R., THOMSON, K. & UNDERHILL, J.R. 1994. Quantification of Tertiary erosion in the Inner Moray Firth by sonic velocity data from the Chalk and Kimmeridge Clay. *Marine and Petroleum Geology*, **11**, 283–293.

HOSPERS, J. & EDIRIWEERA, K.K. 1991. Depth and configuration of the crystalline basement in the Viking Graben area, Northern North Sea. *Journal of the Geological Society, London*, **148**, 261–265, https://doi.org/10.1144/gsjgs.148.2.0261

HUSMO, T., HAMAR, G., HØILAND, O., JOHANNESSEN, E.P., RØMULUND, A., SPENCER, A. & TITTERTON, R. 2003. Lower and Middle Jurassic. *In*: EVANS, D., GRAHAM, C., ARMOUR, A. & BATHURST, P. (eds) *The Millennium Atlas: Petroleum Geology of the Central and Northern North Sea*. Geological Society, London, 129–155.

IHS EDIN 2017. https://my.ihs.com/energy [last accessed January 2017].

JEREMIAH, J.M. 2000. Lower Cretaceous turbidites of the Moray Firth: sequence stratigraphical framework and reservoir distribution. *Petroleum Geoscience*, **6**, 309–328, https://doi.org/10.1144/petgeo.6.4.309

JOHNSON, H., WARRINGTON, G. & STOKER, S.J. 1994. Permian and Triassic of the Southern North Sea. *In*: KNOX, R.W.O'B. & CORDEY, W.G. (eds) *Lithostrigraphic Nomenclature of the UK North Sea*. British Geological Survey (on behalf of the UK Offshore Operators Association), Keyworth.

JOHNSON, H., LESLIE, A.B., WILSON, C.K., ANDREWS, I.J. & COOPER, R.M. 2005. *Middle Jurassic, Upper Jurassic and Lower Cretaceous of the UK Central and Northern North Sea*. British Geological Survey Research Report **RR/03/001, 42**. British Geological Survey, Keyworth.

JOHNSON, R.J. & DINGWALL, R.G. 1981. The Caledonides: their influence on the stratigraphy of the northwest European Shelf. *In*: ILLING, L.V. & HOBSON, G.P. (eds) *Petroleum Geology of the Continental Shelf of North-West Europe*. Heyden, London, 88–97.

JONES, E., JONES, R. *ET AL*. 2003. Eocene. *In*: EVANS, D., GRAHAM, C., ARMOUR, A. & BATHURST, P. (eds) *The Millenium Atlas: Petroleum Geology of the Central and Northern North Sea*. Geological Society, London, 261–277.

KARNIN, W.D., IDIZ, E., MERKEL, D. & RUPRECHT, E. 1996. The Zechstein Strassfurt carbonate hydrocarbon system of the Thuringian Basin, Germany. *Petroleum Geoscience*, **2**, 53–58, https://doi.org/10.1144/petgeo.2.1.53

KARSTENS, J. 2015. *Focused fluid conduits in the Southern Viking Graben and their implications for the Sleipner CO_2 storage project*. PhD thesis, Christian-Albrechts-University.

KARSTENS, J. & BERNDT, C. 2015. Seismic chimneys in the Southern Viking Graben – implications for palaeo fluid migration and overpressure evolution. *Earth and Planetary Science Letters*, **412**, 88–100.

KEARSEY, T., ELLEN, R., MILLWARD, D. & MONAGHAN, A.A. 2015. *Devonian and Carboniferous Stratigraphical Correlation and Interpretations in the Central North Sea, Quadrants 25-44, 21CXRM Palaeozoic Project*. British Geological Survey, Energy and Marine Geoscience Programme Commissioned Report **CR/15/117**, Draft 0.1. British Geological Survey, Keyworth.

KIM, Y.S. & SANDERSON, D.J. 2005. The relationship between displacement and length of faults: a review. *Earth Science Reviews*, **68**, 317–334.

KOMBRINK, H., BESLY, B.M. *ET AL*. 2010. Carboniferous. *In*: DOORNENBAL, J.C. & STEVENSON, A.G. (eds) *Petroleum Geological Atlas of the Southern Permian Basin Area*. European Association of Petroleum Geologists, Houten, 81–99.

KUBALA, M., BASTOW, M., THOMPSON, S., SCOTCHMAN, I. & OYGARD, K. 2003. Geothermal regime, petroleum generation and migration. *In*: EVANS, D., GRAHAM, C., ARMOUR, A. & BATHURST, P. (eds) *The Millennium Atlas: Petroleum Geology of the Central and Northern North Sea*. Geological Society, London, 289–315.

LEEDER, M.R. & BOLDY, S.R. 1990. The Carboniferous of the outer Moray Firth basin, Quadrants 14 and 15, Central North Sea. *Marine and Petroleum Geology*, **7**, 30–37.

LEEDER, M.R. & HARDMAN, M. 1990. Carboniferous geology of the Southern North Sea Basin and controls on hydrocarbon prospectivity. *In*: HARDMAN, R.F.P. & BROOKS, J. (eds) *Tectonic Events Responsible for Britain's Oil and Gas Reserves*. Geological Society, London, Special Publications, **55**, 87–105, https://doi.org/10.1144/GSL.SP.1990.055.01.04122S

LESLIE, A.G., MILLWARD, D., PHARAOH, T., MONAGHAN, A., ARSENIKOS, A. & QUINN, M. 2015. *Tectonic Synthesis and Contextual Setting for the Central North Sea and Adjacent Onshore Areas, 21CXRM Palaeozoic Project*. British Geological Survey Commissioned Report **CR/15/125**. British Geological Survey, Keyworth.

LORENZ, V. & NICHOLLS, L.A. 1984. Plate and intra-plate processes of Hercynian Europe during the Late Paleozoic. *Tectonophysics*, **107**, 25–56.

MARK, D.F., GREEN, P.F., PARNELL, J., KELLEY, S.P., LEE, M.R. & SHERLOCK, S.C. 2008. Late Paleozoic hydrocarbon migration through the Clair field, west of Shetland, UK Atlantic margin. *Geochimica et Cosmochimica Acta*, **72**, 2510–2533.

MARSHALL, J.A.E. 1998. The recognition of multiple hydrocarbon generation episodes: an example from Devonian lacustrine sedimentary rocks in the Inner Moray Firth, Scotland. *Journal of the Geological Society, London*, **155**, 335–352, https://doi.org/10.1144/gsjgs.155.2.0335

MARSHALL, J.E.A. & HEWETT, A.J. 2003. Devonian. *In*: EVANS, D., GRAHAM, C., ARMOUR, A. & BATHURST, P. (eds) *The Millennium Atlas: Petroleum Geology of the*

Central and Northern North Sea. Geological Society, London, 65–81.

Marshall, J.E.A., Rogers, D.A. & Whitely, M.J. 1996. Devonian marine incursions into the Orcadian Basin, Scotland. *Journal of the Geological Society, London*, **153**, 451–466, https://doi.org/10.1144/gsjgs.153.3.0451

McClay, K.R., Norton, M.G., Coney, P. & Davis, G.H. 1986. Collapse of the Caledonian orogen and the Old Red Sandstone. *Nature*, **323**, 147–149.

Mendum, J. & Noble, S.R. 2010. Mid-Devonian sinistral transpressional movements on the Great Glen Fault: the rise of the Rosemarkie Inlier and the Acadian Event in Scotland. *In*: Law, R.D., Butler, R.W.H., Holdsworth, R.E., Krabbendam, M. & Strachan, R.A. (eds) *Continental Tectonics and Mountain Building: The Legacy of Peach and Horne*. Geological Society, London, Special Publications, **335**, 161–187, https://doi.org/10.1144/SP335.8

Monaghan, A.A., Arsenikos, S. *et al.* 2015. *Paleozoic Petroleum Systems of the Central North Sea/Mid North Sea High, 21CXRM Palaeozoic Project*. British Geological Survey, Energy and Marine Geoscience Programme Commissioned Report **CR/15/124**, Draft 0.1. British Geological Survey, Keyworth, http://nora.nerc.ac.uk/516766/

Monaghan, A.A., Johnson, K. *et al.* 2016. *Palaeozoic Petroleum Systems of the Orcadian Basin to Forth Approaches, Quadrants 6 - 21, UK*. British Geological Survey, Energy and Marine Geoscience Programme Commissioned Report **CR/16/038N** British Geological Survey, Keyworth, http://nora.nerc.ac.uk/516781/

Monaghan, A.A., Arsenikos, S. *et al.* 2017. Carboniferous petroleum systems around the Mid North Sea High, UK. *Marine and Petroleum Geology*, **88**, 282–302.

Mudge, D.C. & Copestake, P. 1992. Lower Palaeogene stratigraphy of the northern North Sea. *Marine and Petroleum Geology*, **9**, 287–301.

Løseth, H., Wensaas, L., Arntsen, B., Hanken, N.-M., Basire, C. & Graue, K. 2011. 1000 m long gas blow-out pipes. *Marine and Petroleum Geology*, **28**, 1047–1060.

Partington, M.A., Mitchener, B.C., Milton, N.J. & Fraser, A.J. 1993. Genetic sequence stratigraphy for the North Sea Late Jurassic and Early Cretaceous: distribution and prediction of Kimmeridgian–Late Ryazanian reservoirs in the North Sea and adjacent areas. *In*: Parker, J.R. (ed.) *Petroleum Geology of Northwest Europe: Proceedings of the 4th Conference*. Geological Society, London, 347–370, https://doi.org/10.1144/0040347

Patruno, S. & Reid, W. 2016*a*. New plays on the Greater East Shetland Platform (UKCS Quadrants 3, 8-9, 14-16) – Part 1: Regional setting and a working petroleum system. *First Break*, **34**, 33–45.

Patruno, S. & Reid, W. 2016*b*. An introduction to the plays of the East Shetland Platform and Mid North Sea High (UK 29th Frontier Licensing Round). GeoExpro, **13**, June 2016, www.geoexpro.com/articles/2016/07/the-east-shetland-platform-and-mid-north-sea-high

Patruno, S. & Reid, W. 2016*c*. Chronostratigraphic panels for the PESGB structural framework of the North Sea and Atlantic margin map, 2017 version. Poster published by the *Petroleum Exploration Society of Great Britain*, www.researchgate.net/publication/311317399_North_Sea_and_West_of_Shetlands_chronostratigraphic_framework_from_the_PESGB_Structural_Framework_of_the_North_Sea_and_Atlantic_Margin_-_2017_Edition

Patruno, S. 2017. A new look at the geology and prospectivity of a North Sea frontier area with modern seismic: the East Shetland Platform. *In*: *Oral Presentation Session: Exploration Plays, Prospects and Prospects Evaluation II*. 79th EAGE Conference and Exhibition 2017, 15 June 2017, Paris, France.

Patruno, S. & Reid, W. 2017. New plays on the Greater East Shetland Platform (UKCS Quadrants 3, 8-9, 14-16) – Part 2: Newly reported Permo-Triassic intra-platform basins and their influence on the Devonian–Paleogene prospectivity of the area. *First Break*, **35**, 59–69.

Patruno, S. & Reid, W. 2018. Complex multiphase inversion tectonics in the southern East Shetland Platform, offshore United Kingdom. *In*: Misra, A.A. & Mukherjee, S. (eds) *Atlas of Structural Geological Interpretation from Seismic Images*. Wiley-Blackwell, Oxford, 73–76.

Patruno, S., Hampson, G.J. & Jackson, C.A-L. 2015*a*. Quantitative characterisation of deltaic and subaqueous clinoforms. *Earth-Science Reviews*, **142**, 79–119.

Patruno, S., Hampson, G.J., Jackson, C.A-L. & Dreyer, T. 2015*b*. Clinoform geometry, geomorphology, facies character and stratigraphic architecture of ancient sand-prone subaqueous delta: Upper Jurassic Sognefjord Formation, Troll Field, Offshore Norway. *Sedimentology*, **62**, 350–388.

Patruno, S., Hampson, G.J., Jackson, C.A-L. & Whipp, P.S. 2015*c*. Quantitative progradation dynamics and stratigraphic architecture of ancient shallow-marine clinoform sets: a new method and its application to the Upper Jurassic Sognefjord Formation, Troll Field, offshore Norway. *Basin Research*, **27**, 412–452.

Patruno, S., Triantaphyllou, M.V., Erba, E., Dimiza, M.D., Bottini, & Kaminski, M.A. 2015*d*. The Barremian and Aptian stepwise development of the 'Oceanic Anoxic Event 1a' (OAE 1a) crisis: integrated benthic and planktic high-resolution palaeoecology along the Gorgo a Cerbara stratotype section (Umbria-Marche Basin, Italy). *Palaeogeography, Palaeocology, Palaeoecology*, **424**, 147–182.

Patruno, S. & Lampart, V. 2018. Newly-observed post-Variscan extensional mini-basins: the key to the prospectivity of the under-explored platform areas of the North Sea? *Paper to be presented at the 80th EAGE Conference and Exhibition 2018*, 11–14 June 2018, Copenhagen, Denmark.

Patruno, S., Reid, W., Jackson, C.A. & Davies, C. 2018. New insights into the unexploited reservoir potential of the Mid North Sea High (UKCS Quadrants 35-38, 41-43): a newly-described intra Zechstein sulphate-carbonate platform complex. *In*: Bowman, M. & Levell, B. (eds) *Petroleum Geology of Northwest Europe: 50 Years of Learning – Proceedings of the 8th Petroleum Geology Conference*. Geological Society, London, 87–124, https://doi.org/10.1144/PGC8.9 http://pgc.lyellcollection.org/content/early/2017/06/07/PGC8.9.full.pdf+html

Pegrum, R.M. & Spencer, A.M. 1990. Hydrocarbon plays in the northern North Sea. *In*: Brooks, J. (ed.) *Classic Petroleum Provinces*. Geological Society, London,

Special Publications, **50**, 441–470, https://doi.org/10.1144/GSL.SP.1990.050.01.27

Peryt, T.M., Geluk, M., Mathiesen, A., Paul, A. & Smith, K. 2010. Zechstein. *In*: Doornebal, H. & Stevenson, A. (eds) *Petroleum Geological Atlas of the Southern Permian Basin Area*. European Association of Petroleum Geologists, Houten, 123–147.

Pharaoh, T., Dusar, M. *et al.* 2010. Tectonic evolution. *In*: Doornenbal, J.C. & Stevenson, A.G. (eds) *Petroleum Geological Atlas of the Southern Permian Basin Area*. European Association of Petroleum Geologists, Houten, 25–57.

Phillips, T., Jackson, C.A.L., Bell, R., Duffy, O.B. & Fossen, H. 2016. Reactivation of intrabasement structures during rifting: a case study from offshore southern Norway. *Journal of Structural Geology*, **91**, 54–73.

PGS. 2017. *East Shetland Platform OGA 2D Interpretation Workflow Report*. PGS Reservoir (for UK Oil & Gas Authority), March **2017/16.1931**, available online at: www.ukoilandgasdata.com/dp/pages/apptab/ITabManager.jsp

Platt, N.H. 1995. Structure and tectonics of the northern North Sea: new insights from deep penetration regifonal seismic data. *In*: Lambiase, J.J. (ed.) *Hydrocarbon Habitats in Rift Basins*. Geological Society, London, Special Publications, **80**, 103–113, https://doi.org/10.1144/GSL.SP.1995.080.01.05

Platt, N.H. & Cartwright, J.A. 1998. Structure of the East Shetland Platform, northern North Sea. *Petroleum Geoscience*, **4**, 353–362, https://doi.org/10.1144/petgeo.4.4.353

Plaza-Faverola, A., Westbrook, G.K., Ker, S., Exley, R.J.K., Gailler, A., Minshull, T.A. & Broto, K. 2010. Evidence from three-dimensional seismic tomography for a substantial accumulation of gas hydrate in a fluid-escape chimney in the Nyegga pockmark field, offshore Norway. *Journal of Geophysical Research*, **115**, http://doi.org/10.1029/2009JB007078

Plaza-Faverola, A., Bünz, S. & Mienert, J. 2011. Repeated fluid expulsion through sub-seabed chimneys offshore Norway in response to glacial cycles. *Earth and Planetary Science Letters*, **305**, 297–308.

Rattey, R.P. & Hayward, A.B. 1993. Sequence stratigraphy of a failed rift system: the Middle Jurassic to Early Cretaceous basin evolution of the Central and Northern North Sea. *In*: Parker, J.R. (ed.) *Petroleum Geology of Northwest Europe: Proceedings of the 4th Conference*. Geological Society, London, 215–249, https://doi.org/10.1144/0040215

Reid, W. & Patruno, S. 2015. The East Shetland Platform: unlocking the platform potential. With significant advancements in seismic acquisition technology, it is time to re-visit the East Shetland Platform. *GeoExpro*, **12**, 41–46.

Richardson, N.J., Allen, M.R. & Underhill, J.R. 2005. Role of Cenozoic fault reactivation in controlling pre-rift plays, and the recognition of Zechstein Group evaporite–carbonate lateral facies transitions in the East Orkney and Dutch Bank basins, East Shetland Platform, UK North Sea. *In*: Doré, A.G. & Vining, B.A. (eds) *Petroleum Geology: North-West Europe and Global Perspectives – Proceedings of the 6th Petroleum Geology Conference*. Geological Society, London, 337–348, https://doi.org/10.1144/0060337

Richter-Bernburg, G. 1955*a*. Über salinare Sedimentation. *Zeitschrift der Deutschen Geologischen Gesellschaft*, **105**, 593–645.

Richter-Bernburg, G. 1955*b*. Stratigraphische Gliederung des deutschen Zechsteins. *Zeitschrift der Deutschen Geologischen Gesellschaft*, **105**, 843–854.

Roberts, A.M., Yielding, G., Kuznir, N.J., Walker, & Land Dorn Lopez, D. 1995. Quantitative analysis of Triassic extension in the northern Viking Graben. 1. *Journal of the Geological Society, London*, **152**, 15–26, https://doi.org/10.1144/gsjgs.152.1.0015

Scheck, M., Bayer, U. & Lewerenz, B. 2003. Salt redistribution during extension and inversion inferred from 3D backstripping. *Tectonophysics*, **373**, 55–73.

Schmid, S.M., Bernoulli, D. *et al.* 2008. The Alpine-Carpathian-Dinaridic orogenic system: correlation and evolution of tectonic units. *Swiss Journal of Geosciences*, **101**, 139–183.

Scisciani, V., Patruno, S., Tavarnelli, E., Calamita, F., Pace, & Iacopini, D. In press. Modes of reactivation of Late Paleozoic–Mesozoic extensional basins during the Wilson Cycle: case studies from the North Sea (UK) and central Apennines (Italy). *In*: Wilson, R.W., Houseman, G.A., McCaffrey, K.J.W., Dore, A.G. & Buiter, S.J.H. (eds) *Tectonic Evolution: 50 Years of the Wilson Cycle Concept*. Geological Society, London, Special Publications, **470**.

Seranne, M. 1992. Devonian extensional tectonics v. Carboniferous inversion in the northern Orcadian basin. *Journal of the Geological Society, London*, **149**, 27–37, https://doi:10.1144/gsjgs.149.1.0027

Sibuet, J.C., Srivastava, S.P. & Spakman, W. 2004. Pyrenean orogeny and plate kinematics. *Journal of Geophysical Research: Solid Earth*, **109**, B08104.

Skipper, K. 1977. Offshore petroleum developments in Northwest Europe – an update. *Geoscience Canada*, **4**, 31–40.

Sneider, J.S., de Clarens, P. & Vail, P.R. 1995. Sequence stratigraphy of the Middle to Upper Jurassic, Viking Graben, North Sea. *In*: Steel, R.J., Felt, V.L., Johannesson, E.P. & Mathieu, C. (eds.) *Sequence Stratigraphy on the Northwest European Margin*. Norwegian Petroleum Society, Special Publications, **5**. Elsevier, Amsterdam, 167–197.

Snyder, D.B., England, R.W. & McBride, J.R. 1997. Linkage between mantle and crustal structures and its bearing on inherited structures in northwestern Scotland. *Journal of the Geological Society, London*, **154**, 79–82.

Stampfli, G.M. & Borel, G.D. 2002. A plate tectonic model for the Paleozoic and Mesozoic constrained by dynamic plate boundaries and restored synthetic oceanic isochrons. *Earth and Planetary Science Letters*, **196**, 17–33.

Stampfli, G.M. & Borel, G.D. 2004. The TRASNSMED transects in space and time: constraints on the Paleotectonic evolution of the Mediterranean domain. *In*: Cavazza, W., Roure, F.M., Stampfli, G.M. & Ziegler, P.A. (eds) *The TRANSMED Atlas, The Mediterranean Region from Crust to Mantle*. Springer, Berlin, 52–80.

Statoil 2017. Johan Sverdrup—the North Sea giant, www.statoil.com/en/what-we-do/johan-sverdrup.html

Steel, R.J. 1993. Triassic–Jurassic megasequence stratigraphy in the Northern North Sea. rift to post-rift evolution. *In*: Parker, J.R. (ed.) *Petroleum Geology of North-West*

Europe. Proceedings of the 4th Conference. Geological Society, London, 299–315, https://doi.org/10.1144/0040299

Stemmerik, L., Ineson, J.R. & Mitchell, J.G. 2000. Stratigraphy of the Rotliegend group in the Danish part of the Northern Permian Basin, North Sea. *Journal of the Geological Society, London*, **157**, 1127–1136, https://doi.org/10.1144/jgs.157.6.1127

Stone, P. 2012. The demise of the Iapetus Ocean as recorded in the rocks of southern Scotland. *Journal of the Open University Geological Society*, **33**, 29–36.

Taylor, J.C.M. 1980. Origin of the Werraanhydrit in the U.K. Southern North Sea – a reappraisal. *In*: Fuchtbauer, H. & Peryt, T. (eds) *The Zechstein Basin with Emphasis on Carbonate Sequences. Contributions to Sedimentology*, **9**, 91–113.

Taylor, J.C.M. 1998. Upper Permian – Zechstein. *In*: Glennie, K.W. (ed). *Petroleum Geology of the North Sea: Basic Concepts and Recent Advances*, 4th edn. Blackwell Science, Oxford, 174–211.

Taylor, M.H., Dillon, W.P. & Pecher, I.A. 2000. Trapping and migration of methane associated with the gas hydrate stability zone at the Blake Ridge Diapir: new insights from seismic data. *Marine Geology*, **164**, 79–89.

Thomson, K. & Underhill, J.R. 1993. Controls on the development and evolution of structural styles in the Inner Moray Firth Basin. *In*: Parker, J.R. (ed.) *Petroleum Geology of Northwest Europe: Proceedings of the 4th Conference*. Geological Society, London, 1167–1178, https://doi.org/10.1144/0041167

Torsvik, T.H., Carlos, D., Mosar, J., Cocks, L.R.M. & Malme, T.N.M. 2002. Global reconstructions and North Atlantic paleogeography 440 Ma to Recent. *In*: Eide, E.A. (ed.) *BATLAS – Mid Norway Plate Reconstruction Atlas with Global and Atlantic Perspectives*. Geological Survey of Norway, Trondheim, 18–39.

Tucker, M.E. 1991. Sequence stratigraphy of carbonate–evaporite basins: models and application to the Upper Permian (Zechstein) of northeast England and adjoining North Sea. *Journal of the Geological Society, London*, **148**, 1019–1036.

Turner, C.C., Richards, P.C., Swallow, J.L. & Grimshaw, S.P. 1984. Upper Jurassic stratigraphy and sedimentary facies in the Central Outer Moray Firth Basin, North Sea. *Marine and Petroleum Geology*, **1**, 105–117.

Turner, C.C., Cronin, B.T. *et al.* 2018. The South Viking Graben: overview of Upper Jurassic rift geometry, biostratigraphy and extent of Brae Play submarine fan systems. *In*: Turner, C.C. & Cronin, B.T. (eds) *Rift-related coarse-grained submarine fan reservoirs; the Brae Play, South Viking Graben, North Sea*. American Association of Petroleum Geologists, Memoirs, **115**.

Unida, S. & Patruno, S. 2016. The palynostratigraphy of the upper Maiolica, Selli Level and the lower Marne a Fucoidi units in the proposed Barremian/Aptian (Lower Cretaceous) GSSP stratotype at Gorgo a Cerbara, Umbria-Marche Basin, Italy. *Palynology*, **40**, 230–246.

Underhill, J.R. 1991. Implications of Mesozoic–Recent basin development in the western Inner Moray Firth, UK. *Marine & Petroleum Geology*, **8**, 359–369.

Underhill, J.R. & Partington, M.A. 1993. Jurassic thermal doming and deflation in the North Sea: implication of the sequence stratigraphic evidence. *In*: Parker, J.R. (ed.) *Petroleum Geology of Northwest Europe: Proceedings of the 4th Conference*. Geological Society, London, 337–346, https://doi.org/10.1144/0040337

Underhill, J.R. & Partington, M.A. 1994. Use of maximum flooding surfaces in determining a regional control on the Intra-Aalenian Mid Cimmerian sequence boundary: implications of North Sea basin development and Exxon's sea-level chart. *In*: Posamentier, H.W. & Wiemer, P.J. (eds) *Recent Advances in Siliciclastic Sequence Stratigraphy*. American Association of Petroleum Geologists, Memoirs, **58**, 449–484.

Underhill, J.R., Monaghan, A. & Browne, M.A. 2008. Controls on structural styles, basin development and petroleum prospectivity in the Midland Valley of Scotland. *Marine and Petroleum Geology*, **25**, 1000–1022.

United Kingdom Discovery Digest. 1999. Vol. 2, Awards and Objectives 18th Round. Agip, British Borneo, Statoil: 8/19, 20, 25, 7–11, http://ukdigest.canadiandiscovery.com/?q=node/58247&flipbook_view=1

United Kingdom Discovery Digest. 2001. Vol. 5, Offshore Ocean Acreage Evaluations: the East Shetland Platform, 19–40, http://ukdigest.canadiandiscovery.com/?q=node/58244&flipbook_view=1

Van Wees, J.D., Stephenson, R.A. *et al.* 2000. On the origin of the Southern Permian Basin, Central Europe. *Marine and Petroleum Geology*, **17**, 43–59.

Vane, C.H., Uguna, C., Kim, A. & Monaghan, A.A. 2016. *Organic Geochemistry of Palaeozoic Source Rocks, Orcadian Study Area, North Sea, UK*. British Geological Survey Commissioned Report **CR/16/037**, http://nora.nerc.ac.uk/id/eprint/516754

Vincent, C.J. 2015. *Maturity Modelling of Selected Wells in the Central North Sea*. British Geological Survey Commissioned Report **CR/15/122**, http://nora.nerc.ac.uk/516764/

Waltham, D. 1997. Why does salt start to move? *Tectonophysics*, **282**, 117–128.

Whipp, P.S., Jackson, C.A-L., Gawthorpe, R.L., Dreyer, T. & Quinn, D. 2014. Normal fault array evolution above a reactivated rift fabric; a subsurface example from the northern Horda Platform, Norwegian North Sea. *Basin Research*, **26**, 523–549.

Whitehead, M. & Pinnock, S.J. 1991. The Highlander Field, Block 14/20b, UK North Sea. *In*: Abbotts, I.L. (ed.) *United Kingdom Oil and Gas Fields, 25 Years Commemorative Volume*. Geological Society, London, Memoirs, **14**, 323–329, https://doi.org/10.1144/GSL.MEM.1991.014.01.40

Wilson, R.W., Holdsworth, R.E., Wild, L.E., McCaffrey, K.J.W., England, R.W., Imber, J. & Strachan, R.A. 2010. Basement-influenced rifting and basin development: a reappraisal of post-Caledonian faulting patterns from the North Coast Transfer Zone, Scotland. *In*: Law, R.D., Butler, R.W.H., Holdsworth, R.E., Krabbendam, M. & Strachan, R.A. (eds) *Continental Tectonics and Mountain Building: The Legacy of Peach and Horne*. Geological Society, London, Special Publications, **335**, 795–826, https://doi.org/10.1144/SP335.32

Zanella, E. & Coward, M.P. 2003. Structural framework. *In*: Evans, D., Graham, C., Armour, A. & Bathurst, P. (eds) *The Millennium Atlas: Petroleum Geology of the*

Central and Northern North Sea. Geological Society, London, 42–59.

Ziegler, P.A. 1988. *Evolution of the Arctic–North Atlantic and the Western Tethys*. American Association of Petroleum Geologists, Memoirs, **43**.

Ziegler, P.A. 1989. North German facies patterns in relation to their substrate. *Geologische Rundschau*, **78**, 105–127.

Ziegler, P.A. 1990. *Geological Atlas of Western and Central Europe*. 2nd edn. Shell Internationale Petroleum, The Hague.

Ziegler, P.A. 1992. North Sea rift system. *Tectonophysics*, **208**, 55–75.

Ziegler, P.A., Schumacher, M.E., Dèzes, P., van Wees, J.D. & Cloetingh, S.A.P.L. 2004. Post-Variscan evolution of the lithosphere in the Rhine Graben area; constraints from subsidence modelling. *In*: Wilson, M., Neumann, E.R., Davies, G.R., Timmerman, M.J., Heeremans, M. & Larsen, B.T. (eds) *Permo-Carboniferous Magmatism and Rifting in Europe*. Geological Society, London, Special Publications, **223**, 289–317, https://doi.org/10.1144/GSL.SP.2004.223.01.13

Ziegler, P.A., Schumacher, M., Van Wees, J.-D.&, Cloetingh, S. 2006. Post-Variscan evolution of the lithosphere in the area of the European Cenozoic Rift System. *In*: Gee, D.G. & Stephenson, R.A. (eds) *European Lithosphere Dynamics*. Geological Society, London, Memoirs, **32**, 97–112, https://doi.org/10.1144/GSL.MEM.2006.032.01.06

The Old Red Group (Devonian) – Rotliegend Group (Permian) Unconformity in the Inner Moray Firth

J. E. A. MARSHALL[1]*, K. W. GLENNIE[2], T. R. ASTIN[3] & A. J. HEWETT[4]

[1]*Ocean & Earth Science, University of Southampton, National Oceanography Centre, European Way, Southampton, SO14 3ZH, UK*

[2]*4 Morven Way, Ballater, Aberdeenshire, AB35 5SF, UK*

[3]*The Vicarage, 1 The Avenue, Tadworth, Surrey, KT20 5AS, UK*

[4]*10 Murrells Walk, Great Bookham, Surrey, KT23 3LP, UK*

J.E.A.M., 0000-0002-9242-3646

**Correspondence: jeam@soton.ac.uk*

Abstract: A major stratigraphical problem in the offshore Paleozoic of the Inner Moray Firth is the identification of the top of the Devonian Old Red Sandstone Group beneath the lithologically similar Permian Rotliegend Group. Wireline log criteria for a revised Old Red Group to Permian boundary are given for the Inner Moray Firth. Section lines drawn using these criteria and flattened on the overlying Triassic Smith Bank or Permian Kupferschiefer formations show a relatively thin development of Rotliegend with two depocentres. The underlying Devonian when flattened on the Eday Marl shows a systematic subcrop pattern. There is currently no exposed onshore Old Red to Rotliegend boundary, but the possibility remains that a Permian section is present in the 'Upper Old Red Sandstone' of Tarbat in Easter Ross. An exposed Permian–Devonian boundary is present in East Greenland and provides an analogue for the Inner Moray Firth.

A major problem (e.g. Andrews *et al.* 1990) in producing stratigraphic breakdowns in the Paleozoic of the North Sea area is the separation of the Upper Old Red Group (Devonian, Old Red Sandstone) from the Permian Rotliegend Group (Findhorn Formation). Both intervals are sand rich, lithologically very similar and in contact at an unconformity surface. In terms of mineral composition and texture they are also very similar, with the Rotliegend Group being provenanced directly from the Old Red Group. This problem is not new, as there was always difficulty in recognizing this boundary (Peacock *et al.* 1968; Benton 1995; Diemer 1996) in the onshore successions of the southern Moray Firth. The boundary is poorly exposed and initially went unrecognized with the Permian succession placed within the Devonian based on fossil fish. Subsequently the discovery of both footprints and reptiles in what we now know to be Permian caused a major controversy (Benton 1995, 2018) with claims for them being either Mesozoic based on the fossils or Devonian based on their lithostratigraphic position. At the time this was highly significant because if they had proved to be Devonian then they would be the earliest known reptile fauna. The problem was solved (see discussion in Benton 1995) with the recognition of an unconformity in the quarry section at Cutties Hillock, near Elgin (Fig. 1; also known as Quarry Hill, but see Benton & Walker 1985 for an exhaustive discussion of the locality and its nomenclature/etymology). Here an excavation in *c.* 1882 (see Peacock *et al.* 1968) revealed a barely perceptible unconformity between sandstones with Permian reptiles and near identical sandstones with the Devonian fish *Holoptychius*.

This problem was also very evident in offshore wells drilled during the main phase of exploration (late 1970s to the mid 1990s) in the Inner Moray Firth. It was exacerbated by the requirement, from the then Department of Trade and Industry for all wells to reach economic basement (Devonian or below) before being permitted to TD. This gave rise to a situation where following the drilling of a dry hole the operator would want to TD in 'proven' Devonian as soon as practicably possible so as to release the rig. However, as this required the operator to pick a boundary in two sequences that were lithologically very similar and both devoid of palynomorphs, it led to many ambiguous and hurried well site determinations. These were based (pers. obs.) on mud logger experience, sandstone discrimination or the 'recognition' of fish scales in cuttings. Various laboratory analyses were also used, including clay mineral assemblages, heavy mineral suites and recognition of the barren palynological assemblage, i.e. *it's barren therefore it's Rotliegend* or equally

From: Monaghan, A. A., Underhill, J. R., Hewett, A. J. & Marshall, J. E. A. (eds) 2018. *Paleozoic Plays of NW Europe*. Geological Society, London, Special Publications, **471**, 237–252.
First published online June 22, 2018, https://doi.org/10.1144/SP471.12

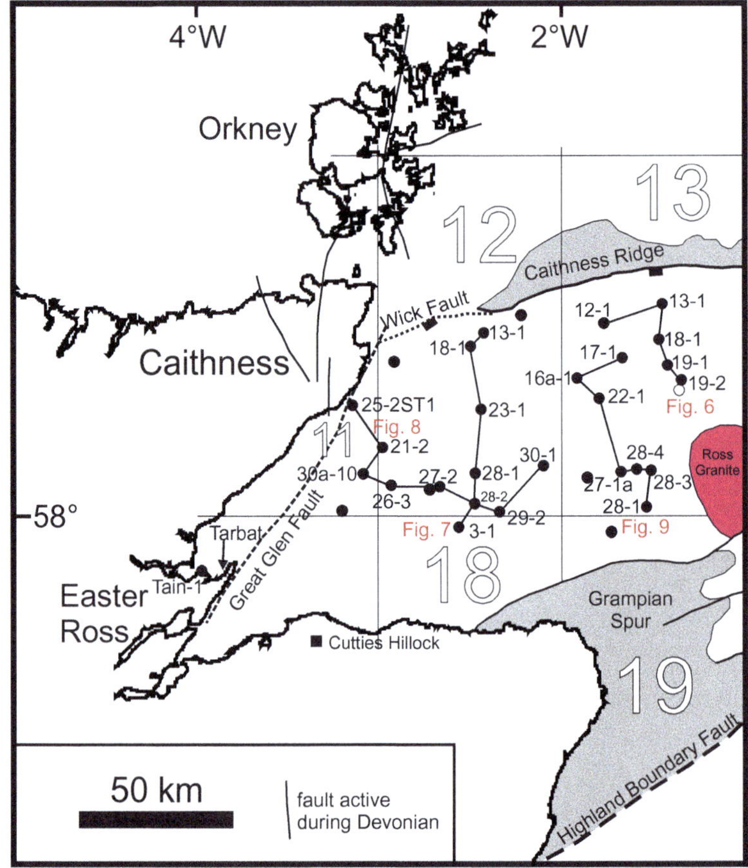

Fig. 1. Map of the Inner Moray Firth showing location of wells with an Old Red–Rotliegend Group contact and the reinterpreted section lines in Figures 6–9. Also shown in the location of Cutties Hillock and the uppermost Old Red Sandstone section on Tarbat.

it's barren therefore it must be Devonian. Many of these analyses were not based on robust criteria since, at that time, the clay or heavy mineral assemblages that were typical for the Rotliegend or Devonian were not well understood. In addition, mass spectrometers that were equipped with a laser ablation system suitable for age dating single grains of heavy minerals were highly specialized instruments and certainly not available for commercial analyses. In contrast, heavy mineral studies are now a sophisticated tool that is routinely applied (e.g. Schmidt *et al.* 2012; Lundmark *et al.* 2013) for subsurface sandstone discrimination. These operationally determined top Devonian picks were then added, without any expression of confidence, review or quality audit to the composite logs and have become the referenced stratigraphic ties and enshrined in a variety of data bases and compilations.

During the technical work for the Millennium Atlas (Marshall & Hewett 2003), all of the Devonian well sections in the North Sea were reviewed and stratigraphic assignments tested with focused palynological sampling and analysis. It then became evident that there were significant inconsistencies in lithostratigraphical correlation between well sections in the upper part of the Devonian interval. This then led us to review and compare all of the sub-Kupferschiefer well sections. To this end we were particularly aided by the advances in knowledge on the onshore stratigraphy (Fig. 2) of the upper parts of the Orcadian Basin (Astin 1985; Marshall 1996) and had completed (Marshall *et al.* 1996) or were compiling studies of the Eday Marl and correlatives (Marshall *et al.* 2011) in Orkney, Easter Ross and the Moray Firth.

This review showed that the Findhorn Siltstone (Cameron 1993) attributed to the Permian was, in fact, the Devonian Eday Marl and that it occurred widely across the Moray Firth in addition to onshore Orkney. This identification was based on its

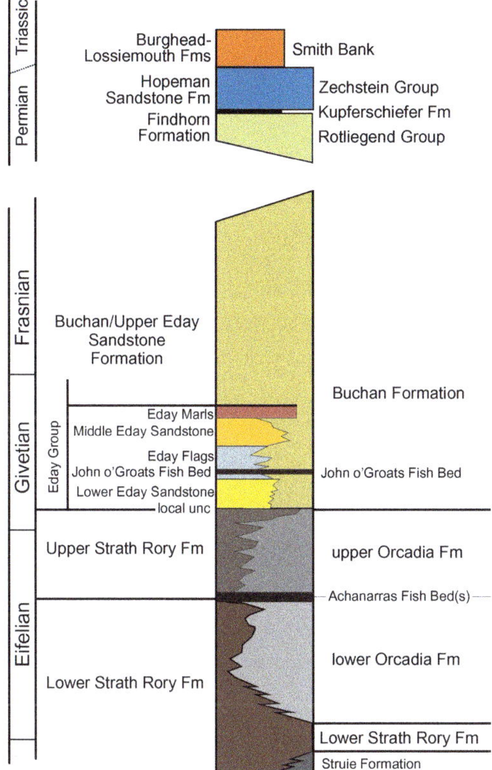

Fig. 2. Onshore–offshore lithostratigraphic scheme for the Old Red Group, Permian and lowermost Triassic of the Inner Moray Firth area. Modified from Marshall & Hewett (2003) with the addition of selected onshore correlative stratigraphy. This scheme has not been unified and creates some ambiguities as the Upper Eday Sandstone is part of the Eday Group and also a correlative of the offshore Buchan Formation.

distinctive lithology as an interval of homogenous red calcareous mudstone, its position within a well understood succession (Fig. 2) of Lower to Upper Eday Sandstone, which could be palynologically dated (Eday Flagstone Formation) and correlated directly to onshore sequences. This assignment was then tested using palynology on selected cuttings samples. To acquire the palynological samples complete sets of 10 foot (3 m) cuttings were scanned by eye or with a hand lens and any fine-grained dark-coloured fragments were picked out using tweezers. This slowly built up enough material for a palynological sample of maybe 1 or 2 g. Essential to this process were complete sets of cuttings from key wells donated by oil companies where there was sufficient volume of material to generate palynological samples that were adequate in size. In contrast, the cuttings samples in the national archive have been subsampled many times with little material left. When insufficient material was left, the 10 ft (3 m) cuttings samples were composited across a greater depth interval. Palynological processing was by standard methods but on a small scale using small beakers and small-diameter sieves to conserve the residue. An initial treatment of of 30% HCl was followed by 60% HF with decant washing after each stage. Following sieving at 15 μm they were then briefly boiled in 30% HCl to remove neoformed fluorides with a final sieving into a storage vial before mounting as strew slides in Elvacite 2044. The resulting residues were very lean with few spores and a high level of contamination by both cavings and extraneous material originating from the cuttings samples.

The impact of this revision is shown in Figure 3 with two wells (18/3-1 and 12/27-2) that are 15 km apart. Well 12/27-2 (Fig. 3a) was reported with a thick sequence of Rotliegend Group, including a Middle Rotliegend Conglomerate Formation, and this was overlying a Lower Devonian sequence that included drab-coloured mudstone that was presumably the rationale for the top Devonian boundary pick. Well 18/3-1 (Fig. 3b) had a not dissimilar Rotliegend sequence that overlaid a thin Middle Devonian shale that was above a thick Middle Devonian conglomerate. Palynological assemblages, in conjunction with log correlation, show that this is a Middle Devonian shale in 18/3-1 (Fig. 3b) and the Achanarras Fish Bed equivalent. The same Achanarras level can be recognized in 12/27-2 (Fig. 3b) where it occurs immediately above the 'Middle Rotliegend Conglomerate Formation'. It can now be seen that the two conglomerates identified as separate Permian and Devonian conglomerates are, in fact, the same conglomerate and both are Eifelian in age. This stratigraphic relationship of a lake (e.g. the Achanarras Fish Bed) immediately overlying a conglomerate (e.g. the Lower Strath Rory Formation) is common in the Orcadian and East Greenland Basins (e.g. Astin 1985; Marshall & Stephenson 1997; Marshall & Fletcher 2002). Following this reinterpretation the base Rotliegend was re-picked at a higher stratigraphic level in both well sections using the consistent set of criteria outlined below. This pick was then confirmed palynologically as the Givetian spore *Geminospora lemurata* (Table 1, Fig. 4b) was recovered from the cuttings sample in 12/27-2 at 2458-2467 m, which is immediately below the gamma ray spike at 2453 m that was picked as the John o'Groats Fish Bed.

Picking the Old Red–Rotliegend Group boundary

The recognition that the Findhorn Siltstone was the Eday Marl has moved the position of the Old Red–

Rotliegend Group boundary significantly up section and made the latter a much thinner interval. As outlined in the Millennium Atlas and evidenced here by the 13/19-1, 12/26-3 and 12/29-2 well correlation (Fig. 5), the boundary was then picked using the following criteria:

- The interval immediately beneath the Kupferschiefer and attributed to the Rotliegend has a consistently flat gamma ray log character with only a gentle ripple in the value.
- This Rotliegend interval has a small but consistent increase in interval transit time (*c.* 5 µs/foot).
- This interval is distinct from the underlying Buchan Formation which has the continuous presence of small gamma ray spikes some 10–20 API above baseline together with rare single or groups of larger magnitude spikes.
- At the interface between the two formations there is often a high-velocity feature (at × on Fig. 5) present in the sonic log of the Buchan Formation and presumed to represent a diagenetically cemented residual weathering surface. This is particularly well defined in 12/13-1, where a compact microcrystalline limestone was identified at the boundary. However, the top of the Buchan Formation can more rarely be a low-velocity feature (e.g. 11/30-A10) that is presumed to represent a less compact weathered zone.

Table 1. *List of palynomorphs recovered in two Buchan Formation samples from wells 12/27-2 and 12/29-2*

Spore	12/29-2	12/27-2
Depth	4780	8200
Geminospora lemurata	+	+
Rhabdosporites langii	+	+
Auroraspora micromanifestus	+	
Ancyrospora spp	+	
Verrucosisporites scurrus	+	
Calamospora atava		+

Two of the spores are illustrated in Figure 4.

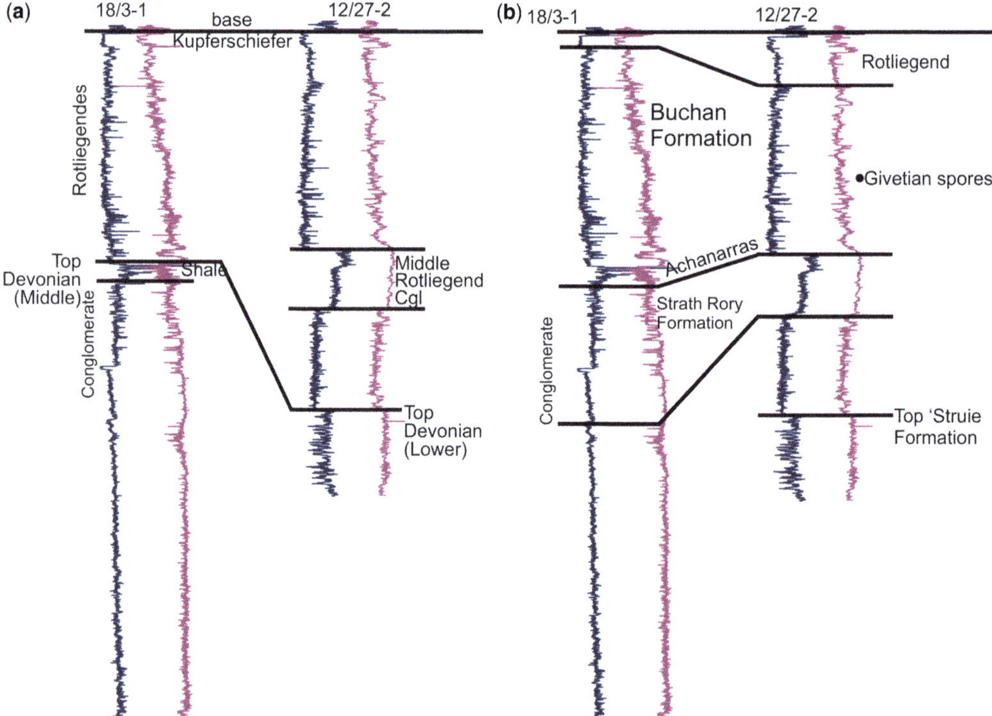

Fig. 3. Comparison of 12/27-2 and 18/3-1 using the composite log stratigraphic picks versus the revised stratigraphy. (**a**) The original picks for 12/27-2 and 18/3-1 with a thick interval of Rotliegend including a single conglomerate. The top of a dark-coloured mudstone unit was picked as Middle Devonian in 18/3-1 but Lower Devonian in 12/27-2. (**b**) Revised correlation that uses palynology to show that the Achanarras Fish Bed level is the Middle Devonian shale overlying the Devonian conglomerate in 18/3-1. The same Achanarras level can be identified overlying the 'Middle Rotliegend Conglomerate' in 12/27-2. The revised Buchan–Rotliegend Formation boundary is picked at a high level and confirmed by palynology from 12/27-2 with the Givetian spore *Geminospora lemurata* present at 2467 m.

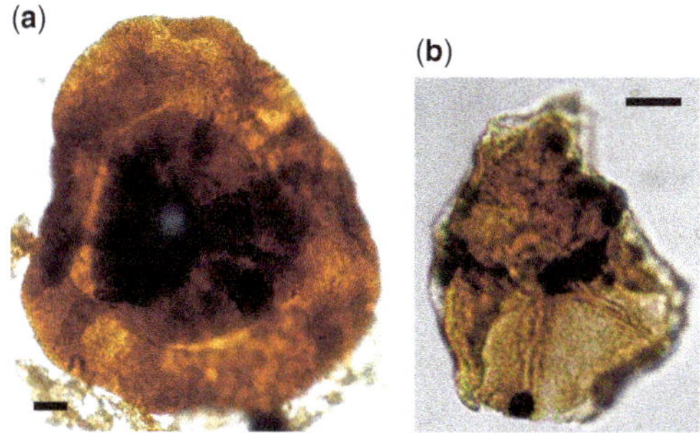

Fig. 4. Illustrative spores from 2467 m in 12/27-2. (**a**) is the Middle Devonian spore *Rhabdosporites langii* abundant in the Eifelian and early Givetian of the Orcadian Basin. (**b**) is *Geminospora lemurata*, which has a Givetian inception and was the microspore of the progymnosperm *Archaeopteris*. Scale bars are 10 µm.

The rationale for accepting the velocity change as the Rotliegend–Old Red Group boundary is that the Moray Firth Buchan Formation interval has a proven greater pre-Permian burial depth (e.g. Marshall 1998) and hence a slight increase in sonic velocity. In addition, the Buchan Formation is dominantly composed of fluvial sandstones. Therefore, the small gamma ray spikes probably represent concentrations of more radioactive minerals (presumably feldspar, mica and heavy minerals). In contrast, the Rotliegend is of aeolian origin, texturally more mature and more likely to give a flat gamma ray log profile. These picks were tested with palynological sampling and rare specimens of *Geminospora lemurata* were identified at 1688.6 m in 13/19-1, at 1421.6 m in 12/29-2 and at 2467.4 m in 12/27-2 (Fig. 4a, b, Table 1). This spore is restricted to the Givetian of Orkney and Shetland (Marshall 1996, 2000). It is not known to occur in the onshore Devonian of the Moray Firth area and is hence unlikely to be reworked.

During compilation of the Millennium Atlas these new picks and unified stratigraphy were used in the Devonian chapter (Marshall & Hewett 2003) with the revised Rotliegend Group thicknesses also reported by Glennie *et al.* (2003; fig. 8.2). However,

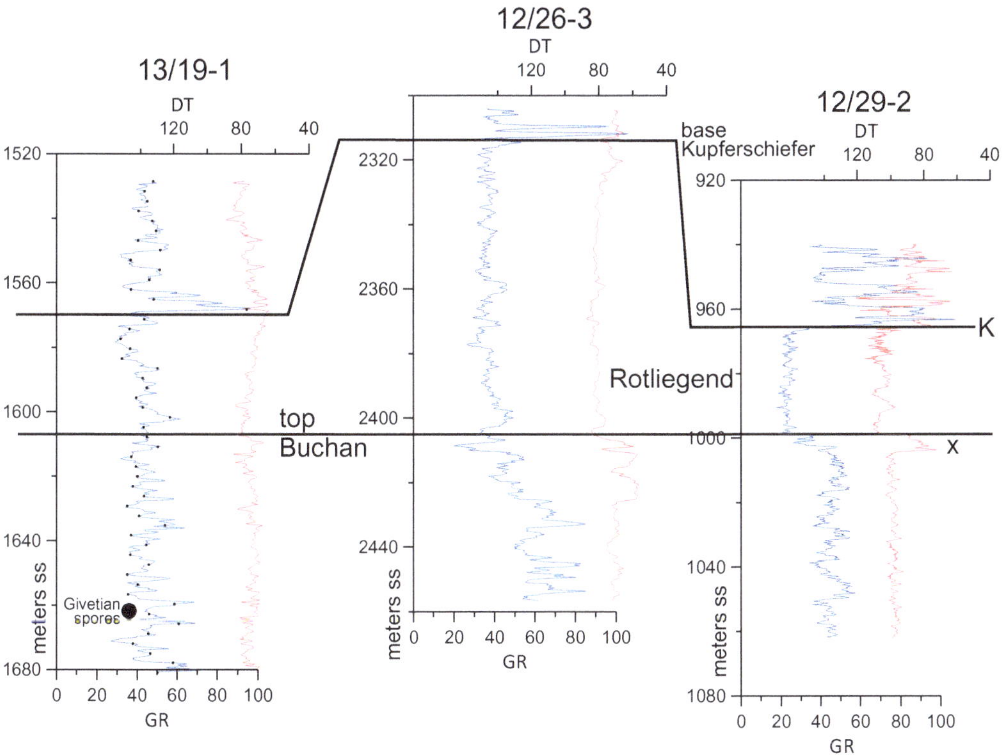

Fig. 5. Example logs from 13/19-1, 12/26-3 and 12/29-2 across the revised Old Red–Rotliegend Group boundary showing how the combination of flatter gamma ray log and increased sonic velocity separates the two formations. These features are not present in all well sections as shown by some higher gamma ray values in 13/19-1. The Buchan Formation can be proved from the presence of rare Givetian spores present in cuttings samples at 1688.6 m in 13/19-1 and 1421.6 m in 12/29-2. A higher velocity feature is often present (e.g. at × in 12/29-2) in many well sections and represents a carbonate-cemented top to the Buchan Formation.

at that time there was not the opportunity to develop an understanding of the implications of the revised stratigraphic boundary. Subsequent to publication of the Millennium Atlas only minor revision has been made to the picks. However, what this contribution will show is how the application of the new boundary as defined in the Millennium Atlas gives a better understanding of the relationship between the Rotliegend Group, its subcropping Old Red Sandstone and the controls on their distribution. It also significantly changes our understanding of the thickness, occurrence and origin of the Rotliegend Group. This new boundary pick is also shown to be internally consistent and self-checking. In short, by achieving a better stratigraphical breakdown it enables a proper understanding of the Rotliegend Group in the Inner Moray Firth for the first time.

The application of this new boundary is demonstrated through a series of cross-sections (Figs 6–9) in the Inner Moray Firth that utilize well logs which were split at the Permian–Devonian boundary and then internally flattened separately on the Smith Bank Formation (Triassic), Kupferschiefer (Permian) and the base Eday Marl or Achanarras/John o'Groats horizons for the Devonian. The position of the section lines is shown in Figure 1. All formation picks are shown in Table 2 together with additional wells not included in the cross-sections. This distribution is deliberately not mapped onto a Moray Firth structural map as these features generally represent Mesozoic and younger structures. It is more important to attempt to use the thickness and distribution of the Rotliegend Group to understand what structures were active at that time.

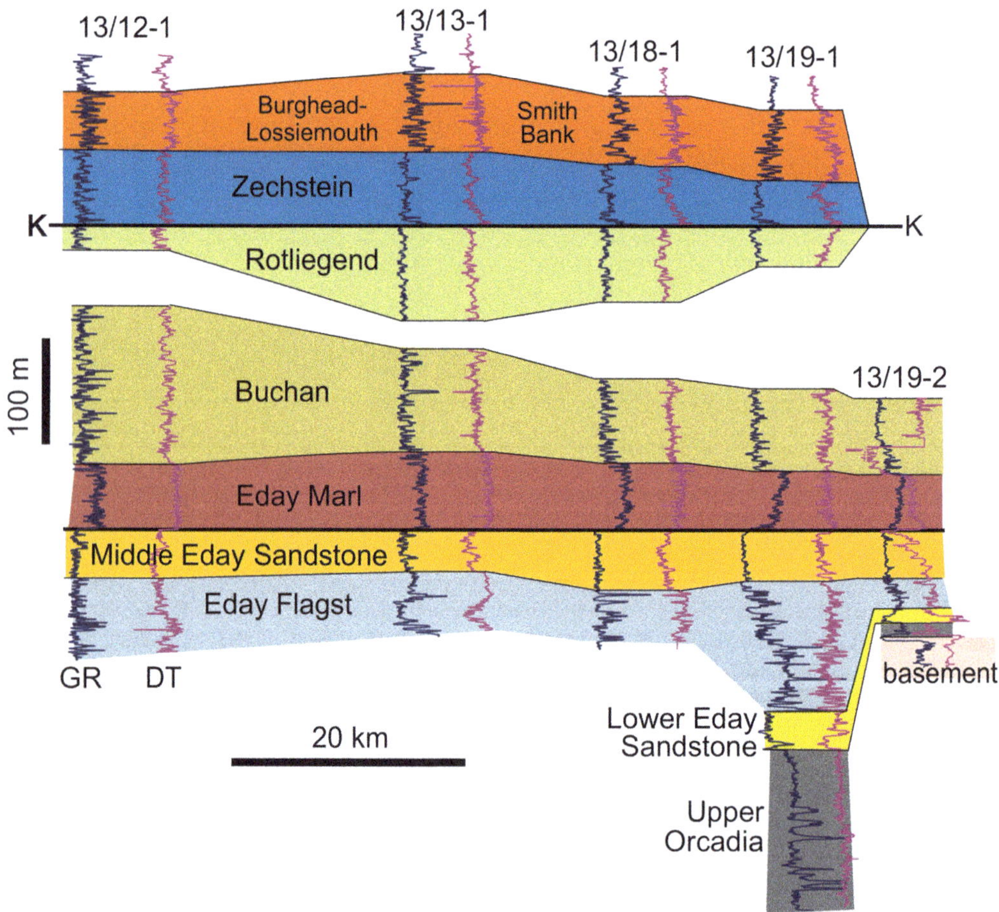

Fig. 6. A cross-section from the Halibut Shelf area that shows a thin simple basin development of Rotliegend Group thinning to the NW and SE. The Buchan Formation shows progressive truncation towards the basement high penetrated in 13/19-2 that has been intruded by the Ross Granite and formed a long-lived positive high (Fig. 1). The Eday Marl also progressively thins onto the same high. GR is gamma ray log and DT is sonic log.

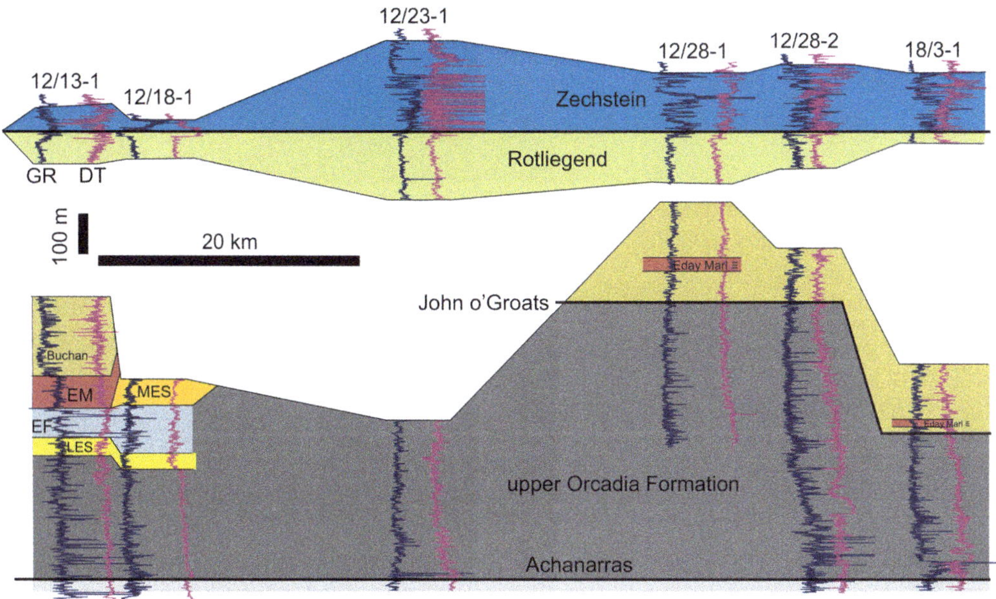

Fig. 7. North–south section across Quadrant 12 showing the same thickness and orientation as the Rotliegend Group as shown by Figure 6. Here the truncation depth has cut deep into the main distribution of lacustrine sediments. A typical development of Eday Marl is present at the southern end of the section line in 12/13-1. At the northern end a ?marginal aeolian facies is present (Eday Marl ≡ equivalent) in 12/28-1 and 18/3-1. GR is gamma ray log and DT is sonic log.

The origin and development of the Rotliegend Basin

The structure of the Rotliegend depositional system can be clearly seen from the three subparallel cross-sections (Figs 6–8) that run approximately NW–SE to north–south across the basin. This shows that the thickness of the Rotliegend Group defines a simple depositional system with a central maximum that thins to a zero edge on its margins. There are no appreciable thickness increases towards major structures such as the Great Glen Fault or Caithness Ridge, implying these were not active structures controlling deposition. However, there are positive areas such as that shown at the southern end of Figure 6 which attenuates in 13/19-2 on the basement high that was intruded by the Ross Granite. When the thicknesses of the Rotliegend are plotted spatially (Fig. 10), these show a relatively simple basin with separate depocentres in the southern part of Quadrant 12 and then a more minor development in Quadrant 13 south of the Caithness Ridge. An attempt was made to digitally contour the thickness distribution but it was limited by the irregular spacing, sharp fault boundaries and limited number of data points. The maximum thickness excepting 12/16-1 is 177 m in 12/23-1. This is now significantly less than the maximum of 700 m reported in 12/23-1 by Andrews *et al.* (1990) and Cameron (1993).

There might appear to be an increase in Rotliegend thickness towards the Wick Fault as shown by 12/16-1. However, this well section is very atypical in having a preserved interval of Carboniferous (?Namurian–Tournaisian) sediments. The Rotliegend is difficult to separate and contains atypical carbonate beds. The overlying 'Fringe' Zechstein is again somewhat atypical and lacks the important Kupferschiefer at its base. The latter has a distinctive lithology, log character and palynological assemblage and acts as a key stratigraphic marker that can be uniquely identified. In the absence of any comparable Carboniferous–Rotliegend sections in the Inner Moray Firth (Bruce & Stemmerik 2003) the established formations are very difficult to pick. Hence, it is possible that this thick interval of apparent Rotliegend Group is a palynologically barren younger Carboniferous section.

It has been shown that the Rotliegend was a relatively thin cover of aeolian and fluvial sediment deposited on the pre-Permian subcrop and accumulating in any topographic lows within it. This thin sequence was then preserved by the Kupferschiefer flooding. The distribution and stratigraphic relationships of the Rotliegend Group shown here demonstrate that is was not deposited in an active

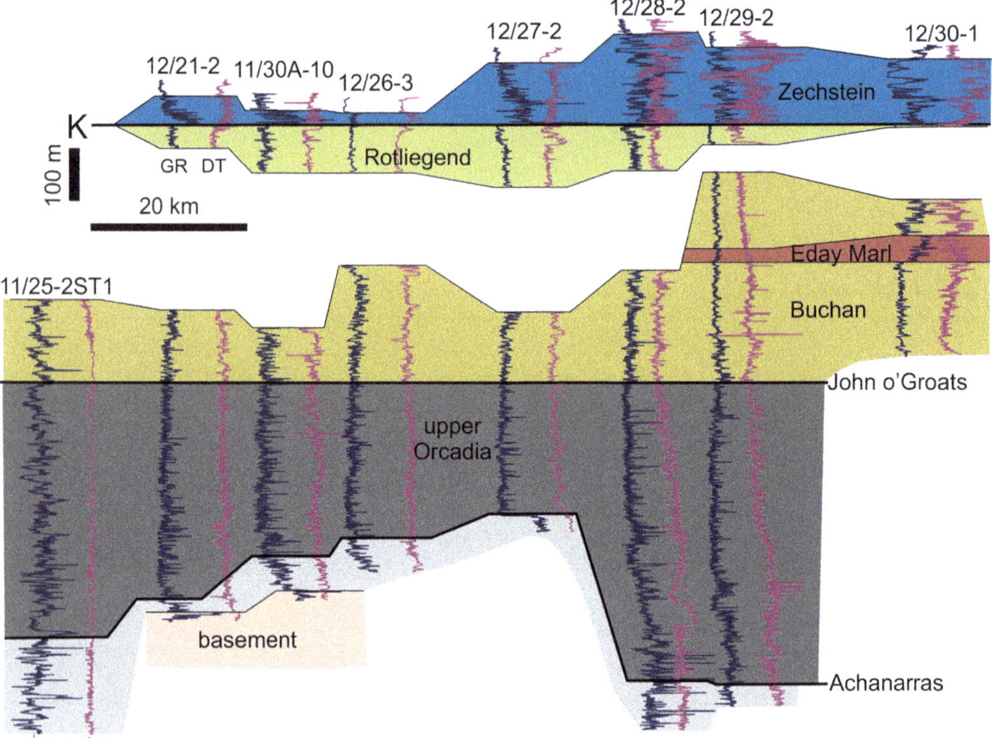

Fig. 8. Section again showing the NE–SW-orientated development of the Rotliegend Basin. The Eday Marl is only present in the SE part of the section line. The John o'Groats horizon represents the third lake development within the Eday Flagstone Formation and using it as an internal datum demonstrates the degree of truncation of the Buchan Formation. The Devonian sediments accumulated within active half-graben with 11/30-A10 and 12/21-2 being thinner sections on a footwall high. The thickest section is present in 12/29-2 with most of the penetration omitted from the section line. GR is gamma ray log and DT is sonic log.

extensional basin. It should also be noted that, although the Rotliegend Group is often referred to the Lower Permian it is in fact Late Permian in age. As a porous red bed sandstone sequence there are no preserved palynological assemblages in the Inner Moray Firth area. However, it has now become clear that the approximately correlative Cutties Hillock Sandstone near Elgin contains reptiles regarded as latest Permian in age (Benton & Walker 1985). In addition, there is palynological and geochronological evidence (reviewed in Glennie 1998) from the Central North Sea and continental Europe that shows it to be Tatarian (Late Permian) in age and contiguous with the overlying Zechstein.

The Zechstein Group

The thickest development of the Zechstein Group is shown on Figures 7 and 8 in Quadrant 12. Here its thickness generally correlates with that of the underlying Rotliegend Group. It thins to the NW towards Quadrant 11 and the trace of the Great Glen Fault, again confirming that this major fault system was not controlling deposition. The maximum thickness is in the southern part of Quadrant 12 where again it, in part, parallels that of the Rotliegend. There is a second separate depocentre in the central part of Quadrant 13 (Fig. 9) that again matches that of the Rotliegend Group.

The accepted scenario for the Zechstein marine transgression in the Permian basins of the North Sea area (Glennie 1998) was the rapid flooding of a low-lying deflationary basin. The Inner Moray Firth was at the periphery of this basin with the revised and now much thinner interval of Rotliegend Group overlain by a thin interval of Kupferschiefer. The latter has a high organic matter content (including amorphous organic matter) and represents microbial production under a stratified water column, i.e. circulation was restricted and the water depth deep enough to prevent wind-driven mixing.

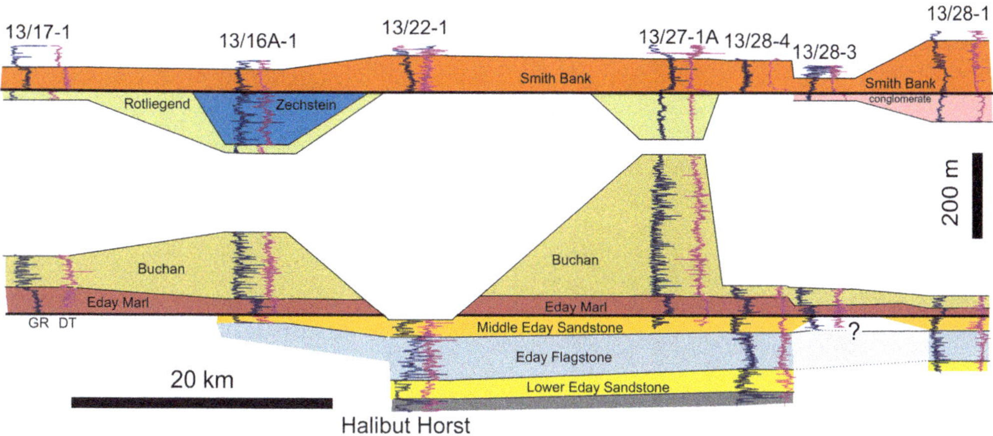

Fig. 9. Cross-section showing a more complex stratigraphy with an interval of homogenous red mudstone at the base of the Smith Bank Formation. This can have a log signature similar to the Eday Marl, but they are distinct and separable when they appear together in the same well. This section crosses the Halibut Horst. Importantly the Middle Eday Sandstone and Eday Flagstone formations do not thin onto this horst. The Buchan Formation is truncated at the unconformity beneath the Triassic Smith Bank Formation, which is also continuous across the structure. This implies that the Halibut Horst had a stage of uplift in the Variscan which caused the downcutting that removed the Buchan and Eday Marl formations. The horst then remained a positive feature during the Rotliegend and Zechstein (Late Permian) but was submerged in sediment by the time of Smith Bank Formation deposition. GR is gamma ray log and DT is sonic log.

The stratigraphy and structure of the Devonian

The uppermost part of the Devonian section is the Buchan Formation, a sandstone interval generally devoid of distinguishing characters. Beneath this there is a sequence of key stratigraphic markers. The only one of these that can always be dated palynologically is the Eday Flagstone Formation, which contains a distinctive Givetian spore assemblage (Marshall 1996) in organic rich lacustrine beds, which also gives the interval its distinctive spiky gamma ray log profile. Higher in the section is the Eday Marl Formation, which is a bright red homogenous calcareous mudstone with a distinctive wireline log signature where both the gamma ray and sonic logs step strongly to the right in contrast to the overlying Buchan and underlying Middle Eday Sandstone Formations. The Eday Marl marks the Taghanic aridity event (Marshall *et al.* 2011) and is an excellent time-constrained late Givetian stratigraphic marker. Figure 6 shows the distribution of the Eday Marl in five wells from the Halibut Shelf area. Using the Eday Marls as a datum demonstrates that the eroded top of the Buchan Formation has a consistent truncation and thickness reduction as it down-steps to the SE onto the high. In addition, the Eday Marl also shows a slight but consistent thickness reduction onto the high. These trends together with that of the overlying Rotliegend Group demonstrate that the stratigraphic pick for the Devonian–Rotliegend boundary is internally consistent.

Figure 9 is a section line that crosses the Halibut Horst and demonstrates a number of key stratigraphic relationships between the wells which can only be elucidated by comparing the different sections. The lacustrine flagstones of the Eday Flagstone Formation can be recognized and constrain the lower part of the well sections. In addition, the Zechstein Group was identified in only one well and is somewhat atypical in character as it developed within a more clastic facies with a poorly defined Kupferschiefer. The Triassic Smith Bank Formation is also present which for its lower part (13/16A-1) or in its entirety is a homogenous interval of bright red mudstone somewhat similar to the Eday Marl. In 13/17-1 and 13/27-1A both the Smith Bank and Eday Marl formations are present and can be identified in stratigraphic sequence and discriminated. The Smith Bank Formation can then be traced across the section line and in 13/28-3 and 13/28-1 has a locally developed conglomerate beneath it that is attributed here to the Triassic.

In the area of 13/22-1 (Fig. 9) the Halibut Horst is present and the Devonian sequence above it has been truncated to a lower level with a very variable thickness of Buchan Formation preserved on its flanks. As the mudstones of the Triassic Smith Bank and the Devonian Lower Eday Sandstone, Eday Flagstone and Middle Eday Sandstone formations

Table 2. *Compilation of reinterpreted log picks for Moray Firth wells*

Well	kb	Top Smith Bank mdst and equivalents	Top Smith Bank Conglomerate	Top Zechstein	Top Kupferschiefer	Top Rotliegend	Thickness Rotliegend	Quality of top Buchan pick	Top Buchan	Top Eday Marl	Top Middle Eday Sandstone base Eday Marl	Top John o'Groats fish bed	Top Eday Flagstone	Top Lower Eday Sandstone	Top upper Orcadia Formation	Top Achanarras	Top basement	TD
11/30-6	32.6			1858.7	1878.5	1880.0	55	2	1934.6							2411.9		2457.0
11/25-2 ST1	36.0							1	2860.5			2996.8				3486.9		3713.1
11/30A-10	36.8			2529.6	2560.1	2560.1	91	2	2651.5			2864.9				3095.0	3306.9	3463.8
12/13-1	25.0			2386.6	2451.8	2453.6	80	2	2534.1	2719.4			2824.9	2902.6	2946.8	3303.4		3368.0
12/14-1	25.9			2736.8	2831.6	2833.7	11	2							2844.4			3208.3
12/18-1	24.4			2097.0	2124.8	2128.1	66	3			2193.6		2264.7	2393.9	2433.5	2716.1		2830.1
12/21-2	32.0			2770.3	2814.8	2829.8	43	2			2872.7	3128.8				3315.9	3486.6	3486.6
12/23-1	29.9			1330.1	1475.8	1478.9	177	2							1655.7	2048.9	2562.8	2614.6
12/26-3	42.7			2288.7	2308.9	2313.4	91	2	2404.9			2635.0				2937.7		3109.6
12/27-1	37.8			1625.8	1663.0	1664.2	60	2	1723.9									3238.8
12/27-2	32.0			2009.5	2119.9	2127.5	122	1	2249.4			2388.1				2644.1		3244.6
12/28-1	23.2			2731.6	2878.5	2881.6	137	2	3018.7	3158.9	3198.6	3279.3						3659.7
12/28-2	25.9			929.9	1097.3	1101.2	93	2	1194.2			1492.0				2055.9		2439.6
12/29-2	35.4			814.1	959.2	965.3	33	1	997.9	1199.1	1240.2	1415.5				2003.8	3457.7	3530.8
12/30-1	33.8			2121.6	2142.1	2244.1	5	2	2249.1	2357.1	2377.1							2575.1
13/12-1	23.5	3472.6		3530.5	3597.6	3603.7	21	2	3625.0	3777.4	3841.4		3888.6					3973.7
13/13-1	25.3	2437.5		2474.1	2589.0	2593.8	94	1	2687.4	2785.0	2864.2		2903.8					2975.8
13/16A-1	25.9	1467.6		1533.8	1684.0	1687.1	21	2	1708.4	1894.3	1943.7		?					2012.6
13/17-1	26.5	2219.9				2281.7	72	2	2354.0	2439.3								2499.4
13/18-1	25.0	1907.4		2008.6	2129.6	2134.2	73	2	2207.7	2288.1	2353.4		2415.5					2473.1
13/19-1	25.9	1452.4		1525.5	1565.1	1571.2	37	1	1607.8	1687.1	1743.2		1805.9	1922.7	1960.8			2127.2
13/19-2	25.3							1	900.7	975.1	1030.8		1075.0	1111.6	1120.7		1134.5	1307.0
13/22-1	26.2	1062.5						1			1162.5		1205.2	1324.1	1385.0			1430.1
13/26-1	22.3			1931.5	1994.9	1996.4	8	2	2004.7									2623.4
13/27-1A	29.9	2164.7				2259.2	143	3	2402.4	2794.7	2850.2							2895.6
13/28-1	24.1	2883.7	2966.0					1	3033.1	3036.1	3089.5		3130.6	3222.0				3357.7
13/28-3	24.1	2770.6	2815.7					1	2838.9	2885.5								2966.9
13/28-4	25.3	2560.9						1	2652.4	2687.4	2737.4		2788.0	2885.5	2955.6			2988.0
18/3-1	25.6			1060.1	1211.9	1213.4	29	2	1242.4	1385.6	1407.0	1422.2				1755.0		3025.4
19/2-1	25.3	2641.1		2831.9	2961.7	2963.3	14	2	2977.0									3303.1

These picks were used in the compilation of Figures 6–9. They update and replace the existing composite log picks. The top Buchan picks are ranked as to quality with 1 being where there is palynological age control in the formation or it is from a section lacking Rotliegend Group, 2 is using solely the wireline log criteria and 3 is where it has been difficult to apply these criteria. Picks are also provided for additional wells that were reviewed but are not included on the cross-section figures. Triassic picks are only given where the interval has been plotted on a cross-section. All picks are in converted to metres, subsea. If comparing with the picks from the original composite log for 13/19-2 note that unusually these were already corrected to subsea rather than being relative to kelly bushing or rotary table elevation as would be expected. This only becomes evident when comparing the same peaks on the digital las file and the composite log. 12/18-1 is significantly deviated and the true vertical – depths are given, all other depths are measured depth.

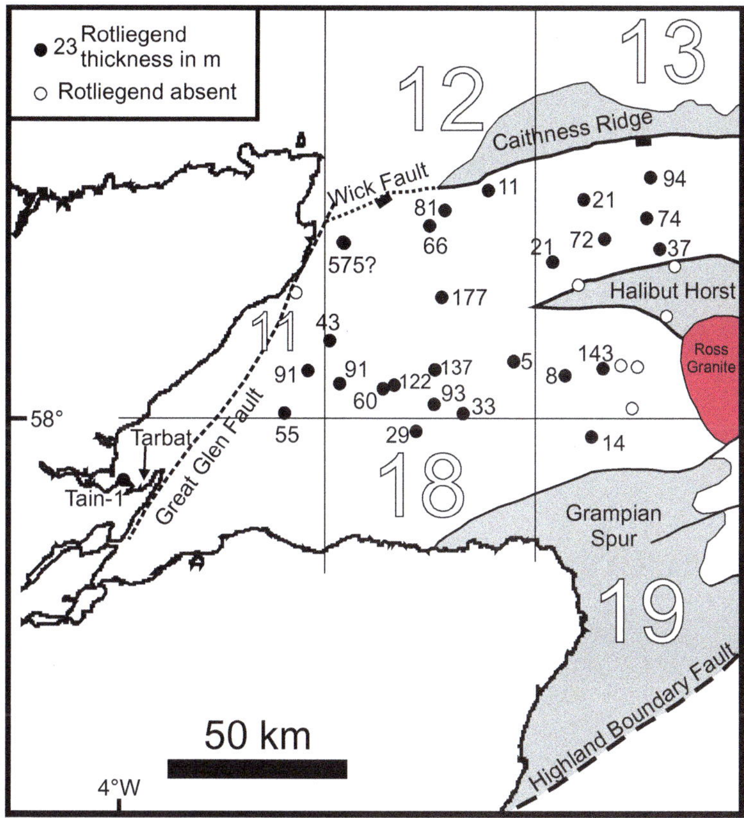

Fig. 10. Thickness of the Rotliegend Group in the Inner Moray Firth. There is a minor development south of the Caithness Ridge with the maximum thickness in the southern part of Quadrant 12. There is a secondary development in the central part of Quadrant 13. The 575 m maximum thickness reported from 12/16-1 is regarded as questionable. It is present above an interval of Carboniferous and there is no proof that it is Rotliegend Group.

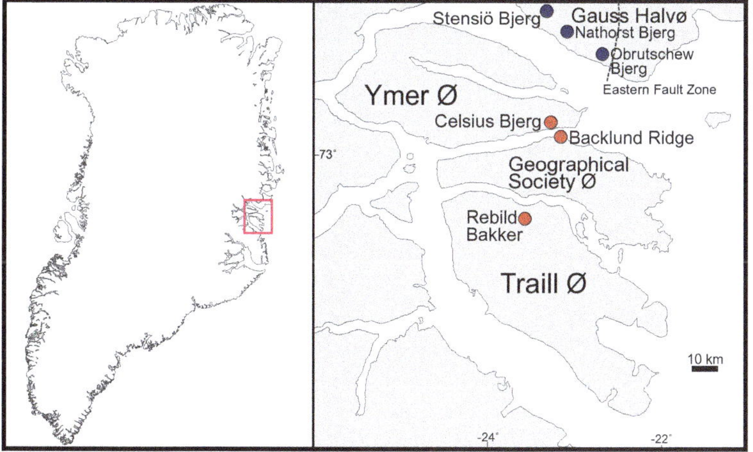

Fig. 11. Map of East Greenland showing location (red circles) of the Harder Bjerg Formation on Traill Ø, Geographical Society Ø and Ymer Ø. The exposed Permian–Devonian boundary sections (blue circles) are all on Gauss Halvø and on the summits of Stensiö Bjerg, Nathorst Bjerg and Obrutschew Bjerg. The Eastern Fault Zone is a major syn-Devonian and younger extensional fault. East of this fault zone there is a development of Permian marginal marine facies.

Fig. 12. (**a**) The summit of Nathorst Bjerg showing two developments (red arrows) of yellow Permian sediment unconformable on the Devonian Stensiö Bjerg Formation. This is photographed from the summit of Stensiö Berg to the west. The base of the Permian is at different heights from north to south. The local development of Stensiö Bjerg Formation (black arrow) is topographically lower than the Permian immediately adjacent to it but has been disrupted by faulting. The distance between the red arrows is 1.3 km. (**b**) The summit of Nathorst Bjerg (UTM 27X 424055 8146050, 1502 m, left-hand arrowed peak in (a)) showing the contrast of yellow Permian sandstones overlying the

are evenly developed across the area, this implies that any relief associated with the Halibut Horst was only present during the Permian and the result of Variscan inversion. The Zechstein and Rotliegend groups both attenuate to the SE against the Halibut Horst. A thicker interval of Buchan and Rotliegend units has been variably preserved to the SE in the area of 13/27-1A.

The Devonian succession in the area of Quadrant 12 is somewhat different in that the Eday Marl is only present in the eastern part in 12/29-2 and 12/30-1 (Fig. 8) and 12/13-1 (Fig. 7). It has also been tentatively identified as a marginal ?aeolian facies (Fig. 7, Eday Marl≡) in 12/28-1 and 18/3-1 (Marshall *et al.* 2011) together with more certain correlations to aeolian facies in the onshore outcrop at Port Tarsuinn and in the Tain-1 well (Marshall *et al.* 2011). It should be noted that in 12/29-2 there was only a weak log response for the Eday Marl. However, all wireline log traces found for this part of the well have been run through casing. Therefore, this pick for the Eday Marl is based on visual recognition of a thick homogenous calcareous red mudstone interval in cuttings samples and as reported on the original mudlog. It should also be noted that the gamma ray responses in wells such as 12/23-1 and 12/28-1 appear to show attenuated and sandier sections but these are older penetrations and may not have accurately calibrated logging tools.

There is no distinctive interval of Eday Flagstones in these Quadrant 12 wells but instead it is represented within a fluvial sandstone sequence by a single or group of gamma ray log spikes that are interpreted as representing the John o'Groats Fish Bed. This is a correlative of the third lake cycle in the Eday Flagstone Formation and represents the deepest and widest extension of this Eday Flagstone lakes with the most diverse fish fauna. Beneath this gamma ray spike the sequence is upper Orcadia Formation with its well-developed multiple gamma ray spikes representing the deep permanent lake facies. In more distal settings there is an interval of sandstone, the Lower Eday Sandstone between the Eday Flagstones and the top of the upper Orcadia Formation. In the more proximal Buchan Formation it can be difficult to discriminate this interval below John o'Groats Fish Bed. Therefore, the top of the upper Orcadia Formation is placed, for simplicity, immediately beneath the John o'Groats pick. This avoids the problem of separating a poorly defined interval of sandstone in the upper part of the Orcadia Formation as shown by comparing 12/28-1 with 18/3-1. This variable development of lacustrine beds in the upper Orcadia Formation is a consequence of drier conditions and hence less extensive lakes at this time (Astin 1985). The upper Orcadia Formation lakes reach their maximum development at the Achanarras Fish Bed level with its distinctive single or pair of high-amplitude gamma ray spikes representing the deepest, widest development of lacustrine facies in the Orcadian Basin (Marshall *et al.* 2007). In several wells (e.g. 12/27-2, 12/29-2, 18/3-1) the Achanarras Fish Bed(s) are developed immediately overlying the conglomerate fans of Lower Strath Rory Formation. The basement penetrations in 12/21-2 and 11/30a-10 are into the footwall of a Devonian half-graben structure. There is a much thicker development of deeper Middle and Lower Devonian fluvial and lacustrine sediments in the deeper sections of other half-graben, for example, 12/29-2, that are omitted from Figure 8.

An exposed Old Red Group: Rotliegend Group unconformity surface?

The physical Old Red–Rotliegend Group unconformity in the onshore Moray Firth was only ever exposed in the quarry at Cutties Hillock (Fig. 1), which has been long disused. Initially the unconformity was revealed by trial pit excavation and then during the continued quarrying operation (Peacock *et al.* 1968). It has not been reported as visibly exposed since the 1880s. The other possible location for an exposed unconformity is in the Portmahomack area on Tarbat. Here there is a thick Devonian section including fish beds and an Achanarras correlative at Hilton of Cadboll (Marshall *et al.* 2007). The

Fig. 12. (*Continued*) Stensiö Bjerg Formation. (**c**) 'Harder Bjerg' Formation on Obrutschew Bjerg showing another of the Gauss Halvø summits capped by an overstepping deposit of poorly consolidated yellow Permian sandstone. The white arrow marks the interpreted unconformity surface. Note that the Harder Bjerg Formation takes its name from a mountain on Gauss Halvø but the type section is on Celsius Bjerg, Ymer Ø. (**d**) Outcrop of the Permian–Devonian unconformity on Nathorst Bjerg at UTM 27X 0433856 8145916 (1231 m). The unconformity is indicated by white arrow heads. The red arrow marks the location of the level with eroded clasts of local origin. Person for scale. (**e**) Clasts from the bed immediately above the Permian–Devonian unconformity level. These are all locally derived. Photo by Simon Johnson. Field notebook 20 × 13 cm. (**f**) Exposed Devonian unconformity surface from Nathorst Bjerg showing altered sediment. There are numerous large clasts (*c.* 10 cm, and indicated by white arrows) of locally derived Devonian sandstone on the surface. Field notebook 20 × 13 cm. (**g**) Permian–Devonian unconformity on the summit of Stensiö Bjerg, the position of the unconformity is marked by the white triangle. UTM 27X 0429046 8148563, 1197 m. Field note book (circled) and rucksack for scale. (**h**) Eroded and altered top surface of Devonian from Stensiö Bjerg locality shown in (g), interpreted as an altered weathering crust. Hammer for scale, length 28 cm.

overlying section includes a John o'Groats correlative at Bindal (NH 933840) and a proximal Eday Marl correlative at Port Tarsuinn (Marshall *et al.* 2011). Above this there is a thick sequence (*c.* 500 m) of sandstone and conglomerate (Tarbat Ness Formation), but at Camas Solais (NH 939877) the succession reverts to soft yellow sandstones that includes a thick intercalation of aeolian and sandy sabka sediment (Fig. 9 in Marshall *et al.* 1996). It has long been known that this sequence includes records of 'reptile' tracks discovered by Campbell and Joass (see review in Diemer 1995) with one of these original specimens relocated by Trewin (2012) which has proved to be an arthropod. Importantly Rogers (1990) discovered a new *in situ* occurrence of the tracks and interpreted them as both tetrapod in origin and Devonian in age. No body fossils or palynomorphs have been reported from this sequence. However, given the occurrence of tetrapod (reptile) tracks in the Permian of the Elgin area from what is lithologically a very similar sandstone sequence then there remains the possibility that this uppermost interval is Permian in age. Along with many others (Gordon & Joass 1863; Harkness 1864), we have looked for a subtle unconformity within the sequence but found only transitional contacts. We have also examined the seismic lines across the putative contact plus the mud log, well log and cuttings from the Tain-1 well, but have failed to detect any unconformity. Clearly the challenge is to now find evidence from body fossils within the interval.

The Permian–Devonian contact in East Greenland

There is a very extensive Devonian Basin in East Greenland (Fig. 11) with a contiguous fill in excess of 7 km of Eifelian to Viséan sediments (Olsen & Larsen 1993). The Devonian–Carboniferous boundary can be recognized as lying within the thin Obrutschew Formation (Marshall *et al.* 1999; Streel & Marshall 2006), which represents a deep and wide, stratified lake that filled the basin to capacity during the end Devonian deglaciation. On Traill Ø, Geographical Society Ø and Ymer Ø (Vigran *et al.* 1999) these sections are contiguous through the boundary into palynologically dated Tournaisian fluvial sediments of the Harder Bjerg Formation. However, on Gauss Halvø there are sections on Stensiö Bjerg and Nathorst Bjerg where the Obrutschew Formation is succeeded or replaced by a contrasting white, red and yellow, poorly cemented sandstones of aeolian and fluvial origin that has also been referred to the Harder Bjerg Formation. On the summit of Nathorst Bjerg there is an obvious thick section of distinctive white-yellow sandstone (Fig. 12a, b, d–f). Importantly the contact can also be located at outcrop (Fig. 12d). This shows the truncation and erosion of Stensiö Bjerg Formation fluvial sediments by an initial thin fluvial unit and then coarse aeolian sandstones. A layer of reworked sandstone clasts is present at the unconformity surface (Fig. 12e) together with a distinct altered top that is also covered with sandstone clasts (Fig. 12f) to the Stensiö Bjerg Formation. The alteration takes the form of mottling, minor concretion development and haematization, all subparallel to the unconformity, rather than the fluvial cross-bedding.

The boundary section on Stensiö Bjerg is less clear (Fig. 12g) with the contact largely obscured by summit block crop debris. However, beneath the contact there is a clear surface (Fig. 12h) that shows both significant alteration with ochreous weathering of a character not found within the normal-bedded Devonian sequence. These are interpreted as the recent weathering of a nodular and calcretized top Devonian unconformity surface.

There is a further very extensive outcrop of Harder Bjerg Formation on Obrutschew Bjerg (Fig. 1c) that we have not been able to visit. It is shown as eroding down into the top of the Obrutschew Formation by Olsen & Larsen (1993; fig. 104A) and its general outcrop pattern is highly suggestive of an unconformable sequence.

The age of the Harder Bjerg Formation on Gauss Halvø is not known, being barren of palynomorphs. In addition, its upper age limit can also not be constrained as it is only present on the top of mountains. However, it contrasts with the supposedly correlative Harder Bjerg fluvial sandstone sections of Tournaisian age from Ymer Ø, Geographical Society Ø and Traill Ø that can be palynologically dated (Vigran *et al.* 1999) and are contiguous into the Viséan. The largely aeolian origin and unconformable basal contact means that a Permian age can be inferred for the Harder Bjerg Formation of Gauss Halvø. This is because its deposition above an unconformity surface means that it is post Variscan (i.e. younger than Carboniferous). The Permian was also the main interval of aridity (i.e. aeolian sediments) in contrast to the overall more humid palaeoclimates in both the Carboniferous and Triassic of East Greenland. It appears to be a continental Permian equivalent to the marginal marine Permian that occurs to the east (Maync 1942; Haller 1971) across the Eastern Fault Zone.

This section gives us a clear analogue for the Old Red–Rotliegend Group boundary in the Moray Firth. It is not an obvious unconformity and is largely concealed by the tendency of the unit to collapse to give a block scree. As a boundary it has been overlooked since the systematic mapping of East Greenland in the 1930s, although its location on the summits of 1200 m-high mountains does act as a deterrent to study.

Acknowledgements The support of numerous companies and data repositories during compilation of the Millennium Atlas is gratefully acknowledged, particularly Dick Sutherland and the Gilmerton Core Store. The donation of cuttings sets by Mobil, Oxy, Kerr-McGee, Burmah and Fina is gratefully acknowledged. Shir Akbari prepared the palynological samples. The support of CASP for the East Greenland fieldwork was essential. We were accompanied onto the summits of Gauss Halvø by Simon Johnson, Clive Johnson and Henning Blom.

Funding This research received no specific grant from any funding agency in the public, commercial, or not-for-profit sectors.

References

Andrews, I.J., Long, D., Richards, P.C., Thomson, A.R., Brown, S., Chesher, J.A. & McCormac, M. 1990. *The Geology of the Moray Firth.* United Kingdom Offshore Regional Report British Geological Survey.

Astin, T.R. 1985. The palaeogeography of the Middle Devonian Lower Eday Sandstone, Orkney. *Scottish Journal of Geology*, **21**, 353–375, https://doi.org/10.1144/sjg21030353

Benton, M.J. 1995. The Elgin reptiles. *In*: Smith, J.S. (ed.) *George Gordon: Man of Science.* Centre for Scottish Studies, University of Aberdeen, Aberdeen, 51–82.

Benton, M.J. 2018. Archibald Geikie and the Elgin reptiles. *In*: Betterton, J., Craig, J., Mendum, J.R., Neller, R. & Tanner, J. (eds) *Aspects of the Life and Works of Archibald Geikie.* Geological Society, London, Special Publications, **480**, https://doi.org/10.1144/SP480.4

Benton, M.J. & Walker, A.D. 1985. Palaeoecology, taphonomy and dating of the Permo-Triassic reptiles from Elgin, north-east Scotland. *Palaeontology*, **28**, 207–234.

Bruce, D. & Stemmerik, L. 2003. Permian. *In*: Evans, D., Graham, C., Armour, A. & Bathurst, P. (eds) *The Millennium Atlas: Petroleum Geology of the Central and Northern North Sea.* 83–89.

Cameron, T.D.J. 1993. Triassic, Permian and pre-Permian of the Central and Northern North Sea. *In*: Knox, R.W.O'B. & Cordey, W.G. (eds) *Lithostratigraphic Nomenclature of the U.K. North Sea.* British Geological Survey, Nottingham, 163.

Diemer, J.A. 1995. The geological work of the Reverend Dr James Maxwell Joass. *In*: Smith, J.S. (ed.) *George Gordon: Man of Science.* Centre for Scottish Studies, University of Aberdeen, Aberdeen, 135–183.

Diemer, J.A. 1996. Old or New Red Sandstone? Evolution of a nineteenth century stratigraphic debate, northern Scotland. *Earth Sciences History*, **15**, 151–166.

Glennie, K.W. 1998. *Petroleum Geology of the North Sea.* 4th edn, Blackwell Science, Oxford.

Glennie, K.W., Higham, J. & Stemmerik, L. 2003. Permian. *In*: Evans, D., Graham, C., Armour, A. & Bathurst, P. (eds) *The Millennium Atlas: Petroleum Geology of the Central and Northern North Sea.* 91–103.

Gordon, G. & Joass, J.M. 1863. On the relations of the Ross-shire sandstones containing reptilian footprints. *Quarterly Journal of the Geological Society of London*, **19**, 506–509, https://doi.org/10.1144/GSL.JGS.1863.019.01-02.45

Haller, J. 1971. *Geology of the East Greenland Caledonides.* Interscience, London.

Harkness, R. 1864. On the Reptiliferous Rocks and the footprint-bearing strata of the north-east of Scotland. *Quarterly Journal of the Geological Society of London*, **20**, 429–443, https://doi.org/10.1144/GSL.JGS.1864.020.01-02.52

Lundmark, A.M., Bue, E.P., Gabrielsen, R.H., Flaat, K., Strand, T. & Ohm, S.E. 2013. Provenance of late Palaeozoic terrestrial sediments on the northern flank of the Mid North Sea High: detrital zircon geochronology and rutile geochemical constraints. *In*: Scott, R.A., Smyth, H.R., Morton, A.C. & Richardson, N. (eds) *Sediment Provenance Studies in Hydrocarbon Exploration and Production.* Geological Society, London, Special Publications, **386**, 243–259.

Marshall, J.E.A. 1996. *Rhabdosporites langii, Geminospora lemurata* and *Contagisporites optivus*: an origin for heterospory in the progymnosperms. *Review of Palaeobotany and Palynology*, **93**, 159–189.

Marshall, J.E.A. 1998. The recognition of multiple hydrocarbon generation episodes: an example from Devonian lacustrine sedimentary rocks in the Inner Moray Firth, Scotland. *Journal of the Geological Society*, **155**, 335–352, https://doi.org/10.1144/gsjgs.155.2.0335

Marshall, J.E.A. 2000. Devonian miospores from the Walls Group, Shetland. *In*: Friend, P.F. & Williams, B.P.J. (eds) *New Perspectives on the Old Red Sandstone.* Geological Society, London, Special Publications, **180**, 473–483, https://doi.org/10.1144/GSL.SP.2000.180.01.25

Marshall, J.E.A. & Fletcher, T.P. 2002. Middle Devonian (Eifelian) spores from a fluvial dominated lake margin in the Orcadian Basin, Scotland. *Review of Palaeobotany & Palynology*, **118**, 195–209.

Marshall, J.E.A. & Hewett, A.J. 2003. Devonian. *In*: Evans, D., Graham, C., Armour, A. & Bathurst, P. (eds) *The Millennium Atlas: Petroleum Geology of the Central and Northern North Sea.* 65–81.

Marshall, J.E.A. & Stephenson, B.J. 1997. Sedimentological responses to basin initiation in the Devonian of East Greenland. *Sedimentology*, **44**, 407–419.

Marshall, J.E.A., Rogers, D.A. & Whiteley, M.J. 1996. Devonian marine incursions into the Orcadian Basin, Scotland. *Journal of the Geological Society*, **153**, 451–466, https://doi.org/10.1144/gsjgs.153.3.0451

Marshall, J.E.A., Astin, T.R. & Clack, J.A. 1999. The East Greenland tetrapods are Devonian in age. *Geology*, **27**, 637–640.

Marshall, J.E.A., Astin, T.R., Brown, J.F., Mark-Kurik, E. & Lazauskiene, J. 2007. Recognizing the Kačák Event in the Devonian terrestrial environment and its implications for understanding land–sea interactions. *In*: Becker, R.T. & Kirchgasser, W.T. (eds) *Devonian Events and Correlations.* Geological Society, London, Special Publications, **278**, 133–155, https://doi.org/10.1144/SP278.6

Marshall, J.E.A., Brown, J.F. & Astin, T.R. 2011. Recognising the Taghanic Crisis in the Devonian terrestrial environment; its implications for understanding land–sea interactions. *Palaeogeography, Palaeoclimatology and Palaeoecology*, **304**, 165–183.

MAYNC, W. 1942. Stratigraphie und faziesverhältnisse der Oberpermischen ablagerungen Ostgrönlands. *Meddeleser om Grønlands*, **115**, 1–128.

OLSEN, H. & LARSEN, P-H. 1993. Lithostratigraphy of the continental Devonian sediments in North-East Greenland. *Grønlands Geologiske Undersøgelse*, **165**, 1–108.

PEACOCK, J.D., BERRIDGE, N.G., HARRIS, A.L. & MAY, F. 1968. *The Geology of the Elgin District*. Memoirs of the Geological Survey Scotland. Explanation of One-inch Geological Sheet, **95**.

ROGERS, D.A. 1990. Probable tetrapod tracks rediscovered in the Devonian of N. Scotland. *Journal of the Geological Society*, **147**, 746–774, https://doi.org/10.1144/gsjgs.147.5.0746

SCHMIDT, A.S., MORTON, A.C., NICHOLS, G.J. & FANNING, C.M. 2012. Interplay of proximal and distal sources in Devonian-Carboniferous sandstones of the Clair Basin, west of Shetland, revealed by detrital zircon U–Pb ages. *Journal of the Geological Society*, **169**, 691–702, https://doi.org/10.1144/jgs2011-148

STREEL, M. & MARSHALL, J.E.A. 2006. Devonian–Carboniferous boundary global correlations and their paleogeographic implications for assembly of Pangaea. *In*: WONG, TH.E. (ed.) *Proceedings of the XVth International Congress on Carboniferous and Permian Stratigraphy*. Utrecht, 10–16 August 2003. Royal Netherlands Academy of Arts and Sciences, 481–496.

TREWIN, N.H. 2012. Historic specimen is not a tetrapod trackway with tail-drag from the Upper Old Red Sandstone of Tarbat Ness, Easter Ross, Scotland. *Scottish Journal of Geology*, **48**, 143–145, https://doi.org/10.1144/sjg2012-451

VIGRAN, J.O., STEMMERIK, L. & PIASECKI, S. 1999. Stratigraphy and depositional evolution of the uppermost Devonian-Carboniferous (Tournaisian- Westphalian) non-marine deposits in North-East Greenland. *Palynology*, **23**, 115–152.

The Paleozoic petroleum system in the north of Scotland – outcrop analogues

JOHN FLETT BROWN[1]*, TIM R. ASTIN[2] & JOHN E. A. MARSHALL[3]

[1]*The Park, Hillside, Stromness, Orkney, KW16 3AH, UK*

[2]*The Vicarage, 1 The Avenue, Tadworth, Surrey, KT20 5AS, UK*

[3]*Ocean & Earth Science, University of Southampton, National Oceanography Centre, Southampton, SO14 3ZH, UK*

J.E.A.M., 0000-0002-9242-3646

**Correspondence: jfletbrown@btinternet.com*

Abstract: All of the components of an exhumed Devonian petroleum system occur in Orkney. These include a good quality mature source rock to bitumen-bearing sandstone reservoirs, with several separate accumulations that could have held about 1.88 billion barrels of oil. The exhumed system presents an excellent analogue for deeply buried petroleum systems offshore. Whilst lighter oils are now absent, reported oil shows occur, commonly associated with faults cutting the Eday Group. On Orkney, Middle Devonian source rocks (750 m thick) were thick lacustrine laminites (fish beds) representing some 30% of the sequence. RockEval and vitrinite analyses show the organic matter is good quality Type I and II and within the early oil window. These source rocks underwent burial until Permian inversion. Several exhumed reservoirs occur on Orkney in aeolian and fluvial sandstones with porosities from 15 to 25%. These reservoirs have been 'breached', losing the light-end hydrocarbons, leaving pore space oil stain and bitumen residues. Thin fluvial and sheet-floods sands found within the lake cycles have bitumen residues and provided connectivity between the thicker reservoir units. All types of trap are found including a major, broad anticline running north–south on Mainland Orkney and an unconformity with fault traps and pinchouts.

In the Devonian Old Red Sandstone (ORS) of Scotland, there is a distinctive northern accumulation of lacustrine, fluvial and aeolian sediments that is known as the Orcadian Basin (Trewin & Thirwall 2002). These sediments were deposited in a series of generally north–south-aligned half-graben that formed in response to extensional collapse of the Caledonian Orogen. The southern limit to the Orcadian Basin (Fig. 1) is the Highland Boundary Fault (Marshall & Hewett 2003), with a series of related half-graben extending northwards to East Greenland. Some of the Early and Mid-Devonian sediments that infilled these half-grabens are organic-rich lacustrine mudrocks and the proven source rock for the Beatrice Field (Marshall *et al.* 1996; Marshall & Hewett 2003) in the Inner Moray Firth, which produced nearly 0.5 billion barrels of oil. The onshore Orcadian Basin has excellent sections of both exposed source rocks and directly linked reservoired oils, although these are now breached and degraded following surface exposure. There are oil seeps within the system, mostly associated with faults (e.g. in the 'red mudstone' core of a fault within the Lower Stromness Flagstone Formation in the Bay of Creekland, south of Bu on North Hoy, HY 237043). Although poorly documented at the time, a 'gas blowout' was reported (Sandy Firth pers. comm.) in World War II when a deep-water well was drilled on the island of Shapinsay. The 'blowout' lasted for three days and is interpreted as representing the trapped gas component of a hydrocarbon system.

We take the opportunity to review the onshore geology of this petroleum system including detailed mapping of a breached, giant oil accumulation. This review will emphasize how recent developments in our understanding of Orcadian Basin structure, stratigraphy, sedimentology and geochronology enable more refined petroleum system models to be developed.

Structural history

The structural history of the Orcadian Basin area started in the late Silurian to early Devonian as a consequence of the collision between the Baltic and Laurentian Plates (Fig. 1). This oblique collision between the two continents created sinistral strike-slip movement on the Great Glen Fault (GGF) system. Early Devonian rebound and extensional tectonics formed sets of north–south-trending half-graben systems that generally dip to the east on the

From: MONAGHAN, A. A., UNDERHILL, J. R., HEWETT, A. J. & MARSHALL, J. E. A. (eds) 2018. *Paleozoic Plays of NW Europe*. Geological Society, London, Special Publications, **471**, 253–280.
First published online December 19, 2018, https://doi.org/10.1144/SP471.14

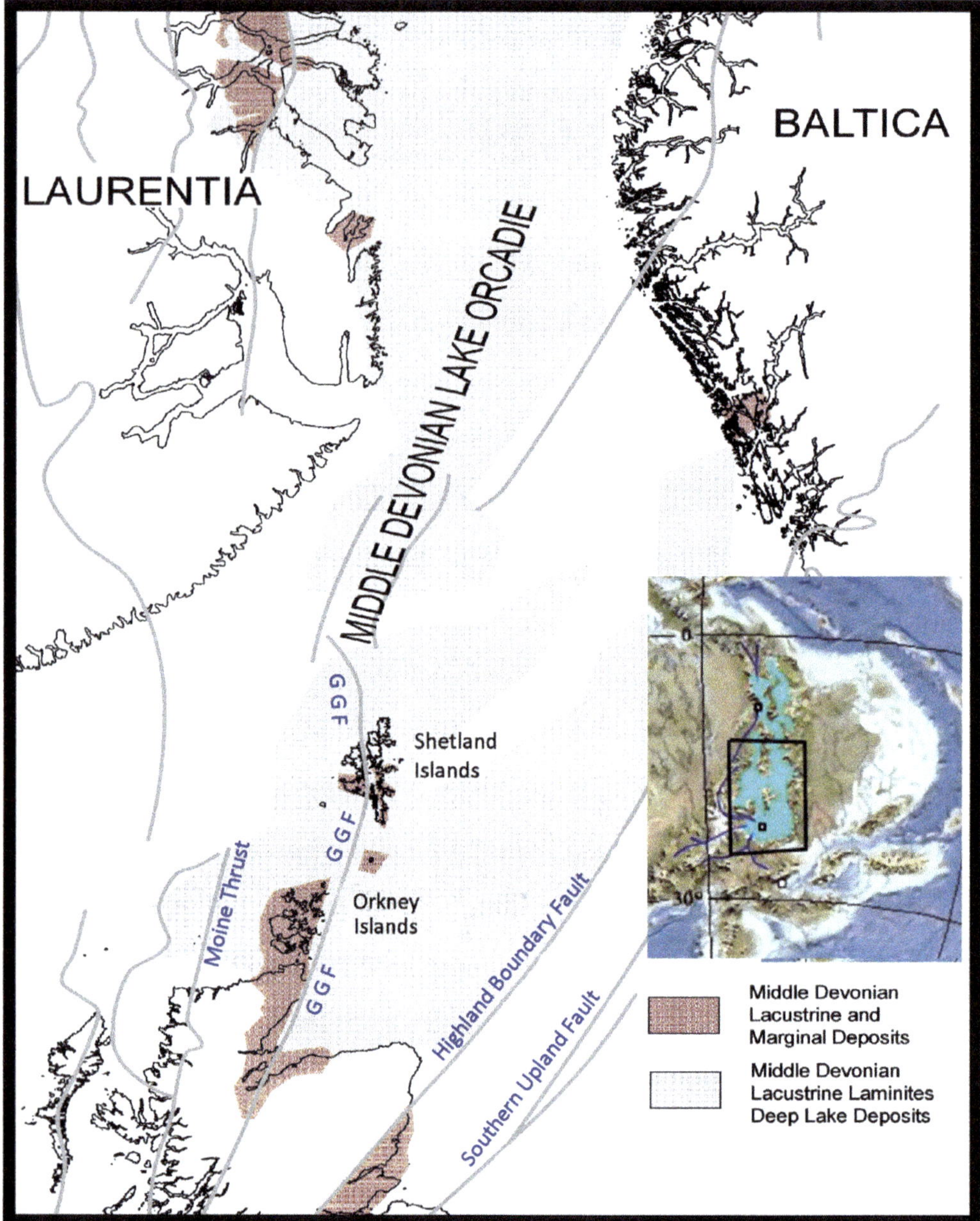

Fig. 1. Palaeogeographical reconstruction of Devonian Lake Orcadie. GGF, Great Glen Fault.

west side of the basin and dip west on the east side. The centre of the basin following along the original suture lay to the east of Orkney and Shetland, which was located (Torsvik & Cocks 2017) about 25° south of the equator at this time. Rapid infill of these original half-graben basins with coarse sediments and boulder conglomerates created a number of essentially isolated lake basins that eventually coalesced to form the larger Orcadian Basin. At maximum extent this Orcadian Basin was about 2000 km long and 250 km wide with subdued topography apart from footwall remnants of basement that were emergent above the basin floor as islands. During the Mid Devonian (approximately early Eifelian to early Givetian), this shallow lake basin was filled with about 1 km of grey and black thinly bedded flagstones that represent the rhythmic lake sediments of Lake Orcadie. The Lake Orcadie waters reached their maximum extent (Marshall *et al.* 2007) in the latest Eifelian during deposition of the Sandwick Fish Bed. During the Early Mid Devonian, episodes of magmatic extrusion occurred particularly in West Shetland (Melby and Eshaness; Mykura *et al.* 1976; Marshall 2000), and are represented in Orkney by an ash layer in the Sandwick Fish Bed. Extrusive volcanic rocks were only erupted extensively in Orkney after deposition of the Lower Eday Sandstone (Fig. 2). The sedimentary style at this time (late Givetian to early Late Devonian) was of mainly fluvial and aeolian sediments.

During the Early Carboniferous, extensional rifting occurred in southern Scotland. In the latest Carboniferous, additional movement on the wrench fault systems (Figs 3 & 4) created broad folding cut by the Late Carboniferous dykes dated at 313 ± 3 Ma at Garthna Geo, Yesnaby, Orkney (Lundmark *et al.* 2011) and perhaps associated with late Variscan regional extension. In northern Scotland, Late Permian sediments (Marshall *et al.* 2018) were deposited unconformably on the Carboniferous erosion surface. This late Carboniferous to early Permian inversion has been related to continuing Variscan shortening (Woodcock & Strachan 2012).

Subsequent dextral movements on the Great Glen Fault and related reactivation of normal faults as reversed faults with mineralization and fault breccias have been dated at 264 ± 3 Ma (Dichiarante *et al.* 2016). Late Permian rifting on the western margin of the Orcadian Basin has been related to early breakup of the north Atlantic, while Triassic extensional rifting in the North Sea area between 251 and 200 Ma led to gentle uplift in the Orcadian Basin area (Goldsmith *et al.* 2003).

During the Late Cretaceous, the central part of the Orcadian Basin, along with the Highlands of Scotland, would have been covered by chalk seas. In the latest Cretaceous and Paleogene apatite fission track analysis shows that the area underwent a major basin inversion (Thomson *et al.* 1999) and was exhumed to form an eroding land mass. This uplift was driven by the mantle plume that was responsible for the final opening of the North Atlantic. The effects of this uplift can be shown in the East Orkney and Dutch Bank basins directly to the east of the Orkney Islands, where there is a preserved Cenozoic infill (Richardson *et al.* 2005). The deformation of this infill shows that there was a general inversion on the faults. The presence of the infill is, in itself, significant as the main sediment source can only have been the emerging and eroding land areas to the west, i.e. Orkney and generally across the Moray Firth area (Guariguata-Rojas & Underhill 2017). In Orkney itself, this breakup of the North Atlantic and North Sea led to compressional tectonics.

The West Orkney Basin

Recent hydrocarbon exploration (Premier Oil 2013) involved the acquisition and reprocessing of 6000 line km of seismic data. This seismic data has provided a clearer picture of the West Orkney Basin structure (Fig. 4). Two wells were drilled in the basin, in 1984 (202/19-1) and 1991 (202/18-1). Both wells intersected a Permo-Triassic succession over 3 km thick (Hitchen *et al.* 1995) but failed to find hydrocarbons. The presence of a Devonian section in the basin centre became a major question. The new data and correlation of the offshore seismic to the onshore geology (Bird *et al.* 2015) has suggested that the Devonian section does indeed extend out into the basin centre.

Bird *et al.* (2015) noted that rift architecture in this area was typically more linear than previously thought. Major faults, as seismically mapped at top basement level, appear to follow rather short linear trajectories. These structures are organized into more complex, segmented fault arrays linked at relay zones higher in the section. It was concluded that these Devonian extensional structures, particularly major detachments, have had a significant influence on later Mesozoic basin geometries.

Onshore faults

Onshore, in Orkney all of the structural elements (Figs 2 & 3; folds, faults, shear zones) show a mean ENE–WSW direction. Similarly, the major strike-slip Great Glen Fault system is compatible with this (Fig. 1). The total offset of post Devonian dextral strike-slip along the Great Glen-Walls Boundary Fault system has been a matter of much discussion (Rogers *et al.* 1989). Estimates range from 300 km of dextral offset through to only 25 km dextral offset. However, it is clear that there was no evidence for strike-slip movements contemporaneous with the deposition of the main Early and Mid Devonian sedimentary sequence in the Orcadian Basin.

The kinematics and timing of the fault reactivation episodes and basin inversion in Orkney have been determined by fieldwork (Hippler 1993). In particular, evidence for strike-slip reactivation along large faults on Orkney suggested early sinistral displacement, and two episodes of later dextral displacements on north–south faults. Late Permian dykes (*c.* 250 Ma as reviewed in Lundmark *et al.* 2011) date these relative movement on major faults. The dykes cross-cut folds associated with the sinistral movements, and intrude the planes of dextral strike-slip faults and therefore post-date the sinistral and earliest dextral deformation.

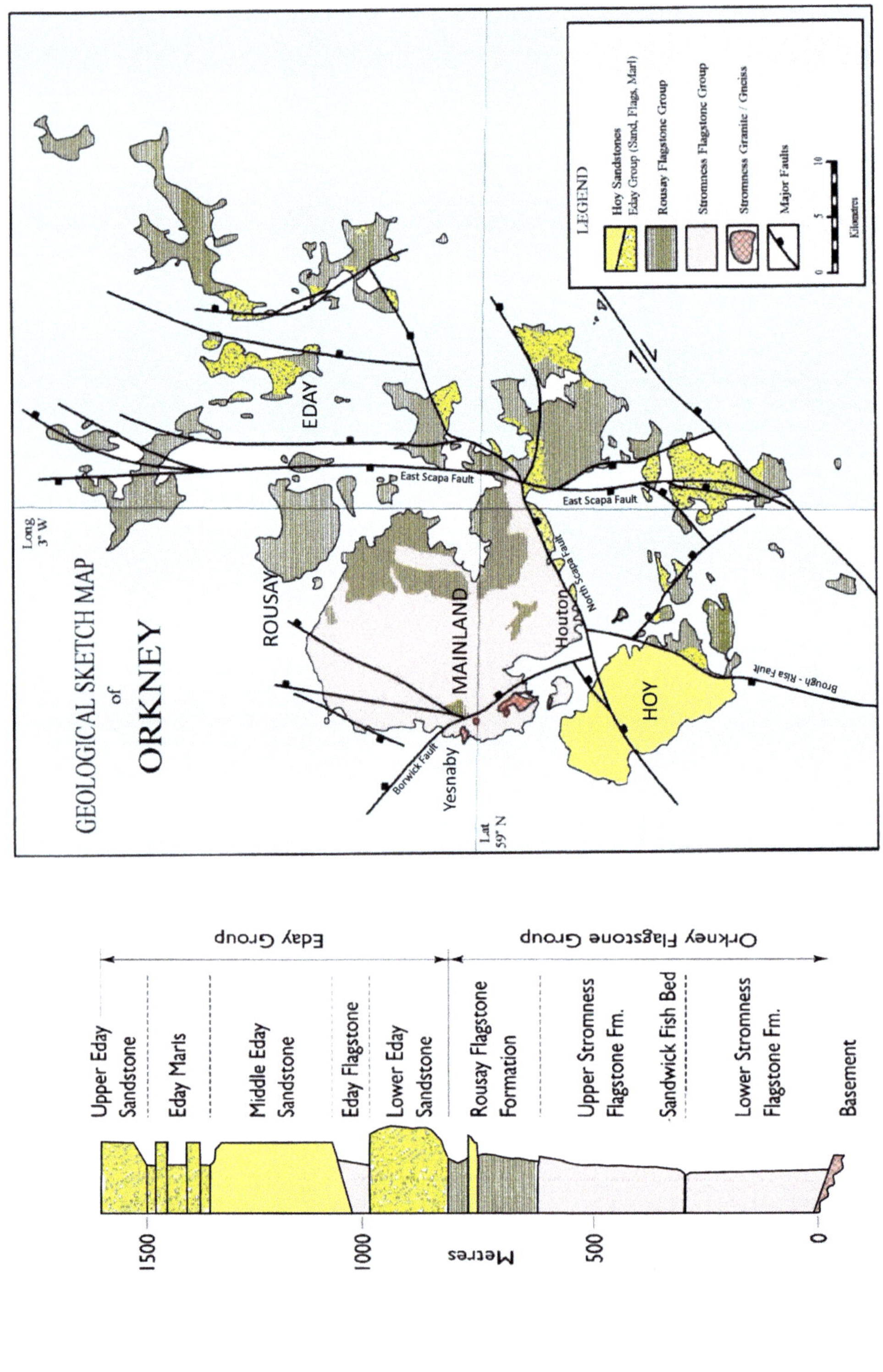

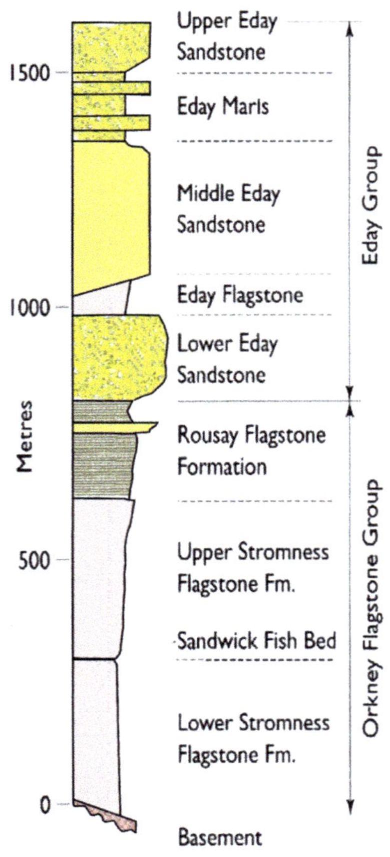

Fig. 2. Geology and stratigraphy of Orkney. From Brown (2003).

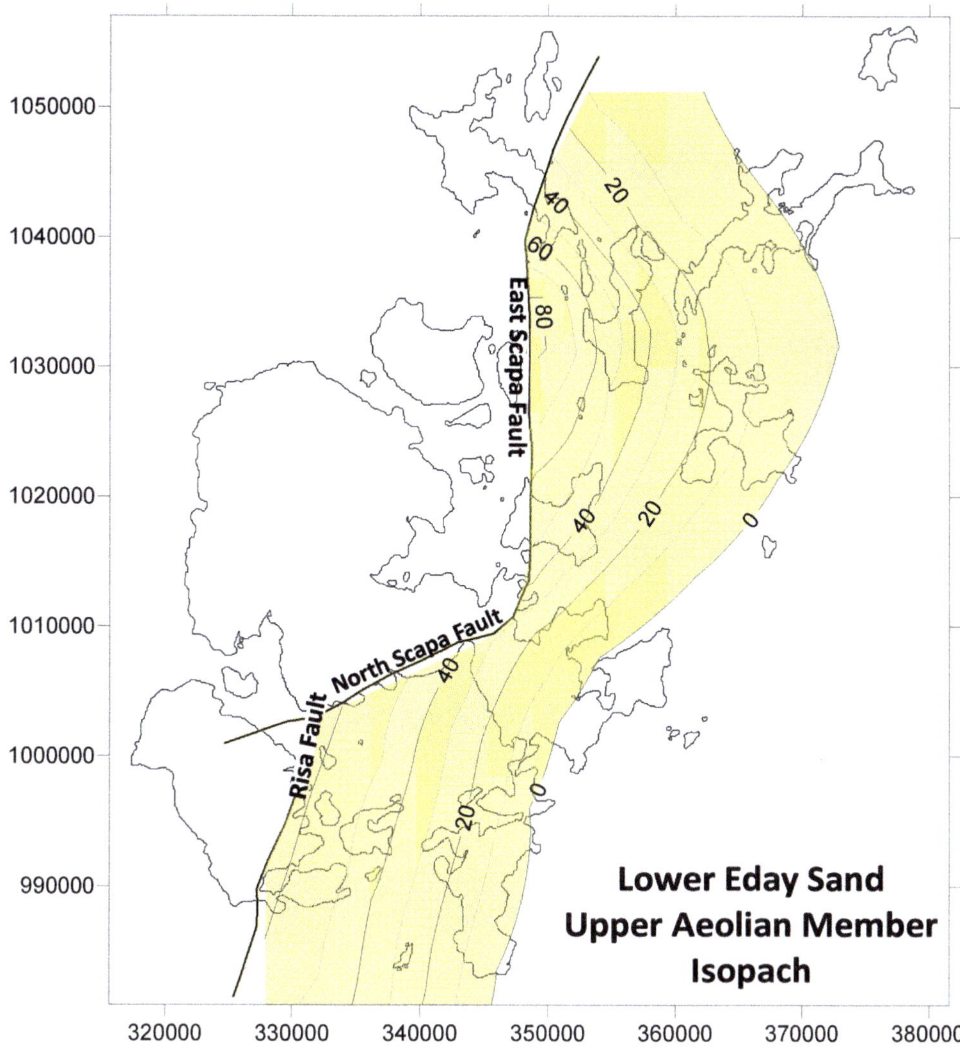

Fig. 3. Isopach thickness of the upper aeolian unit of the Lower Eday Sandstone after Astin (1986). The Risa, North Scapa and East Scapa faults are part of a series of long lived structures including a phase of Mid Devonian extension. Components such as the Risa and East Scapa faults were reactivated with later strike-slip movements.

Significance of basin inversion

Post-Carboniferous uplift of the Orcadian Basin has resulted in inversion geometries (Hippler 1993). In the main Eifelian sequence of lacustrine facies rocks, thrusts have exploited bedding parallel zones forming detachment horizons. Folds and reverse faults also developed because of buttressing against the earlier normal faults. The presence of vein arrays associated with these later reverse faults suggests the existence of high pore fluid pressures. Bitumen in these veins also suggests that the system was oil-wet at this time. The fracture arrays and narrow cataclastic zones provided pathways for these migrating mature hydrocarbons (Hippler 1993; Parnell *et al.* 1998). Rogers *et al.* (1989) show that, although the offshore data (Coward *et al.* 1989) may be interpreted to suggest a Mid Devonian inversion event, the onshore data shows no evidence for this since the folding noted in the onshore Old Red Sandstone sedimentary rocks formed during a Permian inversion event (Hippler 1993). The contractional structures were most likely related to the later tectonic inversion of the basin, just after hydrocarbon maturation was reached. Evidence for the mobility of carbonate and hydrocarbon fluids is observed in the field and in

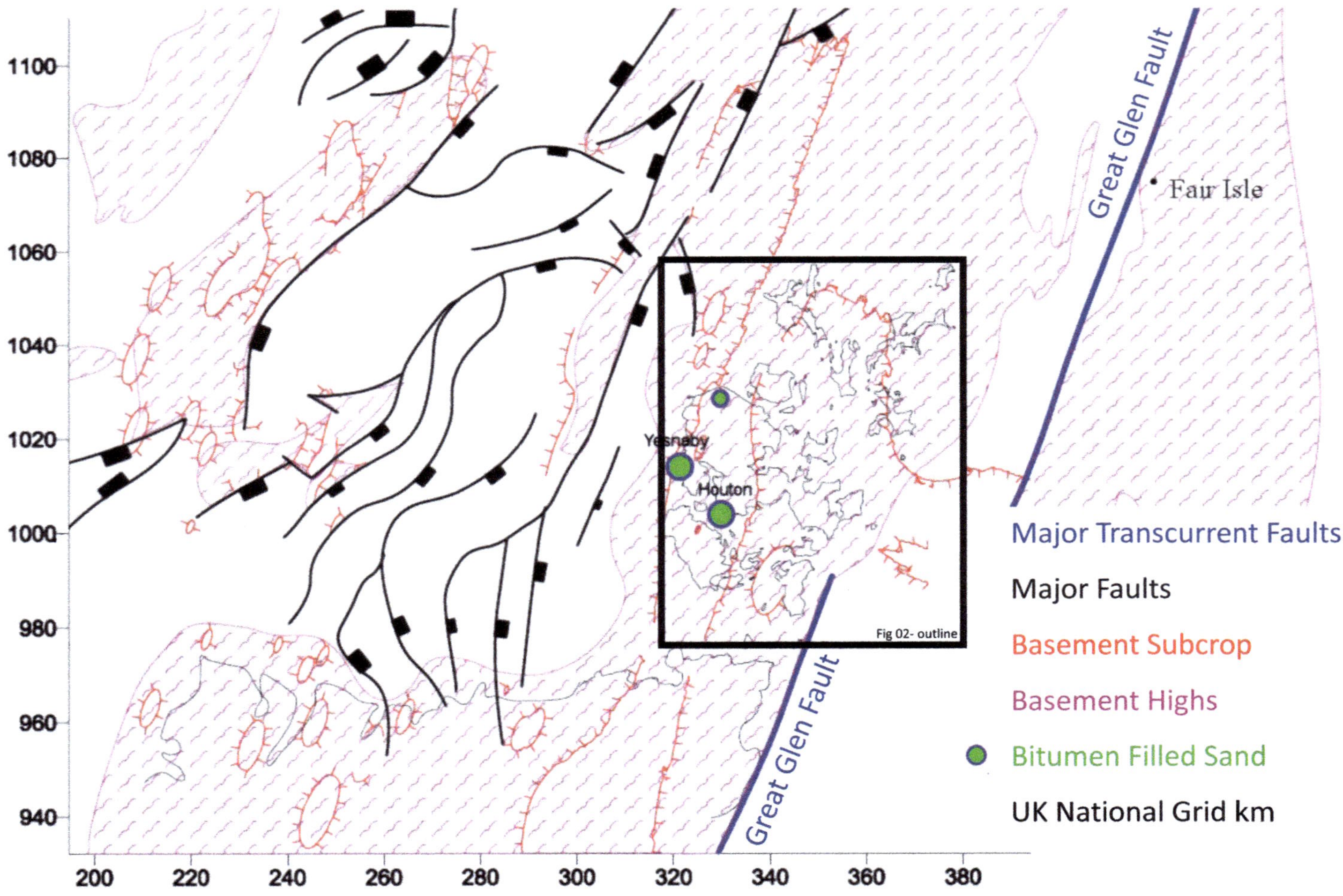

Fig. 4. West Orkney Basin showing location of Figure 2. Basement highs and subcrop faults are shown together with key localities described after Stoker *et al.* (1993).

thin section. Fluids released because of hydrocarbon maturation seem to have contributed to creating the high pore fluid pressures.

Post-Permian extensional structures are seen at Birsay, deforming a Permian dyke. The later extension occurred in a NE–SW direction, but resulted in only minor reactivation of the Mid Devonian extensional faults (Hippler 1993).

Fault movement

Fault timing determined from outcrop and microstructural evidence (Hippler 1993) suggests that two distinct episodes of movement occurred along the North Scapa Fault (Figs 2 & 3). The first domain was deformed and cemented because of earlier-faulting. The later event incorporated casts of the cemented sandstone into the breccia zone. Hydrocarbon staining was observed only in the second domain. The Mid Devonian source rocks in the Orcadian Basin reached peak maturation just before Permian basin uplift. The Eday Sandstone is thicker in the hanging wall of the North Scapa Fault during deposition of the lower half of the formation. This thickening suggests that the initial movements on the North Scapa Fault were syn-depositional and occurred during basin extension in Eday Group times (Astin 1986). This extensional faulting probably continued as the basin subsided and sediments became lithified while buried to depths of up to 2.5 km.

Many of the faults in the Orcadian Basin reactivated during the Permian uplift. Several undeformed Permian dykes exposed at this locality cross-cut deformation features associated with the North Scapa Fault. Thus, the fault reactivation probably occurred during the Permian just after hydrocarbon expulsion from the underlying source rocks and before dyke intrusion.

Stratigraphy and depositional environment

Basement

The oldest rocks found in Orkney (Fig. 2) are granitic-gneiss, migmatite and schist exposed in the West Mainland at Yesnaby, Stromness and Graemsay. They represent part of the metamorphic core of an orogen. These basement rocks in Orkney bear a strong resemblance to migmatized Moine rocks found within the Kirtomy Nappe above the Swordly Thrust in eastern Sutherland (Strachan pers. comm. 2003). These rocks, and by inference the Orkney inliers, with mica schists, psammites and subordinate semipelites, belong to the Loch Eil Group of Moine sedimentary rocks with a depositional age of about 950 Ma. Rims of zircon crystals, which grew during migmatization, have given ages of 461 ± 13 Ma for the Kirtomy assemblage and 467 ± 10 Ma for the Naver assemblage (Strachan 2003). This shows the presence of a Mid Ordovician (Taconic) tectonothermal event. This is in contrast to the age of migmatization found in Moine rocks of the central Highlands where the main tectonothermal event is 840 ± 11 Ma (Woodcock & Strachan 2012). Detrital zircons within the sedimentary protolith of these migmatites fall in the age range of 1850–1000 Ma showing that deposition of the Moinian sediments in Sutherland probably occurred later than 1000 Ma (Kinny *et al.* 1999, 2003; Strachan 2003). However, new structural and geochronological data suggest that a pre-Caledonian basement similar to Moine schist was intruded by syn-tectonic Scandian granites in an extensional setting (Lundmark pers. comm.).

Lacustrine sediments

After uplift, exhumation and erosion, these basement hills of metamorphic rocks formed islands in the Mid Devonian (*c.* 390 million years) Lake Orcadie. Rivers flowed into the lake, bringing mud and silt. This settled on the lake bottom, forming the distinctive grey and black flagstone succession. These characteristic Orkney Flagstone Group (Astin 1990) lacustrine sediments developed as a series of monotonous cycles alternating from deep permanent lacustrine laminites (±2 m thick) to more sand-rich shallow playa lake sediments with ripple marks and mud cracks. These flagstones have a distinctive architecture of lacustrine, fluvial and aeolian sediments. The lack of substantive post-Devonian regional tilting gives stratigraphic continuity from the Eifelian to Givetian. The Flagstone Group is formed from Lower and Upper Stromness flagstones along with the Rousay Flagstone Formation and is 752 m thick. 'Lake Orcadie' has abundant evidence of emergence and non-deposition within this interval. The wide variety of desiccation features that are present, such as mud cracks and evaporite pseudomorphs (Rogers & Astin 1991), supports this.

In the arid climate (25° palaeosouth) evaporation was high and rainfall seasonal. Extensive sand and mudflats formed at the margins of the permanent lake that expanded and contracted across the area in response to climate change. This repetition of permanent lake deposition (laminites) and desert (playa lake) environment continued for several million years. New data on 108 lacustrine cycles in the Mid Devonian is presented below.

A typical cycle within the flagstones

At the base of the cycle (Fig. 5) there is dark-coloured grey flagstone representing deep-water lake sediments, ranging from about 0.2 to 5.4 m thick. When the lake dried out, the sediment

Fig. 5. A representative 5 m laminite cycle, West Shore, Stromness. Photo JFB, HY 251 077.

deposition passed through shallow water to subaerial siltstones and mudstones. These sedimentary layers were constantly being dried out and desiccated, giving rise to a series of mudcrack types, ranging from large to small and including 'synaeresis' cracks (Astin & Rogers 1991). The majority of the synaeresis cracks contain an infilling of sand-sized clastic particles unconnected to any overlying bed of similar grains. These were interpreted (Astin & Rogers 1991) as having formed when sand was transported across a dry lake surface under arid conditions, as opposed to acquiring such an infill under a permanent water column. Shallow-water siltstones and mudstones pass up into ripple marked siltstones and sandstones deposited in a sandflat environment. The water level was probably very shallow as this interval was still intermittently drying out. With deepening water the system passed into another laminated mudstone, again representing the deep lake facies and concluding the cycle (Fig. 5).

Lower Stromness Flagstone Formation

A basal breccia beach deposit fringing the granite gneiss Basement Complex in the Stromness area and reworked Yesnaby Sandstone at Yesnaby passes upwards into 277 m (52 cycles) of lacustrine flagstones. The top of the Formation is marked by the distinctive Sandwick Fish Bed, a 20 m-thick laminite (Marshall *et al.* 2007). This unit can be easily mapped around the western part of the Mainland of Orkney and can be extended offshore as a key stratigraphic marker, as shown in the sections (Fig. 6).

Upper Stromness Flagstone Formation

This unit comprises 285 m (25 cycles) of lake cycle deposits with a higher content of fluvial river sand and sheet flood deposits than the Lower Stromness Flagstones. The clastic sediment input was derived from the NW (Marshall *et al.* 2007). The top of the Formation is less distinct than the base. Recent stratigraphic revision (Leather 2017) places the top at the transition from a series of very sand-rich and thick cycles to a distinct set of much thinner cycles with current ripple marks in the sheet flood sands, indicating flow to the south and SW.

Rousay Flagstone Formation

The third Formation of the Orkney Flagstone Group is the Rousay Flagstone Formation of Givetian age. The Rousay Flagstone Formation is the most

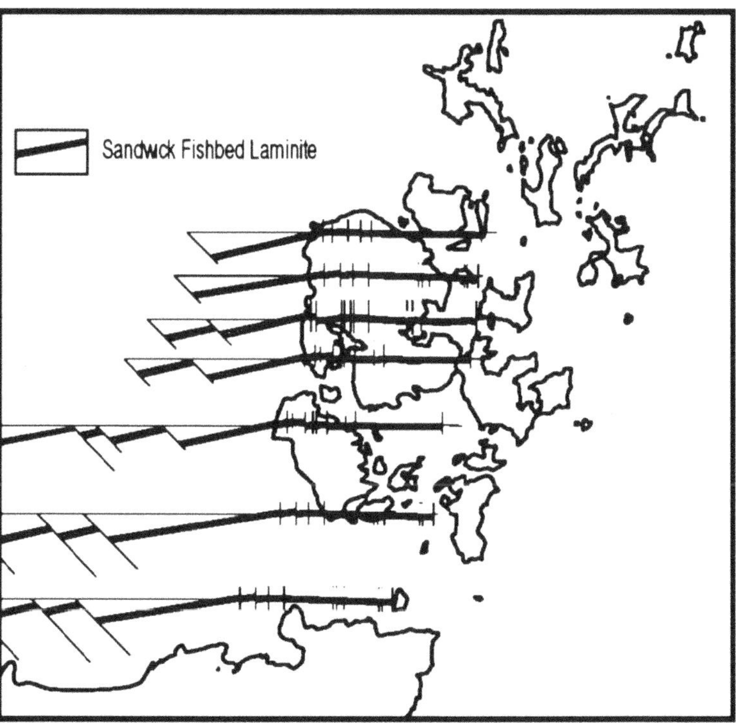

Fig. 6. East–west cross-sections across the West Orkney Basin and Mainland of Orkney showing position of the Sandwick Fish Bed.

widespread exposed unit found in Orkney, occurring over most of the Northern Isles, East Mainland, SE Hoy and parts of the southern islands. The base of the Rousay Flagstone Formation was originally placed at the first occurrence of the fossil branchiopod *Asmussia* (formerly *Estheria*, Wilson *et al.* 1935; Mykura *et al.* 1976). Subsequently, a lithostratigraphic framework was constructed (Astin 1990), which redefined the Orkney flagstones in terms of correlated lake cycles. In that framework, the Rousay Flagstone Formation was redefined as occurring between the 25th lakes cycle above the Sandwick Fish Bed cycle and the overlying Eday Group.

This basal fish bed of this formation is relatively thick and rich in fossil fish and stromatolites. This fish bed with the first occurrence of *Osteolepis panderi* (Michie *et al.* 2015) occurs above an especially thick cycle with poorly developed lake facies. This is followed by 150–230 m (18 cycles) of lake cycle deposits similar to the underlying Upper Stromness Flagstone Formation. While this definition is adequate for the north and western part of Orkney, further south the character of the base of the Rousay Flagstone Formation changes and is more difficult to define.

Near the top of the Formation, at Sacquoy Head, Rousay, is a distinctive pebbly sandstone, of considerable stratigraphic value (Astin 1990). This is known as the Sacquoy Sandstone Member, which thins from 17 m in Rousay to 4 m in the east of Orkney.

Lake laminite deposition timescale

Time constraints on Orcadian Basin Lake laminite deposition have been discussed by many authors with no real consensus (Andrews & Trewin 2010). Most agree, however, that it is a Milankovitch cyclicity that is detected in the flagstone sequences (as reviewed in Andrews & Trewin 2010); however, it is unclear whether the precession (20 000 years), axial tilt (41 000 years) or eccentricity (100 000 years) cycles are recorded. This timing of deposition is important for constraining the burial history of the lacustrine source rocks.

In order to constrain this problem, knowledge of the absolute timeframe of the Mid to Late Devonian is required along with the stratigraphic divisions. It is now generally accepted that the Givetian–Frasnian boundary in the Orcadian Basin (Fig. 2) is to be found no lower than the upper part of the Eday Group (including the Hoy Sandstones) in Orkney and John o'Groats Sandstone in Caithness (Marshall *et al.* 2011). This is because the Taghanic Onlap of latest Givetian age can be identified in the uppermost part of the Eday Marl sequence in Orkney and defines the base of the latest Givetian (Marshall *et al.* 2011).

Although the lower boundary of the latest Givetian is well defined in Orkney, the position of the Eifelian/Givetian boundary is less well known. Marshall *et al.* (2007) placed this boundary at the twenty-first cycle above the Sandwick Fish Bed based on the inception of the Givetian zone fossil *Geminospora lemurata*. The cycle 20 laminite fish beds, *c.* 1.5 m thick, is full of disarticulated plates of the placoderm fish *Dickosteus threiplandi* and marks its last occurrence. This is followed by five relatively barren cycles until the inception of *Osteolepis panderi* and the base of the Rousay Flagstone sequence.

Trewin & Thirwall (2002) placed an unconformity between the Lower and Middle Old Red Sandstone in the Lower Caithness Flagstone Group as the base of the Eifelian. In Orkney, the lowermost flagstones of the Lower Stromness Group rest on a basal breccia which, although not biostratigraphically constrained, is thought (Marshall 1996) to be close to, but above, the base of the Eifelian.

Absolute age dating in the Orcadian Basin

Very little absolute age dating has been done on the rocks of the Orcadian Basin. Most has related to the dyke rocks (Baxter & Mitchell 1984; Brown 1975; Mykura *et al.* 1976; Halliday *et al.* 1977; Lundmark *et al.* 2011; Macintyre *et al.* 1981) and fault minerals (Dichiarante *et al.* 2016) associated with the Permian inversion (250–313 Ma).

Annual varve measurement in laminites

The flagstones, Lower Stromness, Upper Stromness and Rousay Formations represent, on Orkney, a continuous period of cyclical lacustrine sedimentation. The cycles were punctuated by wet periods of laminite (Figs 7 & 8) formation in relatively deep-water conditions. The majority of these laminites have annual varves (Marshall *et al.* 2007) that range in thickness from 50 to 600 μm as measured from the Sandwick and Achanarras Fish Beds. New measurements (Fig. 8) have been made of varve thicknesses from different locations within the Sandwick, Achanarras and Mey (Caithness) fish beds and several other laminites throughout the sequence, making a total of 7598 individual measurements. These all have the same varve thickness range as previously reported. This yields an average deposition rate of 160 μm per year (6250 years per metre). Occasionally thicker intervals are observed consisting of fine silt to sand grain size, possibly indicative of localized sandstorms that brought windblown silt into the lake. Within the Lower Stromness flagstones the lake bed laminites vary in thickness from 20 cm to 5.4 m (Fig. 9), averaging about 1.5 m. A simple calculation, assuming no

Fig. 7. Varves and cycles in the Sandwick Fish Bed, Noust of Netherton, Stromness. (**a**) The general outcrop with location of (b) marked. (**b**) The subdecimetre-scale cycles and location of (c). (**c**) An enlargement to show the annual cycles. Note the prominent annual bands every 6 years. Photo JFB, HY 242 081.

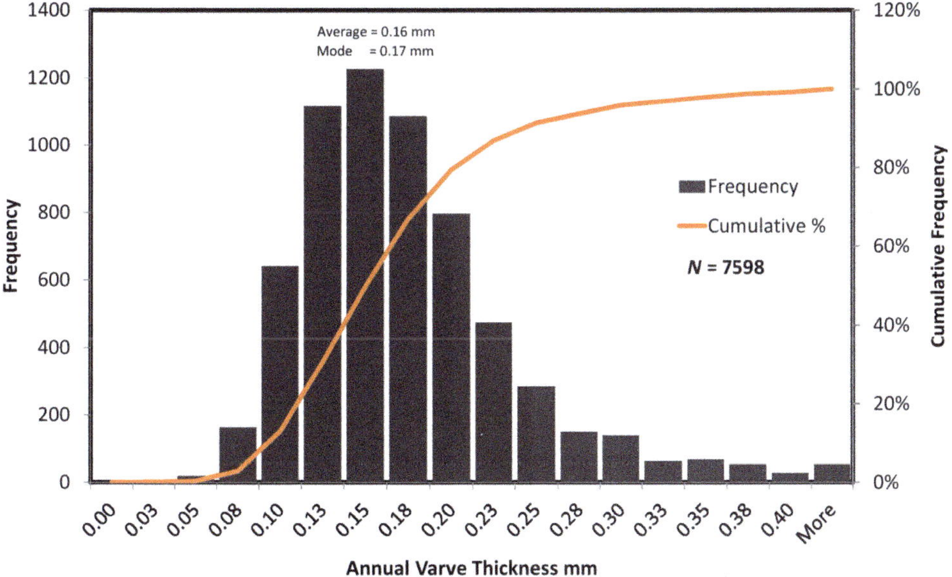

Fig. 8. A compilation of lacustrine varve thicknesses from Orkney. Based on 7598 individual measurements that give an average lamination thickness of 0.16 mm.

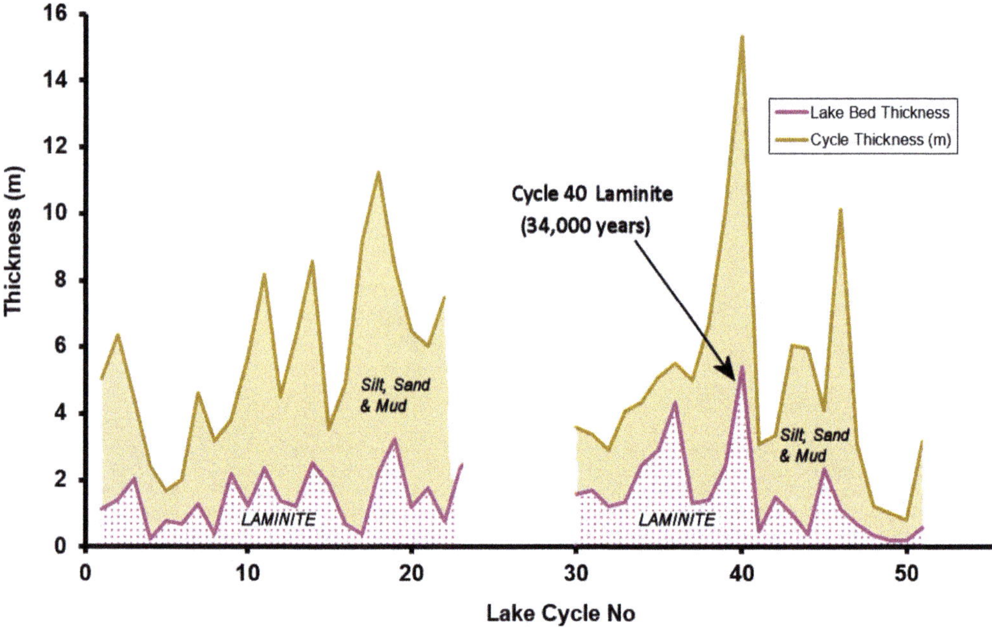

Fig. 9. Lower Stromness Flagstone Formation laminite thickness and cycle thickness plotted against cycle number. Note that approximately every fourth cycle is thicker. The cycles appear to become thicker through time. The gap is section cut out by a fault. The missing thickness was determined by correlation with the nearby Warebeth borehole.

depositional hiatuses within the laminite, gives an average period for the presence of a deep lake of 9500 years.

Only the deep-water lake laminites in each cycle display continuous deposition. The remainder, about two-thirds of the cycle, consists of mudstone, sandstone and siltstone, the deposition of which was discontinuous. The thickness distribution of the individual laminae or varves, as well as of the laminites and the whole cycle thickness, is controlled by the position within the lake basin varying from marginal 'near lake' to deep-water 'distal lake'. Using only the laminite thicknesses and assuming continuous deposition of annual varves, simple calculations based on 20 cm to 5.4 m thicknesses demonstrate deep-water permanent lake durations of 1250–34 000 years.

In common with previous authors (e.g. Andrews & Trewin 2010), we consider that each of these lacustrine cycles represents one Milankovitch cycle and the thickness differences are only related to the position within the topography of the lake during each cycle. Since the thickest 5.4 m laminite of cycle 40 represents 34 000 years, this eliminates the 20 000-year-old precession Milankovitch cycle as a candidate. It is observed (Fig. 9) that there are *c.* 11 peaks in the 51 cycles that represent about 417 000 years.

It is noted that both the cycle thickness and the laminites thickness steadily increase up the stratigraphic section until the final major flooding event of the 20 m-thick Sandwick Fish Bed estimated as lasting for some 125 000 years using the annual varve assumption and depositional rates summarized above. Whilst some sections have missing cycles, e.g. the gap in the Lower Stromness Flagstone Formation related to the presence of a reverse fault, equivalent sections and an adjacent BGS borehole (Brown 2003) have been used to form a complete succession for analysis.

Which Milankovitch cycle controls principal lacustrine cyclicity?

If the absolute age interval for these Orkney flagstone cycles was known, then it would be a simple matter to divide the time interval by the number of cycles measured, to gain an understanding of the control of Milankovitch cyclity on the lacustrine deposition. The present authors have measured the thickness of 108 cycles within the Orcadian Basin principally on the west coast of Orkney and in the North Isles for the Rousay Formation. Since the Eifelian/Givetian boundary is 20 cycles above the Sandwick Fish Bed (Marshall 1996), we have measured 72 cycles within the Eifelian. As the basal unconformity of the Orcadian lake is probably close to the base of the Eifelian (Marshall 1996), it is reasonable to say that there are 72 Eifelian lake cycles in Orkney.

The Mid Devonian stratigraphy has been poorly constrained in terms of absolute ages but values from GTS 2012 (Becker *et al.* 2012) give a value of 5.6 myr for the Eifelian. An order of magnitude answer to the question of which Milankovitch cycle controls cyclicity is given by dividing the 5.6 myr as the average length of the Eifelian by the 72 cycles measured in the Orkney flagstones, which yields an average age of 78 000 years for each cycle length. Therefore, since the maximum measured deep-water laminite as inferred from cycle 40 represents 34 000 years and the order of magnitude calculation indicates a cycle length of 78 000 years, then it can be conjectured that we are looking at an eccentricity signal between 90 000 and 100 000 years.

Lake laminite source rock distribution, quantity, quality and maturity

Source rock facies

The source rocks in the Orcadian basin are organic-rich, lacustrine laminites averaging 2 m thick. These laminites represent some 30% by thickness of the total flagstone cycles within the 'Lake Orcadie' lacustrine basin. They were deposited under standing water and below the wave base. Palynological examination reveals the presence of amorphous organic matter representing the product of microbial production being the major contributor to the internal lamination, probably indicative of seasonal blooms.

Seismic studies in the west Orkney Basin recognized a sequence overlying the basement that can be correlated with the Devonian onshore lacustrine source rocks (Bird *et al.* 2015). This sequence appears to be truncated at the unconformity associated with the late Carboniferous/early Permian inversion. Continued extension during the Mid Devonian lacustrine phase saw some reactivation along the onshore north–south faults and presumably also in the West Orkney Basin. This contemporaneous fault movement served to create a number of sub-basins (Fig. 12) that have local facies patterns including thickening of the individual cycles across such faults and differing source rock richness. This has been noted with particular reference to the East Scapa Fault (Astin 1986; Hippler 1993; Speed 1999). Much younger movement, post-maturation, on the north–south wrench fault systems separates sub-basins that have much higher thermal maturity (Hillier & Marshall 1992).

Geochemical analysis of the lake laminites (Karlsen pers. comm. 2016) shows pristane (Pr) to phytane (Ph) ratios ranging from 0.2 to 0.8 from deep-water lake laminites from the Lower Stromness Flagstone Formation (Fig. 9 – lake cycle numbers: 1–4, 6, 20, 21 and 52). These data strongly support the idea of lake bottom waters being anoxic. However Ghazwani *et al.* (2016) on the basis of 19 somewhat unrepresentative 'dolomitic and calcareous siltstone' samples from across the basin showed Pr/Ph ratios between 1.2 and 1.95 indicative of deposition taking place in oxygen-depleted but not completely anoxic environments, thus suggesting deposition in fluctuating oxic/anoxic bottom waters that were not necessarily representative of the deep-water laminites. Ferrous sulphate levels within the water column are interpreted to have been elevated compared with normal freshwater but lower than seawater (Rønningen 2015; Karlsen *et al.* 2016). High gammacerane indices, as observed in some Orkney rocks, are indicative of saline and stratified water columns and interestingly the highest value recorded in Orkney (Karlsen, pers. comm.) relates to one of the stromatolite-rich locations near the base of the Lower Stromness Flagstone Formation. Based on the lack of fish remains, this was considered to represent highly saline proximal environments favouring the extensive development of stromatolites free from grazing.

Source rock quantity and quality

Marshall *et al.* (1985) presented RockEval Oil Show Analyser and vitrinite reflectance measurements from over 600 samples across the basin, showing that the organic matter from onshore Mid Devonian source rock intervals comprises good-quality Type I and Type II kerogens. The samples were initially collected for palynological work and were biased towards the 70% of the lacustrine sediments that lacked amorphous kerogen. Subsequently many more samples were collected (Fig. 13), particularly from the laminites of the Orkney Flagstone Group, which had total organic carbon (TOC) values greater than 0.5%. Samples with less than this value of TOC tended to be highly mature, dominated by phytoclasts and spores of land plant origin and/or contain bitumen, which skewed the pyrolysis measurements (Marshall *et al.* 1985) and were thus excluded.

Compilation of available data for TOC for a range of Orcadian Basin lithologies ranges from 0.25% up to 4.5% while 166 samples from the laminites/fish beds (Fig. 10) ranged from 0.7 to 4.5% with an average of 1.96%. The modified van Krevelen diagram (Fig. 13) shows clearly that the laminites in the

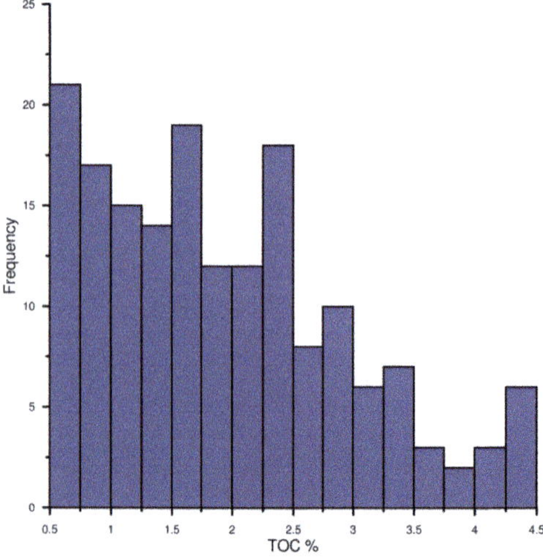

Fig. 10. Orcadian Basin Devonian lacustrine TOC (%) of laminites and siltstones with values greater than 0.5%, Upper Stromness Flagstones, Orkney.

Orkney area range from Type III to Type I kerogen with excellent oil-generative potential and they are all early to mid-mature for oil on Mainland Orkney.

Source rock maturity

Hillier & Marshall (1992; see also Marshall & Hewett 2003) investigated the maturity distribution from all of the the rocks in the onshore Orcadian Basin using spore colour and vitrinite reflectance. They noted that all of the rocks east of the Melby Fault in Shetland and south of Wick in Caithness were over-mature. The presence of the Mid Devonian Sandsting Granite in Shetland they considered potentially responsible and suggested the possibility that a similar unexposed Devonian granite south of Wick could have caused the elevated temperatures and thus over-maturity. However, their Figure 6 shows potential relationships to the set of north–south-trending wrench faults subparallel to the Great Glen Fault, which as noted above, could suggest a potential structural control on the maturity.

Various authors have published calculated burial histories for the Orcadian Basin. The representation (Fig. 14) is based on a number of these where it is considered that at least over the Caithness–Orkney–Shetland post-Carboniferous structural high, no more than 2000 m of sediment had been deposited on the Middle Devonian Flagstone units. The green hatched area depicts the calculated Time Temperature Index of 7–30, i.e. the early mature zone, and clearly shows that the majority of the Eifelian–Givetian flagstones throughout this long period of burial do not become over-mature, at least on Orkney.

This diagram of the burial history is only representing a single location centred on the structural high. The cross-sections are generated by the maturity program *Basin2* showing the evolution of maturity from west to east across the basin at different time intervals. The blue colour is the Eifelian/Givetian lacustrine flagstones while the red represents the calculated maturity at any point and time. These sections clearly show that the laminites within the Orkney Flagstone Group to the west and east of Orkney do not start to mature until the late Carboniferous to early Permian. The deposition of sediments in a passive continental margin environment continued until the Permian within the Orcadian Basin, thickening towards the marginal growth faults. Structural and microstructural analyses combined with Re–Os geochronology (Dichiarante *et al.* 2016) have dated syn-deformational fault infills (pyrite), suggesting that faulting, brecciation and fluid flow events are likely to have occurred during the Permian (267.5 Ma).

This event was associated with widespread carbonate-base metal sulphide mineralization (Dichiarante *et al.* 2016) and this was followed shortly afterwards by the influx of regionally persistent fracture-hosted hydrocarbons that were probably related to the regional phase of hydrocarbon migration. It is thought that this marks the time of uplift and inversion of the Orcadian Basin area and is probably synchronous with the main phase of lamprophyre dyke emplacement. This Permian age inversion event created mineralized thrust fault zones and formed regional north–south folding, and is responsible for anticline trap formation on Mainland Orkney (Fig. 11). Furthermore, the presence of bitumen in faults and sandstone porosity cut by the faults shows that hydrocarbon generation and migration bracket this time range and are consistent with the modelled burial history and maturity.

K/Ar dating of the regional suite of lamprophyre dyke rocks (252 ± 12 Ma and 249–268 Ma; Brown 1975; Baxter & Mitchell 1984 respectively) interacts with the faults confirming a later Permian age for some of these structures. Astin (1990) recognized hydrocarbons within porosity in the Lower Eday Sandstone in the baked alteration zone of a Permian dyke at Houton Head. As the contact metamorphism seals the porosity these hydrocarbons were present as a reservoired hydrocarbon prior to emplacement of the Permian dyke.

In summary, the source rocks underwent burial until the Permian structural inversion brought them to the surface (Fig. 14). Earlier generation and greater maturity would have existed offshore in the half-graben basins to the west and east.

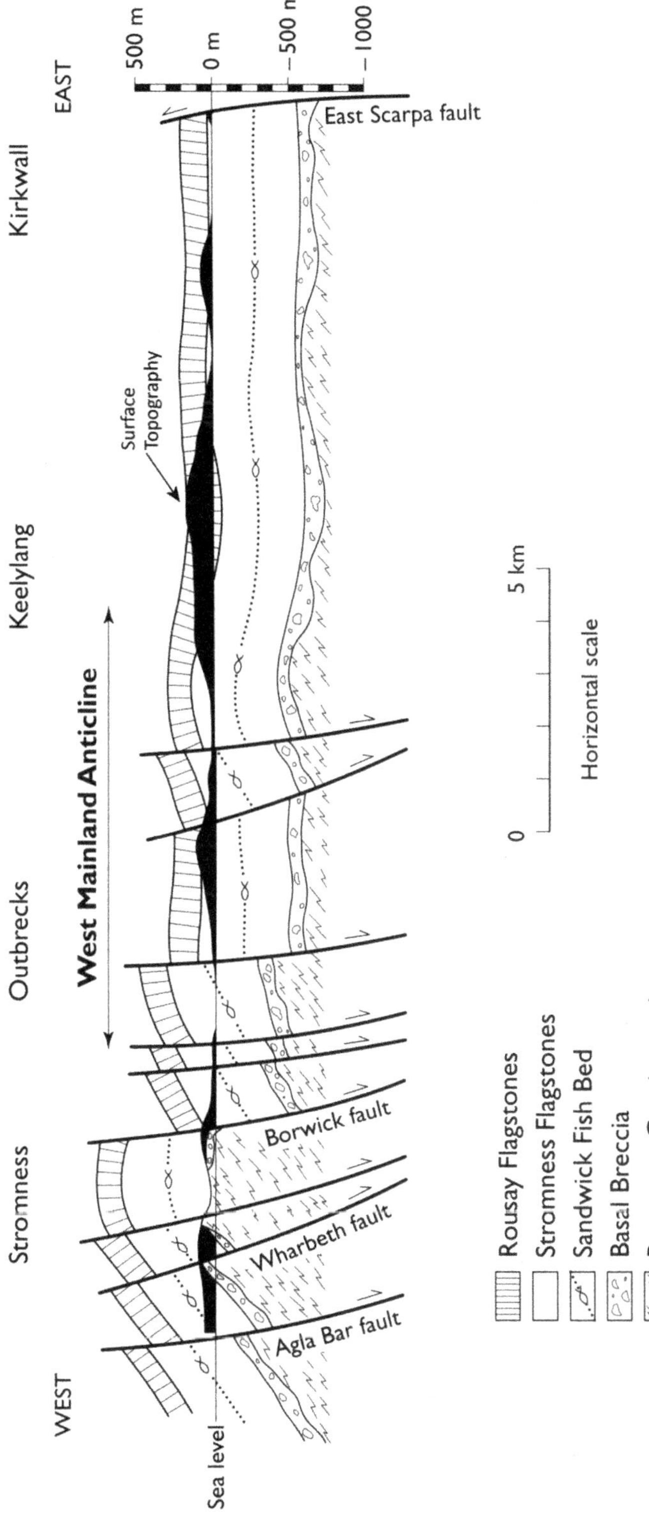

Fig. 11. Schematic geological cross-section and topographic profile, West Mainland, Orkney.

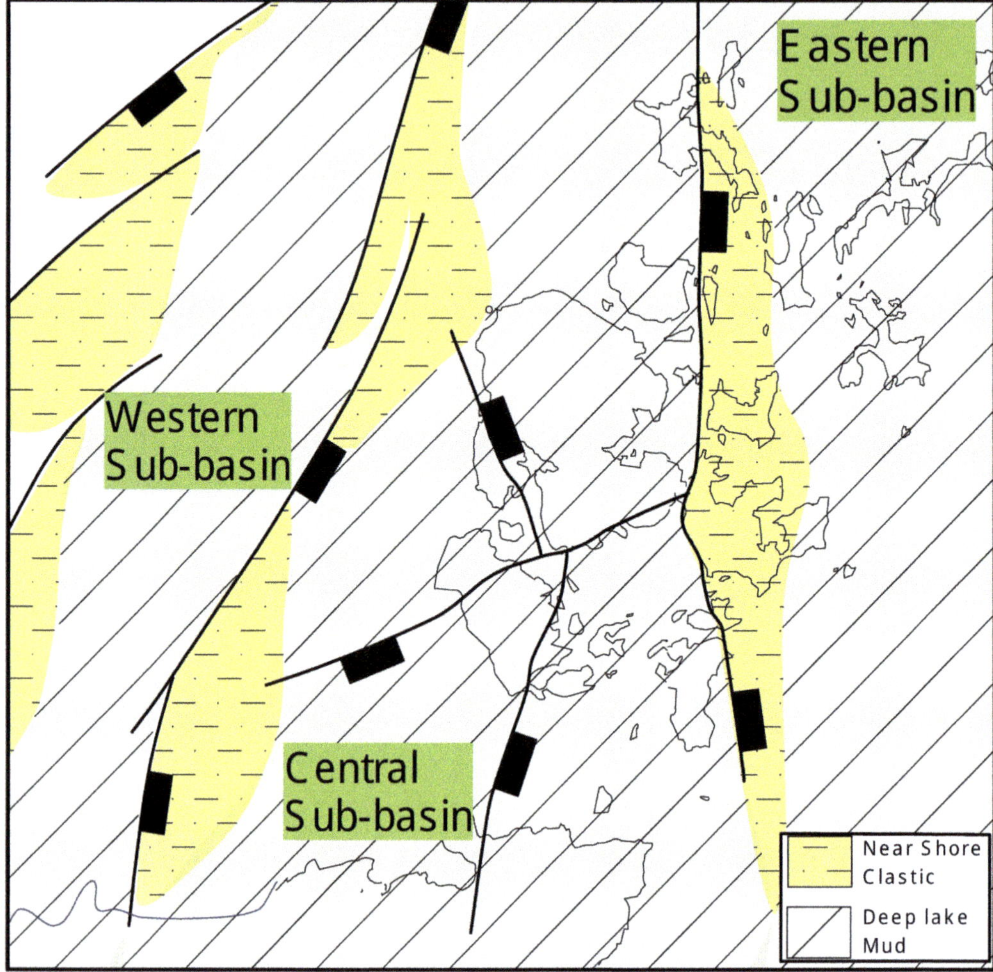

Fig. 12. Possible position of sub-basins within Lake Orcadie, maps based on position of probable faults active in the Devonian during Stromness Group lake deposition.

Apart from extensive occurrences of bitumen as degraded oil in onshore Devonian sandstones in Orkney, there is a viable source rock system in the Inner Moray Firth that has charged the Beatrice Field and satellites (Jacky). This is a very heavy and degraded oil which gives significant technical issues for the successful separation and identification of biomarkers. This attribution has not been without controversy with the oilfield initial claimed as entirely sourced from the Jurassic, a sequence that we now know to be thermally immature (Marshall 1998). It then was attributed to a mixed source (Peters *et al.* 1989, 1999; Bailey *et al.* 1990) before a final compilation by Duncan in Marshall & Hewett (2003) that used stable isotopes to prove a Devonian lacustrine source.

Reservoir outcrops, onshore Orkney

Reservoirs properties

The reservoir rocks consist of fluvial and aeolian sandstones, the latter showing the best reservoir properties. The sandstones are generally fine- to coarse-grained and porosity ranges from 15 to 25% and up to 2700 mD permeability (Owen 1994). Compaction of the sand grains can reduce porosity, which for some rocks can be up to 8% reduction per kilometre of burial.

These reservoirs have been breached (probably by Quaternary glacial erosion), releasing all of the light-end hydrocarbons, leaving pore space bitumen residues. We describe two thick aeolian

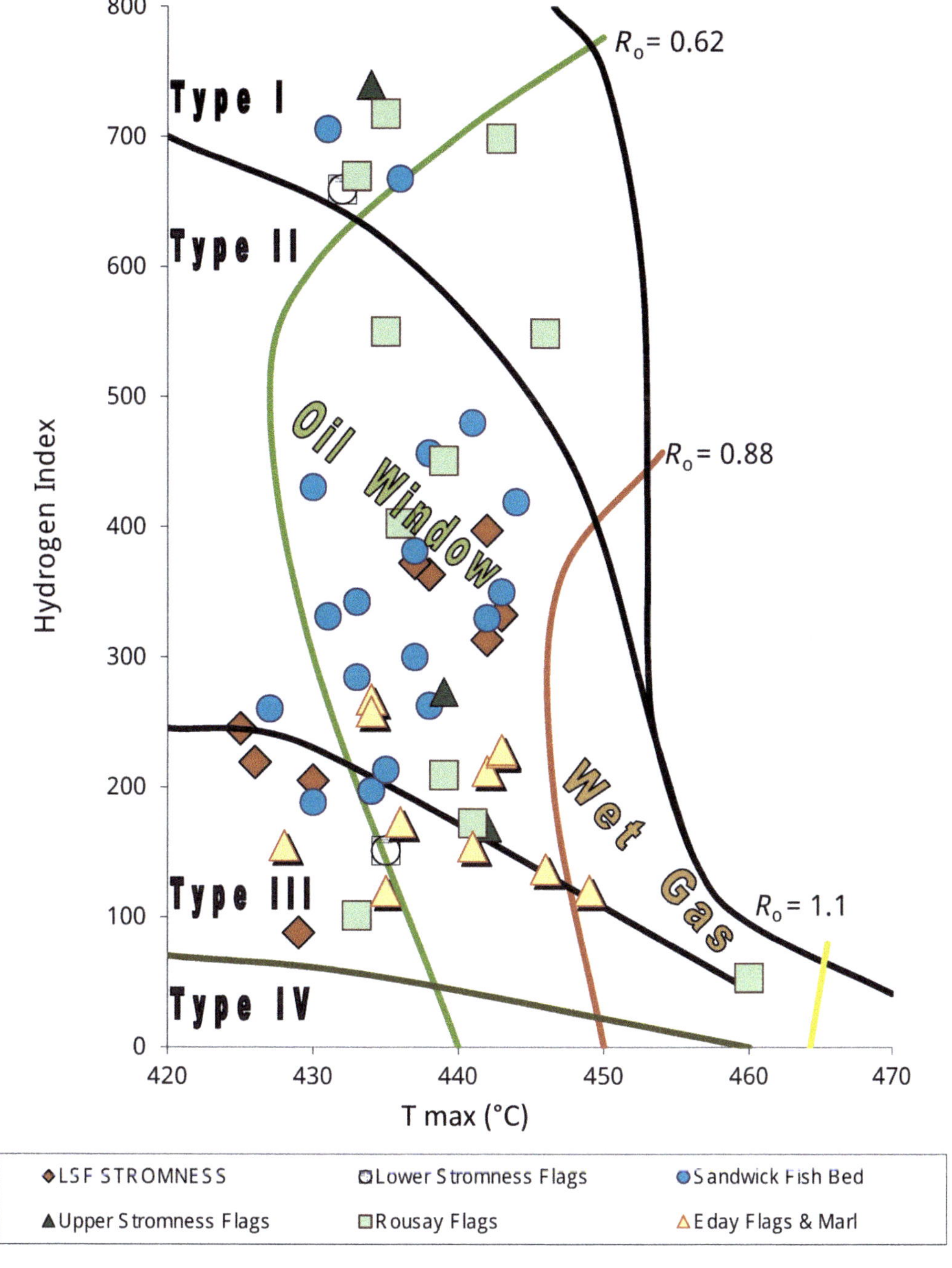

Fig. 13. Modified van Krevelen diagram showing Rock Eval results for the Orcadian flagstones and laminites with TOCs greater than 0.5%. Equivalent vitrinite reflectivity lines are shown (R_o) together with the pre-maturation position for Type I–IV kerogens.

reservoirs on the Orkney Mainland, at Houton Head (Lower Eday Sandstone) and Yesnaby (Yesnaby Sandstone). Thin fluvial sands and sheet-flood deposits 10–50 cm thick within the flagstone cycles also have bitumen residues in the pore space where simple traps existed along the West Mainland

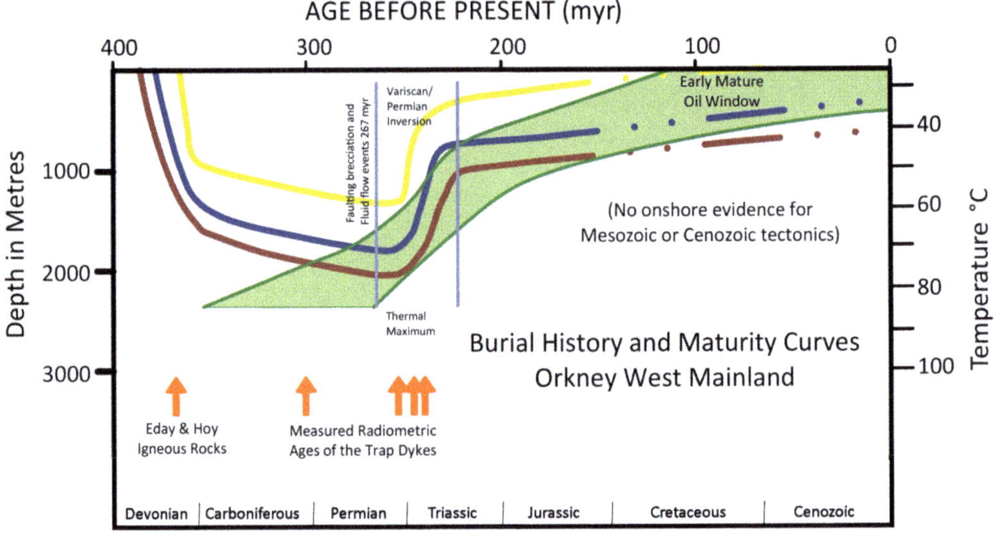

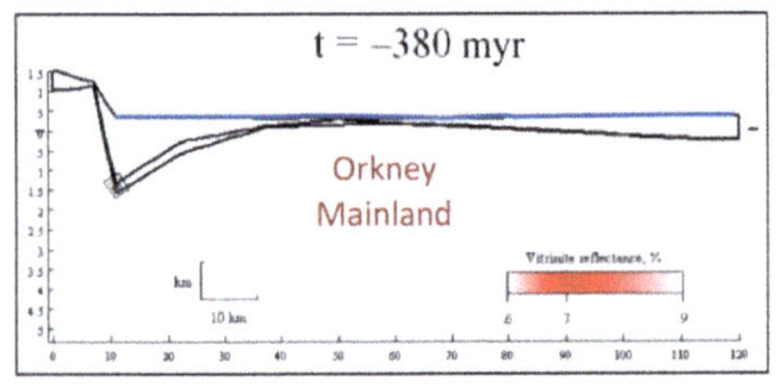

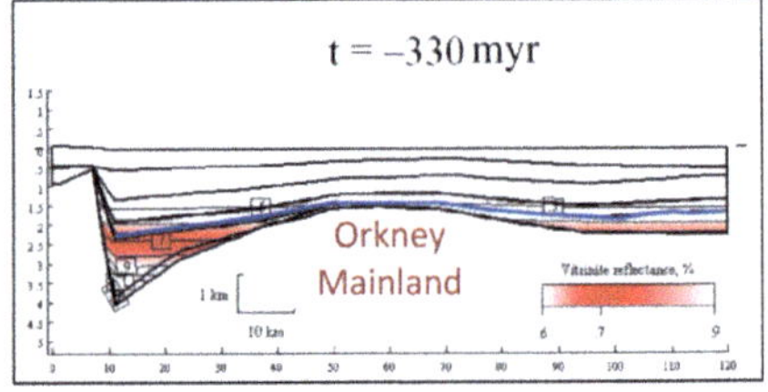

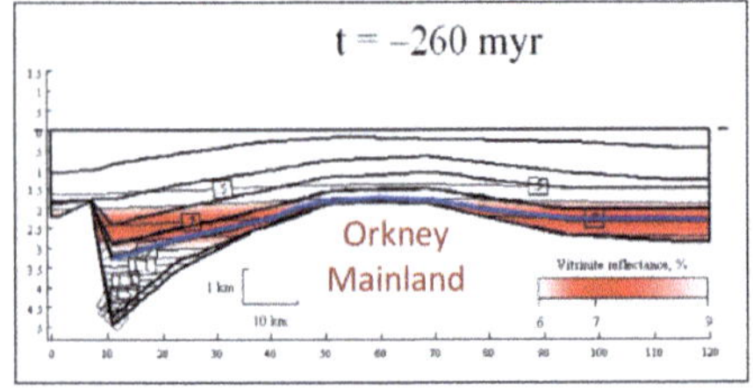

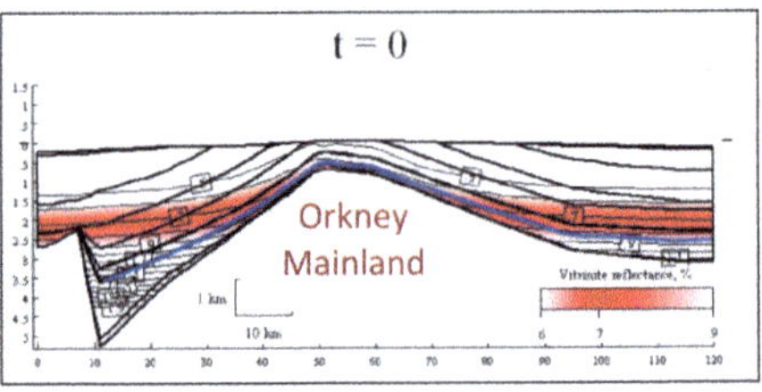

Fig. 14. Orkney Burial history curve from the high point of the West Mainland Anticline. The lower cross-sections are derived using the Basin2 maturity program from the West to East Orkney basins. They are at the times shown ($t = -380, -330, -260, 0$; myr, million years). The blue colour is the Eifelian/Givetian lacustrine flagstones while the red represents the calculated maturity at any point and time. Note that the 260 Ma section indicates maximum regional maturity and the potential migration is similar to that concluded by Dichiarante *et al.* (2016).

Anticline (Fig. 11). Also present are former traps based on unconformities, faults and pinchouts. Sealing formations consist of the lake laminite, mudstones (both source and seal), flagstones and volcanic rocks.

The large variations in porosity and permeability observed in Orkney (Owen 1994) relate principally to the presence of calcite cement, quartz overgrowth and detrital grain dissolution. Replacement by residual bitumen (indicative of hydrocarbon migration) is common at Yesnaby and Houton Head. Residual bitumen in sandstone porosity is ubiquitous throughout the Devonian of northern Scotland. Besides being found in the pore space, bitumen occurs in tension gashes, joints and mineral veins (Parnell 1983); it also occurs as blebs and stringers within the sandstones and within the organic-rich laminites. This bitumen is sometimes associated with metallic sulphides dated at 267.5 ± 3.4 Ma (Dichiarante *et al.* 2016). This bitumen is also found across Orkney and in the past was called 'cloustonite' (Heddle 1880, 1901) after the Orkney naturalist Rev. Charles Clouston of Sandwick, who collected and described specimens at Inganess Bay. The bitumen/cloustonite is dead hydrocarbon as it does not fluoresce. There is no consensus on the relative position of the residual bitumen in the pore-filling sequence but it is noted as post-dating the barite veins at Yesnaby. In some areas the bitumen filling pores can reach up to 22%, particularly at Houton Head.

Liquid hydrocarbons as a component of the fluid phases in fluid inclusions on quartz overgrowths from Devonian sandstones from Yesnaby do fluoresce a green/yellow in UV light (Owen 1994). Monte Carlo simulation of the homogenization temperatures of the oil-bearing fluid inclusions yields an average of 2.4 ± 1.0 km for the sedimentary overburden on the Caithness–Orkney–Shetland structural high (Owen 1994). At the present time, it is very common for sandstone outcrops in Orkney to contain degraded hydrocarbon in the form of bitumen.

The Houton Head Lower Eday Sandstone Formation reservoir

The Lower Eday Sandstone Formation varies in thickness across Orkney from about 30 to 180 m (Figs 2 & 3). Three interacting environments are recognized (Astin 1986), sandy braided rivers, aeolian dunes and lake margins. Two river systems were recognized, a northern, southeasterly-flowing alluvial fan originating from the Eday area and a southern, northeasterly-flowing alluvial fan from Caithness. These rivers prograded eastwards across the lacustrine basin. The East Scapa Fault was interpreted to be active at the time of deposition and dominantly north and NW winds reworked the alluvial plain sands, building up a thick dune field (Fig. 3) on the western part against the fault scarp (Astin 1986). Stratigraphically the Lower Eday Sandstone Formation fluvial sandstone unit lies above a transition zone from the top of the Rousay Flagstone Formation through 'passage beds' into the amalgamated sandstones of the Eday Group. Contemporaneous haematite staining forming the red sandstones probably relates to localized uplift with early diagenesis in the vadose zone (Astin 1986).

The aeolian dune deposits in the Lower Eday Sandstone Formation form an excellent potential reservoir with porosity in the range 16–26% and permeability in the range 1–2700 mD (Owen 1994). Its distribution, up to 80 m thick, exposed on the eastern side of Orkney and linked to a proven basin-wide distribution of similar sandstones, makes this an attractive target. The bitumen staining can be seen both in outcrop (Fig. 15 and see Fig. 16c for a surface polished hand specimen) and in thin section (Fig. 16b), where it is pore filling with clear evidence for degradation in the coarser-grained layers.

At Houton Head the cross-section (Fig. 17b) illustrates three potential traps for hydrocarbon accumulation: northern anticline, central dipping fault

Fig. 15. Visible bitumen in porosity, aeolian sandstone, Lower Eday Sandstone, Houton Head. Scale bar is 25 cm. Photo JFB, HY 304 035.

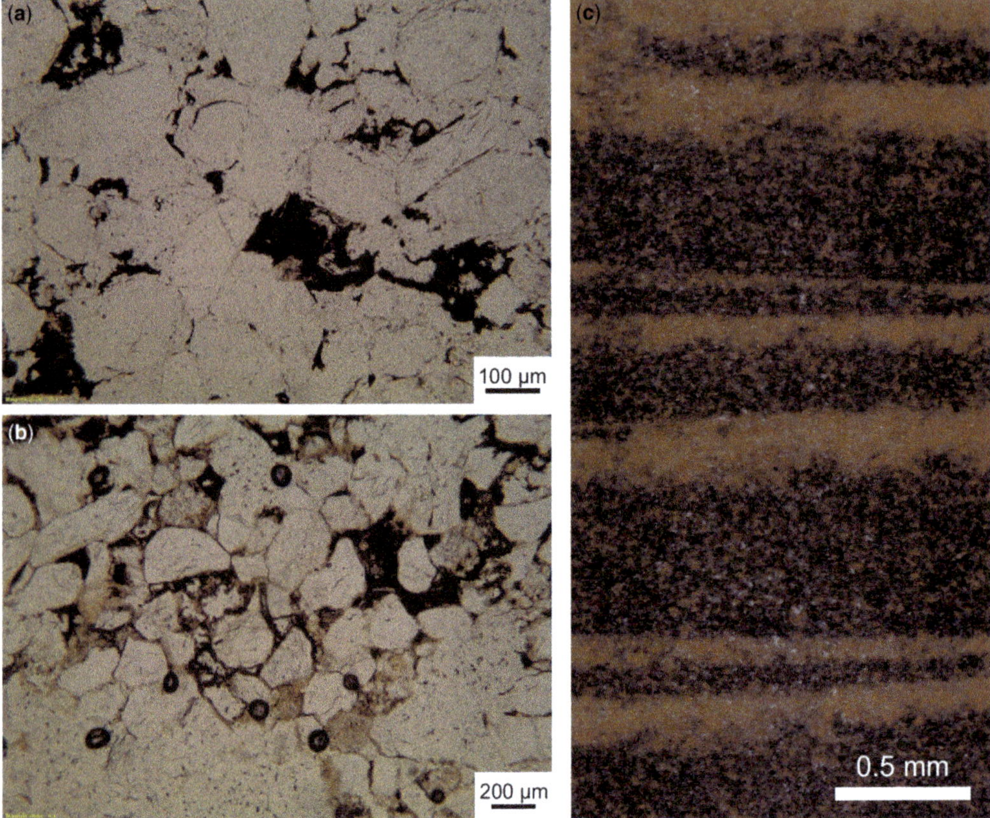

Fig. 16. (**a**) Thin section of aeolian sandstone from Yesnaby with residual bitumen retained within porosity. (**b**) Thin section of Houton Head reservoir sandstone showing bitumen (brown-black) filled porosity separated by areas where the bitumen has been removed by surface degradation. (**c**) Polished surface of a hand specimen of aeolian sandstone from Houton Head. There is a contrast between the bitumen-filled and -stained layers (black) and coarse-grained winnowed sand (white) where the fluid hydrocarbon has been removed by surface degradation. Scale bars as shown.

trap and a southern anticline sealed against the North Scapa Fault. Direct evidence for the original presence of hydrocarbon is only observed at the 'Houton Outcrop' (Fig. 17a) while the other potential traps are speculative.

The central former trap at Houton was shaped (Fig. 17a, b) by the reservoir sandstones dipping south away from the faults and bottom sealed by the Rousay Flagstone Formation. The top seal (cap rock) was not observed in the area. As the possible seal of the Eday Flagstones outcrops to the SE and east of Houton, the facies affinities of the former cap rock are likely to be analogous to the North Hoy facies. In Hoy there is a local unconformity on top of the Lower Eday Sandstone overlain by impermeable cemented volcanic ashes and lava flows.

With a spill point alongside the North Scapa Fault and assuming 23% porosity, 25% water saturation, 225 ft height of oil column and 1.2 volume factor, the accumulation covering a minimum 728 acres could have contained 184 MMBOOIP (Table 1).

The two other potential reservoir intervals in the vicinity with similar properties – a northern anticline trap and an anticline next to the North Scapa Fault – could each speculatively hold over 354 and 218 MMBOIP (Table 1).

Yesnaby Sandstone Reservoir

In Yesnaby around Crua Breck there is an exhumed hill of granite gneiss and schist (Fig. 18). The palaeogeomorphology is interpreted as an exposed feature in the early Eifelian with an estimated palaeorelief of *c.* 200 m (Fig. 19). Deposited against this basement high are conglomerates, mudstone and pebbly arkosic sandstone, forming a scree deposit known as the Hara Ebb Formation. Overlying this and pinching out up the hill are the sands of the Yesnaby

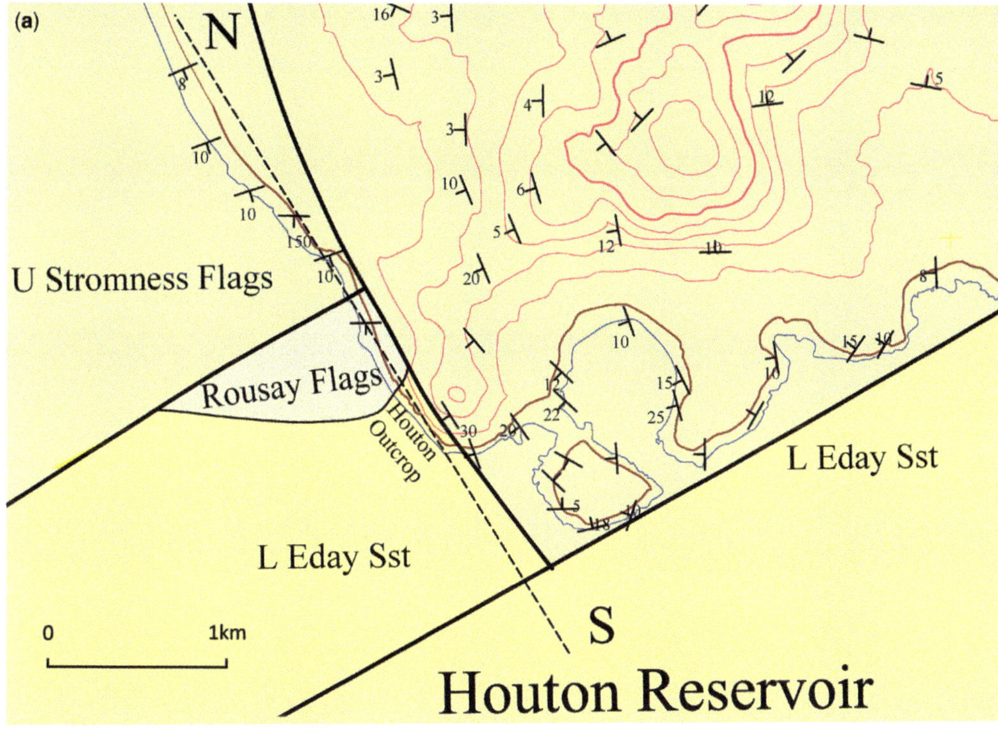

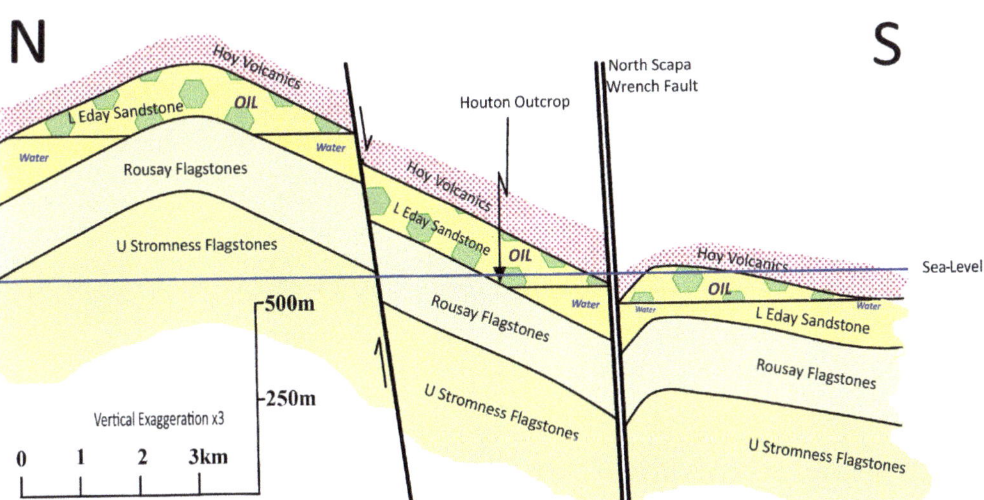

Fig. 17. (**a**) Geological map of the Houton Head Lower Eday Sandstone Formation reservoir. (**b**). Schematic cross-section of the Houton Head hydrocarbon reservoir.

Sandstone Formation. These are then unconformably overlain by the Lower Stromness Flagstone Formation. There is some uncertainty about the relative age of these formations with respect to the lacustrine sedimentary rocks of the Lower Stromness Flagstone Formation, but we interpret the Hara Ebb and Yesnaby Sandstone as differing facies equivalents of the basal breccia outcropping in the Stromness

Table 1. *Potential hydrocarbon volume estimates for Houton sandstone reservoir traps where best estimates are made for extent, porosity, hydrocarbon column height and water saturation*

	Oil column height (ft)	Volume of contour interval (acre/ft)	Original oil in place, stock tank barrels (STB) per reservoir layer	
Houton area				
North Anticline	225	317 800	354 433 236	0.23 Porosity
				0.25 Water saturation
Central Monocline	225	164 575	183 545 783	0.75 Oil saturation
South Anticline	225	195 787	218 356 190	1.2 Formation factor
		Total	**756 MMBOIP**	

Table 2. *Potential hydrocarbon volume estimates for Yesnaby sandstone reservoir traps where best estimates are made for extent, porosity, hydrocarbon column height and water saturation*

Contour height (ft)	Volume of contour interval (acre/ft)	Original oil in place, stock tank barrels per reservoir layer	
Yesnaby main area			
164	201 057	214 472 605	0.22 Porosity
164	153 225	163 448 881	0.25 Water saturation
164	100 529	107 236 303	0.75 Oil saturation
164	76 612	81 724 440	1.2 Formation factor
164	56 345	60 104 218	
131	19 781	21 101 337	
	Total	**648 MMBOIP**	
Yesnaby speculative NE area			
328	265 914	283 657 316	0.22 Porosity
328	154 036	164 313 689	0.25 Water saturation
164	22 700	24 214 649	0.75 Oil saturation
131	4216	4 497 006	1.2 Formation factor
	Total	**477 MMBOIP**	

area. The Geological Survey (Wilson *et al.* 1935) have regarded the Hara Ebb and Yesnaby Sandstone formations as Lower Devonian (i.e. beneath both an unconformity and dated Mid Devonian sediments). Therefore, they have generally been regarded as pre-Eifelian in age (e.g. Mykura *et al.* 1976) in common with similar but palynologically dated occurrences in Caithness (Collins & Donovan 1977) and the Loch Ness (Mykura & Owens 1983) areas. This was predicated on the recognition of a local 10° angular unconformity and a quartz pebble lag deposit in the Old Millstone Quarry at the Point of Qui Ayre.

Two facies are recognized in the Yesnaby Sandstone Formation. The lower facies lying above the Hara Ebb Formation is a carbonate-cemented fine- to medium-grained well-sorted sandstone (Figs 16a & 20) with rounded grains which has developed clear cross-bedding indicative of aeolian deposition. The orientation of the cross bedding suggests a wind direction coming from the west or NW. Lying above the aeolian sediments, the ripple-bedded sandstones in the upper fluvial facies are about 10 m thick and contain numerous worm burrows and larger infilled burrows of arthropods. The upper part of the fluvial sandstone passes into reworked lacustrine sedimentary rocks, eventually becoming flagstone and directly correlatable with the Stromness section.

The Yesnaby Sandstone Formation consists of over 200 m of aeolian sandstone on-lapping basement hills. The low porosity facies of the sandstone still holds degraded bitumen clearly visible both at outcrop (Figs 20 & 21) and in thin section (Fig. 16a). The higher porosity sandstones were similarly charged with oil, now lost. The reservoir thickness is zero at the pinchout in the south and thickens to over 200 m northwards. Porosity is 13–25% with permeability of 3–2000 mD (Owen 1994) – this quality of reservoir could give high recovery factors. The source and seal are interpreted as the overlying Lower Stromness Flagstone Formation (Fig. 21).

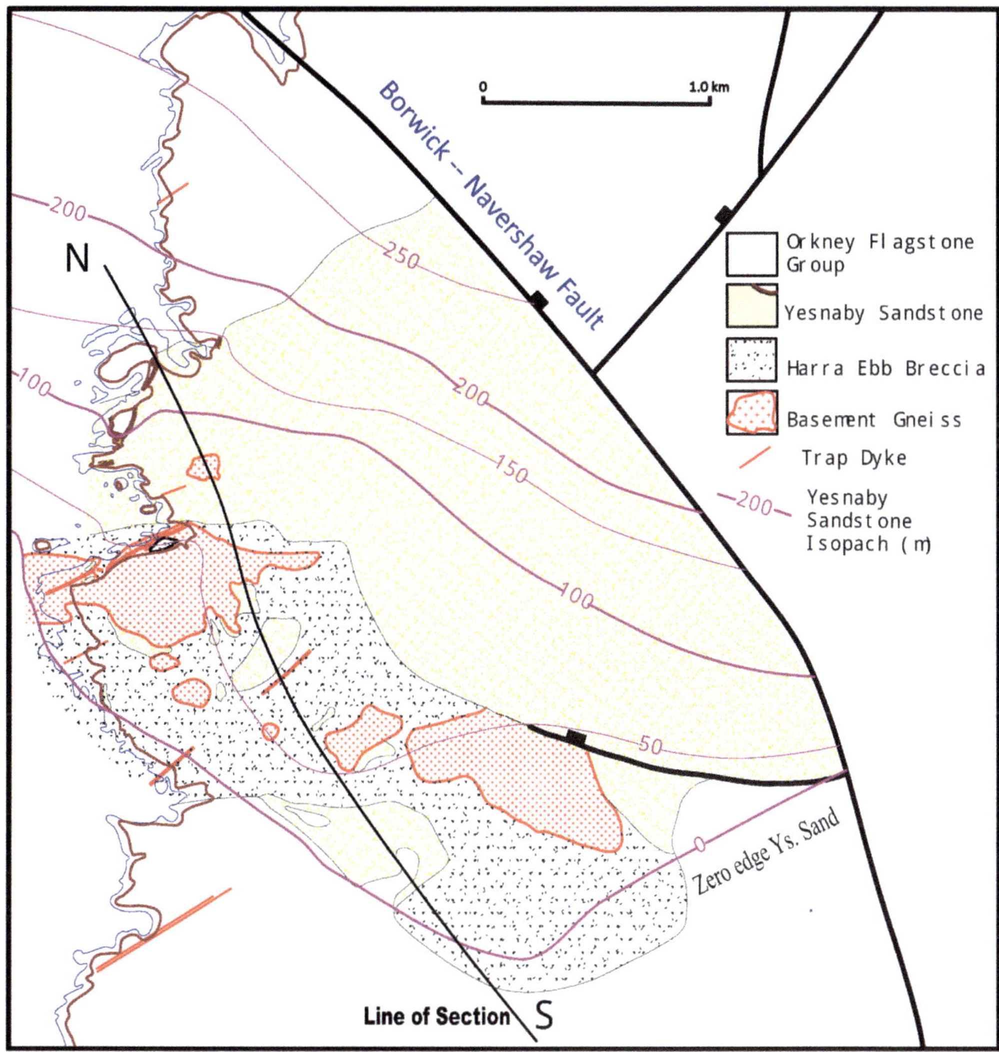

Fig. 18. Geological sketch map of the Yesnaby aeolian sandstone reservoir. Bottom sealed by Harra Ebb Formation, top sealed by the fluvial sands and the Lower Stromness Flagstone. Also shown is the isopach of the oil stained sandstone reservoir thickness up to 280 m as measured in the BGS Yesnaby borehole (HY 2241 1599). These sands thin out over the granite gneiss hillside/island to the zero edge in the south.

The potential reservoir volume for this area shown on the map (Fig. 18) was calculated from the isopach map constructed by subtracting the Harra Ebb structure from the Base Stromness Flagstone structure. Confirmation of this isopach comes from the direct measurement of over 200 m of sandstone containing traces of bitumen in a BGS 1973 borehole. This was drilled near the car park in the northern part of the area (Uisdean Michie pers. comm.).

Simple pyramid rule volume calculations (Table 2) with assumed 22% porosity, 25% water saturation and 1.2 volume factor yield 648 MMBOOIP. We can carry the contouring, with less control, towards the NW beyond the map boundary (Fig. 18). This gives a speculative extension of the structure, which could add at least another 477 MMBOOIP.

The boundary fault on the east side of the Yesnaby area (the Borwick Navershaw Fault) has a post Middle Devonian throw of over 250 m with flagstone forming a lateral seal. By analogy with other faults in the area, the movement is synchronous with early oil generation and migration (Fig. 14).

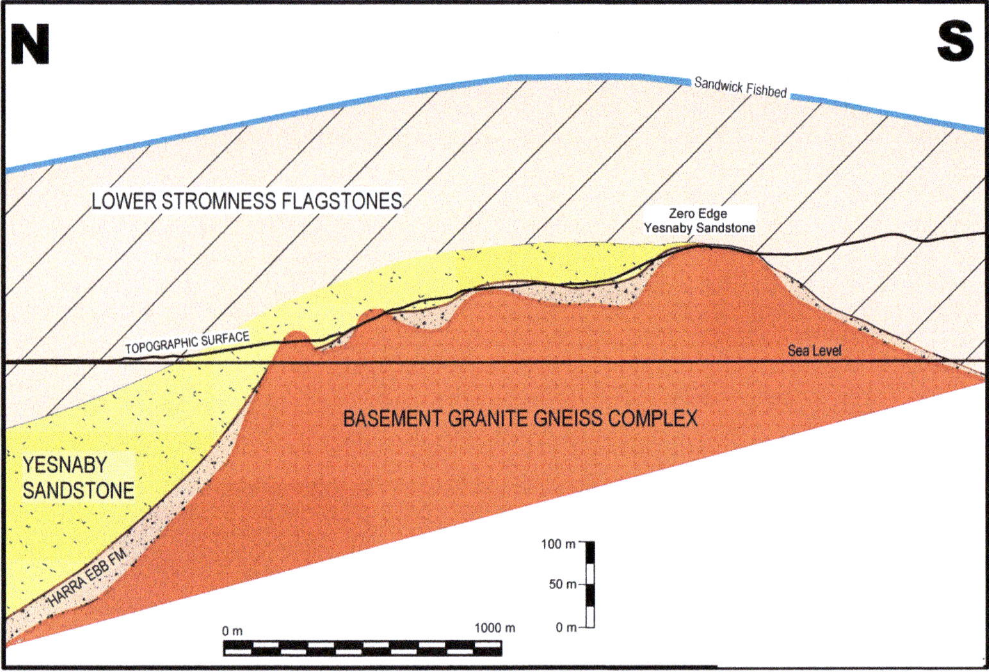

Fig. 19. Cross-section of the Yesnaby area. The block has been tilted to the NW by about 7° and differential compaction is observed over the basement granite gneiss hill to seal the trap.

Fig. 20. Yesnaby oil-stained aeolian sandstone at outcrop. Photo JFB, HY 219 154.

Fig. 21. The Castle of Yesnaby showing the bottom seal of the Harra Ebb Formation, oil-stained aeolian Yesnaby Sandstone (reservoir facies) with the top comprising marginal fluvial sandstones that are the transition to the overlying lacustrine flagstones that seal the system. Height of sea stack, 35 m. Photo JFB, HY 220 153.

One can speculate that hydrocarbon-charged reservoir rocks of the Yesnaby Sandstone Formation exist at depth on the eastern, downthrown, side of the fault. If the Borwick Fault is sealing and splits the potential reservoir and it is also full of hydrocarbon, then this would yield a further speculative 1125 MMBOOIP.

With the exclusion of this speculative trap adjacent to the Borwick Fault, taken together the separate Houton and Yesnaby volumes give a potential total of 1.88 billion barrels of oil formerly in place on Orkney.

Conclusions

In Orkney the evidence for an exhumed Paleozoic petroleum system has existed since the middle of the nineteenth century when the Rev. Charles Clouston collected and described samples of bitumen (the cloustonite of Heddle 1901). These were from Inganess Bay near Kirkwall and in the sandstones at Yesnaby. Since that time, numerous authors have mentioned the presence of this hydrocarbon residue in various Devonian sandstones. These sandstone reservoirs undoubtedly contained liquid hydrocarbons, as evidenced by fluorescent liquid in fluid inclusions. We have demonstrated that about 30% of the 750 m of lacustrine sediments of Eifelian and Givetian age consist of deep-water lake laminites with an average TOC of about 2%, and contain Type I and Type II kerogens, representing good to excellent potential source rocks. Over Orkney, NW Caithness and western Shetland (west of the wrench faults subparallel to the Great Glen Fault), Tmax, spore colour and vitrinite reflectance all confirm that these source rocks are early- to mid-mature.

Dichiarante *et al.* (2016), by establishing their Re–Os age of 267.5 ± 3.4 Ma, give a fixed point in time for major reversed faults movements, fault brecciation and mineralization. Bitumen present in these faults is indicative of hydrocarbon migration concurrent with this faulting episode, which is also probably concurrent with uplift and inversion of the Devonian Basin. The late Permian lamprophyre dyke emplacement (Brown 1975) is post-maturation and at 252 ± 12 Ma gives a fairly tight Permian time bracket for faulting, inversion, maturation, migration and dyke intrusion in this area.

The main anticlinal structure in the West Mainland of Orkney formed during the inversion episode, thus this principal trap formation was concurrent with migration, with few later tectonic movements disturbing the integrity of the reservoirs over a long period of time. The Yesnaby Sandstone Reservoir is a combination stratigraphic–structural trap. No evidence as yet exists for timing of breaching of the reservoirs and consequent expulsion of the light hydrocarbon fluids, but the extensive bitumen-stained sandstones provide significant insight into an exhumed Paleozoic petroleum system of direct relevance to equivalent strata deeply buried in adjacent offshore basins.

Acknowledgements We thank numerous research collaborators and colleagues who have contributed to Orcadian Basin studies during the past 40 years. Particular acknowledgements are due to Steve Hillier, Reuben Speed, Dave Rogers and Mark Owen. The efforts of Alison Monaghan as editor are particularly appreciated.

Funding This research received no specific grant from any funding agency in the public, commercial, or not-for-profit sectors.

References

Andrews, S.D. & Trewin, N.H. 2010. Periodicity determination of lacustrine cycles from the Devonian of Northern Scotland. *Scottish Journal of Geology*, **46**, 143–155, https://doi.org/10.1144/0036-9276/01-395

Astin, T.R. 1986. The palaeogeography of the Middle Devonian Lower Eday Sandstone, Orkney. *Scottish Journal of Geology*, **21**, 353–375, https://doi.org/10.1144/sjg21030353

Astin, T.R. 1990. The Devonian lacustrine sediments of Orkney, Scotland; implications for climate cyclicity, basin structure and maturation history. *Journal of the Geological Society*, **147**, 141–151, https://doi.org/10.1144/gsjgs.147.1.0141

Astin, T.R. & Rogers, D.A. 1991. 'Subaqueous shrinkage cracks' in the Devonian of Scotland reinterpreted. *Journal of Sedimentary Petrology*, **61**, 850–859.

Bailey, N.J.L., Burwood, R. & Harriman, G.E. 1990. Application of pyrolysate carbon isotope and biomarker technology to organofacies definition and oil correlation problems in North Sea basins. *Organic Geochemistry*, **16**, 1157–1172.

Baxter, A.N. & Mitchell, J.G. 1984. Camptonite–Monchiquite dyke swarms of Northern Scotland; age relationships and their implications. *Scottish Journal of Geology*, **20**, 297–308, https://doi.org/10.1144/sjg20030297

Becker, R.T., Gradstein, F.M. & Hammer, Ø. 2012. The Devonian Period. *In*: Gradstein, F.M., Ogg, J.G., Schmitz, M. & Ogg, G. (eds) *The Geologic Time Scale 2012*. Elsevier, Amsterdam, 559–601.

Bird, P.C., Cartwright, J.A. & Davies, T.L. 2015. Basement reactivation in the development of rift basins: an example of reactivated Caledonide structures in the West Orkney Basin. *Journal of the Geological Society*, **172**, 77–85, https://doi.org/10.1144/jgs2013-098

Brown, J.F. 1975. Potassium–argon evidence of a Permian age for the camptonite dykes: Orkney. *Scottish Journal of Geology*, **11**, 259–262, https://doi.org/10.1144/sjg11030259

Brown, J.F. 2003. The geology and landscape of Orkney. *In*: Omand, D. (ed.) *The Orkney Book*, Berlinn, Edinburgh, 1–24.

Collins, A.G. & Donovan, R.N. 1977. The age of two Old Red Sandstone sequences in southern Caithness. *Scottish Journal of Geology*, **13**, 53–57, https://doi.org/10.1144/sjg13010053

Coward, M.P., Enfield, M.A. & Fischer, M.W. 1989. Devonian basins of Northern Scotland. *In*: Couper, M.A. & Williams, G.D. (eds) *Inversion Tectonics*. Geological Society, London, Special Publications, **44**, 275–307, https://doi.org/10.1144/GSL.SP.1989.044.01.16

Dichiarante, A., Holdsworth, R. *et al.* 2016. New structural and Re–Os geochronological evidence constraining the age of faulting and associated mineralization in the Devonian Orcadian Basin, Scotland. *Journal of the Geological Society*, **173**, 457–473, https://doi.org/10.1144/jgs2015-118

Ghazwani, A., Littke, R., Sachse, V., Fink, R., Mahlstedt, N. & Hartkopf-Fröder, C. 2016. Organic geochemistry, petrology and palynofacies of Middle Devonian lacustrine flagstones in the Orcadian Basin, Scotland: depositional environment, thermal history and petroleum generation potential. *Geological Magazine*, **155**, 773–796.

Goldsmith, P.J., Hudson, G. & van Veen, P. 2003. Triassic. *In*: Evans, D., Graham, C., Armour, A. & Bathurst, P. (eds) *The Millennium Atlas: Petroleum Geology of the Central and Northern North Sea*. 105–128.

Guariguata-Rojas, G.J. & Underhill, J.R. 2017. Implications of Early Cenozoic uplift and fault reactivation for carbon storage in the Moray Firth Basin. *Interpretation*, **5**, SS1–SS21.

Halliday, A.N., McAlpine, A. & Mitchell, J.G. 1977. The age of the Hoy Lavas, Orkney. *Scottish Journal of Geology*, **13**, 43–52, https://doi.org/10.1144/sjg13010043

Heddle, M.F. 1880. Geognosy and mineralogy of Scotland. The Orkney Islands. Part II. *The Mineralogical Magazine and Journal*, **3**, 219–251.

Heddle, M.F. 1901. *The Mineralogy of Scotland*. D. Douglas, 148.

Hillier, S. & Marshall, J.E.A. 1992. Organic maturation, thermal history and hydrocarbon generation in the Orcadian Basin, Scotland. *Journal of the Geological Society*, **149**, 491–502, https://doi.org/10.1144/gsjgs.149.4.0491

Hippler, S. 1993. Deformation microstructures and diagenesis in sandstone adjacent to an extensional fault: implications for the flow and entrapment of hydrocarbons. *AAPG Bulletin*, **77**, 625–637.

Hitchen, K., Stoker, M., Evans, D. & Beddoe-Stephens, B. 1995. Permo-Triassic sedimentary and volcanic rocks in basins to the north and west of Scotland. *In*: Boldy, S.A.R. (ed.) *Permian and Triassic Rifting in Northwest Europe*. Geological Society of London, Special Publications, **91**, 87–102, https://doi.org/10.1144/GSL.SP.1995.091.01.05

Karlsen, D.A., Matapour, Z., Rønningen, A., Abay, T., Lerch, B., Flett Brown, J. & Backer-Owe, K. 2016. Evidences for Palaeozoic sourced oils on the Norwegian continental shelf, with links to Palaeozoic bitumen and source rocks from onshore Scandinavia and the Orkneys (sic). *In*: *Palaeozoic Plays of Northwest Europe*. Geological Society, London, Petroleum Group Abstracts.

Kinny, P., Friend, C., Strachan, R., Watt, G. & Burns, I. 1999. U–Pb geochronology of regional migmatites in East Sutherland, Scotland: evidence for crustal melting during the Caledonian orogeny. *Journal of the Geological Society*, **156**, 1143–1152, https://doi.org/10.1144/gsjgs.156.6.1143

Kinny, P., Strachan, R., Friend, C., Kocks, H., Rogers, G. & Paterson, B. 2003. U–Pb geochronology of deformed metagranites in central Sutherland, Scotland: evidence for widespread late Silurian metamorphism and ductile deformation of the Moine Supergroup during the Caledonian orogeny. *Journal of the Geological Society*, **160**, 259–269, https://doi.org/10.1144/0016-764901-087

Leather, D. 2017. Demarcation of the boundary between Middle Devonian Upper Stromness Flagstone and Rousay Flagstone formations in Westray, Orkney. *Scottish Journal of Geology*, **53**, 53–61, https://doi.org/10.1144/sjg2017-007

Lundmark, A.M., Gabrielsen, R.H. & Flett Brown, J. 2011. Zircon U–Pb age for the Orkney lamprophyre dyke swarm, Scotland, and relations to Permo-Carboniferous magmatism in northwestern Europe. *Journal of the Geological Society*, **168**, 1233–1236, https://doi.org/10.1144/0016-76492011-017

MacIntyre, R., Cliff, R. & Chapman, N. 1981. Geochronological evidence for phased volcanic activity in Fife and Caithness necks, Scotland. *Transactions of the Royal Society of Edinburgh: Earth Sciences*, **72**, 1–7.

Marshall, J.E.A. 1996. *Rhabdosporites langii*, *Geminospora lemurata* and *Contagisporites optivus*: an origin for heterospory in the progymnosperms. *Review of Palaeobotany and Palynology*, **93**, 159–189.

Marshall, J.E.A. 1998. The recognition of multiple hydrocarbon generation episodes: an example from Devonian lacustrine sedimentary rocks in the Inner Moray Firth, Scotland. *Journal of the Geological Society, London*, **155**, 335–352, https://doi.org/10.1144/gsjgs.155.2.0335

Marshall, J.E.A. 2000. Devonian miospores from the Walls Group, Shetland. *In*: Friend, P.F. & Williams, B.P.J. (eds) *New Perspectives on the Old Red Sandstone*. Geological Society of London, Special Publications, **180**, 473–483.

Marshall, J.E.A. & Hewett, A.J. 2003. Devonian. *In*: Evans, D., Graham, C., Armour, A. & Bathurst, P. (eds) *The Millennium Atlas: Petroleum Geology of the Central and Northern North Sea*. 65–81.

Marshall, J.E.A., Brown, J.F. & Hindmarsh, S. 1985. Hydrocarbon source rock potential of the Devonian rocks of the Orcadian Basin. *Scottish Journal of Geology*, **21**, 301–320, https://doi.org/10.1144/sjg21030301

Marshall, J.E.A., Rogers, D.A. & Whiteley, M.J. 1996. Devonian marine incursions into the Orcadian Basin, Scotland. *Journal of the Geological Society, London*, **153**, 451–466, https://doi.org/10.1144/gsjgs.153.3.0451

Marshall, J.E.A., Astin, T.R., Brown, J.F., Mark-Kurik, E. & Lazauskiene, J. 2007. Recognising the Kačák Event in the Devonian terrestrial environment and its implications for understanding land–sea interactions. *In*: Becker, R.T. & Kirchgasser, W.T. (eds) *Devonian*

Events and Correlations. Geological Society, London, Special Publications, **278**, 133–155, https://doi.org/10.1144/SP278.6

Marshall, J.E.A., Brown, J.F. & Astin, T.R. 2011. Recognising the Taghanic Crisis in the Devonian terrestrial environment and its implications for understanding land–sea interactions. *Palaeogeography, Palaeoclimatology, Palaeoecology*, **304**, 165–183.

Marshall, J.E.A., Glennie, K.W., Astin, T.R. & Hewett, A.J. 2018. The Old Red Group (Devonian) – Rotliegend (Permian) unconformity in the Inner Moray Firth. *In*: Monaghan, A.A., Underhill, J.R., Hewett, A.J. & Marshall, J.E.A. (eds) *Paleozoic Plays of NW Europe*. Geological Society, London, Special Publications, **471**, https://doi.org/10.1144/SP471.12

Michie, U., Newman, M. & Den Blaauwen, J. 2015. The vertebrate biostratigraphy of the Rousay sequence in the Middle Devonian of Orkney, Scotland. *Scottish Journal of Geology*, **51**, 149–156, https://doi.org/10.1144/sjg2014-011

Mykura, W. & Owens, B. 1983. The Old Red Sandstone of the Mealfuarvonie Outlier, west of Loch Ness, Inverness-shire. *Institute of Geological Sciences, Report*, **83/7**, 1–17.

Mykura, W., Flinn, D. & May, F. 1976. *British Regional Geology: Orkney and Shetland*. Stationery Office Books, UK.

Owen, M.A. 1994. *The controls on reservoir properties of devonian Sandstones in the Orcadian Basin, North East Scotland*. University of Reading.

Parnell, J. 1983. The distribution of hydrocarbon minerals in the Orcadian Basin. *Scottish Journal of Geology*, **19**, 205–213, https://doi.org/10.1144/sjg19020205

Parnell, J., Carey, P. & Monson, B. 1998. Timing and temperature of decollement on hydrocarbon source rock beds in cyclic lacustrine successions. *Palaeogeography, Palaeoclimatology, Palaeoecology*, **140**, 121–134.

Peters, K.E., Moldowan, J.M., Driscole, A.R. & Demaison, G.J. 1989. Origin of Beatrice oil by co-sourcing from Devonian and Middle Jurassic source rocks, Inner Moray Firth, UK. *American Association of Petroleum Geologists Bulletin*, **73**, 454–471.

Peters, K.E., Clutson, M.J. & Robertson, G. 1999. Mixed marine and lacustrine input to an oil cemented sandstone breccia from Brora, Scotland. *Organic Geochemistry*, **30**, 237–248.

Premier Oil, N.S.B.U. 2013. UK P.1577 Licence Relinquishment.

Richardson, N.J., Allen, M.R. & Underhill, J.R. 2005. Role of Cenozoic fault reactivation in controlling pre-rift plays, and the recognition of Zechstein Group evaporite–carbonate lateral facies transitions in the East Orkney and Dutch Bank basins, East Shetland Platform, UK North Sea. *In*: Doré, A.G. & Vining, B.A. (eds) *Geological Society, London, Petroleum Geology Conference Series*, **6**, 337–348, https://doi.org/10.1144/0060337

Rogers, D.A. & Astin, T.R. 1991. Ephemeral lakes, mud pellet dunes and wind-blown sand and silt: reinterpretations of Devonian lacustrine cycles in north Scotland. *In*: Anadon, P., Cabrera, L. & Kelts, K. (eds) *Lacustrine Facies Analysis*. Blackwell Scientific, Oxford, Special Publications International Association of Sedimentologists, **13**, 199–221.

Rogers, D.A., Marshall, J.E.A. & Astin, T.R. 1989. Devonian and Later movements on the Great Glen Fault System, Scotland. *Journal of the Geological Society*, **146**, 369–372, https://doi.org/10.1144/gsjgs.146.3.0369

Rønningen, A. 2015. *The first attempt to correlate the migrated bitumen from the Helgeland Basin cores to Devonian source rocks and oils from the UK Orcadian Basin-Is there a Devonian Orcadian type basin offshore Norway?* Masters thesis, University of Oslo.

Speed, R.G. 1999. *Kerogen Variation in a Devonian Half Graben System*. PhD thesis, University of Southampton.

Stoker, M., Hitchen, K. & Graham, C. 1993. *The Geology of the Hebrides and West Shetland Shelves, and Adjacent Deep-Water Areas*. The Stationery Office, UK.

Strachan, R.A. 2003. The metamorphic basement geology of Mainland Orkney and Graemsay. *Scottish Journal of Geology*, **39**, 145–149, https://doi.org/10.1144/sjg39020145

Thomson, K., Underhill, J.R., Green, P.F., Bray, R.J. & Gibson, H.J. 1999. Evidence from apatite fission track analysis for the post-Devonian burial and exhumation history of the northern Highlands, Scotland. *Marine and Petroleum Geology*, **16**, 27–39.

Torsvik, T.H. & Cocks, L.R.M. 2017. *Earth History and Palaeogeography*. Cambridge University Press, Cambridge.

Trewin, N. & Thirwall, M. 2002. Old red sandstone. *In*: Trewin, N. (ed.). *The Geology of Scotland*. The Geological Society, London, 213–249, https://doi.org/10.1144/GOS4P.8

Wilson, G., Edwards, W., Knox, J., Jones, R.C.B. & Stephens, J. 1935. *The Geology of the Orkneys*. HMSO, Edinburgh, Memoirs of the Geological Survey of Scotland.

Woodcock, N.H. & Strachan, R.A. 2012. *Geological History of Britain and Ireland*. Wiley-Blackwell, Chichester.

An overlooked play? Structure, stratigraphy and hydrocarbon prospectivity of the Carboniferous in the East Irish Sea–North Channel basin complex

T. C. PHARAOH[1]*, C. M. A. GENT[1], S. D. HANNIS[2], K. L. KIRK[1], A. A. MONAGHAN[2], M. F. QUINN[2], N. J. P. SMITH[1,3], C. H. VANE[1], O. WAKEFIELD[1] & C. N. WATERS[1]

[1]*British Geological Survey, Environmental Science Centre, Keyworth, Nottingham NG12 5GG, UK*

[2]*British Geological Survey, The Lyell Centre, Research Avenue South, Edinburgh EH14 4AP, UK*

[3]*2 Downsview Villas, Camp Road, Freshwater, Isle of Wight PO40 9HR, UK*

T.C.P., 0000-0002-0452-5088

**Correspondence: tcp@bgs.ac.uk*

Abstract: Seismic mapping of key Paleozoic surfaces in the East Irish Sea–North Channel region has been incorporated into a review of hydrocarbon prospectivity. The major Carboniferous basinal and inversion elements are identified, allowing an assessment of the principal kitchens for hydrocarbon generation and possible migration paths. A Carboniferous tilt-block is identified beneath the central part of the (Permian–Mesozoic) East Irish Sea Basin (EISB), bounded by carbonate platforms to the south and north. The importance of the Bowland Shale Formation as the key source rock is reaffirmed, the Pennine Coal Measures having been extensively excised following Variscan inversion and pre-Permian erosion. Peak generation from the Bowland source coincided with maximum burial of the system in late Jurassic–early Cretaceous time. Multiphase Variscan inversion generated numerous structural traps whose potential remains underexplored. Leakage of hydrocarbons from these into the overlying Triassic Ormskirk Sandstone reservoirs is likely to have occurred on a number of occasions, but currently unknown is how much resource remains in place below the Base Permian Unconformity. Poor permeability in the Pennsylvanian strata beneath the Triassic fields is a significant risk; the same may not be true in the less deeply buried marginal areas of the EISB, where additional potential plays are present in Mississippian carbonate platforms and latest Pennsylvanian clastic sedimentary rocks. Outside the EISB, the North Channel, Solway and Peel basins also contain Devonian and/or Carboniferous rocks. There have, however, been no discoveries, largely a consequence of the absence of a high-quality source rock and a regional seal comparable to the Mercia Mudstone Group and Permian evaporites of the Cumbrian Coast Group in the EISB.

The productive oilfields and gasfields of the East Irish Sea Basin (EISB) evidence a working, Carboniferous-sourced petroleum system. Whilst a great deal may be known of the Triassic reservoir and seal (Meadows *et al.* 1997), little is known about Carboniferous and Permian petroleum systems at depth and in adjacent basins that may offer significant additional potential. The presence of a Paleozoic hydrocarbon system in the East Midlands and southern North Sea is well documented, however (Fraser *et al.* 1990; Besly 1998; Fraser & Gawthorpe 2003). Following the Wood (2014) review, Paleozoic plays, including those of the greater Irish Sea area, were identified as priority for building regional digital datasets and stimulating exploration. In response, the '21st Century Exploration Roadmap (21CXRM) Palaeozoic Project', running from 2014 to 2016 and openly released in 2017, undertook regional-scale seismic and well interpretation, source and reservoir screening studies, and basin modelling. This paper provides a reinterpretation of the structural history of the greater Irish Sea, and its influence on potential Carboniferous and Permian prospectivity including the marginal basins.

The Carboniferous structure and stratigraphy of the UK sector of the East Irish Sea–North Channel region has been reviewed using all available well and seismic reflection data. The project interpreted about 40 000 km of 2D seismic data of many vintages from 1980 to 2000, with local infill from 3D data, to generate time and depth-converted surfaces for key Paleozoic horizons (Pharaoh *et al.* 2016*a*). Priority was given to the interpretation of long regional speculative lines, with infill from licence- and prospect-scale surveys. These surfaces were then used as the basis for an assessment of Paleozoic

From: Monaghan, A. A., Underhill, J. R., Hewett, A. J. & Marshall, J. E. A. (eds) 2018. *Paleozoic Plays of NW Europe*. Geological Society, London, Special Publications, **471**, 281–316.
First published online March 22, 2018, https://doi.org/10.1144/SP471.7

hydrocarbon prospectivity (Pharaoh *et al.* 2016*b*), which forms the core of this paper. For brevity, the seismic interpretations are summarized using synoptic diagrams ('cartoons'). The present economic focus of the hydrocarbon province is the Morecambe Bay Gasfield and its satellites, located within the EISB, a basin complex of Permian–Mesozoic age comprising a number of mainly north–south-orientated graben and intervening platforms (Jackson *et al.* 1987, 1995, 1997; BGS 1994). Together with the Worcester Graben and Cheshire Basin of the UK onshore, it forms part of the major Permian rift system extending through the north European province of the Pangaea supercontinent (Ziegler 1990; Chadwick & Evans 1995; Coward 1995; Scheck-Wenderoth *et al.* 2008). The principal structures of the EISB are strongly discordant to those in the pre-Permian substrate, which bear the imprint of a long and complex evolution culminating in the Variscan Orogeny in latest Carboniferous time. For these Devonian and Carboniferous tectonic elements, a new terminology is presented here, and the lithostratigraphical nomenclature of Jackson *et al.* (2011) and Waters *et al.* (2011) is used to integrate onshore and offshore successions, allowing more precise correlation than the scheme introduced by Jackson *et al.* (1997).

The Bowland Shale Formation is recognized as a prolific source of gas for the Permo-Triassic reservoirs (Armstrong *et al.* 1997), but potential Namurian and Westphalian reservoirs suffer from low porosity and permeability due to the combined effects of Variscan inversion, deep burial in a Permian–Mesozoic rift, Cenozoic inversion, magmatism and thermal effects associated with the rifting of the North Atlantic (Meadows *et al.* 1997; Quirk & Kimbell 1997). Several areas on the margins of the EISB (the Manx-Furness Ridge, Cumbrian margin, Fylde margin and Cambrian margin) are underlain by the offshore extensions of onshore coalfields or Namurian strata. These areas are covered in some detail by seismic data, and the availability of onshore analogues allows a more realistic assessment in terms of potential for development of non-conventional resources, perhaps from coastal locations.

Methodology and datasets

The exploration datasets used in the regional interpretation are depicted in Figure 1. The 2D seismic datasets include regional speculative data supplied by geophysical companies (CGG, IHS and Western-Geco); licence- and prospect-level datasets provided by the Common Data Access Initiative (CDA) offshore, and United Kingdom Onshore Geophysical Library (UKOGL) nearshore and onshore; and data supplied directly by participating companies (Centrica plc). The 3D dataset used was supplied by CDA, augmented by data from the 3D Terracube supplied by CGG. The well picks were supplied from the Department of Energy and Climate Change (DECC) well database at BGS Edinburgh, with further interpretation during the project.

Pre-Carboniferous structural evolution

The crust of the southern part of the region (North Wales, Anglesey and adjacent offshore areas (Fig. 2) was generated as volcanic and sedimentary complexes in magmatic arc–trench systems during late Proterozoic time. Many early tectonic lineaments (e.g. the Menai Strait Fault Zone: Gibbons 1987) are associated with the accretion and dispersal of various terranes along the margins of Gondwana in Neoproterozoic–Tremadocian time. Many of the lineaments (Dinorwic, Berw) have a SW–NE trend, are relatively straight (implying steep upper-crustal geometry), and have been serially reactivated in Acadian sinistral transpression, Devono-Carboniferous extension, etc. The crust of the northern part of the area (Midland Valley, Scottish Highlands) was generated throughout Proterozoic time. A Neoproterozoic supracrustal metasedimentary sequence, the Dalradian Supergroup, was strongly deformed during the Grampian phase of the Caledonian Orogeny (Smith *et al.* 1999; Chew & Strachan 2014). Its southern limit is marked by the Highland Boundary Fault, which forms the northern boundary of the area of investigation.

The crust in the central part of the region comprises early Paleozoic sedimentary complexes belonging to several different terranes forming part of the Avalonian (Monian, Lakesman) and Laurentian (Southern Uplands, Midland Valley) margins of the Iapetus Ocean, and accreted during the Caledonian Orogeny (Bluck 2002; Barnes *et al.* 2006; Chew & Strachan 2014). Numerous major tectonic lineaments have a typical SW–NE 'Caledonide' trend. These include the Carmel Head Thrust of northern Anglesey, and reactivations of the earlier Monian lineaments; the Ribblesdale Foldbelt (Kirby *et al.* 2000); the Causey Pike Thrust and Southern Borrowdale Lineament of the Lake District (Barnes *et al.* 2006); the numerous accretionary tracts of the Southern Uplands Massif (Bluck 2002); and numerous faults with this trend within the Southern Highlands terrane (Chew & Strachan 2014).

In this study, a NW-dipping zone of enhanced reflectivity in pre-Carboniferous 'basement', previously referred to as the Barrule Thrust (Chadwick *et al.* 2001), was mapped over a large area to NW of the Isle of Man. The analysis of the deep seismic reflection data presented by England & Soper (1997) suggests that this structure lies within the Avalonian

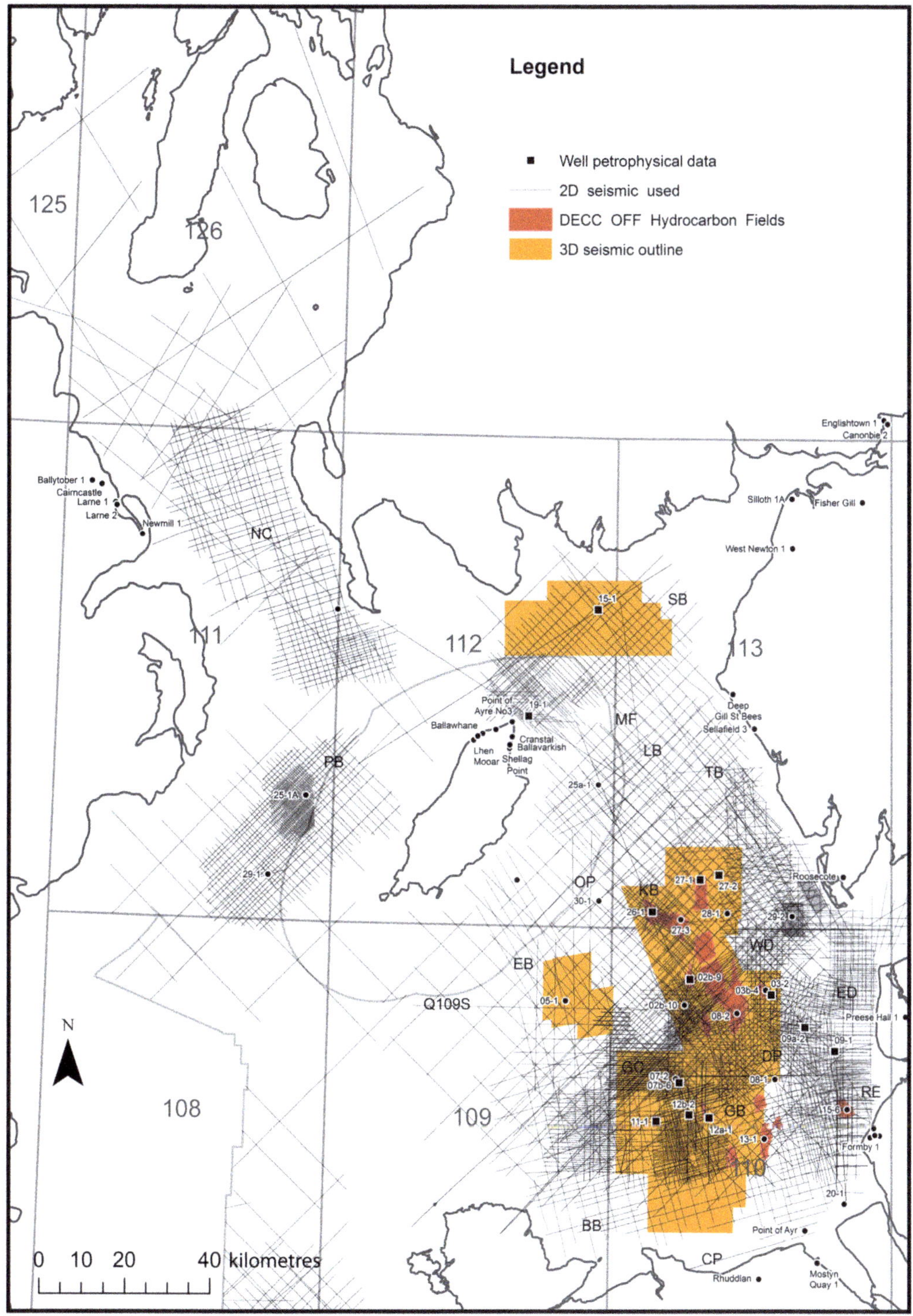

Fig. 1. Key data evaluated during the study (2D seismic, black; 3D seismic outline, orange; wells proving Carboniferous strata, black dot; wells used in the petrophysical study, black square).

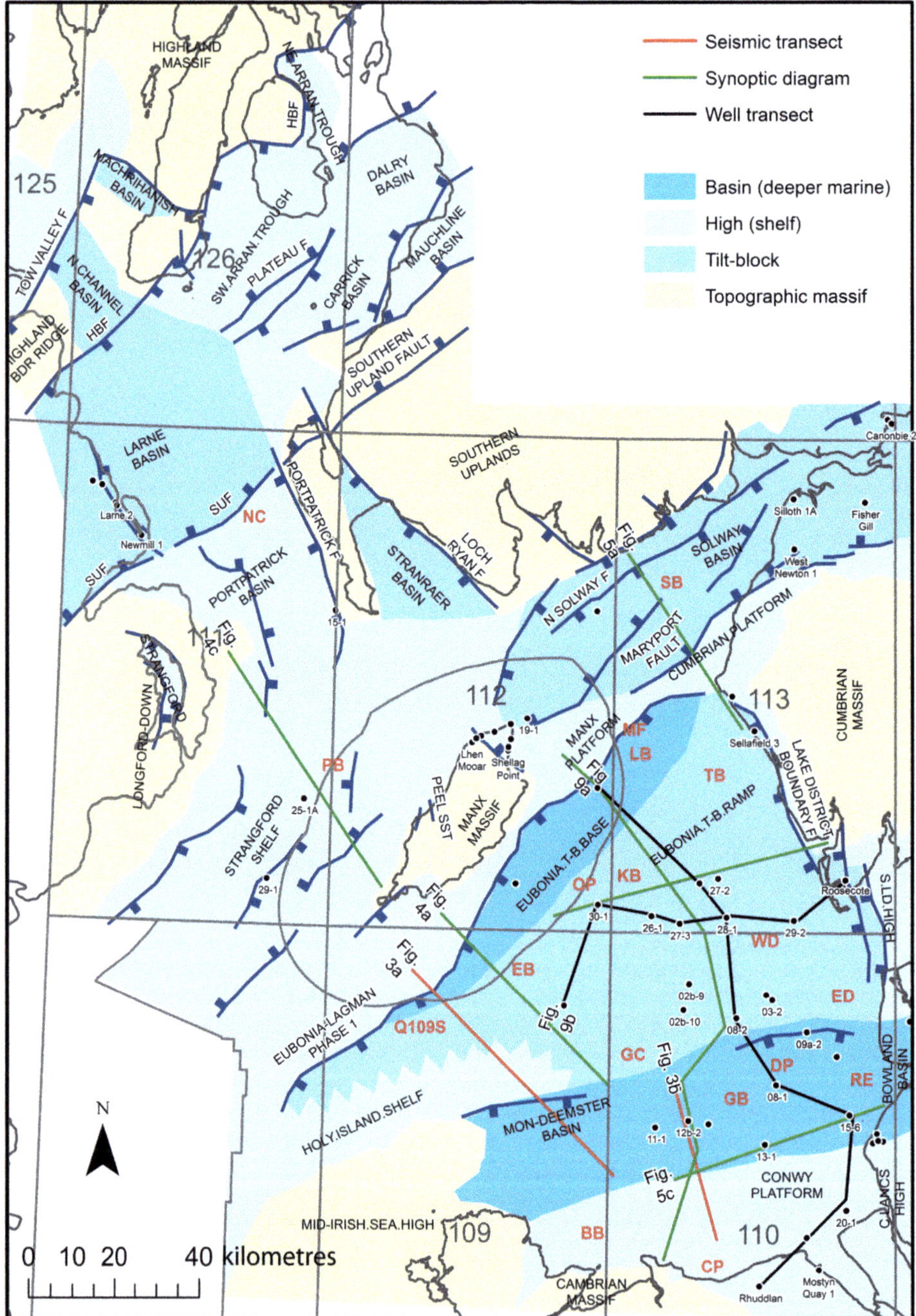

Fig. 2. Key Mississippian structural elements of the greater Irish Sea province. Incorporates information from Maddox *et al.* (1997), Parnell (1997) and Shelton (1997). Location of Permo-Triassic basinal features, following Jackson & Mulholland (1993) and BGS (1994), for reference purposes, in red: BB, Berw Basin; CP, Conwy Platform; DP, Deemster Platform; EB, Eubonia Basin; ED, East Deemster Basin; GB, Gogarth Basin; GC, Godred Croven Basin; KB, Keys Basin; LB, Lagman Basin; MF, Manx–Furness Ridge; NC, North Channel Basin; OP, Ogham Platform; PB, Peel Basin; Q109S, Quadrant 109 Syncline; SB, Solway Basin; TB, Tynwald Basin; WD, West Deemster Basin. Location of sections depicted in later figures: red line, seismic profiles (Fig. 3); green line, synoptic diagrams (Figs 4 & 5); black line, well transects (Fig. 9).

footwall of the Iapetus Suture, rather than representing the suture itself. A further zone of NNW-dipping basement reflectivity underlies the southern part of the EISB (Jackson & Mulholland 1993; Pharaoh *et al.* 2016*a*, *b*), being particularly prominent beneath the Conwy Platform, just off the north coast of Wales (Fig. 2). The dip of this zone steepens as it approaches the coast, and it is inferred to correlate with the southernmost strands of the Menai Strait Lineament (i.e. the Menai and Dinorwic fault zones). Although the seismic coverage is relatively poor in this area, the available data suggest that this zone represents the deepest regional detachment, with all subsequent extensional faulting (of Carboniferous and Permian–Mesozoic age) penetrating no deeper into the crust.

During the Acadian phase of the Caledonian Orogeny, most of the lineaments identified above were reactivated within a sinistrally transpressive regime, associated with the late orogenic collapse of the Caledonian mountains chain, stretching from the Appalachians through Ireland and Scotland to Greenland and Norway (Woodcock & Soper 2006; Chew & Strachan 2014). The most obvious element of this regime is the Great Glen–Walls Boundary Fault System. Devonian strata are thickest in the north of the study area, in the Midland Valley and form the molasse to the Caledonian Orogen (Trewin & Thirlwall 2002). In the south (Anglesey), Devonian strata are more limited in development and related to local faulted basin margins (Hillier & Williams 2006). In this tectonic regime, west–east extension is anticipated (Coward 1993). Basins related to such an orientation are tentatively identified within the Orcadian Basin (Leslie *et al.* 2015) but are less clearly identified in the study area, except, perhaps, in the rift basins (North Channel, Stranraer, Carlingford Lough) within the Southern Uplands Massif and the Peel Sandstone Graben of the Isle of Man (Maddox *et al.* 1997; Parnell 1997; Quirk & Kimbell 1997).

Carboniferous structural and stratigraphic evolution

An extensional–transtensional tectonic regime persisted into Carboniferous time (Leslie *et al.* 2015). Although a general west–east extensional regime has been invoked in Mississippian time (Coward 1993), extension occurred on faults with a diversity of orientations, but with reactivation of earlier basement structures (of various trends) being a common feature: for example, in the Northumberland Basin (De Paola *et al.* 2005). This reflects partitioning of the tectonic regime (Leslie *et al.* 2015). East of the study area, in Lancashire, the Bowland Basin reflects deeper-water deposition in a basin bounded by SW–NE-trending faults (e.g. Pendle Monocline) representing reactivations of earlier basement structures (Kirby *et al.* 2000). The Solway Basin is the offshore continuation of the Northumberland Basin (Chadwick *et al.* 1995), and is controlled by major bounding faults on a SW–NE trend. The Peel Basin, along strike to the SW, has a similar trend but opposite structural polarity and a very different basin setting in the Carboniferous (Fig. 2). However, the evolution of both basins appears to have been strongly influenced by the extensional reactivation of underlying structures in the Caledonide basement. The Midland Valley (and Firth of Clyde basins) also exhibit a SW–NE trend, which persists up to the Highland Boundary Fault.

Carboniferous extensional basins

The Carboniferous substrate of the EISB comprises a number of basin elements, comparable to that of the UK onshore. Figure 2 presents a speculative reconstruction of the principal tectonic elements in Mississippian time. It is based heavily on seismostratigraphic and structural interpretation, as only five offshore boreholes penetrate Visean strata in the whole of the province (112/25a-1 and 113/27-2 in the EISB; 111/25-1A and 111/29-1 in the Peel Basin; and 112/19-1 in the Solway Basin: Fig. 2). In the centre of the EISB, a major basin, here referred to as the Eubonia Tilt-block (Fig. 2), is inferred to extend from the Quadrant 109 (Q109) Syncline in the SW (BGS 1994) to the Ogham Platform (Fig. 2). Extension farther east, beneath the Lagman and Tynwald (Permian–Mesozoic) basins of the EISB, towards the western edge of the Lake District, is also inferred. The presence of a major half-graben (tilt-block), controlled by a major syndepositional bounding fault on its NW margin, the Eubonia–Lagman Fault System (Figs 2 & 3a), is indicated by the seismic reflection data. The structure was not identified as a tilt-block by Jackson & Mulholland (1993, p. 800), but they did recognize the marked asymmetry of the northern limb of the Q109 Syncline/Basin and the presence of up to 7.5 km of Visean–late Westphalian (and, possibly, Stephanian) strata. Figure 3a shows a seismic line extending SE from the Isle of Man towards Anglesey (Fig. 2). It demonstrates the presence of over 2.5 s two-way travel time (TWTT) of Carboniferous strata east of the Eubonia Fault, in what is referred to as the Eubonia Tilt-block (Pharaoh *et al.* 2016*b*). Poor well control is provided by a few distant wells on the western edge of the EISB (Fig. 2) and the picks are not well constrained.

Towards the top of the tilt-block in the south, on the Holy Island Shelf, brighter reflectivity in the upper Visean interval may represent the development of reefal carbonates. The southern end of the

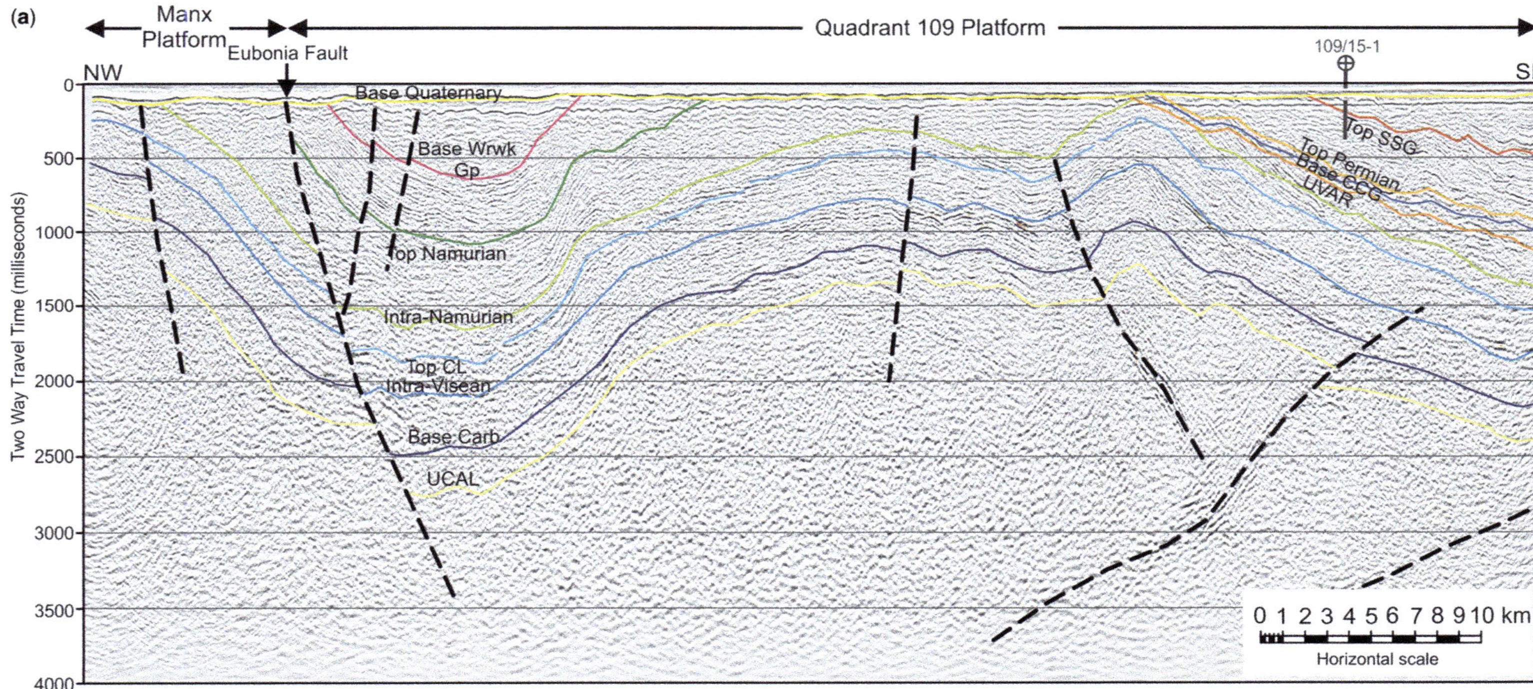

Fig. 3. Seismic reflection data. Locations are shown in Figures 2 and 6. Note: vertical scales in s TWTT. (**a**) Migrated seismic reflection line NW–SE across Quadrant 109: JEBCO JS-MANX-138. Includes content supplied by IHS Global Ltd. Copyright © IHS Global Ltd (2016). All rights reserved. Note the considerable thickness of Carboniferous strata in the Eubonia Tilt-block, here exceeding 2.5 s TWTT; brighter reflectivity towards the top of the tilt-block below the Intra-Visean Unconformity, possibly reflecting reefal development; and in the south, a northwards-vergent anticline–thrust inversion couple, defining the northern edge of the Môn–Deemster Basin. The presence of Warwickshire Group strata is inferred from seismostratigraphic principles and has not yet been confirmed by drilling.

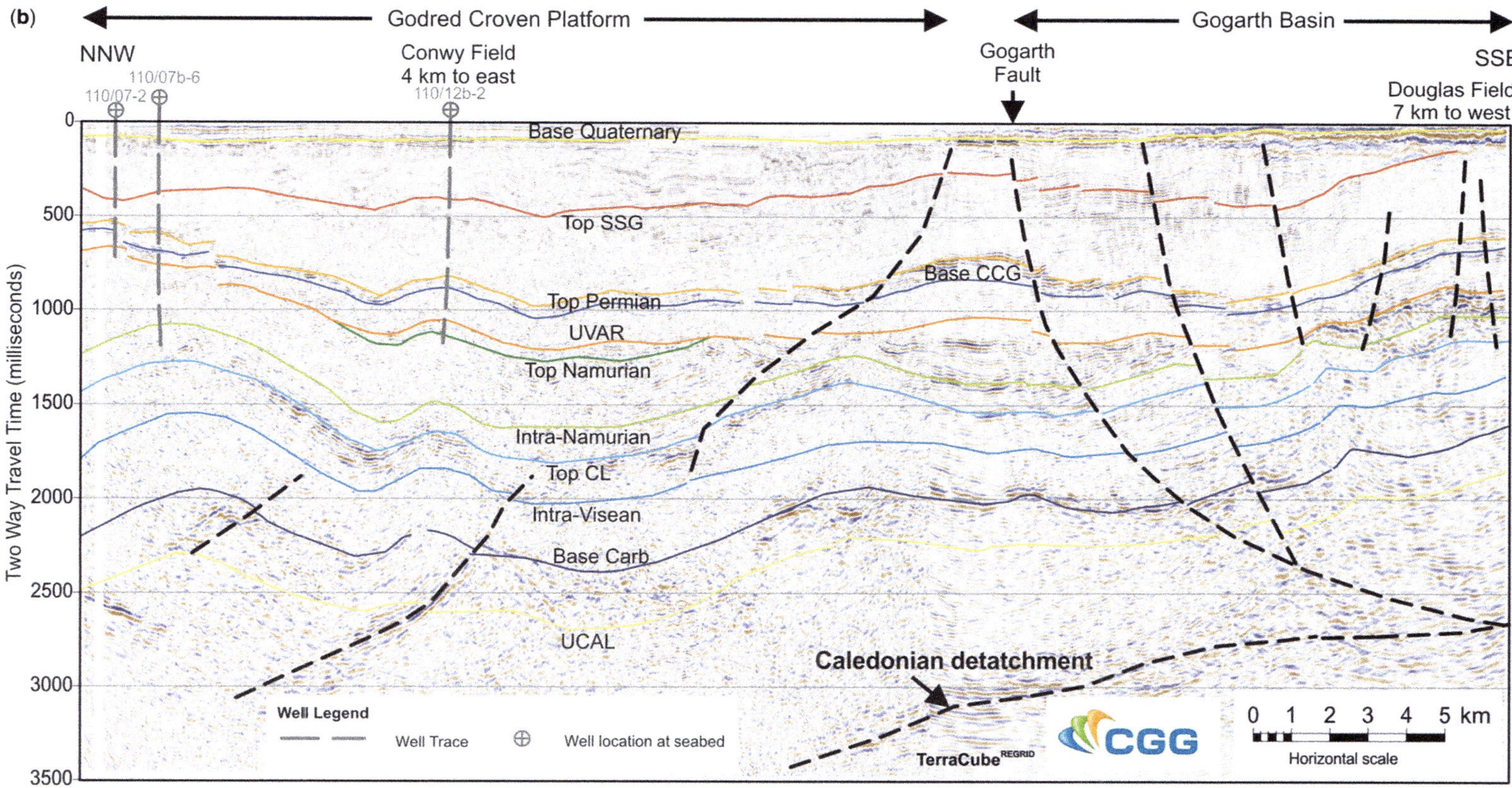

Fig. 3. (*Continued*) Seismic reflection data. Locations are shown in Figures 2 and 6. Note: vertical scales in s TWTT. (**b**) Arbitrary NNW–SSE line through the migrated 3D TerraCube® dataset, supplied courtesy of CGG GeoSpec. Note the presence of a series of inversion anticlines (Môn–Deemster Foldbelt) in the Carboniferous sequence, associated with thrusts (fault-plane reflections) which penetrate into the Caledonian basement. A less steeply dipping detachment is present at depth. The Bowland Shale Formation is inferred to occupy a rather transparent zone, sandwiched between more reflective Carboniferous Limestone Group (below) and Millstone Grit Group (above). The mildly deviated well 110/07b-6 proved 450 m of Bowland Shale Formation before terminating in strata of Pendleian age (unbottomed). Westphalian strata have been almost completely eroded following strong inversion during the earliest Variscan phase. Inversion anticlines of this generation were reactivated 'posthumously' by further compression during the Alpine Orogeny in Miocene time, producing more gentle anticlines in the Permo-Triassic cover, including the traps in the Ormskirk Sandstone Formation (Top SSG pick) hosting the Conwy, Douglas and other fields.

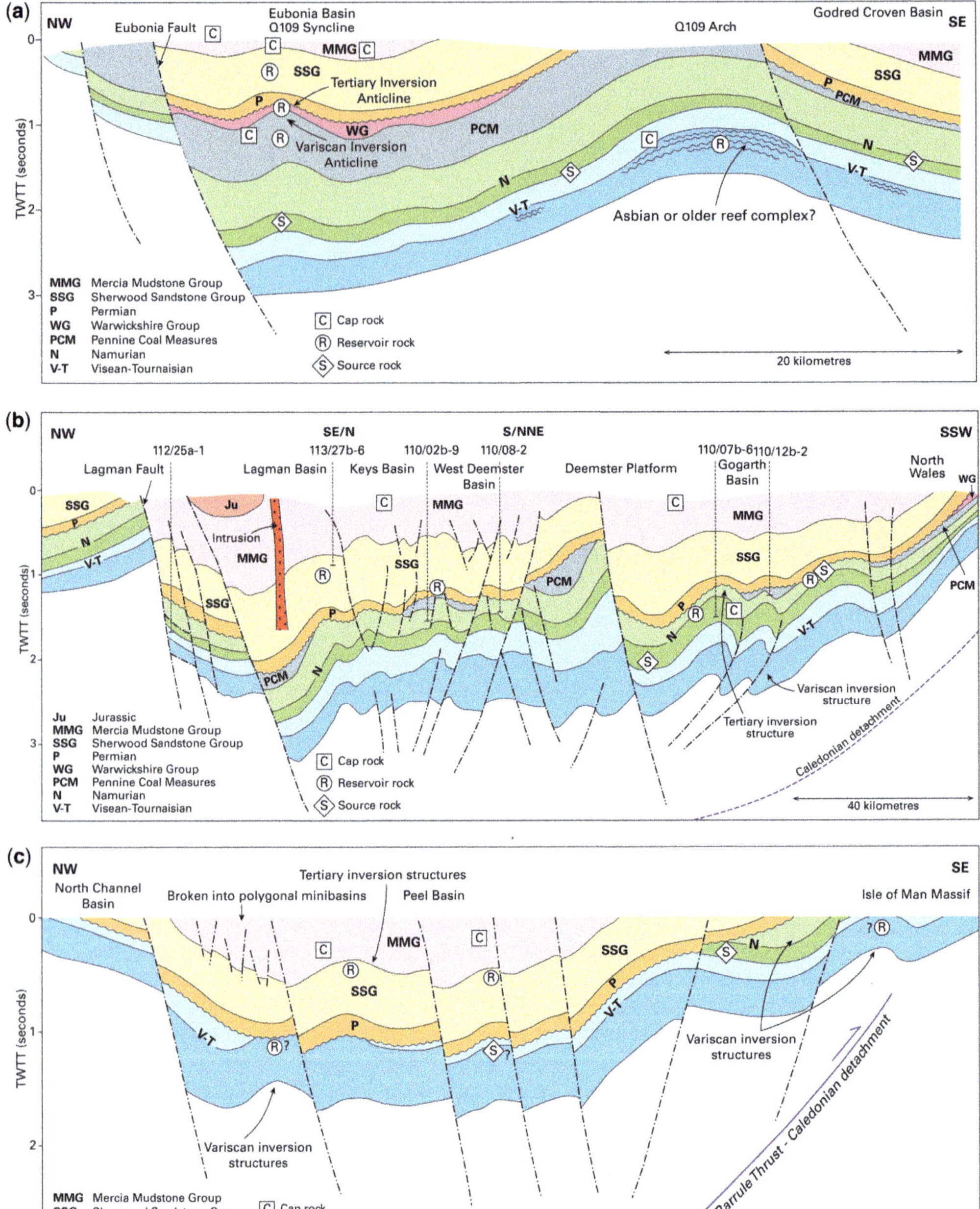

Fig. 4. Synoptic diagrams ('cartoons') to illustrate principal elements of the hydrocarbon system of the greater Irish Sea province. Locations are shown in Figures 2 and 6. Note: vertical scales in s TWTT. The basin names used are principally those of the Permo-Triassic EISB and contemporary basins (Jackson & Mulholland 1993; BGS 1994), rather than those of the Carboniferous elements newly named here. (**a**) Principal elements of the hydrocarbon system in the Eubonia Basin and Q109 Arch. The transect is parallel to Figure 3a and farther east, into the Eubonia Basin. (**b**) Principal elements of the EISB from the Lagman Fault (north) to the Welsh margin (south). The northern part crosses from the Ogham Platform, to the Lagman and Keys basins, where Westphalian strata are almost completely removed, the West Deemster Basin and Deemster Platform. Note the thickening of the Bowland Shale Formation beneath the Deemster Platform, associated with the offshore extension of the Bowland Basin. The southern part

section crosses a northward-vergent inversion anticline–thrust couple, defining the northern limit of the Môn–Deemster Fold Belt. This is a 25–30 km-wide belt of strong Variscan inversion, extending ENE from the north coast of Anglesey, from the Q109 Arch to the Deemster Platform (Fig. 2). The internal structure of this belt is imaged on numerous north–south profiles crossing the Godred Croven Basin, and Figure 3b, an arbitrary line through 3D data in this area, is representative. A schematic profile is presented in Figure 4b. A series of parallel WSW–ENE-trending anticlinal folds has been mapped through the area. The internal structure of this inversion belt is complex, comprising a fan-like array of anticlines and synclines with associated thrusts, SSE-vergent in the south and NNW-vergent in the north (Fig. 4b). Figure 3b clearly shows discordant reflections in the Visean sequence, extending down into the Caledonian basement, interpreted here as fault-plane reflections. Below 3 s TWTT, a further zone of intra-basement reflectivity is interpreted as a deeper Caledonian detachment surface, as recognized by Jackson & Mulholland (1993, p. 805). Well 110/07b- 6 was clearly a test of the structure with the greatest amplitude, at the northern end of the profile. This slightly deviated well proved 450 m of (presumed) Namurian Bowland Shale Formation (Pendleian unbottomed) beneath 550 m of Millstone Grit Group, Westphalian strata being absent beneath the Base Permian Unconformity (Fig. 3b). As noted above, northward-vergent structures have been identified on the northern edge of the Q109 Arch (Fig. 3a), and they have also been mapped beneath the northern part of the Deemster Platform. Several NNW–SSE- to north–south-trending graben of the EISB (Godred Croven, Gogarth and East Deemster basins) discordantly overlie this Carboniferous hinge zone. The inversion belt is very similar in its structure and orientation to the Ribblesdale Foldbelt of the Lancashire onshore, representing the Variscan-inverted Bowland Basin (Corfield *et al.* 1996; Kirby *et al.* 2000). It seems logical to infer connection of the two, via the Fylde coast of Lancashire, as proposed by Corfield *et al.* (1996). If this inference is true, then the southern edge of the zone may represent a reactivated extensional fault, analogous to the Pendle Lineament of Lancashire; and the Visean carbonate platform (Holy Island and Conwy platforms) to the south, with a thin or absent Namurian cover, are the equivalent of the Central Lancashire High (Kirby *et al.* 2000). Also, by analogy with the Bowland Basin/Ribblesdale Fold belt onshore, the greatest thickness of Bowland Shale offshore was probably deposited within a rift basin ancestral to the presently observed Môn–Deemster inversion zone. Further seismic mapping is required to confirm this, however.

That part of the Eubonia Tilt-block lying east of the Keys Fault was subsequently almost obliterated by the combined effects of latest Variscan inversion and pre-Permian erosion. The original eastern limit of the tilt-block is uncertain. It is likely to have continued beyond the Tynwald Basin, where the en echelon faults of the Lake District Boundary Fault System may have acted as transfer faults, offsetting extensional subsidence farther south into the Craven Basin. On the northern margin of the tilt-block, to the NW of the Eubonia–Lagman Fault System, an extensive shallow-marine carbonate platform developed in Visean time. This is well represented by outcrop in the south of the Isle of Man (Chadwick *et al.* 2001), the northern edge of the Lake District and adjacent offshore (Ramsey–Whitehaven Ridge) (Fig. 2). Because of significant pre-Permian uplift and erosion, it is not possible to determine the subsidence regime in which Westphalian strata were deposited, but it was probably dominated by post-extensional thermal subsidence, as elsewhere in southern Britain, the depocentre lying near Manchester (Fraser *et al.* 1990; Fraser & Gawthorpe 2003).

A few wells penetrate the Carboniferous sequence beneath the Peel Basin (Fig. 2) and demonstrate that an extensive carbonate platform (Manx Platform and Strangford Shelf) extends west to Ireland and north towards the North Channel. The present study revealed that the undifferentiated Carboniferous strata on BGS (1994) mapping are principally of Visean age, Namurian strata being largely eroded (Pharaoh *et al.* 2016*a*). The Permo-Triassic Peel Basin has the form of an asymmetrical graben controlled by a major bounding fault on the northern side (Fig. 4c), and extensional faults with smaller throws on the southern side, developed in the hanging wall of the Barrule Thrust (Chadwick *et al.* 2001). Lack of evidence for significant Carboniferous syndepositional throw, and the larger Permo-Triassic throws, suggests that there was probably not a significant basin here in Visean time, although the poor quality of the seismic data allows some uncertainty. Faulting at the top of the Appleby Group (Permian) has a predominantly NW–SE trend (Quirk *et al.* 1999), akin to that of the North Channel Basin.

In contrast, the Solway Basin, underlying the Permian–Mesozoic Carlisle Basin along strike to

Fig. 4. (*Continued*) crosses the Môn–Deemster Foldbelt and is virtually colinear with Figure 3b. (**c**) Principal elements of the hydrocarbon system in the Peel Basin. Note that fault displacements appear to be largely of post-Permian age, indicating little if any syndepositional thickening across the Visean carbonate platform. Post-Visean strata were only preserved on the Manx margin.

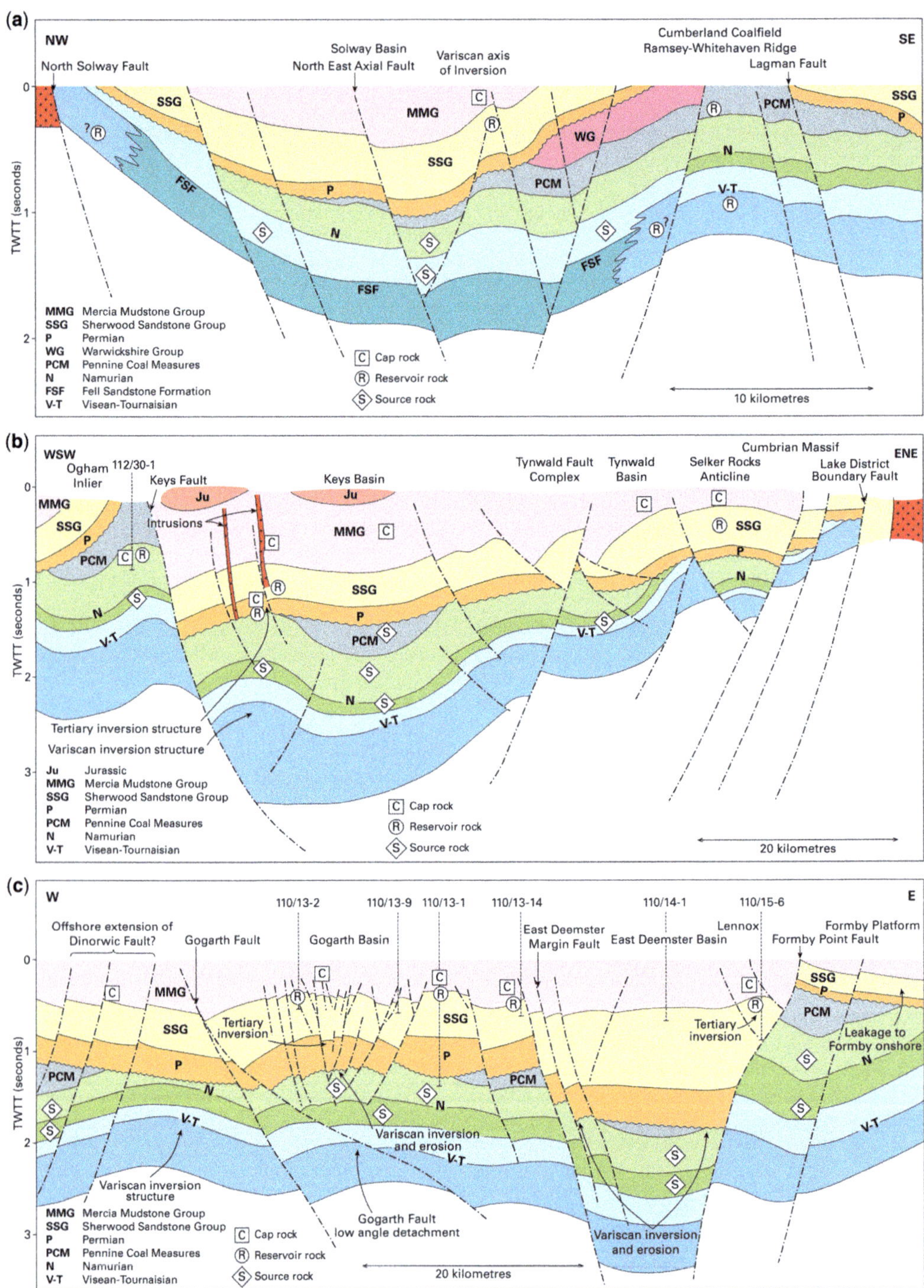

Fig. 5. Synoptic diagrams ('cartoons') to illustrate principal elements of the hydrocarbon system of the greater Irish Sea province. Locations are shown in Figures 2 and 6. Note: vertical scales in s TWTT. The basin names used are principally those of the Permo-Triassic EISB and contemporary basins (Jackson & Mulholland 1993; BGS 1994), rather than those of the Carboniferous elements newly named here. (**a**) Principal elements of the hydrocarbon system in the eastern part of the Solway Firth Basin. Note the preservation of the late Westphalian Warwickshire Group

the NE of the Peel Basin, is asymmetrical with a principal controlling fault on the southern side (Ramsey–Whitehaven Ridge) (Fig. 5a). The Carboniferous basin fill comprises fluvio-deltaic Border and Yoredale Group strata with greater affinity to the Northumberland Trough sedimentary sequence than the carbonate platforms of the southern Irish Sea (Chadwick *et al.* 1995), together with a greater thickness of preserved Pennsylvanian strata.

The present study found no convincing evidence for the presence of significant thicknesses of Carboniferous strata beneath Permo-Trias in the Portpatrick Basin, the southern part of the North Channel basin complex: the only well to penetrate Permian in this basin (111/15-1) unfortunately terminated in early Paleozoic rocks, having passed through the marginal fault. The absence of Carboniferous strata may be a consequence of erosion following late Variscan inversion on the NNW-trend (see below). However, they are present within re-entrants at the northern edge of the Southern Upland Massif (Stranraer, Strangford Lough), and are certainly present to north of the Southern Upland Fault (Larne and Rathlin basins, and SW Arran Trough). All of these basins are very poorly explored by deep boreholes and only very general conclusions can be made about their Mississippian evolution, largely by inference from nearby analogues onshore (Read *et al.* 2002).

Early phase of Variscan inversion

Through Pennsylvanian time, the impact of the Variscan Orogeny resulting from the collision of numerous Gondwana-derived terranes (e.g. Armorica, Central Massif, Bohemian Massif) with the southern margin of Laurussia (Ziegler 1990; Pharaoh *et al.* 2006) became increasingly evident in Britain. Large-scale northwards thrust and nappe emplacement occurred in southern Britain, South Wales and southern Ireland, but the region lay in the northern foreland of the Variscan Foldbelt (Besly 1988; Ziegler 1990; Pharaoh *et al.* 2010). In late Pennsylvanian (Westphalian C) time, an early phase of inversion was followed by deposition of strata of the Warwickshire Group, above a regional unconformity (Eastwood *et al.* 1931; Akhurst *et al.* 1997; Dean *et al.* 2011; Jones *et al.* 2011; Waters *et al.* 2011). The Whitehaven Sandstone Formation (equivalent to the Warwickshire Group and of latest Westphalian–?Stephanian age) has divergent palaeocurrents to the south in Cumbria, and to the north at Canonbie, reflecting penecontemporaneous growth of the Solway inversion anticline (Jones *et al.* 2011). In the EISB, this study has identified SSW–ENE-trending inversion structures parallel to the Eubonia–Lagman Fault System in the north (Fig. 6), as well as in the Môn–Deemster inversion belt described above. The study has shown that the early phase of Variscan inversion structures are cut discordantly by the NNW–SSE- to north–south-trending faults of the Permian–Mesozoic main graben structures of the EISB, such as the Keys Fault, Godred Croven Fault and western marginal fault of the East Deemster Basin. North of the Ramsey–Whitehaven Ridge, both the Solway and Peel basins suffered strong inversion on SSW–NNE ‘Caledonoid’ trends, with uplift and erosion of most of the post-rift (Namurian–Westphalian) successions, prior to deposition of Warwickshire Group strata (Jackson *et al.* 1995; Newman 1999). Variscan reversal of the Maryport Fault is demonstrated by the preservation of a much more complete post-rift sequence on its footwall block (Ramsey–Whitehaven Ridge) than in the Solway Basin, its hanging-wall block (Chadwick *et al.* 1993).

Later phase of Variscan inversion

In late Pennsylvanian time, the final deformation phases of the Variscan Orogeny are associated with the closure of the Uralian Ocean basin, and the collision of the Kazakhstan and Siberian plates (Zonenshain *et al.* 1984; Puchkov 1997; Brown *et al.* 2002), resulting in west–east-orientated compressional stress (Coward 1993, 1995). In the study area, inversion occurred along NNW–SSE- to north–south-trending faults, such as the Keys Fault, the Gogarth Fault, the western marginal fault of the East Deemster Basin and the Formby Point Fault System. Evidence for this is provided by the Carboniferous subcrop pattern presented by BGS (1994). The pre-Permian subcrop inset in the marginalia of this map clearly shows erosion of Westphalian strata in NNW- to north–south-trending belts associated

Fig. 5. (*Continued*) (Whitehaven Sandstone Formation) in the Cumbrian Coalfield adjacent to the Maryport Fault, and the axis of Alpine inversion significantly offset from the Variscan one. The Visean strata here are Border Group and Yoredale facies, with uncertain source potential. (**b**) Principal elements of the hydrocarbon system in the northern part of the EISB. The WSW–ENE transect crosses Variscan second-phase inversion structures obliquely in the Ogham Inlier, western Keys Basin and the Cumbrian margin. (**c**) Principal elements of the hydrocarbon system in the southern part of the EISB. The west–east transect is located parallel to the southern edge of the Môn–Deemster Foldbelt, crossing some of the Variscan first inversion phase structures obliquely. North–south-trending inversion structures of the second Variscan phase at the margins of the East Deemster Basin and on the Formby Platform are crossed obliquely. Note the excision of Westphalian strata on these inversion systems. Modified from Yaliz (1997, fig. 4).

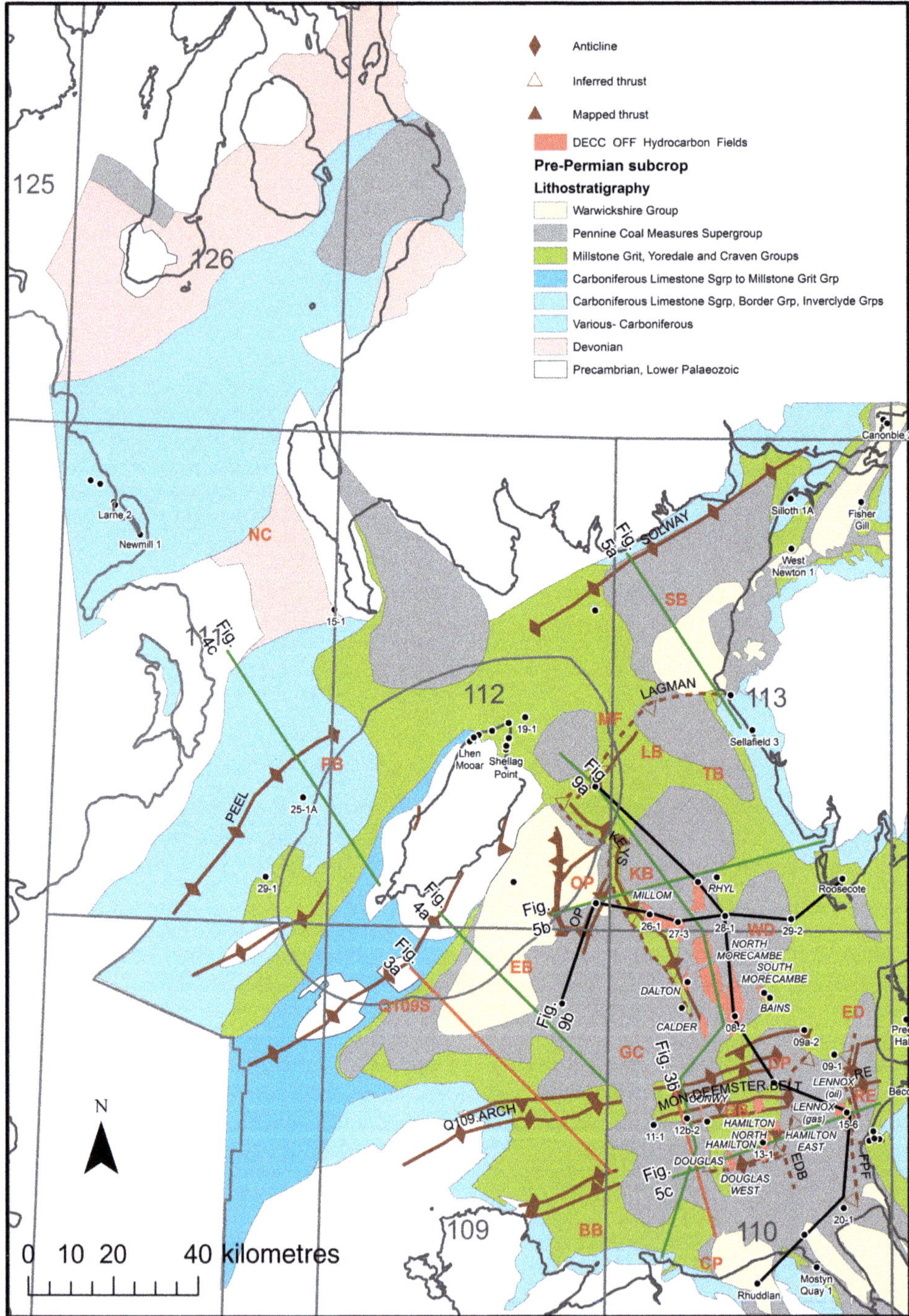

Fig. 6. Pre-Permian subcrop map showing key Variscan inversion structures (after Pharaoh *et al.* 2016*b*). Variscan inversion structures in the Ogham Platform after Quirk & Kimbell (1997). Abbreviated structure names: DP, Deemster Platform; EDB, East Deemster Western Boundary Fault; FPF, Formby Point Fault; OP, Ogham Platform; RE, Ribble Estuary Inlier. The location of sections depicted in other figures: red line, seismic profiles (Fig. 3); green line, synoptic diagrams (Figs 4 & 5); black line, well transects (Fig. 9). Hydrocarbon fields are from the OGA website: http://data.ogauthority.opendata.arcgis.com/datasets/

with the hanging walls of the Keys Fault (Fig. 5b), Gogarth Fault (Fig. 4a) and the Lake District marginal faults (Fig. 5c). By contrast, Westphalian strata are well preserved on the footwall of these structures. The seismic data indicate the presence of north–south-trending anticlinal folds cored by Namurian strata, dissected by faulting on their overturned limbs. Similar subcrop patterns, with Namurian subcrops in the cores of Variscan inversion anticlines (e.g. the Murdoch Anticline), are observed in quadrants 43 and 44 in the Southern North Sea (Corfield *et al.* 1996), and, indeed, the two basins exhibit a similar degree of inversion. At present, the NNW–SSE- to north–south-trending faults are extensional structures of Permian and younger age; but these are here inferred to have initiated as thrusts or positive flower structures ('ancestral faults') on the overturned limb of the anticlines during Variscan inversion, as reported in the Ogham Inlier by Quirk & Kimbell (1997). A component of sinistral shear is likely from the observed relationship of the folds in the Ogham Inlier to the ancestral Keys Fault. Seismic mapping in the present study (Fig. 6) confirms the pre-Permian subcrop pattern presented by BGS (1994) and has identified a possible interference structure between the two trends in the Ribble Estuary Inlier. Although it is conceivable that inversion on faults with both WSW–ESE and NNW–SSE trends could have occurred in one Variscan phase of inversion, comparable to the partitioned deformation system advocated for the Northumberland Basin by De Paola *et al.* (2005), the above evidence would appear to suggest that two, separate, nearly orthogonal phases of Variscan inversion are more likely. Extensional reactivation of the ancestral late Variscan structures in west–east extension during Permian–Mesozoic time facilitated the development of NNW–SSE- to north–south-trending graben of the EISB, strongly discordant to the strong SW–NE structural grain established by Caledonian compression, Mississippian extension and early Variscan inversion. Strong uplift and erosion during the Variscan inversion led to the complete removal of the Pennine Coal Measures strata underlying the Lagman Basin. The ancestral Keys Fault played a key role in partitioning the former Eubonia Tilt-block into western and eastern segments, the latter being almost obliterated by post-Variscan events. Inversion on the same trend may have led to uplift and erosion of Carboniferous strata deposited within basins on the North Channel basin complex.

Post-Variscan structural evolution

The post-Variscan structural evolution of the EISB has been thoroughly described in numerous previous publications (Jackson *et al.* 1987, 1995, 1997; Jackson & Mulholland 1993; BGS 1994). As a result, only a generalized account, focusing on those elements where the Paleozoic structure has a bearing, will be presented here. Following the Variscan basin inversion and regional uplift described above, there is clear evidence on seismic profiles for the erosion of Pennine Coal Measures strata from the crests of inversion anticlines, and tectonic dissection of the latter adjacent to the Keys, Lagman, Lake District Boundary and Formby Point faults prior to deposition of Permian strata (BGS 1994). Jackson & Mulholland (1993, p. 793) and Jackson *et al.* (1997, fig. 2) recognized significant thickening of the Appleby Group (Lower Permian), possibly to as much as 1150 m (Jackson & Mulholland 1993), in a belt extending from the Berw Basin to the Formby Oilfield. For example, the well 110/11-1 proved 763 m of Collyhurst Sandstone Formation (Appleby Group), while 110/7-2 12 km to the north proved only 40 m, and none is present in the vicinity of the Morecambe fields. The belt of thick Appleby Group strata directly overlies the Môn–Deemster Foldbelt, providing strong evidence for significant early Permian penecontemporaneous relief within, and deep erosion of, the tectonically weakened inversion belt. The area must have had a substantial topography in early Permian time. It is interesting to note that significant pre-Permian palaeotopography was described at Formby by Falcon & Kent (1960).

A series of NNW–SSE- to north–south-trending rifts began to develop in response to west–east extension affecting the crust of the Pangaea supercontinent that was established during the Variscan Orogeny (Whittaker 1985; Chadwick & Evans 1995; Coward 1995). In the Worcester and Knowle basins onshore, rifting was able to exploit the north–south ('Malvernoid') grain previously established by the late Precambrian orogeny (Pharaoh *et al.* 1987; Barclay *et al.* 1997) and subsequent Variscan inversion (Chadwick 1993). The rifts propagated with stepwise, en echelon offsets through the province, from the Stafford and Cheshire basins and EISB through the Portpatrick and Larne basins and the North Channel to the western Scottish offshore basins (Ziegler 1990). The Solway and Peel basins subsided less than the EISB, and are elongated SW–NE, reflecting structural control by the extensionally reactivated Caledonide basement structure within the Iapetus Convergence Zone. Nevertheless, it is notable that the majority of small- to medium-sized intrabasinal normal faults (Chadwick *et al.* 2001) take up the new north–south trend, as in the Cheshire Basin (Chadwick 1997). By Triassic time, the EISB was a mature component of the Central European Basin System (Scheck-Wenderoth *et al.* 2008; Pharaoh *et al.* 2010), receiving up to 5 km fill of Sherwood Sandstone Group clastic sedimentary rocks and Mercia Mudstone Group mudstones and evaporites (Jackson & Mulholland 1993). Small relict outliers of Lias (early Jurassic) strata in the Carlisle Basin (Warrington 1997), Peel Basin (Chadwick *et al.* 2001) and EISB (Jackson & Mulholland 1993) indicate that subsidence continued into Jurassic time. Evidence for mid- and late Jurassic subsidence has been removed subsequent to Cenozoic inversion, uplift and erosion. The magnitude of post-Triassic displacement is difficult to estimate due to this erosion, but it is likely that the Lagman and Keys faults, together with the Maryport, Portpatrick, Loch Ryan and St Patrick faults, suffered significant normal movement (Jackson & Mulholland 1993; Quirk *et al.* 1999). Apatite fission-track analysis indicates that for parts of the Ramsey–Whitehaven Ridge, maximum post-Variscan burial was achieved in early Cretaceous time (Green *et al.* 1997). This was associated with peak generation of hydrocarbons from Carboniferous source rocks throughout the region. Soon after this, a fall in relative sea level and erosion resulted in the Late Cimmerian Unconformity, found throughout the British Isles (Whittaker 1985). The reduction in confining pressure may have been enough to allow early formed hydrocarbons, principally oil, to escape early reservoir structures in gentle rollover anticlines associated with the shallow detachment tectonics in the centre of the Main Graben, towards rollover traps at the marginal faults (Pharaoh *et al.* 2016*b*).

Opening of the Atlantic Ocean east of Greenland by Paleocene times associated with putative Icelandic Plume activity (e.g. Brodie & White 1994; Nadin & Kuznir 1995) resulted in voluminous magmatism in the Inner Hebrides and in Northern Ireland, just to the west of the study area. The Fleetwood Dyke Complex (Kirton & Donato 1985; Arter & Fagin 1993) was intruded en echelon across the main graben of the EISB. Magmatic and thermal processes on a lithospheric scale resulted in regional thermal doming of the crust below the EISB (White 1988) in the Paleogene or, possibly, late Cretaceous (Cope 1994, 1997). Across the study area, the combination of enhanced regional and local heat flow led to a further phase of hydrocarbon generation (Cowan & Bradney 1997; Meadows *et al.* 1997). Superimposed on the regional, thermal uplift described above were the effects of later crustal shortening, associated with the developing Alpine Orogeny in southern Europe. Apatite fission-track data indicate a second Cenozoic phase of cooling at 25–20 Ma (Newman 1999), compatible with the region being affected by the Oligo-Miocene phase of inversion found in southern Britain and the Southern North Sea (Van Hoorn 1987; Badley *et al.* 1989; Chadwick 1993). Inversion of the Solway Basin led to development of a major anticlinal structure in the hanging-wall block of the Maryport Fault (Chadwick *et al.* 1993) on the northern side of the Ramsey–Whitehaven Ridge. On the southern side of the ridge, the reversal of the Lagman Fault led to the generation of small hanging-wall anticlines (Chadwick *et al.* 2001). Flower structures and 'pop-up' structures are found along the Keys Fault and Formby Point Fault (e.g. the Rhyl and Lennox fields) (Haig *et al.* 1997), reflecting the 'buttressing' effect of the margins of the EISB (Pharaoh *et al.* 2016*b*). Throughout the EISB, seismic data indicate the presence of gentle Cenozoic inversion anticlines (Fig. 4a–c) superimposed on an earlier generation of Variscan inversion anticlines (Pharaoh *et al.* 2016*a*, *b*), the 'posthumous' tectonic style recognized by Jackson & Mulholland (1993). Further tightening of the Variscan inversion anticlines during Cenozoic (Alpine) crustal compression resulted in the development of more open structures in the Permo-Triassic cover. This was likely to have been an important process in the generation of the traps in the Hamilton fields (posthumous upon the Môn–Deemster inversion belt), and the Millom, Dalton and Calder fields (posthumous on the Keys trend of latest Variscan inversion). The Cenozoic inversion history is thus complex, involving contractional reactivation of precursor normal faults, posthumous folding, and regional arching and uplift of basin depocentres, upon which various thermal effects due to magmatic intrusion and possible underplating have been superimposed. A detailed treatment of these potential Cenozoic impacts upon the Paleozoic hydrocarbon system is beyond the scope of this paper.

Petroleum systems of the Carboniferous basins of the EISB

In the EISB, a proven petroleum system is present involving a Carboniferous source (Colter & Barr 1975; Stuart & Cowan 1991; Stuart 1993; Armstrong *et al.* 1997), reservoirs of the Ormskirk Sandstone, locally the uppermost formation of the Triassic Sherwood Sandstone Group, and halite seals (Fig. 7). A substantial number of exploration wells have been drilled but few penetrate the Permian, and the potential pre-Permian resource underlying the EISB fields is poorly known. The North and South Morecambe gasfields (Fig. 6), with a combined in place recoverable of 5.2 Tcf (trillion cubic feet) (Cowan 1996), were discovered in the 1970s and lie in large regional anticlines associated with rollover and salt-facilitated low-angle detachment faulting of Triassic–Jurassic age (Knipe *et al.* 1993). A further modification of trap geometry occurred in Miocene times as a result of Alpine inversion. An initial charge of hydrocarbons (probably mostly oil) in Jurassic times was originally thought to have been derived from Pennine Coal Measures source rocks, as in the Southern North Sea (Bushell

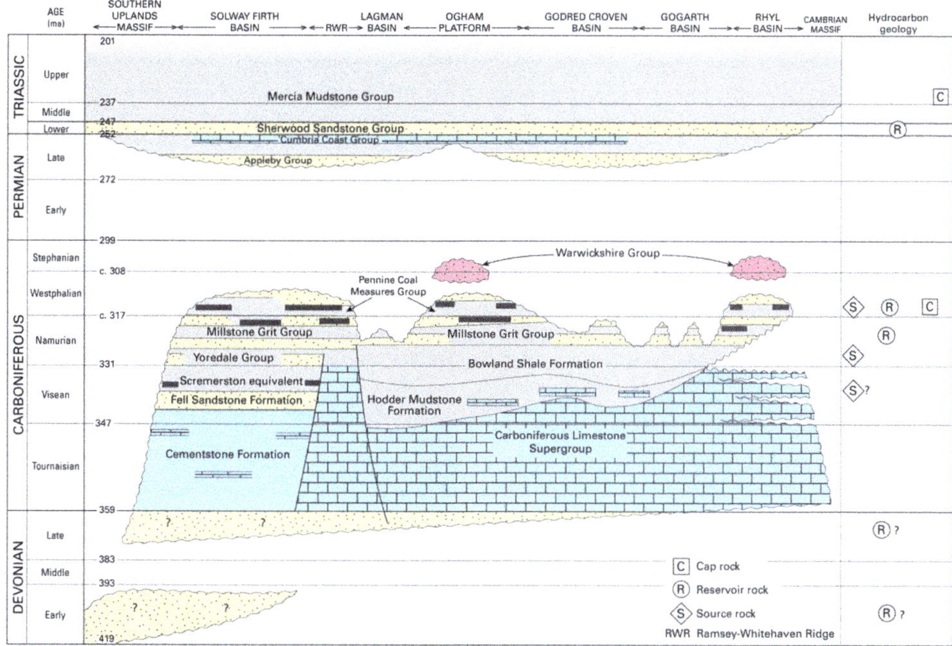

Fig. 7. Petroleum system elements in a north–south transect across the central part of the region.

1986). Subsequently, the Bowland Shale Formation was confirmed as the source (Armstrong *et al.* 1997). This early charge, associated with the formation (at *c.* 180 Ma) of a 'platy-illite' layer and interpreted as a palaeo-hydrocarbon–water contact (Bushell 1986; Woodward & Curtis 1987; Knipe *et al.* 1993), was lost during the early Cretaceous. The present (mostly) gas charge is believed to result from a further cycle of hydrocarbon generation (also from the Bowland Shale Formation?) associated with an elevated geothermal gradient during the early Cenozoic (Cowan & Bradney 1997). Hydrocarbon migration continues in the basin to the present day, as witnessed by the seepage of oil into Quaternary sands and peats at Formby, on the Lancashire coast.

In the 1990s, the Hamilton, Douglas, and Lennox fields, with a mixture of oil and gas, were discovered parallel with the North Wales coast in the southern part of the EISB (Fig. 6). Most of the deep wells of these fields encountered the Millstone Grit Group below the Variscan Unconformity, as at Formby. Using isotopes, the sampled oils (from 110/15-6 and 110/13-10, the Lennox and Douglas fields, respectively) were correlated with each other, and the Holywell bitumen and the Holywell shales (correlative of the Bowland Shale Formation) of NE Wales, thereby proving the Bowland Shale source (Armstrong *et al.* 1997). These were isotopically lighter (more negative) than Westphalian cannel coals of Type I kerogen: for example, those formerly mined and used to make oil at Leeswood in North Wales (Falcon & Kent 1960). Waxy crude shows in the Millstone Grit Group in well 110/07b-6 (1510–1675 m: released geochemical report) showed an isotopically similar source to shows in wells 110/07-2, 110/08-3 and Formby. The API of the Irish Sea oils range from 40° to 45° at Lennox and Douglas (Hardman *et al.* 1993), to 37° at Formby (Armstrong *et al.* 1997), perhaps suggesting a less mature source in the onshore field. Many additional small fields have been discovered subsequently, mostly in the centre of the EISB and mostly containing gas, culminating with the Rhyl discovery in 2009. In the Irish Sea, no significant Carboniferous reservoirs or good shows have been reported but there has been at least one discovery (113/27-2) in the Collyhurst Sandstone (Appleby Group).

Stratigraphy of the petroleum system

Carboniferous source rocks are shown, in Figure 7, as covering the lower part of the Namurian and highest part of the Visean where shales are developed; Pennine Coal Measures may make a contribution where preserved. The lithostratigraphical terminology used here is that introduced by Waters *et al.* (2011) to better integrate the offshore with the onshore geology than previous schemes (e.g. Jackson *et al.* 1997). The Carboniferous source rocks are separated from the Triassic Ormskirk Sandstone reservoir rocks by the Millstone Grit Group and, where present, by Pennine Coal Measures and Warwickshire groups. Above the Variscan Unconformity, the Permian Appleby and Cumbrian Coast groups, and the lower tight part of the Triassic Sherwood Sandstone Group, also intervene. A Pendleian time slice (Fig. 8) highlights the persistence of the relatively deep-marine hemipelagic successions (Bowland Shale Formation) across the central part of the British Isles, including the Craven Basin, the EISB and westwards towards the Dublin Basin (Ramsbottom 1969; Cope *et al.* 1992; Jackson & Mulholland 1993; Andrews 2013; Wakefield *et al.* 2016). The late Pendleian saw the first major influx of thick fluvial and deltaic sandstones into the Craven Basin, both from the north and from the south. The northern basin fill is characterized by thick prodeltaic ramp turbidites, overlain by a siltstone-dominated slope succession, in turn overlain by a fluvio-deltaic, delta-top sandstone (Collinson 1988; Wakefield *et al.* 2016). The hemi-pelagic successions have gamma values which suggest potential as source rocks. The overlying successions of the Pennine Coal Measures and Millstone Grit groups have potential as a combined source–reservoir unit, with secondary sources from marine influxes and coaliferous sediments.

Clastic intervals within the Carboniferous and Permian successions that are evaluated for reservoir

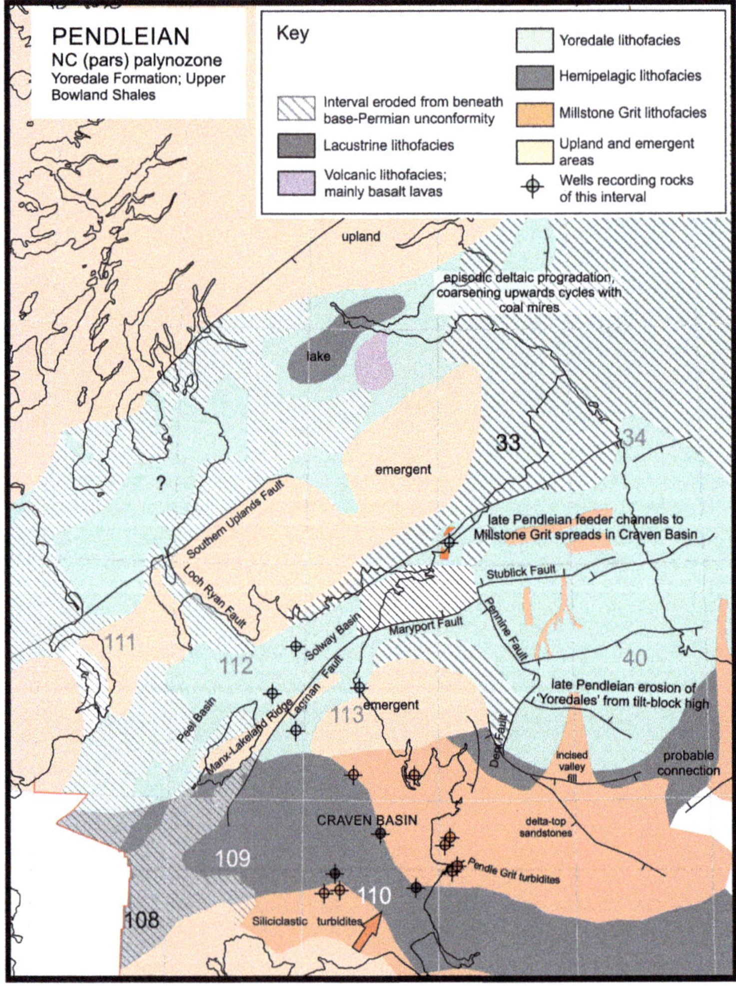

Fig. 8. Pendleian palaeogeography showing the Bowland Shale source-rock distribution and lateral variation with Millstone Grit facies (from Wakefield *et al.* 2016).

potential include the Appleby Group, the Warwickshire Group, the Pennine Coal Measures Group, the Millstone Grit Group and the Bowland Shale Formation. The Carboniferous Limestone Supergroup has been assessed as a potential reservoir, although the effect of secondary, karstified and fracture porosity has not been analysed. The preservation and thickness of the possible reservoir units is variable, particularly the Carboniferous units beneath the Variscan Unconformity (Fig. 6). Interpretation of well logs and associated core analyses (e.g. biostratigraphy, poroperm) frequently provide alternative stratigraphic interpretations to those shown on the well composite log, and have been carried out in this study (Fig. 9a, b). Many authors have referred to the problems in identification that result from secondary reddening of the Carboniferous strata below the Variscan Unconformity (Trotter 1954; Falcon & Kent 1960; Jackson *et al.* 1995) in both the adjacent onshore and within the EISB. In the south of the basin, thick Appleby Group strata overlie the Variscan Unconformity and stratigraphic interpretation is straightforward. However, in the Morecambe fields area, the Appleby Group is absent and the Cumbrian Coast Group is interpreted to overlie the Variscan Unconformity (Fig. 9b). This is important because it shows the probable topography of the Carboniferous surface, deformed and uplifted by the Variscan Orogeny, and the extent of erosion and eventual burial. The Cumbrian Coast Group comprises a varied sequence of thin sandstones, anhydrites, limestones, halites and mudstones, mostly red in colour. Underlying red beds have therefore been interpreted either as a mudstone facies of the Appleby Group or as Warwickshire Group strata, on well composite logs. The favoured interpretation, combining all the seismic and well evidence, is that the red beds directly underlying the Cumbrian Coast Group are secondarily reddened. They often include thin sandstones and high-gamma shales, and rarely contain coals, and are believed to be mostly of Namurian depositional age.

Source rocks

One of the key risks in the Paleozoic of the greater Irish Sea province is the quality, extent and maturity of source-rock intervals (Fig. 10). Potential source rocks include coals of the Pennine Coal Measures (Westphalian) and the upper Millstone Grit (Namurian) groups; shales of the Bowland Shale Formation and the Millstone Grit Group (Pendleian and Arnsbergian); and older Visean shales (unproven by sample data): for example, in the lower part of the Yoredale Group. Compilation of the Rock-Eval source-rock geochemical data from released legacy reports revealed a small dataset (264 samples), limiting the analysis that could be undertaken (Vane *et al.* 2016). Where penetrated, the Pennine Lower Coal Measures Formation, Millstone Grit Group and Bowland Shale Formation are mainly gas-prone strata with poor–fair remaining generative potential, and are mature to the gas window at the sampled intervals in quadrants 110 and 113 (Vane *et al.* 2016). Some shales within the Millstone Grit Group have total organic content (TOC) values (Fig. 11f) and S1 hydrocarbon values (Fig. 11a) greater than the Bowland Shale Formation. Given the maturity levels, source-rock potential in these wells is likely to have been depleted by hydrocarbon generation, or the original quality of these source rocks was poor–fair. The Cumbrian Coast Group, Appleby Group and Carboniferous Limestone Supergroup sampled in two wells in Quadrant 111 are oil to gas window mature, but have low TOC values and low residual (S1) hydrocarbon generative potential. Data are generally lacking to characterize kerogen types using a Van Krevelen plot: however, data from well 110/02b-10 (Fig. 11b) suggest a kerogen mix between Type II and III for the Millstone Grit Group and Pennine Coal Measures. A similar mixed system can also be expected for the Bowland Shale Formation but with a higher proportion of Type II kerogens. The high TOC and widespread extent of the Bowland Shale Formation favour it as the primary source rock, at least in the southern part of the Irish Sea. The other potential sources are ranked as secondary to this.

Hydrocarbon maturation and generation

Vitrinite reflectance (VR) data (Fig. 11d) shows that the Bowland Shale source rocks in wells are mature for oil and gas generation (Corcoran & Clayton 1999; Vane *et al.* 2016). EISB oils were considered to have been derived from the source in the range 0.75–0.85%Ro maturity, and the condensate from >1.0%Ro (Armstrong *et al.* 1997). Given the structural complexity for the area of interest, a singular burial trend and maturity profile cannot be defined. Cowan *et al.* (1999) gave examples of varying thermal and burial history at the basin margins changing over tens of kilometres. Three wells show a correlation of maturity increase with depth within the T_{max} dataset: 110/07b-6, 110/02b-10 and, to a lesser extent, 113/27-1, indicating progressive oil-window into gas-window maturity with depth. Some of the T_{max} data indicate a wide spread of temperatures at the same depth, perhaps reflecting reworked and caved material in addition to *in situ* measurements or possibly due to T_{max} suppression caused by variable kerogen and free oil composition (Fig. 11c). Onshore Isle of Man boreholes (Shellag, Ballavarkish and Black Marble Quarry: Fig. 5) show a similar range of T_{max}, albeit with few samples (Racey 1999).

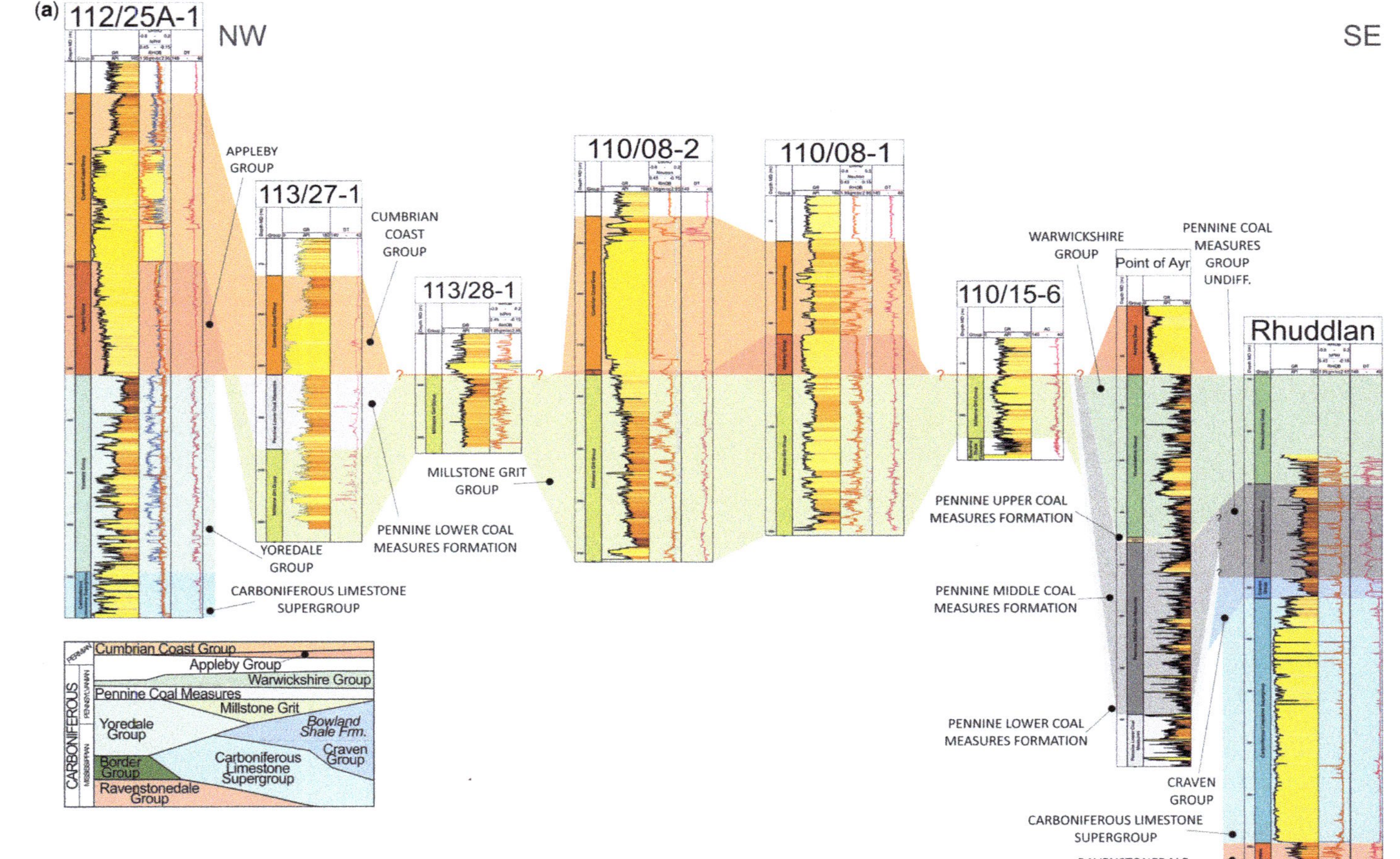

Fig. 9. Well transects (from Wakefield *et al.* 2016). Locations are shown in Figures 2 and 6. (**a**) North–south transect across the EISB from the Lagman Basin to Rhuddlan, onshore North Wales. Note the truncation of the Warwickshire Group north of Point of Ayr and the condensation of the underlying Westphalian strata, southwards onto the Cambrian margin. Also note the variation in the thickness of the Appleby Group.

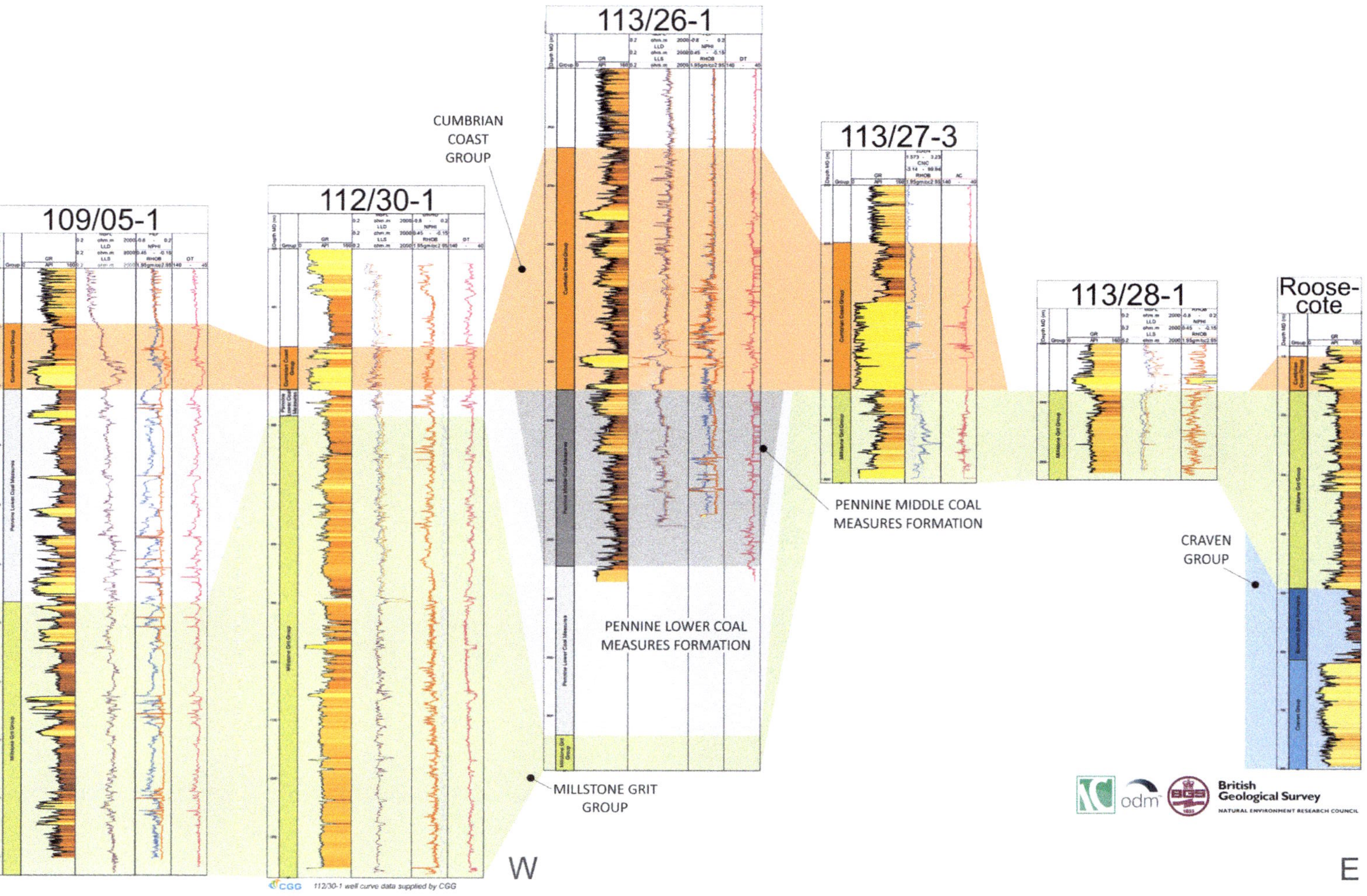

Fig. 9. (*Continued*) Well transects (from Wakefield *et al.* 2016). Locations are shown in Figures 2 and 6. **(b)** West–east transect across the centre of the EISB from 109/5-1 in the Eubonia Basin to Roosecote, onshore north Cumbria.

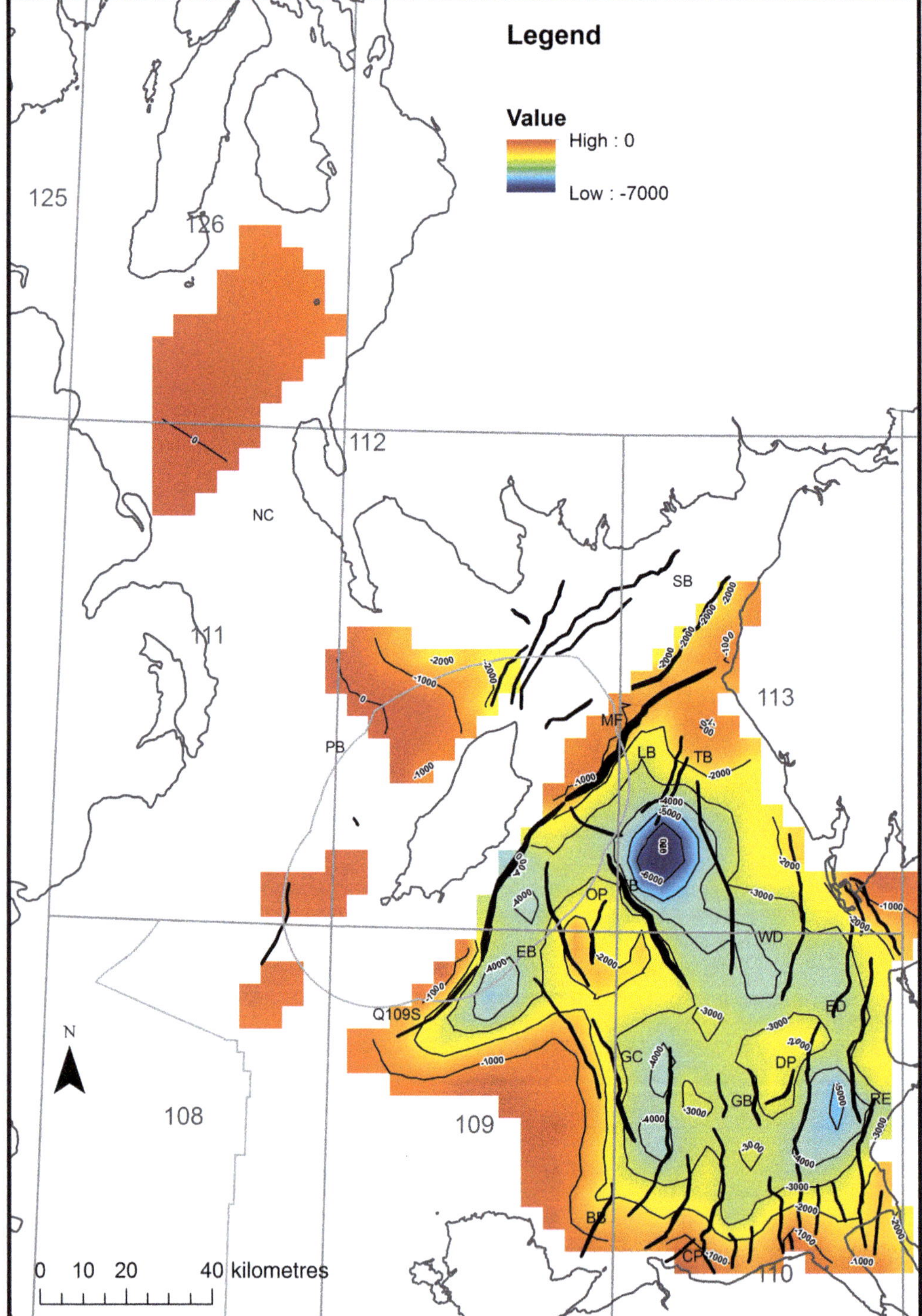

Fig. 10. Seismic structure map in depth (metres subsea level) for the Intra-Namurian pick, equated with the base of the Millstone Grit Group. For the location of abbreviated Permo-Triassic basinal features (for reference purposes), refer to the key in Figure 2.

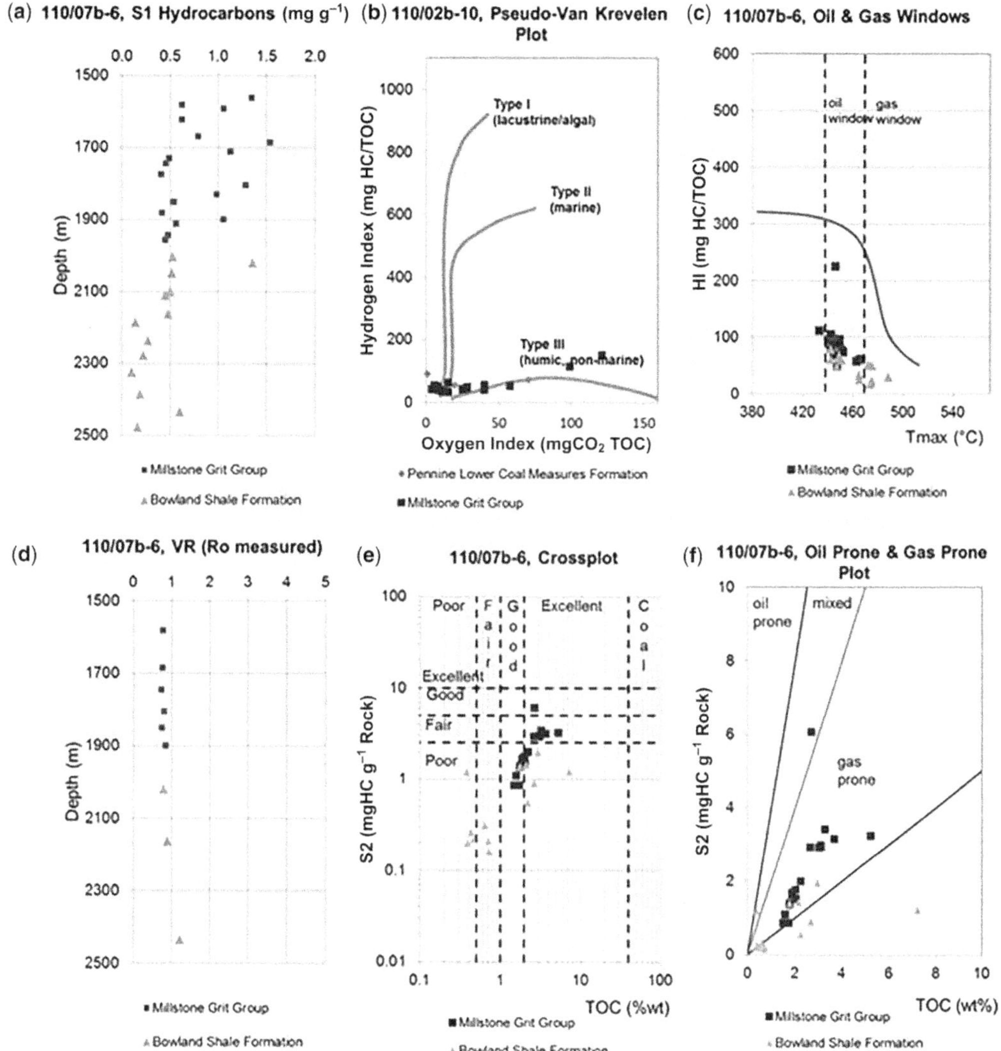

Fig. 11. A summary of the available geochemical data for Bowland Shale Formation lithologies in wells 110/07b-6 and 110/02b-10. Data are sourced from released legacy reports. Note that no oxygen index (OI) data are available for well 110/07b-6, so the data in the pseudo-Van Krevelen plot (Fig. 11b) are from well 110/02b-10 (Millstone Grit and Pennine Coal Measures groups).

Basin modelling

A lack of preserved post-Jurassic strata has resulted in a range of burial and thermal models for the EISB: for example, Cenozoic uplift estimates ranging from <1 km to up to 3 km (Cowan *et al.* 1999; Quirk *et al.* 1999 and references therein). In this study, well 110/07b-6 was chosen for burial and thermal modelling as it had the most complete geochemical profile and thick Carboniferous section (Gent 2016) (Fig. 12). The well is situated on a minor Variscan structural high, and is considered reasonably representative of the more marginal areas of the basin. The burial model was matched to the measured VR profile and the calculated VR profile (from T_{max}) (Fig. 12). Using published studies (Cowan *et al.* 1999; Quirk *et al.* 1999) and seismotectonic interpretations from this study, a 700 m uplift event in the late Carboniferous, followed by a minor 150 m uplift during development of the Late Cimmerian Unconformity, and a final 1100 m uplift and increase in palaeo-heat-flow in the Cenozoic were included. The modelling shows that burial of the Bowland Shale Formation source rock in the

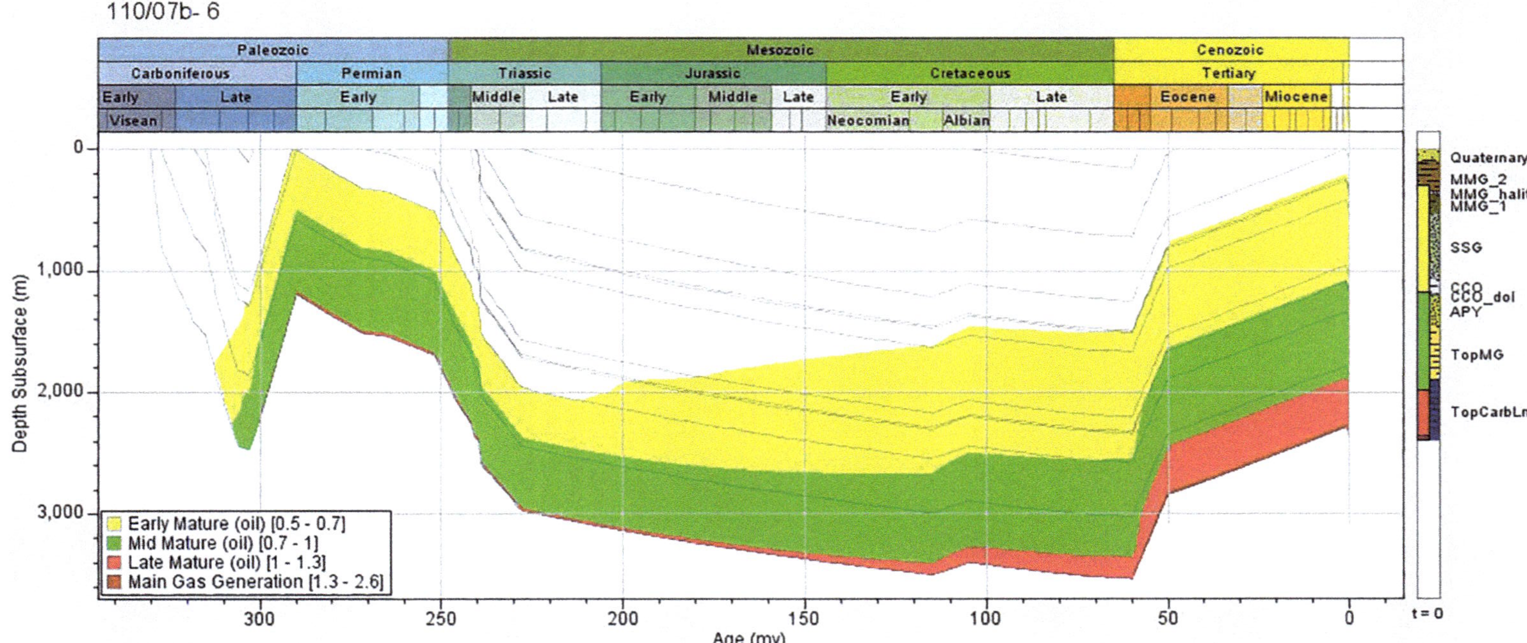

Fig. 12. Modelled burial history for well 110/07b- 6 showing that the Bowland Shale source rock entered the main gas generation window in the late Cretaceous–early Cenozoic. The well terminates within the Bowland Shale Formation.

Carboniferous resulted in the early–mid mature-oil window being reached, before uplift and subsequent deeper burial in the early Cenozoic, just reaching main gas generation in the base of the drilled strata (Fig. 12). This is consistent with the oil shows documented in the well geochemical report (Geochem Laboratories Ltd 1988). Carboniferous trap formation, migration and generation were all likely to have occurred during the Variscan Orogeny. However, subsequent uplift would have almost certainly breached the traps. Migration and trap formation was renewed in the Mesozoic and Cenozoic, with any modern-day hydrocarbon accumulations required to have survived the potential structural breach as a result of Cenozoic inversion.

Migration

Migration of hydrocarbons into Triassic reservoirs and traps has clearly been successful, as evidenced by the producing oilfields and gasfields of the EISB. Oil migration to the Triassic Hamilton fields may have occurred, vertically along faults, in Jurassic and Cretaceous times (Haig *et al.* 1997; Yaliz 1997; Yaliz & Taylor 2003). This study has highlighted how these fields overlie the Môn–Deemster inversion belt described above (Fig. 6), the structures of which may have acted as first-stage reservoirs and were subsequently breached to allow migration into overlying Triassic traps, formed posthumously as late as Cenozoic times, on a template created by the Variscan inversion structures. In a similar way, the Millom, Dalton and Calder fields, lying close to the Keys Fault, and the Lennox Field, close to the Formby Point Fault, are Cenozoic-age traps formed posthumously on a template provided by the second phase of Variscan inversion structures. As the basin depocentre widened and new areas came into the oil window, additional hydrocarbons may have been generated and continued to migrate southwards. The basin depocentre within the dismembered Eubonia Tilt-block entered the gas window, and gas migrated into the Morecambe and other fields. This may have occurred both pre- and post-Late Cimmerian uplift/sea-level fall (Bushell 1986). In a conceptual Carboniferous petroleum system model, migration is away from the steadily deepening and expanding hydrocarbon kitchen towards the margins of the basin, where these strata fail by thinning and overlap. In the north, the boundary is strongly faulted (Lagman, Eubonia and Lake District boundary faults).

Characteristics of potential reservoirs

A reservoir evaluation of Permian and Carboniferous intervals, designed as a quick-look regional overview, was based on legacy core-plug-measured porosity and permeability data and continuous petrophysical interpretations for eight wells (Hannis 2016). Net-to-gross, porosity and basic permeability estimates were calculated for each formation, and are summarized in Table 1. In general, the results illustrate fairly low net-to-gross values of <10% (except in the Permian-aged Appleby Group where net-to-gross was 79%), low porosities (highest formation averages mostly around 10% but up to 19% in the Appleby Group) and mainly poor average permeabilities (highest formation averages mostly less than 10 mD). Further examination of the distribution of potentially higher permeabilities within the Millstone Grit sandstone intervals could be worthwhile (Table 1). The core-plug-measured porosity v. permeability data by formation are exhibited in Figure 13.

The aeolian-dominated Permian Appleby Group strata that include the Collyhurst Sandstone are a prospective reservoir interval. The group, as proven in well data, is commonly defined by a basal breccia, overlain by a thick clean sequence of aeolian sandstones, culminating in an upper sequence of breccias (Wakefield *et al.* 2016). Based on six wells in Quadrant 110 in the depth range 1300–00 m, maximum measured core porosity is 21% with a highest formation average in all wells of 13%. Permeability is, however, poor, with a maximum measured permeability of 71.5 mD (vertical, k_v), and a highest formation average of 0.8 mD (horizontal, k_h) and 7.90 mD (vertically). Petrophysical analysis has confirmed the group to be a sandstone-dominated interval with an average net-to-gross ratio of 79%. Petrophysical porosity and permeability calculations match with the core-measured values, with the highest average porosity calculated at 19% and highest average permeability estimates of 6.89 mD, with some estimates in the 50–100 mD range for several wells (Table 1) (Hannis 2016).

The Warwickshire Group is the equivalent of the Ketch and Boulton formations of the southern North Sea in Quadrant 53 and quadrants 43 and 44 (Waters *et al.* 2011). Onshore, the Warwickshire Group of North Wales and Cheshire Basin comprises predominantly red, brown, purple-grey mudstones and sandstones, and locally green-grey siltstones and mudstones with thin coals. However, potential reservoir sandstones can be locally significant. The amount of sandstone relative to mudstone and siltstones within constituent formations of the Warwickshire Group varies considerably. In West Cumbria, the Whitehaven Sandstone Formation, at least 280 m thick (Akhurst *et al.* 1997; Dean *et al.* 2011), is mainly a red to deep purple or purplish brown, cross-bedded, micaceous, medium- to coarse-grained sandstone (Wakefield *et al.* 2016). The Halesowen Formation was productive in the small mined Coalport Tar Tunnel ‘field’ in Shropshire during the eighteenth and early nineteenth

Table 1. *Synthesis of petrophysical results by formation*

Stratigraphic unit name	Code	Log derived							Core measured							Comments
		Net thickness (m)	NTG	Highest average porosity	Porosity range	Highest average permeability estimate	Permeability estimate range	*Metres of log*	Highest average porosity	Porosity range	Highest average permeability estimate		Permeability estimate range		*Samples*	
											Horizontal (k_h)	Vertical (k_v)	Horizontal (k_h)	Vertical (k_v)		
Cumbrian Coast Group	CCO	81	0.07	0.14	0.05–0.43	0.17	0.02–1.39	*1197*	0.04	0.02–0.07	3.06		0.01–15.20		*6*	
Appleby Group	APY	936	0.79	0.19	0.05–0.40	6.89	0.17–82.27	*1191*	0.13	0.05–0.21	0.80	7.90	0.00–1.72	0.17–71.5	*154*	Highest net-to-gross, highest porosity. Highest permeabilities values in the 50–100 mD range for several wells
Pennine Coal Measures Group	PCM	62	0.09	0.11	0.05–0.26	0.79	0.02–61.45	*795*	0.06	0.01–0.10	1.07	0.01	0.00–9.43	0.01–0.13	*55*	Low NTG (although third highest of the units examined). Reasonable average porosity. Permeabilities appear low. (Highest values of 61.4 mD in one well, but with no core data over that interval)
Millstone Grit Group	MG	293	0.10	0.11	0.05–0.31	367.74	0.17–10 000	*2971*	0.06	0–0.10	0.04	0.05	0.00–0.37	0.00–0.13	*49*	Highest permeability (although estimated with low confidence: seen in only one of three wells (113/27-2), with a relatively poor core-log data fit). Low NTG (although second highest of the units examined)
Yoredale Group	YORE	16	0.02	0.07	0.05–0.30			*783*	0.01	0.00	0.00	0.00	0.00	0.00	*9*	
Bowland Shale Formation	BSG	16	0.03	0.07	0.05–0.23	0.75	0.15–16.19	*551*							*0*	
Carboniferous Limestone Supergroup	CL	0	0.00	0.05	0.05–0.05			*246*							*0*	Matrix porosities are less than 5%, therefore the unit is not considered to have any 'net' using the cut-offs applied

NTG, net reservoir thickness to gross formation thickness. Porosity and net-to-gross are expressed as a fraction. Minimum porosity in the log-derived porosity range is 0.05, the net reservoir porosity cut-off value. Permeability figures are in mD. Core porosity and permeability data are synthesized from legacy reports.
From Hannis (2016).

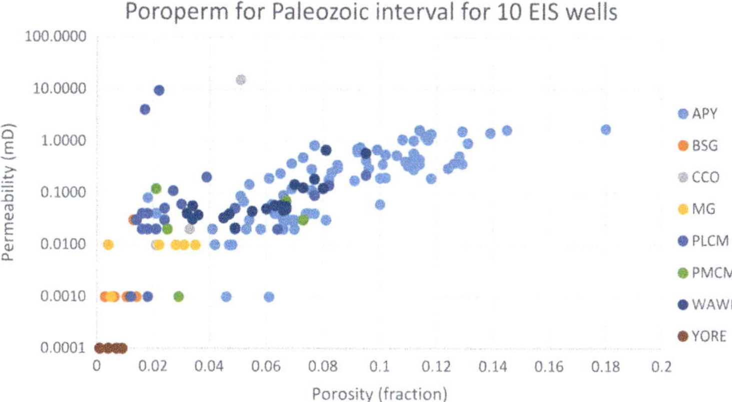

Fig. 13. Cross-plot of core porosity and permeability for East Irish Sea Basin samples. For the key to abbreviations see Table 1, except for: PLC, Pennine Lower Coal Measures; PMCM, Pennine Middle Coal Measures; WAWK, Warwickshire Group.

centuries (Smith *et al.* 2005). In the East Midlands, the Warwickshire Group has been documented as having better reservoir characteristics than productive older late Carboniferous strata, but was spatially confined to the synclines (BGS 1984; Pharaoh *et al.* 2011). Data from Quadrant 53 and the English Midlands shows that an average porosity of 16% is likely, with a permeability of several hundred millidarcies (mD), although the bulk of the data were from above 600 m depth. Therefore, an investigation of the Warwickshire Group as a reservoir interval offshore was considered, although seismic mapping indicated a limited extent in the greater Irish Sea province (Fig. 6) and there are no well penetrations and, therefore, no reservoir data for the group. However, Bolsovian–Asturian (Westphalian C–D)-age strata are recorded in well 33/22-1 along strike in the Kish Bank Basin (Jenner 1981).

In the EISB, the Pennine Coal Measures Group comprises interbedded grey mudstone, siltstone and pale grey sandstone, commonly with mudstones containing marine fossils in the lower part of the lower and upper part of the middle subdivisions, and more numerous and thicker coal seams in the intervening interval. The group shows an overall blocky to erratic log response, with thick high-gamma mudstone and siltstone intervals, and relatively thin (3–15 m) low-gamma sandstones. The sandstones show considerable variation in wireline log character, including 'boxcar' motifs in thick, distributary channel sandstones (Wakefield *et al.* 2016). Onshore, sandstones are also frequently encountered (e.g. Cefn Rock and Hollin Rock of the NE Wales coalfields, Worsley Delf Rock, Prestwich Rock and Newton Rock of the Lancashire Coalfield) and are approximate equivalents to the productive sandstones in basinwards East Midlands fields (e.g. Oak Rock, Crawshaw Sandstone and Wingfield Flags). Based on five wells in quadrants 110 and 113, in the depth range 1400–50 m, maximum measured core porosity is 10% with a highest formation average in all wells of 6%. Permeability is generally poor, with a maximum measured horizontal permeability (k_h) of 9.43 mD and a highest formation average for k_h of 1.07 mD. Petrophysical analysis of the Pennine Coal Measures Group provides a similar outlook, with an average net-to-gross of 9%. Net intervals have reasonable porosities, with the highest average porosity at 11%. Permeability is generally poor, with the highest average permeability estimated at 0.8 mD. However, permeability up to 61 mD was estimated in one well (110/02b-9: Table 1) (Hannis 2016).

The Namurian-aged Millstone Grit Group comprises cyclic sequences of quartzo-feldspathic sandstone, grey mudstone, thin coal and prominent seatearths, resulting from deposition by repeated progradational deltas (Collinson 1988). Common marine bands are present and represent discrete flooding events (Waters & Condon 2012). Thick reservoir intervals are uncommon, with initial turbidite lobes passing into delta-top deposits with thin sandstones typically contained within sheetfloods, overbank deposits and stacked channels. Onshore and, potentially, offshore, thicker sand bodies (up to 50 m thick) occupy incised valleys (Waters & Condon 2012; Wakefield *et al.* 2016). Jackson *et al.* (1997) (Fig. 6) identified a Kinderscoutian sandstone unit up to 90 m thick in the Liverpool Bay region (111/20-1), which can be correlated with wells farther north (112/30-1 and 113/27-2), although considerably reduced in thickness. Onshore, Millstone Grit sandstones are encountered in NE Wales (e.g. Cefn-y-Fedw, Gwespyr Sandstone and Aqueduct

Grit), Lancashire (e.g. Fletcherbank Grit, Pendle Grit and Warley Wise Grit) and in producing East Midland fields (e.g. the Rempstone Oilfield). The Namurian (Marsdenian) depocentre extends from the Staffordshire Gulf, probably to Preston and thins to SW under the Cheshire Basin (Collinson 1988; Smith *et al.* 1995). This pattern continues into the offshore of the EISB, with Namurian absent at the Rhuddlan well on the North Wales coast (Figs 5 & 9b). Based on samples from four wells in quadrants 110 and 113, at 1950–50 m depth, maximum measured core porosity is 10%, with a highest formation average in all wells of 6% (Table 1). Permeability is poor, the maximum measured was 0.37 mD (k_h), and the highest formation averages for k_h and k_v were 0.04 and 0.05 mD, respectively. Petrophysical analysis provides a more promising outlook for the group, although the average net-to-gross is 10%. Net intervals have a reasonable porosity, the highest average porosity is 11%. Permeability is poor, with an average estimate of 0.2–2.1 mD, apart from one well, 113/27-2, which shows an average of 367.7 mD (Table 1). Further analysis of these sandstones could therefore be beneficial (Hannis 2016).

The Bowland Shale Formation is only examined in wells 110/11-1 and 110/07b-6; however, the formation broadly shows an upwards decrease in carbonate turbidites and an increase in siliciclastic sandstone turbidites (Wakefield *et al.* 2016). Potential thin reservoir sandstones may be present. Well 110/07b-6 encounters a total of 16 m of these sandstones, giving a net-to-gross of 3%. (The other well examined, 113/27-2, contained no net intervals.) No core samples were taken, but petrophysical interpretation revealed that the net intervals had porosities up to 23%, although the average porosity was 7%. Permeability estimates appear poor, with an average of 0.7 mD and maximum of 16.2 mD (Table 1) (Hannis 2016).

Carboniferous Limestone Supergroup sequences are interpreted to be widespread over the EISB and thus worthy of investigation as a reservoir. Petrophysical analysis of the limestones encountered in two wells (112/25a-1 and 111/25A-1) appear clean, but have matrix porosities that are too low (<5%) to be considered as a reservoir (Table 1), but accumulations could be hosted in secondary porosity as a result of karstfication or fracturing. Onshore, the Hardstoft Oilfield in Derbyshire (Craig *et al.* 2013) produced from the top of the Carboniferous Limestone but, despite numerous shows, no further production was established from this reservoir in the East Midlands fields (Falcon & Kent 1960). Karstified limestones, such as those known from Anglesey (Walkden & Davies 1983), and apron reefs like those which crop out at Castleton, Derbyshire might be present in the offshore. Seismic evidence for the possible presence of reefs towards the top of the ramp of the Eubonia Tilt-block in Quadrant 109 (Fig. 4a) was described above and is indicated schematically in Figure 4a. Waulsortian mud-mounds of pre-Asbian age may also be possible reservoirs. They are seen at outcrop in the south of the Isle of Man (Dickson *et al.* 1987) and in the Craven Basin. In the prolific Williston Basin of Canada, collapsed mud-mounds up to 100 m tall provide excellent porosity but were initially hard to identify on seismic data (Kupecz *et al.* 1996).

Seal rocks

The Cumbrian Coast Group, which includes the Manchester Marls (Fig. 7), provides the most extensive potential seal to Permian or Carboniferous rocks across the whole of the greater Irish Sea area. The unit consists of thick evaporites in the north and central East Irish Sea, thinning southwards, passing laterally into dolomitic mudstones (Jackson & Mulholland 1993; Wakefield *et al.* 2016), and is encountered in wells in surrounding sub-basins. This seal has been proven to trap hydrocarbons in well 113/27-2 and the sealing potential is proven in 112/25a-1, with minor gas shows in the tight Appleby Group. In the producing EISB fields, any Cumbrian Coast Group seals were breached as the fluids migrated out of the Carboniferous and Permian into the Triassic Ormskirk Sandstone reservoir (Colter 1997). Carboniferous intraformational mudstone seals have proved adequate in all the onshore fields of the East Midlands (Pharaoh *et al.* 2011), Cousland in Scotland (Hallett *et al.* 1985), various fields in the Silver Pit and Cleaver Bank basins of the southern North Sea, and numerous fields in The Netherlands and Germany (Pletsch *et al.* 2010), and could be expected to work in Carboniferous basins of the Irish Sea.

Hydrocarbon prospectivity of the Carboniferous basins outside the EISB

Whilst basins of the greater Irish Sea province outside the EISB have extensive seismic coverage of variable quality, there are few wells. Data are therefore lacking to constrain their hydrocarbon systems and is heavily dependent on onshore analogues.

Solway Basin

The Permian–Jurassic Solway Basin, linked NE to the Carlisle Basin and SW to the Peel Basin, is underlain by a Carboniferous basin of the same trend, an extension of the Northumberland Trough (Chadwick *et al.* 1995) (Fig. 2). Two well penetrations (112/15-1 and 112/19-1) prove a Visean–

Namurian Yoredale Group, distinguished from the Carboniferous Limestone Supergroup by the presence of fewer carbonates (Fig. 7). The Yoredale Group sandstones, limestones and siltstones represent a fluvio-deltaic depositional environment (see Wakefield *et al.* 2016) that is a northwards lateral equivalent of the basinal Bowland Shale Formation; that is, the Bowland Shale facies is not proven and may not be present. The presence of delta-top lacustrine facies is a possibility, but has not been demonstrated. In the onshore Cumberland Coalfield, the coals are gassy (Colter 1997), but the Pennine Coal Measures Group have not been penetrated offshore in the Solway Basin. Potential Carboniferous reservoir intervals include a relatively small area of Warwickshire Group on both sides of the Maryport Fault (Figs 5a & 6) and the Fell Sandstone Formation in the main part of the basin.

Peel Basin

The Peel Basin is a Permian–Jurassic basin lying between the Isle of Man and Northern Ireland, underlain by a Carboniferous carbonate platform. Wells 111/25a-1 and 111/15-1 penetrated the Mississippian-age Carboniferous Limestone Supergroup, in contrast to the time-equivalent Yoredale Group encountered in the along-strike Solway Basin. The lack of a clastic, fluvio-deltaic system may enhance the likelihood of the Bowland Shale (source rock) equivalent being present in younger strata between 111/25a-1 and the Isle of Man coast, but there are no data to test this hypothesis. The seismic reflection data are generally of poor quality, but allow the presence of a small outlier of Namurian strata to the NW of the Isle of Man. The Peel Basin may extend to the Carlingford Lough area near the Irish border, south of the Mourne Mountains (Fig. 2). BGS boreholes (in Quadrant 112, near the Irish coast) 73/65 and 73/67 are of probable Visean age, and form a rim to the Lower Paleozoic Longford-Down Massif. BGS borehole 71/43 near the Isle of Man coast was dated as Namurian. The data available preclude evidence of a working Paleozoic petroleum system in the Peel Basin, a conclusion previously reached by both Newman (1999), Quirk *et al.* (1999) and Floodpage *et al.* (2001).

North Channel Basin

The North Channel Basin is a NW-trending Permo-Triassic basin complex lying between the Southern Uplands and the Longford-Down Massif of Northern Ireland (Quinn 2008) and forms the main rift through the massif. Two tilt-blocks, the east-dipping Portpatrick and the west-dipping Larne sub-basins, recognized by Maddox *et al.* (1997), are separated by the Southern Upland Fault (Fig. 2). Several smaller basins lie parallel in Scotland (Stranraer and Lochmaben) and Ireland (Strangford Lough). In the Portpatrick Sub-basin, the underlying strata are possibly Devonian, although the seismic is poorly resolved because the only well (111/15-1) passed through a fault adjacent to the Southern Uplands, and did not prove a Carboniferous section. Data are lacking for the presence of source, reservoir and seal in this area (Maddox *et al.* 1997). Permo-Triassic and underlying Devonian and Carboniferous strata are present onshore in the Larne (Penn *et al.* 1983) and Lough Neagh basins of Northern Ireland (Mitchell 2004). Onshore in the Midland Valley of Scotland and in Northern Ireland, a range of potential Carboniferous source rocks (coals, carbonaceous mudstones) and sandstone reservoir intervals are documented, although there is considerable spatial variability (Browne *et al.* 1999; Read *et al.* 2002; Reay 2004, 2012; Underhill *et al.* 2008; Monaghan 2014). Onshore in Northern Ireland, a Carboniferous prospect was drilled by Infrastrata plc in Woodburn Forest in 2016, without success (www.infrastrata.co.uk/index.php?option=com_content&task=view&id=413\l). Seismic interpretation offshore (Pharaoh *et al.* 2016*a*) has included a Carboniferous succession in the Larne Basin buried to 5000 m, and with faulting and folding observed offering potential for structural traps. However, the interpretation is poorly constrained by data, precluding a detailed assessment of petroleum system elements.

Brief mention can be made of the Rathlin Trough, which lies outside the study area and for which only limited seismic data, covering the offshore extension of the Machrihanish Coalfield, have been studied. The source rocks include coals and oil shales (Murlough Bay Formation) of early Carboniferous age that have excellent TOC values and which are mostly in the oil window, with smaller areas in the gas window (Reay 2012). This sequence, together with volcanic rocks, invites comparison with the Lothian part of the Midland Valley of Scotland (Read *et al.* 2002). Drilling took place at Magilligan, in the west of the basin, and at Ballinlea in 2008. In the latter well, oil was produced from the Carrickmore Formation sandstones (Providence Resources plc 2013) of the wide Visean subcrop (Smith 1985).

Petroleum system knowns and risks

The distribution of the principal Carboniferous source rock (Bowland Shale Formation) as inferred from the seismic interpretation is constrained by a few borehole penetrations in the EISB, but the absence of boreholes in the deepest part of the basin (Keys and Lagman basins) and onto the Manx–Furness Ridge means that the northern limit is poorly constrained. The nature of the transition to the Solway Firth and Northumberland basins, where boreholes prove time-equivalent Yoredale facies, is therefore poorly known. The lack of any offshore well data requires analogy with the adjacent onshore Carboniferous. In the northern part of the EISB, the very deep burial of the source (now at >7 km depth despite Cenozoic inversion) and the strong thermal impact from the Fleetwood Dyke (Arter & Fagin 1993) means it is probably overmature, compatible with high CO_2 and nitrogen levels observed in Rhyl and neighbouring fields (Cowan 1996; Centrica pers. comm. 2015). Leakage of hydrocarbons in the Cenozoic following fault reactivation and degassing, consequent upon regional uplift, are further risks throughout the region. Source rocks may also be present in the Clyde basins and the adjacent North Channel Basin, but are unlikely to be present in the southern part of the latter or beneath the Peel Basin. Attenuation of the Carboniferous sequence southwards towards the Welsh Massif (Figs 4b & 9a) also increases the source risk in this direction. The paucity of data on the maturity of the source means that this parameter cannot be mapped in detail. Similarly, the reservoir porosity–permeability characteristics are poorly known over large parts of the region studied. The petrophysical analyses presented here suggest that the Carboniferous sandstones beneath the Morecambe fields have very poor porosity and permeability, confirming information provided by Centrica (pers. comm. 2015). This is, no doubt, a consequence of their deep burial, and processes such as platy-illite development and silica cementation that severely affect even the overlying Triassic formations (Bushell 1986; Colter 1997; Woodward & Curtis 1987; Cowan & Bradney 1997; Stuart 1993). The Carboniferous tight gas play may work if hydraulic fracturing can be applied, as is presently being attempted at Kirby Misperton (Cleveland Basin) and in the Ravenspurn Deep (Southern North Sea). Extensive carbonate platforms surrounding the Isle of Man (Manx Platform) and off North Wales (Colwyn Platform) also have unknown porosity–permeability (poro–perm) characteristics. Until more is known about possible secondary porosity (following dedolomitization) and fracture density, the reservoir properties of these areas are ranked as high risk.

The Mercia Mudstone Group is a proven cap rock to Sherwood reservoirs and is present throughout the EISB but is absent across the margins of the basin complex. The potential seal of the Permian Cumbrian Coast Group sequence thins and fails in the same directions. In the EISB, a relatively thick shale and evaporite (St Bees Evaporites, Cumbrian Coast Group) may be developed. The same is true in the Portpatrick and Larne basins, where several Triassic halites are present (Quirk *et al.* 1999; Quinn 2008).

Analysis of seismic data, integrated with, for example, well and core data, indicates that the marginal areas of the EISB hold the greatest potential for undiscovered hydrocarbon resources in the Carboniferous, although the geochemical, petrophysical and other essential data are scant. In general, the presence of an effective seal is considered to represent the biggest risk in the hydrocarbon system at the margins of the EISB. Yet-to-find prospects are anticipated to be relatively small in volume and with shallow column heights supported by Carboniferous intra-formational seals. The most prospective parts of the region, outside the Triassic play, are considered to be:

- Thick Westphalian combined reservoir and source-rock sequences preserved in the Eubonia Tilt-block in Quadrant 109 (Fig. 4a), located outside the main Permian–Mesozoic graben system and less affected by Cenozoic inversion. The presence and quality of seals form a major risk as the Cumbrian Coast Group seal is thin or absent and Carboniferous intraformational seals are required but untested. Based on the limited dataset available in adjacent basins, reservoir quality is also a significant risk.
- A belt of Variscan inversion structures (the Môn–Deemster Foldbelt: Fig. 4b) correlated with structures on the Formby Platform, and the onshore Ribbledale Foldbelt, from which hydrocarbons sourced by a thick Bowland Shale sequence have leaked into the overlying Triassic-hosted Hamilton fields (Block 110/13). The biggest risk here is whether reservoirs exist and remain unbreached at the pre-Permian level, and retain good poro–perm characteristics at depths of about 2500 m.
- A more speculative play lies in the extensive carbonate platform in Quadrant 109 and surrounding the Isle of Man (Fig. 4a), in Asbian reefal facies with enhanced secondary porosity. Here, source-rock presence and migration pathways, reservoir properties, and seal quality are major risks.
- The Ribble Estuary Inlier east of the Formby Point Fault (Figs 5c & 6) may contain a working petroleum play. It lies adjacent to the deep Deemster Basin where there is a thick sequence of Upper Carboniferous sedimentary rocks preserved, and between the Formby and Lennox fields. Well 110/9-1, within the Deemster Basin, was dry but appears to have good porosity in the Ormskirk Sandstone, although no shows. Fluorescence was recorded in the Appleby Group.
- A potential play exists sourced from the Bowland Shale Formation in the deep Godred Croven Basin drilled by well 110/11-1 migrated into the Carboniferous reservoir on the faulted highs of

its flanks. The Ormskirk Sandstone is very shallow in these locations but the Carboniferous strata might be securely sealed by the Cumbrian Coast Group.

Discussion

The pre-Permian structural synthesis presented here is speculative in view of the limited number of offshore well penetrations of Carboniferous strata. For example, further tectonic partitions may exist within the inferred Eubonia Tilt-block. It is possible, for example, that the eastern part of the structure (underlying the Keys and Tynwald basins of the EISB) may represent a separate tilt-block with a hinge in the Lake District Boundary Fault System, and a master controlling fault in the west (ancestral Keys system). The presence of such a basin, referred to as the Lancaster Fells Basin, was inferred by Cowan *et al.* (1999). However, the available evidence suggests that the NNW structural trend did not play a significant role until latest Carboniferous time, so that an ancestral Keys Fault is regarded an unlikely Visean structural element. The nature of the link between the structures of Quadrant 109 and onshore Lancashire has been much speculated on in the past (e.g. Ramsbottom *et al.* 1978). Jackson & Mulholland (1993) recognized the Menai Strait–Pendle Line link, but preferred to link the Q109 Arch to the High Haume Anticline of the Furness Inlier in the southern Lake District. This paper shows that the Ribblesdale Foldbelt does extend west of the Leyland Basin and Formby Point Fault (cf. Jackson & Mulholland 1993, fig. 4, p. 797) and links to the Q109 Arch, via the Môn–Deemster Foldbelt. More detailed seismic mapping of the Upper Carboniferous interval will be required to elucidate what is probably an intricately folded subcrop pattern here. We support the proposed continuity of the Bowland Basin towards the SW into the offshore area, as inferred by Corfield *et al.* (1996) and Cowan *et al.* (1999). From the perspective of hydrocarbon prospectivity, the presence of the prolific Bowland Shale Formation source rock interpreted across much of the EISB has been a key element in the hydrocarbon system of the overlying Permian–Mesozoic basins. Prospective reservoir intervals with moderate porosity are likely to exist in the Warwickshire Group and Pennine Coal Measures Group in the marginal parts of the EISB, although the permeability is likely to be poor. The EISB lay to west (Fig. 8) of the main Pennine deltaic and fluviatile fairway in the onshore (Fraser *et al.* 1990), and consequently shows a lower net-to-gross sand ratio. Evidence from geophysical logs indicates that the EISB was less influenced by the repeated deltaic incursion so evident onshore, and a sharp transition between deltaic and basinal shale facies, of the type seen at Mam Tor and Edale, is neither observed nor to be expected there.

In a review of the deep reflection seismic data for the Irish Sea, principally BIRPS' WINCH lines and some deep data from JEBCO, England & Soper (1997) stated that, from this limited dataset, there is no clear evidence for reactivation of earlier structures during either Carboniferous sedimentation or Variscan inversion. Using the exploration seismic data, this study describes the presence of fold–thrust structures in the pre-Carboniferous basement and, in the Môn–Deemster Foldbelt, demonstrates their role in controlling both Carboniferous extensional and inversion structures. The interpretation presented supports the view of England & Soper (1997) that the faults controlling Permian and Mesozoic basin development are discordant to the Caledonian, Acadian and early Carboniferous structural grain (as exemplified by the Q109 structures), and are therefore juvenile structures developed in late Westphalian–early Stephanian time. Evidence presented here suggests that these were initiated as a result of a late phase of Variscan inversion, reflecting west–east Uralide compression, superimposed on an earlier phase produced by north–south compression. The timing of these two inversion phases is imprecisely defined in the Irish Sea due to the significant missing stratigraphic section. However, in late Variscide intramontane basins in central France, north–south compression in Stephanian B time is followed by inferred phases of compression on NW–SE (late Stephanian B) and west–east (mid-Stephanian C) principal stress axes (Gélard *et al.* 1986), the 'Bourbonnaise phase' of Grolier (1971). Although interpreted in terms of systematic rotation of the principal horizontal compressive stress axis (Gélard *et al.* 1986; Blès *et al.* 1989; Ziegler 1990), Faure (1995) considered these deformations to be a consequence of late Variscan orogenic collapse. In the UK region, other manifestations of the late west–east compressive phase may include the west–east-orientated basaltic dykes, and a component of growth of north–south-trending folds, within the Midland Valley of Scotland (Monaghan & Pringle 2004; Timmerman 2004); and the west–east-directed transport of fold nappes on the eastern margin of the Worcester Graben (Peace & Besly 1997).

The observed variation in Variscan structural orientation in the Variscan Foreland of Britain is currently explained in terms either of one resolved compressional vector (Corfield *et al.* 1996), or of strain-partitioning across a heterogeneous basement template (e.g. De Paola *et al.* 2005). In the Irish Sea, it is difficult to argue for a strong control by a north–south-orientated basement grain, as identified, for example, within the Midlands Microcraton (Corfield *et al.* 1996), and the presence of two discrete late Variscan deformation phases is regarded as a more likely scenario. Another expression of the multiple inversion history, and very significant for the formation of Ormskirk traps in the cover, is the impact of posthumous folding. This process, first recognized and described by Suess (1904), is very clearly demonstrated in the Irish Sea, where a template of Variscan inversion anticlines in the Carboniferous sequence underlies structures with a similar trend but lower amplitude in the Permo-Triassic cover.

Conclusions

The study has demonstrated that the basins of the Irish Sea preserve a Phanerozoic geological history as complex as that of the UK onshore. A strong SW–NE structural grain was imprinted on the crust during late Precambrian and Caledonian accretion and orogenic deformation. Dipping zones of strong reflectivity in seismic sections are interpreted as major thrusts and shear zones, some of which can be correlated with known examples onshore. Mississippian rifting on SW–NE-trending faults resulted in depocentres that accumulated marine shale source rocks, preceding regional thermal subsidence. The Eubonia Tilt-block is a major Carboniferous syndepositional element beneath the northern part of the EISB, but was partially dismembered by the formation of the ancestral Keys Fault System. The Eubonia–Lagman Fault System formed the syndepositional bounding fault to the tilt-block. The Bowland Shale Formation forms the main source-rock interval, with inferred thickest development likely to be within the Môn–Deemster Foldbelt, the offshore correlative of the Bowland Basin, and its inversion, the Ribblesdale Foldbelt. This source rock is buried to depths of >7 km under the Lagman and Keys basins, and is probably post-mature there at the present day.

The Millstone Grit Group and Bowland Shale Formation contain thin clean sandstones locally up to 90 m thick which could be considered potential reservoirs. Prospective areas at these stratigraphic levels may exist at depth adjacent to the Keys Basin, and west of the Keys Fault. The Millstone Grit Group also has the potential to act as a secondary source rock, as do the Pennine Coal Measures Group when buried deep enough to achieve maturity. However, the latter were stripped from a large area of the EISB following Variscan inversion. Pennsylvanian strata exhibit marked thinning to the south onto the Conwy Platform. Burial by Upper Carboniferous sediments is likely to have resulted in early maturation of kerogen in source rocks within the deepest basins, but in the destruction of reservoir porosity and permeability in the depocentres. Warwickshire Group sedimentary rocks were not so deeply buried, and are likely to retain better reservoir characteristics.

The Variscan Orogeny, in late Carboniferous time, caused uplift, folding and thrusting on both WSW–ENE (Môn–Deemster) and NNW–SSE to north–south (Keys–Gogarth) trends, probably in two phases, corresponding to well-documented main compressional phases of the Variscan–Uralian Orogen. The later inversion phase occurred on NNW–SSE- to north–south- trending zones of deformation which would subsequently become localized as the main synsedimentary bounding faults of the EISB in Permian–Mesozoic time. Corfield *et al.* (1996) provided a definition of inversion intensity. In the greater Irish Sea region, the intensity ranges from moderate (in the EISB, Solway and Clyde basins) to strong, with almost complete removal of the post-rift fill (in the North Channel basins and Peel Basin). The timing of these events is poorly constrained in the Irish Sea due to a significant missing stratigraphic section but, by comparison with intramontane basins in France, is likely to be of intra-Stephanian age. The Variscan inversion structures have not yet been adequately tested as targets. They form both first-stage hydrocarbon reservoirs and the structural template for more gentle, 'posthumous' folds produced by Alpine inversion which form traps in the Triassic cover (e.g. the Hamilton fields). Deposition of the Permian Appleby Group and Cumbrian Coast Group strata resulted in a potential reservoir–seal combination overlying the Carboniferous source rocks. Permian–Mesozoic rifting is along NNW–SSE and north–south trends. These faults cut discordantly across the early Carboniferous structures and have allowed late Cretaceous–early Cenozoic vertical migration of Carboniferous-sourced hydrocarbons into Triassic reservoirs. There is a migration route to Triassic reservoirs in the centre of the EISB because the Warwickshire Group and Appleby Group strata have been removed from that area, and the thin Cumbrian Coast Group seal breached, where the producing hydrocarbon fields are located. The Clyde–North Channel basin complex, and the Solway and Peel basins also contain Devonian and/or Carboniferous rocks beneath Permo-Triassic strata, but are likely to have been buried less deeply than those in the EISB. The North Channel basins may also have suffered significant Variscan inversion. Extensive 2D seismic datasets cover the latter areas but there are only four well penetrations. There have been no discoveries, which is interpreted to be largely a consequence of the absence of a regional seal comparable in quality to the Mercia Mudstone Group in the EISB. The prolific Bowland Shale source is also absent in these basins, being replaced by fluvio-deltaic sedimentation of the Yoredale Group. Very limited well penetrations do not presently allow a realistic assessment of the prospectivity of the Carboniferous strata underlying these poorly drilled basins.

Acknowledgements This work was undertaken as a joint-industry project for the former Department of Energy and Climate Change (DECC) now Oil and Gas Authority (OGA), Oil and Gas UK, and 49 oil and gas company sponsors as part of the 21CXRM (21st Century Exploration Roadmap) Palaeozoic Project. Thanks to Richard Milton-Worsell for his enthusiastic guidance, and to the Technical Steering Committee of the project. The role of CGG Geospec, IHS and Western-Geco in supplying regional speculative data for this study is gratefully acknowledged, likewise the provision of data from the CDA and UKOGL data libraries. The staff of Centrica plc are thanked for very helpful and productive discussions. The assistance of the CCS team at BGS, in particular John Williams, in pre-conditioning the well database was of great help. Mike Sankey is thanked for supplying seismic data from the DECC-OGA digital datastore. Keith Henderson and Sandy Henderson improved the quality of many of the diagrams. Finally, we acknowledge the contribution of the reviewers (Ian Andrews and two anonymous referees) and publication editor (John Underhill) in improving the manuscript. This paper is published with the permission of the Executive Director, British Geological Survey.

References

Akhurst, M.C., Chadwick, R.A. *et al.* 1997. *Geology of the West Cumbria District. Memoir for 1:50 000 Geological Sheets 28 (Whitehaven), 37 (Gosforth) and 47 (Bootle).* Memoirs of the Geological Survey of Great Britain, England and Wales (Sheet – New Series). British Geological Survey, Keyworth, Nottingham, UK.

Andrews, I.J. 2013. *The Carboniferous Bowland Shale Gas Study: Geology and Resource Estimation.* British Geological Survey for the Department of Energy and Climate Change, London.

Armstrong, J.P., Smith, J., D'Elia, V.A.A. & Trueblood, S.P. 1997. The occurrence and correlation of oils and Namurian source rocks in the Liverpool Bay–North Wales area. *In*: Meadows, N.S., Trueblood, S.P., Hardman, M. & Cowan, G. (eds) *Petroleum Geology of the East Irish Sea and Adjacent Areas.* Geological Society, London, Special Publications, **124**, 195–211, https://doi.org/10.1144/GSL.SP.1997.124.01.12

Arter, G. & Fagin, S.W. 1993. The Fleetwood Dyke and the Tynwald fault zone, Block 113/27, East Irish Sea Basin. *In*: Parker, J.R. (ed.) *Petroleum Geology of Northwest Europe: Proceedings of the 4th Conference.* Geological Society, London, 835–843, https://doi.org/10.1144/0040835

Badley, M.E., Price, J.D. & Backshall, L.C. 1989. Inversion, reactivated faults and related structures: seismic examples from the southern North Sea. *In*: Cooper, M.A. & Williams, G.D. (eds) *Inversion Tectonics.* Geological Society, London, Special Publications, **44**, 201–219, https://doi.org/10.1144/GSL.SP.1989.044.01.12

Barclay, W.J., Ambrose, K. *et al.* 1997. *Geology of the Country around Worcester. Memoir for 50 000 Geological Sheet 199 (England and Wales).* Memoirs of the Geological Survey of Great Britain, England and Wales (Sheet – New Series). HMSO, London.

Barnes, R.P., Branney, M.J., Stone, P. & Woodcock, N.H. 2006. The Lakesman Terrane: the Lower Palaeozoic record of the deep marine Lakesman Basin, a volcanic arc and foreland basin. *In*: Brenchley, P.J. & Rawson, P.F. (eds) *The Geology of England and Wales.* 2nd edn. Geological Society, London, 103–129, https://doi.org/10.1144/GOEWP.5

Besly, B.M. 1988. Palaeogeographic implications of late Westphalian to early Permian red-beds. *In*: Besly, B. & Kelling, G. (eds) *Sedimentation in a Synorogenic Basin Complex: The Upper Carboniferous of Northwest Europe.* Blackie, Glasgow, 200–221.

Besly, B.M. 1998. Carboniferous. *In*: Glennie, K.W. (ed.) *Petroleum Geology of the North Sea: Basic Concepts and Recent Advances.* Blackwell Science, Oxford, 104–136.

BGS 1984. *East Irish Sea. Special Sheet Edition. Solid Geology.1:250 000.* British Geological Survey, Edinburgh.

BGS 1994. *East Midlands Hydrocarbon Prospectivity Report Confidential for DTI.* British Geological Survey, Edinburgh.

Blès, J.L., Bonijoly, D., Castaing, C. & Gros, Y. 1989. Successive post-Variscan stress fields in the European plate (Massif Central and its borders): comparison with geodynamic data. *Tectonophysics*, **169**, 79–111.

Bluck, B.J. 2002. The Midland Valley Terrane. *In*: Trewin, N.H. (ed.) *Geology of Scotland.* 4th edn. Geological Society, London, 149–166, https://doi.org/10.1144/0036-9276/01-383

Brodie, J. & White, N. 1994. Sedimentary basin inversion caused by igneous underplating: northwest European continental shelf. *Geology*, **22**, 147–150.

Brown, D., Juhlin, C. & Puchkov, V. (eds). 2002. *Mountain Building in the Uralides: Pangea to the Present.* American Geophysical Union, Geophysical Monograph, **132**.

Browne, M.A.E., Dean, M.T., Hall, I.H.S., McAdam, A.D., Monro, S.K. & Chisholm, J.I. 1999. *A Lithostratigraphical Framework for the Carboniferous Rocks of the Midland Valley of Scotland.* British Geological Survey, Research Report RR/99/07. British Geological Survey, Keyworth, Nottingham, UK.

Bushell, T.P. 1986. Reservoir geology of the Morecambe Field. *In*: Brooks, J., Goff, J.C. & van Hoorn, B. (eds) *Habitat of Palaeozoic Gas in N.W. Europe.* Geological Society, London, Special Publications, **23**, 189–208, https://doi.org/10.1144/GSL.SP.1986.023.01.12

Chadwick, R.A. 1993. Aspects of basin inversion in southern Britain. *Journal of the Geological Society, London*, **150**, 311–322, https://doi.org/10.1144/gsjgs.150.2.0311

Chadwick, R.A. 1997. Fault analysis of the Cheshire Basin, NW England. *In*: Meadows, N.S., Trueblood, S., Hardman, M. & Cowan, G. (eds) *Petroleum Geology of the Irish Sea and Adjacent Areas.* Geological Society, London, Special Publications, **124**, 297–313, https://doi.org/10.1144/GSL.SP.1997.124.01.18

Chadwick, R.A. & Evans, D.J. 1995. The timing and direction of Permo-Triassic extension in southern Britain. *In*: Boldy, S.A.R. & Hardman, R.F.P. (eds) *Permian and*

Triassic Rifting in NW Europe. Geological Society, London, Special Publications, **91**, 161–192, https://doi.org/10.1144/GSL.SP.1995.091.01.09

Chadwick, R.A., Evans, D.J. & Holliday, D.W. 1993. The Maryport Fault: the post-Caledonian tectonic history of southern Britain in microcosm. *Journal of the Geological Society, London*, **150**, 247–250, https://doi.org/10.1144/gsjgs.150.2.0247

Chadwick, R.A., Holliday, D.W., Holloway, S. & Hulbert, A.G. 1995. *The Northumberland–Solway Basin and Adjacent Areas: Subsurface Memoir*. British Geological Survey, Keyworth, Nottingham, UK.

Chadwick, R.A., Jackson, D.I. *et al.* 2001. *The Geology of the Isle of Man and Its Offshore Area*. British Geological Survey Research Report RR/01/06. British Geological Survey, Keyworth, Nottingham, UK.

Chew, D.M. & Strachan, R.A. 2014. The Laurentian Caledonides of Scotland and Ireland. *In*: Corfu, F., Gasser, D.Q. & Chew, D.M. (eds) *New Perspectives of the Caledonides of Scandinavia and Related Areas*. Geological Society, London, Special Publications, **390**, 45–91, https://doi.org/10.1144/SP390.16

Collinson, J. 1988. Controls on Namurian sedimentation in the Central Province basins of northern England. *In*: Besly, B. & Kelling, G. (eds) *Sedimentation in a Synorogenic Basin Complex: The Upper Carboniferous of Northwest Europe*. Blackie, Glasgow, 200–221.

Colter, V.S. 1997. The East Irish Sea Basin – from caterpillar to butterfly, a thirty-year metamorphosis. *In*: Meadows, N.S., Trueblood, S.P., Hardman, M. & Cowan, G. (eds) *Petroleum Geology of the Irish Sea and Adjacent Areas*. Geological Society, London, Special Publications, **124**, 1–10, https://doi.org/10.1144/GSL.SP.1997.124.01.01

Colter, V.S. & Barr, K.W. 1975. Recent developments in the geology of the Irish Sea and Cheshire basins. *In*: Woodland, A.W. (ed.) *Petroleum and the Continental Shelf of North West Europe. Volume 1: Geology*. Applied Science, London, 61–75.

Cope, J.C.W. 1994. A latest Cretaceous hotspot and the southeasterly tilt of Britain. *Journal of the Geological Society, London*, **151**, 905–908, https://doi.org/10.1144/gsjgs.151.6.0905

Cope, J.C.W. 1997. The Mesozoic and Tertiary history of the Irish Sea. *In*: Meadows, N.S., Trueblood, S.P., Hardman, M. & Cowan, G. (eds) *Petroleum Geology of the Irish Sea and Adjacent Areas*. Geological Society, London, Special Publications, **124**, 47–59, https://doi.org/10.1144/GSL.SP.1997.124.01.04

Cope, J.C.W., Ingham, J.K. & Rawson, P.F. 1992. *Atlas of Palaeogeography and Lithofacies*. Geological Society, London, Memoirs, **13**, https://doi.org/10.1144/GSL.MEM.1992.012.01.16

Corcoran, D. & Clayton, G. 1999. Interpretation of vitrinite reflectance profiles in the Central Irish Sea area: implications for the timing of organic maturation. *Journal of Petroleum Geology*, **22**, 261–286.

Corfield, S.M., Gawthorpe, R.L., Gage, M., Fraser, A.J. & Besly, B.M. 1996. Inversion tectonics of the Variscan foreland of the British Isles. *Journal of the Geological Society, London*, **153**, 17–32, https://doi.org/10.1144/gsjgs.153.1.0017

Cowan, G. 1996. The development of the North Morecambe gas field, East Irish Sea Basin, UK. *Petroleum Geoscience*, **2**, 43–52, https://doi.org/10.1144/petgeo.2.1.43

Cowan, G. & Bradney, J. 1997. Regional diagenetic controls on reservoir properties in the Millom accumulation: implications for field development. *In*: Meadows, N.S., Trueblood, S.P., Hardman, M. & Cowan, G. (eds) *Petroleum Geology of the Irish Sea and Adjacent Areas*. Geological Society, London, Special Publications, **124**, 373–386, https://doi.org/10.1144/GSL.SP.1997.124.01.22

Cowan, G., Burley, S. *et al.* 1999. Oil and gas migration in the Sherwood Sandstone of the East Irish Sea Basin. *In*: Fleet, A.J. & Boldy, S.A.R. (eds) *Petroleum Geology of Northwest Europe: Proceedings of the 5th Conference*. Geological Society, London, 1383–1398, https://doi.org/10.1144/0051383

Coward, M.P. 1993. The effect of Late Caledonian and Variscan continental escape tectonics on basement structure, Paleozoic basin kinematics and subsequent Mesozoic basin development in NW Europe. *In*: Parker, J.R. (ed.) *Petroleum Geology of Northwest Europe: Proceedings of the 4th Conference*. Geological Society, London, 1095–1108, https://doi.org/10.1144/0041095

Coward, M.P. 1995. Structural and tectonic setting of the Permo-Triassic basins of northwest Europe. *In*: Boldy, S.A.R. & Hardman, R.F.P. (eds) *Permian and Triassic Rifting in Northwest Europe*. Geological Society, London, Special Publications, **91**, 7–40, https://doi.org/10.1144/GSL.SP.1995.091.01.02

Craig, J., Gluyas, J., Laing, C. & Schofield, P. 2013. Hardstoft – Britain's first oil field. *Oil Industry History*, **14**, 97–116.

De Paola, N., Holdsworth, R.E., McCaffrey, K.J. & Barchi, M.R. 2005. Partitioned transtension: an alternative to basin inversion models. *Journal of Structural Geology*, **27**, 607–625.

Dean, M.T., Browne, M.A.E., Waters, C.N. & Powell, J.H. 2011. *A Lithostratigraphical Framework for the Carboniferous Successions of Northern Great Britain (Onshore)*. British Geological Survey Research Report RR/10/07. British Geological Survey, Keyworth, Nottingham, UK.

Dickson, J.A.D., Ford, T.D. & Swift, A. 1987. The stratigraphy of the Carboniferous rocks around Castletown, Isle of Man. *Proceedings of the Yorkshire Geological Society*, **46**, 203–229, https://doi.org/10.1144/pygs.46.3.203

Eastwood, T., Dixon, E.E.L., Hollingworth, S.E. & Smith, B. 1931. *The geology of the Whitehaven and Workington district: Explanation of Sheet 28*. Memoirs of the Geological Survey of Great Britain, England and Wales (Sheet – New Series). HMSO, London.

England, R.W. & Soper, N.J. 1997. Lower crustal structure of the East Irish Sea from deep seismic reflection data. *In*: Meadows, N.S., Trueblood, S.P., Hardman, M. & Cowan, G. (eds) *Petroleum Geology of the Irish Sea and Adjacent Areas*. Geological Society, London, Special Publications, **124**, 61–72, https://doi.org/10.1144/GSL.SP.1997.124.01.05

Falcon, N.L. & Kent, P.E. 1960. *Geological Results of Petroleum Exploration in Britain 1945–1957*. Geological Society, London, Memoirs, **2**, https://doi.org/10.1144/GSL.MEM.1960.002.01.01

Faure, M. 1995. Late orogenic Carboniferous extensions in the Variscan French Massif Central. *Tectonics*, **14**, 132–153.

Floodpage, J., Newman, P. & White, J. 2001. Hydrocarbon prospectivity in the Irish Sea area: insights from recent exploration of the Central Irish Sea, Peel and Solway basins. *In*: Shannon, P.M., Haughton, P.D.W. & Corcoran, D.V. (eds) *The Petroleum Exploration of Ireland's Offshore Basins*. Geological Society, London, Special Publications, **188**, 107–134, https://doi.org/10.1144/GSL.SP.2001.188.01.06

Fraser, A.J. & Gawthorpe, R.L. 2003. *An Atlas of Carboniferous Basin Evolution in Northern England*. Geological Society, London, Memoirs, **28**, https://doi.org/10.1144/GSL.MEM.2003.028.01.08

Fraser, A.J., Nash, D.F., Steele, R.P. & Ebdon, C.C. 1990. A regional assessment of the intra-Carboniferous play of Northern England. *In*: Brooks, J. (ed.) *Classic Petroleum Provinces*. Geological Society, London, Special Publications, **50**, 417–440, https://doi.org/10.1144/GSL.SP.1990.050.01.26

Gélard, J.-P., Castaing, C., Bonijoly, D. & Grolier, J. 1986. Structure et dynamique de quelques bassins houilliers limniques du Massif Central. *Memoire Société Géologique de France*, **149**, 576–572.

Gent, C.M.A. 2016. *Maturity Modelling of Well 110/07b-6*. British Geological Survey Commissioned Report CR/16/043. British Geological Survey, Keyworth, Nottingham, UK, http://nora.nerc.ac.uk/id/eprint/516793

Geochem Laboratories Ltd 1988. *Geochemical Evaluation and Correlation Study Morecambe Bay 110/7b-6 Well*. Prepared for Kelt UK Ltd, London (110_07b_6_rep_GEOL_CHEM_105249616.pdf released on CDA). Geochem Laboratories Ltd, Chester, UK.

Gibbons, W. 1987. The Menai Strait fault system: an early Caledonian terrane boundary in North Wales. *Geology*, **15**, 744–747.

Green, P.F., Duddy, I.R. & Bray, R.J. 1997. Variation in thermal history styles around the Irish Sea and adjacent areas: implications for hydrocarbon occurrence and tectonic evolution. *In*: Meadows, N., Trueblood, S., Hardman, M. & Cowan, G. (eds) *Petroleum Geology of the Irish Sea and Adjacent Areas*. Geological Society, London, Special Publications, **124**, 73–93, https://doi.org/10.1144/GSL.SP.1997.124.01.06

Grolier, J. 1971. Le tectonique du socle hercynien dans le Massif Central. *In*: Jung, J. (ed.) *Symposium.: Géologie, géomorphologie et structure profonde du Massif Central français*. Plein Air Service, Clermont-Ferrand, France, 215–260.

Haig, D.B., Pickering, S.C. & Probert, R. 1997. The Lennox oil and gas field. *In*: Meadows, N.S., Trueblood, S.P., Hardman, M. & Cowan, G. (eds) *Petroleum Geology of the Irish Sea and Adjacent Areas*. Geological Society, London, Special Publications, **124**, 417–436, https://doi.org/10.1144/GSL.SP.1997.124.01.25

Hallett, D., Durant, G.P. & Farrow, G.E. 1985. Oil exploration and production in Scotland. *Scottish Journal of Geology*, **21**, 547–570, https://doi.org/10.1144/sjg21040547

Hannis, S. 2016. *Reservoir Evaluation of 8 Wells in the Palaeozoic of the Irish Sea: Petrophysical Interpretations of Clay Volume, Porosity and Permeability Estimations*. British Geological Survey Commissioned Report CR/16/042. British Geological Survey, Keyworth, Nottingham, UK, http://nora.nerc.ac.uk/516774/

Hardman, M., Buchanan, J., Herrington, P. & Carr, A. 1993. Geochemical modelling of the East Irish Sea Basin: its influence on predicting hydrocarbon type and quality. *In*: Parker, J.R. (ed.) *Petroleum Geology of Northwest Europe: Proceedings of the 4th Conference*. Geological Society, London, 809–821, https://doi.org/10.1144/0040809

Hillier, R.D. & Williams, B.P.J. 2006. The alluvial Old Red Sandstone: fluvial basins. *In*: Brenchley, P.J. & Rawson, P.F. (eds) *The Geology of England and Wales*. 2nd edn. Geological Society, London, 155–172, https://doi.org/10.1144/GOEWP.8

Jackson, D.I. & Mulholland, P. 1993. Tectonic and stratigraphical aspects of the East Irish Sea Basin and adjacent areas: contrasts in their post-Carboniferous structural styles. *In*: Parker, J.R. (ed.) *Petroleum Geology of Northwest Europe: Proceedings of the 4th Conference*. Geological Society, London, 791–808, https://doi.org/10.1144/0040791

Jackson, D.I., Mulholland, P., Jones, S.M. & Warrington, G. 1987. The geological framework of the East Irish Sea Basin. *In*: Brooks, J. & Glennie, K. (eds) *Petroleum Geology of North West Europe*. Graham & Trotman, London, 191–203.

Jackson, D.I., Jackson, A.A., Evans, D., Wingfield, R.T.R., Barnes, R.P. & Arthur, M.J. 1995. *United Kingdom Offshore Regional Report: The Geology of the Irish Sea*. HMSO, London.

Jackson, D.I., Johnson, H. & Smith, N.J.P. 1997. Stratigraphical relationships and a revised lithostratigraphical nomenclature for the Carboniferous, Permian and Triassic rocks of the offshore East Irish Sea Basin. *In*: Meadows, N.S., Trueblood, S.P., Hardman, M. & Cowan, G. (eds) *Petroleum Geology of the Irish Sea and Adjacent Areas*. Geological Society, London, Special Publications, **124**, 11–32, https://doi.org/10.1144/GSL.SP.1997.124.01.02

Jackson, D.I., Jones, N.S. & Waters, C.N. 2011. Chapter 16: Irish Sea (including Kish Bank). *In*: Waters, C.N., Somerville, I.D. *et al.* (eds) *A Revised Correlation of Carboniferous Rocks in the British Isles*. Geological Society, London, Special Reports, **26**, 110–116.

Jenner, J.K. 1981. The structure and stratigraphy of the Kish Bank Basin. *In*: Illing, L.V. & Hobson, G.D. (eds) *Petroleum Geology of the Continental Shelf of North West Europe*. Heyden, London, 426–431.

Jones, N.S., Holliday, D.W. & McKervey, J.A. 2011. Warwickshire Group (Pennsylvanian) red-beds of the Canonbie Coalfield, England–Scotland border, and their regional palaeogeographical implications. *Geological Magazine*, **148**, 50–77.

Kirby, G.A., Baily, H.E. *et al.* 2000. *The Structure and Evolution of the Craven Basin and Adjacent Areas*. Subsurface Geology Memoir. British Geological Survey, Keyworth, Nottingham, UK.

Kirton, S.R. & Donato, J.A. 1985. Some buried Tertiary dykes of Britain and surrounding waters deduced by magnetic modelling and seismic reflection methods.

Journal of the Geological Society, London, **142**, 1047–1057, https://doi.org/10.1144/gsjgs.142.6.1047

KNIPE, R.J., COWAN, G. & BALENDRAN, V.S. 1993. The tectonic history of the East Irish Sea Basin with reference to the Morecambe Fields. *In*: PARKER, J.R. (ed.) *Petroleum Geology of Northwest Europe: Proceedings of the 4th Conference*. Geological Society, London, 857–866, https://doi.org/10.1144/0040857

KUPECZ, J.A., ARESTAD, J.F. & BLOTT, J.E. 1996. Integrated study of Mississippian Lodgepole Waulsortian Mounds, Williston Basin, USA. Paper SEG-1996-0401 presented at the 1996 SEG Annual Meeting, 10–15 November 1996, Denver, Colorado, USA.

LESLIE, A.G., MILLWARD, D., PHARAOH, T., MONAGHAN, A.A., ARESENIKOS, S. & QUINN, M. 2015. *Tectonic Synthesis and Contextual setting for the Central North Sea and Adjacent Onshore Areas, 21CXRM Palaeozoic Project*. British Geological Survey Commissioned Report CR/15/125. British Geological Survey, Keyworth, Nottingham, UK, http://nora.nerc.ac.uk/516757/

MADDOX, S.J., BLOW, R.A. & O'BRIEN, S.R. 1997. The geology and hydrocarbon prospectivity of the North Channel Basin. *In*: MEADOWS, N.S., TRUEBLOOD, S.P., HARDMAN, M. & COWAN, G. (eds) *Petroleum Geology of the Irish Sea and Adjacent Areas*. Geological Society, London, Special Publications, **124**, 95–111, https://doi.org/10.1144/GSL.SP.1997.124.01.07

MEADOWS, N.S., TRUEBLOOD, S.P., HARDMAN, M. & COWAN, G. (eds) 1997. *Petroleum Geology of the Irish Sea and Adjacent Areas*. Geological Society, London, Special Publications, **124**, https://doi.org/10.1144/GSL.SP.1997.124.01.26

MITCHELL, W.I. (ed.). 2004. *The Geology of Northern Ireland-Our Natural Foundation*. Geological Survey of Northern Ireland, Belfast.

MONAGHAN, A.A. 2014. *The Carboniferous Shales of the Midland Valley of Scotland: Geology and Resource Estimation*. British Geological Survey for the Department of Energy and Climate Change, London.

MONAGHAN, A.A. & PRINGLE, M.S. 2004. ^{40}Ar/^{39}Ar geochronology of Carboniferous–Permian volcanism in the Midland Valley, Scotland. *In*: WILSON, M., NEUMANN, E.-R., DAVIES, G.R., TIMMERMANN, M.J., HEEREMANS, M. & LARSEN, B.T. (eds) *Permo-Carboniferous Magmatism and Rifting in Europe*. Geological Society, London, Special Publications, **223**, 219–241, https://doi.org/10.1144/GSL.SP.2004.223.01.10

NADIN, P.A. & KUZNIR, N.J. 1995. Palaeocene uplift and Eocene subsidence in the northern North Sea Basin from 2D forward and reverse stratigraphic modelling. *Journal of the Geological Society, London*, **152**, 833–848, https://doi.org/10.1144/gsjgs.152.5.0833

NEWMAN, P.J. 1999. The geology and hydrocarbon potential of the Peel and Solway Basins, East Irish Sea. *Journal of Petroleum Geology*, **22**, 305–324.

PARNELL, 1997. Fluid migration history in the North Irish Sea–North Channel region. *In*: MEADOWS, N.S., TRUEBLOOD, S.P., HARDMAN, M. & COWAN, G. (eds) *Petroleum Geology of the Irish Sea and Adjacent Areas*. Geological Society, London, Special Publications, **124**, 213–228, https://doi.org/10.1144/GSL.SP.1997.124.01.13

PEACE, G.R. & BESLY, B.M. 1997. End-Carboniferous fold-thrust structures, Oxfordshire, UK: implications for the structural evolution of the late Variscan foreland of south-central England. *Journal of the Geological Society, London*, **154**, 225–237, https://doi.org/10.1144/gsjgs.154.2.0225

PENN, I., HOLLIDAY, D.W. *ET AL*. 1983. The Larne No. 2 Borehole: discovery of a new Permian volcanic centre. *Scottish Journal of Geology*, **19**, 333–334, https://doi.org/10.1144/sjg19030333

PHARAOH, T.C., WEBB, P.C., THORPE, R.S. & BECKINSALE, R. D. 1987. Geochemical evidence for the tectonic setting of late Proterozoic volcanic suites in central England. *In*: PHAROAH, T.C., BECKINSALE, R.D. & RICKARD, D.T. (eds) *Geochemistry and Mineralisation of Proterozoic Volcanic Suites*. Geological Society, London, Special Publications, **33**, 541–552, https://doi.org/10.1144/GSL.SP.1987.033.01.36

PHARAOH, T.C., WINCHESTER, J.A., VERNIERS, J., LASSEN, A. & SEGHEDI, A. 2006. The western accretionary margin of the east European craton: an overview. *In*: GEE, D.G. & STEPHENSON, R.A. (eds) *European Lithosphere Dynamics*. Geological Society, London, Memoirs, **32**, 291–311, https://doi.org/10.1144/GSL.MEM.2006.032.01.17

PHARAOH, T.C., DUSAR, M. *ET AL*. 2010. Chapter 3: Tectonic evolution. *In*: DOORNENBAL, J.H. & STEVENSON, A.G. (eds) *Petroleum Geological Atlas of the Southern Permian Basin Area*. European Association of Geoscientists and Engineers (EAGE), Houten, The Netherlands, 25–58.

PHARAOH, T.C., VINCENT, C.J., BENTHAM, M.S., HULBERT, A.G., WATERS, C.N. & SMITH, N.J. 2011. *Structure and Evolution of the East Midlands Region of the Pennine Basin*. Subsurface Geology Memoir. British Geological Survey, Keyworth, Nottingham, UK.

PHARAOH, T.C., KIRK, K., QUINN, M., SANKEY, M. & MONAGHAN, A.A. 2016*a*. *Seismic Interpretation and Generation of Depth Surfaces for Late Palaeozoic Strata in the Irish Sea Region*. British Geological Survey Commissioned Report CR/16/041. British Geological Survey, Keyworth, Nottingham, UK, http://nora.nerc.ac.uk/cgi/search/archive/simple?screen=Search&dataset=archive&order=&q=21CXRM&_action_search=Search

PHARAOH, T.C., SMITH, N.J.P., KIRK, K., KIMBELL, G.S., GENT, C., QUINN, M. & MONAGHAN, A.A. 2016*b*. *Palaeozoic Petroleum Systems of the Irish Sea*. British Geological Survey Commissioned Report CR/16/045. British Geological Survey, Keyworth, Nottingham, UK.

PLETSCH, T., APPEL, J. *ET AL*. 2010. Chapter 12. Petroleum generation and migration. *In*: DOORNENBAL, J.C. & STEVENSON, A.G. (eds) *Petroleum Geological Atlas of the Southern Permian Basin Area*. European Association of Geoscientists and Engineers (EAGE), Houten, The Netherlands, 225–253.

PROVIDENCE RESOURCES PLC 2013. Frontier exploration opportunities Rathlin Trough, offshore Northern Ireland. Presented at the Prospex-2013 Conference and Exhibition, 11–12 December 2013, London.

PUCHKOV, V.N. 1997. Structure and geodynamics of the Uralian orogeny. *In*: BURG, J.-P. & FORD, M. (eds) *Orogeny Through Time*. Geological Society, London, Special Publications, **121**, 201–236, https://doi.org/10.1144/GSL.SP.1997.121.01.09

Quinn, M.F. 2008. *A Geological Interpretation of the Larne and Portpatrick Sub-Basins, Offshore Northern Ireland, with an Evaluation of an Area Proposed for Gas Storage in Salt Caverns.* British Geological Survey Commissioned Report, CR/08/064 (Commercial-in-Confidence). British Geological Survey, Keyworth, Nottingham, UK.

Quirk, D.G. & Kimbell, G.S. 1997. Structural evolution of the Isle of Man and central part of the Irish Sea. *In*: Meadows, N.S., Trueblood, S., Hardman, M. & Cowan, G. (eds) *Petroleum Geology of the Irish Sea and Adjacent Areas.* Geological Society, London, Special Publications, **124**, 135–159, https://doi.org/10.1144/GSL.SP.1997.124.01.09

Quirk, D.G., Roy, S., Knott, I., Redfern, J. & Hill, L. 1999. Petroleum geology and future hydrocarbon potential of the Irish Sea. *Journal of Petroleum Geology*, **22**, 243–260.

Racey, A. 1999. Palynolgical and geochemical analysis of Carboniferous borehole and outcrop samples from the Isle of Man. *Journal of Petroleum Geology*, **22**, 349–362.

Ramsbottom, W.H.C. 1969. Reef distribution in the British Lower Carboniferous. *Nature*, **222**, 765–766.

Ramsbottom, W.H.C., Calver, M.A., Eagar, R.M.C., Hodson, F., Holliday, D.W., Stubblefield, C.J. & Wilson, R.B. 1978. *A Correlation of Silesian Rocks in the British Isles.* Geological Society, London, Special Reports, **10**.

Read, W.A., Browne, M.A., Stephenson, D. & Upton, B.G.J. 2002. Carboniferous. Chapter 9. *In*: Trewin, N.H. (ed.) *Geology of Scotland.* 4th edn. Geological Society, London, 149–166, https://doi.org/10.1144/GOS4P.9

Reay, D.M. 2004. Oil and gas. *In*: Mitchell, W.I. (ed.) *The Geology of Northern Ireland – Our Natural Foundation.* Geological Survey of Northern Ireland, Belfast, 273–290.

Reay, D.M. 2012. Geology and gas in Northern Ireland. Presented at the Future of Natural Gas Seminar, 27 March 2012, Dundalk, Ireland.

Scheck-Wenderoth, M., Krzywiec, P., Zühlke, R., Maystrenko, Y. & Froitzheim, N. 2008. Permian to Cretaceous tectonics. *In*: McCann, T. (ed.) *The Geology of Central Europe. Volume 2: Mesozoic and Cenozoic.* Geological Society, London, 999–1030, https://doi.org/10.1144/CEV2P.4

Shelton, R. 1997. Tectonic evolution of the Larne Basin. *In*: Meadows, N.S., Trueblood, S.P., Hardman, M. & Cowan, G. (eds) *Petroleum Geology of the Irish Sea and Adjacent Areas.* Geological Society, London, Special Publications, **124**, 113–134, https://doi.org/10.1144/GSL.SP.1997.124.01.08

Smith, M., Robertson, S. & Rollin, K.E. 1999. Rift basin architecture and stratigraphical implications for basement–cover relationships in the Neoproterozoic Grampian Group of the Scottish Caledonides. *Journal of the Geological Society, London*, **156**, 1163–1173, https://doi.org/10.1144/gsjgs.156.6.1163

Smith, N.J.P. (compiler) 1985. *Map 1: Pre-Permian Geology of the United Kingdom (South), 1:1 000 000 Scale.* British Geological Survey, Keyworth, Nottingham, UK.

Smith, N.J.P., Chadwick, R.A., Warrington, G., Kirby, G.A. & Jones, D. 1995. *The Hydrocarbon Prospectivity of the Cheshire Basin and Surrounding Areas.* BGS Technical Report WA/94/95C. British Geological Survey, Keyworth, Nottingham, UK.

Smith, N.J.P., Kirby, G.A. & Pharaoh, T.C. 2005. *Structure and Evolution of the South-West Pennine Basin and Adjacent Area.* Subsurface Geology Memoir. British Geological Survey, Keyworth, Nottingham, UK.

Stuart, I.A. 1993. The geology of the North Morecambe Gas Field, East Irish Sea Basin. *In*: Parker, J.R. (ed.) *Petroleum Geology of Northwest Europe: Proceedings of the 4th Conference.* Geological Society, London, 883–895, https://doi.org/10.1144/0040883

Stuart, I.A. & Cowan, G. 1991. The south Morecambe Field, blocks 110/2a, 110/3a, 110/8a, UK East Irish Sea. *In*: Abbotts, I.L. (ed.) *United Kingdom Oil and Gas Fields, 25 Years Commemorative Volume.* Geological Society, London, Memoirs, **14**, 527–541, https://doi.org/10.1144/GSL.MEM.1991.014.01.66

Suess, E. 1904. *The Face of the Earth.* Clarendon Press, Oxford.

Timmerman, M.J. 2004. Timing, geodynamic setting and character of Permo-Carboniferous magmatism in the foreland of the Variscan Orogen, NW Europe. *In*: Wilson, M., Neumann, E.-R., Davies, G.R., Timmermann, M.J., Heeremans, M. & Larsen, B.T. (eds) *Permo-Carboniferous Magmatism and Rifting in Europe.* Geological Society, London, Special Publications, **223**, 41–74, https://doi.org/10.1144/GSL.SP.2004.223.01.03

Trewin, N.H. & Thirlwall, M.F. 2002. Chapter 8. Old red sandstone. *In*: Trewin, N.H. (ed.) *Geology of Scotland.* 4th edn. Geological Society, London, 149–166, https://doi.org/10.1144/GOS4P.8

Trotter, F.M. 1954. Reddened beds in the Coal Measures of Lancashire. *Bulletin of the Geological Survey*, **5**, 61–80.

Underhill, J.R., Monaghan, A.A. & Browne, M.A.E. 2008. Controls on structural styles, basin development and petroleum prospectivity in the Midland Valley of Scotland. *Journal of Marine and Petroleum Geology*, **25**, 1000–1022.

Vane, C.H., Uguna, C., Kim, A.W. & Monaghan, A.A. 2016. *Organic Geochemistry of Palaeozoic Source Rocks of the Irish Sea, UK. British Geological Survey.* Commissioned Report CR/16/044. British Geological Survey, Keyworth, Nottingham, UK, http://nora.nerc.ac.uk/516753/

Van Hoorn, B. 1987. Structural evolution, timing and tectonic style of the Sole Pit inversion. *Tectonophysics*, **137**, 239–284.

Wakefield, O., Waters, C.N. & Smith, N.J.P. 2016. *Carboniferous stratigraphical correlation and interpretation in the Irish Sea.* British Geological Survey Commissioned Report CR/16/040. British Geological Survey, Keyworth, Nottingham, UK.

Walkden, G.M. & Davies, J.R. 1983. Polyphase erosion of subaerial omission surfaces in the late Dinantian of Anglesey. *Sedimentology*, **30**, 861–878.

Warrington, G. 1997. The Penarth Group–Lias Group succession (Late Triassic-Early Jurassic) in the East Irish Sea Basin and neighbouring areas: a stratigraphical review. *In*: Meadows, N.S., Trueblood, S.P., Hardman, M. & Cowan, G. (eds) *Petroleum Geology of the Irish Sea and Adjacent Areas.* Geological Society, London,

Special Publications, **124**, 33–46, https://doi.org/10.1144/GSL.SP.1997.124.01.03

Waters, C.N. & Condon, D.J. 2012. Nature and timing of Late Mississippian to Mid-Pennsylvanian glacio-eustatic sea-level changes of the Pennine Basin, UK. *Journal of the Geological Society, London*, **169**, 37–51, https://doi.org/10.1144/0016-76492011-047

Waters, C.N., Somerville, I.D. *et al.* (eds). 2011. *A Revised Correlation of Carboniferous Rocks in the British Isles*. Geological Society, London, Special Reports, **26**.

White, R.S. 1988. A hot-spot model for early Cenozoic volcanism in the N Atlantic. *In*: Parson, L.M. & Morton, A.C. (eds) *Early Cenozoic Volcanism and the Opening of the North Atlantic*. Geological Society, London, Special Publications, **39**, 3–13, https://doi.org/10.1144/GSL.SP.1988.039.01.02

Whittaker, A. (ed.) 1985. *Atlas of Onshore Sedimentary Basins*. Blackie, Glasgow.

Wood, I. 2014. *UKCS Maximising Recovery Review: Final Report*. HMSO, London.

Woodcock, N.H. & Soper, N.J. 2006. The Acadian Orogeny: the mid-Devonian phase of deformation that formed slate belts in England and Wales. *In*: Brenchley, P.J. & Rawson, P.F. (eds) *The Geology of England and Wales*. 2nd edn. Geological Society, London, 103–129, https://doi.org/10.1144/GOEWP.6

Woodward, K. & Curtis, C.D. 1987. Predictive modelling of the distribution of production constraining illites – Morecambe Gas Field, Irish Sea, Offshore UK. *In*: Brooks, J. & Glennie, K. (eds) *Petroleum Geology of North West Europe*. Graham & Trotman, London, 205–215.

Yaliz, A. & Taylor, P. 2003. The Hamilton and Hamilton North Gas Fields, Block 110/13a, East Irish Sea. *In*: Gluyas, J.G. & Hitchens, H.M. (eds) *United Kingdom Oil and Gas Fields, Commemorative Millennium Volume*. Geological Society, London, Memoirs, **20**, 77–86, https://doi.org/10.1144/GSL.MEM.2003.020.01.06

Yaliz, A.M. 1997. The Douglas oil field. *In*: Meadows, N.S., Trueblood, S.P., Hardman, M. & Cowan, G. (eds) *Petroleum Geology of the Irish Sea and Adjacent Areas*. Geological Society, London, Special Publications, **124**, 399–416, https://doi.org/10.1144/GSL.SP.1997.124.01.24

Ziegler, P.A. 1990. *Geological Atlas of Western and Central Europe*. 2nd edn. Shell Internationale Petroleum Maatschappij, The Hague.

Zonenshain, L.P., Korinevsky, V.G., Kazmin, V.G., Pechersky, D.M., Khain, V.V. & Mateveenkov, V.V. 1984. Plate tectonic model of the south Urals development. *Tectonophysics*, **109**, 95–135.

Seismostratigraphic analysis of Paleozoic sequences of the Midlands Microcraton

MALCOLM BUTLER

UK Onshore Geophysical Library, 46–50 Coombe Road, New Malden, Surrey KT3 4QF, UK, malcolm@butlermail.org

M.B., 0000-0002-0462-9135

Abstract: A regional review of publicly available seismic reflection lines and wells enables the identification across the Midlands Microcraton (MMC) of four Paleozoic seismostratigraphic megasequences, bounded by the Shelveian, Acadian, Symon and Variscan unconformities. The southern boundary of the MMC is drawn at a line of major change in pre-Permian subcrop, and most of central southern England is considered underlain by Paleozoic rocks of MMC character. The Lower Silurian Shelveian Unconformity cuts down through Ordovician and Cambrian rocks to the Precambrian and largely defines the distribution of these rocks in the region. However, more than 2500 m of Tremadoc shales are preserved SW of Swindon. Above the Shelveian, a characteristic shallow-marine Silurian is overlain by up to 3000 m of Old Red Sandstone facies, preserved in a north–south-orientated syncline. The Acadian Unconformity cuts down through folded Lower Paleozoic rocks, with the Frasnian transgression overlying Precambrian in places. Little Carboniferous was deposited across the MMC until uppermost Westphalian–Stephanian Warwickshire Group sandstones and coals were laid down across the erosion surface of the Symon Unconformity. The northern boundary of thin-skinned Variscan thrusting can be interpreted on seismic data but appears to have had little effect on the regional pre-Permian subcrop.

This paper presents the results of a regional review of the area of the Midlands Microcraton (MMC) using the digital, post-stack seismic lines, and well and borehole formation tops and times (Fig. 1), freely available to academic users through the UK Onshore Geophysical Library (UKOGL), together with lithological descriptions from released wells. This area has been of little interest for oil and gas exploration over the past 25 years, and the details of the Paleozoic rocks are not widely known. Few seismic lines have been recorded over the eastern and westernmost parts of the MMC, and lines recorded over the rest of the area vary greatly in vintage and quality. However, many of the lines in the central area show remarkably good detail of both Lower and Upper Paleozoic sequences and tectonic events. Interpretation has enabled the identification of four major unconformities, which define the distribution of Paleozoic rocks beneath the Permian and Triassic cover across the MMC. It has also enabled the recognition of a series of 'packages' of seismic character that can be used to identify the four seismostratigraphic megasequences separated by these unconformities. It is hoped that the information presented here will stimulate interest in the area and lead to more detailed studies for both commercial and academic purposes.

Definition of the Midlands Microcraton

The Midlands Microcraton (MMC) (Pharaoh *et al.* 1987) is a wedge-shaped area, the NE and NW margins of which were defined as bounding an area comparatively undisturbed by Caledonide folds, the Midlands Massif, by Turner (1949). It was relatively unaffected by major crustal deformation from at least Llandovery times until the present day, although it was not entirely stable. It is bounded to the NW by the Pontesford Lineament (Woodcock 1984), a series of faults that form part of the boundary of the Lower Paleozoic Welsh Basin, and to the NE by a series of faults that form the boundary to the Anglo-Brabant Acadian deformation zone (Smith *et al.* 2005). The plate movements that defined these boundaries are described in detail in Smith *et al.* (2005) and will not be addressed here. The boundaries to the NE and NW are postulated to have been accentuated by the northwards movement of the MMC at the time of the Acadian deformation, creating thrusting and sinistral strike-slip along the NW boundary, with thrusting and dextral strike-slip along the NE boundary (Soper *et al.* 1987). The southern boundary of the MMC has previously been taken at the inferred northern margin of Variscan foreland deformation, referred to as the 'Variscan Front' and here located after Chadwick *et al.* (1989).

From: MONAGHAN, A. A., UNDERHILL, J. R., HEWETT, A. J. & MARSHALL, J. E. A. (eds) 2018. *Paleozoic Plays of NW Europe*. Geological Society, London, Special Publications, **471**, 317–332.
First published online April 30, 2018, https://doi.org/10.1144/SP471.6

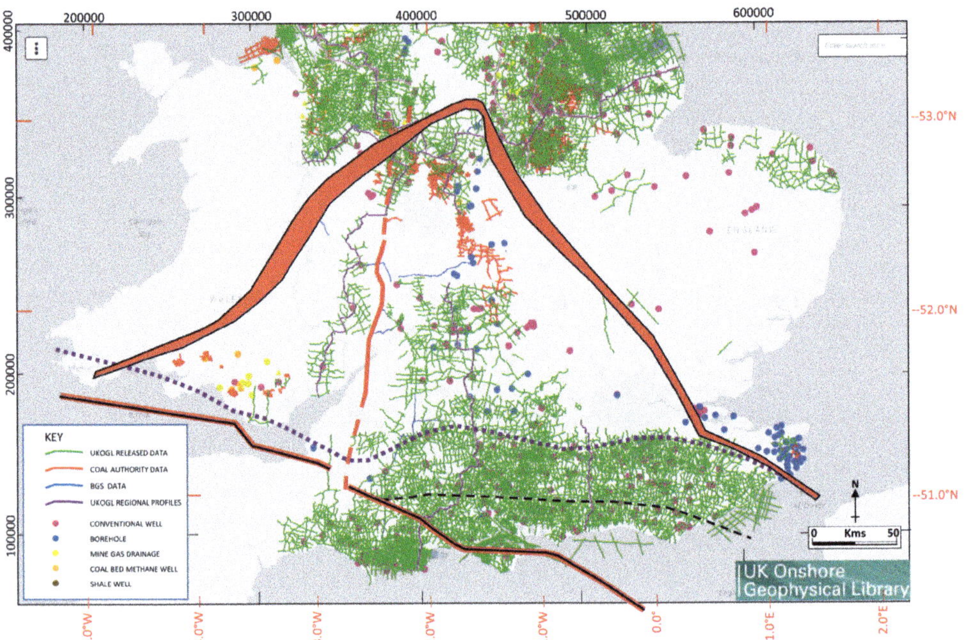

Fig. 1. Database of seismic lines and basic well information freely available through UKOGL (www.ukogl.org.uk) and used in this study. Structural elements are labelled in Figure 2.

Although the line of the Variscan Front (Fig. 2) does coincide with the northern margins of the Mesozoic Pewsey and Weald basins, thin-skinned thrusting occurs well north of this line. However, the interpreted subcrop maps indicate no significant change in the Paleozoic sequences over most of central southern England from those of the MMC further north. There are two alternative lines along which to take the southern boundary of the MMC: the first is the line south of which the Old Red Sandstone facies seen in core and cuttings samples from released wells appears to be more indurated and is possibly equivalent to the Devonian sandstones of north Devon; and the second is the better-defined northern boundary of the subcrop of highly deformed Devonian and Carboniferous shales identified by well and borehole information beneath the Permo-Trias. In Dorset, this boundary also marks the northern extent of the thick Permian Aylesbeare Mudstone sequence (Butler 1998). This second alternative is taken as the boundary of the Variscide Rhenohercynian Terrane and the southern margin of the MMC.

Seismic megasequences in the Paleozoic of the MMC

Figure 3 illustrates the four megasequences here recognized in the Paleozoic of the MMC. These have been simplified because they are primarily based on the recognition of 'packages' of seismic events, rather than detailed stratigraphy. Nevertheless, the bases of the packages bear some resemblance to the 'carpets' identified over the MMC by Wills (1978). In general, the underlying Precambrian includes few events that can be followed on the seismic, although traces of major faults and indications of strong dips can be seen. A marked, angular unconformity between the Cambrian and Precambrian is seen at outcrop in Shropshire (Greig *et al.* 1968) and can sometimes be identified on seismic line data in parts of the Welsh Borderlands (Fig. 4). However, it cannot be mapped over any significant area.

The Cambrian sequence is poorly represented in outcrop across the MMC but appears to consist of a series of sandstones and shales, probably age-equivalent to the sequence seen in outcrop at Comley (Smith *et al.* 2005), forming stacked, relatively high-amplitude events, and its appearance is quite characteristic. However, the upper part of the Cambrian, equivalent to the White Leaved Oak Shale of Merioneth age in the Malvern area, is indistinguishable from the Tremadoc, and these shales together form a very characteristic opaque seismic character (combined as 'Tremadoc' on the interpreted seismic lines). Smith & Rushton (1993) interpreted the thick Tremadoc seen on isolated seismic lines in the Worcester Graben area as the remnants

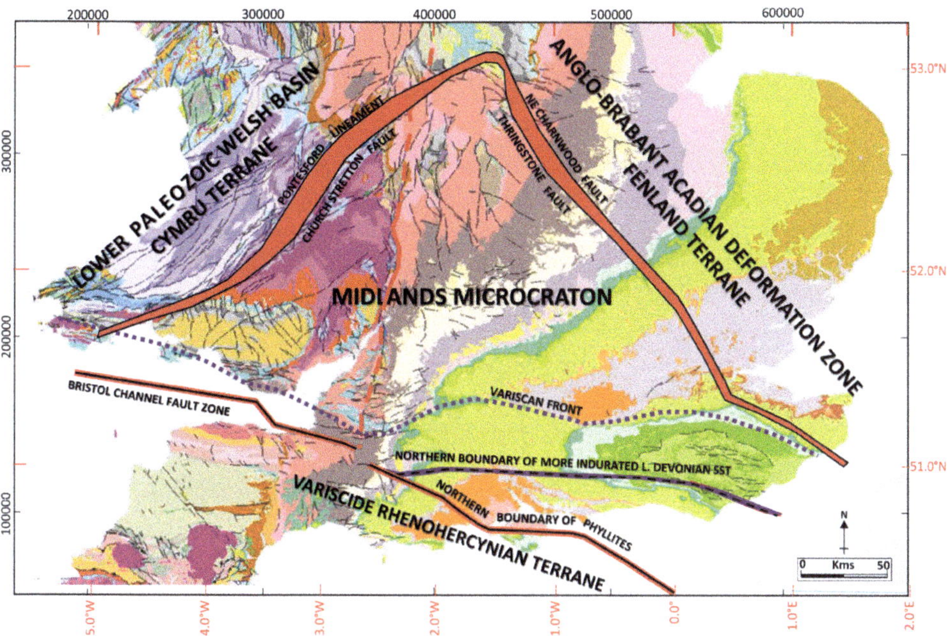

Fig. 2. The Midlands Microcraton: terrane boundaries, adapted from Pharaoh *et al.* (1987), and possible southern limits. The base map was taken from the published 1:625 000 geological map with the permission of the British Geological Survey.

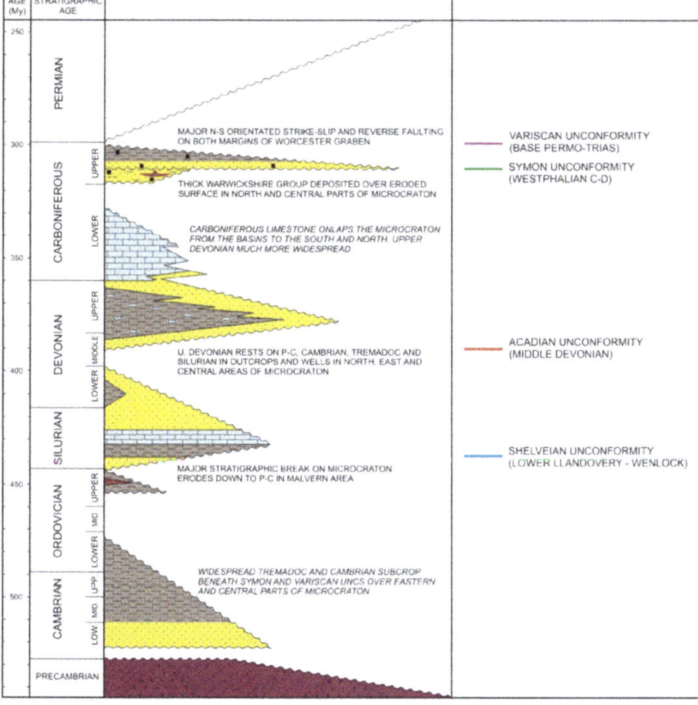

Fig. 3. Paleozoic stratigraphy of the MMC, showing the ages of the four megasequences and the unconformities that separate them.

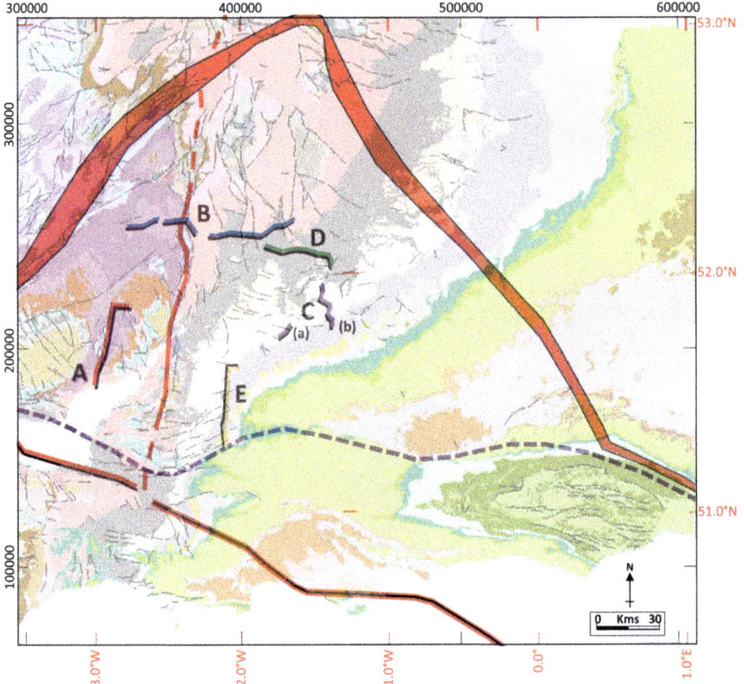

Fig. 4. The location of seismic lines used to illustrate this paper. The base map was taken from the published 1:625 000 geological map with the permission of the British Geological Survey.

of deposition in a series of extensional graben. The current study suggests it is more likely that a relatively uniform, passive margin sequence of Tremadoc was deposited over most of the central part of the MMC but was uplifted and eroded by the Shelveian Unconformity. It is interesting to note that well-preserved Tremadocian acritarchs were identified at two levels within Emsian-age Lower Old Red Sandstone of the Apley Barn borehole (Richardson & Rasul 1979). Further erosion took place at the Acadian and Variscan unconformities to give rise to the currently preserved thicknesses. Above the eroded Ordovician and older rocks, the banded amplitudes created by the limestones, shales and volcanics of the Llandovery–Ludlow sequence form a recognizable package below the seismically opaque Old Red Sandstone of Přídolí–Early Devonian age. The contrast between the Cambrian, Tremadoc and Silurian seismic character was used by Smith *et al.* (2005, fig. 8) as a guide to interpret the BGS Duffield seismic line, in the northern apex of the MMC. The Upper Devonian rocks have a similar seismic character to the Lower Devonian and the two can rarely be differentiated away from well control, while the Lower Carboniferous limestones can be recognized where they subcrop the Triassic in the southern part of the MMC by characteristic remnant topographical highs, often underlain by small thrusts. Over most of the north central part of the MMC, the youngest pre-Permian package is formed by the Warwickshire Group, which has a very characteristic series of high-amplitude events at its base.

The Shelveian Unconformity

The unconformity seen near the base of the Llandovery in the Welsh Borderlands appears to have had a major impact across the MMC. This unconformity was named Shelveian by Toghill (1992), who mapped the Shelve Inlier in Shropshire. Review of seismic and borehole data, combined with previous work by Smith (1987) and Smith *et al.* (2005), indicate that the Shelveian Unconformity removes the upper and middle parts of the Ordovician over most of the MMC, leaving just the Tremadoc, and cuts down to the Precambrian in parts of the NW region of the MMC. However, the Caradoc seems to be present beneath it in parts of Shropshire and, possibly, Herefordshire (Fig. 5), and Arenig rocks were recorded in well Strat A-1 on the northern margin of the Weald Basin. The age of the outcropping rocks above the unconformity ranges from Early Llandovery (Rhuddanian) in Gwent and Herefordshire to

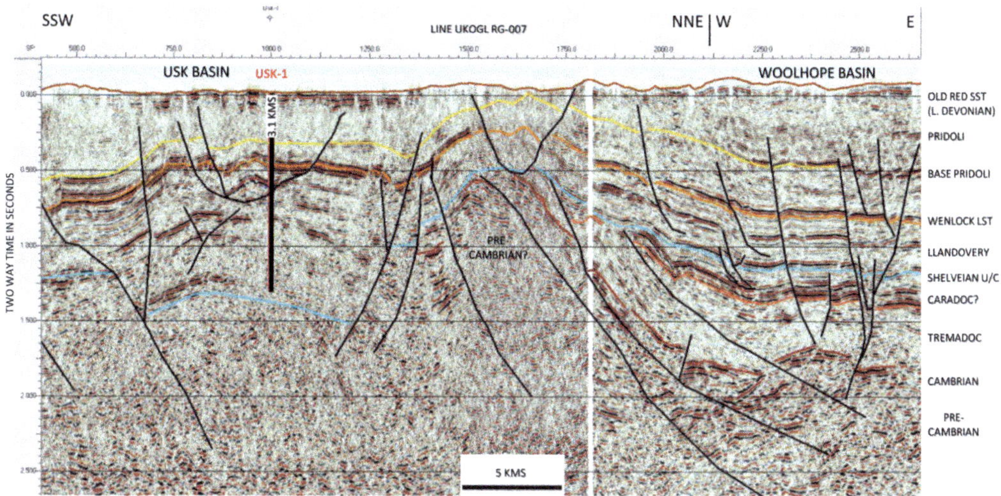

Fig. 5. Seismic Line A, through the southern Welsh Borderlands, showing the anomalous Usk Basin and tight folding below the Shelveian Unconformity.

Wenlock in parts of Shropshire (Greig *et al.* 1968). The Llandovery–Ludlow rocks form a characteristic banded package on seismic lines, which enables them to be mapped across the MMC, possibly as far as the Weald Basin, despite a general lack of well control. Woodcock & Pharaoh (1993) compared the facies of the limited penetrations of Silurian rocks in East Anglia with those of the Welsh Basin and concluded that they were very similar: two primarily deep-water, shale-dominated sequences separated by the thin shelf sequence postulated to have been deposited over the MMC. However, there is little information available from wells to determine the lithologies of this shelf sequence or the exact age of the subsurface rocks overlying the Shelveian Unconformity, which probably range in age from latest Ordovician to Wenlock.

There is widespread evidence of folding and minor thrusting beneath the Shelveian Unconformity (Fig. 5), although the opaque nature of the Tremadoc makes it difficult to see evidence of deformation within it. However, well evidence indicates the presence of very variable dips and reverse faulting within this sequence. Figure 5 is part of regional profile UKOGL RG-007 (available from the UKOGL website: www.ukogl.org.uk) through an area studied in detail by Butler *et al.* (1997). Part of the line was interpreted by Barclay & Smith (2002), who postulated the presence of Caradoc above the Tremadoc in the northern part of Figure 5. Unfortunately, the next well control is Fownhope-1, 17 km to the NNE, which is on the other side of the Woolhope Fault, a thrust with a strike-slip component, and it is difficult to correlate the horizons across it. The line demonstrates tight folding and thrusting below what is interpreted to be the Shelveian Unconformity in a WSW–ENE-orientated, unnamed high that separates the Usk Basin from the Woolhope Basin. Well Usk-1 is anomalous, because it penetrated a very thick sequence of clastics of Llandovery age and was still within rocks dated as Rhuddanian at total depth. The Usk Basin appears to have formed as an extensional graben that filled with the products of erosion around its northern and eastern margins, and was apparently no longer active by the end of the Llandovery. Seismic lines to the west of Usk-1 indicate the presence of a thick wedge of seismically opaque sediments that pinch out before reaching the well. These may be shales forming an extension of the Welsh Basin sequence that opens up off the margin of the MMC to the NW (Woodcock *et al.* 1996; Butler *et al.* 1997), although the Usk Basin still lies 50 km or more from the Pontesford Lineament.

In the area to the west and north of the Malvern Hills, the Shelveian Unconformity appears to cut down to the Precambrian, as it does on the SE side of the Church Stretton Fault (Greig *et al.* 1968) and in the Telford area (Smith *et al.* 2005). Figure 6 presents a compiled seismic cross-section from west to east across the Malvern Line. The interpretation indicates that the Precambrian basement encountered below the Variscan Unconformity in Kempsey-1 (Whittaker 1980*a*), which lies some 3.5 km south of the compiled line in the gap between BGS84-02 and BGS86-03, is part of a major Shelveian high that extended over a large area of Herefordshire and Worcestershire. The Silurian sequences above the unconformity show no sign of thinning towards the Malvern Line (Fig. 7). It is postulated that the occurrence of Precambrian subcrop beneath the

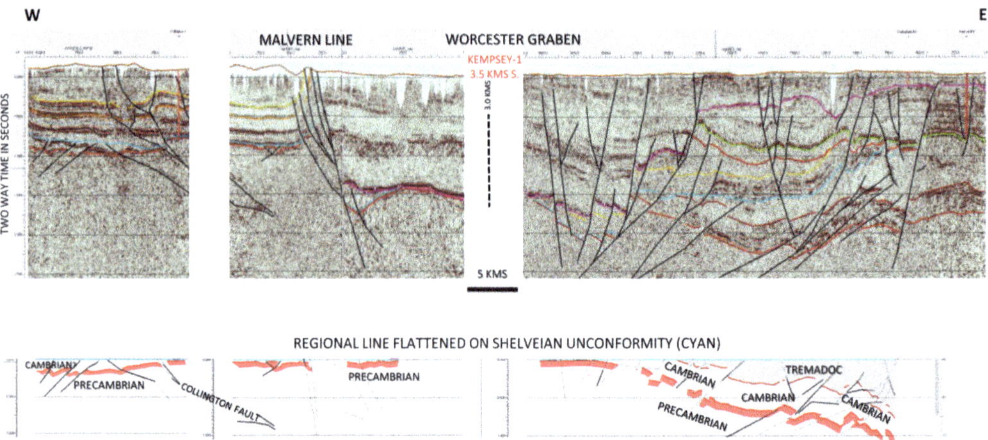

Fig. 6. Seismic Line B: regional compilation from north Herefordshire across the Malvern Line to Stratford-on-Avon, together with events flattened on Shelveian Unconformity to demonstrate the interpreted regional Shelveian high. Lines BGS84 -01, BGS84-02 and BGS86-03 were reprocessed by UKOGL in 2012 and used with the permission of the British Geological Survey. ©NERC 2012.

Permian in the centre of the Worcester Graben is the result of erosion of this extensive high zone by unconformities at Shelveian, Acadian and two Variscan episodes, rather than a simple Variscan uplift and later inversion of what is now the Permian Basin by Variscan thrusts reaching the pre-Permian surface at the Malverns in the west and the zone of the Inkberrow Fault in the east, as proposed by Chadwick (1993) and Peace & Besly (1997). This is in agreement with Smith (1987), who inferred a Precambrian subcrop beneath the Devonian in the Worcester Graben area. The steep nature of the faulting on the Malvern Line suggests strike-slip-induced reversed faulting, although there is clear deep-seated evidence of thrusting on the Collington Fault, which may be of Acadian age but unfortunately reaches the surface (Brandon 1989) in an area where there is no outcrop younger than Early Devonian.

The eastern half of Seismic Line B (Fig. 8) demonstrates the increased preservation of Tremadoc and

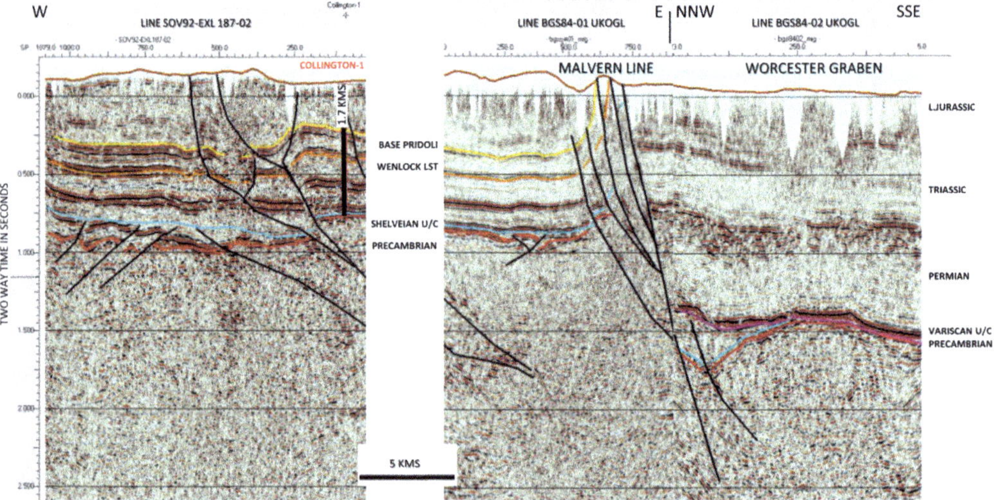

Fig. 7. Western half of Seismic Line B, showing detail of the Collington Fault and the Malvern Line. Lines BGS84-01 and BGS84-02 were reprocessed by UKOGL in 2012 and used with the permission of the British Geological Survey. ©NERC 2012.

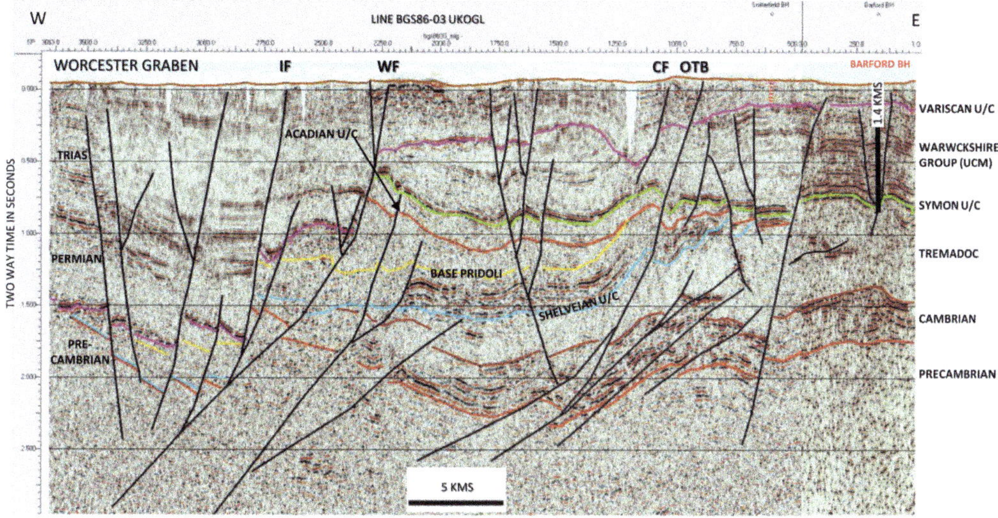

Fig. 8. Eastern half of Seismic Line B, showing the Shelveian, Acadian, Symon and Variscan unconformities. The line was reprocessed by UKOGL in 2012 and used with the permission of the British Geological Survey. ©NERC 2012. IF, Inkberrow Fault; WF, Weethley Fault; CF, Clopton Fault; OTB, Oxfordshire Thrust Belt.

Cambrian rocks towards the east, and also shows the characteristic seismic response of the events at these levels (cf. the northern part of Fig. 5). The Cambrian events appear to show tight folding and reverse faulting similar to that seen in Figure 5, on the other side of the Malvern Line, and there is some evidence of similar deformation within the Tremadoc in the eastern part of the line. This predates the Shelveian Unconformity and presumably took place during the Late Ordovician.

The concealed Llandovery–Lower Devonian megasequence

As noted earlier, the Silurian beds seen to the west of the Malverns on Figures 6 and 7 show no evidence of thinning towards the present-day Worcester Graben, and it seems reasonable to suppose that they continued across to link up with the thinner sequence of the same age interpreted in Figure 8. It is also likely that rocks of Přídolí and Early Devonian age were deposited across this area. On the eastern side of the Worcester Graben, it is postulated that the Shelveian erosion becomes less extreme away from the centre of the graben, and that progressively younger rocks are preserved beneath a thin Llandovery, Wenlock and Ludlow sequence, defined by characteristic banded amplitudes on the seismic data.

To the east of the Worcester Graben, beneath the Oxfordshire Coalfield, lies a syncline in which is preserved a thick sequence of Lower Devonian and Přídolí Old Red Sandstone (ORS) facies. Figure 9 shows the approximate outline of the thick ORS, although the western boundary is difficult to interpret because it is caught up in the tight folding and faulting associated with Variscan movements along the eastern margin of the present-day Worcester Graben. Although the ORS has a generally opaque seismic character, with few continuous seismic markers, there is little evidence from Figure 10 and other lines in the area that it thins towards its truncated margin below the Acadian Unconformity. There are few thick penetrations of this sequence in the region, and the two wells with velocity surveys available, Faringdon-1 and the Steeple Aston BH (Poole 1977), indicate interval velocities of approximately 3900 m s^{-1} for their short penetrations of both Upper and Lower Devonian. This is slower than the 4343 m s^{-1} interval velocity for the 695 m thickness of the ORS in Usk-1 and may indicate less burial. In estimating the maximum thickness beneath the Oxfordshire Basin, an interval velocity of 3900 m s^{-1} has been used, which equates to around 3000 m of ORS.

The Acadian Unconformity

There is good evidence from wells across the eastern and central parts of the MMC that the Acadian Unconformity cuts down deeply into the Lower Paleozoic rocks. The Frasnian marine transgression and Famennian regression can be recognized in wells and outcrop from Brightling-1 to Willesden-1, across to Steeple

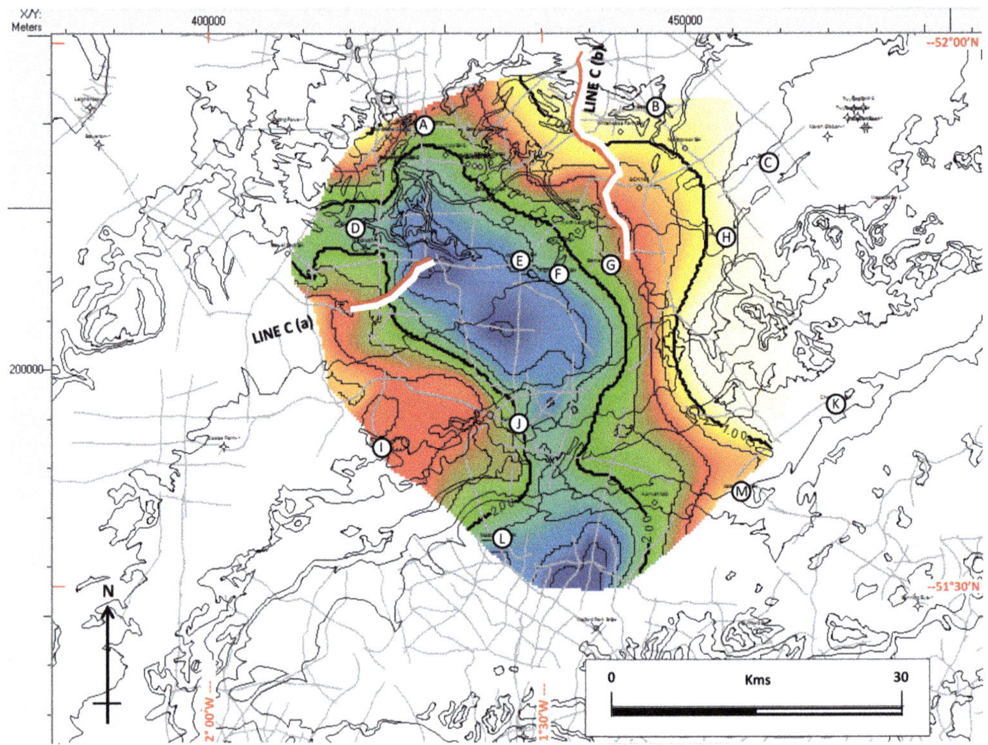

Fig. 9. Form-line time–structure map on the Shelveian Unconformity, indicating the shape of the overlying thick Lower Devonian–Přídolí ORS basin (C.I. 250 ms). Key wells are indicated by the following letters: A, Ash Farm-1; B, Steeple Aston BH; C, Bicester-1; D, Sherbourne-1; E, Apley Barn-1; F, High Cogges BH; G, Barnard Gate BH; H, Noke Hill G-1; I, Highworth-1; J, Faringdon-1; K, Chalgrove BH; L, Maddle Farm BH; M, Aston Tirrold BH. Grey lines show seismic coverage.

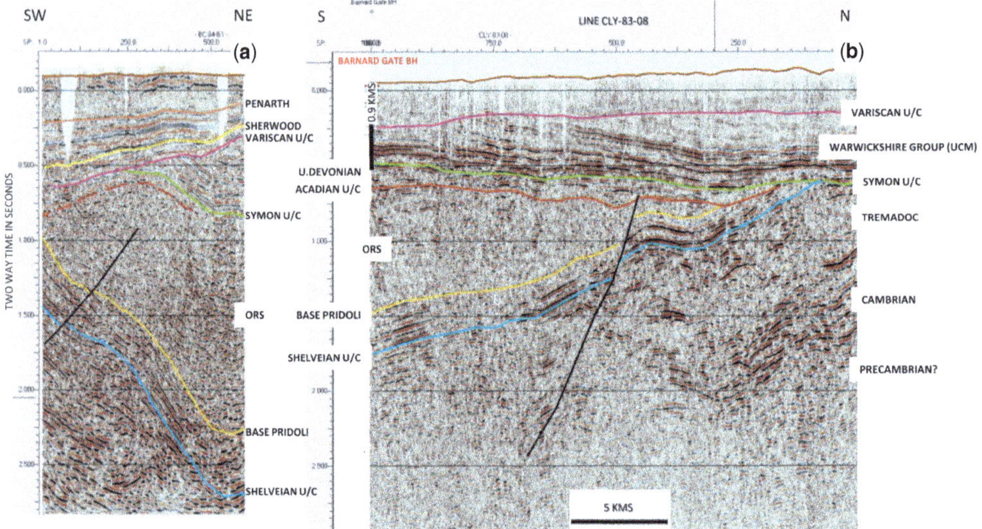

Fig. 10. Seismic Line C: (**a**) & (**b**) illustrating the thick ORS preserved in a syncline beneath the Oxfordshire Coal Basin. The truncation beneath the Acadian Unconformity can be seen in (b), where the presence of Upper Devonian is tied to the Barnard Gate BH, but is more speculative in (a).

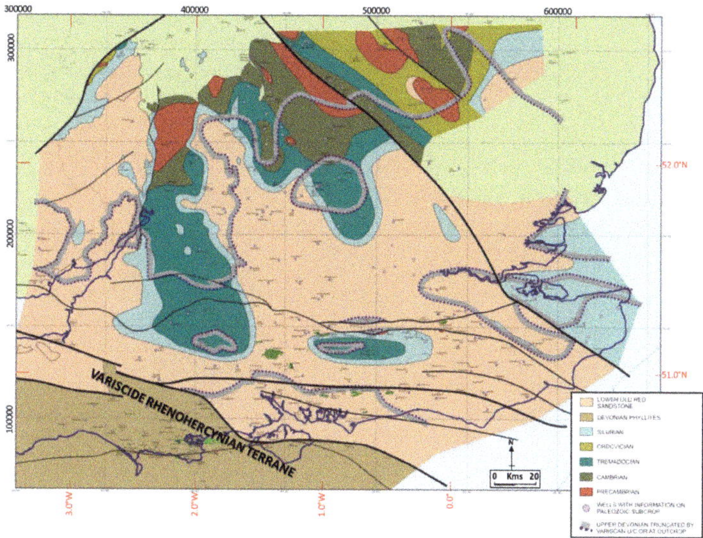

Fig. 11. Map showing the interpreted subcrop to the Acadian Unconformity, based on well, seismic and outcrop information and relying heavily on previous work by Smith (1987), Smith *et al.* (2005) and Pharaoh *et al.* (2011) for the NE part.

Aston BH and over the Malvern Line as far as Powys (Wills 1952; Butler 1981), Similarly, the late Famennian transgression and transition to the Carboniferous Lower Limestone Shale can be seen in wells and outcrop from SE England to South Wales. However, this unconformity is not often resolved on seismic data. Interpretation of the position of the Acadian Unconformity in Figure 8 is based on evidence of truncation on this and other seismic lines to the south (Fig. 10b), combined with nearby well evidence. The Acadian Unconformity is interpreted to truncate the Lower Devonian and Přídolí ORS facies, with its characteristic opaque seismic character, and to cut down as far as the Tremadoc in Figure 8.

It is difficult to determine when the Silurian–Upper Devonian beds were removed from the floor of the present-day Worcester Graben. Work by Barclay *et al.* (2013) has demonstrated that the Halesowen Beds (Warwickshire Group) unconformably overlie both the Precambrian and the Přídolí in trenches excavated across the Malvern Line fault complex close to line BGS84-01, which would place the uplift and erosion on this margin of the Worcester Graben no later than the Symon Unconformity. Outcrop control in the Tortworth Inlier (Cave *et al.* 1977) and eastern Mendips (Green *et al.* 1965), although the mapped unconformity here may be a thrust boundary (Hancock 1982), indicates that the Acadian cuts down to Wenlock or Llandovery to the south of the Malverns, and outcrops and boreholes west of Nuneaton indicate that the Acadian cuts down to the Tremadoc and Cambrian to the NE of the Malverns. Interpretation of seismic data in the Devizes area of Wiltshire suggests that the Acadian Unconformity eroded down to the Tremadoc south of the Worcester Graben. On balance, it seems likely that much of the Silurian and Lower Devonian sequence was removed from the Worcester Graben area during Acadian uplift. The map of interpreted Acadian subcrop (Fig. 11) has borrowed heavily from the previous work of Pharaoh *et al.* (1991, 2011), Smith (1987) and Smith *et al.* (2005) for the northern and eastern parts of the MMC, and is subject to the limitations of sparse borehole data points over much of the area. An attempt has been made to define the present-day limits of the Acadian Unconformity (it is present on the dotted side of the grey shaded line in Fig. 11), beyond which the Upper Devonian has been removed by a combination of the Symon and Variscan unconformities, and the map is even more speculative.

The Symon Unconformity

The Symon Unconformity, at the base of the Warwickshire Group (UCM), is very clearly seen in Figures 8 and 10. There is considerable angular truncation at this unconformity on seismic lines across the area but, where the Acadian Unconformity can be seen, it appears that much of the erosion may have taken place at this older level. Figure 10b demonstrates this on one of the few seismic lines that shows both unconformities.

The Warwickshire Group

The sequence making up the Oxfordshire Coalfield has been described in detail by Dunham & Poole (1974). The beds are age-equivalent to the Halesowen, Keele and Enville groups of the Warwickshire and South Staffordshire Coalfields, the Coal Measures of the Forest of Dean, and the upper part of the Coal Measures of the Kent Coalfield; the uppermost parts being of Stephanian age (Waters *et al.* 2011). Across the MMC, they overlie a surface already greatly eroded by the Acadian Unconformity but also overstep the Lower Coal Measures and the feather edge of the Carboniferous Limestone in the Reading Coal Basin. The preserved area of the Warwickshire Group subcrop beneath the Variscan Unconformity has been mapped from boreholes by Dunham & Poole (1974) and Foster *et al.* (1989), but examination of seismic data indicates that it once covered a much larger area and, although there appears to be regional thinning towards the NE (Poole 1977) (Fig. 10b) there is little evidence of a depositional margin to the north or east. Away from the complexities of the Oxfordshire Thrust Zone, the Warwickshire Group shows regional tilting prior to Variscan erosion. Olivine-bearing basalts and dolerite, with dolerite sills, have been described from the Lower and Middle Coal Measures of the Reading Coal Basin (Foster *et al.* 1989) but there is seismic evidence of an apparent igneous intrusion uplifting and breaking through the Warwickshire Group on the NE margin of the Oxfordshire Coalfield at about National Grid Reference [45200 24600]. This shows as a pronounced negative on both the published reduced-to-pole magnetic field −2 and 10 km continuation residual maps (Busby *et al.* 2006) but does not appear to have any significant effect on the residual gravity maps. It is truncated by the Variscan Unconformity, which is here overlain by Triassic Sherwood Sandstone.

Late Variscan tectonics and the Variscan Unconformity

The Oxfordshire Thrust Belt (OTB) was named by Peace & Besly (1997) to encompass the intense late Variscan deformation seen in parts of the eastern boundary of the Worcester Graben. These authors had access to good-quality seismic lines acquired by Clyde Petroleum in the 1980s over the area between Stow-on-the-Wold and just north of Chipping Camden, where the late Variscan faulting and folding is most intense. Further seismic data now available demonstrate that the intense structuration dies out very rapidly south of this area (see line RG-004 in Butler & Jamieson 2013), although Peace & Besly (1997) demonstrated reverse faulting in the Upton borehole, some 12 km south of Stow-on-the-Wold. Comparison between Figures 8 and 11, which are about 10 km apart at the OTB location, shows that it also dies out northwards. It is possible that this intense deformation zone is bounded by strike-slip faults: a lineation orientated NE–SW was identified from gravity information in the BGS Worcester Sheet Memoir (Barclay *et al.* 1997) and may define the northern boundary of this zone, although the actual track of this lineation intersects Figure 8 at about the location of the Inkberrow Fault and no seismic data are available for some 50 km to the north. Barclay *et al.* (1997) noted that the projection of this lineation coincides with the southern limit of exposed Malvern Complex rocks, although it appears to have little effect on the Permo-Trias strata. It also has the same orientation as the Severn Estuary Fault Zone of Wilson *et al.* (1988). Note the differences in data quality and consequent interpretation of line BGS86-03 (Fig. 8) between the reprocessed version published here and the original version published by Chadwick & Smith (1988). Note also that the Clopton Fault now appears to sole out into the Warwickshire Group.

Figure 12 demonstrates the nature of the tight folding and thrusting seen in the Oxfordshire Thrust Zone and also the reversal of the overall sense of fault movement caused by extension at the beginning of Permian times to create the Worcester Graben. Figure 13 shows a reconstruction of the horizons interpreted in Figure 8, by flattening on the Variscan Unconformity, demonstrating the persistent nature of the Precambrian cored high beneath the Worcester Graben. Although the central part of the Worcester Graben includes a thick Permian section, in its southern extension and on the margins of the basin the Variscan Unconformity is overlain by Triassic Sherwood Sandstone or Mercia Mudstone. Erosion of these areas continued for some considerable time and was presumably responsible for recycling the Warwickshire Group sediments into the Permian and lower Triassic sequence.

The Gloucestershire and Wiltshire Tremadoc Basin, and the Variscan Front

South of the Permian depocentre of the Worcester Graben, there is a large area in which rocks of Tremadoc seismic character subcrop beneath the base Permo-Trias. The main age control for this part of the MMC was provided by the drilling of Cooles Farm-1 in 1976. This well penetrated a 2281 m-thick section of Tremadoc and Upper Cambrian White Leaved Oak shales below the Triassic Bunter Pebble Beds, before reaching a total depth of 3513 m in what was assumed to be Hollybush Sandstone (Shell 1976). Smith (1993) published an interpretation of

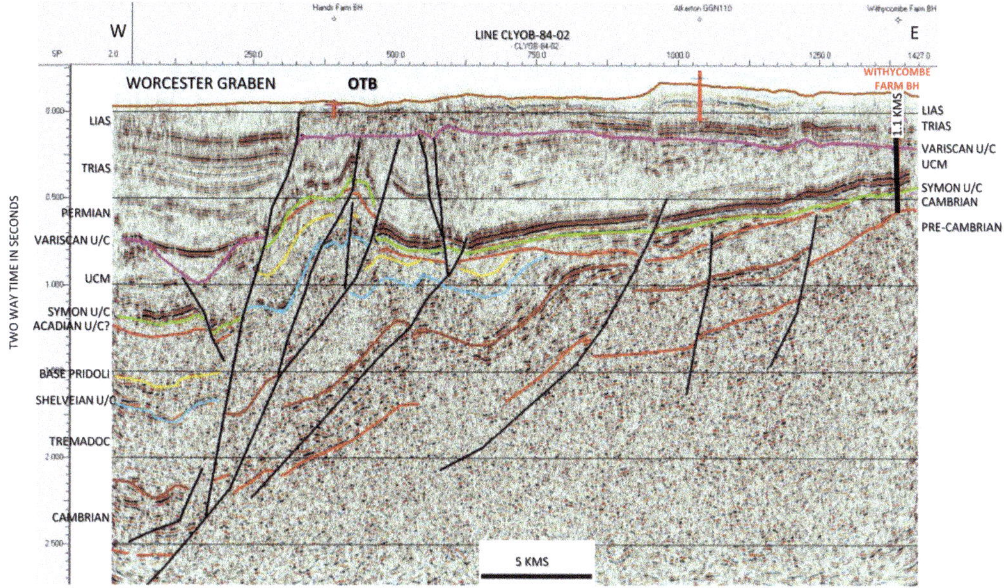

Fig. 12. Line D through the zone of Variscan deformation along the eastern side of the Worcester Graben. Note that the main Variscan deformation post-dates the deposition of the Warwickshire Group (UCM).

seismic line CLY-83-10, running eastwards from Cooles farm-1 to Faringdon-1. Line E (Fig. 14) runs through Cooles Farm-1 and then turns south along IGS79-01, reprocessed by UKOGL, to Devizes-1, which lies close to the line of the Variscan Front, and encountered the transition from post-Acadian Upper Devonian sandstones and shales to basal Carboniferous Limestone. The northern part of the line shows the top Hollybush Sand ('Cambrian') reflector and the package beneath it to be affected by the tight folding and thrusting characteristic of the Cambrian elsewhere in the MMC. The opaque nature of the Tremadoc rarely allows recognition of deformation on the seismic data, although well data provide evidence of this. A reverse fault was interpreted in Cooles Farm-1 by Shell (1976, based on the dipmeter) and is shown in Figure 14. The beds above it dip at between 8° and 23° to the SE, those below dip at between 13° and 24° to the SW, although the lower part of the Tremadoc and the Upper Cambrian change back to variable dips to the south and SE. Dips within the 1251 m-thick Tremadoc section of Shrewton-1, south of Devizes, range from 20° to 43° to the SW, although dips of 35°–50° to the

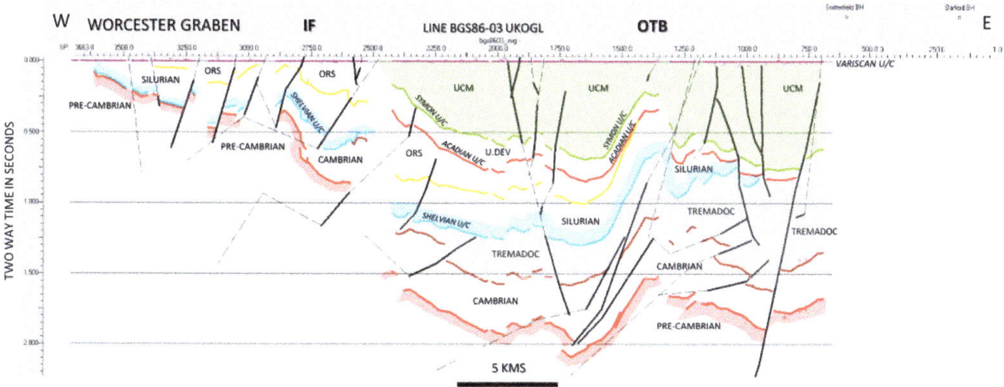

Fig. 13. Interpretation of the eastern half of Line B, flattened on the Variscan Unconformity. No attempt has been made to redraft the fault planes. The TWT scale is in seconds.

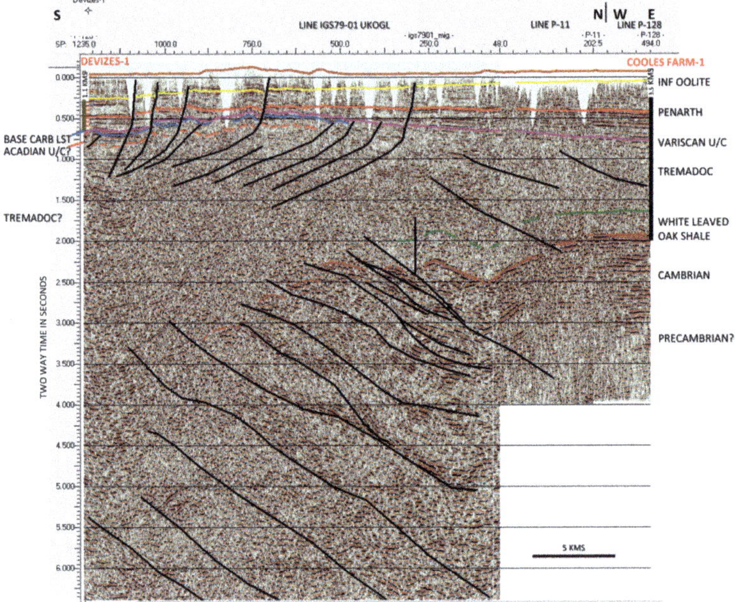

Fig. 14. Line E: a composite line linking Devizes-1 to Cooles Farm-1, providing ties to the Tremadoc and Cambrian in the north, and the Lower Carboniferous and Upper Devonian in the south. Line IGS79-01 was reprocessed by UKOGL in 2012 and is included by permission of the British Geological Survey. ©NERC 2012.

NNW were recorded from the uppermost 40 m of the Tremadoc section penetrated by the well, which did not reach the bottom of this stage (Whittaker 1980*b*).

Figure 14 (Line E) seems to show a change in structural style from a thick package of Cambrian and Ordovician (and probably Precambrian) in the north, with reverse faulting generally southerly directed (although there is no good quality east–west data available to indicate the true direction) to thin-skinned thrust faulting directed to the north. The southern portion of the line appears to show several palaeotopographical highs at the base Triassic (Variscan) Unconformity, similar to those seen on seismic data south of the Mendip Hills where Variscan thrusts affect the Carboniferous Limestone (e.g. see Holloway 1982). The northern limit of these base Triassic perturbations occurs at about CDP 300 on line IGS79-01 and this is probably the local limit of Variscan deformation, although it is well to the north of the generally accepted Variscan Front (Fig. 2). Although Devizes-1 encountered the basal Carboniferous and Upper Devonian, evidence from seismic data linking Devizes-1 to Shrewton-1 suggests that the Tremadoc is not far below the Upper Devonian at this location due to Acadian erosion. The Tremadoc and Cambrian sequences are affected by a series of broad folds in this area but line IGS79-01 probably runs sub-parallel to the fold axes. Although the deep data quality worsens on the southern end of the line, it is still possible to map events suggestive of the Cambrian at around 3–4 s two-way time (TWT). These suggest that thicker Tremadoc is preserved beneath a combination of the Acadian and Variscan unconformities to the south. Immediately to the west of the northern end of line IGS79-01 (Fig. 14), the Cambrian package at the base of the Tremadoc can be seen on several east–west lines to change rapidly from almost horizontal to steeply dipping, the fold limb rising westwards with a strike of about N10°E up to about 12 km west of line IGS79-01, when data quality becomes too poor. However, Cave *et al.* (1977) estimated the minimum thickness for the entire Tremadoc in the Tortworth Inlier, some 25 km west of the northern end of IGS79-01, to be 2286 m, without seeing either the top or base of the sequence, which might suggest the structure becomes synclinal again beyond the limit of the seismic data. The rocks in this inlier of poorly exposed outcrops show some tight folding and slickensides, with dips in various directions measured at between 4° and 70° to vertical, suggesting that measured thicknesses may be exaggerated by structural complications (Cave *et al.* 1977). The overall impression of the Cambrian and Tremadoc seismic package in this area of NW Wiltshire and SE Gloucestershire is that it has been compressed

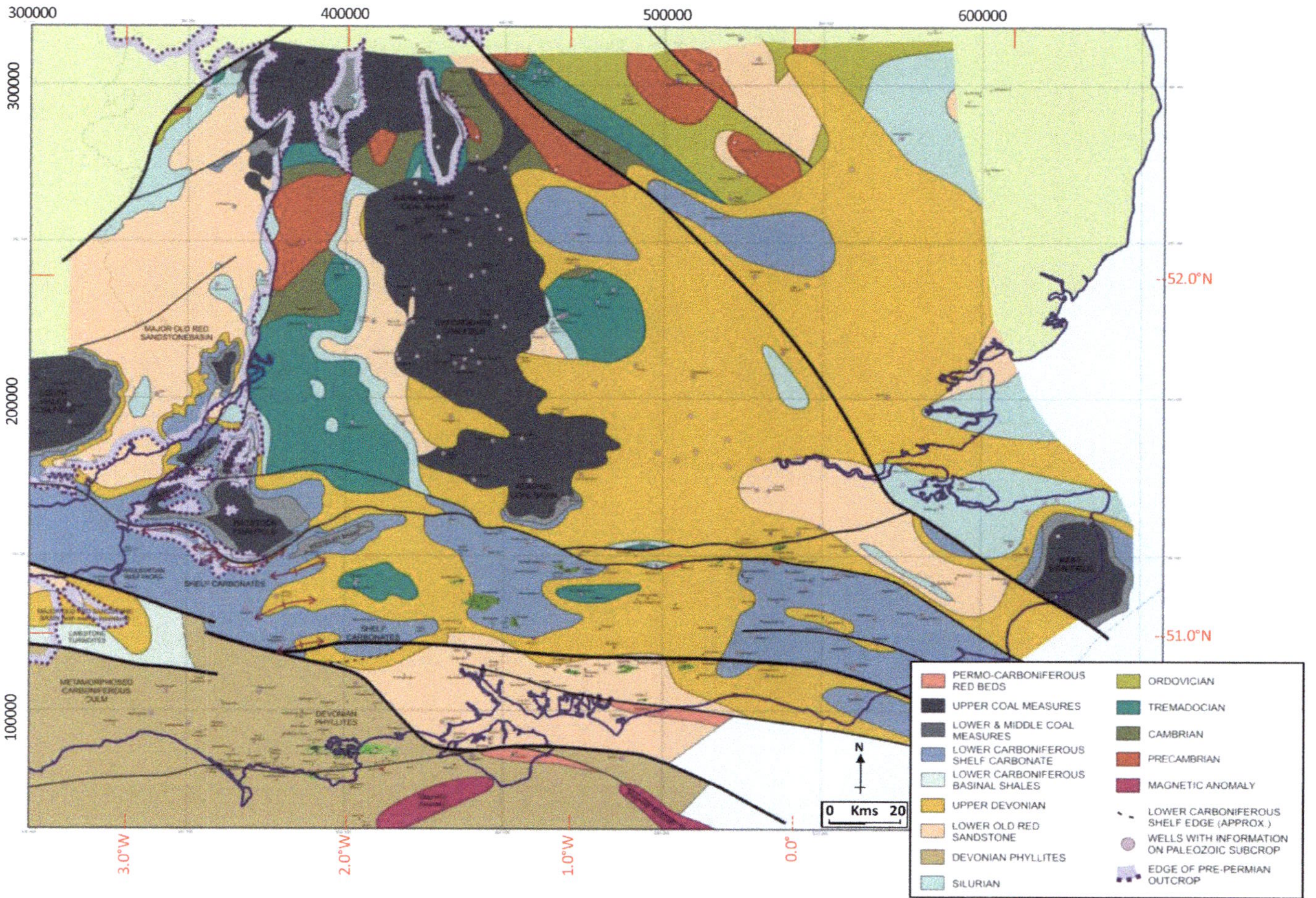

Fig. 15. Map showing the interpreted subcrop to the Variscan Unconformity, based on well, seismic and outcrop information and relying on previous work by Smith (1985), Pharaoh *et al.* (1991, 2011) and Smith *et al.* (2005) for the NE parts.

into a series of broad folds orientated almost north–south. Changes in thickness of the Tremadoc sequence identified on seismic lines in the area are likely to be due, in large part, to later erosion rather than original deposition, since the top of the unit is always an unconformity surface. The Silurian package characteristic of the post-Shelveian Unconformity sequence can be seen to be folded in the same way where it occurs in small outliers beneath the Variscan Unconformity just to the NW of Cooles Farm-1. The orientation of the syncline in the ORS of Figures 9 and 10 has a similar north–south trend, which would suggest this regional-scale folding to be of Acadian age.

Unfortunately, lines IGS79-01 and IGS79-02, at right angles to it and with poor data quality, provide the only deep data in the area, being recorded to 12 s TWT, but there are several long lines immediately east of it that were recorded to 4 s TWT in the mid-1980s and reprocessing of these might help in understanding the structural picture.

Variscan Unconformity subcrop map

Interpretation of the available seismic, well and outcrop data has enabled the construction of a map of the subcrop underlying the Variscan Unconformity (Fig. 15), which is a diachronous erosional surface overlain by rocks ranging in age from early Permian to late Trias, and, in the eastern Mendips, as young as Inferior Oolite (Woodward 1894), although older Mesozoic rocks may have been removed post-Trias. This mapping has, again, made use of the previous work undertaken by Smith (1985), Smith *et al.* (2005) and Pharaoh *et al.* (2011), and the data points of Molyneux (1991), on the areas to the NE of the seismic coverage. This is considered to be a more accurate representation than the Acadian subcrop map (Fig. 11) because of the availability of considerably more data. However, it is subject to change in the Oxfordshire Thrust Belt, because of the complexity of the structures, and in areas where there is little seismic coverage. Based on well penetrations and limited seismic evidence, the subcrop in the area south of the Variscan Front but north of the inferred boundary of the Variscide Rhenohercynian Terrane is interpreted as consisting of mainly Upper Devonian and Carboniferous Limestone, with a number of erosional windows through to the pre-Acadian Lower Paleozoic rocks. Indications are that these rocks were subjected to open folding and thin-skinned thrusting similar to that seen in outcrop in the Mendip Hills. The present-day limits of the Variscan Unconformity are shown on the map (it is present on the dotted side of the grey shaded line in Fig. 15), beyond which the basal Permo-Trias has been eroded and pre-Permian rocks outcrop.

Conclusions

Study of the released seismic and well data over the MMC shows that it is possible to correlate seismostratigraphic sequences over large parts of the area. These sequences appear to continue southwards across much of southern England and suggest that the southern margin of the MMC should be taken at a line running from north Dorset to the Isle of Wight, rather than at the northern margins of the Wessex and Weald basins (the Variscan Front). Where available, seismic data reveal considerable tectonic activity across the MMC prior to the Shelveian, Acadian, Symon and Variscan unconformities, and indicates that the MMC has not always been a stable block. This review gives a general indication of the history of the area during Paleozoic times but it is hoped that the availability of released seismic and well data through UKOGL will encourage more detailed studies.

Acknowledgements Special thanks are due to Nigel Smith and Chris Pullan for many hours of enthusiastic discussion and advice during the interpretation of this challenging dataset. The British Geological Survey have kindly agreed to allow the publication of their geological base maps and the reprocessed seismic lines. The trustees of the UK Onshore Geophysical Library continue to stress the importance of disseminating the information in its free archive as widely as possible and have kindly agreed to contribute towards the cost of publishing in this paper. Thanks are due to Eric Stuckey for his help in drafting the figures. Comments and suggestions from John Marshall, Nigel Woodcock and an anonymous reviewer are greatly appreciated.

References

Barclay, W.J. & Smith, N.J.P. 2002. *Geology of the Country between Hereford and Ross-on-Wye: A Brief Explanation of the Geological Sheet 215.* Sheet Explanations England & Wales (1:50 000). British Geological Survey, Keyworth, Nottingham, UK.

Barclay, W.J., Ambrose, K. *et al.* 1997. *Geology of the Country around Worcester. Memoir for 50 000 Geological Sheet 199 (England and Wales).* Memoirs of the Geological Survey of Great Britain, England and Wales (Sheet – New Series). HMSO, London.

Barclay, W.J., Olver, P., Payne, J., Hay, S., Jenkins, M., Watkins, N. & Nicklin, J. 2013. The Precambrian inlier at Martley, Worcestershire: Martley Rock rediscovered. *Transactions of the Woolhope Naturalists' Field Club, Herefordshire*, **60**, 1–15.

Brandon, A. 1989. *Geology of the Country between Hereford and Leominster. Memoir for 50 000 Geological Sheet 198 (England and Wales).* Memoirs of the Geological Survey of Great Britain, England and Wales (Sheet – New Series). HMSO, London.

Busby, J.P., Walker, A.S.D. & Rollin, K.E. 2006. *Regional Geophysics of South-east England.* Version

1.0 on CD-ROM. British Geological Survey, Keyworth, Nottingham, UK.

Butler, A.J., Woodcock, N.H. & Stewart, D.M. 1997. The Woolhope and Usk Basins: Silurian rift basins revealed by subsurface mapping of the southern Welsh Borderlands. *Journal of the Geological Society*, London, **154**, 209–223, https://doi.org/10.1144/gsjgs.154.2.0209

Butler, D.E. 1981. Marine faunas from concealed Devonian rocks of southern England and their reflection of the Frasnian transgression. *Geological Magazine*, **118**, 679–697.

Butler, M. 1998. The geological history of the southern Wessex Basin – a review of new information from oil exploration. *In*: Underhill, J.R. (ed.) *Development, Evolution and Petroleum Geology of the Wessex Basin*. Geological Society, London, Special Publications, **133**, 67–86, https://doi.org/10.1144/GSL.SP.1998.133.01.04

Butler, M. & Jamieson, R. 2013. *Preliminary Interpretation of Six Regional Seismic Profiles Across Onshore Basins of England*. UK Onshore Geophysical Library, digital publication, http://www.ukogl.org.uk

Cave, R., Kellaway, G.A. *et al.* 1977. *Geology of the Malmesbury District: Explanation of One-Inch Geological Sheet 251*. Memoirs of the Geological Survey of Great Britain, England and Wales (Sheet – New Series). HMSO, London.

Chadwick, R.A. 1993. Aspects of basin inversion in southern Britain. *Journal of the Geological Society*, London, **150**, 311–322, https://doi.org/10.1144/gsjgs.150.2.0311

Chadwick, R.A. & Smith, N.J.P. 1988. Evidence of negative structural inversion beneath central England from new seismic reflection data. *Journal of the Geological Society*, London, **145**, 519–522, https://doi.org/10.1144/gsjgs.145.4.0519

Chadwick, R.A., Pharaoh, T.C. & Smith, N.J.P. 1989. Lower crustal heterogeneity beneath Britain from deep seismic reflection data. *Journal of the Geological Society*, London, **146**, 617–630, https://doi.org/10.1144/gsjgs.146.4.0617

Dunham, K.C. & Poole, E.G. 1974. The Oxfordshire Coalfield. *Journal of the Geological Society*, London, **130**, 387–391, https://doi.org/10.1144/gsjgs.130.4.0387

Foster, D.D., Holliday, D.W., Jones, C.M., Owens, B. & Welsh, A. 1989. The concealed Upper Palaeozoic rocks of Berkshire and South Oxfordshire. *Proceedings of the Geologists' Association*, **100**, 395–407.

Green, G.W., Welch, F.B.A. *et al.* 1965. *The Geology of the Country around Wells and Cheddar. Explanation of One-Inch Geological Sheet 280*. Memoirs of the Geological Survey of Great Britain, England and Wales (Sheet – New Series). HMSO, London.

Greig, D.C., Wright, J.E., Hains, B.A. & Mitchell, G.S. 1968. *Geology of the Country around Church Stretton, Craven Arms, Wenlock Edge and Brown Clee. Explation of Sheet 166*. Memoirs of the Geological Survey of Great Britain, England and Wales (Sheet – New Series). HMSO, London.

Hancock, N.J. 1982. Stratigraphy, palaeogeography and structure of the East Mendips Silurian inlier. *Proceedings of the Geologists' Association*, **93**, 247–261.

Holloway, S. 1982. *Completion Report for Bruton Borehole*. Institute of Geological Sciences, London, Deep Geology Unit, Report 82/6.

Molyneux, S.G. 1991. The contribution of palaeontological data to an understanding of the early Palaeozoic framework of eastern England. *Annales de las Société Géologique de Belgique*, **114**, 93–105.

Peace, G.R. & Besly, B.M. 1997. End-Carboniferous fold-thrust structures. Oxfordshire. UK: implications for the structural evolution of the Variscan foreland of south-central England. *Journal of the Geological Society*, London, **154**, 225–237, https://doi.org/10.1144/gsjgs.154.2.0225

Pharaoh, T.C., Merriman, R.J., Webb, P.C. & Beckinsale, R.D. 1987. The concealed Caledonides of eastern England: preliminary results of a multidisciplinary study. *Proceedings of the Yorkshire Geological Society*, **46**, 355–369, https://doi.org/10.1144/pygs.46.4.355

Pharaoh, T.C., Merriman, R.J., Evans, R.J., Brewer, T.S., Webb, P.C. & Smith, N.J.P. 1991. Early Palaeozoic arc-related volcanism in the concealed Caledonides of southern Britain. *Annales de la Société Géologique de Belgique*, **114**, 63–91.

Pharaoh, T.C., Vincent, C.J., Bentham, M.S., Hulbert, A.G., Waters, C.N. & Smith, N.J.P. 2011. *Structure and Evolution of the East Midlands Region of the Pennine Basin: Subsurface Memoir*. British Geological Survey, Keyworth, Nottingham, UK.

Poole, E.G. 1977. *Stratigraphy of the Steeple Aston Borehole, Oxfordshire*. Bulletin of the Geological Survey of Great Britain, **57**.

Richardson, J.B. & Rasul, S.M. 1979. Palynological evidence for the age and provenance of the Lower Old Red Sandstone from the Apley Barn Borehole, Witney, Oxfordshire. *Proceedings of the Geologists' Association*, **90**, 27–42.

Shell. 1976. *Completion Log: Cooles Farm-1*. Released well data. Shell UK Ltd (Exploration and Production), London.

Smith, N.J.P. (compiler) . 1985. *Map 1 Pre-Permian Geology of the United Kingdom (South)*. British Geological Survey, Keyworth, Nottingham, UK.

Smith, N.J.P. 1987. The deep geology of central England: the prospectivity of the Palaeozoic rocks. *In*: Brooks, J. & Glennie, K.W. (eds) *Petroleum Geology of Northwest Europe: Proceedings of the 3rd Conference*. Graham & Trotman, London, 217–224,

Smith, N.J.P. 1993. The case for exploration of deep plays in the Variscan fold belt and its foreland. *In*: Parker, J.R. (ed.) *Petroleum Geology of Northwest Europe: Proceedings of the 4th Conference*. Geological Society, London, 667–675, https://doi.org/10.1144/0040667

Smith, N.J.P. & Rushton, A.W.A. 1993. Cambrian and Ordovician stratigraphy related to structure and seismic profiles in the western part of the English Midlands. *Geological Magazine*, **130**, 665–671.

Smith, N.J.P., Kirby, G.A. & Pharaoh, T.C. 2005. *Structure and Evolution of the South-west Pennine Basin and Adjacent Area: Subsurface Memoir*. British Geological Survey, Keyworth, Nottingham, UK.

Soper, N.J., Webb, B.C. & Woodcock, N.H. 1987. Late Caledonian (Acadian) transpression in north-west England: timing, geometry and geotectonic significance.

Proceedings of the Yorkshire Geological Society, **46**, 175–192, https://doi.org/10.1144/pygs.46.3.175

Toghill, P. 1992. The Shelveian event, a late Ordovician tectonic episode in Southern Britain (Eastern Avalonia). *Proceedings of the Geologists' Association*, **103**, 31–35.

Turner, J.S. 1949. The deeper structure of central and northern England. *Proceedings of the Yorkshire Geological Society*, **27**, 280–297, https://doi.org/10.1144/pygs.27.4.280

Waters, C.N., Jones, N.S. & Besly, B.M. 2011. South Midlands and Kent. *In*: Waters, C.N. (ed.) *A Revised Correlation of Carboniferous Rocks in the British Isles*. Geological Society, London, 44–48.

Whittaker, A. 1980*a*. *Shrewton No.1: Geological Well Completion Report*. Institute of Geological Sciences, London, Deep Geology Unit, Report 80/1.

Whittaker, A. 1980*b*. *Kempsey No.1: Geological Well Completion Report*. Institute of Geological Sciences, London, Deep Geology Unit, Report 80/2.

Wills, L.J. 1952. *A Palaeogeological Atlas of the British Isles and Adjacent Parts of Europe*. Blackie & Son, London.

Wills, L.J. 1978. *A Palaeogeological Map of the Lower Palaeozoic Floor below the Cover of Upper Devonian, Carboniferous and Later Formations*. Geological Society, London, Memoirs, **8**, 7–9, https://doi.org/10.1144/GSL.MEM.1978.008.01.01

Wilson, D., Davies, J.R., Smith, M. & Waters, R.A. 1988. Structural controls on Upper Palaeozoic sedimentation in south-east Wales. *Journal of the Geological Society*, **145**, 901–914, https://doi.org/10.1144/gsjgs.145.6.0901

Woodcock, N.H. 1984. The Pontesford Lineament, Welsh Borderlands. *Journal of the Geological Society*, London, **141**, 1001–1014, https://doi.org/10.1144/gsjgs.141.6.1001

Woodcock, N.H. & Pharaoh, T.C. 1993. Silurian facies beneath East Anglia. *Geological Magazine*, **130**, 681–690.

Woodcock, N.H., Butler, A.J., Davies, J.R. & Waters, R. A. 1996. Sequence stratigraphical analysis of late Ordovician and early Silurian depositional systems in the Welsh Basin: a critical assessment. *In*: Hesselbo, S.P. & Parkinson, D.N. (eds) *Sequence Stratigraphy in British Geology*. Geological Society, London, Special Publications, **103**, 197–208, https://doi.org/10.1144/GSL.SP.1996.103.01.11

Woodward, H.B. 1894. *The Jurassic Rocks of Britain, Volume 4: The Lower Oolitic rocks of England (Yorkshire excepted)*. Memoirs of the Geological Survey (Stratigraphical Monographs). HMSO, London.

Paleozoic gas potential in the Weald Basin of southern England

CHRISTOPHER P. PULLAN[1]* & MALCOLM BUTLER[2]

[1]*CP Exploration, Potterne, Devizes, Wiltshire, SN10 5LR, UK*

[2]*UK Onshore Geophysical Library, 46–50 Coombe Road, New Malden, Surrey, KT3 4QF*

M.B., 0000-0002-0462-9135

**Correspondence: chris.pullan1@btinternet.com*

Abstract: Gas has been found in Mesozoic reservoirs in the Weald Basin, particularly along the northern margin. Most of the gas is dry, with a high methane content and often associated nitrogen. Isotopic evidence indicates that the gas is from a thermogenically mature marine source. Although there is evidence of some shallow, biogenic gas, only the lowermost Lias is projected to have reached the thermogenic gas window before Tertiary uplift. Estimated maturities from isotopic data from the main gas accumulations indicate significantly greater levels than those projected for Liassic shales: thus, the gas is thought to have originated from Paleozoic rocks.

Data on the distribution of Paleozoic rocks subcropping the Variscan unconformity is limited. However, available data suggest that their distribution owes more to Acadian erosion than to Variscan. It is thought that the Upper Devonian and Lower Carboniferous transgressed over a thick, folded Tremadocian shale sequence in the west, and over folded Silurian and Lower–Middle Devonian rocks in the central Weald. There is some evidence for the presence of isolated late Carboniferous or early Permian clastics but no significant coals have been encountered to date. Regional source studies suggest that the only Paleozoic rocks with potential are post-Acadian-aged Devonian shales.

The Weald Basin is the easternmost sub-basin of the southern England Mesozoic Basin and is located south of London (Fig. 1). Exploration began in the area in the late nineteenth century and the first significant discovery of gas was made at Heathfield in 1895 (Dawson 1898). Sporadic exploration subsequently was largely unsuccessful, although a gas field was discovered at Bletchingley in 1965 by Esso. The basin experienced a major phase of exploration during the late 1970s through to the oil-price collapse of the late 1980s. Oil fields were discovered on the southern and western edges of the basin, whilst both oil and gas discoveries were made along the northern edge of the area.

Numerous studies have shown that the oil in these discoveries is sourced from Jurassic shales, and, in particular, the Oxford Clay and Kimmeridge Clay intervals (Ebukanson & Kinghorn 1986; Burwood *et al.* 1991; Harriman 2010). Gas has been found in a number of places in the Weald, particularly along the northern margin of the basin. Several theories have been proposed for the source of the gas in the area but none have been definitively demonstrated (Andrews 2014); in this study, evidence is presented for a Paleozoic source. Although Jurassic reservoir development is generally poor, shows and accumulations have been encountered within porous zones in Middle–Upper Jurassic reservoir intervals (Trueman 2003). Both Tertiary anticlinal and pre-Albian structural traps have been found to be hydrocarbon bearing (Fig. 2) (Hancock & Mithen 1987; Butler & Pullan 1990; Trueman 2003).

Basin development in southern England began during the Lower Permian, following the Variscan Orogeny, and development of the Weald Basin appears to have begun in the Triassic (Hawkes *et al.* 1998). Two phases of rifting are recognized: the first commenced in the Triassic and ended in the Middle Lias; the second, related to the opening of the Bay of Biscay and the North Atlantic, began in the Upper Jurassic (Kimmeridgian), continued into the Lower Cretaceous and ended with the Aptian unconformity, which marked drift onset in the Bay of Biscay (Fig. 3) (Butler & Pullan 1990; Hawkes *et al.* 1998). Tertiary compressional movements are recognized, and are considered to have had two phases of movement: the first, in the Paleogene, resulted in regional uplift and is thought to be associated with the North Atlantic opening; whilst the second, in the Miocene, appears to be related to Alpine tectonism (Jones 1999). These movements resulted in fault reactivation, as well as in regional uplift of the entire basin centred to the SE. Various studies have attempted to calculate the amount of uplift, with estimates ranging from 3700 to 6800 ft (1128–2073 m) (Butler & Pullan 1990; Jones 1999; Andrews 2014).

From: MONAGHAN, A. A., UNDERHILL, J. R., HEWETT, A. J. & MARSHALL, J. E. A. (eds) 2018. *Paleozoic Plays of NW Europe*. Geological Society, London, Special Publications, **471**, 333–363.
First published online March 22, 2018, https://doi.org/10.1144/SP471.1

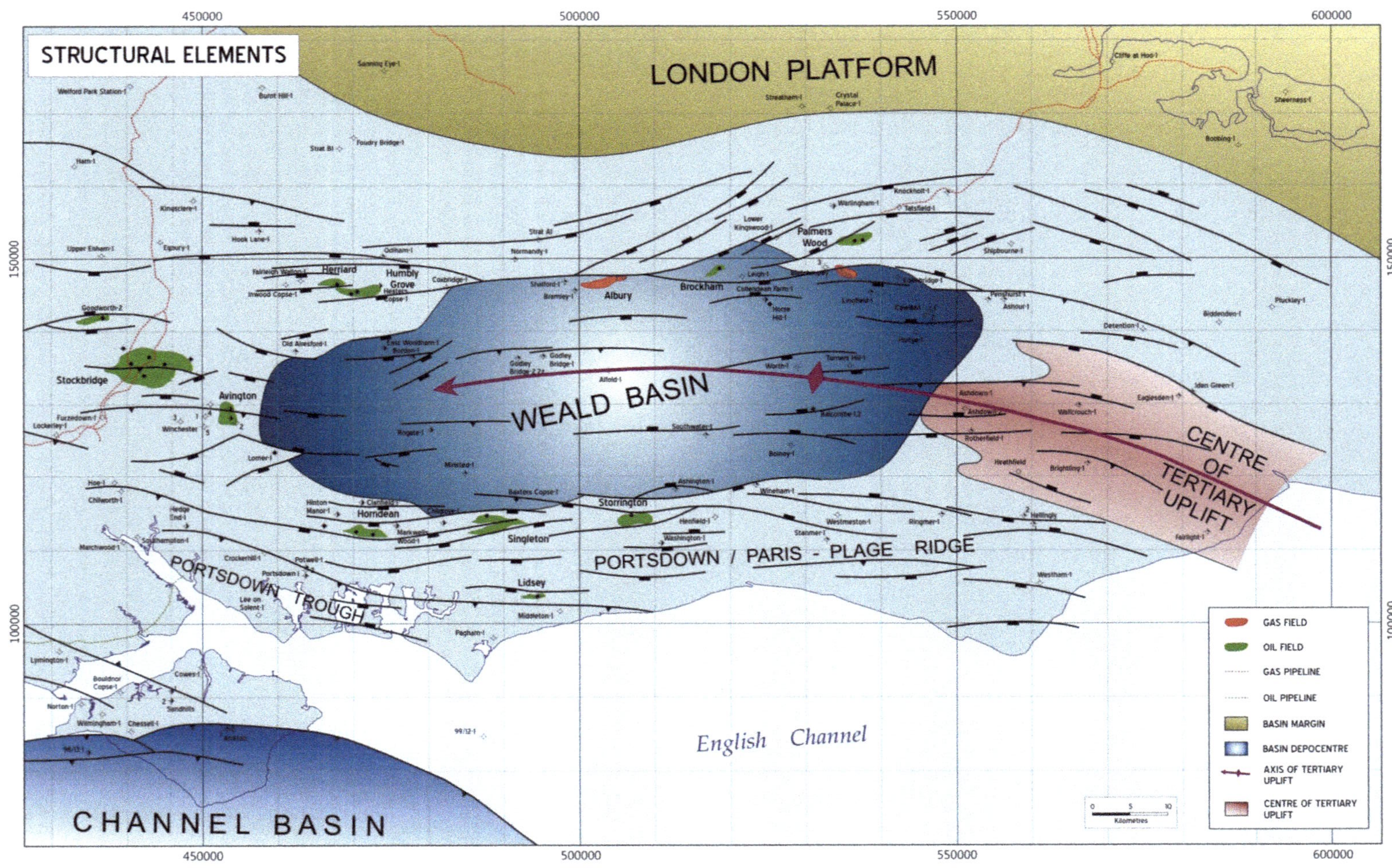

Fig. 1. Weald Basin location, with structural elements. Adapted from Magellan Petroleum (UK) Ltd (2012).

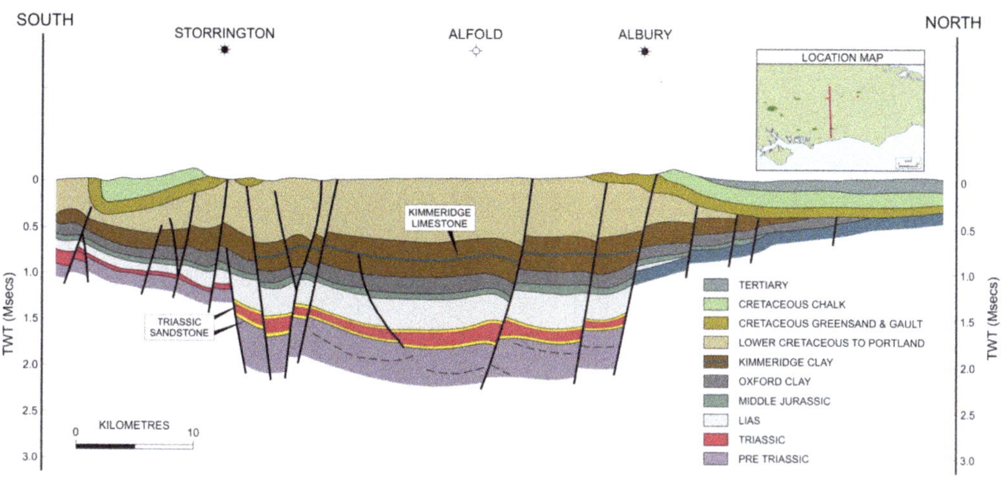

Fig. 2. Geological cross-section of the Weald Basin. After Butler & Jamieson (2013, plate 1).

Subsequently, the area has undergone major erosion, so that Lower Cretaceous rocks crop out in the basin centre, with Upper Cretaceous and Lower Tertiary strata cropping out around the flanks.

All the relevant available released well and regional reports, together with the comprehensive seismic database held in the UK Onshore Geophysical Library (UKOGL) database, over the area of interest have been used in this study for mapping and analysis.

Gas type

Results of the analysis of gas recovered during testing of Weald Basin wells are shown in Table 1 (see also Fig. 4). These data have been compiled from operators' final well reports and Annex B submissions. Only the final results are available, so there may be some questions over reliability in places.

The gas analysis shows that the majority of gas encountered is dry, with a high methane content, particularly when the gas is non-associated. In addition, much of the gas has a high nitrogen content (2.5–12.5%). Nitrogen is considered to be the product of source rocks at an advanced stage of thermal maturity (Tissot & Welte 1984). This amount of nitrogen is consistent with generation from a source in the dry-gas window (Ro >1.4%). No major compositional variations appear to exist between the shallow and deeper gas pools in the same field, as seen in the Godley Bridge or Humbly Grove fields.

The oils typically found in the area are light (39–41° API), low sulphur, waxy crudes. The gas/oil ratios (GORs) of the oils vary, being generally low to the south and west, but higher along the north flank. Most fields do not have a gas cap. Only one analysed oil demonstrates evidence of post-emplacement modification and, therefore, biodegradation is not thought to be a widespread phenomenon. This is in keeping with the theory that the entire basin was widely 'pasteurized' prior to the Tertiary inversion (Wilhelms *et al.* 2001).

Isotope work on the gases by Conoco showed that the gas associated with the oil found at Godley Bridge, Baxters Copse and Palmers Wood has a mainly thermogenic origin (Figs 5 & 6) (Conoco (UK) Ltd 1986). Some of the methane at Godley Bridge may be derived from biodegradation of the associated oil. The data also suggest that all the gases were derived from a marine source (Type II/III), which is in the wet-gas window (vitrinite reflectance (VR) of 1.1–1.2%) (Stahl 1977).

Timing of gas charge

The available seismic data across known gas accumulations were studied and available diagenetic studies reviewed in an attempt to estimate the timing of the gas charge.

Gas was found at Godley Bridge at three stratigraphic levels: Inferior Oolite, Great Oolite and Portland Sandstone. The structure is a simple downthrown closure against an earlier extensional fault (Fig. 7). Although there is no post-Lower Cretaceous cover, the Godley Bridge structure lies on trend to the Hindhead surface anticline (Thurrell *et al.* 1968; Butler & Pullan 1990), which is thought to be of Tertiary age by analogy with structures of similar style where a Tertiary age can be assigned with confidence: for example, the Winchester Anticline (Sterling Resources (UK) Ltd 2001). This suggests some of the gas migration occurred during the

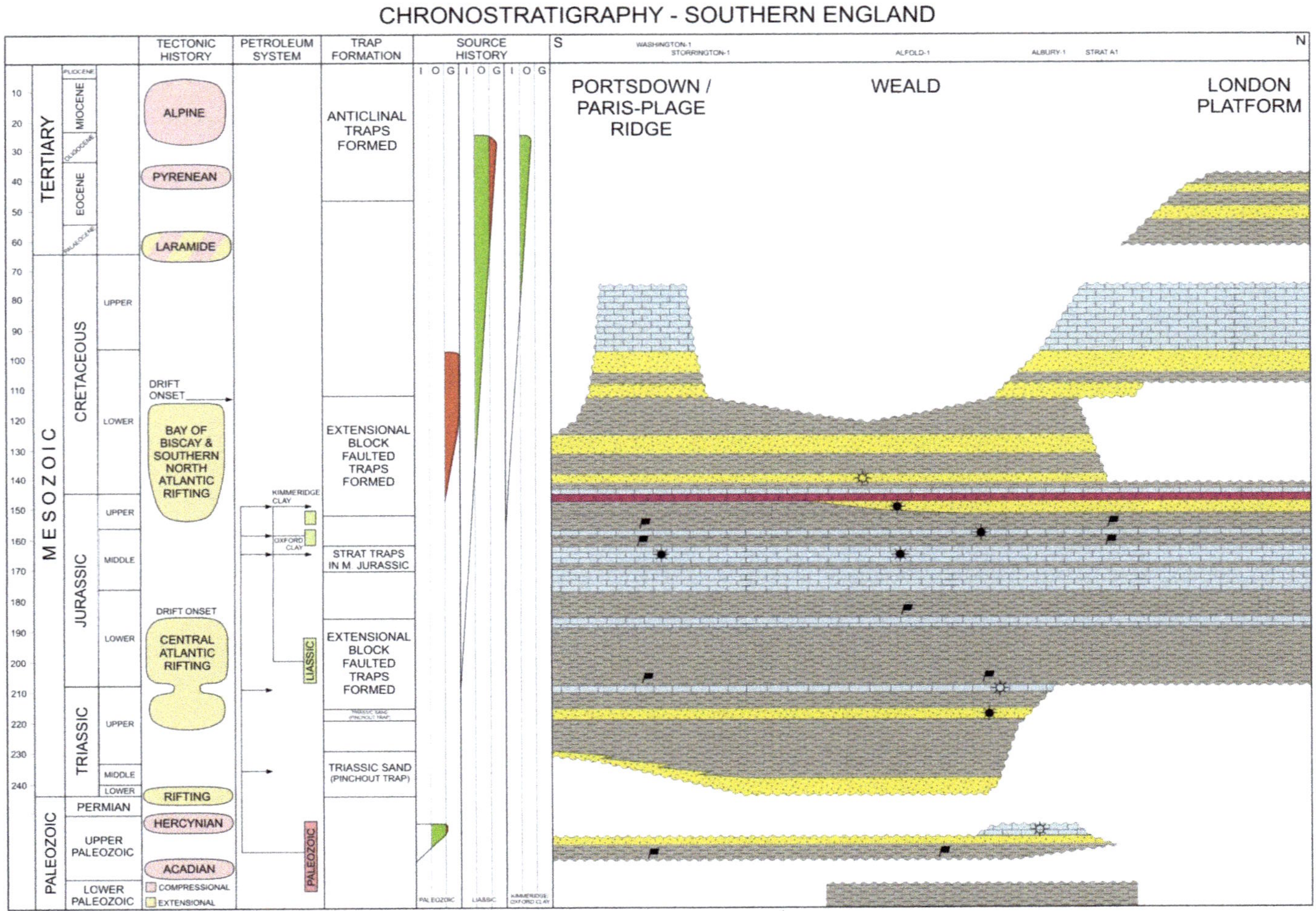

Fig. 3. Chronostratigraphy of the Weald Basin, making use of numerous sources including Butler & Pullan (1990) and Hawkes *et al.* (1998).

Table 1. *Gas composition (the details are shown in bold type when associated with oil)*

	C1 (%)	C2 (%)	C3 (%)	C4 (%)	C5 (%)	N_2 (%)	CO_2 (%)
Albury	97.6	0.6	0.4	0.4	0.2	0.9	
Ashdown	92.2	0.2				7.6	
Bletchingley	79.5	3.5	2.3	1.6		12.8	
Bolney	86.5					10.7	
Godley Bridge – Port	**85.6**	**5.1**	**2.8**	**1.2**	**0.4**	**4.2**	**0.2**
Godley Bridge – Great Oolite	84.3	6.7	2.7	1.2	0.5	3.5	0.4
Godley Bridge – Inferior Oolite	82.9	6.9	3.2	1.5	0.7	4.1	0.3
Heathfield	93.4	2.3	1.6			2.7	
Henfield	73.8	6.9				19.3	
Humbly Grove – Great Oolite	**81.5**	**5.6**	**1.9**	**0.7**	**0.3**	**9.6**	**0.1**
Humbly Grove – Rh	82.0	6.5	3.8	1.2	0.3	5.4	0.4
Palmers Wood	**83.7**	**4.4**	**5.2**	**1.2**		**2.2**	**0.1**
Storrington	**69.4**	**14.5**	**9.0**	**1.0**		**2.7**	**4.0**

Tertiary, possibly as a result of remigration from existing traps during the uplift.

The gas found at Humbly Grove is in a structure that predates the Aptian unconformity (see Fig. 8) and has not undergone any Tertiary reactivation (Hancock & Mithen 1987; Trueman 2003). Unlike most of the Weald Basin Middle Jurassic Great Oolite fields, at Humbly Grove a gas cap exists at the Great Oolite level, but gas is also trapped in an underlying Triassic Penarth (Rhaetic) reservoir. Since the trap seems to have been formed by fault movement during the Wealden and early Lower Greensand (Fig. 8), the gas could have migrated into the structure from mid Cretaceous times on. This does not rule out a Tertiary charge; however, it would imply a very limited charge volume at that time since the trap is not gas filled to spill. In order to better constrain this timing, the diagenetic history of the gas reservoirs was reviewed.

In the Humbly Grove Oil Field, the Great Oolite carbonate reservoir is characterized by an upper, high-permeability oil- and gas-bearing zone, which overlies a poor-permeability oil-bearing reservoir (Fig. 9). This interface is sharp and occurs at a depth of 3395 ft (1035 m) subsea. Extensive cementation occurs below, with porosity reduced, on average, by 2.6%, but average permeability is reduced from 64 mD in the upper good zone to 0.6 mD in the lower poor zone (see Fig. 10). In addition, the contact is marked by a zone of sphalerite enrichment (Sellwood & Evans 1986; Heasley *et al.* 2000).

As can be seen from the structure map (Fig. 9), which is constrained by seismic and well control, the permeability contrast surface is coincident with the current structure and is almost flat. It has therefore undergone little later structural modification. The permeability contrast surface is thought to represent a palaeo-gas–water contact (Carless Exploration Ltd 1983).

In an attempt to gain an understanding of the controls on these reservoir qualities, the then operator (Carless Exploration) undertook a series of detailed diagenetic studies (Sellwood & Evans 1986, 1987; Sellwood *et al.* 1989*a*, *b*). These studies showed several diagenetic phases.

The first phase was an early, near-surface diagenesis, which lithified the sediments, and this was followed by a second phase characterized by the deposition of cements from hot, saline fluids (95–110°C, 18–19.5% NaCl), whose origin is unclear. These cements include non-ferroan calcite and saddle dolomite, with sphalerite and barytes being deposited in fractures and fissures and at the palaeo-hydrocarbon contact (Heasley *et al.* 2000). Studies on other wells in the basin (Sellwood & Evans 1986; McLeod 2002) show that this hot water flush was a regional phenomenon. The high temperatures (based on the present-day geothermal gradient of 2.2°F/100 ft (4.0°C/100 m) suggest an origin from depths in excess of 7500 ft (2286 m). Well and seismic data over the field show that the Variscan surface lies currently at around 4770 ft (1454 m). Subsea depth reconstructions suggest that during the Albian, the Variscan surface would lie at depths of 3750 ft (1143 m); this would indicate that the fluids originated from sediments below the Variscan unconformity in the Humbly Grove area or that the geothermal gradients were significantly greater in the past, for which the source-rock maturity data show no evidence. An origin from below the Mesozoic requires fault migration pathways due to the presence of Jurassic shale aquacludes, and this suggests fluid movement from early Cretaceous times onwards.

Subsequently, there was a change from waters with high salinities and temperatures to waters characterized by lower temperatures and salinities (40–80°C, 6.5–15% NaCl) and a third diagenetic phase,

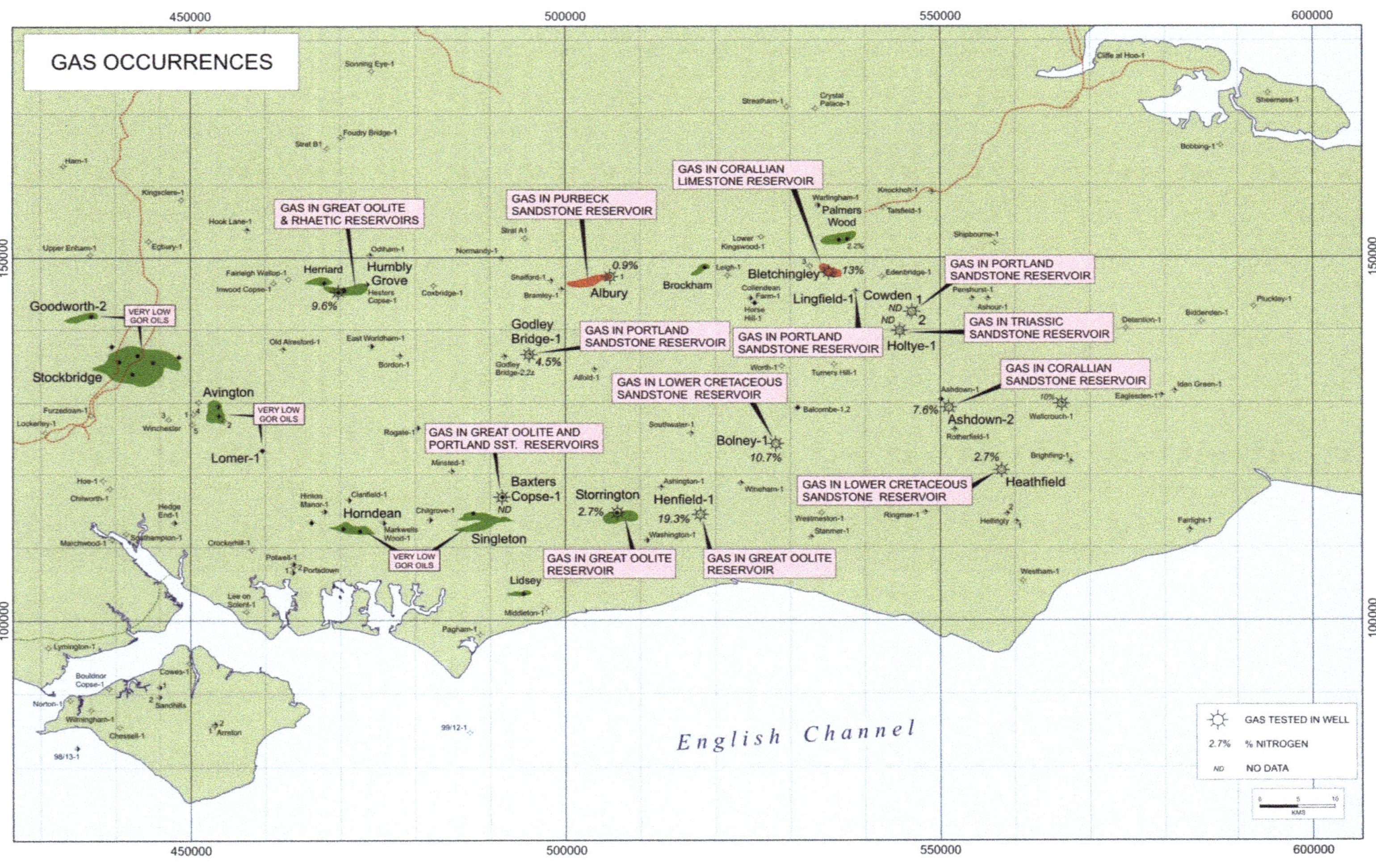

Fig. 4. Gas occurrences. Compiled from released well data.

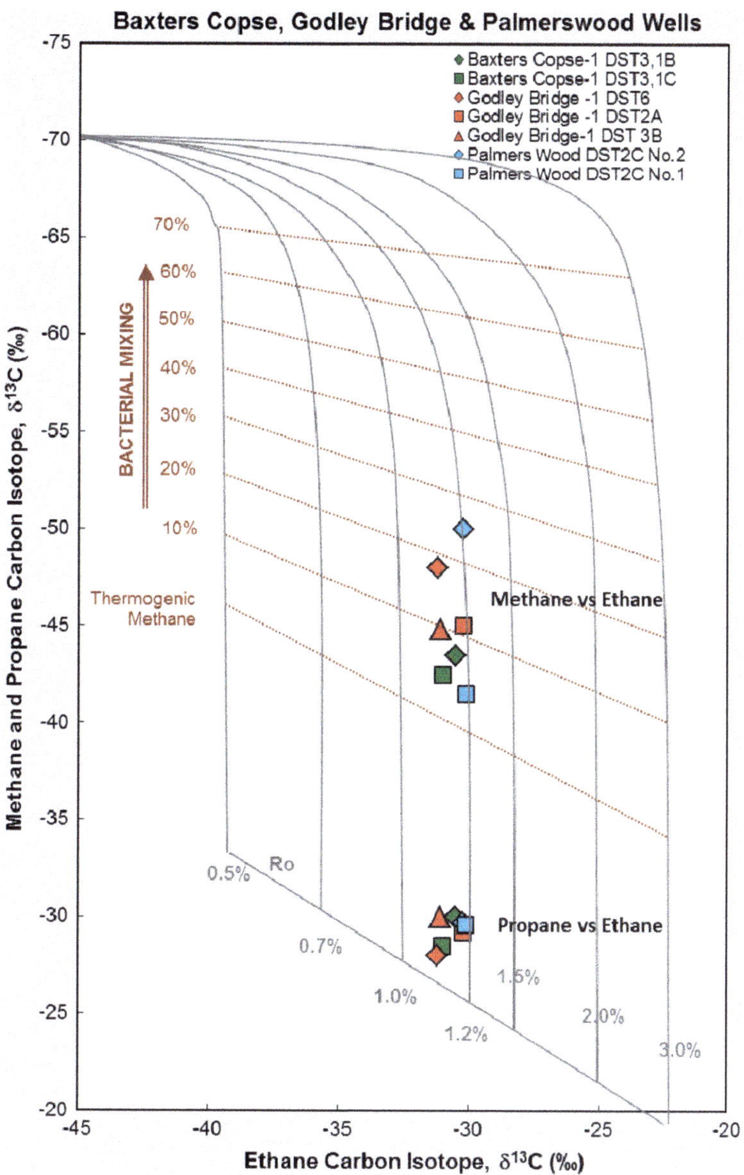

Fig. 5. Schoell plot for Weald Basin gases. Based on Conoco (UK) Ltd (1986).

which led to the deposition of mildly ferroan, coarse spar calcite cements. This diagenetic phase was the main Great Oolite porosity destruction event. Fluid-inclusion studies reveal that these cements are rich in oil inclusions, so it is suggested that the Jurassic-sourced oil was emplaced at the same time (Sellwood *et al.* 1989*a*; Heasley *et al.* 2000). Regional porosity mapping of the Middle Jurassic Great Oolite in the Weald (Butler & Pullan 1990) indicates that porosity loss is related to the depth of burial, and the area of maximum loss is coincident with the zone of maximum burial during the late Cretaceous. Since the oil emplacement is associated with the second cementation phase and this appears to be coincident with burial in the Upper Cretaceous, it would imply that the oil charge was coincident with Upper Cretaceous burial and predated the Tertiary basin inversion.

The Rhaetic gas reservoir is a hydrothermally altered limestone conglomerate and calcarenite that has been chertified. A study by the field operator

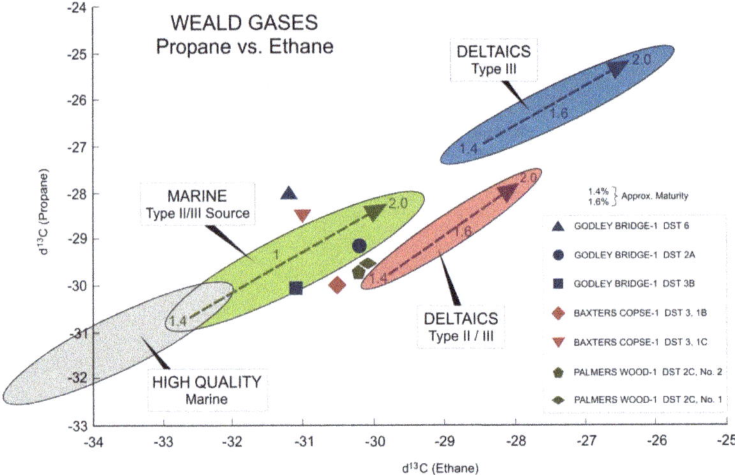

Fig. 6. Isotope plot of ethane v. propane. Based on Conoco (UK) Ltd (1986).

(Carless Exploration 1983) showed that, where gas bearing, the reservoir appears to have undergone a similar high-temperature diagenetic episode with sphalerite deposition, as seen in the Great Oolite. On the basis of these studies, it is postulated that the mineralization in the Rhaetic and Great Oolite reservoirs was coeval. A similar diagenetic alteration is found in rocks at outcrop in the Lower Lias- and Inferior Oolite-aged Harptree Beds, on the flanks of the Mendip Hills, where it occurs in association with the Biddle Fault, and is possibly post-Albian in age (Green & Welch 1965).

The gas charge is thought to have occurred before the deposition of the mildly ferroan, coarse spar calcite cements (third diagenetic phase). It is considered that this early gas charge reached the Humbly Grove trap, filling the Great Oolite reservoir with gas, preserving the permeability and also charging

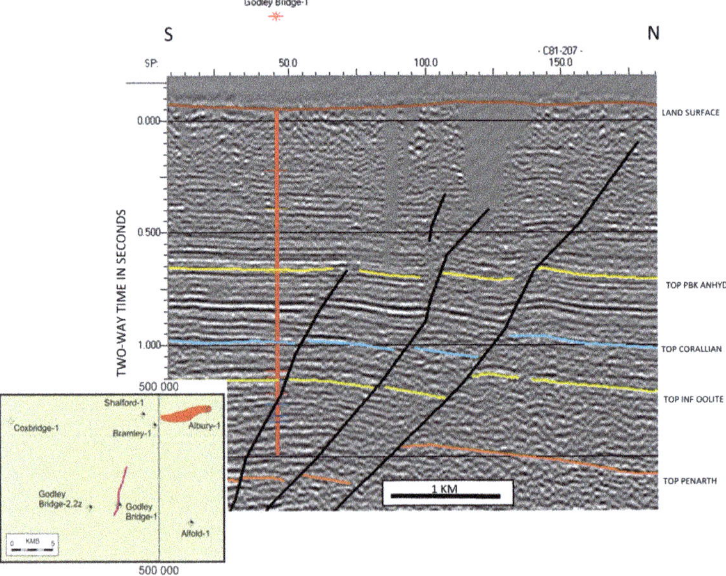

Fig. 7. North–south seismic line across the Godley Bridge Field.

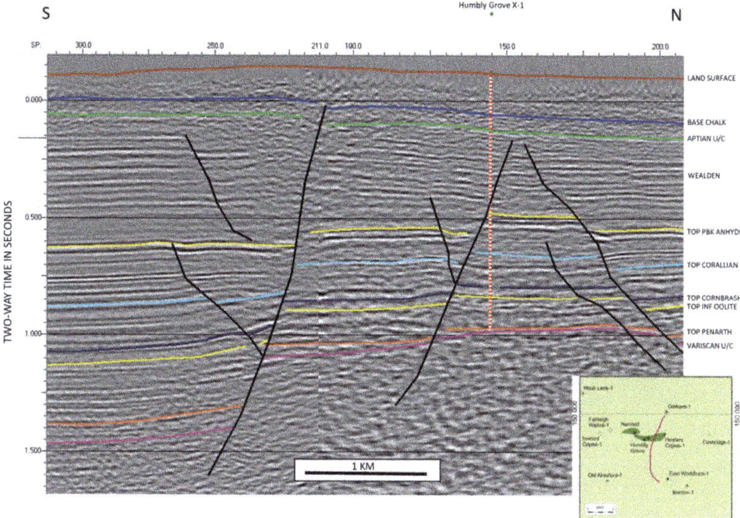

Fig. 8. North–south seismic line across the Humbly Grove Field. (U/C, unconformity.)

the Rhaetic reservoir. Later burial and compression is thought to have confined the gas to the crestal areas of the reservoir, allowing space for an oil charge into the preserved high-permeability reservoir. A later Upper Cretaceous hydrocarbon charge from Jurassic source rocks then created the main oil accumulation.

Interpretation suggests that this gas charge occurred after the Aptian, associated with the main regional faulting event which occurred before the Aptian unconformity (Chadwick & Evans 2005). Most Weald Basin Great Oolite fields do not have a gas cap but also do not have a high-permeability reservoir present. The Herriard Field, just to the NW of Humbly Grove, does not have a gas cap or high-permeability reservoir despite an apparently identical structural history.

Interpretation suggests that the basin had an early Cretaceous gas charge, followed by an oil charge later in the Cretaceous, with remigration of both oil and gas in the Tertiary. This remigration is demonstrated not only at Godley Bridge but also at the Stockbridge, Singleton and Storrington fields (Trueman 2003).

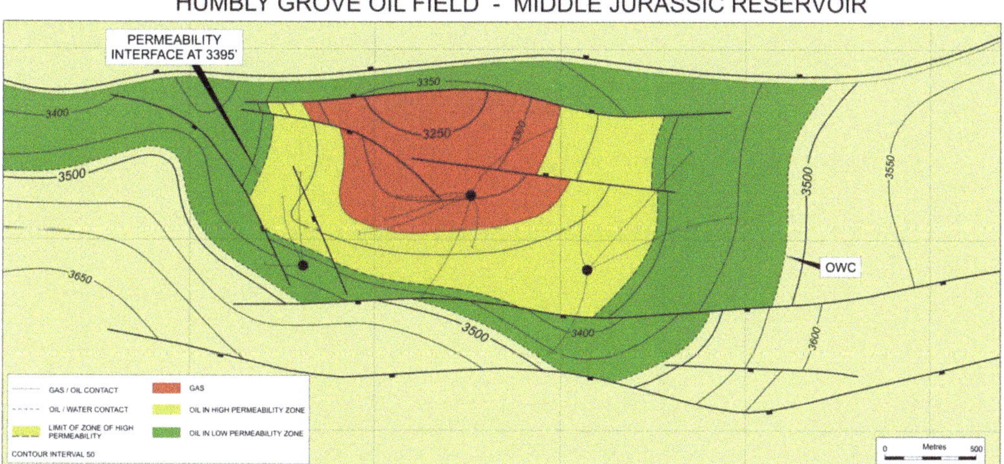

Fig. 9. Humbly Grove: depth structure map at the Top Great Oolite Reservoir. Adapted from Hancock & Mithen (1987).

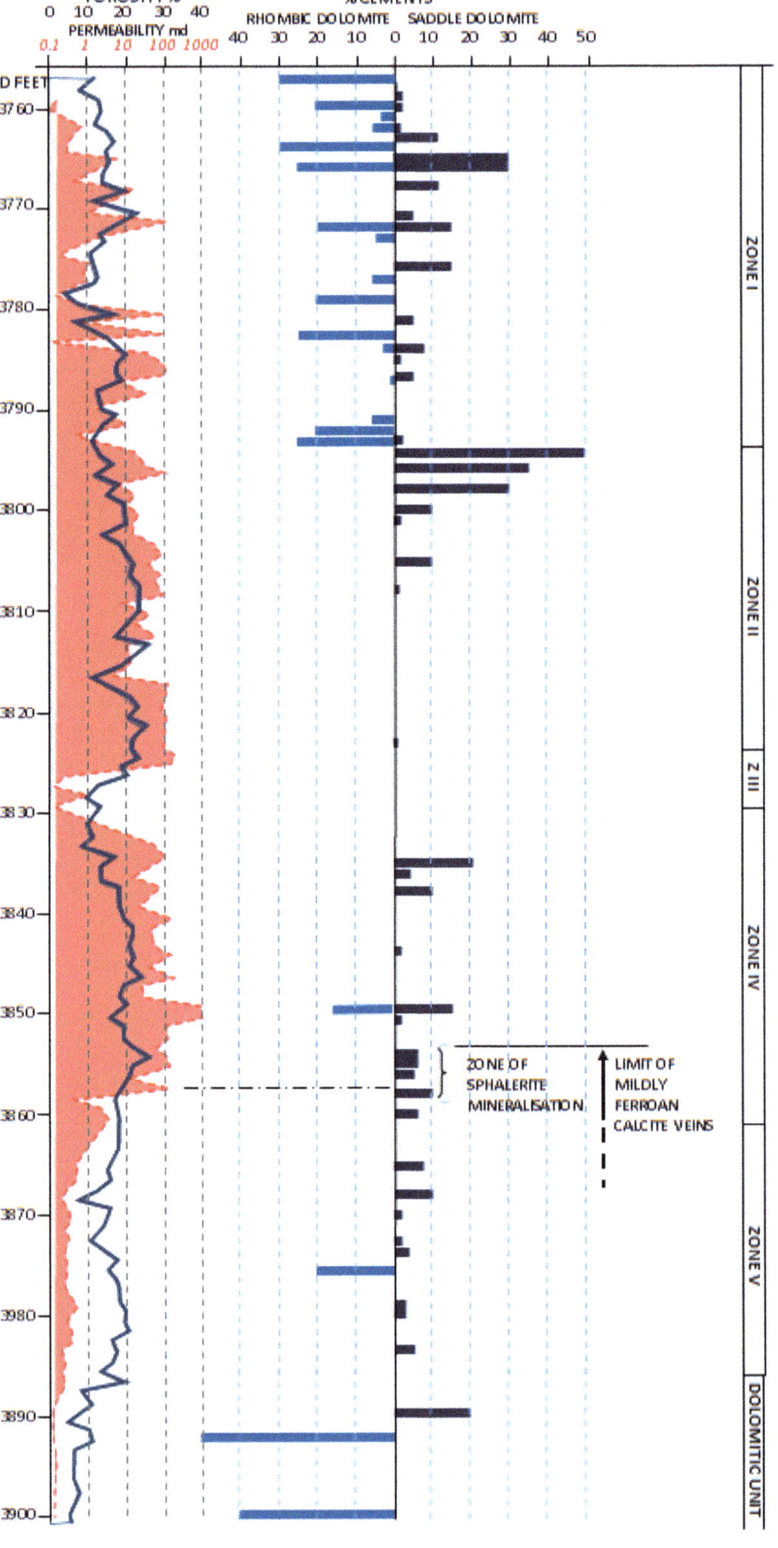

Fig. 10. Humbly Grove – Great Oolite: core porosity–permeability plot. Redrawn from Sellwood & Evans (1986).

Origin of gas

Several theories have been put forward for the origin of this gas (Ebukanson & Kinghorn 1986; Conoco (UK) Ltd 1986; Andrews 2014):

- the gas comes from mature Lower Jurassic shales, deeply buried in the basin centre;
- the gas encountered is biogenic rather than thermogenic;
- the gas has been released by ex-solution caused by uplift in the Tertiary;
- gas reserves in the Weald Basin were not sourced from the Jurassic but, rather, from a deeper Paleozoic source rock interval.

These theories are not mutually exclusive: the gases may have had contributions from different source rocks and may be mixed.

Jurassic source

Thermogenic gas being derived from a Jurassic source is compatible with the isotope data. However, Figure 11 shows that the gas occurrences are not coincident with areas where the Lower Jurassic shales are most mature.

The maturity modelling of the Liassic in Figure 11 is based on regional seismic mapping of the top Penarth event. In the Celtique Energie study, an unpublished interpretation was used in which synsedimentary thickening of the Liassic across the bounding faults in the basin centre had the effect of deepening the Liassic in this area. In order to compensate for the Tertiary inversion effect, the Top Cretaceous was flattened using the Top Chalk structure map of Butler & Pullan (1990). Individual wells were modelled using Platt River® BasinMod software and verified using existing well maturity studies in order to check the assumed temperature gradients. A series of dummy wells based on seismically controlled sections were used to calibrate the untested basin centre (Fig. 12). This work shows that only in the very centre of the basin have the Lower Liassic shales reached the gas window. An alternative interpretation by the British Geological Survey (BGS) (Andrews 2014) does not have the Liassic sediments thickening across the bounding faults, so the Penarth is shallower in this area and the Lower Lias does not reach the gas window. Maturity modelling by the Central Weald Group (Magellan Petroleum (UK) Ltd 2012) shows that the Lower Jurassic interval only reached the gas window in the latest Cretaceous or Tertiary, which does not agree with the timing derived from the Humbly Grove data. This modelling work also suggests that the source has not been buried sufficiently to have started the generation of nitrogen, although it could be argued that this was caused by the mixing and contamination of the hydrocarbon gas pool by a nitrogen-rich flush from a deeper source.

The model for the gas to have been derived from mature Jurassic shales faces two additional problems. The distribution of the gas away from the potential gas kitchen area implies widespread lateral migration of over 15 km for the gas discoveries on the north flank. This is problematic since the Liassic interval lacks any regional carrier beds and has extremely poor reservoir development. An additional problem is that where seen in the wells in the area, the Liassic shales are generally of fair hydrocarbon-generation quality at best. It can be argued that better source quality will exist in the basin centre, which is undrilled. Work by the BGS has concluded that the Lower Lias has a minimal source potential (Andrews 2014).

Biogenic origin

The available isotopic data (Conoco (UK) Ltd 1986) suggest that most of the gas is thermogenic and therefore does not have a biogenic origin. A biogenic origin would also not explain the high nitrogen content seen in the gas.

Shallow gas has been discovered in the basin centre at Heathfield and Bolney, in Lower Cretaceous reservoirs; but, unfortunately, no samples are available to ascertain whether these gases have a biogenic origin.

Peak biogenic methane production occurs at temperatures of 40°C (Clayton 2009), which equates to a burial depth of approximately 3000 ft (914 m). Analysis suggests that only the Lower Cretaceous in the centre of the basin was in the zone of peak biogenic gas productivity at end Cretaceous times, prior to Tertiary uplift.

Isotopic data indicate that the Godley Bridge-1 gas discovery, also located in the basin centre, has gas that has been contaminated with a biogenic component (see Fig. 5). Figure 13 shows the age of the strata which lie at the base of the zone of pasteurization (70°C) at the end of the Cretaceous, prior to uplift. Analysis of the Portlandian reservoir in Godley Bridge-1 indicates that it most likely did not reach pasteurization temperatures of 70°C. The temperature of the reservoir at the present day is approximately 39°C. In Brockham-2, the analysis indicates that the Portlandian reservoir only reached temperatures of approximately 50°C prior to uplift, which would explain the slight biodegradation of the oil at that locality.

Exsolution

An alternative explanation for the occurrence of gas in the Weald Basin is that, during and following the Tertiary uplift, gas was released from porewaters

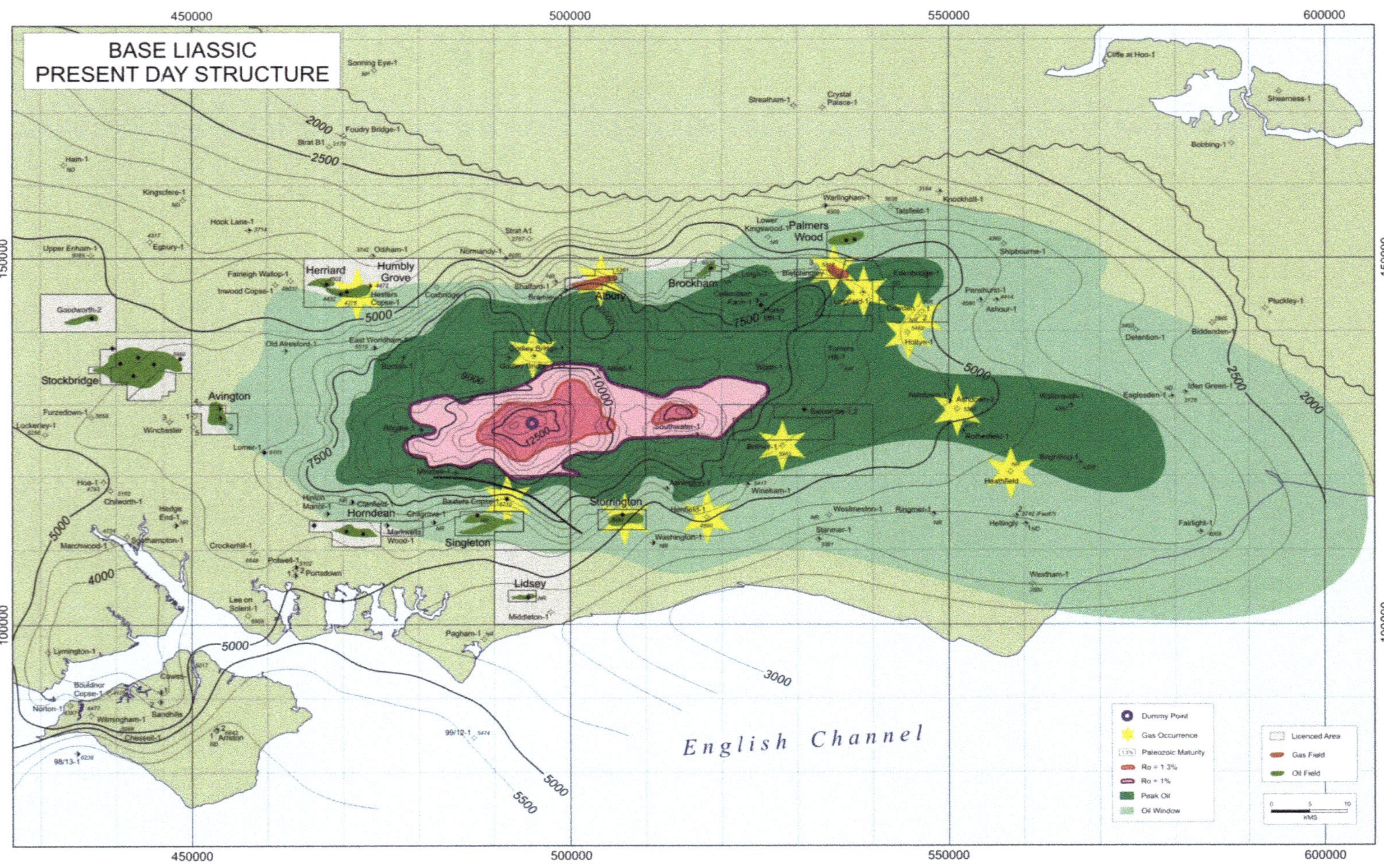

Fig. 11. Gas occurrences placed on a map showing the maturity at the base of the Lower Jurassic. Based on unpublished maturity modelling by Celtique Energie.

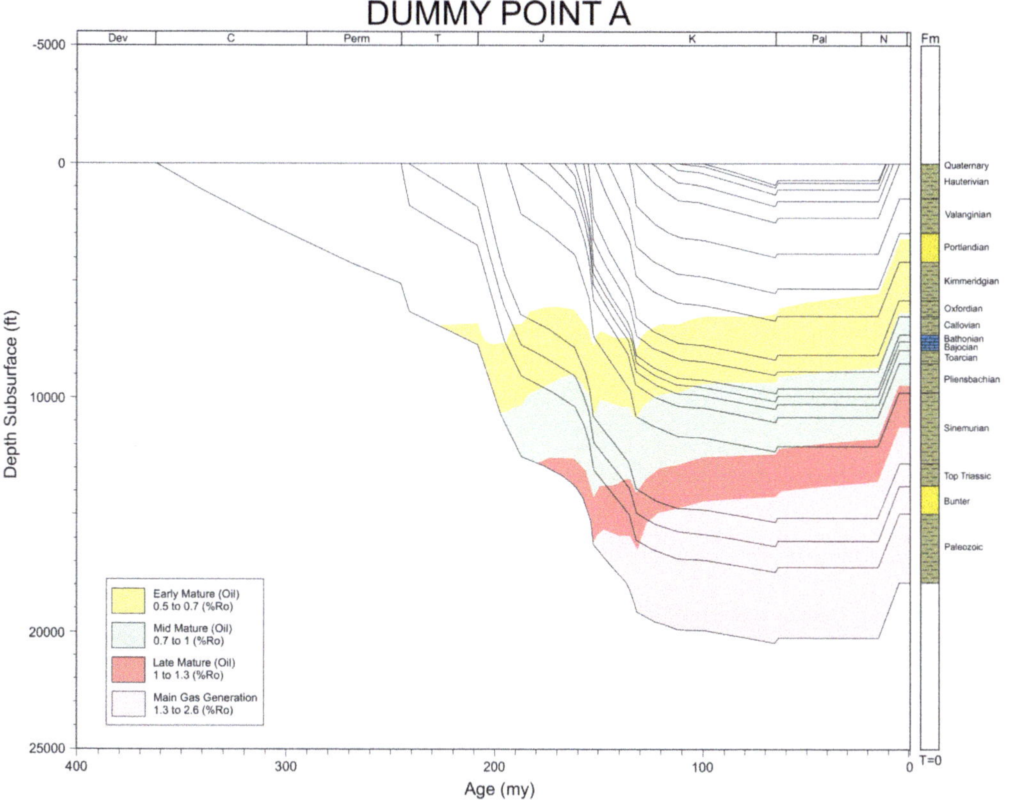

Fig. 12. Burial depth plot for the deepest part of the Weald Basin. Based on unpublished maturity modelling by Celtique Energie.

through ex-solution of dissolved gas. This was a model proposed by Conoco (UK) Ltd (1986), in which it was thought that, during burial, Jurassic strata would have been generating wet gas, some of which would have been dissolved in the pore-waters. During inversion, the associated reduction in confining pressure and temperature would significantly decrease the solubility of methane and the dry gas in solution would ex-solve. According to the Conoco modelling, gas ex-solved from pore-waters is estimated to be more than 8 Tcf (trillion cubic ft). This gas could have displaced previously trapped oil to the basin peripheries. This ex-solved gas would be much drier and preferentially rich in methane due to its higher solubility in water in comparison to ethane/propane. Furthermore, all the gases would have a carbon isotope signature indicative of thermogenic generation associated with the oil. It was believed by Conoco that the pore-waters were saturated with methane at an early stage, which would explain the difference in maturity compared to the ethane and propane.

The ex-solution theory does not account for all the observations made. If this were the mechanism, one would expect to see more gas occurrences in the area of greatest uplift. This is not the case, with the majority of gas occurring at the peripheries of the basin. A further problem is that the oils seen are generally low GOR, non-gassy oils. This may suggest that little gas was generated in the first place, although it is possible that these low GOR oils have already ex-solved volumes of gas originally in solution. However, although the gas is dry, consistent with the ex-solution model, the high nitrogen content cannot be explained since the source rocks did not reach sufficient maturity to generate nitrogen.

Paleozoic origin

Most oil and gas exploration companies have taken the view that the Paleozoic of southern England does not have any petroleum potential due to overmaturity caused by the Variscan Orogeny. Figure 14 demonstrates the oil and gas shows encountered in Paleozoic rocks of southern England. Paleozoic oil and gas shows also occur to the north of the Weald Basin, on the London Brabant Massif. Such shows are located over 30 km away from mature

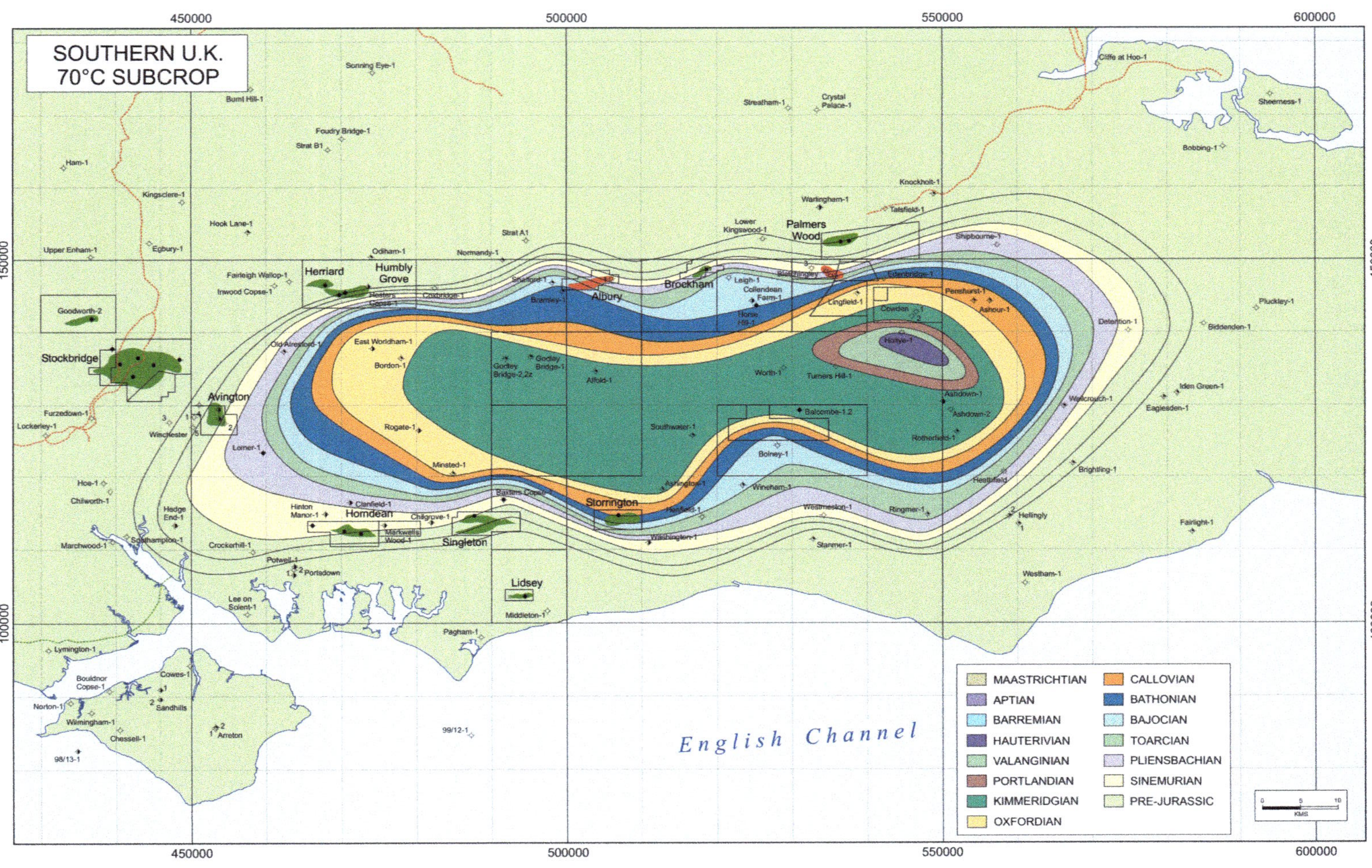

Fig. 13. Formations subcropping the pasteurization floor. Based on unpublished maturity modelling by Celtique Energie.

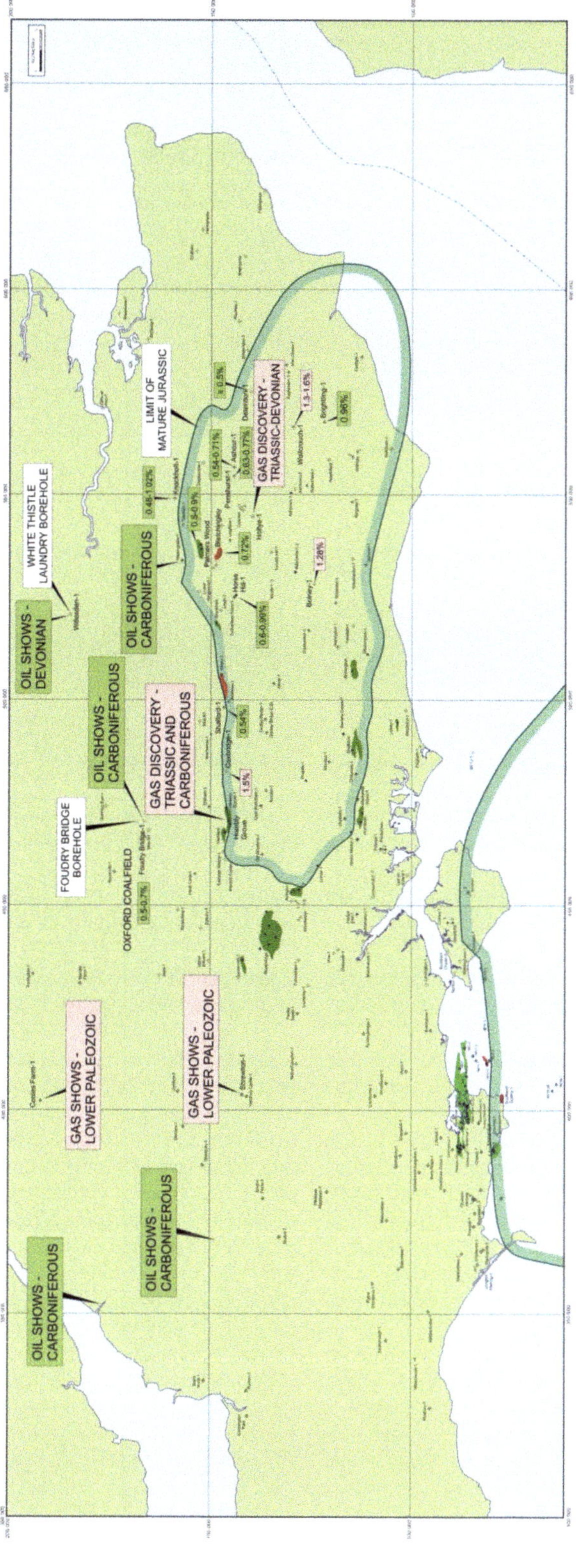

Fig. 14. Southern England Paleozoic hydrocarbon occurrences. Based on reports from released wells.

Jurassic source rocks, whilst those to the west occur over 100 km from the nearest mature Jurassic source, and it is therefore difficult to explain all the occurrences as a result of a charge from the Jurassic. Furthermore, since the oil occurrences at Foudry Bridge-1 occur 470 ft (143 m) below the Variscan unconformity surface and in the White Thistle Laundry borehole, near Willesden in London, the shows occur 620 ft (189 m) below the Variscan unconformity surface, it is difficult to explain using a Jurassic source and downwards migration. The presence of a Paleozoic source system is implied. Unfortunately, most of these wells are old and no samples are available, so no attempt could be made to type the occurrences to a Paleozoic source interval.

The source potential of the Paleozoic section in the Weald is poorly known, since of the wells that penetrated the Paleozoic most only reached the Carboniferous and many only penetrated a thin (<100 ft) (30 m) interval of the Paleozoic. Many of the intervals were defined on lithological grounds (Fig. 15), since dating is limited and often difficult due to widespread caving.

Paleozoic source rocks

In order to gain an understanding of the source potential of the Paleozoic, a review of geochemical data from wells and outcrops in southern and central England was undertaken. This study reveals that a number of potential source rock horizons exist in the Cambrian, Devonian and Upper Carboniferous intervals (Fig. 16).

The Cambrian shales are known as the White Leaved Oak or Bentley Foot Shales at outcrop in the Midlands and the Welsh Borderlands, and have total organic carbon (TOC) values of 4–7% (Fig. 17). The interval is a proven source rock in the Baltic Sea area, North America and North Africa. Although a potential source to the NW, no shales of this age have been encountered in wells in the Weald Basin.

Two wells to the west, Shrewton-1 and Cooles Farm-1, penetrated a very thick section of early Ordovician (Tremadocian) age but, although these rocks had minor hydrocarbon shows, geochemical analysis reveals that the interval has little to no source potential with only lean shales encountered.

Silurian shales have been encountered in wells in the east and north of the Weald. Analysis by Celtique Energie (Harriman 2010) and other authors (Lamb 1983; Andrews 2014) have failed to find any significant source potential within the intervals, which are of post-Lower Llandovery age. However, work in North Africa and the rest of Europe has shown that the best source development occurs in the basal Silurian Lower Llandovery (Boote *et al.* 1998; Lüning *et al.* 2003), an interval not encountered in the drilling to date. Therefore, the possibility exists of encountering these older shales with source potential in the Weald.

The Devonian section can be subdivided into pre- and post-Acadian intervals. Unfortunately, few wells penetrate a significant section of the Devonian, except those to the north on the London Brabant Massif. Other than outcrops to the west of the Weald Basin, the most complete sections have been penetrated in the eastern part of the basin. Shipborne-1 encountered a 600 ft (188 m) section of Old Red Sandstone facies, the upper parts of which Shell identified as being of Eifelian–Fammenian age, which is considered to be post-Acadian Middle–Upper Devonian, overlying Lower Devonian, Pridoli and Ludlovian rocks. Brightling-1, also in the eastern part of the Basin, penetrated a 602 ft (183 m) section of Old Red Sandstone facies dated as probably Lower Devonian (Falcon & Kent 1960). Devonian marine shales were deposited during the Givetian–Frasnian transgression, following the Acadian unconformity, and inner–outer shelf conditions existed to the north with deeper marine conditions in the south. Butler (1981) shows marine Frasnian to extend from the SE up to a line running from north Essex to southern Pembrokeshire, with deeper marine shales over much of the Weald. As part of the current study, these shales have been dated and analysed in a number of wells and were found to have source potential in Coxbridge-1 (Molyneux 2010*a*), where the section was reliably dated for the first time, and Detention-1 (Molyneux 2010*b*) (Figs 18 & 19). Uppermost Devonian sands and shales have been encountered in the Weald Basin either below the Variscan unconformity surface or conformably below the Lower Carboniferous limestones. These shales are often reddened and, where analysed, have no source potential.

In the Carboniferous, black shales are known from near the base of the Lower Carboniferous at outcrop: for example, in the Mendips (Black Rock Limestone). Numerous studies have shown that these shales do not have any source potential (Geochem Laboratories Ltd 1980; Lamb 1983; PetraChem Ltd 1986). To the SW, in the Wessex Basin, a change from shallow-water carbonates to deeper-water shales is seen within the Lower Carboniferous sequence. However, in this area, these rocks are overmature and their potential cannot be established (Cornford *et al.* 1987).

Coals are developed in the Upper Carboniferous Coal Measures, and are well known to the east in the Kent Coalfield and to the north in the Berkshire–Oxfordshire Coalfield (Foster *et al.* 1989). These coals would represent an excellent gas source (Kettel 1989). No Upper Carboniferous coals have been seen in the Weald Basin south of

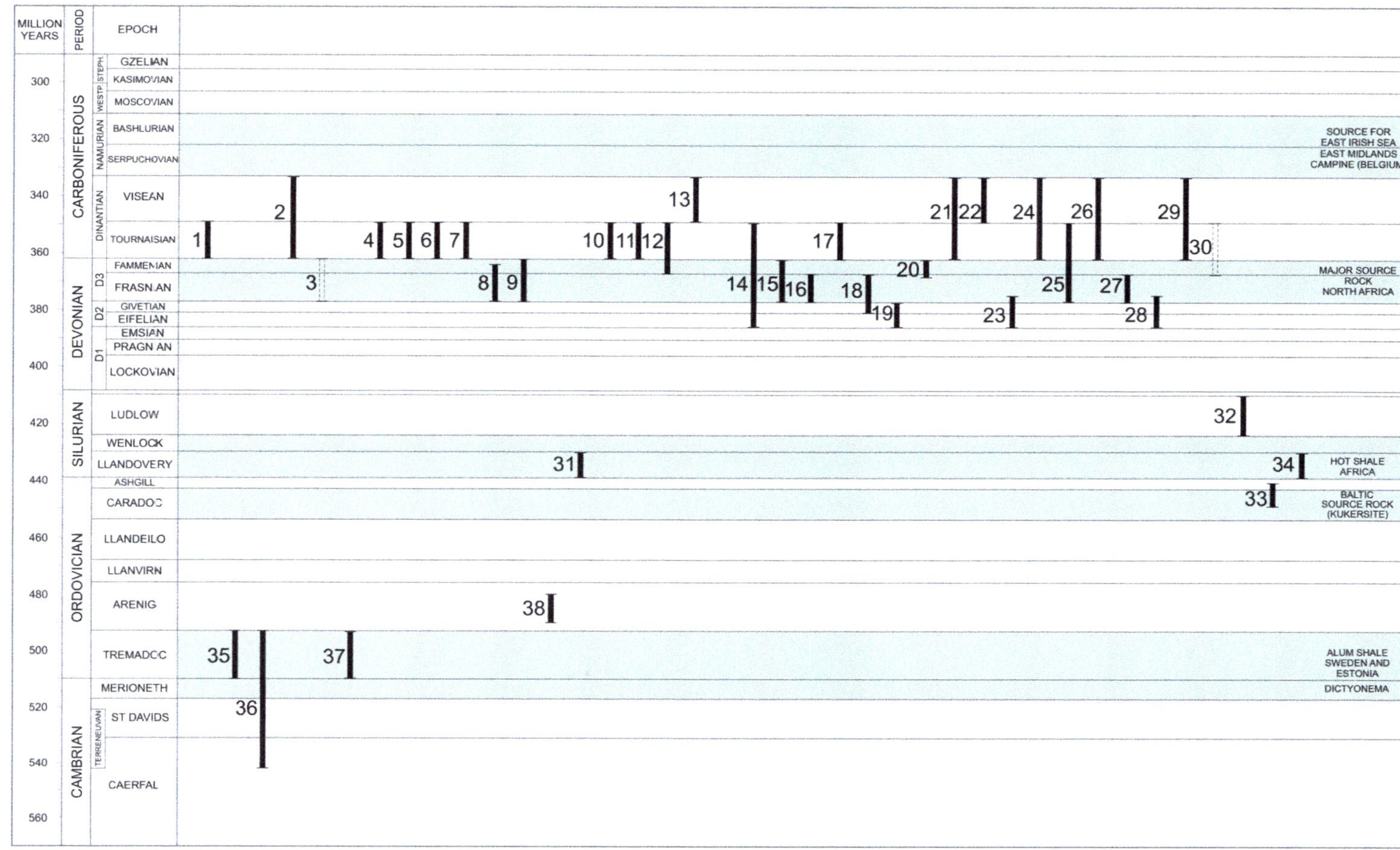

Fig. 15. Paleozoic well penetrations: 1, Devizes 1; 2, Farleigh Wallop 1; 3, Lomer 1; 4, Herriard 1; 5, Humbly Grove 1; 6, Humbly Grove 2; 7, Hesters Copse 1; 8, Coxbridge 1; 9, Baxters Copse 1; 10, Albury 1; 11, Brockham 1; 12, Horse Hill 1; 13, Wineham 1; 14, Bolney 1; 15, Stanmer 1; 16, Bletchingley 1; 17, Warlingham 1; 18, Palmers Wood 1; 19, Tatsfield 1; 20, Holtye 1; 21, Penshurst 1; 22, Ashour 1; 23, Shipbourne 1; 24, Hellingly 2; 25, Westham 1; 26, Wallcrouch 1; 27, Brightling 1; 28, Detention 1; 29, Iden Green 1; 30, Fairlight 1; 31, Shalford 1; 32, Biddenden 1; 33, Bobbing 1; 34, Chilham 1; 35, Shrewton 1; 36, Cooles Farm 1; 37, East Worldham 1; 38, Strat A1.

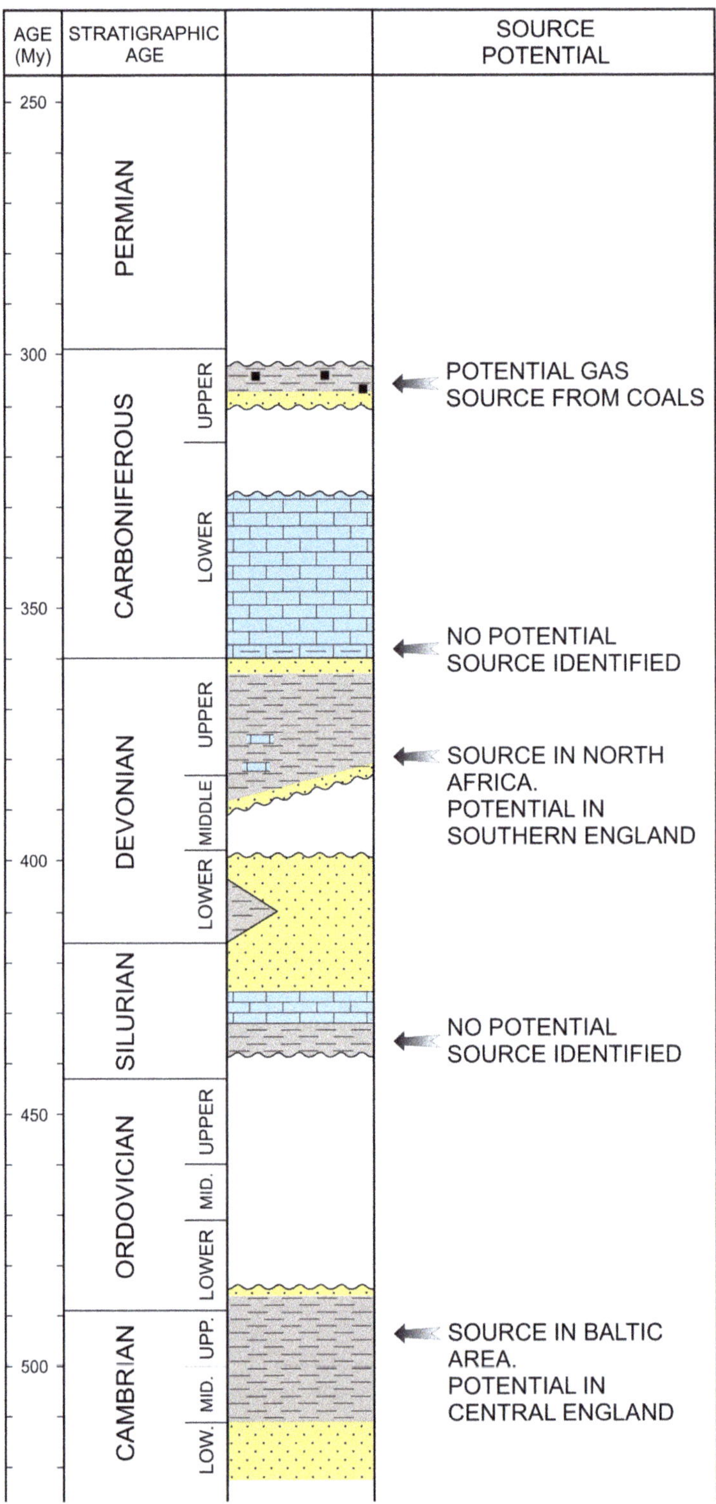

Fig. 16. Potential Paleozoic source rock intervals.

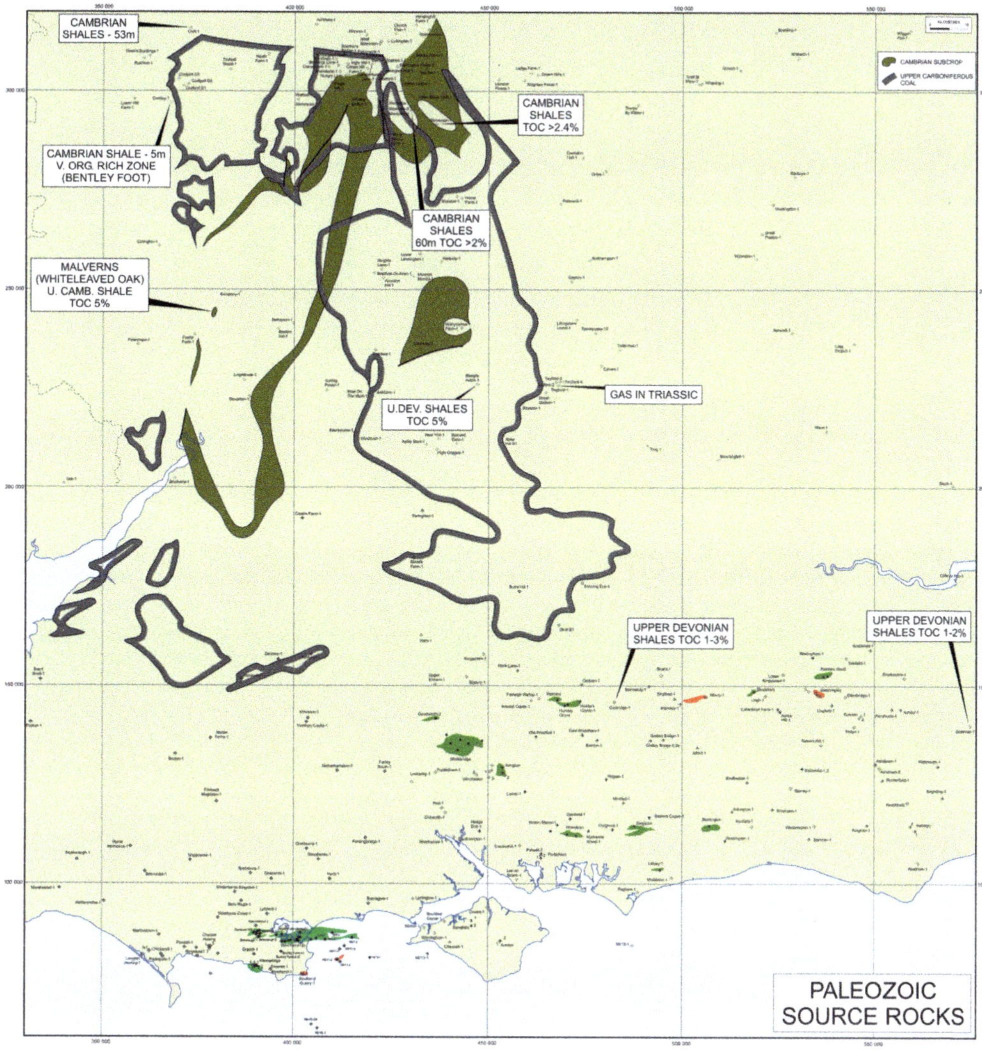

CAMBRIAN SHALES - 53m
CAMBRIAN SHALE - 5m V. ORG. RICH ZONE (BENTLEY FOOT)
CAMBRIAN SHALES TOC >2.4%
CAMBRIAN SHALES 60m TOC >2%
MALVERNS (WHITELEAVED OAK) U. CAMB. SHALE TOC 5%
GAS IN TRIASSIC
U.DEV. SHALES TOC 5%
UPPER DEVONIAN SHALES TOC 1-3%
UPPER DEVONIAN SHALES TOC 1-2%
CAMBRIAN SUBCROP
UPPER CARBONIFEROUS COAL
PALEOZOIC SOURCE ROCKS

Fig. 17. Paleozoic source rock occurrences.

Coxbridge 1 (L.Famm-Frasnian)

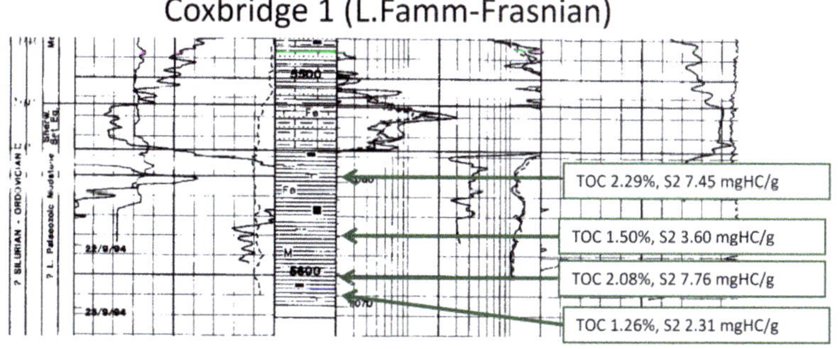

Fig. 18. Devonian section: Coxbridge 1. Released well data.

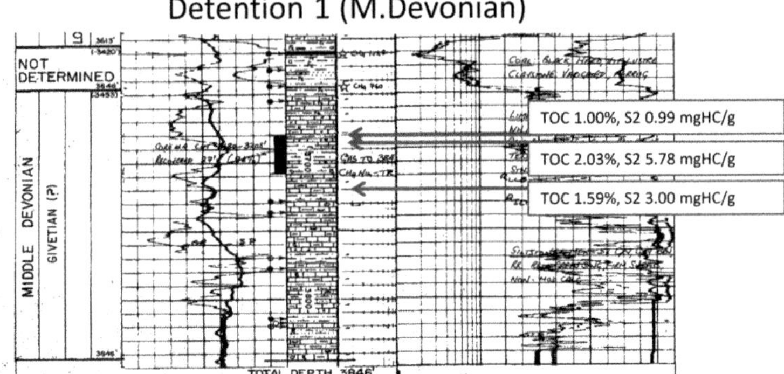

Fig. 19. Devonian section: Detention 1. Released well data.

the northern bounding fault. This may be a sampling problem, since most wells are located on palaeo-highs; but with over 37 well penetrations, the absence of coals suggests that these rocks are not present and were either not deposited or were eroded prior to Triassic deposition. In a number of wells to the east, thin coals of possibly early Jurassic or Upper Triassic age have been encountered. In the Southern England Basin only one well, the Westbury borehole located south of the Bristol–Radstock Coal Basin, encountered Coal Measures south of the northern basin-bounding fault, often referred to as the Variscan Front. There is some evidence for the development of isolated basins of possible late Carboniferous or early Permian red-bed clastics in the southern Weald (e.g. Middleton-1).

Regional distribution

In order to get an understanding of the distribution of potential Paleozoic rocks, it is necessary to define the Paleozoic subcrop to the Variscan unconformity surface.

In southern England, the effects of several tectonic episodes during the Paleozoic control the resultant subcrop. The Caledonian Orogeny was characterized by deformation phases caused by two plate-docking episodes (Fig. 20). The first of these was the closure of the Tornquist Ocean (between Avalonia and Baltica) in the east in late Ordovician–early Silurian times, which was marked by the Shelvian unconformity when erosion down to the Precambrian took place in parts of the Midlands Microcraton. The second was the closure of the Iapetus Ocean (between Avalonia and Laurentia) to the NW in the late Silurian–early Devonian. The latter, Acadian, orogenic phase is marked by a pronounced unconformity below the Upper Devonian across southern England.

The first phase of the Variscan Orogeny seen in southern England began in the Middle Carboniferous and ended in the late Carboniferous–early Permian. It marked the progressive closure of the Rheic Ocean as the Avalonian and Armorican plates moved together. The deformation in southern England is thought to represent a north-verging east–west-trending thrust belt, with Paleozoic rocks being thrust northwards towards the foreland of the Midlands Microcraton–London Brabant Massif. The activity was periodic and reflected in the development of a series of unconformities but, overall, the deformation decreases in age northwards.

The new subcrop map of southern England (Fig. 21) is based on well data, seismic interpretation and coalfield mapping, and makes use of previous work carried out by Busby & Smith (2001). Seismic lines have been used where events can be mapped at Paleozoic levels outside the Mesozoic depocentres. Unfortunately, over much of the Weald Basin, the subcrop map is reliant on limited well control because of poor seismic penetration and, therefore, poor quality, and there are few deep well penetrations in the basin centre.

The problems in determining the Paleozoic structure beneath the Weald Basin are well illustrated at Horse Hill. The seismic data show beds dipping to the south, which depth conversion would tend to accentuate slightly; however, the dip-meter in the well demonstrates that the interval has reasonably consistent dips of 20°–35° to the north (shown in white in Fig. 22; see also Fig. 23).

In the NW of the Wessex Basin (the Western Sub-basin of the Southern England Basin: Hawkes *et al.* 1998), the Paleozoic section can sometimes be resolved on the seismic data and can be tied to deep well control. Figure 24 links the Yarnbury-1 and Shrewton-1 boreholes to the south, which encountered Lower Ordovician Tremadocian sequences below the Mesozoic (Whittaker 1980),

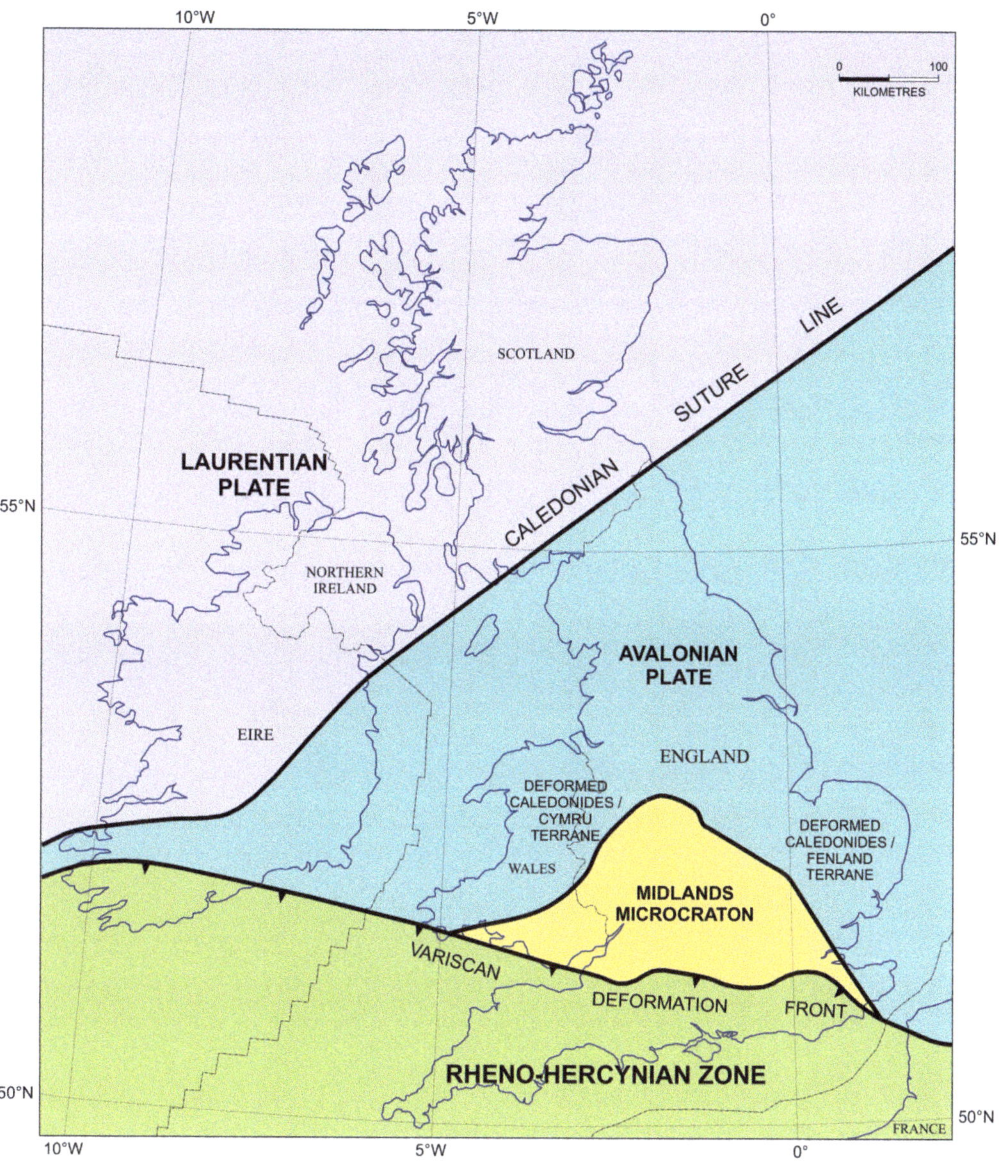

Fig. 20. Regional structural setting. Simplified from Chadwick & Evans (2005).

with the Devizes-1 borehole (*c.* 20 km to the north), which encountered Lower Carboniferous limestones below the Mesozoic, with an apparently conformable transition from the presumed Upper Devonian beneath. The line gives no indication of a major dislocation between the two subcrop areas and suggests that the Upper Devonian lies unconformably on the Tremadocian in this area. This would be consistent with outcrop information in the eastern Mendips and Tortworth inlier, where the Acadian unconformity has cut down at least to the Llandovery (Green & Welch 1965; Cave 1977), and with the gravity interpretation of the Vale of Wardour (Chadwick *et al.* 1981). The Tremadocian subcrops below the Variscan unconformity in East Worldham 1 and the Ordovician subcrops in Strat A 1, in the central part of the Weald Basin. Whereas the presence of the Tremadocian subcrop beneath the centre of the Weald might be considered to be due to deep erosion of younger beds during Variscan movements and the consequent loss of potential late Devonian source rocks, it seems more likely that Acadian erosion of

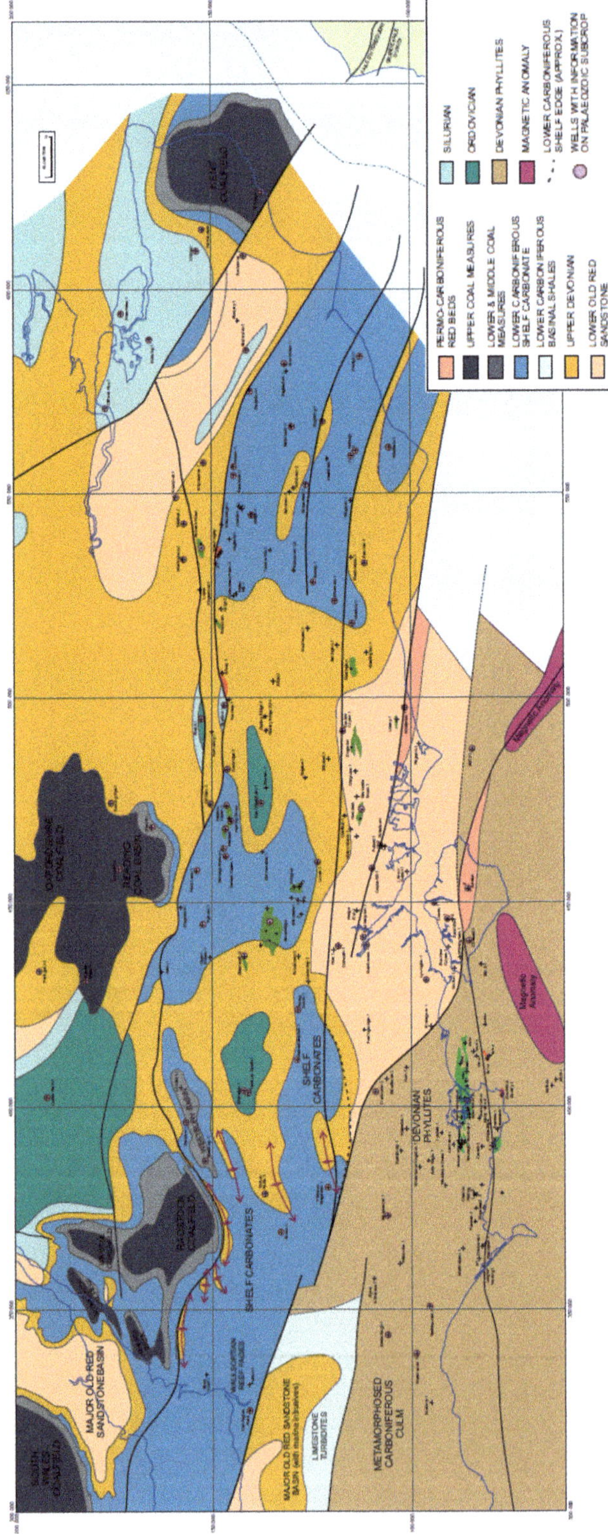

Fig. 21. Variscan subcrop map based on seismic mapping, well and coalfield data, and making use of Busby & Smith (2001).

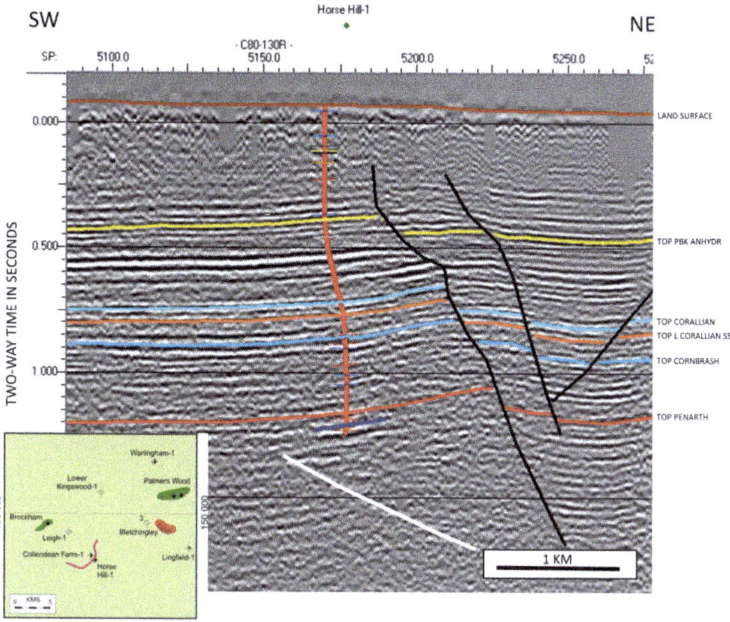

Fig. 22. Seismic line through the Horse Hill structure.

early Devonian and Silurian beds led the post-Acadian Frasnian unconformity to lie directly on Tremadocian rocks that were later exposed to the surface by minor Variscan erosion. This leads to the conclusion that there may be a wide distribution of potential source rocks in the late Devonian across east and central parts of the Weald Basin (Fig. 21).

The subcrop data in the Weald Basin suggest that the Upper Devonian and Lower Carboniferous transgressed over thick, folded but unmetamorphosed Lower Devonian, and that older Lower Paleozoic rocks deformed during Acadian movements. The entire section, including Upper Paleozoic post-Acadian sediments, then deformed during the Variscan movements. To the east, the metamorphosed Lower Paleozoic rocks of the Acadian-age Anglo-Brabant Deformation Belt underlie the Upper Devonian and Lower Carboniferous in the Kent Coalfield.

Paleozoic maturity

The available maturity data for the Paleozoic of the Weald Basin are shown in Figure 25 and demonstrate that the Paleozoic petroleum system is a valid source objective since it is not overmature in the Weald Basin. Also shown on this map is the extent of the Mesozoic depocentre, where burial is calculated to have allowed the restarting of hydrocarbon generation from the deeper buried Paleozoic source rocks during the Cretaceous.

The data in Figure 25 have been compiled from well samples near or at the top of the Paleozoic sequences, just below the Hercynian unconformity surface. Although of variable quality, the data demonstrate that the sediments in the north of the area are still in the oil window (VR of 0.5–1.1%), while those from more basinal wells indicate an increase in maturity and a move into the gas window (VR of 1.3–1.6%). There is no significant break in the maturity profile with respect to the overlying Mesozoic in the north and centre of the basin. However, to the south and SW, the effects of Variscan metamorphism are seen (Cornford *et al.* 1987). Since these values are measured at the top of the interval, they do not represent the maturity of the deeper Paleozoic, which would be higher and can be projected to reach the gas window. The majority of the data come from the post-Acadian Devonian and Carboniferous sequences. The maturity levels are commensurate with the results of the isotope analyses, as well as the nitrogen content of the gases.

Due to the absence of vitrinite, maturity of the Lower Paleozoic has been measured (Molyneux 2010*a*) using the acritarch alteration index (AAI), as well as spore coloration. These measurements were then converted to vitrinite equivalents (Legall *et al.* 1981).

In order to put these data into a regional perspective, the available data from the rest of southern England have been compiled in Figure 26.

HORSE HILL-1

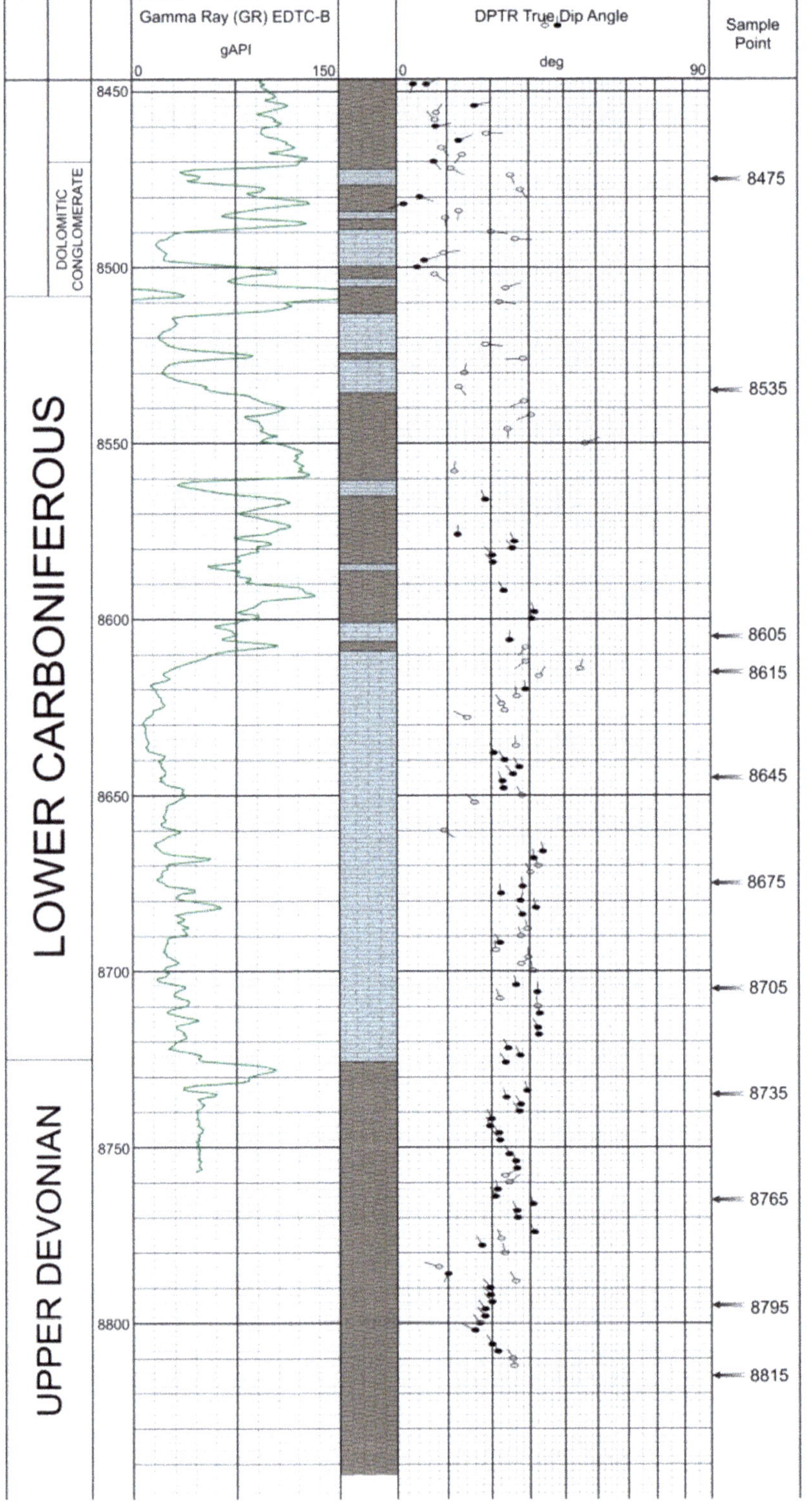

Fig. 23. Well Horse Hill 1 dipmeter. Published with the permission of UKOG plc.

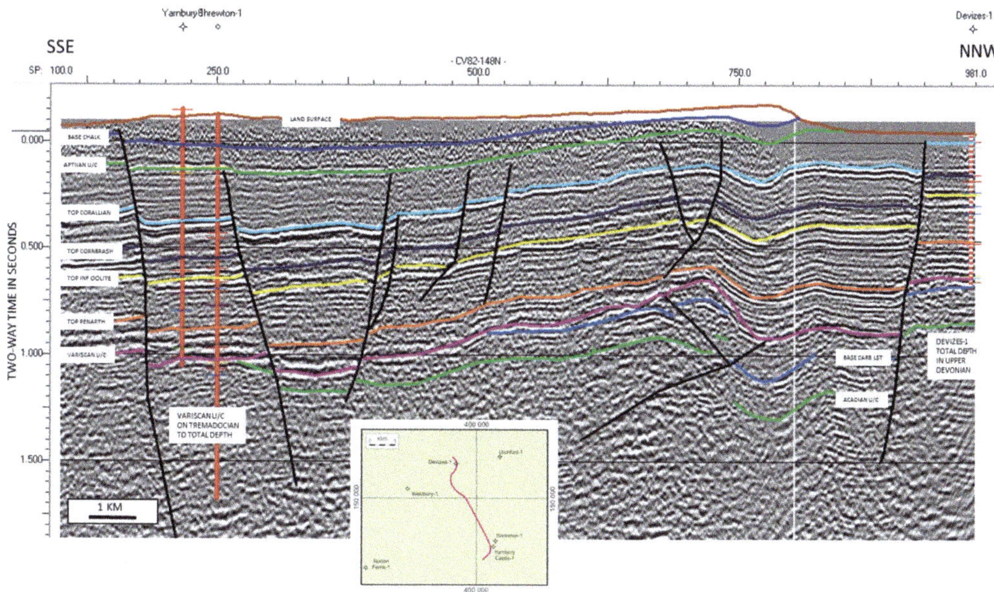

Fig. 24. SSE–NNW seismic line: NW Wessex Basin.

These data show that the Paleozoic in the southern Wessex Basin is at a post-gas generation stage. The rocks reach metamorphic grades, suggesting a position in the internal part of the Hercynian Deformation Belt. This higher grade of maturity appears to extend to the east of the southern flank of the Weald Basin, often called the Portsdown–Paris-Plage Ridge (see Fig. 1) in the literature (Butler & Pullan 1990; Andrews 2014). Although there is no maturity data from the wells in this area, the log and lithological data indicate that the rocks are more indurated, deformed and are likely to be mildly metamorphosed. In this area, the seismic shows no response in the Paleozoic, suggesting strong deformation, whilst the overlying Mesozoic thins markedly and is cut by numerous extensional faults. This change in structural character, compared to the relatively undeformed and unfaulted Mesozoic of the Central Weald, suggests a different basement fabric.

The maturity data from the Tremadocian and Cambrian of the Shrewton 1 and Cooles Farm 1 wells (Whittaker 1980; Harriman 2010) show that these rocks are in the wet-gas window. Since the post-Acadian and Mesozoic burial in the vicinity of these wells has not been high, it is believed that this maturity was reached as a result of pre-Acadian burial.

Discussion

A vexed question in the literature has been the position of the northern limit of intense Variscan deformation, the ‘Variscan Front’, in southern England since it does not crop out but lies beneath the Mesozoic cover. With the realization that much of the deformation of the Paleozoic sequences may result from Acadian movements, the limit of Variscan deformation has become more difficult to define.

To the SW, in Devon and Cornwall at outcrop, the Upper Paleozoic rocks are folded and metamorphosed, and the deformation front has been taken to lie in southernmost Wales where outcrop information allows the limit of intense deformation to be more clearly defined, although Variscan folding and thrusting continues well north of this line (Chadwick & Evans 2005). Previous workers, including Chadwick (1986) and Busby & Smith (2001), have extrapolated this front as a line running eastwards from the Mendip Hills across southern England using the most northerly prominent Mesozoic and Tertiary reactivated faults, such as the Pewsey Fault and the Hog’s Back Fault, as the surface trace. It is clear that Upper Paleozoic rocks in the Weald were deformed and folded prior to Variscan erosion (as demonstrated in the Horse Hill well). Although to the north of this ‘Variscan Front’ line the Paleozoic section appears less deformed, there are still indications of east–west-orientated Variscan structures on seismic lines across from Wiltshire into north Kent, the most important of which are shown on Figure 21. Chadwick & Evans (2005, fig. 32) projected that the main Variscan thrust subcrops the Permo-Triassic some way north of the Pewsey Fault, and Butler & Jamieson (2013, plates 1 & 4) show

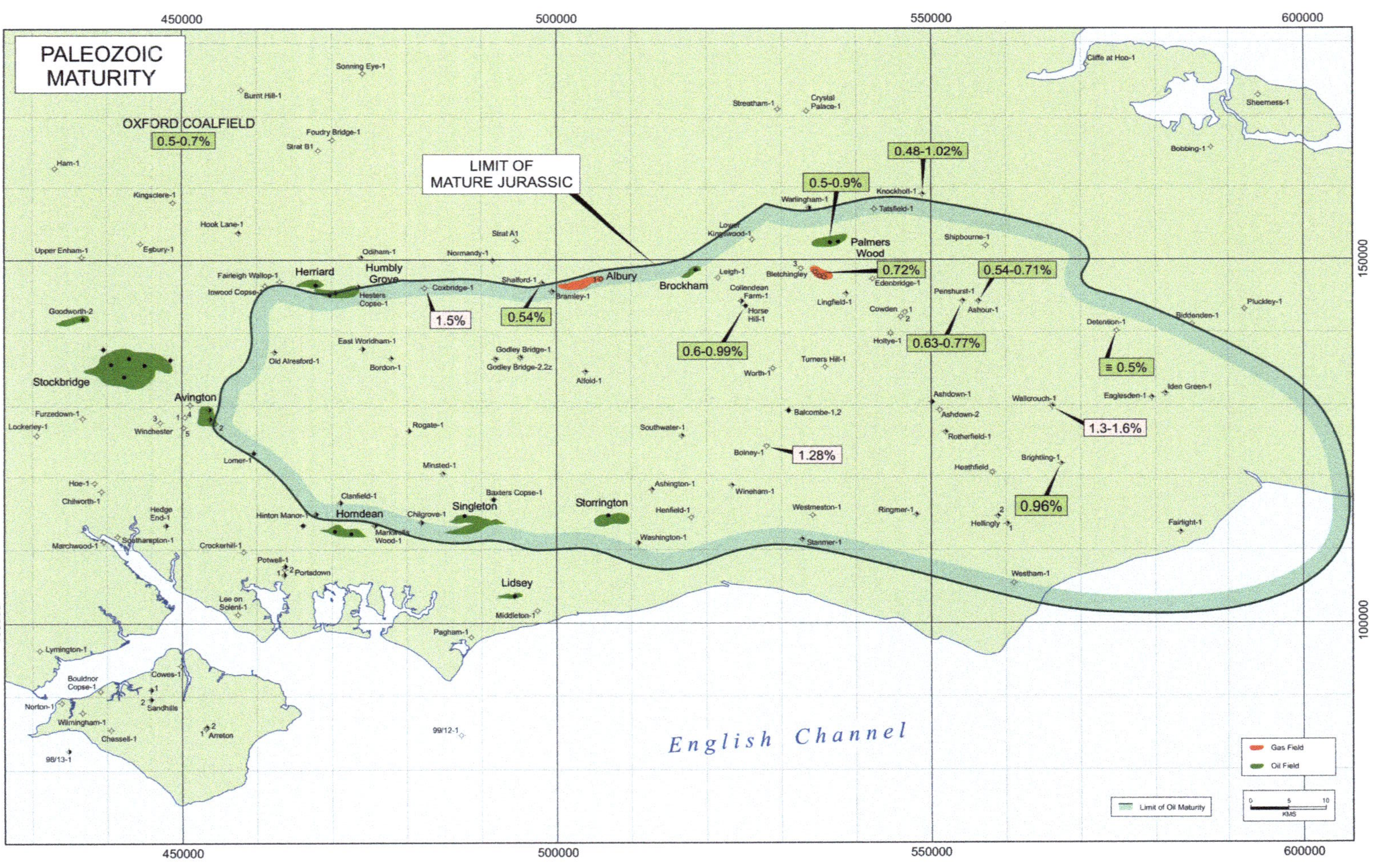

Fig. 25. Top Paleozoic maturity. From released well and industry reports.

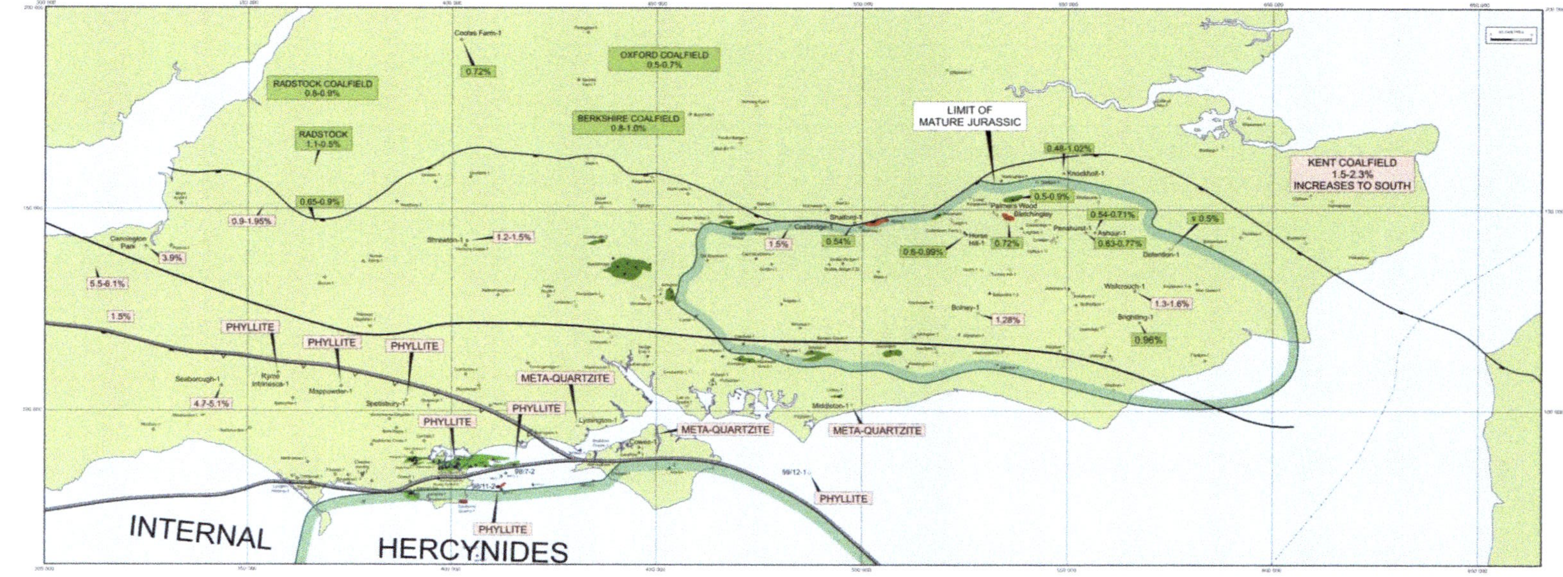

Fig. 26. Southern England: Top Paleozoic maturity. Compiled from various sources, including Taylor (1986) and Smith (1993).

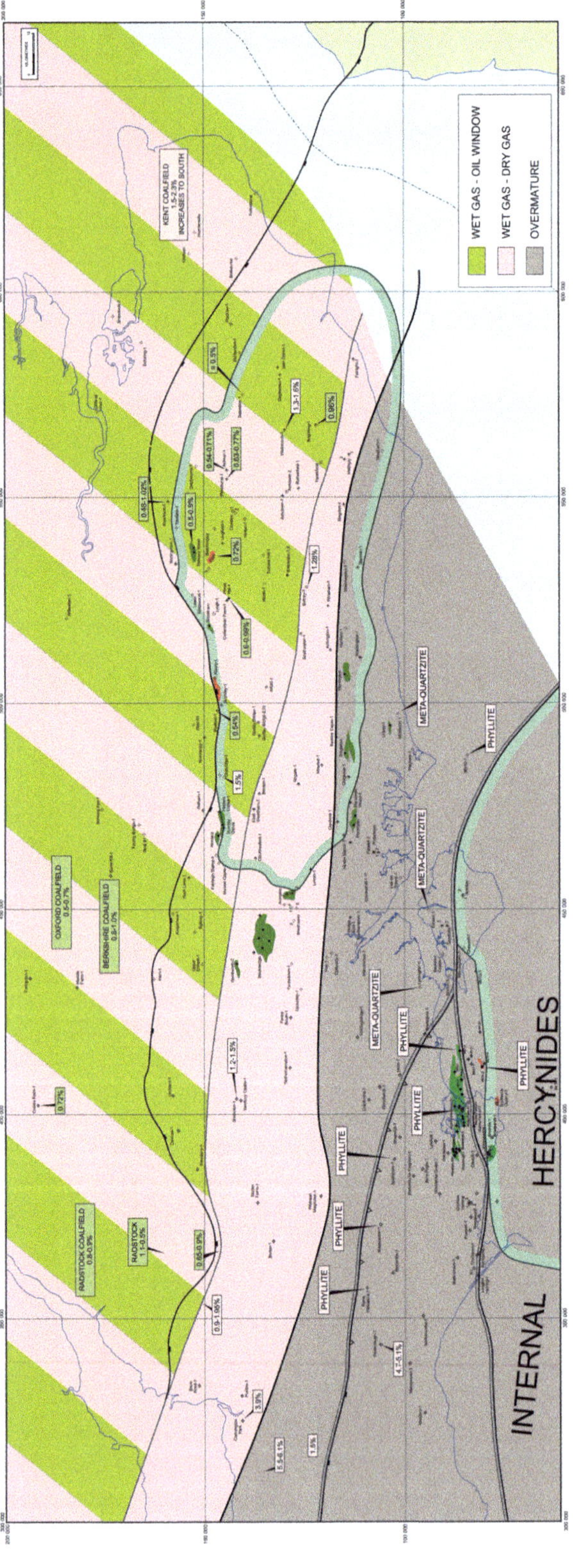

Fig. 27. Southern England: Top Paleozoic hydrocarbon-generation zones.

apparent Variscan thrusting and folding of the Paleozoic well north of the Pewsey and Hog's Back faults. On the basis of the scattered evidence from seismic data and well penetrations, it seems likely that the structure of the Paleozoic rocks beneath the Weald itself is similar to that postulated by Williams & Brooks (1984) for the area SE of the Mendips. However, the rocks are not metamorphosed. In the absence of good seismic definition of pre-Mesozoic structures, it seems sensible to define the limit of intense Variscan deformation by the onset of low-grade metamorphism in the Paleozoic subcrop, which lies on the southern edge of the Weald Basin (Fig. 27).

In the Wytch Farm Oil Field in Dorset, England, the largest onshore oil field in NW Europe, a conventional play reliant on Triassic Sherwood sandstone reservoirs sealed by Triassic and Lower Jurassic shales and sourced by Lower Jurassic through fault juxtaposition has been proven. An extension of this play into the Weald Basin has suffered from being unable to demonstrate a hydrocarbon charge to the deeper reservoir targets. These indications of a deeper gas source open up the opportunity for plays beneath the established Jurassic reservoirs of the Weald Basin, with hydrocarbons trapped in Paleozoic and Triassic reservoirs sourced by the underlying Paleozoic shales. Since Triassic river systems are expected to have followed the depositional axis of the basin, flowing from east to west downdip of structural highs, the thickest Triassic sands may be expected in the undrilled basin centre. This play remains untested.

Acknowledgements The authors would like to thank Celtique Energie and Magellan Petroleum for permission to publish this work and to UKOG plc for permission to publish information from Horse Hill-1. Thanks also go to the technical team at Celtique Energie who assisted with the geological, geophysical and petrophysical interpretation; in particular, Martin Berry, Ken Dups and Sarah Shirt. Thanks also go to Gareth Harriman (GH Geochem) for his geochemical analyses and advice. Very helpful discussions were had with Bernard Cooper of Horus Petroleum, who had thought deeply about the same issue. Thanks also go to Eric Stuckey of Windmill Graphics for drafting the figures and maps.

Funding The authors would like to thank the UK Onshore Geophysical Library (UKOGL) for providing financial support for this publication.

References

Andrews, I.J. 2014. *The Jurassic Shales of the Weald Basin: Geology and Shale oil and Shale Gas Resource Estimate*. British Geological Survey for Department of Energy and Climate Change, London.

Boote, D.R.D., Clark-Lowes, D.D. & Traut, M.W. 1998. Palaeozoic petroleum systems of North Africa. *In*: MacGregor, D.S., Moody, R.T. & Clarke-Lowes, D.D. (eds) *Petroleum Geology of North Africa*. Geological Society, London, Special Publications, **132**, 7–68, https://doi.org/10.1144/GSL.SP.1998.132.01.02

Burwood, R., Staffurth, J., de Walque, L. & de Witte, S.M. 1991. Petroleum geochemistry of the Weald–Wessex Basin of southern england: a problem in source-oil correlation. *In*: Manning, D. (ed.) *Organic Geochemistry: Advances and Applications in Energy and the Natural Environment. Extended Abstracts. 15th Meeting of the European Association of Organic Geochemists*. Manchester University Press, Manchester, UK, 22–27.

Busby, J.P. & Smith, N.J. 2001. The nature of the Variscan basement in southeast England: evidence from integrated potential field modelling. *Geological Magazine*, **138**, 669–685.

Butler, D.E. 1981. Marine faunas from concealed Devonian rocks of southern England and their reflection of the Frasnian transgression. *Geological Magazine*, **118**, 679–697.

Butler, M. & Jamieson, R. 2013. *Preliminary Interpretation of Six Regional Seismic Lines Across Onshore Basins of England*. UK Onshore Geophysical Library, London.

Butler, M. & Pullan, C.P. 1990. Tertiary structures and hydrocarbon entrapment in the Weald Basin of southern England. *In*: Hardman, R.F. & Brooks, J. (eds) *Tectonic Events Responsible for Britain's Oil and Gas Reserves*. Geological Society, London, Special Publications, **55**, 371–391, https://doi.org/10.1144/GSL.SP.1990.055.01.19

Carless Exploration Ltd 1983. *The Humbly Grove Oilfield – Field Development and Production Programme, Annex B*. Available online through UKOGL, https://ukogl.org.uk/industry-reports

Cave, R. 1977. *Geology of the Malmesbury District*. HMSO, London.

Chadwick, R. 1986. Extension tectonics in the Wessex Basin, southern England. *Journal of the Geological Society, London*, **143**, 465–488, https://doi.org/10.1144/gsjgs.143.3.0465

Chadwick, R.A. & Evans, D.J. 2005. *A Seismic Atlas of Southern Britain – Images of Subsurface Structure*. British Geological Survey, Occasional Publications, **7**. British Geological Survey, Keyworth, Nottingham, UK.

Chadwick, R.A., Kirby, G.A., Kubala, M., Sobey, R.A., Whittaker, A., Kenolty, N. & Penn, I.E. 1981. *A Hydrocarbon Prospectivity Study of the Vale of Wardour Area*. Deep Geology Report DP 81/1. Institute of Geological Sciences, London.

Clayton, C. 2009. *A New Understanding of Biogenic Gas: An Explorers Guide*. Petroleum Exploration Society of Great Britain (PESGB), Croydon, UK.

Conoco (UK) Ltd 1986. *Weald Basin Gas Hydrocarbon Generation Montage*. Available online through UKOGL, https://ukogl.org.uk/industry-reports

Cornford, C., Yarnell, L. & Murchison, D.G. 1987. Initial vitrinite reflectance results from the Carboniferous of north Devon and north Cornwall. *Proceedings of the Ussher Society*, **6**, 461–467.

DAWSON, C. 1898. On the discovery of natural gas in East Sussex. *Quarterly Journal of the Geological Society, London*, **54**, 564–574, https://doi.org/10.1144/GSL.JGS.1898.054.01-04.38

EBUKANSON, E.J. & KINGHORN, R.R. 1986. Oil and gas accumulations and their possible source rocks in southern England. *Journal of Petroleum Geology*, **9**, 413–428.

FALCON, N.L. & KENT, P.E. 1960. *Geological Results of Petroleum Exploration in Britain 1945–1957*. Geological Society, London, Memoirs, **2**, https://doi.org/10.1144/GSL.MEM.1960.002.01.01

FOSTER, D., HOLLIDAY, D.W., JONES, C.M., OWENS, B. & WELSH, A. 1989. The concealed Upper Palaeozoic rocks of Berkshire and South Oxfordshire. *Proceedings of the Geologists' Association*, **100**, 395–407.

GEOCHEM LABORATORIES LTD 1980. *Geochemical Evaluation of Carboniferous Outcrop Samples from Maesbury and Windsor Hill, Mendips. Report for St. Joe Petroleum Ltd.* Available online through UKOGL, https://ukogl.org.uk/industry-reports

GREEN, G.W. &. WELCH, F.B.A. 1965. *Geology of the Country around Wells and Cheddar*. HMSO, London.

HANCOCK, F.R. & MITHEN, D.P. 1987. The geology of the Humbly Grove Oilfield, Hampshire, U.K. *In*: BROOKS, J. & GLENNIE, K.W. (eds) *Petroleum Geology of Northwest Europe: Proceedings of the 3rd Conference*. Graham & Trotman, London, 119–128.

HARRIMAN, G.H. 2010. *Geochemical Evaluation of Oils and Samples in the Weald Basin*. Unpublished report for Magellan Petroleum (UK) Ltd

HAWKES, P.W., FRASER, A.J. & EINCHCOMB, C.C. 1998. The tectono-stratigraphic development and exploration history of the Weald and Wessex basins, Southern England, UK. *In*: UNDERHILL, J.R. (ed.) *Development, Evolution and Petroleum Geology of the Wessex Basin*. Geological Society, London, Special Publications, **133**, 39–65, https://doi.org/10.1144/GSL.SP.1998.133.01.03

HEASLEY, E.C., WORDEN, R.H. & HENDRY, J.P. 2000. Cement distribution in a carbonate reservoir: recognition of a palaeo oil–water contact and its relationship to reservoir quality in the Humbly Grove Field. *Marine and Petroleum Geology*, **17**, 639–654.

JONES, D.K. 1999. Uplift and denudation of the Weald. *In*: SMITH, B.J. & WORKE, P.A. (eds) *Uplift, Erosion, and Stability: Perspectives on Long-term Landscape Development*. Geological Society, London, Special Publications, **162**, 25–43, https://doi.org/10.1144/GSL.SP.1999.162.01.03

KETTEL, D. 1989. Upper Carboniferous source rocks north and south of the Variscan Front (NW and Central Europe). *Marine and Petroleum Geology*, **6**, 170–181.

LAMB, R.C. 1983. *Hydrocarbon Prospectivity of the Weald and Eastern English Channel. Volume 5: Source Rock Potential and Maturity*. IGS Deep Geology Unit Report to Department of Energy 83/3/5.

LEGALL, F.D., BARNES, C.R. & MACQUEEN, R.W. 1981. Thermal maturation, burial history and hotspot development, Palaezoic strata of southern Ontario–Quebec, from conodont and acritarch alteration studies. *Bulletin of Canadian Petroleum Geology*, **29**, 492–539.

LÜNING, S., ADAMSON, K. & CRAIG, J. 2003. Frasnian organic-rich shales in North Africa: regional distribution and depositional model. *In*: ARTHUR, T.J., MACGREOGOR, D.S. & CAMERON, N.R. (eds) *Petroleum Geology of Africa: New Themes and Developing Technologies*. Geological Society, London, Special Publications, **207**, 165–184, https://doi.org/10.1144/GSL.SP.2003.207.9

MAGELLAN PETROLEUM (UK) LTD 2012. *Relinquishment Report PEDLs 135 and 242*. Available online through UKOGL, https://ukogl.org.uk/industry-reports/

MCLEOD, R. 2002. *A reservoir study of the Great Oolite Limestone Formation in the Wiltshire–Hampshire high, Weald Basin*. MSc thesis, University of Reading, Reading, UK.

MOLYNEUX, S.G. 2010*a*. *Palynology Report on Palaeozoic Samples from Bobbing, Cooles Farm 1, Coxbridge 1, East Worldham 1, Harmansole, Shalford 1, Shrewton 1 and Strat A1 Boreholes*. British Geological Survey Commissioned Report CR/10/123. UKOGL, London.

MOLYNEUX, S.G. 2010*b*. *Palynology Report on Palaeozoic Samples from Baxters Copse 1, Brockham 1, Coxbridge 1, Detention 1, Goodworth 1, Knockholt 1, Lomer 1 and Palmers Wood 1 Boreholes*. British Geological Survey Commissioned Report CR/11/054. UKOGL, London.

PETRA-CHEM LTD 1986. *Geochemical Source Rock Evaluation of Carboniferous and Lower Jurassic Sediments from Selected Wells/Locations Adjacent to XL130, Somerset, UK*. Available online through UKOGL, https://ukogl.org.uk/industry-reports

SELLWOOD, B.W. & EVANS, M.R. 1986. *Regional Sedimentology and Diagenesis of the Great Oolite in the Wessex and Weald Basins. First Report, Carless Exploration Ltd.* Available online through UKOGL, https://ukogl.org.uk/industry-reports

SELLWOOD, B.W. & EVANS, M.R. 1987. *Regional Sedimentology and Diagenesis of the Great Oolite in the Wessex and Weald Basins. Second Report, Carless Exploration Ltd.* Available online through UKOGL, https://ukogl.org.uk/industry-reports

SELLWOOD, B.W., SHEPERD, T.J., EVANS, M.R. & JAMES, B. 1989*a*. Origin of late cements in oolitic reservoir facies: a fluid inclusion and isotopic study (Mid-Jurassic), southern England. *Sedimentology Geology*, **61**, 223–237.

SELLWOOD, B.W., WILKES, M. & JAMES, B. 1989*b*. Hydrocarbon inclusions in late calcite cements: migration indicators in the Great Oolite Group, Weald Basin, southern England). *Sedimentology Geology*, **84**, 51–55.

SMITH, N.J. 1993. The case for exploration of deep plays in the Variscan fold belt and its foreland. *In*: PARKER, J.R. (ed.) *Petroleum Geology of Northwest Europe: Proceedings of the 4th Conference*. Geological Society, London, 667–675, https://doi.org/10.1144/0040667

STAHL, W.J. 1977. Carbon and nitrogen isotopes in hydrocarbon research and exploration. *Chemical Geology*, **20**, 121–149.

STERLING RESOURCES (UK) LTD 2001. *UK Onshore 10th Round Application Weald*. Available online through UKOGL, https://ukogl.org.uk/industry-reports

TAYLOR, J.C.M. 1986. Gas prospects in the Variscan thrust province of southern England. *In*: BROOKS, J., GOFF, J.C. & VAN HOORN, B. (eds.) *Habitat of Palaeozoic Gas in NW Europe*. Geological Society, London, Special Publications, **23**, 37–53, https://doi.org/10.1144/GSL.SP.1986.023.01.03

THURRELL, R.G., WORSSAM, B.C. & EDMONDS, E.A. 1968. *Geology of the Country Around Haslemere*. HMSO, London.

TISSOT, B.P. & WELTE, D.H. 1984. *Petroleum Formation and Occurrence*. 2nd edn. Springer, Berlin.

TRUEMAN, S. 2003. The Humbly Grove, Herriard, Storrington, Singleton, Stockbridge, Goodworth, Horndean, Palmers Wood, Bletchingley and Albury Fields, Hampshire, Surrey and Sussex, UK. *In*: GLUYAS, J.G. & HICHENS, H.M. (eds) *United Kingdom Oil and Gas Fields Commemorative Millennium Volume*. Geological Society, London, Memoirs, **20**, 929–941, https://doi.org/10.1144/GSL.MEM.2003.020.01.79

WHITTAKER, A. 1980. *Shrewton No. 1 Geological Well Completion Report*. Report of the Deep Geology Unit Institute of Geological Sciences 80/1. Institute of Geological Sciences, London.

WILHELMS, A., LARTER, S.R., HEAD, I., FARRIMOND, P., DI-PRIMIO, R. & ZWACH, C. 2001. Biodegradation of oil in uplifted basins prevented by deep-burial sterilization. *Nature*, **411**, 1034–1037.

WILLIAMS, G.D. & BROOKS, M. 1984. A reinterpretation of the concealed Variscan structure beneath southern England by section balancing. *Journal of the Geological Society, London*, **142**, 689–695, https://doi.org/10.1144/gsjgs.142.4.0689

A Paleozoic-sourced oil play in the Jura Mountains of France and Switzerland

C. P. PULLAN[1]* & M. BERRY[2]

[1]*CP Exploration, Potterne, Devizes, Wiltshire SN10 5LR, UK*

[2]*Icon Energy Limited, 4 Miami Key, Broadbeach, QLD 4218, Australia*

M.B., 0000-0002-0462-9135

**Correspondence: chris.pullan1@btinternet.com*

Abstract: The Jura region of France and Switzerland has significant hydrocarbon potential. This arcuate fold-and-thrust belt, located in front of the Western Alps, is composed of Mesozoic rocks, which overlie the North Swiss Basin Complex filled with Permo-Carboniferous sediments of Late Westphalian–Lower Permian (Autunian) age. The existence of three small producing gas fields, and the presence of recorded oil and gas shows in a number of petroleum wells, indicate the presence of an active petroleum system. Shows occur in stratigraphic horizons that lie beneath the regional Triassic evaporite seal. Very effective Permo-Carboniferous oil-prone source rocks include algal organically rich Autunian shale beds and Stephanian-aged bituminous shales. Carboniferous and Permian coals are also present, which would be excellent sources of gas. The primary reservoir target is the fluviatile Triassic Bunter Sandstone Formation, which is widely developed across the area and thickens to the NE. Reservoir quality is best developed in the cleaner channel sands. The La Chandelière oil field, discovered in 1989 by Exxon, is crucial evidence of the oil potential and trapping mechanism. Geochemical analysis demonstrates that the 41° API oil at La Chandelière is derived from Lower Permian Autunian shales. The trapping mechanism, interpreted from seismic and confirmed by structural modelling, is that of an unreactivated Mesozoic-aged structural high bounded by thrust faults, resulting in a large-scale rollover within the core of the structure. This combination of critical elements sets up the Paleozoic-sourced oil play within the Jura Mountains of France and Switzerland.

The Jura is not generally regarded as a major petroleum province. It is a region of complex structuring and mountainous terrain. Switzerland, a country renowned for its beautiful mountain scenery, has no major oil and gas discoveries. However, this poorly explored region has active petroleum systems with good hydrocarbon shows and a number of small gas fields demonstrating hydrocarbon generation. Good-quality source rocks and excellent reservoir development have been proven by drilling, and seismic interpretation has demonstrated the existence of large closures at reservoir level. The La Chandelière oil discovery, in the French Jura, proves the presence of a Paleozoic-sourced oil play.

Structural history and associated stratigraphy

The Jura is an arcuate-shaped fold-and-thrust belt (Fig. 1) that extends eastwards from eastern France, through NW Switzerland for 370 km, and has a maximum width of 75 km. The belt comprises Mesozoic rocks that were mainly developed during the Mio-Pliocene Alpine Orogeny at the front of the Western Alps (Sommaruga 1997). It is separated from the Alps by the Oligo-Miocene Molasse Foredeep Basin. It is surrounded by the Tertiary depocentres of the Bresse graben to the west and the Rhine graben to the north, which are associated with the Oligocene West European Rift System (Sommaruga *et al.* 2012). The location of the Jura Mountains appears to be intimately linked with the presence and distribution of the Triassic salt, which acts as a major décollement horizon (Laubscher 1961; Rigassi 1977). Work by a number of authors suggest that the tectonics of the Jura and Molasse Basin are closely linked (Laubscher 1961; Burkhard & Sommaruga 1998).

In the Jura region, the first major phase of deformation that is recognized occurred during the Hercynian Orogeny in the Lower Carboniferous and established the predominantly NE–SW structural fabric in the area (Nagra 1991, 2015). The later phases of this compression (an example of 'escape tectonics') resulted in movement along major strike-slip faults, with the creation of a series of NE–SW-trending extensional Permo-Carboniferous basins (McCann 2008*a*).

The Upper Carboniferous–Permian sequences that fill the North Swiss Basin are not exposed at the surface in the Jura, but seismic and well data show that the interval is unconformably overlain by Triassic and younger aged sediments, which

From: Monaghan, A. A., Underhill, J. R., Hewett, A. J. & Marshall, J. E. A. (eds) 2018. *Paleozoic Plays of NW Europe*. Geological Society, London, Special Publications, **471**, 365–387.
First published online March 22, 2018, https://doi.org/10.1144/SP471.2

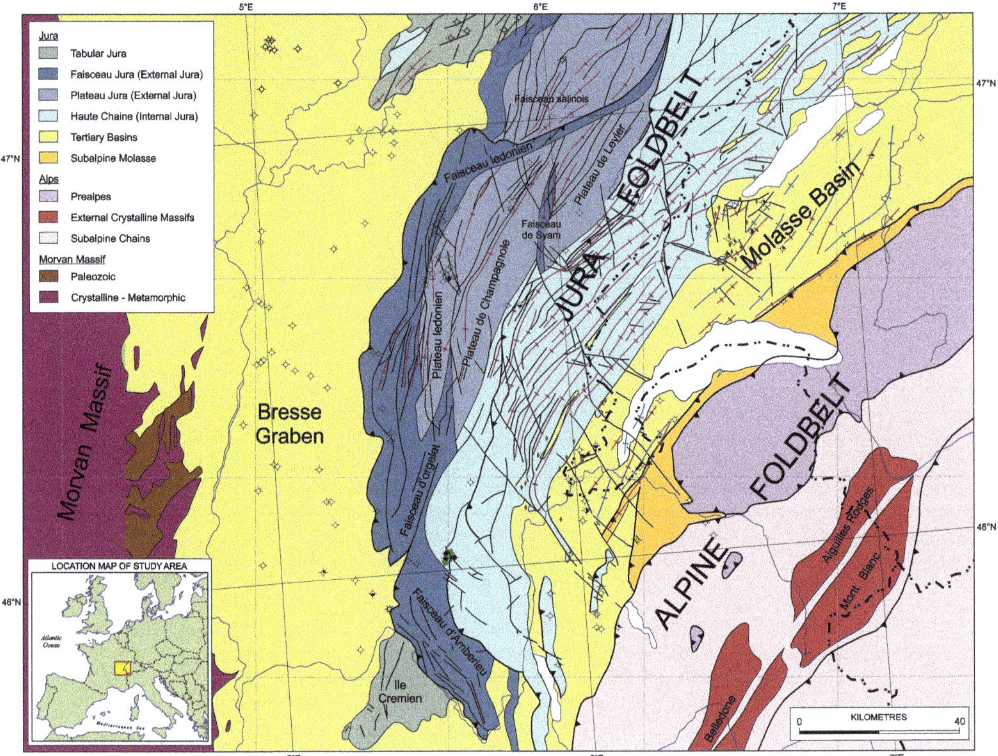

Fig. 1. Regional structural elements of the Jura.

thicken from 1.5 km in the north to 3 km in the south (Sommaruga 1997).

The Permo-Carboniferous sequences are also exposed around the basin's edges to the west, on the east flank of the Morvan Massif in the Autun, Blanzy and St Etienne basins; whilst to the north, the sediments are seen in the Ronchamp Basin on the south flank of the Vosges Massif (Chateauneuf & Farjanel 1989). In the Autun Basin, the Permian (Autunian) interval contains a series of continental argillaceous-dominated sequences deposited in a fluvial or lacustrine setting (Doubinger 1974).

During the Mesozoic, the Jura lay on the northern margin of the Tethys Ocean (Stampfli & Borel 2002). The evolution of the area reflects regional extension caused by the break-up of this margin associated with the opening of the Central Atlantic and its eastern prolongation into the Tethys Ocean, the Ligurian Ocean (Ziegler 1990). The first evidence of faulting is seen in the Triassic and the formation of NE–SW-trending extensional basins. Synsedimentary fault movement resulted in the localized thickening of evaporites within the rift complexes. This phase of rifting appears to have ended in the late Lower Jurassic.

A three-fold differentiation of the Triassic sedimentary section is possible (McCann 2008*b*). The basal section, or Bunter, is a fluvial sequence which fines upwards overall and unconformably overlies the Permo-Carboniferous. The overlying Muschelkalk sequence is carbonate rich and is capped by the Lettenkohle dolomites. The Muschelkalk interval is salt bearing to the NW, but elsewhere it is shale and anhydrite dominated. The overlying Keuper interval is dominated by two evaporite sequences that contain salt. These two units are separated by a dolomite interval. The Lower Jurassic section is characterized by thick marine shales that vary in thickness locally, reflecting synsedimentary fault movement.

Following cessation of the faulting in the Lower Jurassic, passive margin deposition dominated. The Middle and Upper Jurassic section is represented by a thick carbonate sequence containing thinner intervening shale beds with little evidence of deformation. Renewed extension in the Late Jurassic, caused by the continued opening of the Atlantic, resulted in fault reactivation (Wetzel *et al.* 2003).

The Cretaceous cover is absent in large areas of the Jura due to subsequent Tertiary uplift and erosion, and therefore the timing of fault movement during the Cretaceous cannot be determined accurately. The Cretaceous section that is preserved is carbonate dominated.

The first phase of Alpine compression was associated with the closure of the Ligurian Ocean, and with the collision of the Austro-Alpine and Briançonnais continental fragments, which began in eastern Switzerland in the Upper Cretaceous (80 myr ago) (Stampfli 2001). These movements finally ended in the Lower Eocene (50 myr ago). The main phase of the Alpine Orogeny began in the earliest Oligocene (35 myr ago) with the emplacement of the Austro-Alpine–Briançonnais continental fragments onto the Swiss margin. Deformation progressed with the accretion of the Apulian Plate onto the North European margin. In the Upper Miocene (Serravallian – 12 myr ago), the deformation front stepped further forwards and incorporated the Molasse and Jura regions, with the Alpine Front then marked by the western leading edge of the Jura Fold Belt (Becker 2000). Studies indicate that the area did have a full cover of Tertiary sediments, with the Molasse Basin originally extending to the north (Moss 1992). These sediments in the Jura are now only preserved as relicts, and their deformation indicates that the folding and thrusting post-dated Molasse deposition (Becker 2000). The area underwent deformation into the Plio-Pleistocene, and continues to undergo significant uplift and erosion, with the removal of much of the Tertiary and Cretaceous sediment cover.

Exploration history

There have been several phases of exploration within the Jura Fold Belt. Asphalt and surface hydrocarbon occurrences have been known in the area since antiquity, and asphalt mining has taken place in three locations in Switzerland since the eighteenth century. Hydrocarbon exploration began in the Swiss area of the Jura in 1912, with drilling located close to the surface seeps in the adjacent Molasse Basin (Schardt 1911; Breynaert 1912; Spycher 1994).

The main phase of exploration in the Jura Fold Belt region began in the 1950s and continued into the 1960s. A total of 48 exploration wells have been drilled in the Jura, 24 of which reached the Bunter Sandstone. Of these, 15 wells (62.5%) were drilled prior to 1965 when the available seismic were both limited and of poor quality. Drilling targeted the Muschelkalk carbonate reservoir objective below the Triassic evaporite seal. The traps were the compressional anticlinal structures that had been recognized and mapped at the surface. Exploration success occurred with the first gas discovery in the Muschelkalk carbonates in 1957 at Lons Le Saunier (2.5 Bcf (billion cubic ft)) (Elf Aquitaine 1991). This was followed by the gas discovery at Valempoulieres (3.8 Bcf) in 1961. Unfortunately, work showed that the reservoir quality was poor and relied on fracturing. Oil was also discovered in the Muschelkalk at Briod 101 in 1954. During this period, many of the wells were drilled prior to the acquisition of usable seismic data, resulting in many wells being located off-structure at the Bunter level.

This phase of exploration drew to a close due to the small size of the reserves discovered, and the unpredictability of the quality and extent of the Triassic Muschelkalk reservoir. However, the work showed that an active petroleum system existed in the area.

Exploration resumed in the 1980s with BP and Shell licensing acreage and acquiring modern seismic data in Switzerland (Shell 1977; BP 1986*a*, *b*, 1989), while Exxon was active in the French Jura. Exxon developed a thin–thick-skinned structural model, using much improved seismic data, invoking a pre-salt structural high bounded on its flanks by overthrusted post-salt section, with faults soling into the Triassic evaporates. The reservoir target was the Triassic Bunter sands, charged from the underlying Permo-Carboniferous source rocks and sealed by the overlying Triassic evaporites, in structural traps below the Triassic salt. Exxon's programme met with success with the Chaleyriat/La Chandelière oil discovery in 1989.

In contrast, exploration in Switzerland during the 1980s was unsuccessful. BP applied a thin-skinned tectonic model, based on a Triassic salt décollement horizon (BP 1986*a*, 1989). This approach did not allow them to identify pre-salt structuring.

From the early 1990s to the mid-2000s, the area had not been explored due to the shift of companies' exploration focus away from onshore Europe, the low oil-price environment and the fact that most companies overlooked a Triassic Bunter play in the Jura.

The most recent phase of exploration began in 2006, when Celtique Energie Ltd began acquiring exploration acreage in France and Switzerland over the Jura region. Recognizing the hydrocarbon potential of the area while working for Celtique, the authors assembled the most comprehensive database over the region, carried out detailed geological and geophysical evaluation, and generated multiple prospects along this Paleozoic-sourced hydrocarbon play.

Database

A digital database integrating information from both France and Switzerland was compiled for the Jura and surrounding region, utilizing all available well and seismic data. The often-incomplete well datasets varied in content and existed in hard copy only. Wireline logs from all key modern wells were digitized, which allowed petrophysical interpretation, accurate correlation and the generation of synthetic seismograms. In Switzerland, previous operators' reports were available for review in the Vaud Canton Archive in Lausanne. In France, interpretational data were not available, and so the work and reports of previous operators could not be accessed. However, all the basic well data and reports were available from the French Government. Seismic data in Switzerland were recovered from the Canton Vaud Archive; whilst in France, field data were purchased from Bureau de Recherches Géologiques et Minières (BRGM). All the data were reprocessed, which resulted in a significant improvement in data quality. The seismic interpretation was carried out using Kingdom software. However, to generate regional maps of the Jura covering both France and Switzerland, a combined dataset was created using the Swiss Coordinate (CH1903/LV03) system. This was the first time that a comprehensive digital well and seismic database had been created covering the Jura of France and Switzerland, enabling both sides of the border to be worked simultaneously, providing a better understanding of the hydrocarbon potential of the area.

Source potential

Surface hydrocarbon seeps and the presence of multiple oil and gas shows in exploration wells drilled in the Jura region indicate the presence of an active petroleum system (Shell 1977; Zweidler 1981, 1985; BP 1986*b*; Celtique Energie 2009, 2010, 2012). These hydrocarbon occurrences across the study area are detailed in Figure 2. The presence of shows in stratigraphic horizons beneath the regional Triassic evaporite seal, including the gas discoveries at Valempoulieres and Lons Le Saunier, indicate that the reservoirs are fed by very effective

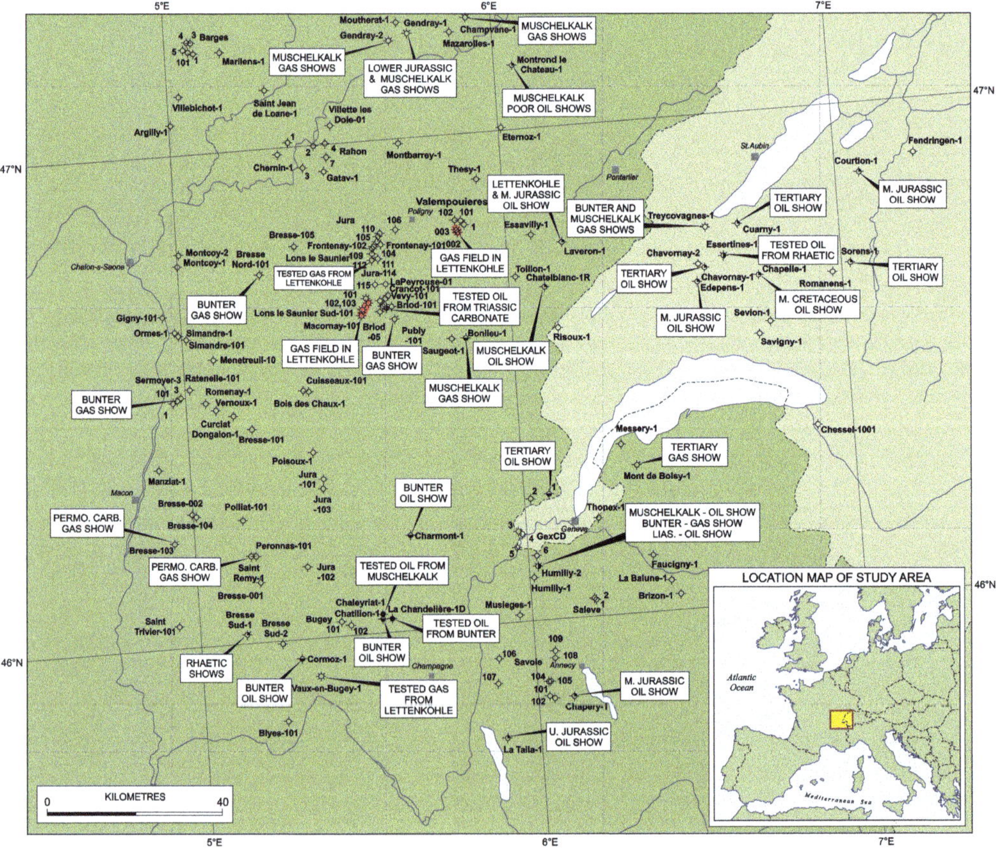

Fig. 2. Hydrocarbon occurrences in the Jura.

Permo-Carboniferous source rocks that have generated both oil and gas. At La Chandelière, the oil is a light (41° API), low sulphur (0.1%) crude with a gas/oil ratio (GOR) of 70 scft/bbl (standard cubic ft/barrel). Geochemical analysis carried out as part of this work reveals that the oil has lacustrine markers (including carotene), and is characterized by the absence of sterane biomarkers and has a light isotopic signature (Celtique Energie 2009; Harriman 2009). Biomarkers suggest that anoxia was achieved through water stratification. The oil found in the Briod-101 well, as well as the numerous oil shows seen in various wells in the area, are believed to be from this same source. Compositional data reveal that the gas found at Valempoulieres and Lons Le Saunier is dry and from a mature pre-salt source.

Two potential oil-prone source rocks are developed within the pre-evaporite source rock interval: Lower Permian (Autunian) shales; and Upper Carboniferous (Stephanian) shales and coals (Celtique Energie 2009).

The Autunian section of Lower Permian age contains rich, oil-prone source rocks which are well developed both in the Upper and Lower Autunian-aged sequences (Chateauneuf & Farjanel 1989). The source rocks are algal rich and were deposited in a lacustrine setting where water stratification caused anoxic conditions.

These lacustrine sequences are seen in all the Permian basins around the Jura to the east (Autun and St Blazny) and to the north in Germany (Fig. 3). In the Autun Basin, the individual beds of organically very rich shale are 0.5–10 m thick within an overall 250 m-thick shale section. The shales are laterally pervasive and extend over the entire basin area (Kettel & Herzog 1988; BRGM 1989; IFP 1994, 1996). The source rocks are oil prone (Type I), algal rich with total organic carbon (TOC) values of 3–25% and with pyrolysis yields of 12–135 kg/tonne (averaging 75 kg t^{-1}). Autun Basin coals have measured TOCs ranging from 27 to 47% (Table 1). The source-rock analysis carried out as part of this work confirms the results of more extensive studies undertaken by the French Institute of Petroleum (IFP).

The North Swiss Basin complex extends to the NE to the Bodensee area of Switzerland/Germany (McCann *et al.* 2006). Well control in Germany has shown that a thick Permian section exists, with lacustrine sediments forming 30% of the overall section. In the Weiach-1 well, at the northern end of the basin, both Autunian and Stephanian sections

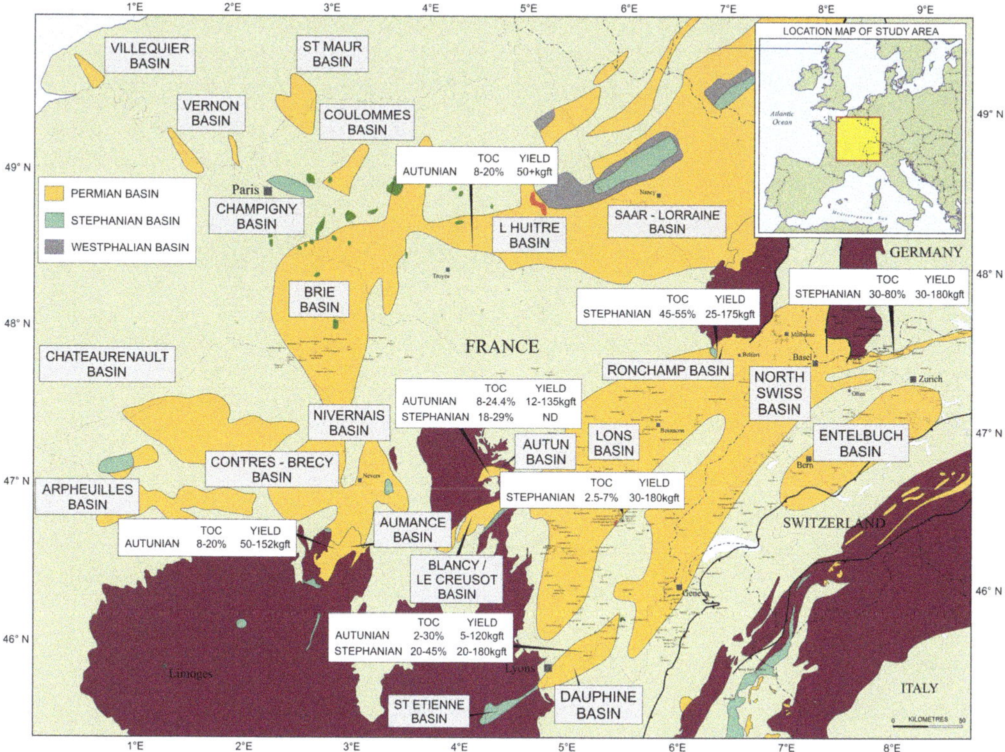

Fig. 3. Regional source potential in the pre-Triassic.

Table 1. *Lower Permian (Autunian) source potential*

Basin	Lithology	Thickness (m)	TOC (%)	Yield (kg t^{-1})
Autun	Coal		27–47	
Dauphine	Shale	10	2–30	5–120
Autun	Shale – 10 beds	200–250	3–24.4	12–135
Jura (Frick Bodensee)	Shale	up to 35	1.7–10.3	

Source: IFP (1994, 1996, 2001), Celtique Energie (2009) and Harriman (2009).

have been proven. The Autunian shales here are oil prone, up to 35 m thick, with TOCs of 1.7–10.3% (Spycher 1994).

Analysis of the Autunian shales shows that, when mature, a light, low-sulphur (0.1%) oil would be generated (Harriman 2009). There are clear facies variations within the Autunian, and this gives rise to a range of isotopic signatures and variable concentrations of some of the biomarkers. However, the geochemical analysis carried out as part of this work suggests that the oil found in La Chandelière is derived from a Permian Autunian source rock. This oil is very mature, generated at the base of the oil window (Ro equivalent of 1%) from more mature shales than from those penetrated in the wells, thought to be present at greater depths on the flanks of the regional high.

Additional source rocks are developed in the shales and coals of the Stephanian in the Upper Carboniferous. Regional studies have shown that these shales and coals are an effective source, are widely developed, and appear to have generated both oil and gas (Courel & Paquette 1981). To the SE of the Jura, IFP (IFP 1994, 1996) have reported that coals and good-quality bituminous shales are developed in the Dauphine Basin area with TOCs of 20–45% and yields of 20–180 kg t^{-1} (Table 2). To the north in the Ronchamp Basin, on the south flank of the Vosges Massif, the Stephanian interval contains both coals and bituminous shales within a section of over 200 m (Chateauneuf & Farjanel 1989), with the shale unit being over 150 m thick. The coals have TOCs of 45–55% with yields of 25–175 kg t^{-1}. The coal seams are up to 15 m thick. In the Autun Basin, the Stephanian interval unconformably underlies an Autunian section and is over 500 m thick. The section contains coals with TOCs of 18–29%. The coals appear to be best developed within the Middle Stephanian section, and constitute 3–5% of the overall section and thicken eastwards.

In the Jura, the extent and thickness of the coals in the subsurface is not well known. The best control on the Upper Carboniferous interval in the Jura is provided by Charmont-1, where the coals are 52.5 m thick (out of a penetrated section of 496 m, 10.5%) with TOCs of 8–46% and yields of 7.3–55.7 kg t^{-1} (Celtique Energie 2009). In the Lons Le Saunier area, to the NW, Lower–Middle Stephanian coals exist at several levels with individual seams, up to 7 m thick, that have yields of 30–180 kg t^{-1}. The associated shales have TOCs in the range 2.5–7% (IFP 1994, 1996). The coals within the interval are thought to be more gas prone, while the shales originally had oil potential. The composition of the gas found in wells drilled in the Jura suggests a thermogenic source at a maturity equivalent to Ro 1.5–2.1% (Harriman 2009), which is supported by isotopic data. This would suggest an origin from the Carboniferous coals. The gas in the seep at Cuarny and at Essertines is thermogenic and partly originates from a source with a maturity of Ro equivalent of 1.5% (Celtique Energie 2009); only a Paleozoic source satisfies this maturity constraint.

Table 2. *Upper Carboniferous (Stephanian) source potential*

Basin	Lithology	Thickness (m)	TOC (%)	Yield (kg t^{-1})	Seam thickness (m)
Ronchamp	Coal		45–55	25–175	6–15
Ronchamp	Shale	50	2–30	5–200	
Dauphine	Coal		20–45	20–180	0.5–4
Lons	Coal	145	2.5–7	30–180	
Autun	Coal	50–250	18–29		0.7–2
Jura (Frick Bodensee)	Coal		30–80	30–180	

Source: Nagra (1988), IFP (1994, 1996, 2001), Celtique Energie (2009) and Harriman (2009).

Maturity modelling

Extensive maturity modelling has been carried out as part of this study to define where the Mesozoic and Paleozoic source rocks are mature for the generation of both oil and gas and the timing of that generation. Information from 12 wells have been modelled (together with several 'dummy' locations where the stratigraphic section was derived from seismic control) in order to obtain reasonable spatial sampling. Well temperature information was used to obtain average present-day geothermal gradients for the modelled wells. The results showed a variation, from 25.7°C km^{-1} in Chatelblanc 1 to 39.1°C km^{-1} in Essertines 1. In order to determine the heat flow, the lithological information from the well was loaded. Previous work in the area (Leu *et al.* 1997; Schegg & Leu 1998; Mazurek *et al.* 2006; Timar-Geng *et al.* 2006; Strobl 2007) suggested that the heat flow varied over time because of the variability of the tectonic regimes. In their models, the highest gradients were achieved in the late Paleozoic with a gradual decline until a heat flow anomaly in the early Cretaceous (130–95 Ma). Schegg & Leu (1998) decreased the heat flow into the Tertiary, whilst Strobl (2007) and Mazurek *et al.* (2006) increased the heat flow into the Tertiary based on their apatite fission-track work. In Celtique's models, a similar changing heat flow with time was used, with an increase in the mid Cretaceous and again in the Tertiary.

A key uncertainty to maturity modelling in the area is the amount of uplift and erosion that has occurred. Three major unconformities are recognized: Late Paleozoic, Base Tertiary and Base Quaternary, and these represent major phases of uplift and erosion in the pre-Permian, Eocene and late Mio-Pliocene. Work by Schegg & Leu (1998) show that eastern Switzerland was uplifted in the Late Tertiary by 500 m, but by up to 2500 m in western Switzerland. These estimates of the amount of uplift are constrained by stratigraphic data, vitrinite information and velocity data, as well as work undertaken on the clay mineralogy. Their work also demonstrated that the uplift in the late Cretaceous and early Tertiary was mainly centred to the east, as shown by the progressive older subcrop to the Base Tertiary in that direction (Schegg & Leu 1998). A relatively minor phase of uplift is identified in the western Jura based on field relationships. The timing and amount of uplift of these events were incorporated into the uplift assumptions used in the Celtique modelling work.

Individual wells were modelled using Platte River Associates Inc. BasinMod software, and the models have been constrained by maturity data measured in wells (Nagra 1988; Strobl 2007). An example burial history plot is shown for the Chatelblanc 1 well (Fig. 4).

The work shows that the maturity of the Permo-Carboniferous section was greatest in the north and east; with maturity progressively decreasing to the

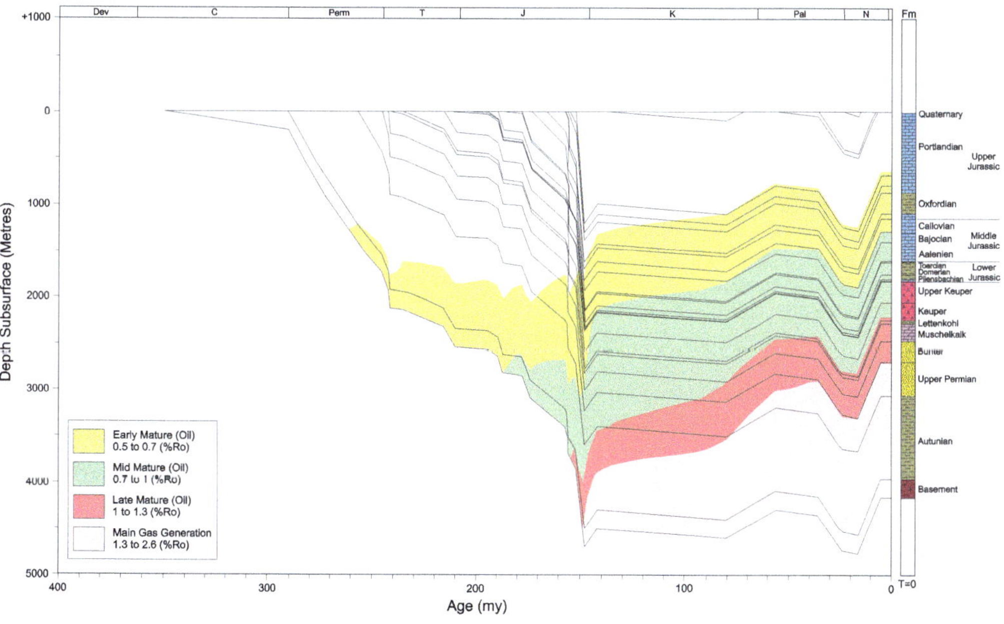

Fig. 4. Burial history plot of Chatelblanc 1.

west and SW. The onset of oil generation is reached at a depth of 1500–2000 m during the Jurassic in the north and in the Lower Cretaceous in the south. Peak generation began in the Lower Cretaceous in the north and in the Late Cretaceous in the south. The onset of gas generation is reached at 3000 m in the Upper Cretaceous in the north; however, the modelling indicates that the gas-generation window was not reached at all in the south. These conclusions agree well with the conclusions of previous studies (Strobl 2007).

Reservoir

The primary reservoir target within the Jura is the Triassic Bunter Sandstone, which is the reservoir at the La Chandelière Field. A detailed reservoir study was carried out on core material from La Chandelière 1D, Chaleyriat 1, Chatillon 1D and Charmont 1. This work included descriptions of the depositional facies within the Bunter Sandstone and the underlying Permo-Carboniferous, a lithostratigraphic scheme for the Bunter Sandstone reservoir and detailed petrological analyses (Macchi 2008; Celtique Energie 2009, 2012). This work showed that the Bunter Sandstone is a fluviatile deposit consisting of cross-bedded sandstones with interbedded mudstones forming a series of individual fining-upwards sequences, within an overall fining-upwards sequence. The sandstones were deposited by perennial, laterally shifting, low-sinuosity braided streams, whilst the mudstones represent flood (low-flow) conditions or, possibly, channel abandonment. The nature of the fluvial channels, having a semi-random stacking pattern, means that they amalgamate laterally and the thin intervening shales are not laterally extensive. This would suggest that permeability barriers are not laterally continuous, which infers that vertical communication will exist within the reservoir (Macchi 2008; Celtique Energie 2009, 2012). A further consequence is that detailed stratigraphic correlation within the Bunter Sandstone is virtually impossible.

The Triassic fluvial systems flowed to the NE towards the basin centre (Ziegler 1990). The Bunter Sandstone is thickest along a NE–SW-trending depositional axis, with an overall thickening to the NE (refer to the Bunter Sandstone isopach map: Fig. 5a). These sandstones were deposited over a pre-existing palaeotopographical surface with basement highs and intervening areas of lower relief. The topographical relief contrast would have been greater to the south where the basement was exposed.

The petrological analysis of the Bunter sands shows that they are medium to coarse grained, moderately sorted and contain sub-angular quartz grains, suggesting relatively short transport distances. Framework compositions show that most lithologies can be classed as subarkosic with a generally high feldspar content (Fig. 5b). Clays are dominated by illite and illite–smectite (smectite <5%) and subordinate volumes of kaolinite. The full range of mineralogical and textural data acquired from thin section microscopy are presented in Appendix A. The source for the sandstones was the Hercynian basement to the south, SW and west, which was dominated by metamorphics with a minor igneous component.

This new core study (Macchi 2008; Celtique Energie 2012) has helped in the construction of a new geological model for the Bunter Sandstone, modified from the classic point-bar model for a meandering stream system (Allen 1964, 1970). Integrating both the lithological data from core and cuttings and the depositional facies interpreted from the core study, with wireline log character, has enabled the identification of five primary genetic facies over the Triassic Bunter Sandstone and Permian intervals. Although a range of wireline logs were used (SP, gamma-ray, sonic, density, neutron and caliper) to help with lithology interpretation and borehole condition, it is the gamma-ray log character that is the best representation of the depositional facies. These electro-facies, once established in wells with core data, can then be extrapolated to wells without available core data (Celtique Energie 2009). The primary genetic facies identified within the sequence are as follows:

- fluvial channel facies (FC): sandstones with thin siltstones and/or shales, generally fining upwards, stacked channels, laterally continuous.
- crevasse splay channel facies (CSC): interbedded sandstones and siltstones, several metres thick, change vertically into floodplain or fluvial facies, laterally discontinuous.
- floodplain crevasse splay facies (FP-CS): interbedded fine-grained sandstones and siltstones, generally thin (<2 m), change laterally into fluvial or crevasse splay channel facies.
- floodplain peat mire facies (FP-PM): thick coal beds, and interbedded coals, siltstones and thin sandstones, change laterally and vertically into FP-CS facies.
- delta floodplain facies (DFP): siltstones and/or shales deposited in a low-energy floodplain, lacustrine and often shallow-marine environment can change laterally into FP-CS or FP-PM facies.

An example of the application of this facies interpretation can be seen in the Bunter Sandstone and Permian intervals in Charmont 1, located 19 km NE of the La Chandelière Field (Fig. 6a). A cross-section based on these facies between the three wells in the La Chandelière Field (La Chandelière 1D, Chatillon 1D and Chaleyriat 1) and Charmont 1 illustrates clearly the nature of the stacked channels and discontinuous permeability barriers within the Bunter Sandstone (Fig. 6b). The cross-section from La Chandelière to Treycovagnes 1 in Switzerland (Fig. 6c) (Macchi 2008; Celtique Energie 2012), representing a distance of approximately 120 km, further illustrates the complexity in correlation within the Bunter Sandstone and the stacked nature of the channel facies.

In the La Chandelière Field, the sands are 14–17 m thick with average porosities of 12% (Fig. 5c). The cleaner channel sands have the best reservoir development, which core analysis shows can be as high as 24% porosity and 1894 mD permeability. La Chandelière is located slightly to the SE of the depositional axis, resulting in a general NW increase in gross thickness of the Bunter Sandstone in the vicinity of the field.

In addition to thickening towards the NE, the Bunter Sandstone also appears to show an improvement in reservoir quality in the same direction; a result of better sorting and increasing mineralogical maturity. At Chatelblanc 1, located approximately 80 km NE of the La Chandelière Field, the Bunter Sandstones reach a gross thickness of 106 m, with a net reservoir thickness of 40 m. The reservoir quality within these sands is very good, with core analyses showing porosity in the range 7–21.3% (averaging 13.7%), and permeabilities in the range 0.3–827 mD (averaging 193 mD).

The petrological analysis (Macchi 2008; Celtique Energie 2012) has, for the first time, provided detailed information on the primary mineral components, diagenetic alteration, visual porosity and on the clay fraction within the Bunter Sandstone reservoir (Appendix A). The sandstones also contain microporosity, partly resulting from feldspar dissolution and partly associated with the presence of clays. These petrological data were incorporated into a new petrophysical evaluation of the wireline log data from the La Chandelière Field, using the Dual Water Model, and taking into account the clay percentages, the macroporosity and the microporosity within the reservoir rock. An example output plot for the La Chandelière 1D well is shown in Figure 5c.

Reservoir quality distribution is extremely variable and very difficult to predict; however, there is a general relationship between porosity and permeability that reflects the volume of clay. The best reservoirs coincide with the cleanest intervals. The clays reduce pore space and promote ductile deformation. Porosity loss can be related to grain size, clay enrichment, cementation and mechanical compaction; however, there does not appear to be one consistent set of controls on porosity reduction or compaction within the sandstones.

Microporosity (from potassium feldspar dissolution) is a significant contributor (visual estimates range up to 6%) to total pore volume (Macchi 2008; Celtique Energie 2012). Clay volume will also contribute to microporosity, which is non-effective; this is the reason why some zones with high total porosity can have low permeability.

The sands have undergone complex diagenesis with quartz, calcite, anhydrite, clay cements, and minor hematite and halite. The diagenetic sequence is complex with no single diagenetic control on porosity reduction. There is a lot of precipitation but little dissolution, although much of the pore system may be secondary in origin.

Secondary reservoir potential

Secondary reservoir objectives, identified during the early exploration in the Jura, were carbonates in the Muschelkalk and Lettenkohle formations, which were gas bearing at Lons Le Saunier and Valempoulieres (Elf Aquitaine 1991). Shows were often recorded in these formations associated with the limited fracturing of the carbonates since the reservoir properties of the matrix are poor. These carbonates are poor reservoirs at La Chandelière.

Other potential reservoir horizons include:

- Middle Keuper dolomite: the reservoir development is generally poor, and is likely to be reliant on secondary porosity and permeability from natural fractures.
- Permo-Carboniferous: variable sands deposited in alluvial-fan complexes have low porosity and low permeability due to poor sorting and low compositional maturity.
- Rhaetic sandstones: sands are thin, and reservoir quality can vary over short distances.

Seal

The Bunter Sandstone reservoir interval is sealed by the overlying shales and evaporates of the Triassic Muschelkalk and Keuper intervals (Rigassi 1977). Regional mapping of these evaporites shows that the interval generally thickens to the NE, but locally thickens into NE–SW-trending troughs (Sommaruga 1997). Three potential seal zones are developed in the Triassic interval:

- The Upper Keuper is anhydritic and is most thickly developed to the north.
- The Keuper evaporite sequence is the most widely developed seal and accounts for the majority of the salt in the Triassic. It passes southwards into anhydrite before pinching out to the south and west. It is locally halokinetic and reaches a thickness of up to 720 m.
- The Muschelkalk salt is only developed to the NE. The interval is up to 375 m thick. Where it is thin, the underlying Muschelkalk carbonates may act as a waste zone to the Bunter sands.

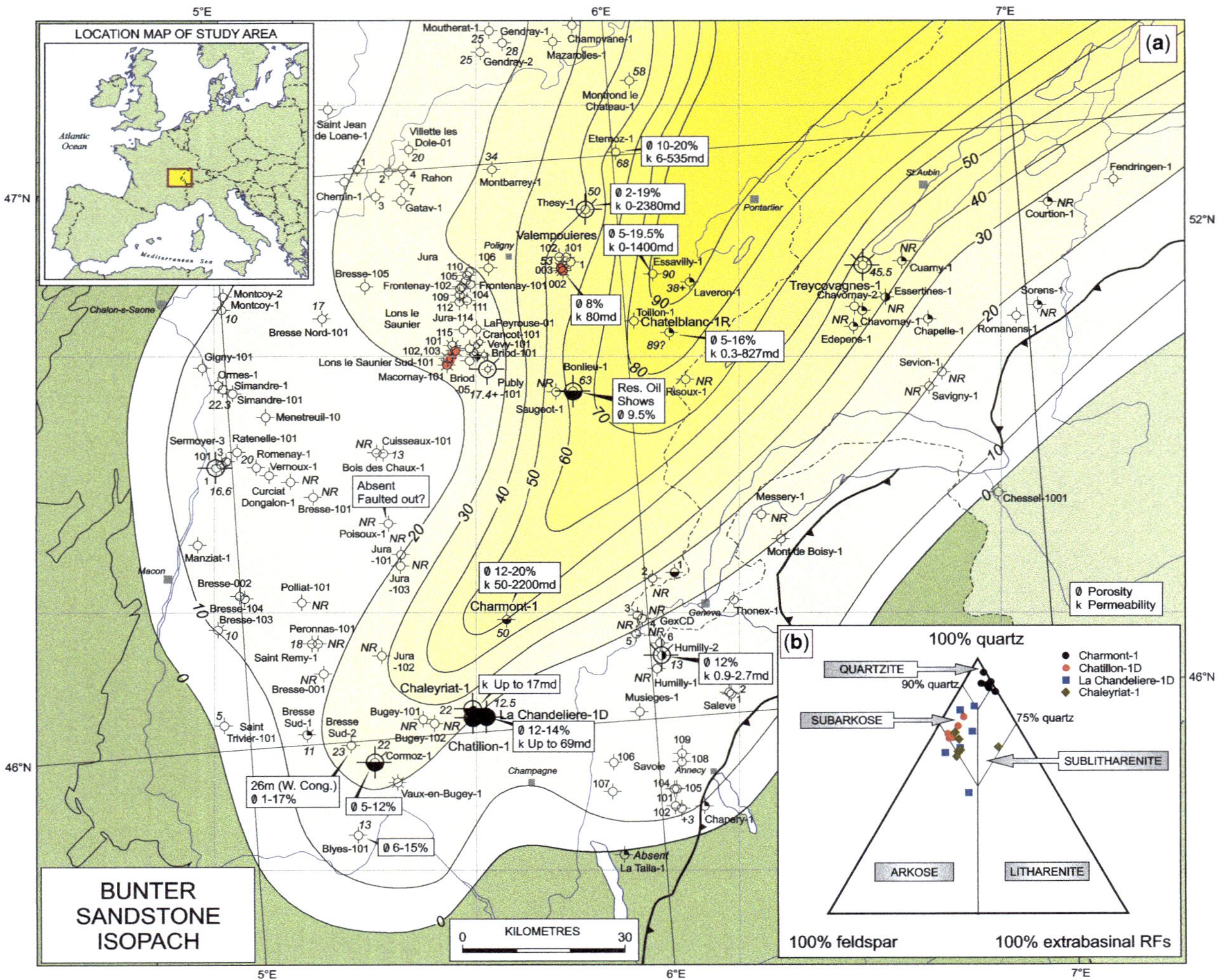

Fig. 5. (**a**) Bunter Sandstone isopach map (contour interval 10 m). (**b**) Normalized framework compositions for the Bunter Sandstone in Chaleyriat 1, La Chandelière 1D, Chatillon 1D and Charmont 1 (Macchi 2008). Particle types exclude intra-formational rock fragments.

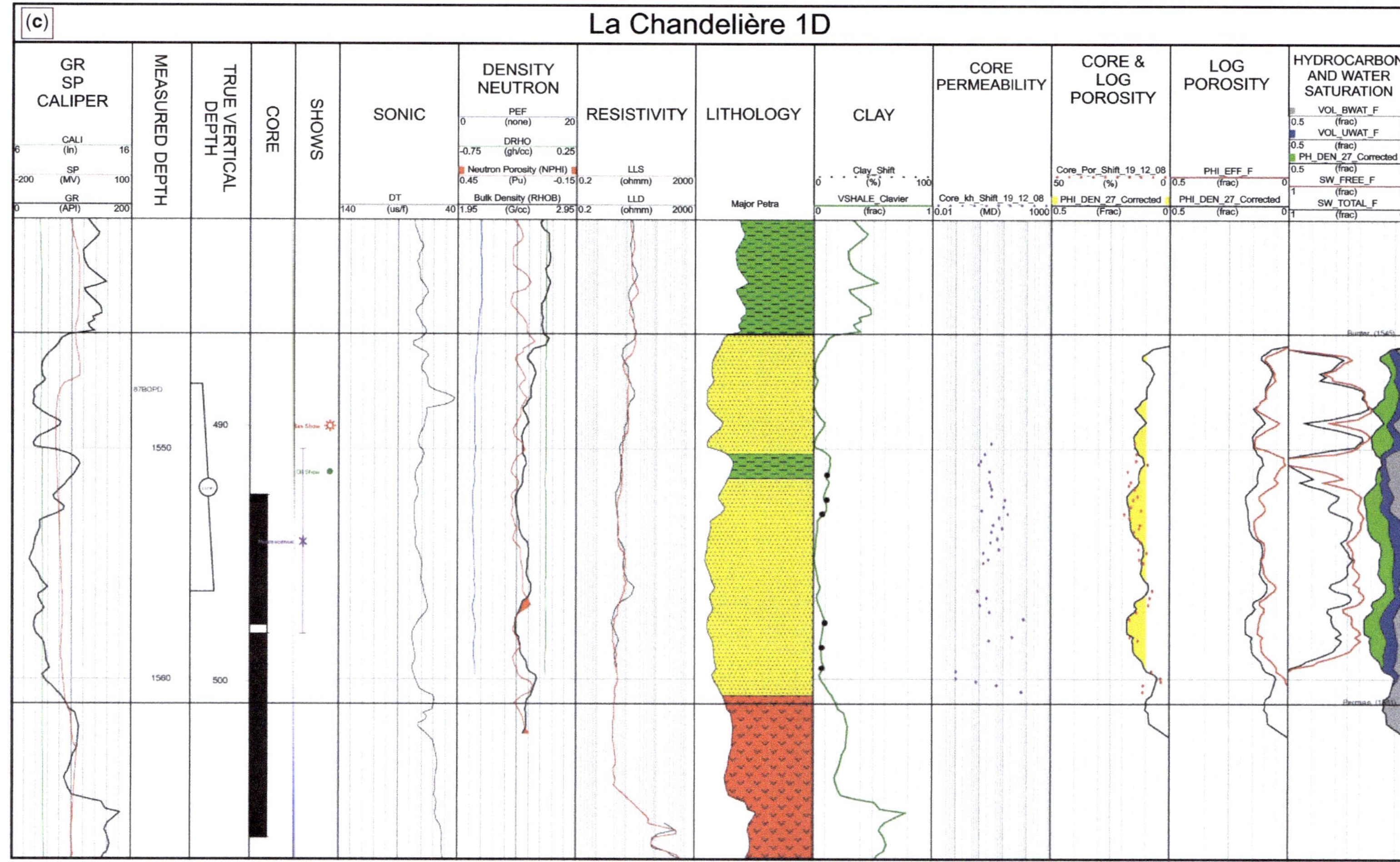

Fig. 5. (*Continued*) (**c**) La Chandelière 1D Bunter Sandstone petrophysical interpretation.

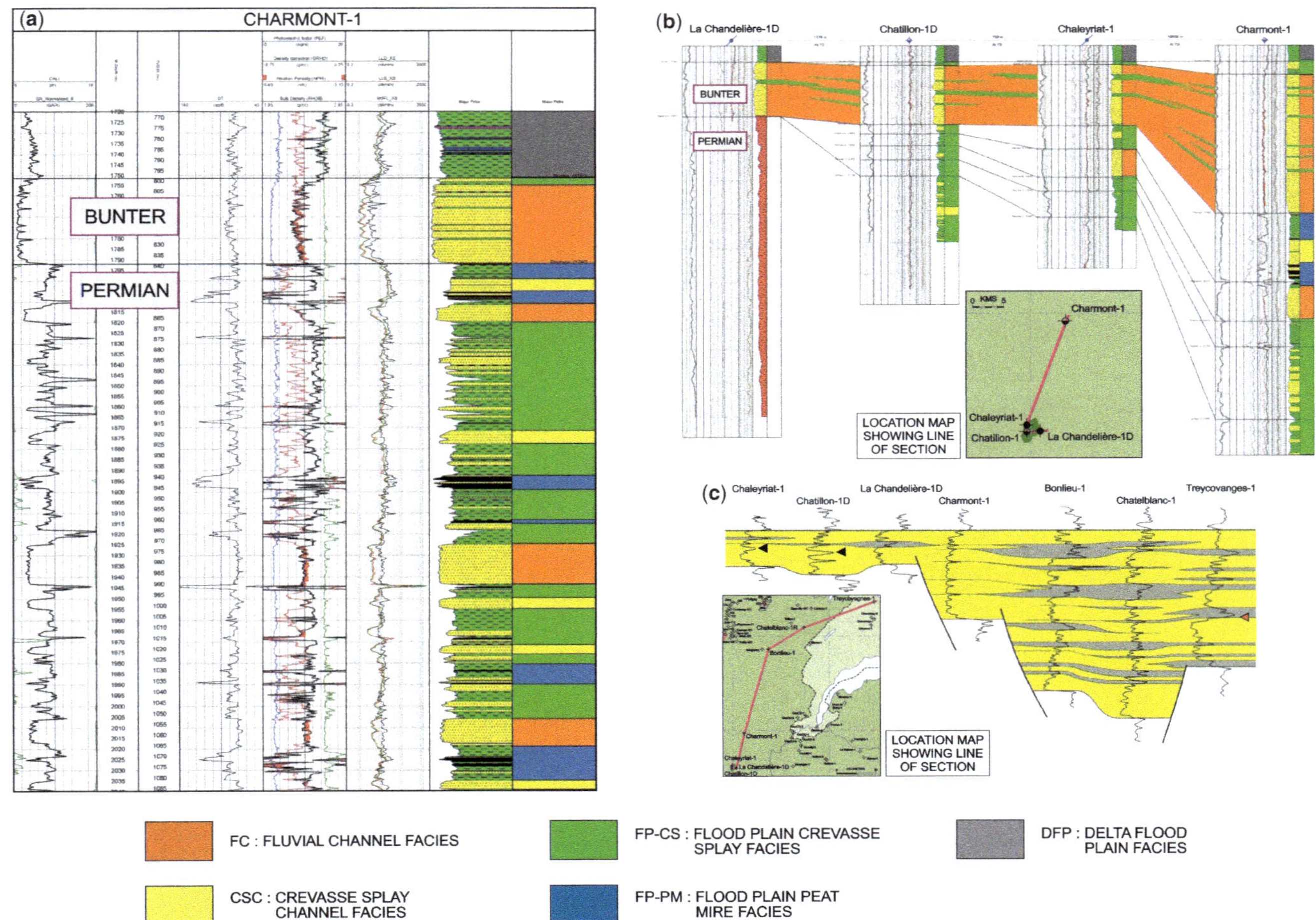

Fig. 6. (**a**) Genetic facies interpreted from Charmont 1 (19 km NE of La Chandelière) based on core study, lithological data and wireline log character. (**b**) The facies cross-section between the La Chandelière Field wells and Charmont 1 illustrates the stacked channels and discontinuous permeability barriers in the Bunter Sandstone. (**c**) The cross-section from La Chandelière to Treycovagnes in Switzerland further illustrates the complexity in correlation within the Bunter Sandstone and the stacked nature of the channel facies (Macchi 2008).

La Chandelière structure

The La Chandelière Field is located within the southern Jura in a major north–south-orientated thrust-faulted horst complex. It was discovered in June 1989 by Exxon and has been penetrated by three wells: Chaleyriat 1 in June 1989, La Chandelière 1D in December 1989 and Chatillon 1D in September 1991. A new interpretation (Fig. 7a), based on successful reprocessing of the existing 2D seismic dataset, shows the La Chandelière structure to be a horst that was subsequently deformed creating a NE–SW-trending anticline covering an area of 6.35 km^2. At reservoir level, the structure is bounded to the NW by a normal fault zone (Fig. 7b) that was reactivated during compression, resulting in an overthrusted package of post-Upper Triassic sediments to the NW.

Structural modelling

Two- and three-dimensional structural modelling was carried out as part of this work to validate the pre-Triassic exploration play identified by seismic interpretation at the La Chandelière Field, to determine the geometry and depth of the target structure and to understand the structural evolution and timing (Turrini *et al.* 2009).

All the available geophysical and geological data for the region were used to interpret key horizons and faults (using Kingdom software). The outcrop geology was used to constrain the seismic geometries, and a number of seismic lines were converted to depth. The depth sections were integrated to provide the final structural contour maps and the related 3D model (using Midland Valley software) of the potential exploration targets.

In the Jura Fold Belt, the structures show mild (external Jura to the north) to strong (internal Jura to the south) folding and thrusting, which is generally considered to form above a regional detachment level hosted in the Triassic salt layers (Philippe 1994; Sommaruga 1997; Turrini *et al.* 2009). The modelling suggested that the variable distribution (presence/absence, thickness) of the Triassic evaporates provides control to the deformation of the shallow post-Triassic units, which are structurally discordant with respect to the deep pre-Triassic units. Thick-skinned tectonics deforming the pre-Triassic basement was consistently modelled at depth below the shallow, thin-skinned post-Triassic structures.

The review of the seismic and outcrop data resulted in the creation of a top Bunter Sandstone map. Although the fold belt appears to be continuous from a regional perspective at the surface, the modelling suggests that it is more likely to be segmented, being offset by NNW–SSE-trending deformation zones, and creating a series of tectonic domains below the salt and the shallow structures (Fig. 8). Within each domain, an overall asymmetrical wedge-like morphology exists with a deepening to the east and to the south, towards the Alps. In each domain, the position of the change from intense to relatively mild deformation varies. A tectonic assemblage can be recognized within each domain, based on the association of three constituent parts (Fig. 9a). These are:

- A thrust-related structural high with basement involvement (Area A) which passively transports the earlier pre-Alpine palaeostructures. These pre-Alpine structures are the trapping mechanism for the hydrocarbon play in the Jura.
- At the basement fault tip (Area B to the west) highly complex, shallow thrusted-folds which are most likely detached at the Triassic salt level.
- On the backlimb of the basement anticline units (Area C to the east) intense thrusting and tectonic stacking occurs.

Three tectonic domains (Southern, Central and Eastern) have been recognized as part of this modelling (Figs 8 & 9b–d). The dimensions and orientations of these domains show control from the pre-Alpine basement grain. The southern domain (Les Moussières domain) is the narrowest and most deformed region (refer to the map in Fig. 8 and the line of section A–A′ in Fig. 9b). Further north, the central domain (Pontarlier-Vallorbes domain) is a broad, less-deformed SSE-dipping gentle monocline that dips below the Jura Deformation Belt (refer to the map in Fig. 8 and the line of section B–B′ in Fig. 9c). However, on its SE flank, the deformation is intense with tectonic stacking occurring. The eastern domain (Neuchâtel domain) is a broad zone that is a mirror image of the Les Moussières domain, but where the deformation is less intense (refer to the map in Fig. 8 and the line of section C–C′ in Fig. 9d).

In each domain, the position of the palaeostructures is critical from a petroleum prospectivity standpoint. The prospective structures are early formed horst blocks, with associated gentle, large-scale rollovers. As a result of the Tertiary compressional movements, the flanks of these horst blocks were reactivated with thrusting from both the east and west along thrust faults that sole out into the Triassic evaporite décollement zone. The overthrusted units result in a significant topographical expression, whilst the crestal anticlinal feature is expressed at the surface by a valley complex. As demonstrated in the La Chandelière Field, the hydrocarbon trap is an independent, four-way dip closure with minimal faulting at the Bunter Sandstone level (Fig. 7b).

The structural modelling carried out as part of this work successfully validated the geometry and

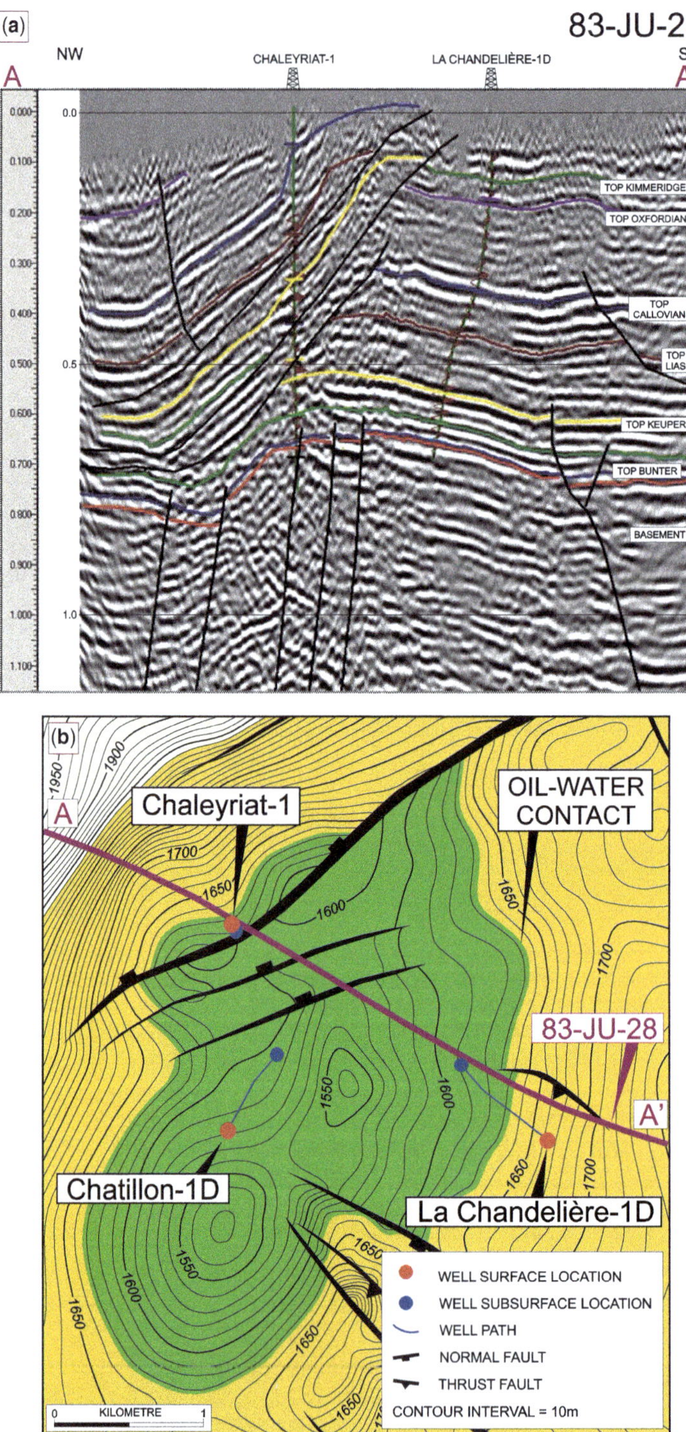

Fig. 7. (**a**) Interpretation of the reprocessed seismic line 83-JU-28 over the La Chandelière Field shows the faulting complexity in the vicinity of Chaleyriat 1 (seismic datum 1125 m above mean sea level (amsl)). (**b**) Top Bunter Sandstone depth structure map of the La Chandelière Field (seismic contour interval 10 m). The oil field extent is indicated in green down to a lowest closing contour level (as marked).

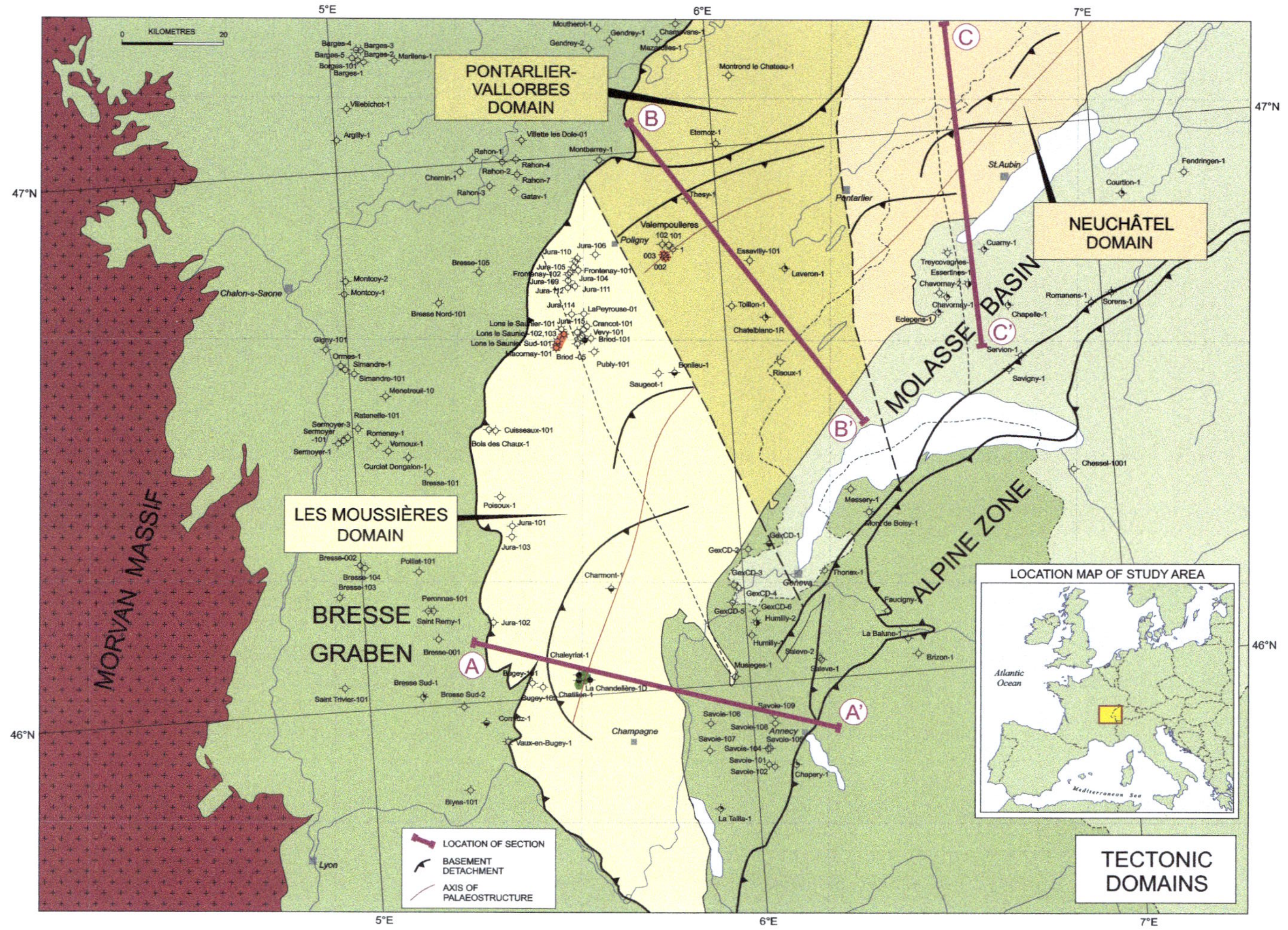

Fig. 8. Jura Fold Belt tectonic domains.

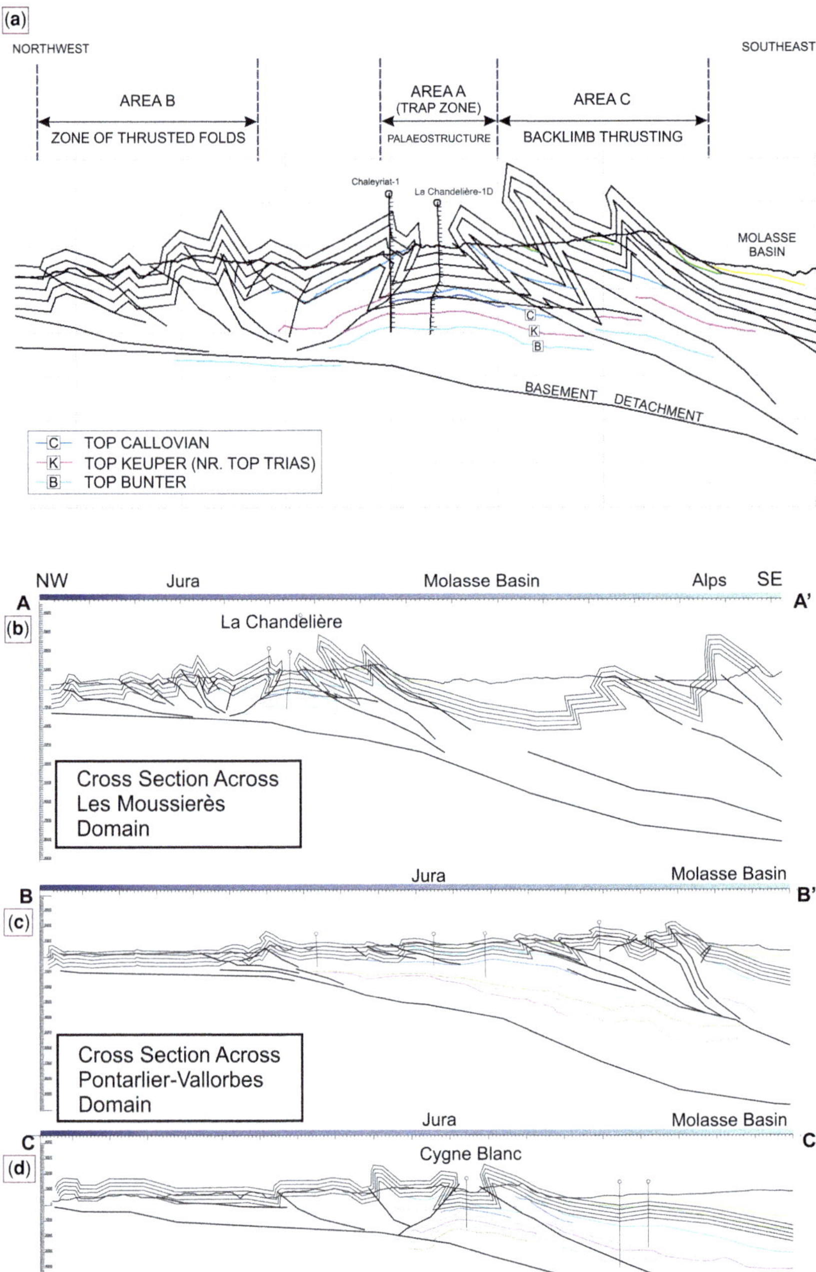

Fig. 9. Schematic cross-sections from 2D structural modelling (output from Midland Valley software) of the Jura carried out to validate the pre-Triassic exploration play (Turrini *et al.* 2009). Coloured lines represent horizons picked from 2D seismic; black lines are output from Midland Valley structural modelling software. (**a**) Schematic cross-section illustrating the tectonic assemblage. (**b**) Schematic cross-section across the southern (Les Moussières) domain (A–A′ in Fig. 8). (**c**) Schematic cross-section across the central (Pontarlier-Vallorbes) domain (B–B′ in Fig. 8). (**d**) Schematic cross-section across the eastern (Neuchâtel) domain (C–C′ in Fig. 8).

kinematics of the targeted structures interpreted from seismic data. This methodology demonstrated its significance in contributing to an understanding of this complex region, and confirmation of an unexplored exploration play within the Jura of France and Switzerland.

Conclusions

- A Paleozoic-sourced oil play exists in the Jura Mountains of France and Switzerland.
- The existence of three, small producing gas fields, and the presence of recorded oil and gas shows in a number of petroleum wells in stratigraphic horizons that lie beneath the regional Triassic evaporite seal, indicate the presence of an active petroleum system.
- Geochemical analysis suggests that hydrocarbons are sourced by very effective Permo-Carboniferous source rocks.
- The source rocks include oil-prone, algal (Type I), organically rich Autunian shale beds, Stephanian-aged bituminous shales, and both Carboniferous and Permian coals, which would be an excellent source of gas.
- Geochemical analysis demonstrates that the 41° API oil at La Chandelière is derived from Lower Permian Autunian shales.
- The primary reservoir target is the fluviatile Triassic Bunter Sandstone Formation. Reservoir quality is best developed in the cleaner channel sands, with porosities as high as 24% and permeabilities of up to 1894 mD.
- The trapping mechanism, interpreted from seismic and confirmed by structural modelling, is that of an unreactivated Mesozoic-aged structural high bounded by thrust faults, resulting in a large-scale rollover within the core of the structure.
- The La Chandelière oil field, discovered in 1989 by Exxon, validates the petroleum system and trapping mechanism.
- This model of the Triassic Bunter Sandstone reservoirs, charged by Paleozoic source rocks and sealed by Triassic salt, is an unexplored exploration play within the Jura fold and thrust belt through France and Switzerland.

Acknowledgements The authors would like to dedicate this paper to Lou Macchi (Resource Associates International), who passed away in June 2017. Lou's work on the Bunter Sandstone was extremely important to our understanding of the reservoir potential of the Jura. He was a good friend and will be sadly missed.

The authors would like to thank Celtique Energie Ltd for permission to publish this work. Thanks also go to the technical team at Celtique who contributed to the effort of assembling the extensive database from multiple sources across France and Switzerland, and for assisting with the geological, geophysical and petrophysical interpretation; in particular, Geoff Davies, Ken Dups and Sarah Shirt. Thanks also go to Gareth Harriman (GH Geochem) for his geochemical analyses and to Claudio Turrini (CT Geological Consulting) for his structural modelling work of the Jura Fold Belt. Our understanding of the area was considerably enhanced after discussions with Gilles Borel, Anna Sommaruga and Jon Mosar. Thanks also go to Eric Stuckey of Windmill Graphics for drafting the figures and maps.

Appendix A

Mineralogical and textural data from thin section microscopy (L. Macchi 2008)

Table A1. *Mineralogical data for Chaleyriat 1*

Well: Chaleyriat 1	Chlyt-1	Chlyt-1	Chlyt-1	Chlyt-1	Chlyt-1	Chlyt-1
Depth (mDDRKB)	1328.30	1327.00	1324.63	1322.80	1320.80	1319.75
Depth (mMDRKB)	1328.93	1327.55	1325.04	1323.10	1320.97	1319.86
Median grain-size estimate (mm)	1.40	0.55	0.50	0.50	0.60	0.30
Sorting estimate (σ SD)	0.80	0.70	0.65	0.75	0.90	0.50
Powers angularity estimate (quartz)	0.36	0.50	0.55	0.45	0.45	0.57
Median primary pore-size estimate (mm)	0.040	0.035	0.020	n.a.	0.001	0.015
Sedimentary texture	Stratified	Uniform	Aligned	Uniform	Massive	Aligned
Unicrystalline quartz (%)	23.33	26.33	28.33	24.67	33.00	35.00
Polycrystalline quartz (%)	26.67	20.33	26.67	18.00	12.33	18.00
Chert/chalcedony (%)		Trace	Trace	Trace		Trace
Potassium feldspar (%)	8.00	17.00	21.00	18.00	14.67	16.33
Plagioclase feldspar (%)			Trace	Trace		Trace
Volcanic rock fragments (%)	Trace					

(*Continued*)

Table A1. *Mineralogical data for Chaleyriat 1* (*Continued*)

Well: Chaleyriat 1	Chlyt-1	Chlyt-1	Chlyt-1	Chlyt-1	Chlyt-1	Chlyt-1
Metamorphic rock fragments (%)	18.33	8.00	9.67	8.33	6.33	5.00
Indeterminate rock fragments (%)		1.00	0.33			0.33
Intraformational rock fragments/ soil peds (%)	0.33	0.33	0.33	0.33		
Mica (%)	Trace		Trace	Trace		Trace
Ore minerals (%)	Trace	Trace	Trace	0.67	Trace	Trace
Other heavy minerals (%)		Trace	0.33			Trace
Syntaxial quartz (%)	1.00	2.67	2.33	Trace	Trace	8.67
Epitaxial sanidine (%)			Trace			
Non-ferroan calcite (intergranular) (%)	11.00	0.67	Trace	0.33	22.67	
Non-ferroan calcite (replacive) (%)	0.67	0.33	Trace	Trace	5.00	
Non-ferroan dolomite (intergranular) (%)				2.00		
Non-ferroan dolomite (replacive) (%)				2.33		
Ferroan dolomite (intergranular) (%)				2.00		
Ferroan dolomite (replacive) (%)				0.67		
Anhydrite (intergranular) (%)						
Anhydrite (replacive) (%)						
Barite (intergranular) (%)			Trace			
Barite (replacive) (%)			Trace			
Halite (intergranular) (%)			Trace			
Halite (replacive) (%)						
Anatase (intergranular) (%)		Trace	Trace			
Hematite (intergranular) (%)						
Hematite (replacive) (%)		Trace				
Pyrite/pyrrhotite (intergranular) (%)		Trace		2.00	Trace	0.67
Pyrite/pyrrhotite (replacive) (%)		Trace	Trace	2.00	0.67	1.67
Kaolinite (intergranular) (%)		0.33	Trace	0.33	0.33	Trace
Kaolinite (replacive) (%)	1.33	0.67	1.00	1.00	Trace	0.67
Illite (intergranular) (%)	0.67	0.33	0.33	2.00		
Illite (replacive) (%)	Trace			0.33		
Amorphous clay (intergranular) (%)	1.67	0.33	2.67	3.00	0.33	2.33
Amorphous clay (replacive) (%)		Trace	0.67	1.67		1.67
Residual oil		Trace				
Intergranular macroporosity (%)	5.67	18.67	3.67	5.33	1.00	6.00
Grain-dissolution macroporosity (%)	0.67	2.33	0.67	2.33	3.33	2.67
Intergranular microporosity (%)	Trace	0.33	1.33	0.67	Trace	0.67
Secondary microporosity (%)	0.67	0.33	0.67	1.00	Trace	0.33
Particle fracture porosity (%)	Trace	Trace	Trace	1.00	0.33	Trace
Total porosity (%)	7.00	21.67	6.33	10.33	4.67	9.67
Quartz–feldspar–rock fragment ratio	66–10–24	64–23–13	64–24–12	62–26–12	68–22–10	71–22–07
Depositional porosity estimate (%)	33.60	34.50	35.00	34.00	33.00	37.40
Compactional porosity loss (%)	17.00	14.60	27.50	18.90	11.10	23.30
Helium porosity (%)	No data	23.63	13.23	14.58	10.43	21.47
Horizontal permeability (mD)	No data	1390.40	24.80	90.00	27.40	425.60
Grain density (g cm^{-3})	No data	2.65	2.64	2.66	2.65	2.67

mDDRKB, metres drillers depth rotary kelly bushing; mMDRKB, metres measured depth rotary kelly brushing.

Table A2. *Mineralogical data for Chatillon 1*

Well: Chatillon 1D	Chtn-1D	Chtn-1D	Chtn-1D	Chtn-1D	Chtn-1D
Depth (mDDRKB)	1556.54	1556.40	1551.65	1551.27	1547.85
Depth (mMDRKB)	1555.44	1555.30	1550.55	1550.17	1546.75
Median grain-size estimate (mm)	0.65	0.40	0.45	0.55	0.85
Sorting estimate (σ SD)	0.55	0.55	0.60	0.85	1.50
Powers angularity estimate (quartz)	0.50	0.45	0.52	0.50	0.45
Median primary pore-size estimate (mm)	0.015	0.003	0.005	n.a	0.010
Sedimentary texture	Uniform	Aligned	Massive	Massive	Uniform
Unicrystalline quartz (%)	27.00	35.33	28.33	36.33	35.67
Polycrystalline quartz (%)	23.67	25.00	21.00	16.33	20.00
Chert/chalcedony (%)	Trace	0.33	0.33	0.33	
Potassium feldspar (%)	17.67	12.67	17.00	14.00	20.00
Plagioclase feldspar (%)	Trace		Trace	0.33	
Volcanic rock fragments (%)					
Metamorphic rock fragments (%)	3.33	4.67	5.00	4.00	4.67
Indeterminate rock fragments (%)		0.33		0.67	0.33
Intraformational rock fragments/soil peds (%)	0.33		1.00	0.67	
Mica (%)	0.33	Trace			Trace
Ore minerals (%)	Trace		Trace	Trace	
Other heavy minerals (%)	Trace	Trace	Trace	Trace	Trace
Syntaxial quartz (%)	8.33	1.33	6.00	0.33	4.67
Epitaxial sanidine (%)	Trace				Trace
Non-ferroan calcite (intergranular) (%)	0.67	Trace	Trace		Trace
Non-ferroan calcite (replacive) (%)	0.67	0.33	Trace		Trace
Non-ferroan dolomite (intergranular) (%)					Trace
Non-ferroan dolomite (replacive) (%)		0.33	Trace	Trace	Trace
Ferroan dolomite (intergranular) (%)	0.33				
Ferroan dolomite (replacive) (%)	0.33		Trace	0.67	
Anhydrite (intergranular) (%)	1.33	0.33	13.67	13.33	2.67
Anhydrite (replacive) (%)	Trace	Trace	2.67	3.00	0.67
Barite (intergranular) (%)			Trace	Trace	Trace
Barite (replacive) (%)			Trace	Trace	Trace
Halite (intergranular) (%)					
Halite (replacive) (%)					
Anatase (intergranular) (%)					
Hematite (intergranular) (%)					
Hematite (replacive) (%)					
Pyrite/pyrrhotite (intergranular) (%)	Trace	Trace		Trace	
Pyrite/pyrrhotite (replacive) (%)	Trace	Trace		Trace	
Kaolinite (intergranular) (%)	1.33	Trace	1.00	1.67	
Kaolinite (replacive) (%)	1.33	2.00	1.00	1.33	0.67
Illite (intergranular) (%)	0.67	1.00		0.33	1.00
Illite (replacive) (%)					Trace
Amorphous clay (intergranular) (%)	2.00	8.00	0.33	0.33	3.00
Amorphous clay (replacive) (%)	0.67	1.67	0.33	1.33	0.33
Residual oil					
Intergranular macroporosity (%)	5.33	1.33	1.33	2.67	3.00
Grain-dissolution macroporosity (%)	2.33	1.00	0.67	1.67	0.33
Intergranular microporosity (%)	1.00	3.67	Trace	Trace	2.33
Secondary microporosity (%)	1.00	0.67	0.33	0.67	0.67
Particle fracture porosity (%)	0.33	Trace	Trace		Trace
Total porosity (%)	10.00	6.67	2.33	5.00	6.33
Quartz–feldspar–rock fragment ratio	70–25–05	77–16–06	69–24–07	74–20–06	69–25–06
Depositional porosity estimate (%)	36.50	36.50	35.80	33.20	30.50
Compactional porosity loss (%)	19.30	24.70	17.30	17.90	16.60
Helium porosity (%)	14.67	14.67	No data	8.41	11.01
Horizontal permeability (mD)	14.70	17.70	No data	2.61	3.19
Grain density (g cm^{-3})	2.63	2.63	No data	2.67	2.65

mDDRKB, metres drillers depth rotary kelly bushing; mMDRKB, metres measured depth rotary kelly brushing.

Table A3. *Mineralogical data for La Chandelière 1D*

Well: La Chandelière 1D	Chdle-1D	Chdle-1D	Chdle-1D	Chdle-1D	Chdle-1D	Chdle-1D
Depth (mDDRKB)	1560.90	1559.96	1558.86	1555.20	1554.65	1553.50
Depth (mMDRKB)	1558.90	1557.96	1556.86	1553.30	1552.75	1551.60
Median grain-size estimate (mm)	0.50	0.54	0.60	0.75	0.55	0.30
Sorting estimate (σ SD)	0.70	0.90	1.50	0.65	0.60	0.70
Powers angularity estimate (quartz)	0.40	0.30	0.25	0.45	0.45	0.50
Median primary pore-size estimate (mm)	n.a.	n.a	0.005	n.a.	n.a	0.002
Sedimentary texture	massive	massive	massive	massive	massive	uniform
Unicrystalline quartz (%)	38.00	27.67	27.33	12.67	31.33	26.67
Polycrystalline quartz (%)	21.67	15.67	27.67	15.33	22.33	20.67
Chert/chalcedony (%)					0.33	0.67
Potassium feldspar (%)	10.67	15.33	12.33	17.33	6.33	21.67
Plagioclase feldspar (%)	1.00				0.33	0.33
Volcanic rock fragments (%)			Trace		Trace	
Metamorphic rock fragments (%)	2.67	7.33	8.67	13.67	5.00	5.00
Indeterminate rock fragments (%)	0.33	0.33	0.67		0.33	0.67
Intraformational rock fragments/soil peds (%)	3.67	22.67	0.67	6.67	6.67	
Mica (%)			0.33			Trace
Ore minerals (%)	Trace	0.33	Trace	Trace	Trace	Trace
Other heavy minerals (%)		Trace	Trace			Trace
Syntaxial quartz (%)	0.33		0.33	0.33	0.33	1.33
Epitaxial sanidine (%)						
Non-ferroan calcite (intergranular) (%)			Trace		Trace	
Non-ferroan calcite (replacive) (%)	1.33		0.67		Trace	
Non-ferroan dolomite (intergranular) (%)	Trace				Trace	
Non-ferroan dolomite (replacive) (%)					0.33	
Ferroan dolomite (intergranular) (%)				0.33	Trace	0.33
Ferroan dolomite (replacive) (%)	0.67	0.33	1.00	1.33	0.33	0.33
Anhydrite (intergranular) (%)	6.33		0.67	2.33	3.33	2.00
Anhydrite (replacive) (%)	2.00	0.33	0.33	7.00	4.00	1.33
Barite (intergranular) (%)						
Barite (replacive) (%)		Trace				
Halite (intergranular) (%)		0.33	1.67	3.33	0.33	
Halite (replacive) (%)		0.33	0.67	1.00		
Anatase (intergranular) (%)						
Hematite (intergranular) (%)		Trace				
Hematite (replacive) (%)	Trace	Trace	Trace			Trace
Pyrite/pyrrhotite (intergranular) (%)					Trace	
Pyrite/pyrrhotite (replacive) (%)		Trace				
Kaolinite (intergranular) (%)	0.33		0.67		Trace	0.67
Kaolinite (replacive) (%)	0.67	1.33	1.33	0.33	0.33	3.00
Illite (intergranular) (%)	0.33		0.33		0.33	1.00
Illite (replacive) (%)	Trace		Trace			Trace
Amorphous clay (intergranular) (%)	3.33	5.33	6.67	6.00	7.33	4.67
Amorphous clay (replacive) (%)	2.33	0.33	0.67	1.00	3.00	2.00
Residual oil						
Intergranular macroporosity (%)	0.67		0.67	9.00	3.00	Trace
Grain-dissolution macroporosity (%)	1.67	0.67	1.00	2.33	1.00	2.00
Intergranular microporosity (%)	0.33	1.00	3.33	Trace	2.33	2.00
Secondary microporosity (%)	1.67	0.67	2.33	Trace	1.33	3.67
Particle fracture porosity (%)	Trace	Trace	Trace	Trace	Trace	
Total porosity (%)	4.33	2.33	7.33	11.33	7.67	7.67
Quartz–feldspar–rock fragment ratio	80–16–04	65–23–12	72–16–12	43–26–31	82–10–08	63–29–08
Depositional porosity estimate (%)	34.50	33.00	30.50	35.00	35.80	34.50
Compactional porosity loss (%)	25.90	28.20	18.90	17.40	22.70	25.60
Helium porosity (%)	7.60	13.56	17.04	12.05	15.41	17.80
Horizontal permeability (mD)	0.10	2.40	69.30	1.20	11.30	2.50
Grain density (g cm^{-3})	2.76	2.64	2.67	2.67	2.67	2.65

mDDRKB, metres drillers depth rotary kelly bushing; mMDRKB, metres measured depth rotary kelly brushing; n.a., not available.

Table A4. *Mineralogical data for Charmont 1*

Well: Charmont 1	Chrmnt-1	Chrmnt-1	Chrmnt-1	Chrmnt-1	Chrmnt-1	Chrmnt-1
Depth (mDDRKB)	1761.95	1759.42	1757.67	1755.06	1753.70	1751.90
Depth (mMDRKB)	1763.05	1760.52	1758.77	1756.16	1754.40	1752.60
Median grain-size estimate (mm)	0.75	0.45	0.80	0.70	0.50	0.28
Sorting estimate (σ SD)	0.65	0.60	0.90	0.80	0.70	0.55
Powers angularity estimate (quartz)	0.52	0.57	0.60	0.52	0.60	0.52
Median primary pore-size estimate (mm)	0.050	0.003	0.060	0.050	0.040	0.010
Sedimentary texture	Uniform	Bedded	Massive	Bedded	Bedded	Aligned
Unicrystalline quartz (%)	27.67	39.00	32.67	32.33	24.67	33.00
Polycrystalline quartz (%)	36.00	24.00	32.33	32.00	35.67	28.33
Chert/chalcedony (%)	Trace			0.33	Trace	0.33
Potassium feldspar (%)	Trace	0.33	0.33	0.67	0.67	1.67
Plagioclase feldspar (%)		Trace		0.33	0.67	0.67
Volcanic rock fragments (%)						
Metamorphic rock fragments (%)	6.00	3.00	8.67	5.33	6.00	4.00
Indeterminate rock fragments (%)				0.33	0.33	
Intraformational rock fragments/soil peds (%)	Trace	7.00	Trace	1.33	2.33	
Mica (%)		Trace	Trace	Trace	Trace	Trace
Ore minerals (%)	Trace	Trace	0.33	0.33	Trace	Trace
Other heavy minerals (%)	Trace	Trace	Trace	Trace	Trace	Trace
Syntaxial quartz (%)	3.67		6.33	3.33	3.67	3.00
Epitaxial sanidine (%)						
Non-ferroan calcite (intergranular) (%)						
Non-ferroan calcite (replacive) (%)						
Non-ferroan dolomite (intergranular) (%)						
Non-ferroan dolomite (replacive) (%)						
Ferroan dolomite (intergranular) (%)	2.00	Trace	1.33	2.33	3.00	1.67
Ferroan dolomite (replacive) (%)	Trace	Trace	1.00	1.67	0.33	1.67
Anhydrite (intergranular) (%)						
Anhydrite (replacive) (%)						
Barite (intergranular) (%)	Trace	Trace	Trace		0.33	
Barite (replacive) (%)		Trace	Trace			
Halite (intergranular) (%)						
Halite (replacive) (%)						
Anatase (intergranular) (%)						
Hematite (intergranular) (%)	Trace					
Hematite (replacive) (%)				Trace	Trace	
Pyrite/pyrrhotite (intergranular) (%)		Trace	Trace	0.67	1.67	2.00
Pyrite/pyrrhotite (replacive) (%)	Trace			Trace	0.67	0.33
Kaolinite (intergranular) (%)	2.33	Trace	1.67	1.67	3.00	2.33
Kaolinite (replacive) (%)	1.67	1.00	1.33	2.33	2.33	4.33
Illite (intergranular) (%)	Trace	3.00		0.33	Trace	
Illite (replacive) (%)		0.67				
Amorphous clay (intergranular) (%)	0.67	14.00	1.00	0.67	0.67	2.00
Amorphous clay (replacive) (%)	Trace	2.00	0.33	0.67	0.33	0.67
Residual oil	Trace	Trace				
Intergranular macroporosity (%)	14.67	Trace	10.00	9.33	9.33	6.00
Grain-dissolution macroporosity (%)	4.67	1.00	2.67	3.33	3.00	6.33
Intergranular microporosity (%)	0.33	4.00	Trace	0.33	1.33	1.00
Secondary microporosity (%)	0.33	1.00	Trace	0.33	Trace	0.67
Particle fracture porosity (%)						Trace
Total porosity (%)	20.00	6.00	12.67	13.33	13.67	14.00
Quartz–feldspar–rock fragment ratio	91–00–09	95–00–05	88–00–12	89–01–10	89–02–09	91–03–6
Depositional porosity estimate (%)	35.00	35.80	33.00	33.60	34.50	36.50
Compactional porosity loss (%)	14.80	10.80	15.90	18.40	14.90	22.60
Helium porosity (%)	20.60	14.60	18.40	17.70	17.50	17.04
Horizontal permeability (mD)	2225.23	12.40	980.00	441.00	202.88	63.76
Grain density (g cm^{-3})	2.64	2.63	2.64	2.64	2.71	2.66

mDDRKB, metres drillers depth rotary kelly bushing; mMDRKB, metres measured depth rotary kelly brushing.

References

ALLEN, J.R.L. 1964. Studies in fluviatile sedimentation: six cyclothems from the Lower Old Red Sandstone, Anglo-Welsh Basin. *Sedimentology*, **3**, 163–198.

ALLEN, J.R.L. 1970. A quantitative model of grain size and sedimentary structures in lateral deposits. *Geological Journal*, **7**, 129–146.

BECKER, A. 2000. The Jura Mountains – an active foreland fold-and-thrust belt? *Tectonophysics*, **321**, 381–406.

BP 1986*a*. *Seismic Interpretation Summary of Vaud, NE (Switzerland)*. BP Exploration Canton Vaud Geological Archives.

BP 1986*b*. *Well Geochemistry*. BP Exploration Canton Vaud Geological Archives.

BP 1989. *Vaud Prospectivity Review*. BP Exploration Canton Vaud Geological Archives.

BREYNAERT, M. 1912. Le gisement d'asphalte du Val-de-Travers [The asphalt field of Val de Travers]. *Annales des Mines*, **2**, 316–347.

BRGM 1989. *Synthese Geologique Des Bassins Permian Francais [Geological Synthesis of French Permian Basins]*. Memoire du Bureau de Recherches Géologiques et Minières, **128**.

BURKHARD, M. & SOMMARUGA, A. 1998. Evolution of the western Swiss Molasse Basin: structural relations with the Alps and the Jura belt. *In*: MASCLE, A., PUIDGEFABREGAS, C., LUTERBACHER, H.P. & FERNANDEZ, M. (eds) *Cenozoic Foreland Basins of Western Europe*. Geological Society, London, Special Publications, **134**, 279–298, https://doi.org/10.1144/GSL.SP.1998.134.01.13

CELTIQUE ENERGIE 2009. *Annual Vallorbes Licence Presentation*. Canton Vaud Geological Archives.

CELTIQUE ENERGIE 2010. *Annual Vallorbes Licence Presentation*. Canton Vaud Geological Archives.

CELTIQUE ENERGIE 2012. *Annual Vallorbes Licence Presentation*. Canton Vaud Geological Archives.

CHATEAUNEUF, J. & FARJANEL, G. (eds). 1989. *Synthese Geologique des Bassins Permiens Francais [Geological Synthesis of French Permian Basins]*. Memoire du Bureau de Recherches Géologiques et Minières, **128**.

COUREL, L. & PAQUETTE, Y. 1981. Place du Charbon Dans Le Remplissage de Trois Bassins Limniques du Massif Central Francais. *Bulletin des Centres de Recherches Exploration–Production Elf Aquitane Pau*, **5**, 473–490.

DOUBINGER, J. 1974. Etudes palynologiques dans l'Autunien. *Review of Palaeobotony and Palynology*, **17**, 21–38.

ELF AQUITAINE 1991. *Monographies des principaux champs petroliers de France [Monograph of the principal oil and gas fields in France]*. Chambre syndicale et Société nationale Elf Aquitaine, Paris.

HARRIMAN, G. 2009. *Geochemical Evaluation of Jura Wells*. GH Geochem Ltd. Unpublished report for Celtique Energie Petroleum Ltd

IFP 1994. Jura septentrional et son avant-pays [Northern Jura and its foreland]. *In*: JACQUART, G., DEVILLE, E. & MASCLE, A. (eds) *Rapport regional d'evaluation petroliere*. Institut Francais du Petrol (IFP), Paris.

IFP 1996. Jura Meridional et region Rhone-Alpes [Southern Jura and the Rhone Alpes Region]. *In*: JACQUART, G. & DEVILLE, E. (eds) *Rapport regional d'evalutaion petroliere*. Institut Francais du Petrol (IFP), Paris.

IFP 2001. Paris basin petroleum potential. *In*: DELMAS, J., HOUEL, P. & VIALLY, R. (eds) *Rapport regional d'evaluation petroliere*. Institut Francais du Petrol (IFP), Paris.

KETTEL, D. & HERZOG, M. 1988. The Permocarboniferous of the South German Molasse basin – a source rock for oil and gas? *Erdol, Erdgas, Kohle*, **104**, 154–157.

LAUBSCHER 1961. Die Fernschubhypothese des Jurafaltung [The remote thrusting hypothesis of the Jura Folding]. *Eclogae geologicae Helvetica*, **54**, 43–72.

LEU, W., SCHEGG, R., GREBER, E., MATTER, A., MAZUREK, M., HURFORD, A. & JONES, M. 1997. Burial and temperature history in NE-Switzerland – implications for hydrocarbon generation. Paper presented at the EAGE 59th Conference and Technical Exhibition, 26–30 May 1997, Geneva, Switzerland.

MACCHI, L. 2008. *The Bunter Sandstone Reservoir Jura Wells (France)*. L. Macchi, Reservoir Associates International. Unpublished report for Celtique Energie Petroleum Ltd.

MAZUREK, M., HURFORD, A. & LEU, W. 2006. Unravelling the multi-stage burial history of the Swiss Molasse Basin: integration of apatite fission track, vitrinite reflectance and biomarker isomerisation analysis. *Basin Research*, **18**, 27–50.

MCCANN, T. (ed.) 2008*a*. *The Geology of Central Europe. Volume 1: Precambrian and Palaeozoic*. Geological Society, London.

MCCANN, T. (ed.) 2008*b*. *The Geology of Central Europe. Volume 2: Mesozoic and Cenozoic*. Geological Society, London.

MCCANN, T., PASCAL, C. *ET AL.* 2006. Post-Variscan (end Carboniferous–Early Permian) basin evolution in Western and Central Europe. *In*: GEE, D.G. & STEPHENSON, R.A. (eds) *European Lithosphere Dynamics*. Geological Society, London, Memoirs, **32**, 355–388, https://doi.org/10.1144/GSL.MEM.2006.032.01.22

MOSS, S. 1992. Organic maturation in the French Subalpine Chains: regional differences in burial history and the size of tectonic loads. *Journal of the Geological Society, London*, **149**, 503–515, https://doi.org/10.1144/gsjgs.149.4.0503

NAGRA 1988. *Sondierbohrung Weiach – Untersuchungsbericht [Report on the Weiach Borehole]*. Nagra Technischer Bericht NTB 86-01.

NAGRA 1991. *Zur Tektonik Der Zentralen Nordschweiz – Untersuchungsbericht [On the tectonics of Central Northern Switzerland]*. Nagra Technischer Bericht NTB 90-04, *Wettingen*.

NAGRA 2015. *Overview of Seismic Exploration of Northern Switzerland*. SASEG Annual Convention, Basel, Switzerland.

PHILIPPE, Y. 1994. Transfer zone in the Southern Jura Thrust Belt (Eastern France): geometry, development, and comparison with analogue modelling experiments. *In*: MASCLE, A. (ed.) *Hydrocarbon and Petroleum Geology of France*. European Association of Petroleum Geoscientists, Special Publications, **4**, 328–346.

RIGASSI, D.A. 1977. Genese tectonique du Jura: une nouvelle hypothese [The tectonic genesis of the Jura: a new hypothesis]. *Paleolab News*, **2**, 1–27.

SCHARDT, H. 1911. Note sur les gisements asphaltiferes du Jura neuchatelois [Note on the asphalt fields of the

Neuchatel Jura]. *Bulletin de la Société neuchâteloise des sciences naturelles*, **37/38**, 398–424.

SCHEGG, R. & LEU, W. 1998. Analysis of erosion events and Palaeogeothermal gradients in the North Alpine Foreland Basin of Switzerland. *In*: DUPPENBECKER, S.J. & ILIFFE, J.E. (eds) *Basin Modelling: Practice and Progress*. Geological Society, London, Special Publications, **141**, 137–155, https://doi.org/10.1144/GSL.SP.1998.141.01.09

SHELL 1977. *Shell Switzerland (Exploration) 1977 Trecovagnes 1 Report*. Canton Vaud Geological Archives.

SOMMARUGA, A. 1997. *Geology of the Central Jura and the Molasse Basin: New Insight into An Evaporite-Based Foreland Fold and Thrust Belt*. Institut de Geologie, Universite de Neuchâtel, Switzerland.

SOMMARUGA, A., EICHENBERGER, U. & MARILLIER, F. 2012. *Seismic Atlas of the Swiss Molasse Basin*. Matériaux pour la Géologie de la Suisse – Géophysique, **44**.

SPYCHER, A. 1994. *Les mines d'asphalte de la Presta/Val-de-Travers [The Presta asphalt mine, Val de Travers]*. Société suisse des traditions populaires, Basle, Switzerland.

STAMPFLI, G. (ed.) 2001. *Geology of the Western Swiss Alps, A Guide-Book*. Memoires de Geologie (Lausanne), **36**.

STAMPFLI, G. & BOREL, G. 2002. A plate tectonic model for the Paleozoic and Mesozoic constrained by dynamic plate boundaries and restored synthetic oceanic isochrones. *Earth and Planetary Science Letters*, **196**, 17–33.

STROBL, C. 2007. *GIS-gestutzte Beckenanalyse am Beispiel des Franzosischen Juragebirges [TA GIS-linked basin analysis of the French Jura Mountains]*. PhD thesis, Ludwigs-Maximilians-Universitat, Munchen, Germany.

TIMAR-GENG, Z., FUGENSCHUH, B., WETZEL, A. & DRESMANN, H. 2006. The low-temperature thermal history of northern Switzerland as revealed by fission track analysis and inverse thermal modelling. *Eclogae geologicae Helvetica*, **99**, 255–270.

TURRINI, C., DUPS, K. & PULLAN, C. 2009. 2D and 3D structural modelling in the Swiss–French Jura Mountains. *First Break*, **27**, 65–71.

WETZEL, A., ALLENBACH, R. & ALLIA, V. 2003. Reactivated basement structures affecting the sedimentary facies in a tectonically 'quiescent' tepicontinental basin. *Sedimentary Geology*, **157**, 153–172.

ZIEGLER, P. (ed.) 1990. *Geological Atlas of Western and Central Europe*. 2nd edn. Shell Internationale Petroleum Maatschappij, The Hague. Geological Society, London.

ZWEIDLER, D. 1981. *Le gisement d'asphalte de La Presta (Val-de-Travers, Suisse) [The Presta asphalt mine (Val de Travers, Switzerland)]. PhD thesis*, l'Universite de Neuchâtel, Neuchâtel, Switzerland.

ZWEIDLER, D. 1985. *Genese des gisements d'asphalte des formations de La Pierre Jaune de Neuchatel et des calcaires urgoniens du Jura (Jura Neuchatelois et Nord Vaudois, Suisse) [The Genesis of the asphalt fields in the Yellowstones of Neuchatel and the Urgonian carbonates of the Jura]*. PhD thesis, l'Universite de Neuchâtel, Neuchâtel, Switzerland.

Index

Page numbers in *italics* refer to Figures. Page numbers in **bold** refer to Tables.